A COMPANION TO GREEK WARFARE

BLACKWELL COMPANIONS TO THE ANCIENT WORLD

This series provides sophisticated and authoritative overviews of periods of ancient history, genres of classical literature, and the most important themes in ancient culture. Each volume comprises approximately twenty-five and forty concise essays written by individual scholars within their area of specialization. The essays are written in a clear, provocative, and lively manner, designed for an international audience of scholars, students, and general readers.

Ancient History

A Companion to the Roman Army
Edited by Paul Erdkamp

A Companion to the Roman Republic
Edited by Nathan Rosenstein and Robert Morstein-Marx

A Companion to the Classical Greek World
Edited by Konrad H. Kinzl

A Companion to the Ancient Near East
Edited by Daniel C. Snell

A Companion to the Hellenistic World
Edited by Andrew Erskine

A Companion to Late Antiquity
Edited by Philip Rousseau

A Companion to Ancient History
Edited by Andrew Erskine

A Companion to Archaic Greece
Edited by Kurt A. Raaflaub and Hans van Wees

A Companion to Julius Caesar
Edited by Miriam Griffin

A Companion to Byzantium
Edited by Liz James

A Companion to Ancient Egypt
Edited by Alan B. Lloyd

A Companion to Ancient Macedonia
Edited by Joseph Roisman and Ian Worthington

A Companion to the Punic Wars
Edited by Dexter Hoyos

A Companion to Augustine
Edited by Mark Vessey

A Companion to Marcus Aurelius
Edited by Marcel van Ackeren

A Companion to Ancient Greek Government
Edited by Hans Beck

A Companion to the Neronian Age
Edited by Emma Buckley and Martin T. Dinter

A Companion to Greek Democracy and the Roman Republic
Edited by Dean Hammer

A Companion to Livy
Edited by Bernard Mineo

A Companion to Ancient Thrace
Edited by Julia Valeva, Emil Nankov, and Denver Graninger

A Companion to Roman Italy
Edited by Alison E. Cooley

A Companion to the Etruscans
Edited by Sinclair Bell and Alexandra A. Carpino

A Companion to the Flavian Age of Imperial Rome
Edited by Andrew Zissos

A Companion to Science, Technology, and Medicine in Ancient Greece and Rome
Edited by Georgia L. Irby

A Companion to the City of Rome
Edited by Amanda Claridge and Claire Holleran

A Companion to Greeks Across the Ancient World
Edited by Franco De Angelis

A Companion to Late Ancient Jews and Judaism—Third Century BCE—Seventh Century CE
Edited by Naomi Koltun-Fromm and Gwynn Kessler

A Companion to the Archaeology of Early Greece and the Mediterranean
Edited by Irene S. Lemos and Antonios Kotsonas

A Companion to Assyria
Edited by Eckart Frahm

A Companion to Sparta
Edited by Anton Powell

A Companion to Greco-Roman and Late Antique Egypt
Edited by Katelijn Vandorpe

Literature and Culture

A Companion to Greek and Roman Music
Edited by Tosca Lynch and Eleonora Rocconi

A Companion to Classical Receptions
Edited by Lorna Hardwick and Christopher Stray

A Companion to Greek and Roman Historiography
Edited by John Marincola

A Companion to Catullus
Edited by Marilyn B. Skinner

A Companion to Roman Religion
Edited by Jörg Rüpke

A Companion to Greek Religion
Edited by Daniel Ogden

A Companion to the Classical Tradition
Edited by Craig W. Kallendorf

A Companion to Roman Rhetoric
Edited by William Dominik and Jon Hall

A Companion to Greek Rhetoric
Edited by Ian Worthington

A Companion to Ancient Epic
Edited by John Miles Foley

A Companion to Greek Tragedy
Edited by Justina Gregory

A Companion to Latin Literature
Edited by Stephen Harrison

A Companion to Greek and Roman Political Thought
Edited by Ryan K. Balot

A Companion to Ovid
Edited by Peter E. Knox

A Companion to the Ancient Greek Language
Edited by Egbert Bakker

A Companion to Hellenistic Literature
Edited by Martine Cuypers and James J. Clauss

A Companion to Vergil's Aeneid and Its Tradition
Edited by Joseph Farrell and Michael C. J. Putnam

A Companion to Horace
Edited by Gregson Davis

A Companion to Families in the Greek and Roman Worlds
Edited by Beryl Rawson

A Companion to Greek Mythology
Edited by Ken Dowden and Niall Livingstone

A Companion to the Latin Language
Edited by James Clackson

A Companion to Tacitus
Edited by Victoria Emma Pagán

A Companion to Women in the Ancient World
Edited by Sharon L. James and Sheila Dillon

A Companion to Sophocles
Edited by Kirk Ormand

A Companion to the Archaeology of the Ancient Near East
Edited by Daniel Potts

A Companion to Roman Love Elegy
Edited by Barbara K. Gold

A Companion to Greek Art
Edited by Tyler Jo Smith and Dimitris Plantzos

A Companion to Persius and Juvenal
Edited by Susanna Braund and Josiah Osgood

A Companion to the Archaeology of the Roman Republic
Edited by Jane DeRose Evans

A Companion to Terence
Edited by Antony Augoustakis and Ariana Traill

A Companion to Roman Architecture
Edited by Roger B. Ulrich and Caroline K. Quenemoen

A Companion to Sport and Spectacle in Greek and Roman Antiquity
Edited by Paul Christesen and Donald G. Kyle

A Companion to Plutarch
Edited by Mark Beck

A Companion to Greek and Roman Sexualities
Edited by Thomas K. Hubbard

A Companion to the Ancient Novel
Edited by Edmund P. Cueva and Shannon N. Byrne

A Companion to Ethnicity in the Ancient Mediterranean
Edited by Jeremy McInerney

A Companion to Ancient Egyptian Art
Edited by Melinda Hartwig

A Companion to the Archaeology of Religion in the Ancient World
Edited by Rubina Raja and Jörg Rüpke

A Companion to Food in the Ancient World
Edited by John Wilkins and Robin Nadeau

A Companion to Ancient Education
Edited by W. Martin Bloomer

A Companion to Ancient Aesthetics
Edited by Pierre Destrée and Penelope Murray

A Companion to Roman Art
Edited by Barbara Borg

A Companion to Greek Literature
Edited by Martin Hose and David Schenker

A Companion to Josephus in His World
Edited by Honora Howell Chapman and Zuleika Rodgers

A Companion to Greek Architecture
Edited by Margaret M. Miles

A Companion to Plautus
Edited by Dorota Dutsch and George Fredric Franko

A Companion to Ancient Near Eastern Languages
Edited by Rebecca Hasselbach-Andee

A Companion to Ancient Greece and Rome on Screen
Edited by Arthur J. Pomeroy

A Companion to Euripedes
Edited by Laura K. McClure

A Companion to Ancient Near Eastern Art
Edited by Ann C. Gunter

A Companion to Ancient Epigram
Edited by Christer Henriksén

A Companion to Late Antique Literature
Edited by Scott McGill and Edward Watts

A Companion to Religion in Late Antiquity
Edited by Josef Lössl and Nicholas Baker-Brian

A COMPANION TO GREEK WARFARE

Edited by

Waldemar Heckel
University of Calgary
Calgary, Alberta

F.S. Naiden
University of North Carolina, Chapel Hill
Chapel Hill, North Carolina

E. Edward Garvin
University of Alberta
Edmonton, Alberta

John Vanderspoel
University of Calgary
Calgary, Alberta

WILEY Blackwell

This edition first published 2021

Registered Office
John Wiley & Sons, Inc., 111 River Street, Hoboken, NJ 07030, USA

Editorial Office
111 River Street, Hoboken, NJ 07030, USA

For details of our global editorial offices, customer services, and more information about Wiley products visit us at www.wiley.com.

Wiley also publishes its books in a variety of electronic formats and by print-on-demand. Some content that appears in standard print versions of this book may not be available in other formats.

Library of Congress Cataloging-in-Publication Data
Names: Heckel, Waldemar, 1949- author. | Garvin, E. Edward (Erin Edward), author. | Vanderspoel, John, 1954- author. | Naiden, F. S., author.
Title: A companion to Greek warfare / edited by Waldemar Heckeln, F. S. Naiden, E. Edward Garvin, John Vanderspoel
Description: Hoboken, NJ: John Wiley & Sons, 2021. | Series: Blackwell companions to the Ancient World | Includes bibliographical references and index. | Summary: "The first part of the book, a "Historical Survey," focuses on changing issues in strategy and grand strategy-on Greek motives, goals, and responses to military success and failure, the Macedonians, Persians, Romans, and Carthaginians being given due weight. These five chapters cover chronological fundamentals, but deal with modalities and patterns more than with campaigns, and with campaigns more than with battles. Johannes Heinrichs, in "Bronze Age and Early Greek Wars," sees Mycenaean warfare in terms of "the centralized organization of life," then deals with Homer as evidence of an ideological rather than sociological character. Among late Archaic Wars, Sparta's efforts to dominate the southern Peloponnesus illustrate a grand strategy that evolves from annexation to domination. The next chapter, Sabine Müller's "The Persian Wars to Alexander," gives a holistic account of the conflicts between the Achaemenids and both the Greeks and Macedonians, one that dismantles propaganda and rationalizations in order to demonstrate interdependence as well as rivalry. Frances Pownall's chapter on "Internal Wars from the 'First Peloponnesian War' to Chaeronea" covers much the same period but centers on alliance systems and spheres of influence; one after another, the Greeks states failed to implement strategies that would that would do more collective good than harm. She emphasizes Persia's role in this sequence of events"-- Provided by publisher.
Identifiers: LCCN 2020047069 (print) | LCCN 2020047070 (ebook) | ISBN 9781119438816 (hardback) | ISBN 9781119438830 (paperback) | ISBN 9781119438823 (pdf) | ISBN 9781119438854 (epub) | ISBN 9781119438847 (ebook)
Subjects: LCSH: Military art and science--Greece--History--To 500. | Greece--History, Military--To 146 B.C. | Military art and science--History--To 500. | Military history, Ancient.
Classification: LCC U33 .H43 2021 (print) | LCC U33 (ebook) | DDC 355.020938--dc23

LC record available at https://lccn.loc.gov/2020047069
LC ebook record available at https://lccn.loc.gov/2020047070

Cover image: © Photo Vault/Alamy Stock Photo
Cover design by Wiley

Set in 10/12.5, GalliardStd by Integra Software Services, Pondicherry, India.

10 9 8 7 6 5 4 3 2 1

Contents

Notes on Contributors x
Abbreviations and Short Titles xvi
List of Maps and Figures xvii

Introduction 1
Waldemar Heckel and F.S. Naiden

Part I Historical Survey **9**

1. Bronze Age and Early Greek Wars 11
Johannes Heinrichs

2. The Persian Wars to Alexander 21
Sabine Müller

3. Internal Wars from the "First Peloponnesian War" to Chaeronea 31
Frances Pownall

4. Hellenistic Land Warfare 42
Edward M. Anson

5. Greek Warfare in Sicily 58
Melanie Jonasch

Part II Military Operations **73**

6. The Nature of Hoplite Warfare 75
Fernando Echeverría

7. Cavalry Battle in Greece and the Hellenistic East 88
Carolyn Willekes88

8. Siege Warfare 99
David Whitehead

Part III Military Personnel **117**

9. The Organization of Greek Armies 119
F.S. Naiden

10. Generalship 137
Joseph Roisman

11. Greek Camps and Camp Followers 148
Mauricio G. Álvarez

12. Greeks in Foreign Service: The Case of the Achaemenid Empire 160
Jeffrey Rop

Part IV War with Non-Greeks **171**

13. The Royal Elite of the Achaemenid Army 173
Michael B. Charles

14. Parthian Warfare Under the Early Arsacids 185
Marek Jan Olbrycht

15. Elephants in Hellenistic Warfare 202
Christopher Epplett

16. Thracian Warfare 214
Emil Nankov

Part V Technical and Economic Context of Greek Warfare **225**

17. Logistics 227
Stephen O'Connor

18. Fortifications 241
Silke Müth

19. Military Intelligence 252
Frank Russell

20. The Economics of War 261
Johannes Heinrichs

21. War and Slavery in the Greek World 271
Peter Hunt

22. Agriculture and Greek Warfare 286
Jeanne Reames and Ann Haverkost

Part VI Social and Political Context of Greek Warfare **299**

23. Battle Trauma in Ancient Greece 301
Lawrence Tritle

24. Religion and Warfare 312
F.S. Naiden

25. Women and War in the Greek World 329
Elizabeth D. Carney

26. Piracy, Brigandage, and Terrorism 339
Sheila L. Ager

27. Civil War in Greece: Forgetting as the Path to Reconciliation 358
Gordon Shrimpton

Part VII Warfare in Art and Literature **367**

28. The Iconography of War 369
Olga Palagia

29. War Monuments and Memorials 384
Lawrence A. Tritle

30. War in Greek Poetry 394
J. Vela-Tejada

31. War and Propaganda 406
Sabine Müller

Bibliography 416
Index 464

Notes on Contributors

Sheila L. Ager is Professor of Classical Studies and currently Dean of Arts at the University of Waterloo, Waterloo, Canada. She is the author of *Interstate Arbitrations in the Greek World, 337–90 BC* (1996), co-editor (with Riemer Faber) of *Belonging and Isolation in the Hellenistic World* (2013), and editor of *A Cultural History of Peace in Antiquity* (2020). Much of her research has focused on international relations in antiquity, in particular peaceful conflict resolution. She is also the author of several papers on Hellenistic monarchy, especially the Hellenistic queens of the Ptolemaic and Seleukid house.

Mauricio G. Álvarez holds a Ph.D. in Ancient History from the Autónoma University, Madrid, and an M.A. in Mediterranean Studies from King´s College London, and he is a former DAAD grant holder. Dr. Álvarez is member of the Research Group "History, Archeology and Documentation" in La Coruña University (Spain) and of the Spanish Association for Military History, and in recent years has served as a professor in Nebrija University, Madrid. His principal fields of interest are Greek and Roman military camps, ancient Greek and Roman armies, Alexander the Great, the culture and history of the Mediterranean basin, and teaching methods for Geography and History.

Edward M. Anson is Professor of Ancient History at the University of Arkansas, Little Rock. His publications include numerous articles and book chapters, and he has authored or edited *Philip II, the Father of Alexander the Great: Themes and Issues* (2020); *Affective Relations and Personal Bonds in Hellenistic Antiquity: Studies in Honor of Elizabeth D. Carney*, with Monica D'Agostini and Frances Pownall (2020); *Alexander's Heirs: The Age of the Successors 323–281 BC* (2014); *Alexander the Great: Themes and Issues* (2013); *After Alexander: The Age of the Diadochi (323–281 BC)* with Victor Alonso Troncoso (2013). He is an associate editor of the *Ancient History Bulletin*, an Assessor for Classics for the Australian Research Council, and a fellow of the University of Waterloo's Institute for Hellenistic Studies.

Elizabeth D. Carney is Professor of History and Carol K. Brown Scholar in the Humanities, Emerita, at Clemson University. Her focus has been on Macedonian and Hellenistic monarchy and the role of royal women in monarchy. She has written *Women and Monarchy in Ancient Macedonia* (2000), *Olympias, Mother of Alexander the Great.* (2006), *Arsinoë of Egypt and Macedon: A Royal Life* (2013), and *Eurydice and the Birth of Macedonian Power* (2019). Some of her articles dealing with monarchy, with new afterwords, are collected in *King and Court in Ancient Macedonia: Rivalry, Treason and Conspiracy* (2015). She and Sabine Müller have co-edited *The Routledge Companion to Women and Monarchy in the Ancient Mediterranean World* (2020).

Michael B. Charles, Associate Professor, is a member of the School of Business, Law and Arts at Southern Cross University, Gold Coast, Australia. Michael has published over 100 refereed journal articles in a variety of subject areas, together with numerous book chapters and refereed conference papers. Michael's Ph.D. (University of Queensland) was on the dating of the late antique military writer Vegetius, but he has moved on to other aspects of ancient history, especially the use of elephants in antiquity, the political role of eunuchs, and Persian military studies. As part of his "day job," Michael also publishes in the area of public policy, with a particular interest in science and technology policy, together with transport and infrastructure studies. Michael also has a Master of International Business Studies from the Queensland University of Technology and teaches in the area of infrastructure management. He currently resides on the Gold Coast and, when he has free time, enjoys gardening/landscaping, antique furniture restoration, reading about popular culture, learning about wine (and, of course, drinking it), and listening to a variety of music.

Fernando Echeverría is lecturer in Ancient History at the Complutense University in Madrid, Spain. His research focuses on the social and political implications of war and military transformations in Archaic and Classical Greece, combining literary and iconographical approaches. He is the author of *Ciudadanos, Campesinos y Soldados. El Nacimiento de la Polis Griega y la Teoría de la Revolución Hoplita* (2008), and currently conducts a project on the nature and dynamics of Greek siege warfare.

Christopher Epplett teaches ancient history in the Department of History at the University of Lethbridge, Alberta, Canada. His research interests include Hellenistic history, the armies of Greece and Rome, and ancient sport and spectacle.

Ann Haverkost is an archaeologist and graduate student in the History Department at the University of Nebraska at Omaha. She is currently researching cultural and political reflections in classical Macedonian funerary imagery. In addition, she is a supervisor with Bethsaida Archaeological Excavations in northern Israel. Her research interests include religion and cultural exchange between the Mediterranean and Ancient Near East during the Iron Age through the Classical periods.

Waldemar Heckel is Professor Emeritus of Greek and Roman Studies and a Research Fellow of the Centre for Military and Strategic Studies. He taught at the University of Calgary from 1977 until his retirement at the end of 2013. He is author of numerous books, including (most recently) *Alexander's Marshals. A Study of the Makedonian Aristocracy and the Politics of Military Leadership* (London, 2016), *In the Path of Conquest. Resistance to Alexander the Great* (Oxford 2000), and *Who's Who in the Age of Alexander and his Successors. From Chaironeia to Ipsos, 338–301 BCE* (London 2021).

Johannes Heinrichs is Professor in Ancient History at the University of Cologne (Germany), currently a researcher in the project *Corpus Inscriptionum Iudaeae Palaestinae*, and one of the editors of the *Lexicon of Argead Makedonia*. His focus is on Achaemenid, Greek, and Latin epigraphy, Persian, Greek, and Roman history and numismatics, Roman imperial prosopography (*PIR*, vols. VIII.1 and 2), and Celtic and Germanic ethnic groups in Western Europe. His recent papers include discussions of a new bronze tablet from the Lykaion in archaic Arkadia, and legal aspects of the Hefzibah inscription that contains letters by the Seleukid king, Antiochos III.

Peter Hunt, Professor of Classics and (courtesy) History at the University of Colorado Boulder, studies warfare and society, slavery, historiography, and oratory with a focus on Classical Greece. His first book, *Slaves, Warfare and Ideology in the Greek Historians* (1998), discerns a conflict between the extent of slave and Helot participation in Greek warfare and the representation of their role in contemporary historians. His second book, *War, Peace, and Alliance in Demosthenes' Athens* (2010), uses the evidence of deliberative oratory as evidence for Athenian thinking and feelings about foreign relations. His third book is a survey on *Ancient Greek and Roman Slavery* (2018). In addition to articles and reviews, he has contributed chapters to *The Oxford Handbook of Demosthenes, Brill's Companion to Thucydides, The Oxford Handbook of Thucydides, The Cambridge History of Greek and Roman Warfare, The Cambridge World History of Slavery, and The Cambridge History of the World.* He is now starting a major project on Plutarch's *Phocion*.

Melanie Jonasch is with the German Archaeological Institute. Her dissertation dealt with *Les pierres du bon buveur. Eine regionale Konvention unter den Grabstelen des römerzeitlichen Gallien* (2011). She is author of *The Fight for Greek Sicily: Society, Politics and Landscape* (2020).

Sabine Müller is Professor of Ancient History at Marburg University. She studied Medieval and Modern History, Art History, and Ancient History. Her research focuses on the Persian Empire, Argead Macedonia, the Hellenistic Empires, Macedonian royal women, Lukian of Samosata, and reception studies. Her publications include monographs on Alexander the Great, Ptolemy II and Arsinoë II, and Perdikkas II. She is co-editor of the *Lexicon of Argead Makedonia* (2020).

Silke Müth is a Classical archaeologist specializing in topography and architecture, particularly in Greek cities and fortifications. She coordinated a research project on the city wall of Messene (Peloponnese) from 2004 to 2008, co-directed the international network "Focus on Fortification. Ancient Fortifications in the Eastern Mediterranean" from 2008 to 2016, and held a fellowship of the German Research Association (DFG) on Symbolic Functions of Greek and Roman Fortifications at the Athens branch of the German Archaeological Institute and the Humboldt University of Berlin from 2012 to 2016. Currently she is a Senior Researcher at the National Museum of Denmark, co-directing a research project on the Archaic-Classical city of Sikyon (Peloponnese).

F.S. Naiden, professor of Greek History at the University of North Carolina at Chapel Hill, studies ancient Greek law, religion, and warfare, including Near Eastern parallels, especially among the Western Semites. Chief periods of interest are the Classical and Hellenistic. Recently completed is a study of Alexander the Great, his officers, and the role of religion in Macedonian conquests, *Soldier, Priest, and God*, combining his interests in warfare, religion, and the Near East. Now underway is a study of war councils and command and control throughout antiquity and into the early modern period. The languages used for his research are Greek, Latin, Hebrew, and Akkadian

Emil Nankov is Assistant Professor at the National Archaeological Institute with Museum at the Bulgarian Academy of Sciences. He received his Ph.D. in Art History and Archaeology at Cornell University (2009). His main scholarly interests focus on urbanization, warfare, and ancient economy of ancient Thrace. He is the editor of *Sandanski and Its Territory During Prehistory, Antiquity and Middle Ages: Current Trends in Archaeological Research: Proceedings of an International Conference at Sandanski, September 17–20, 2015. Papers of the American Research Center in Sofia*, vol. 3 (Veliko Tarnovo). He is also a co-editor of *Heraclea Sintica: from Hellenistic polis to Roman civitas (4th c. BC–6th c. AD). Proceedings of a Conference at Petrich, Bulgaria, September 19–21, 2013,* published in PARCS vol. 2 (2015*),* and *A Companion to Ancient Thrace* (2015). Currently he is a Deputy Director of the archaeological excavations at Emporion Pistiros.

Stephen O'Connor is Associate Professor of History at California State University, Fullerton. His research focuses on the economics and logistics of Classical Greek warfare. He has published articles on the diet and pay of Greek sailors and soldiers and, especially, the markets they participated in while on campaign. He is currently preparing a monograph entitled *Feeding Maritime Empire: Grain Markets and Athenian Imperial Domination in the 5th Century BCE.*

Marek Jan Olbrycht is Professor of History and Ancient Oriental Studies at Rzeszów University, Poland, Member of the Oriental Commission of the Polish Academy of Sciences, Kraków, and Member of the Polish Archaeological Institute in Athens, Greece. Formerly: Member at the School of Historical Studies, Institute for Advanced Study, Princeton and Humboldt Visiting Professor in Germany, University of Münster. His focus has been the history of ancient Iran and Central Asia, Alexander the Great, and ancient warfare. He published more than 100 articles and several books, including *Parthia et Ulteriores gentes* (1998) and *Alexander the Great and the Iranian World* (2004). In 2010, he established the journal *Anabasis. Studia Classica et Orientalia.*

Olga Palagia is Professor of Classical Archaeology, Emerita, National and Kapodistrian University of Athens. She has published widely on Greek sculpture and Macedonian painting and has edited several collective volumes. Recent edited books include *Sailing to Classical Greece in Honour of Petros Themelis* (2011) (co-edited with H.R. Goette), *Naupaktos* (2016), *From Hippias to Kallias: Greek Art in Athens and Beyond 527–449 BC* (2019) (co-edited with E.P. Sioumpara), and *De Gruyter Handbook of Greek Sculpture* (2019).

Frances Pownall is Professor of Classics at the University of Alberta. She has published widely on Greek historiography (particularly the fourth century and the Hellenistic period), the source tradition on Macedonia and the Successors, and the historiographical tradition of Sicily and the Greek West. She has written updated editions, translations, and historical commentaries for a number of important fragmentary historians (including Hecataeus, Hellanicus, Duris, Aristobulus, Eratosthenes, and Philistus) for *Brill's New Jacoby*. She is the author of *Lessons From the Past. The Moral Use of History in Fourth-Century Prose* (2004), *Ancient Macedonians in the Greek and Roman Sources* (co-edited with T. Howe, 2018), *Lexicon of Argead Macedonia* (co-edited with W. Heckel, J. Heinrichs, and S. Müller, 2020), and *Affective Relations & Personal Bonds in Hellenistic Antiquity* (co-edited with E.M. Anson and M. D'Agostini, 2020).

Jeanne Reames is Director of the Ancient Mediterranean Studies Program at the University of Nebraska-Omaha, where she also served as the History Graduate Program Chair until recently. She specializes in Argead Macedonia up through the court of Alexander the Great, including areas of social history (both Macedonian and Greek): religion, ethnicity, military, and sexuality. In addition, she has a long-time interest in Alexander's portrayal in modern pop culture.

Her Ph.D. is from the Pennsylvania State University, 1998.

Joseph Roisman is a Professor Emeritus of Classics at Colby College. His main publications deal with Greek political, social, and military history and the Attic orators. They include *Lycurgus, Against Leocrates. Introduction and Commentary by Joseph Roisman. Translation by Michael Edward* (2019), *The Classical Art of Command: Eight Greek Generals Who Changed the History of Warfare* (2017), *Alexander's Veterans and the Early Wars of the Successors* (2012), and *The General Demosthenes and His Use of Military Surprise* (1993). He edited *Ancient Greece from Homer to Alexander: The Ancient Evidence.* Translation by John Yardley (2011), and co-edited with Ian Worthington *A Companion to Ancient Macedonia* (2010).

Jeffrey Rop is Associate Professor of History at the University of Minnesota, Duluth. His publications include *Greek Military Service in the Ancient Near East, 401–330 BCE* (2019), and articles on the military and political history and historiography of ancient Greece and Persia.

Frank Russell received his B.A. in Classics from Loyola Marymount, and his M.A. and Ph.D. in Classics from UCLA. His areas of research are Greek and Roman intelligence, frontier zones, and counterinsurgency. He teaches at Transylvania University, where he is Professor of History.

Gordon Shrimpton is Professor Emeritus at the University of Victoria, British Columbia. His books include *Theopompus the Historian* (1991) and *History and Memory in Ancient Greece* (1997). The chapter in this collection explores the remarkable way one city, Athens, was able to suppress its bitter memory of civil conflict.

J. Vela-Tejada is Professor of Classics at University of Zaragoza (Spain). He has written extensively on Greek warfare and military literature from his dissertation on the language of Aeneas Tacticus (1991). He is author of articles on "Warfare, History and Literature in the Archaic and Classical Periods," *Historia* 53, 2004, and "Creating Koiné: Aineias Tacticus in the History of the Greek Language" (in M. Pretzler and N. Barley (eds.), *Brill's Companion to Aineias Tacticus*, 2017). His research area also covers the scope of Xenophon (*Post H.R. Breitenbach: tres décadas de estudios sobre Jenofonte (1967–1997)*, Zaragoza, 1998; he is co-author of *Xenophontis operum Concordantiae, vols. I: Hellenica, II: Anabasis*, 2002; *III: Cyrupaedia*, 2003; *V: Opuscula*, 2008), and Plutarch ("Atticism in Plutarch: a μίμησις τῶν ἀρχαίων or diglossia?," *Euphrosyne* 47, 2019).

Lawrence Tritle is Professor Emeritus of History at Loyola Marymount University, Los Angeles, California (1978–2017); in 2012 he was selected as Daum Research Professor of History and in 2014 received the Rains Award for Distinguished Research and Scholarship. An ancient historian by training (Ph.D., University of Chicago, 1978), his publications include *From Melos to My Lai. War and Survival* (2000), *A New History of the Peloponnesian War* (2010), *The Oxford Handbook of Warfare in the Classical World* (edited with B. Campbell, 2013), and most recently *The Many Faces of War* (edited with J.W. Warren, 2018). Tritle's research focuses on comparative war and violence. His investigations center on the experience of war, investigating war's impact on the individuals who fight and the wider consequences of violence on culture and society. A combat veteran of the Vietnam War, Tritle supported South Vietnamese forces in ground operations. His decorations include the Combat Infantrymen's Badge, Bronze Star, and Vietnamese Cross of Gallantry.

David Whitehead, a one-time pupil of M.I. Finley, is Emeritus Professor of Ancient History at Queen's University, Belfast, and an elected Member of the Royal Irish Academy. After monographs on the metics

(1977) and the demes (1986) of Athens, his research output has focused on two main fields: Athenian courtroom oratory—commentaries on Hyperides (2000) and Isocrates (forthcoming); and, as exemplified in the present volume, Greek military history and military writers, with particular reference to siege warfare.

Carolyn Willekes is a Senior Lecturer in the Department of General Education at Mount Royal University. She received her Ph.D. in Greek and Roman Studies from the University of Calgary. Her research focuses primarily on the horse–human relationship in the ancient world, with a particular interest in the use of horses in warfare. She is author of *The Horse in the Ancient World: From Bucephalus to the Hippodrome* (2016) and *Greek Warriors: Hoplites and Heroes* (2017), as well as several chapters and articles, including "A Tale of Two Games: Cirit, Buzkashi, and the Horsemen of the Ancient Asiatic Steppe" (*Nomadic Peoples*, 2017) and "Breeding Success: The Creation of the Racehorse in Antiquity" (*Mouseion*, 2019).

Abbreviations and Short Titles

With a few exceptions, journal titles in the bibliography follow the standard abbreviations specified in *L'Année Philologique*. Abbreviations in the notes follow the *Oxford Classical Dictionary*, ed. S. Hornblower, A. Spawforth, and E. Eidinow. 4th ed., Oxford, 2012.

AOAT = *Alter Orient und Altes Testament.*

Barrington Atlas = *Barrington Atlas of the Greek and Roman World*, ed. R.J.A. Talbert, Princeton, and Oxford, 2000.

CAH = *The Cambridge Ancient History*, multiple editors, volumes and editions, Cambridge, 1970–2005.

JÖAI = *Jahrshefte des Osterreichischen Archaologischen Instituts.*

ML = Russell Meiggs and David Lewis. 1969. *A Selection of Greek Historical Inscriptions to the end of the Fifth Century BC*, Oxford; 2nd revised ed. Oxford 1988.

RE = *Real-Encyclopädie der classischen Altertumswissenschaft.*

SEG = *Supplementum Epigraphicum Graecum.*

List of Maps and Figures

Maps

Map 1 Greece and Anatolia xviii
Map 2 Sicily and Magna Graecia xix
Map 3 The Hellenistic East xx

Figures

14.1 An Arsacid Gorytos 188
14.2 The Prince of Shami 189
14.3 Antiochus I of Commagene 191
14.4 Parthian Leg Armor 193
14.5 The Defenses of Merv 200
28.1 Hoplites on the Chigi Vase 371
28.2 Block from the South Frieze of the Temple of Athena Nike 372
28.3 Block from the South Frieze of the Temple of Athena Nike 373
28.4 The Alexander Mosaic 375
28.5 Alexander as He Appears in the Mosaic 375
28.6 Alexander on the Attack in the Mosaic 376
28.7 Darius as He Appears in the Mosaic 377
28.8 Darius' Horses in the Mosaic 378
28.9 The Pursuit of Darius on an Apulian Vase 379
28.10 A Cavalry Attack According to a Funeral Relief 381
28.11 The Battle of Pydna According to a Relief on a Statue Base 383
29.1 The Funeral Relief in Honor of Dexileus 393

Table

12.1 Greek Military Service in the Persian Empire 161

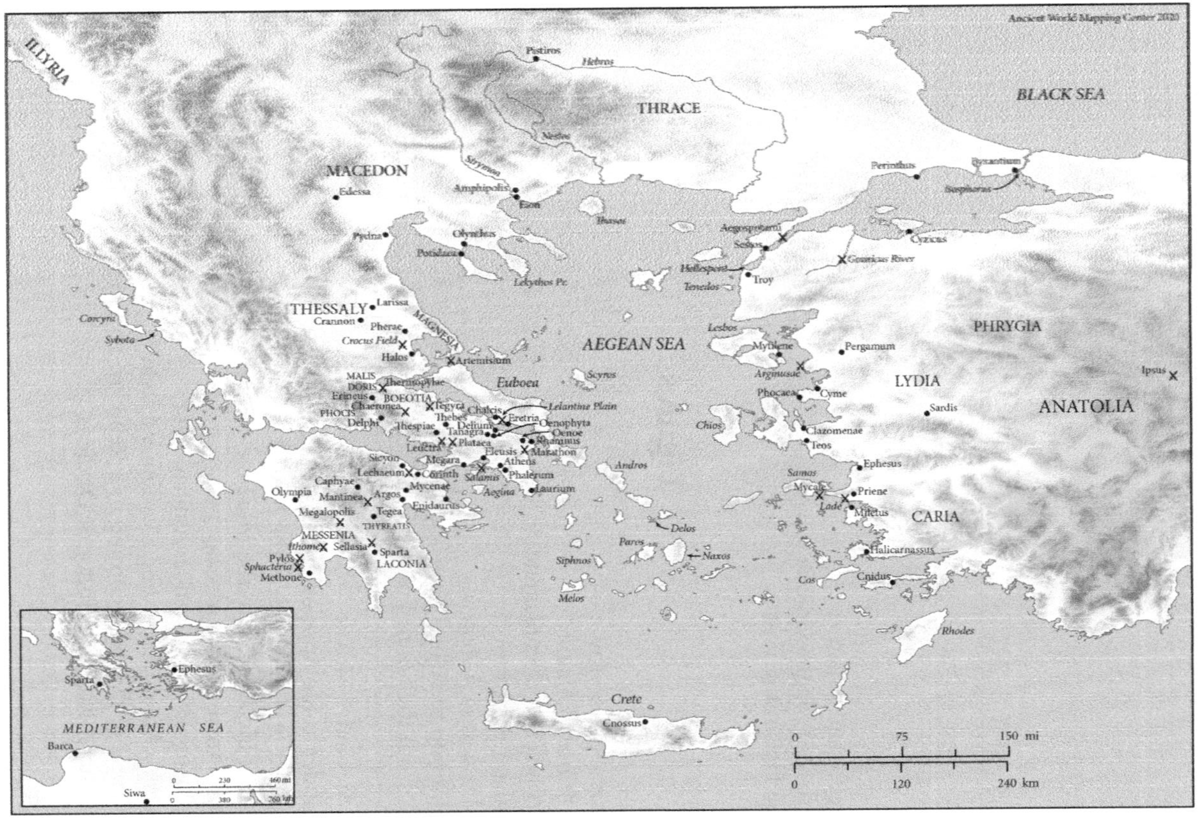

Map 1 Greece and Anatolia.

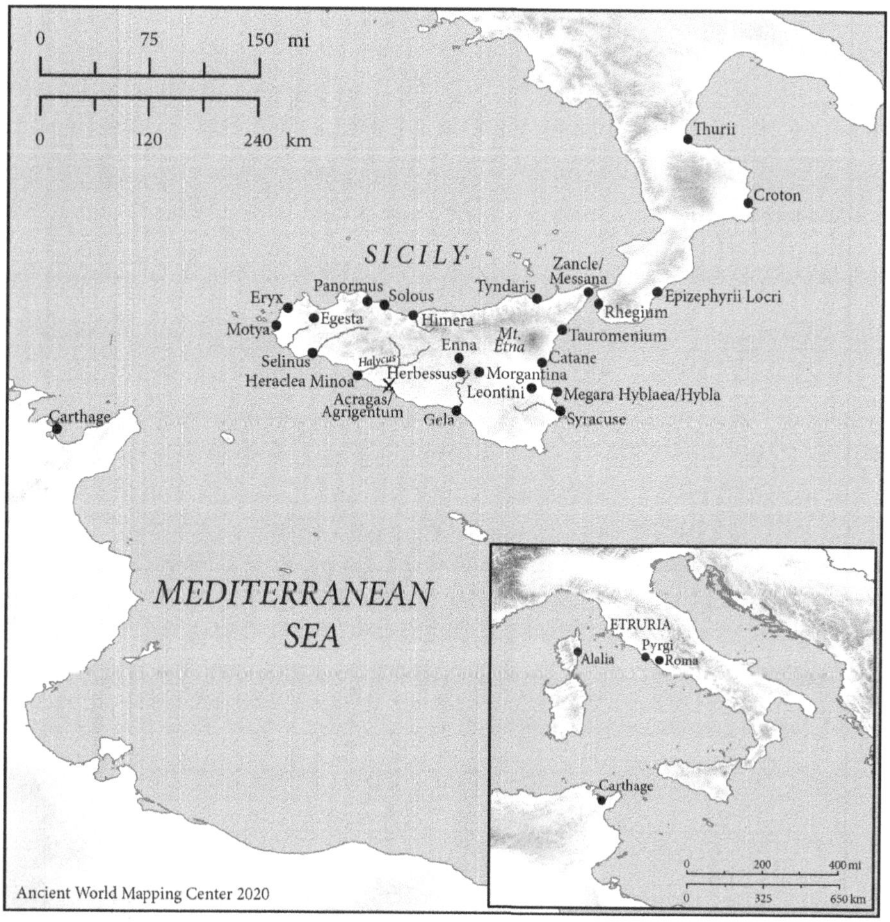

Map 2 Sicily and Magna Graecia.

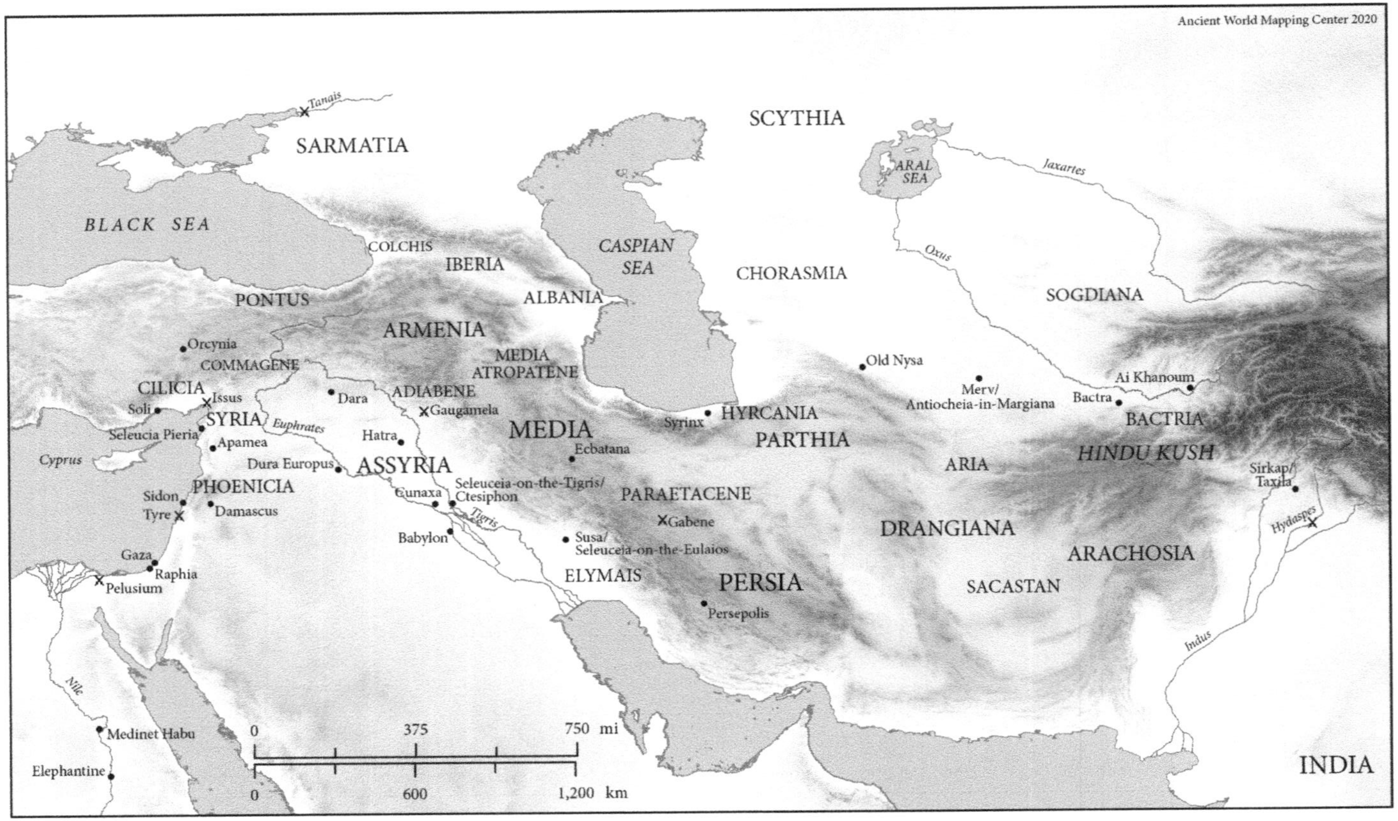

Map 3 The Hellenistic East.

Introduction

Waldemar Heckel and F.S. Naiden

Although the study of ancient Greek military history has flourished in the last few decades, no handbook or survey exclusively about this subject has appeared. *The Wiley Companion to Greek Warfare* fills this intellectual and pedagogical gap. It considers how Greek states prepared themselves, materially, psychologically, and politically, for confrontations with their adversaries; how the techniques used to overcome the enemy on the battlefield—whether on land or sea—were honed through training, tactical innovation, and technological advances; how the logistical demands of maintaining an army on the march or in distant lands were met, and how their warfare was justified in the face of moral and religious prohibitions against murder and other types of violence.

Even though an army moved on its stomach, it was motivated to move by the heart, and directed by the mind. Propaganda was used for justification and condemnation; the noble and manly virtues of the good soldier contrasted with the alleged cowardice and effeminacy, or the inhumanity and brutality, of the enemy, who was on some occasions to be feared and on others to be despised. For Greek imperialists, degraded enemies were easily confected, and invented atrocities promulgated. For those who faced aggression, the defense of family, property, and shrines was sufficient motivation; but even then the accumulated warrior traditions of the past, such as those that inspired the Thebans in 335 (Diod. Sic. 17.11.5), were no less essential than mobilizing troops or digging trenches and strengthening walls.

Equally important were the skills of the general, scout, spy, mechanic, smith, diviner, architect, teamster, and horseman. These and other *technai* let the ancient Greeks invent or make better artillery, improve ship design, create some genres of military literature, advance the use of combined arms, and make strategy, previously a royal and aristocratic monopoly, a matter of public debate. In all these regards, the Greeks were learning from, and competing with, their Near Eastern and European neighbors from Thrace to Persia. Notions that Greek

A Companion to Greek Warfare, First Edition. Edited by Waldemar Heckel, F.S. Naiden,
E. Edward Garvin, and John Vanderspoel.

warfare was unique or seminal, or even that it was superior to the warfare of neighboring peoples, are all exaggerated, the truth being that the Greeks and the Macedonians were distinctive and influential, but superior only for a brief period, one not much longer than the period of German superiority in modern Europe, and much shorter than the period of Roman superiority that would begin in the third century BCE.[1] The Romans came to regard the Greeks as unmilitary, if not effeminate.[2] The Pharaohs who hired Greek mercenaries thought differently.

Waging war was not an isolated event (or even a series of events), but a process that touched upon virtually all aspects of Greek life. And, even though the Greeks did not worship Mars (Ares) as the Romans did—indeed Ares is depicted in a most unsympathetic way by Homer (*Il.* 5.888–898)—war dominated much of their lives or at least hovered on the horizon. Alexander the Great was paradigmatic. From birth, he had been raised for war, as both a combatant and a tactician, a leader and a comrade—a soldier foremost in both the formulation and execution of orders. His first teachers imbued in him a passion for the *aristeia* of the Homeric hero and the strict discipline that had made Sparta the living guidebook of military virtue, and he went on to read not only the *Iliad* and the works of the Athenian tragedians but also Xenophon and, most likely, Herodotus. The written word and its interpretation by some of the Greek world's greatest minds provided the underpinning of his warcraft, while the veteran generals of Macedon were there to advise, their actions clearly visible; for the prince had been thrust at an early age into the thick of battle.

Although arguably the greatest warrior of antiquity, Alexander was not unique. Every man's fame, as the man from Seriphus reminded Themistocles, depended on the prestige of his homeland, and that, in turn, depended on public and especially military service. Even those leaders who were orators rather than generals celebrated military heroes, many of them ancestors, others ordinary citizens. What these men did, and how they acted in what must be one of life's greatest trials, reflected the many arts (and the mundane demands) of war described in this volume, one that probes beyond the familiar if often dubious record found in the works of the historians and orators or in contemporary public documents.

Unlike two other recent books, *The Cambridge History of Greek and Roman Warfare* and *The Oxford Handbook of Warfare in the Classical World*, this volume does not subordinate Greece to Rome, or the entire length of Greek history to whatever part of it is "Classical." So much is clear from the titles of these books, but this one also differs from the other two by covering more topics in a larger number of chapters.[3] The editors have sought to include all major topics in the field of Greek and also Macedonian military history, including not just armies and their operations but also the economic, social, psychological, and cultural context of warfare. We have added neglected operational topics such as strategy and the role of Greeks in foreign service, and four chapters on the armies and methods of the Greeks' and the Macedonians' opponents, but we have also added new contextual topics such as religion, civil

---

[1] For the Greeks as unique and seminal, see Chapter 4 in this volume. A critique of sundry views that Greece is the point of origin for later historical phenomena: Goody 2008, 26–67.

[2] E.g. Liv. 29.1.1–11, 36.11.3–4. A contrast with Romans: 34.3.3–6. The next step in the argument: service in the Greek East renders Romans "lax," as in Wheeler 1996.

[3] Thirty-one chapters vs. 25 in the *Oxford Handbook* edited by B. Campbell and L. Tritle, and vs. 9 and parts of 6 others in the *Cambridge History* edited by P. Sabin, H. van Wees, and M. Whitby. In addition, chs. 2–5 of the former, and 1–3 of the latter, cover ancient military history in general, rather than Greece alone. Both devote more space to Rome than to Greece, granting that the *Cambridge History* does so in the course of two volumes.

war, iconography, propaganda, and the roles of slaves and women. Basics such as generalship and siege warfare receive whole chapters. Compared to other works, this book has more to say about *realia*, less about discourse, and in the latter case, it pays more attention to propaganda and monuments and less to general features of Greek culture.

This book also draws on more diverse contributors, totaling 28 from 11 countries. The majority are not from the United States and the United Kingdom; 11 are not from English-speaking countries. Some are authorities who have written for other military handbooks; others are young scholars working from monographs in neglected fields; most are ancient historians, but some are specialists such as archaeologists, or historians of religion. All in all, these scholars have less investment in *idées reçues* concerning Greek hoplites, Greek models, and Greek distinctiveness.

These differences reflect a bigger one, concerning method. The chapters in this book center on some type of evidence, or some period that written sources establish as distinctive, or some locale that archaeological and artistic sources establish as distinctive. Approaches to this body of evidence vary; they may be biographical, operational, institutional, social, economic, or cultural. Operations do not always predominate, but always matter: why and how wars are won or lost, and made or avoided, is crucial for some topics, important for others, and relevant to all. The prelude and sequel to wars receives less attention; any homology between war and society receives less attention than differences between the two.

Both the Cambridge and Oxford volumes center largely on themes derived from social scientists, especially Weber, Durkheim, and Marx—although the last of these is manifest indirectly, via the *Annales* school of history. Sociology, in a word, outweighs political history. Operations reflect the influences of technological determinism or cultural hegemony. Why and how wars are won and lost thus becomes a misleading question. For combatants and civilians, all wars are lost. War itself is the only victor, and an undeserving one at that.[4] These two books are not pacifist any more than this book is militarist, but an ethical as well as intellectual gap divides the two.

Never acknowledged, but nonetheless felt, is an assumption on the part of the Cambridge authors that Greek warfare is a distinctly Greek, not ancient, phenomenon; in other words, that it owes little or nothing to either neighboring societies or earlier ones, even if it may serve as the origin for later, Euro-American warfare. In this regard, the volume is better termed Finleyite than Marxist, since it adheres to a history of warfare that runs parallel to Finley's notion of ancient Greece as the first slave society.[5]

At the same time, the book does not attempt to supplant the best work done in the Cambridge and Oxford volumes, such as the essays on navies and land battle in the former and the treatment of military medicine and mercenaries in the latter, all of which topics are accordingly omitted. In two cases, intelligence gathering and mental trauma, Frank Russell

---

[4] See Wheeler 2011; for a similar view, but restricted to the Archaic Period, see Chapter 6 in this volume. Like other disputes in Classics, this one has roots in the ancient Greek sources, such as the account given by Herodotus 1.1–4 of the origin of the Persian Wars, and the account given by Thucydides of the origin of the Peloponnesian War, esp. 1.4–5.

[5] For Greece as the first "slave society," as opposed to ancient societies with an economically secondary slave populations, see Finley 1980, ch. 3. Perhaps the best known of the scholars agreeing with Finley: Patterson 1982, App. 3, omitting slavery in the ancient Near East, including the Old Testament. Finley himself may be described as an ex-Marxist, or, as Ernst Badian once said to F. S. Naiden, "an anti-anti-Marxist." See both Naiden 2014 and Shaw 2014, along with Naiden 2017b. The Stalinist view of Greek slavery assimilated it to diverse Near Eastern systems of servitude.

and Lawrence Tritle, who also contributed to (and, in the latter case, co-edited one of) the earlier volumes, provide updated essays focused on certain aspects of their subjects.

This book suffers from several of the same gaps as its predecessors. It does not describe Greek federal armies, and it leaves aside important Near Eastern forces against whom the Greeks and Macedonians did some fighting, including the forces of Carthage, Lydia, and Assyria.

All manuals of ancient warfare derive from the dated but valuable reference works written by German, often Prussian, scholars in the late nineteenth and early twentieth centuries.[6] In the *Cambridge History*, the German legacy influences chapters on "Military Forces," "War," and "Battle," whereas interest in the context of warfare accounts for "Warfare and the State" and "War and Society," the switch from the one to the other being the occasion for introductory chapters on methods and sources. In the *Oxford Handbook*, similar introductory chapters come first, but a social and psychological concern, the common soldier's experience of combat, supplies the title of the next part, "The Face of Battle," as in John Keegan's 1976 book, whereas the other large part, "Impacts and Techniques" reconfigures traditional matter such as operations. This book again follows a different course, for it avoids generalizations about sources and methods, issues left to the authors of the chapters. It also avoids long, broad chapters such as "Warfare and the State," which lead to some subjects being raised briefly in one chapter, and then briefly in another, yet leads to other topics being omitted. As for the context of warfare, this book avoids favoring any one perspective, and with it, any one author—in the case of "The Face of Battle," Keegan's work on infantry combat of the same name. Operational matters still come first, but each topic receives its due, repetition and omission both being avoided. The context of warfare is divided in two parts, on the one hand, the economic and technical context, and, on the other hand, the social and political, a third part being reserved for warfare in art and literature, in other words, for the ancient cultural context, as opposed to any contemporary, twenty-first century intellectual context into which ancient literary and artistic sources should fit.

Part I, a "Historical Survey," focuses on changing issues in strategy and grand strategy—on Greek motives, goals, and responses to military success and failure, the Macedonians, Persians, Romans, and Carthaginians being given due weight. These five chapters cover chronological fundamentals, but deal with modalities and patterns more than with campaigns, and with campaigns more than with battles. Johannes Heinrichs, in Chapter 1 "Bronze Age and Early Greek Wars," sees Mycenaean warfare in terms of "the centralized organization of life," then deals with Homer as evidence of an ideological rather than sociological character. Among late Archaic Wars, Sparta's efforts to dominate the southern Peloponnesus illustrate a grand strategy that evolves from annexation to domination. Chapter 2, Sabine Müller's "The Persian Wars to Alexander," gives a holistic account of the conflicts between the Achaemenids and both the Greeks and Macedonians, one that dismantles propaganda and rationalizations in order to demonstrate interdependence as well as rivalry. Frances Pownall's Chapter 3 on "Internal Wars from the 'First Peloponnesian War' to Chaeronea" covers much the same period but centers on alliance systems and spheres of influence; one after another, the Greeks states failed to implement strategies that would that

[6] Rüstow and Köchly 1852; Bauer 1893; Delbrück 1900; Kromayer and Veith 1928. Pritchett 1971–1991, at vols. 1 and 2 (1971, 1974), is wider ranging, but unsystematic; it does mark the first attempt to include topics outside the ambit of military equipment, operations, and organization.

would do more collective good than harm. She emphasizes Persia's role in this sequence of events.

Edward Anson, in Chapter 4 "Hellenistic Land Warfare," shows how innovations introduced by Philip II of Macedon improved and professionalized armies, and thus enlarged the scale of Greek politics and warfare. Some changes occurred in weaponry, some in tactics, and some in the composition of armies. In this instance, military changes affected politics more than politics affected the military. Yet Melanie Jonasch, in Chapter 5 "Greek Warfare in Sicily," this part's last chapter, points out a major regional exception, in Sicily, where problems of colonization and rivalry with Carthage led to some technical innovations that Philip would adopt, but also to uniquely Sicilian patterns of warfare.

Part II, "Military Operations," deals with the fundamentals of infantry, cavalry, and sieges, but not in uniform fashion. Heavy infantry fighting has received too much attention for too long, and has been used, or misused, to explain too much, from Greek constitutions to Greek farming, and so Fernando Echeverría Rey, in Chapter 6 "The Nature of Hoplite Warfare," sets out to correct these emphases. This chapter is polemical, whereas Chapter 7, Carolyn Willekes' "Cavalry Battle in Greece and the Hellenistic East," builds on recent useful work about cavalry by analyzing the use of horses in combat. She seeks to recapture the experience of the ancient rider in handling his mount, whereas Echeverría Rey seeks to demolish the speculations of scholars who, unlike some German writers of the nineteenth century, did not have the benefit as well as the drawback of being infantry officers. David Whitehead's topic, "Siege Warfare," in Chapter 8 presents an altogether different challenge, that of blending Greek historiography with the long, complex development of ancient literature on the subject of *poliorketika*. He thus begins and ends with the most important siege writer, Philo of Byzantium, but after evaluating evidence from Herodotus and others. Here, as with the innovations of Philip II, changes in materiel did not by themselves confer decisive advantages.

Part III, "Military Personnel," identifies a topic treated in handbooks going back to the nineteenth century, but never emphasized. The four chapters in this part of the book seek to correct the misplaced egalitarianism found in recent hoplite studies by acknowledging the place of both superiors and inferiors in Greek armies. Chapter 9, F.S. Naiden's "The Organization of Greek Armies," assesses the size of Greek and Macedonian forces, identifies large and small units, and specifies the numbers and ranks of Greek and Macedonian officers. Better command and control emerge not only among Spartans as opposed to others, but cavalry as opposed to infantry, and Macedonians as opposed to Greeks, while Hellenistic increases in army size do not accompany ascertainable gains in this regard. In Chapter 10 on "Generalship," Joseph Roisman shows how effective high command, a long-recognized feature of fourth-century warfare, began far earlier, as shown by Homer, and thus was never a Spartan monopoly. Roisman describes the development of strategy through the career of Pericles as well as that of Philip II. The Hellenistic commander Epichares of Athens reveals another aspect of Greek generalship, protecting resources and frontiers in time of peace.

Many ancient authors report camp followers and their doings, but Chapter 11 by Mauricio Álvarez Rico on "Greek Camps and Camp Followers" is the first ever devoted expressly to this subject. Working from his monograph of 2013, he summarizes evidence from Aeneas Tacticus and others that the Greeks practiced castramentation, contrary to clichés about Roman orderliness and discipline and a Greek lack of the same. Camp followers, called "the mob" (*ho ochlos*) in Thucydides, might better be termed the denizens of a city on the move, to which the Macedonians imparted their own distinctive features.[7] Another neglected aspect of

Greek military personnel, Greeks serving in foreign armies—not merely mercenaries, but mercenaries adapting to the needs of foreign rulers—illustrates this book's theme of Greek and Near Eastern interdependence. Jeffrey Rop's Chapter 12 on this subject, "Greeks in Foreign Service: The Case of the Achaemenid Empire," is also adopted from a recent monograph. Rop rebuts the common view that Greeks in Persian service were very largely hoplites who became valued by the Persian kings after Marathon. In fact, they first entered Persian service in 525, and often served as sailors and engineers.

Part IV, "War with Non-Greeks," takes up the theme of Greco–Persian overlap through Michael Charles' Chapter 13 "The Royal Elite of the Achaemenid Army," in which he summarizes his recent articles to show that the famed 10,000 "Immortals" disappeared sometime after Xerxes, to be replaced by mercenary hoplites; two other units, the smaller Kinsmen Cavalry and the Apple-Bearers, an infantry counterpart to this cavalry, both survived. In Chapter 14 "Parthian Warfare under the Early Arsacids," Marek Olbrycht surveys Parthian military organization, weapons, and armor before concentrating on the cataphracts, whom he describes as a heavy cavalry that developed from the confluence of "Central Asian nomadic cavalry tradition, Achaemenid military innovation, and the Macedonian challenge." The long spear was first borrowed from the Macedonians not by the Arsacids, but by steppe peoples, the Massagetae and Dahae. It reached Parthia by an indirect route typical of the cross-connections traced in this volume.

The remaining chapters in this part deal with two less studied enemies, India and Thrace. India provided the Macedonians with elephants, the subject of Chapter 15 by Christopher Epplett, "Elephants in Hellenistic Warfare," which asks whether the Successors and later the Romans made good use of this Indian military export. Epplett answers that at Gaugamela Persian mismanagement of elephants offered the Macedonians a warning that Alexander ignored, especially after he witnessed the success of Porus in using them at the Hydaspes. The Successors (but not Alexander) then put this new military arm to the test in more than a century of battles. The record was mixed: success at Ipsus, but failure on some other occasions. The Successors thus abandoned this exotic weapon, in spite of its shock power.

If Indian elephants were a luxury for Macedonian rulers, Greek and Macedonian equipment and methods were a necessity for Thracian tribes wishing to prevail against each other and against foreign invaders, as described by Emil Nankov in Chapter 16 "Thracian Warfare." Nankov's study explores the methodological problems of reconstructing military history from exiguous literary sources and scattered archaeological remains, especially battlefields, weapons hoards and *militaria* from graves and tombs.

The remaining parts of this book, Parts V–VII, turn from what might be called the foreground of warfare—strategy, operations, personnel, and foes—to the background, which is the general context in which ancient states made war. Part V deals with economic and technical context that directly affected victory and success; Parts VI and VII deal with topics that affected performance indirectly.

The first chapter in Part V, Chapter 17 deals with logistics—the little studied art of keeping soldiers alive and well before battle. The author, Stephen O'Connor, takes advantage of the numerous remarks on this subject in Xenophon, and branches out toward other commanders and other problems. Clichés about Greek improvisation and self-reliance prove untrue, the same as similar clichés about castramentation. The next chapter, Silke Müth-Frederiksen's

[7] Th. 7.78.2. "City on the move": Rostovtzeff 1941, 1.146.

Chapter 18 on "Fortifications," builds on recent scholarship that has shown how fortifications, "besides being military necessities, ... are architectural monuments with multiple purposes and historical witnesses with much information to impart." This chapter provides the material background for Whitehead's on siege warfare, Chapter 8. Third in this part is Frank Russell's "Military Intelligence," Chapter 19, which deals with both strategic concerns and operations, and divides the latter according to defensive and offensive contexts. His pages on towers and fortlets complement Müth-Frederiksen's in Chapter 18.

The remaining chapters in this part deal with economic matters. In Chapter 20 "The Economics of War," Johannes Heinrichs moves from war finance, a much-treated topic, to the actual or expected effects of war on production, distribution, wealth, and growth. Greek states cared more about these issues than scholarship on taxes and budgets has recognized. Another underestimated aspect of Greek warfare is the use of slave labor, the subject of Peter Hunt's Chapter 21 "War and Slavery in the Greek World." Hunt finds the role of slaves to be ambiguous—indispensable as attendants and porters, but prone to desert, valuable as rowers, and occasionally as soldiers, but prone to rebel, easy to replace by taking captives, but only if the would-be captor was not captured and enslaved himself. Slave labor was thus both an asset and a potential liability, as especially illustrated by the Peloponnesian War. The same was true of agricultural land, as described in Chapter 22 "Agriculture and Greek Warfare" by Jeanne Reames and Anne Haverkost. This chapter shows how wars fought to obtain agricultural land gave way to devastating farms and fields as a military aim, a change culminating in the Peloponnesian War, which again is central. Then, in the Hellenistic period, belligerents increasingly relied on farmers' markets to feed larger armies operating farther from home. Instead of providing a socioeconomic basis for armies, as in the work of Victor Davis Hanson, agriculture alternately provided an asset and a target.

Like Part V, Part VI has two sections, the first devoted to the social and psychological context of warfare, and the second to the political and legal. In Chapter 23 on "Battle Trauma in Ancient Greece," Lawrence Tritle narrows the focus of his previous work on suffering in battle and considers psychological trauma, as reported in Herodotus and other sources. The disorientation, or even hallucinations, that Tritle discovers overlap with the military epiphanies reported by F.S. Naiden in Chapter 24 "Religion and Warfare," which covers epiphanies as well as sacrifice, divination, and post-battle supplication, games, and the erection of trophies. This chapter describes all the religious rites that would commonly occur during a campaign. The following piece, Elizabeth Carney's Chapter 25 "Women and War in the Greek World," might appear in any of several parts of this book: the abuse of women captives links it to the subject of trauma, the enslavement of captives links it to slavery, and the role played by women in sieges links it to siege warfare. Most camp followers were women, and just a few of them, all Macedonian or Epirote, were in charge of armies, even if they left operations to a male lieutenant.

The chapters touching on politics begin with Chapter 26 by Sheila Ager on several phenomena that these rules did not always either prevent or recognize, the "Piracy, Brigandage, and Terrorism" of her title, for all of which Greek literature supplies examples from Homer onwards. She draws discrete parallels with modern events, while always mindful of tendentious vocabulary, whether ancient or modern. Last in this part of the volume is Chapter 27 by Gordon Shrimpton on "Civil War: Forgetting as the Path to Reconciliation," for which he provides a range of examples of *stasis*, showing how this term applies both to civil wars and to the disagreements and rivalries leading up to it, mainly by the use of elaborate examples from the history of Athens and Corcyra.

Finally, Part VII on "Warfare in Art and Literature" includes two chapters on visual material and two on literary texts. The first, Chapter 28 "The Iconography of War" by Olga Palagia, deals with the iconography of historical battles in Greek art. She provides a survey from the Persian Wars down through the battle of Pydna some 300 years later. Next, Lawrence Tritle, in Chapter 29, writes on "War Monuments and Memorials," with sections on battlefield memorials, battlefield graveyards, military cemeteries, military inscriptions, and heroic military sculpture. All these forms of expression extolled "civic virtue," and thus civic responsibility. José Vela Tejada finds the same civic propaganda in Chapter 30 "War in Greek Poetry," beginning with Homer, who inaugurates the enduring theme of the "beautiful death" suffered in exchange for imperishable glory. Archaic poets add a civic motive for this sort of death. Some plays deprecate civic martyrdom, but only some, and none of these works, Vela Tejada warns, should be termed pacifist. Civic propaganda, a Greek invention, was never subjected to thorough Greek criticism. "War and Propaganda," in turn, is the title of the last chapter in this volume, Chapter 31. In it, Sabine Müller analyzes the propaganda propounded by Greek assemblies and officials, and later by Macedonian kings, and finds several themes that have lasted until our own times: protecting the homeland, liberating the oppressed, exacting revenge, safeguarding the peace, and fighting in a just cause. All might be accompanied by claims of divine support, even of predestined victory, a conclusion linking this chapter to Naiden's Chapter 24 on religion and warfare, which describes the rituals used to make these claims.

The editors wish to thank both the contributors and the team of Todd Green and Andrew Minton at Wiley for their patience as we assembled and prepared the 31 chapters in this book. Waldemar Heckel conceived this volume and recruited nearly all the contributors; F.S. Naiden recruited the others and joined Heckel in editing it; and Erin Garvin and John Vanderspoel did the indispensable work of line editing and compiling the index and bibliography.

PART I

Historical Survey

CHAPTER 1

Bronze Age and Early Greek Wars

Johannes Heinrichs

This survey begins with the final Mycenaean period, around 1200. For the preceding Mycenaean phases evidence is lacking, and in the yet-older Minoan period, though a part of the Greek mythological tradition, a non-Greek language was spoken and written (an Aegean substrate, in Linear A script, not yet deciphered). Written in Linear B script (expressing the Greek language) they seemingly belong to fourteenth-century strata, but for both linguistic and paleographic reasons must be coeval with the tablets from Pylos (from c. 1220/1200). The geographic theater is restricted to the south west Peloponnese (Pylos) and central Crete (Knossos). Tablets from further residences such as Mycenae or Thebes (the *Kadmeía*) do not shed light on military aspects.

Defense of the Mycenaean Kingdom of Pylos (c. 1220/1200)

Epigraphic evidence[1]—reasonably coherent, albeit demanding interpretation—comes from an archive unearthed at the site of the palace of Pylos. It comprises sun-dried clay tablets covered with lines of script, written in vowels, syllabic units, and ideograms to record stored objects and services due to the palace. Another archive is lost; it was possibly housed in the

1 All pertinent texts (with introduction, discussion, and—if possible—translation) in Ventris and Chadwick 1956 and Palmer 1963 (see the registers). Abbreviations indicate provenance (here PY: Pylos; KN: Knossos), series (e.g An: personnel, men), and number in inventory. Generally, see Ventris and Chadwick 1956, 183–194. A general overview: the papers ed. by Laffineur 1999, Sections 8 (Les sources écrites et l' administration), 9 (Les armes et l'armement), and 10 (La guerre).

A Companion to Greek Warfare, First Edition. Edited by Waldemar Heckel, F.S. Naiden, E. Edward Garvin, and John Vanderspoel.

adjacent residence, probably of the kingdom's highest military official, the *ra-wa-ke-ta* (*lawagétas*, i.e. commander: *hagetas*, of the army: *lawos*).[2] When no longer needed, the clay tablets were regularly brought outside to decay; only the last set survived in the form of ceramics when the palace went up in flames. There must have been information on a threat emanating from the sea: defense posts were organized along the Messenian south west coast. Approximately in the middle, some way off the Bay of Pylos/Navarino, the palace of the king (*wa-na-ka*: *wánax*) was situated.

The actual spot of an attack was not clear.[3] So the whole coast had to be guarded by small "commands" of watchers (*o-ka*: *orché* = *arche*), stationed at ten places, recorded on five tablets,[4] each containing two records.[5] The first starts with a headline for the whole series, the following entries are of a uniform type: "Thus watchers are guarding the coast. Command of Mareus: (4 personal names). At (place name) men from (2 place names): (ideogram for) MAN 50." In this case, the *o-ka* consisted of one officer and four troops, assigned to watch the sea and, in case that enemy ships approached, to alarm both the *lawagétas* and the warships presumably concentrated in the central Navarino Bay.

The second group of 50 men must have been regarded as sufficient to delay landing operations, at first by long distance weapons such as arrows and slings, then by spears, especially where the Messenian coast was rocky,[6] until the bulk of the army had arrived. All detachments add up to 740 + x men (some numbers have perished), probably c. 1,000 in all. Their nine commanders are labeled as *e-qe-ta*: followers (*heq*^{u}*etai*).[7] That they belong to the nobility is indicated by a patronym added in most cases; moreover, they possess land parcels under very favorable terms.[8] Hence, the *e-qe-ta* should be regarded as professional military officers paid indirectly by the palace, with responsibilities for the troops temporarily attached to the *o-ka* on the coast.

Their duties also include the recruitment of the ship crews. PY An 610 contains an incomplete list of these "rowers" (*e-re-ta*: *eretai*); its 19 entries contain local groups of men, 443 in all. If we add the average per group for nine further lost entries (approximately 210) the total is c. 650 "rowers." Since in PY An 12 the crew of one vessel comprises 30 men, 650 men are sufficient for some 20 crews plus a reserve. This might be what the kingdom could muster in times of threat. Obviously, marines were also on board: much later, in 433, Thucydides (1.49) mentions sea-going hoplites, archers, and spearmen, giving the fighting "done in the old fashion" the character of a land battle. The command of the sea crews was evidently committed to the *e-qe-ta*; therefore it included the ships.

---

2 Nikoloudis 2008. For the Spartan *arch-hagétai* see the section on "Sparta's Wars: Immigration and Conquest of Laconia (c. 950/930)".

3 Historical setting: Palaima 1995, cf. Halstead 2007.

4 PY An 657, 519, 654, 656, 661.

5 Lang 1990.

6 Cf. Thuc. 4.11ff.

7 From the Indo-Europ. root (s)equ- (cf. lat. *sequi*): to follow (*scil.* the king). A detailed study has been submitted by Deger-Jalkotzy 1978, cf. Ventris and Chadwick 1956, 121; Killen 2007. At Knossos these officials are also involved in raising tributes (KN As 821).

8 Evident from a document that lists missing rowers (PY An 724, 11f.). We may conclude that some *e-qe-ta* whose responsibility it was to organize personnel had been negligent and were reprimanded by hinting at the privileges they had received, cf. Ventris and Chadwick 1956, 188, s.v. *e-ko-si-qe*, and generally 232ff.

For the land troops under the *lawagétas* we are virtually at a complete loss. His high command will have included training in peace, in cooperation with palace authorities controlling the stocks of bronze, also in sanctuaries. Whether resources sufficed to equip more than some dozens of men can be doubted. In the palace's final year, fewer than 50 corselets are listed (PY Sh 736: *to-ra-ke*: *thorakes*), some with hoops of bronze and probably also some made of linen with metal plates sewn all over.[9] They constitute ensembles with "helmets" (PY Sh 737, 740: *ko-ru-to*: of a *korys*), which appear in ideograms as sets.[10] Armor was probably reserved for charioteers and some high-ranking officers.[11] We may take for granted that such men had not only spears but also swords. However, tablets dealing with standard military "swords" (KN Ra 1458: *pa-ka-na*: *phasgana*) are lost at Pylos; at Knossos they occur in some number.[12]

Like the Egyptians, Hittites, and the north Syrian Kingdom of Mitanni,[13] Greek Mycenaean kingdoms also had cohesive units of chariots, though information is scarce at Pylos, and once more comes substantially from the Knossos tablets.[14] Since a special driver was indispensable to give the charioteer a free hand in fighting (he held on to the chariot with his left hand and could not hold the reins and a whip), a pair of horses—at times even three—were needed to draw two men.[15]

High expenses for chariots could be borne only by the palaces. Along with them, this weaponry disappears, but its memory endured. In the *Iliad*, Nestor arrays the Pylian chariots in front of the infantry (*Il.* 4.297–309) and warns the charioteers not to leave their formation: "Thus the old man exhorted them, from time immemorial an expert in fighting" (310). Nestor is generally styled as a representative of a bygone age; so the way he arranges his Pylian chariots at Troy as a unit, in striking contrast to the individual practise of younger Greek kings, may well be a reflection of the Mycenaean tactics.[16]

---

9 Ventris and Chadwick 1956, 291ff.; cf. the Homeric epithet *Chalcochiton*: "garment covered with bronze" (e.g. *Il.* 1.371, 2.47). For a photo of a suit of armor see Dickinson 1994, 204, pl. 5.21.

10 Ventris and Chadwick 1956, 375.

11 PY Sh 733–740, see Ventris and Chadwick 1956, 376 with nos. 292–295. At Knossos, corslets are combined with pairs of horses and chariots (KN Sc 222 and 226 = Ventris and Chadwick nos. 297–298; no helmets are mentioned there separately): "Such body armour may have been restricted almost entirely to charioteers, for whom the absence of a shield made it a necessity; which would help to explain its absence from chieftains' graves." (ib. 376). Charioteers needed their left hand to hold on to the chariot. Moreover, if there would have been a shield of a prestigious type, elements of it would have survived: at least the boss and the rim will have been made of bronze.

12 KN Ra 1450: the number is incomplete (50 [+x, i.e. between 51 and 59]).

13 On the Egyptian chariot contingents see generally Drews 1993, 104–134 and 209–225. In two great thirteenth-century battles large pharaonic chariot units were involved: in 1274, the defeat of Ramesses II against the Hittites at Kadesh (on the Orontes): ib. 130–134; in c. 1200, the victory of Ramesses III over the "Sea Peoples" (in this case Philistines) in the Delta, illustrated at Medinet Habu, ib. 157ff. (frescoes: ib. pl. 6 and 7 on pp. 159–160). The fictional report in the Old Testament (Ex. 14.7: 600 + x Egyptian chariots pursue the Israelites) can be taken as a further general hint at the Egyptian strength of chariots at that time. On the Hittites generally Kuhrt 1995, 225–282, on Mitanni ib. 289–300.

14 Ventris and Chadwick 1956, 361–375, with a (modern) sketch (362). At Pylos, chariots are indicated indirectly by separately stored wheels: Ventris and Chadwick 1956, 373ff. The practise to take them off the chariots when on stands is testified in *Il.* 5.722.

15 E.g. KN Sd 403; in turn, also one-horse chariots are listed, e.g. KN Sd 409 + 481, see Ventris and Chadwick 1956, 365ff., with these and further texts.

16 Maybe the passage is inspired by an epic older than the *Iliad* (a Nestoris); possibly also a famous episode from Nestor's youth was taken from it (*Il.* 11.668ff.). What is preserved of such epics is collected in the Loeb edn. of *Greek Epic Fragments* (by M.L. West 2003).

At Knossos,[17] chariots have special ideograms; we may assume the same for Pylos where pertinent evidence is largely missing. Military chariots must once more have been the responsibility of the *lawagétas*—palace records will pertain to civilian versions of high officers. Since Greece is a rough country, chariots needed leveled roads and bridges, of the sort partly preserved to this day at some spots near Mycenae.[18] There must have been many others, crumbled now or unknown. The costs for building and upkeep were certainly substantial, a further reason for abandoning chariot contingents in post-Mycenaean times.[19]

At the other end of the hierarchy rank the large majority of the infantry. Certainly, men were lightly armed, fighting as skirmishers with small shields of wicker and "caps of dog's skin" (*kynées*). The bulk served as slingers using stones and as archers. Therefore, the supply of arrows—for both the army and the fleet—must have been immense. The palace, generally responsible for importing raw materials, had to supply the bronze needed for arrowheads.

To judge from archaeological finds, a single arrowhead weighed up to 1.5 g. Information about their production comes from the Jn series.[20] According to PY Jn 829 "those in charge of a region ... and their deputies and ... are to give bronze from sanctuaries for arrows and spears, as heads."[21] The long list of officials is strongly abbreviated here, but their contributions in weights are moderate and uniform. So we may guess that all men had to get the prescribed metal as personal loans, otherwise one person would have sufficed to get the total amount. It can be calculated to 34 talents (of 29 kg each) and 26 double MNA (of 1/30 talents each = 0.97 kg), in all 1,011 kg (PY Ja 449). At a minimum, it sufficed for 674,000 arrowheads or 2,886 spearheads (up to 350 g each). Both items were produced in proportions unknown to us.[22]

Since sufficient amounts of wood were required for fuel, the production of the weapons was outsourced to villages on the land. Men living there as farmers, shepherds, or craftsmen had exactly defined obligations to work for the palace as (bronze)smiths (*ka-ke-we*: *Chalcees*). Their local chiefs (*pa₂-si-re-we*: *basilées*[23]) received portions of the collected bronze, distributed personal allotments,[24] and sent back the heads to the palace to be redistributed to the armed units.[25] In all, a surprisingly detailed system of production.

---

17 Ventris and Chadwick 1956, 365 calculate a total of 400 + x chariots at Knossos: "it was evidently immeasurably smaller than the bodies of chariotry deployed by the Egyptians and Hittites on the more open battlefields of Syria".

18 Hope Simpson 1981, 15ff. (with several plates: 2b, 6a–b, 8a–9c), esp. 17: "there can be little doubt that the main purpose of the roads was military ... as a means of rapid deployment of mobile forces, presumably including chariots." Cf. id. 2002. For a sketch of the suggested Mycenaean road system see Dickinson 1994, 163, fig. 5.34.

19 But some principal routes were kept intact, such as that from the Bay of Pylos via the valleys of Alpheus and Eurotas to Lacedaemon/Pellana, cf. *Od.* 3.481ff. (the Pherae near this route is not identical with Pherae (Pharae)/ Kalamata, but with Aliphera, south of Heraea: *Barr. Atlas*, map 58, B2).

20 Jn series: Ventris and Chadwick 1956, 352ff.; Smith 1992/1993.

21 *jo do-so-si ko-re-te-re ... po-ro-ko-re-te-re-qe ... ka-ko na-wi-jo pa-ta-jo-i-qe e-ke-si-qe ai-ka-sa-ma*, in alphabetic Greek: ὡς δώ{σω}σι χωρηταὶ ... προχωρηταί τε ... χαλκὸν νάϝιον παλτοίς τε ἔγχεσι τ' αἰχμάς. Ventris and Chadwick 1956, 357 translate *na-wi-jo* "for ships" (*naus*: ship, never spelt with *vau/digamma*), but this makes no sense here; actually to be connected with *na(w)os*: temple.

22 Ventris and Chadwick 1956, 361 refer to a tablet from Knossos (KN R 482) "on which the ARROW ideogram is followed by the high numbers 6010 and 2630 (which would require about 13 kg of bronze)".

23 Up to the Homeric *basilées* it was a long way, but it starts just here, cf. Ventris and Chadwick 1956, 121f., 253.

24 E.g. PY Jn 658. For a table see Ventris and Chadwick 1956, 355–356.

25 Killen 2006.

Efforts were not effective: The Pylos palace went up in flames, as did others during the time that followed.[26] Thus ended the centralized organization of life, not by one single catastrophe, but by a series that was probably caused by different groups of Sea Peoples. They also affected Crete, Egypt, the Levant, and the Anatolian Hittite Kingdom. What we read in Egyptian and Hittite texts (and see on the frescoes at Medinet Habu) suggests what also happened to the Mycenaean palaces in Greece proper.

Dark Ages and Archaic Period

The "dark" centuries that followed the age of the Mycenaean palaces (starting at different times in different regions), up to the middle of the eighth century, are characterized by decentralized structures without written evidence. Writing recommenced in the middle of the eighth century, with an alphabet of newly devised letters on Phoenician models. It gave expression to the same Greek language, though now used not for running central archives, but for private dedications and public laws, and for the tradition of poetry of different genres.

Myth to "History": The *Iliad*[27]

The new literacy starts with the *Iliad*, an epic with a mythological plot. A broader concept of authorship and authenticity applies: as a literary work, it widely consists of formulaic passages, verses, and phrases that belong to different periods, in most cases preceding the individuals who fell back on such elements to establish their own versions within epic traditions.[28] Such poetry was transmitted orally, constantly revised from generation to generation. Only in the end of the process, when the new system of writing became available,[29] were the versions then most esteemed noted down and attributed to a legendary poet under the name of Homer.

Because of the process that precedes Homer, the *Iliad* is an amalgam, not a mirror of actual stages of warfare; nor does it describe historical societies. Themes and perspectives of the epics are clearly aristocratic, with a view to the audience the *rhapsodoi* (and *aoidoi*) appealed to.[30] So what we can learn about eighth-century warfare is sparse, if we are looking for the historical situation, whereas examples of aristocratic chivalry (as well as inappropriate behavior) are paramount. But for the Greeks who followed, including Herodotus (7.171.2) and Thucydides (1.9ff.), the Trojan War was an historical event.

To illustrate the transformation of historical recollection into epic poetry by an example, we once more turn to the chariots. In the *Iliad*, the majority of the heroes can afford such vehicles, not for fighting in units, but to be transported to the battle ground and to have a

26 Desborough 1975.
27 Generally Kirk 1975, 820–828 and van Wees 1997
28 Research on oral poetry was initiated as a discipline in the 1930s by Milman Parry (died 1935), during field studies in Bosnia. His papers applying his results on the Homeric epics were collected in Parry 1987; cf. Visser 2006.
29 Powell 1991.
30 Generally, Donlan 1999.

quick escape if necessary;[31] they jump off and fight heroically in individual duels (*aristeiai*),[32] whereas drivers wait at a safe distance—and at times take guard of valuable booty.[33] In Homeric battle, chariots are tantamount to taxis, as in peacetime.[34]

History to Myth: Early Greek Wars

Historiography as a genre started only in the later fifth century, with Herodotus and Thucydides.[35] What they could still learn about early wars had undergone more or less a process of transformation into myth or, in turn, pseudo-historical myths had been invented to explain how actual structures had come about and why certain claims were justified.

Thucydides (1.12ff.) gives a catalog of early Greek wars. Starting from the Trojan War, he distinguishes between migrations involving many groups and conflicts between two neighbors (1.15.2). An exception is the Lelantine War, when further Greek states and groups of allies[36] supported the two opponents (1.15.3); so we have before us more than a neighborhood conflict.[37]

The Lelantine War[38] (Probably c. 710/700)

Authorities from different periods allude to the war as an actual event. Hesiod (*floruit* in the later eighth century) won a prize in the funeral games for the Chalcidean noble Amphidamas (*Erg.* 650ff.).[39] If this man had been killed in action just before, as suggested by the games, the pertinent war took place c. 710/700. This accounts for the still Homeric practice of funeral games reminiscent of those for Patroclus (*Il.* 23), in the version we read from around 750.

Archilochus (c. 680–645) describes a war "on the plain," being fought by "those spear-famed squires of Euboea" (fr. 3 Gerber).[40] The character of the fighting is reminiscent of the *aristeiai* between noble single combatants in the *Iliad*,[41] clearly before the introduction of

31 Drews 1993, 161: "The Homeric description of chariots as battle taxis may be a reminiscence of this twelfth-century development. Possibly in (LH) IIIC Greece the horses and vehicles that survived from the pre-Catastrophe chariot forces became nothing more than prestige vehicles for the professional warriors…".

32 E.g. *Il.* 4.365, 419 with 5.109ff. (Diomedes, Sthenelus); 5.221ff. (Aeneas, Pandarus).

33 The divine and therefore most prestigious horses of Aeneas are a most instructive example. Before attacking the Trojan chariot, Diomedes instructs his own driver to get hold of them in case he should kill or put to flight Aeneas and Pandarus being with him: *Il.* 5.259ff.

34 *Il.* 24.325ff., cf. 440ff. (Priam, Hermes); *Od.* 3.475ff. (Telemachus, Pisistratus).

35 Also the two originators participated in "reconfiguring the past"; see e.g. Dillery 1996 (for Herodotus).

36 Plut. *mor.* 760f.–761a: Cleomachus of Pharsalus.

37 Even more wars were fought by coalitions, e.g. the First Sacred War (c. 595–585, of dubious historicity): Allegedly it was undertaken by the Pylaean-Delphic amphictyony on behalf of the interests of Delphic Apollo, to punish the city of Crisa for having violated the sanctuary's holy claims. But the true intention must have been to eliminate a place situated in the plain below Delphi near the coast since it controlled the ascent from the sea up to Delphi and thence through Phocis into Thessaly. See Forrest 1956; Jeffery 1976, 73–82; Scott 2014, 71ff., 144–145.

38 Stein-Hölkeskamp 1999, 38. Generally Tausend 1987; Parker 1997; Walker 2004, 156–182

39 Tedeschi 1975.

40 Donlan 1970.

41 Most typical in chivalry (albeit a duel is avoided) is the encounter between Diomedes and Glaukos (*Il.* 6.119–236).

the phalanx. In accordance with this, Strabo learned much later about Eretria and Chalcis: When quarrels arose between them "for the Lelantine Plain" they came to terms "under which they were to conduct the 'strife' (*agon*)." They engraved these terms on a stele "that banned long distance missiles ... by archers, slingers and javelin-throwers" (10.1.12).[42] The still entirely aristocratic agreement must belong to a very early time, albeit a stele with clauses of a treaty cannot have existed then; the detail was evidently invented later.

Also Theognis (second half of the sixth century) adopts a contemporary perspective on an Euboean war (v. 891–94 Gerber),[43] the later sixth century being additionally confirmed by the fact that the mythological king Pheidon of Argos (mostly dated by the Greek tradition to this century) plays a role.[44] Finally, the Lelantine Plain was conquered by the Athenians in 506 (Hdt. 5.74–77).

The 200 years from Hesiod to the Athenian conquest are far too much for one single war, even if it had some length. What we find in the Greek tradition must indeed be understood as a series of separate conflicts. The term in its proper sense applies to only one of them, probably that mentioned by Hesiod.[45] It has been proposed to attach it to the series as a whole,[46] but the tradition that "the whole Greek world[47] divided joining the party of one of the two" (Thuc. 1.15.3) runs counter to this view: it is impossible that such a deep division could have escaped the attention of our sources for the archaic period.

Excursus: Naval Battles

Thucydides introduces such conflicts as a further category; few in all, the earliest was that between Corinth and Corcyra/Corfù (1.13.4, around 681 or 664).[48] He regards more recent naval battles as insignificant (1.14.3) and skips entirely over those between Greeks and non-Greeks, such as the defeat of the Phocaeans against the allied fleets of Etruscans and Carthaginians off Alalia (on Cyrnus/Corsica, c. 540: Hdt. 1.166f.),[49] and Hieron's victory against the Etruscans near Campanian Cyme (474: Diod. Sic. 11.51).[50]

42 For a possible explanation see Walker 2004, 159. A contrary view: Wheeler 1987.

43 Jeffery 1976, 66; Parker 1997, 82ff.

44 Bradeen 1947, cf. Ragone 2006.

45 The early option is supported by Thucydides reporting that "the (Lelantine) War occurred in a very early time" (1.15.3: *palai pote genomenos polemos*). This dating also marks the end of the Dark Age settlement near modern Lefkandi, situated on the Lelantian coast halfway between Chalcis and Eretria, see Boardman and Price 1980, 74: "The Lefkandi desertion should have taken place about 710, but possibly later...".

46 Walker 2004, 156ff.

47 Herodotus (5.99.1) reports that the Milesians supported Eretria whereas the Samians took the side of Chalcis, maybe motivated by regional conflicts in Asia Minor. The wide support can be explained by the fact that both opponents were prominent in the colonization. So early, purely economic motives can be excluded.

48 Following Thucydides, "260 years before the end of this (i.e. the Peloponnesian) war" (421, 404?); cf. Salmon 1984, 218–219 and 222–223.

49 Asheri et al. 2007, 186ff., with pertinent scholarly literature; Sachs 2014, 132ff.

50 An Etruscan bronze helmet from the spoils, dedicated at Olympia, bears the inscription: "Hieron, son of Deinomenes | and the Syracusians | for Zeus, from the *Tyrhen(ioi* = Etruscans) at Cyme" (ML 29, with comm.).

Sparta's Wars: Immigration and Conquest of Laconia (c. 950/930)

The most important of the migrations, from the north west into the Peloponnese, was labeled by the Greek tradition as the "Return of the *Herakleidai*," actually often referred to as the "Doric Migration," a term no less misleading. The historical process took much time, in several stages. It was fought by groups belonging to the same *ethnos* but independent of each other. One of them, consisting of several subunits, arrived at the site of the future Sparta not much later than 950.[51] Still in the middle of the fifth century the Spartans knew that their ancestors had immigrated from the (intermediate stage of the) Doris region just north of the Thermopylae (Thuc. 1.107.2: *Lakedaimonion metropolis*, three small towns, among them Erineus). Since 1971, we have had at our disposal a papyrus fragment containing parts of an elegy by the Spartan poet-general Tyrtaeus. Active in the later seventh century, he combines the mythological tradition[52] with historical facts: "For he himself, the son of Cronus and spouse of fair-crowned Hera, Zeus, has given this city (Sparta) to the descendants of Heracles. Leaving windy Erineus with them, we arrived in Pelops' wide island."[53] During the period of migration from Erineus to Sparta the mythological *Herakleidai* had not been commanders, but two *archagetai* ("dukes"),[54] interpreted as "kings" by later Greeks. Under them, two subgroups had arrived in Laconia and established the originally four neighboring villages of Sparta on virgin ground. Starting from the new settlement, they conquered the Laconian Plain down to the sea. The indigenous population was subdued by force to a state of bondage (the *heilotai*) or political dependence (the *perioikoi*), with one substantial exception.

The inhabitants of the small town of Amyclae, situated some 5 km south of Sparta, were integrated as a fifth village.[55] This is even more strange since the Amyclaeans were non-Dorian, and their sanctuary (*Amyklaion*) belonged to *Hyakinthos*, a god with a pre-Greek name;[56] he was later equated with Apollo, as the originally Dorian *Ortheia* with Artemis, brother and sister. The Amyclaeans had evidently successfully resisted the Spartan expansion and received a privileged position so that their military potential was under control.

51 Cartledge 1979, 65–81, who overestimates the gap between late Mycenaean and Dark Age (Proto-Geometric) ceramics at the *Amyklaion*; there was no interval of settlement for some centuries, as Coulson 1985, 63–66 points out after fresh analysis of the (very fragmented) pottery. The earliest sherds unearthed in Sparta, from the sanctuary of (Artemis) *Ortheia* (found in a stratified excavation on virgin ground), from an Heroon near the Eurotas and, not stratified, from the *akropolis* near the temple of Athena *Chalcioikos*, go back to the middle of the tenth century (or soon after). The vessel types are clearly influenced by those from Amyclae which, in turn, depend on the late Mycenaean ceramics; cf. Coulson 1985, 30–31 (with n. 30–31).

52 Collected by Hammond 1975, 690–696.

53 *POxy* 38.2824 = fr. 2 in Gerber's Loeb edn. of "Greek Elegiac Poetry," *ll.* 12–15, with Turner's supplements (in *POxy*) adopted by Gerber.

54 From lat. *ducere*: to lead. The Greek term is derived from *arche*: "command, i.e. body of troops" (*LSJ* II5, cf. the Pylian *o-ka*), and (Doric) *hagéomai* (noun: *hagetas*): "to go before, lead the way" (*LSJ* 1a), here meaning the commander of a migration group organized as a military body. The position became permanent after settling down. The two leaders were regarded as descendants of Heracles; see Hdt. 7.204 and 8.131.

55 Lazenby 1985, 69 with n. 21 is more reserved, but fundamentally agrees. Possibly the Great Rhetra (transmitted by Plut. *Lyc.* 6.2, see Tyrtaeus fr. 4 Gerber) belongs to this context.

56 Ventris and Chadwick 1956, 27 (words in *-inthos*). On his cult feast (*Hyakinthia*) see Pettersson 1992, 9–41.

Expansion Realized and Thwarted: Messenia, Argos, and Arcadia (c. 700/690–550)

Sparta's two Messenian Wars (c. 700/690–680/670 and 640/630–600),[57] fought in a developed phalanx tactic as reported by Tyrtaeus (fr. 11 Gerber),[58] resulted in a state of *heiloteia* of the central Messenian *Stenyklaros-Pamisos* plain (Tyrt. fr. 5f. Gerber); the historical events are completely superseded by myths.[59] The same applies for Sparta's wars against Argos for the control of the intermediate region of the Thyreatis. The final engagement perhaps belongs to the middle of the sixth century. Herodotus (1.82) adopts a strongly reworked tradition: Spartans and Argives had allegedly limited their phalanx to 300 hoplites each—as many served in the guards of the Spartan kings, e.g. at the Thermopylae; but the chivalrous arrangement failed and an ordinary battle ensued, with a Spartan victory.[60]

Against Arcadian Tegea, Sparta's attack failed in the middle of the sixth century. The *polis* transferred her activities to the diplomatic field, by constituting a system of unequal bilateral alliances ("Peloponnesian League") to secure her military supremacy.[61] Also the tradition on Tegea in Herodotus (1.66ff.),[62] written down after a longer process of oral tradition,[63] has undergone transformation into myth.

Abortive Campaign Against Cleisthenic Athens (506)

Sparta's most recent major campaign before the Persian Wars was the final act in a series of events that had started in 510 with the overthrow of the Athenian tyrant Hippias. A city situated so close to the Isthmus of Corinth, the highway into the Peloponnese, was of primary strategic and political interest. Since a non-oligarchic regime implied substantial political risks, Sparta was eager to establish at Athens a person she could trust. After some setbacks, a campaign to achieve this was prepared by a coalition: the Peloponnesian League from the south, the Boeotians and Euboean Chalcis from the north and the north east. The attack was cancelled when the Peloponnesians learned about Athens entering the Persian Empire (Hdt. 5.73.2ff., following a different version). By doing so, Athens thwarted a new Spartan attempt to establish a pro-Spartan government,[64] and indeed became unassailable. But she also undertook the first step on the way straight into the Persian Wars.

57 Meier 1998, 91–99.

58 Loeb edition of *Greek Elegiac Poetry*.

59 Meier 1998, 100–120.

60 Dillery 1996; cf. (generally) Krentz 2007a. Sparta was victorious once more in 494, at Sepeia, when Argos lost practically a whole generation, allegedly by acts of deceit of the Spartan King Cleomenes (Hdt. 6.77ff.); but Herodotus evidently draws from traditions hostile to the king.

61 Cawkwell 1993; Lendon 1994; Cartledge 2002, 118ff.

62 Cartledge 2002, 121–122.

63 On the theory of oral tradition, see Vansina 1961 and 1985 (based on research on African tribal societies), Finnegan 1977, and some papers in Olson and Torrance 1991.

64 Cf. Thuc. 1.19.1: "The Lacedaemonians did not collect tribute from their allies but saw to it that their states were oligarchies...".

Athens Versus Aegina (505/504–481)

For the moment, Sparta's allies paid the price: after the Peloponnesians had retreated without a strike, Athens defeated the Boeotians and the Chalcideans. From the Boeotians she took much ransom money, and from Chalcis additionally the Lelantine Plain. A subsequent Boeotian campaign for vengeance failed, but supporting Thebes the Aeginetans attacked the coast of Attica, taking a large booty and causing great damage (Hdt. 5.79ff.; 6.87ff.).[65] A long conflict ensued that was settled only on the eve of Xerxes' invasion (481: Hdt. 7.145.1). Hinting at this war, Themistocles convinced his fellow citizens to build a strong fleet of modern triremes (Hdt. 7.144.1). It would become essential for resisting Xerxes in 480/479 and, during the following decades, for Athens' ascent to hegemony.

65 Jeffery 1962; Figueira 1977.

CHAPTER 2

The Persian Wars to Alexander

Sabine Müller

Introduction: Persia and Greece, No Archenemies

As this chapter on Greco-Persian wars shows, from the end of the sixth century to the Macedonian conquest of Persia, the Persian Empire under the Achaemenids, Darius I's family, formed an inextricable part of the history of Greek politics and warfare. For Greek politicians, Persia was never out of sight or monolithically perceived as an oppressive force—a symbol of tyranny, despotism, and backwardness—as suggested by the biased images in literary sources.[1] For the Greek *poleis*, variously, the Achaemenid king could be an ally, host, protector, intermediary, supporter, source of hope, and guarantor of peace or an intruder, threat, and enemy. His role was hybrid, depending on the contemporary context, political issues, constellations, and needs of the respective Greek *polis* dealing with Persia. Persian subsidies were never far from Greek politician's minds and played a frequent role in inner-Greek conflicts, often enabling the Great King to partly influence Greek affairs.

From the Ionian Revolt to Xerxes' Greek Campaign

The Persian Empire's founder Cyrus II (c. 559–530) established Persian control over the Greek cities in Ionian Asia Minor in the 540s when he conquered the Lydian Empire.[2] Under Cyrus' family, the Teispids, no conflict with mainland Greece is attested. The

1 In fact, there was a gap between the actual Greek knowledge and attitude concerning the Persian Empire and its stereotypical depiction in political discourse and literature (Madreiter 2012, 180–184).
2 Van de Mieroop 2007, 287.

A Companion to Greek Warfare, First Edition. Edited by Waldemar Heckel, F.S. Naiden, E. Edward Garvin, and John Vanderspoel.

so-called Persian Wars, a series of military campaigns against alliances of certain Greek *poleis*, were fought in the early fifth century under the Teispids' successors, the Achaemenids: the Ionian Revolt, the campaign of Datis and Artaphernes (leading up to Marathon), and Xerxes' invasion and its aftermath. Xerxes' war played a special role in distorted, Western concepts, and triggered everlasting, biased images of an impious tyrant flooding Greece with soldiers but unable to defeat the brave, free-spirited Greeks who were inferior in numbers but superior in manly courage.[3] In Greek literary sources, these campaigns were mostly called the Median War (*Medikos polemos*) or *ta Medika* (Hdt. 9.64). Consequently, Greeks siding with the Persians were accused of *Medismos* or *medizein* with its complex meaning and "specially derogatory and odious connotation."[4] In truth, the Achaemenids did not aim at subjugating the whole of Greece, and not all of the Greek *poleis* fought against the Persians.

The roots of Darius' and Xerxes' campaigns can be traced back to the end of the sixth century. While according to Greek—particularly Athenian—interpretation, the Persian kings were aggressors who planned to enslave all Greece and—in later reception—even the world, in fact Athens' policy was the central factor leading to the wars.[5] In 507/506, in order to receive support against Sparta and her allies, Athenian ambassadors accepted the offer by Darius I's brother Artaphernes, the satrap in Sardis, to give earth and water to the king in exchange for Persian help. By accepting this compact, Athens became a part of the Achaemenid Empire. Engaging in window dressing, the Athenians blamed the ambassadors for this mishap. However, they could not have acted on their own account without any instructions or permission, even in this early stage of (proto-) democracy. It is also incredible that they were unaware of the consequences of their agreement. Probably, quickly shifting alliances and factional strife in Athens caused indignation about the arrangement.[6] Perceiving Athens as a member of his empire, Darius acted in accordance with his royal prerogatives when he ordered Artaphernes to make the Athenians restore their expelled tyrant Hippias, who had taken refuge at his court (Hdt. 5.96). Athens was unwilling to obey. In order to be publicly rid of their inconvenient status as dependent members of the Achaemenid Empire, the Athenians made an ostentatious political statement in the context of the Ionian Revolt (500/499–494) against the Persian dominion.

In accordance with the general outline in Herodotus' report, this uprising is predominantly seen as a consequence of the ambitions of individual Ionian tyrants, namely Histiaeus and Aristagoras of Miletos probably pushed by a widespread discontent with tyrants.[7] The older hypothesis that an (alleged) economic depression led to the revolt is contradicted by archeological evidence. In 494, the uprising in which Cyprus and Caria also became involved was finally crushed at the sea battle of Lade (Hdt. 6.13–14). Miletos, the leading Ionian city, was captured (Hdt. 6.18). When in 492 a peace settlement was imposed on Ionia, Mardonius, Darius' most important general and an Achaemenid himself, reportedly established forms of "democracy" in the cities (Hdt. 6.43).

3 Impious Persians: Briant 1996, 531–533. Countless, too: Hdt. 7.61–99; Aesch. *Pers.* 12, 56–57, 61, 929–930. Brave Greeks: Bridges 2015.

4 Graf 1984a, 15.

5 Hdt. 7.138; Thuc. 1.18.1–2; Aesch. *Pers.* 234.

6 Hdt. 5.73 with Zahrnt 1992, 256–257; Ruberto 2010, 4, 24.

7 Hdt. 5.30, 49–51; cf. Evans 1976 and Manville 1977. Discontent: Chapman 1972; Badian 2004.

Subsequently, Mardonius restored Persian control over Thrace and Macedon weakened during the revolt (Hdt. 7.108.1). Both regions had become Darius' subjects in the course of his campaign against the "European Scythians" (514/513). Amyntas I of Macedon had given earth and water to Darius' ambassadors and become the Persian king's *hyparchos* (Hdt. 5.18.1, 20.4), in this case meaning a client ruler, not a satrap. Amyntas will have recognized the chance to elevate his and his family's status by connecting with a worldwide empire. Presumably, Macedon's short-lived secession from Persia about 496 was initiated not by the Argeads but by influential Macedonian nobles disliking the upgrade of their *primus inter pares* ruler, and also by Thrace.[8]

Athens came to the aid of the rebellious Ionians by sending 20 ships (Hdt. 5.97.3). Coming as it did before Themistocles' naval program, and in view of the permanent Aiginetan threat, this was no small gesture.[9] Euboean Eretria, an old ally of Miletos (Hdt. 5.99), joined Athens by sending five ships. Both parties were involved in the Ionian capture and raid of Sardis (498) during which the local temple of the goddess Cybele (Hdt. 5.102.1) went up in flames. After being defeated, both parties withdrew from the war, but the dispatching of 20 ships made the Athenians rebels in the eyes of the Great King, hence representatives of *drauga*, the evil lie according to the ideology circulated by Darius (DB §§ 10, 52, 63; cf. DPd § 3). This openly hostile act, in Athens' case even a rebellious insubordination, publicly challenged Darius' authority and called for immediate punitive reaction in order to avoid further insubordination by imitators.[10] As Herodotus states, "these ships started the troubles for both the Greeks and the barbarians" (5.97.3). Allegedly, Darius swore vengeance against Athens, ordering one of his servants to remind him of Athens three times every day (5.105.1–2). In 490, as a response to the *poleis*' intervention in the Ionian Revolt, a Persian sea force under the command of Datis and Artaphernes was sent to Greece. Eretria's punishment was accomplished: the city was sacked, the inhabitants enslaved, a payback for Sardis' fate (Hdt. 6.101.1–3). However, when the Persians sailed to Attica to punish Athens, the Athenians under Miltiades, aided by Plataea, won the battle at Marathon, and hence remained unpunished (Hdt. 6.102–117.1). In Athens, Marathon became one of the important *lieux de mémoire*, proof of the city's glorious past and military superiority, especially because of Sparta's failure to come to their aid, something that enhanced Athens' glory.[11]

After Darius' death in 486, his son and successor Xerxes became enmeshed in the unsettled Athenian affair that challenged the Great King's authority. First, he and his advisors aimed at a diplomatic solution. During 483–480, they tried to prevent Athens from going to war while isolating her in Greece by courting other non-involved parties (Hdt. 7.32, 132–133.1). Herodotus reports that they won over Thessaly (at this time dominated by the Aleuads from Larisa), Dolope, Enia, Perrhaebia, Locris, Magnesia, Melos, Phthia, Thebes and the rest of Boeotia except for Thespiae and Plataea (7.132). However, Athens rejected all Persian offers to come to terms with Xerxes. Her response was clear: building up her fleet (Plut. *Them.* 3.5–4.6). Financed by the silver mines of Laurium, somewhere between 100 (Hdt. 7.144) and 200 triremes ([Aristot.] *Ath. Pol.* 22.7) were constructed. Athens' massive reinforcement triggered an arms race, forcing the Persian side to consider an invasion by

8 Müller 2017, 76–77.
9 Tozzi 1978, 161; Zahrnt 1992, 259.
10 Walser 1987, 165.
11 Cf. Thuc. 1.73.4; 2.34.5; Aeschin. 3.181.

both land and sea and on a much wider scale than planned.[12] For good reason, Persian leading circles tried to avoid such a huge logistical task. Ironically, the timber for the Athenian ships and oars probably came from Macedonia. In turn, Athenian payment helped Alexander I, the Argead owner of the timber business and Xerxes' loyal ally, finance his support of the Persian campaign.[13]

Even during the invasion, Xerxes and his general Mardonius strove for a de-escalation, sending Alexander I as an ambassador to the Athenians to persuade them to cancel their war plans (Hdt. 7.143.3; 8.136.1–2; 8.140–142). The Athenians chose to persist, as did the other Greeks who had joined them in a "Hellenic League." The inscription of the serpent column dedicated to Delphic Apollon after the victory of Plataea lists 31 League members.[14]

When in 480 Xerxes' land forces had crossed Macedonia, Alexander I managed to clear a path for the Persian troops through the Vale of Tempe into Thessaly by diplomatic means, but Leonidas of Sparta and his army made a stand at the Thermopylae and thus delayed the Persian advance, presumably a decisive factor with regard to logistics: for, toward the end, the Persians ran short of supplies.[15] Athens was evacuated before being captured and sacked by the Persians. Later on, the Athenian memory of the Persian conflagration of the Acropolis and temple of Athena (in revenge for the destruction at Sardis) was kept alive as an everlasting reminder of the courageous Athenian stand in the Persian Wars. Pausanias (1.27.6) attests that even in the second century CE Athens' damaged and blackened statues of Athena were on public display, allegedly the very artifacts burned by Xerxes' Persians and left as memorials by the Greek warriors. The story that Xerxes stole the statues of Harmodius and Aristogeiton, one first attested by the Alexander historians, is likely to be a propagandistic invention. These historians wished to credit their hero with the return of the statues.[16]

The Persian fleet was defeated off Salamis in 480, in a battle the Athenians claimed as their own victory.[17] Xerxes left Greece, entrusting affairs to Mardonius. In 479, the Persians lost the battles both at Mycale in Asia Minor and Plataea in Boeotia where Mardonius was killed in action (Hdt 9.46.1–63.2; 9.101.3–102.3). Sparta contributed the lion's share to this victory. Hence, in Athenian cultural memory, Plataea played a minor role, particularly in comparison to Salamis (cf. Thuc. 1.73.4–74.3).

For the Persian side, the campaign was an annoying waste of lives, time, finances, and resources—surely a blow to Xerxes' image but neither so serious nor the devastating setback the Greek sources make it out to be. Firmly establishing Achaemenid control over the rich satrapy of Egypt was much more important than the trouble in peripheral Greece.[18] Indubitably, the Persian Wars meant much more to the invaded Greeks, and particularly the Athenians, than to the invading Persians.

The Persians withdrew from northern Greece, attacked by the Thracians and spared by the Macedonians (Hdt. 9.89.4). They left a power vacuum. Thracian *ethnoi*, Argead Macedonia and Athens tried to annex as much territory as possible, apparently based on mutual

12 Wiesehöfer 2013, 281.
13 Heinrichs and Müller 2008, 288; Heinrichs 2017, 88–93.)
14 ML no. 27.
15 Hdt. 7.143.3, with Heinrichs 2017, 91–3; Hdt. 7.205.2–220.4.
16 Arr. *Anab.* 3.16.7–8; 7.19.1–2 with Müller 2016a, 173–187.
17 Cf. Aeschin. 2.172; 3.181; Lyc. *Leocr.* 70, 73.
18 Balcer 1984, 281.

arrangements.[19] The Athenians under Cimon cleared the territory by eliminating the last Persian strongholds at the Thracian coast (such as Eion) and Chersonese.[20] In the Aegean, they gained control of most islands and the Hellespont. In the (early?) 460s, at Eurymedon in Pamphylia, under Cimon, they damaged the Persian fleet severely by destroying the important Phoenician force (Thuc. 1.100.1). Persia lost control of the Ionian coast. Athens followed them in the leadership over the Greeks of Asia Minor. As the Ionian Revolt had made clear, dominion over the Ionian cities depended on the naval supremacy in the coastal areas of Asia Minor.[21]

While the Persians showed no ambition to return to Greece with fresh troops, there was common fear in Greece that they would. As a reaction, and as a way to glorify their own role in the Persian Wars, the Athenians convinced other Greek to join their first naval confederacy, the Delian League, founded in 478/477 (Thuc. 1.96.1–2). They proclaimed they were the only power capable of securing their League's members from future Persian attacks (Hdt. 7.139–140). By developing their hegemonic methods, Athens seems to have learned from the enemy.[22]

The Persians failed to return, and meanwhile the Delian League alienated its members by its rigid hegemonic politics and a strict no-exit clause. The Athenians were deemed oppressive. The consequence was a widespread wish among their *symmachoi* to be set free. In the Peloponnesian War, Athens' opponent Sparta exploited this discontent.

It is a matter of debate whether in the early 440s, after Cimon's campaign against Cyprus (451), the Peace of Callias (named after the Athenian ambassador) was settled between Athens and Artaxerxes I (465–424/423). Reportedly, it restricted the areas of naval action: Whereas the Persian fleet was excluded from the zone between the Bosporus and Lycia, the confederate Greek fleet renounced action in the Levant and in Egypt.[23] In view of the lack of explicit contemporary evidence, some scholars suggest that the Peace was invented in the fourth century to highlight Athens' past glory. If the Peace existed at all, perhaps it was not a formal treaty but an informal arrangement.[24]

Persia and Greece from the Peloponnesian War to the King's Peace

Persia intervened only in the second phase of the Peloponnesian War (from 412). There was common ground with the Spartans who wanted to end the Athenian supremacy in the Aegean. In the name of Darius II (423–405), Tissaphernes, the satrap in Sardis,

19 Thus, Alexander I of Macedon seems to have had an arrangement with Cimon regarding his expansion into parts of Bisaltia in about 477/476 (Heinrichs 2017, 79 n.1). Later on, in the late 460s, when Cimon was tried in Athens, his opponents also relied on his activities in the past, accusing him of having been bribed by Alexander (Plut. *Cim.* 14.3). For Argead Macedon, obviously, the trouble did not start with the Persian presence in northern Greece but with the Athenians as the new dominating force posing a constant threat to Argead autonomy, policy, and freedom of action under Perdiccas II who found his realm surrounded by Athenian foundations, allies, and members of the Athenian Naval Confederacy (Müller 2017, 125–224).

20 Hdt. 7.107; Thuc. 1.98.1; Plut. *Cim.* 14.1.

21 Heinrichs 1989, 85.

22 Cf. Kurt Raaflaub, "with few exceptions, the entire range of the Athenian instruments of empire was derived from Persian models" 2009, 97.

23 Isoc. *Pan.* 117–118; Plut. *Cim.* 13.4–5.

24 Invented: Rhodes 2006, 185. Informal: Heinrichs, loc. cit.

stereotypically depicted as a deceitful liar by Thucydides and Xenophon, negotiated Sparta's consent to the renewed Persian claim to coastal cities in exchange for Persian subsidies to build and maintain a fleet able to match that of the Athenians. Thus, when the Spartan commander Lysander outwitted the Athenian fleet at Aegospotami in 405, he had about 200 triremes at his disposal.[25] Athens' position in the Hellespont was crushed, the Athenian Empire (ἀρχή) ceased to exist, and the Greeks of Asia were once more under Persian control.

The Spartans' victory that had ended the Peloponnesian War confronted them with new requirements difficult to meet. Stepping into the power vacuum left by Athens meant that they had to adopt the role of the hegemon of Greece *and* the eastern Greeks. Consequently, a conflict between Sparta and the Persian king concerning Greek Asia Minor was inevitable.

Factional strife in Sparta and a lack of both manpower and finances increased the problem of coping with Athens' "legacy." Additionally, Karl-Wilhelm Welwei points to the Spartan failure to create a clear concept of how to handle the new situation that discredited Sparta as a stabilizing factor.[26]

The cause of the coastal Greeks put Sparta's relationship with Artaxerxes II (404–359) to the test several times. Early in his reign, Sparta and the Greeks supported the failed attempt of his rebellious brother Cyrus the Younger to overthrow him. Cyrus was on good terms with Lysander and had backed the Spartan establishment of short-term oligarchic *dekarchies* in Asia Minor. After Cyrus' defeat and death at the battle of Cunaxa near Babylon in 401 (Xen. *Anab.* 1.8.23–9.1), Artaxerxes II ordered Tissaphernes to collect tribute from Cyrus' Ionian supporters, openly demonstrating who was in charge. These cities called upon the aid of their Spartan protector (Xen. *Hell.* 3.1.4–8). In 400/399, the Spartan commander Thibron was sent to Asia to "liberate" the Ionian Greeks. However, in 397, Sparta, then represented by Dercylidas, ended up settling an armistice with Artaxerxes II whose claim to the coast remained unchallenged (Xen. *Hell.* 3.2.12–20).

In 396–394, the next Spartan liberator of the Greeks, Agesilaus II, pointed at the Persian naval threat and invaded Asia Minor.[27] In fact, he plundered the rich satrapies Phrygia, Lydia, Paphlagonia, Mysia, and Cappadocia.[28] Agesilaus wanted to acquire finances for further Spartan wars necessary to consolidate the hegemony in Greece. Plutarch's claim that Agesilaus wanted to overthrow the Persian king (*Ages.* 15.3) is an unreliable Second Sophistic retrospective, colored by Plutarch's regret that in the end not a Greek but a Macedonian accomplished this. Agesilaus' actions, number of troops, and lack of siege equipment suggest that he set his mind on plundering and probably on securing a permanent income from Spartan control over western Asia Minor.[29]

The Persian side used the discontent of Sparta's opponents in mainland Greece to get Agesilaus out of Asia Minor and weaken Sparta's hegemony. Artaxerxes II gave subsidies to an alliance of Corinth, Argos, Athens, and Thebes (Diod. Sic. 14.82) that confronted Sparta in the Corinthian War (395–387/386). Agesilaus was forced to return to Greece. Xenophon's comment that only Persian bribery prevented Agesilaus from liberating the Ionians who

25 Xen. *Hell.* 2.1.27–28; Diod. Sic. 13.105–106.

26 Welwei 2006, 535–536.

27 Xen. *Ages.* 1.6–7; *Hell.* 3.4.1, 5; Nep. *Ages.* 2.1–2.

28 Nep. *Ages.* 3.1–2; *Hell. Oxy.* 14.1–16.2; 24.1–25.4.

29 McKechnie and Kern 1988, 181.

"mourned his departure as though they were saying good-bye not only to a ruler, but to a father or a *hetairos*" (*Ages.* 1.38) is biased by his admiration for his hero to whom he might have owed his Peloponnesian rural estate. Reportedly, Agesilaus made a numismatic joke saying that he was driven from Asia by 30,000 archers of the Persian king (Plut. *Art.* 20), referring to the iconography of Achaemenid (gold) coins showing a royal archer on the obverse.

In the same year, 394, Sparta became embroiled in a war on two fronts: Persian ships commanded by the Athenian admiral Conon and the Persian satrap Pharnabazus challenged Sparta's position in the Aegean.[30] Sparta's fleet was crushed at Cnidus, Spartan officials (harmosts) and garrisons were expelled from the Aegean and Ionian cities. Thus, the Spartan sphere of influence was limited to the Hellespont. Persian money even enabled the completion of the rebuilding of Athens' Long Walls torn down in 404.

Hard-pressed on several fronts, in 392, the Spartans sent Antalcidas, who advocated a moderate policy, to negotiate a settlement acknowledging Achaemenid control over the coastal cities. However, negotiations failed until 387/386, when the hostilities of the Corinthian War ended with the so-called Peace of Antalcidas or King's Peace. A new strategy to control Greece as a hegemon was introduced: a panhellenic *koine eirene*.[31] Sparta was the guarantor of the Peace in Greece, Persian was the beneficiary in Asia, claiming the mainland, Cyprus, and Clazomenae, and all other Greek cities were to be autonomous. Sparta lost her position in the Aegean but, bolstered by the King's Peace, carried on her offensive strategy in Greece until the Theban victory near Leuctra in 371 ended her hegemony. According to an anecdote, after Leuctra, Artaxerxes II ignored Antalcidas (who asked him for money) and voiced his opinion that Spartans were the most impudent people of all (Plut. *Art.* 22).

During the following decades, the Greeks respected the clause of the King's Peace placing the Ionians under Achaemenid power. Now and then, minor conflicts occurred. The obscure and scarcely documented so-called Satraps' Revolt in the 360s (Diod. Sic. 15.90.1–92.5) seems to have attracted only minimal Greek interference. In the Social War, which the Athenians fought in 357–355, the local dynast and satrap Mausolus of Caria annoyed Athens by supporting Chios, Rhodes, Byzantium, and Cos—rebellious members of Athens' Second Athenian Confederacy, founded in 377.[32]

Persia, Macedon, and Greece from Philip II to Alexander III

Before the end of the 340s, there are no traces of any tensions between Macedon, rising under the reign of Philip II (360/359–336), and Persia, consolidated under Artaxerxes III (359–338)—not even when Artabazus, the satrap of Hellespontine Phrygia, who had revolted against royal authority (Heckel 2006, 55), found refuge with his huge family at Philip's court. Arrian's information about a *philia kai symmachia* between Philip II and

30 Satrapal coins bearing the legend ΦΑΡΝΑΒΑ and showing his portrait on the obverse, support the view that the naval campaign was financed by Persia.

31 Xen. *Hell.* 5.1.35–36; Diod. Sic. 14.110.3; cf. Cawkwell 1981a. *Koine Eirene*: Jehne 1994.

32 Dem. 15.3; Diod. Sic. 16.21.1–4 with Cawkwell 1981b, 52–55; Rhodes and Osborne no. 22.

Artaxerxes III may be plausible.[33] However, the situation changed with Philip's expansion into eastern Thrace. Artaxerxes III was disturbed by the eastern ambitions foreshadowed by Philip's siege of Byzantium and Perinthus in 341/340. According to Diodorus, "Philip's growth in power had been reported in Asia, and the Persian king, viewing this power with alarm, wrote to his satraps on the coast to give all possible assistance to the Perinthians" (16.75.1–2; trans. C.H. Oldfather). Philip's ambitions regarding the Hellespontine sphere offered the first threat to the Persian–Greek balance of power established by the King's Peace. Additionally, Artaxerxes III will have (rightly) feared that Philip's next step would be to plunder the rich satrapies of Asia Minor, challenge the Persian control over the Greek cities, and make them pay their tribute to him. For a time, however, the Macedonians drew back. Philip imposed Macedonian hegemony on Greece in 338. Then, in spring 337, when Greek cities met in the context of the Corinthian League, war against Persia was declared and Philip was elected leader of the campaign (Diod. Sic. 16.89.3). Possibly, he planned to limit his war to the conquest of the coastal cities in order to secure the permanent income necessary for financing wars to maintain Macedon's hegemony.[34] Very probably, Philip's negotiations with Pixodarus, the Carian satrap who might have participated in blocking the Macedonians at Perinthus and Byzantium, were connected with the Persian campaign and served to form an alliance with the man who was in control of the important naval base of Halicarnassus.[35] However, the Persian king seems to have interfered and sent his own man, Orontobates, to Caria to watch over Pixodarus.[36] A Macedonian force commanded by Parmenion and Attalus went ahead in order to secure the coastline (Diod. Sic. 17.24.1), but their initial conquests in Asia Minor, where several oligarchies and tyrannies favorable to the Persians were toppled and replaced by dependent democracies, were countered by the famous Rhodian mercenary Memnon, son-in-law of Darius' influential courtier Artabazus. Memnon kept the Macedonians confined to a small beachhead at Abydus.[37] In the meantime, Darius III (338–330) offered subsidies to the Greeks to induce them to revolt against the Macedonians and keep them busy in Europe. Demosthenes formed part of the Athenian political faction pleading in favor of anti-Macedonian cooperation with Persia.[38]

However, in the autumn of 336, Philip was assassinated and the Persian campaign, duly propagated as a panhellenic act of avenging Persian war atrocities and liberating the Greeks, fell to his son and successor, Alexander. Strictly speaking, the influential generals Parmenio and his son Philotas led the campaign in the beginning. They remained key figures until their elimination in 330.[39]

The Macedonian army supported by Greek contingents of the Corinthian League crossed the Hellespont undisturbed and defeated the satrapal armies of Asia Minor in 334 at the

33 *Anab.* 2.14.1–2 with Olbrycht 2010c, 350. Arrian, however, problematically refers to a letter written by Darius III to Alexander. Its authenticity is uncertain. But the historical kernel, namely the alliance, might have been trustworthy.

34 Wirth 1985, 148–150; Worthington 2008, 170; Müller 2010, 179.

35 Plut. *Alex.* 10.1–3, Ruzicka 2010, 4–11.

36 Arr. *Anab.* 1.23.6–8; Strab. 14.2.17; cf. Ruzicka 1992, 136–139.

37 Heckel 2006, 162; Panovski and Sarakinski 2011, 8.

38 Revolts: Aesch. 3.239; Din. 1.10; Hyp. 5.17; Diod. Sic. 17.7.2. Demosthenes: Wirth 1999, 75.

39 Philip: Flower 2000; Squillace 2010. The generals: Heckel 2016a, 53–55.

battle of Granicus. The Macedonians conquered Lydia, the leading Greek cities, Caria—but only parts of Halicarnassus—and also Lycia and Pamphylia before entering Cilicia in 333. As the fleet of the Macedonians was inferior in number, experience, and technique, they tried to avoid sea battles and neutralize the threat posed by the Persian fleet by capturing their naval bases. After the fall of Miletos, the expensive Macedonian fleet was disbanded but had to be reconstituted shortly afterwards when the Persian resistance in the Aegean put the Macedonians under pressure.[40]

Aeschines reveals that Persian–Athenian diplomatic exchanges did not cease until the eve of the battle of Issus in 333 (3.132, 164). However, after the Macedonian victory, Darius, who had managed to escape from the battlefield, dropped the plan to stir up revolts in Greece and focused on offering fierce resistance in his own empire. At Damascus, Darius' royal camp including his family and treasure was captured. Curtius comments, "Scarcely any courtier's household was unaffected by the disaster" (3.13.14). Rejecting Darius' attempts to ransom his family members, the Macedonian regime held them as valuable hostages and as tokens of Alexander's legitimization as the would-be king of Asia. The honorable treatment given to Darius' relatives was "a display of domination, and humiliating to the Great King, who was fighting for his life and survival of his empire."[41] When Darius attempted to negotiate with the Macedonians in order to turn Alexander into a vassal confined to the Mediterranean littoral, Alexander responded with theatrical scorn.

The Macedonian generals decided to secure the Levantine coast, aiming at cutting off the Persians from the regional naval bases before pursuing the king. In spring 333, embarking from Cos, Memnon, Darius' commander-in-chief in the Aegean, took the offensive there, having some 300–400 ships at his disposal. After Memnon's early death in about summer 333 during the siege of Mytilene, his brother-in-law Pharnabazus and the admiral Autophradates, an offspring of another old satrapal family, took over and started "a series of aggressive, wide-ranging, and strategically intelligible actions."[42] The crucial base of Chios was in their hands. They seized Lesbos (where the Greek Chares, an old friend of Artabazus' family, commanded 2,000 mercenaries for the Persians), Tenedos, Andros, and Siphnos. They recaptured partly lost Halicarnassus and perhaps Miletos (cf. Curt. 4.1.37), and entered into negotiations with Agis III of Sparta, ambitious to revolt against the Macedonians in Greece. Despite the defeat at Issus and setbacks in the Aegean where the new Macedonian fleet began to strike back, "the grand Persian strategy remained unchanged." The end of the Aegean campaign came only in 332 when the Phoenician fleet defected to the Macedonians, and Tyre, a crucial port, was conquered after seven months of fierce resistance. Gaza and Egypt fell to the Macedonians.[43] They could thus occupy the last Persian naval bases and get their hands on the grain from Cyrene.

In the meantime, Darius assembled another large army supported by the elite of Bactria and Sogdiana. In addition, Indian local dynasts and Scythian ethnics sent troops to him.[44] Following the principle of learning from the enemy, Darius had also ordered military reforms

---

40 Ashley 1998, 91. The fleet disbanded: Diod. Sic. 17.22.5; Arr. *Anab.* 2.20.1. Revived: Curt. 3.1.19; Arr. *Anab.* 3.2.6.

41 Heckel and McLeod 2015, 260.

42 Heckel 2006, 65; the citiation, Ruzicka 1988, 134.

43 The citation, Ruzicka 1988, 144. Levantine events: Ruzicka 2012, 206.

44 Scythians: Diod. Sic. 17.59.5, 8; Curt. 4.14.3, 15.14, 15.18; Arr. *Anab.* 3.8.3, 11.5–6, 13.2, 13.4. Indians: Diod. Sic. 17.59.4; Arr. *Anab.* 3.8.3, 8.6, 11.5–6, 15.1.

in order to be equipped in a manner more suitable to counter the Macedonians.[45] However, his army was defeated again at Gaugamela (Arbela) in late 331. Darius could escape while the Macedonians conquered the abandoned royal residences and plundered the treasuries. Still hoping to rebuild his army, he was assassinated in 330 when his own generals and courtiers conspired against him. For Alexander, the alleged panhellenic mission was accomplished: The conflict, however, went on, triggered by the acclamation of the Bactrian satrap Bessus, probably an Achaemenid adopting the throne name of Artaxerxes V in 330, and by fierce resistance in Bactria and Sogdiana to the Macedonians (329–328). The invaders conquered the eastern parts of the Achaemenid Empire to the river of Hyphasis in India and made their way back. Yet the Indian lands informally connected with the Achaemenid house before could not be ruled directly. Macedonian control faded away as soon as the army left. When Alexander died in Babylon in the summer of 323, his swiftly conquered empire was susceptible to unrest.[46]

In order to cooperate with the indigenous leading circles, Alexander adopted the Achaemenid system of administration, plus elements of court etiquette and royal representation, and integrated indigenous nobles into his court and soldiers into his army. This political continuity was, however, an illusion meant to legitimize the new monarch. Persian rule had been ended—by the army of a descendant of a former Persian *hyparchos*.

45 Heckel 2017a.

46 Bessus: Jacobs 1992. India: Bosworth 2003.

CHAPTER 3

Internal Wars from the "First Peloponnesian War" to Chaeronea

Frances Pownall

Introduction

Although the Persian invasions galvanized many of the Greek city-states in uniting, this new-found harmony was short-lived and shortly afterward the Greeks reverted to their traditional state of perennial internecine warfare, with one important difference—the stakes were now much higher. Instead of relatively small-scale campaigns where citizen armies challenged their neighbors for control of scarce agricultural land (particularly in border territories) in a single pitched infantry battle,[1] the Greek world was now divided into two competing spheres of influence, ensuring that henceforth internal wars were to become lengthy large-scale campaigns.

In response to the Persian threat, the Athenians developed the largest navy in the Greek world. Even more importantly, the Persian Wars instilled into the Athenians the mindset that not only could they protect their fellow Greeks from any possible Persian repercussions, but that as the saviors of Greece they were entitled to exercise hegemony over them (Aesch. *Pers.* 233–234; Hdt. 7.139; Thuc. 1.73–75.1, 5.89, 6.83.2). The foundation of the Delian League (so named after the organization's headquarters on the

1 This "traditional" view of early intra-Greek conflicts (see e.g. Ober 1996; Hanson 2000 and 2009, 27–39; Schwartz 2009, 102–146 and 226–230; cf. Thuc. 1.15) has been challenged, on the grounds that the hoplite phalanx did not develop fully until the Classical period (van Wees 2000), its origins possibly retrojected into the Archaic Period as a nostalgic response to the destructive nature of contemporary warfare (Krentz 2002, 25). For a succinct overview of the controversy, see Lee 2006, 484–486; cf. Chapter 6 in this volume.

A Companion to Greek Warfare, First Edition. Edited by Waldemar Heckel, F.S. Naiden, E. Edward Garvin, and John Vanderspoel.

island of Delos, site of an important pan-Ionian sanctuary) in 478/477 afforded the Athenians leadership over willing allies.[2] Athenian control over the allied naval forces as well as the annual tribute of those allies who did not contribute ships (Thuc. 1.96.2) soon gave them the ability to rival the Spartans' military dominance in Greece through leadership of the Peloponnesian League. As Thucydides observes (1.18.2–3), not long after the Persian Wars the Greek world became divided between two great powers, Athens on the sea and Sparta on land, each heading a coalition of allies. The military polarization between the two soon shaded into the realm of ideology, as the Spartans tended to favor oligarchic governments and the Athenians democratic ones. When the uneasy alliance forged during the Persian Wars began to fray, the stage was set for a major confrontation between Athens and Sparta, each of which now possessed the manpower and the resources to wage extensive year-round campaigns far beyond their borders throughout the Greek world.

The "First Peloponnesian War" (460–446/445)

Although the Spartans at first acquiesced in the Athenian leadership of a maritime naval alliance,[3] it soon became apparent that the Athenians had their own increasingly imperialistic interests in mind at least as much as action against the Persians, the ostensible purpose of the Delian League.[4] According to Thucydides, our main source for the period between the Persian Wars and the Peloponnesian War (the so-called *Pentekontaetia* or "Fifty Years"), the early years of the Delian League were marked not only by naval campaigns against Persia, but also by the increasingly heavy-handed Athenian treatment of their allies, including the forcible subjugation of members of the League who attempted to secede (Thuc. 1.98–100). Thucydides' terse narrative is tendentious, however, and deliberately shaped to support his contention that the "truest cause" (as he put it) of the Peloponnesian War was the growth of Athenian power and the fear it engendered in the Spartans (1.23.6). For what it is worth, the Spartans also experienced difficulty in maintaining control over their Peloponnesian allies in the decades following the Persian Wars.[5] Nevertheless, despite the appearance of cracks in the truce between Sparta and Athens (most notably the Spartan opposition to the Athenian rebuilding of their city walls and the alleged secret promise of the Spartans to support the

2 As Hornblower (2011, 12) observes, Thucydides (1.95.1) offers both a negative and a positive motive for the Ionian approach to the Athenians: resentment of the oppressive behavior of the Spartan regent Pausanias and recognition of the Athenians' traditional status as mother city of the Ionians.

3 As Thucydides implies (1.75.2, 1.95.7; see also Xen. *Hell.* 6.5.34), but cf. 1.92.1 and evidence from other authors suggests otherwise: Hdt. 8.3.2 (on this passage, see Munson 2001, 214–217); [Arist.] *Ath. Pol.* 23.2; Plut. *Arist.* 23.2. Diodorus (11.50) may be correct that the Spartans themselves were divided on this point; see Hornblower 2011, 8–10.

4 Thuc. 1.96.1, who states that the League was founded in order to get revenge on the Persians by ravaging their land; it is generally assumed that the League had a defensive purpose as well as an offensive one (Rhodes 2010, 19), and that it offered the Athenians a pretext (*proschema*) for leadership over the Greeks (Hornblower 2011, 12–13).

5 Herodotos' reference (9.35.2) to Spartan battles in the 470s and 460s at Tegea and Dipaea suggests the presence of anti-Spartan movements in the Peloponnese (cf. Thuc. 1.118.2); see Lewis 1992a, 104–108.

Thasians in their revolt by invading Attica),[6] open hostility did not break out until the Spartans dismissed the Athenian commander Cimon and his troops who had come to assist them in suppressing the helot revolt at Mount Ithome (Thuc. 1.101–102).

This unceremonious dismissal had far-reaching consequences, for the Athenians renounced their existing alliance with the Spartans, and energetically proceeded to make alliances with Sparta's traditional enemy of Argos as well as Thessaly (Thuc. 1.102.3). They also accepted an alliance in 461 with the Megarians, erstwhile members of the Peloponnesian League who were getting the worst of an ongoing border dispute with their Corinthian neighbors (Thuc. 1.103.4). Megara's strategic location on the narrow isthmus joining the Peloponnese to central Greece was not lost on the Athenians, who immediately connected Megara and its port of Nisaea on the Saronic Gulf with protective walls, which they garrisoned with their own troops (Thuc. 1.103.4). Not only was this overtly expansionist activity around the isthmus threatening to the Corinthians in particular (Thucydides 1.103.4 influentially identifies it as the "original cause of the extreme hatred" of Corinth for Athens[7]) but the Athenian control of Megara prevented the Spartans from leading troops out of the Peloponnese by land into central Greece.[8] This blatant Athenian attempt to extend their sphere of influence provoked a series of open conflicts with the Spartans and their Peloponnesian allies (especially Corinth), conventionally but somewhat misleadingly known as the First Peloponnesian War (Lewis 1997, 72).

In the opening engagement, the Athenians attempted to counter Corinthian ambitions in the eastern Peloponnese by attacking the port town of Halieis on the Argolic gulf (possibly on behalf of their new ally Argos, but almost certainly also to consolidate their own control of the Saronic Gulf),[9] where they were defeated in a land battle by the Corinthians, Epidaurians, and Sicyonians (Thuc. 1.105.1, with *SEG* 31.369). Despite this inauspicious beginning, the Athenians defeated the Peloponnesians soon afterward in a sea battle off the small island of Cecryphalea just west of Aegina (Thuc. 1.105.1) and followed up this victory with a decisive defeat of the powerful Aeginetan fleet (along with their Peloponnesian allies), which enabled them to lay siege to Aegina (Thuc. 1.105.2), the large island that was their longtime rival in the Saronic Gulf. The Corinthians attempted to aid the Aeginetans by sending forces to attack Megara, assuming that with so many troops committed overseas in the Egyptian expedition the Athenians would be forced to withdraw from Aegina (Thuc. 1.105.3). The Athenian general Myronides defied expectations by leading the reserve troops (that is, those considered too young or too old for active duty) to Megara, where they defeated the Corinthians first by a narrow margin and then more decisively in a second battle culminating in the massacre of their retreating foes (described vividly by Thucydides 1.106.1–2).[10] Despite their initial success, the Athenians were sufficiently nervous of a

6 On the tension created by the rebuilding of the walls, see Thuc. 1.89.3–93. The tradition that the Spartans were prevented from coming to the Thasians' aid by a devastating earthquake and ensuing helot revolt is probably a later invention designed to explain why they did not seize the opportunity to nip nascent Athenian imperialism in the bud; Rhodes 2018, 47; cf. Rhodes 2010, 31.

7 Lewis 1997; Hornblower 2011, 25–26.

8 Harrison 2006: 518. Hornblower (2011, 26–29) observes that the Spartans were also jostling for position with the Athenians for influence at Delphi.

9 So Lewis 1997, 75–76.

10 An Athenian casualty list (*IG* I3 1147) records the campaigns of Halieis, Aegina, and Megara as occurring in a single year, either 460 or 459 (on the date, see ML no. 33 and Rhodes 2010, 50).

possible Peloponnesian invasion by land that they began to construct the Long Walls linking the city to the Piraeus in order to ensure that Athens could not be cut off from the sea (Thuc. 1.107.1).

It is at this point that Thucydides records the Spartans as openly entering the conflict for the first time,[11] sending an army to liberate their mother city of Doris in central Greece from the Phocians (Thuc. 1.107.2).[12] Although Thucydides (1.107.3) tells us that after their expulsion of the Phocians the Spartan forces were prevented from returning home by land and by sea (thanks to Athenian control of the isthmus and the Corinthian Gulf) and therefore decided to cross into Boeotia instead, it is far more likely that the Spartans' actual intention was to build up the Theban army (in return for support of Theban hegemony over Boeotia) as an effective force against Athens beyond the isthmus.[13] The Athenians, wary of the danger presented by a Spartan–Boeotian *rapprochement* (particularly in light of overtures to the Peloponnesians by political opponents of the democracy in Athens, if Thucydides 1.107.4–6 can be believed), sent a large army to fight at Tanagra, but were no match for the Spartans on land (Thuc. 1.108.1). Two months later, the Athenians under Myronides retaliated by marching north and defeating the Boeotian army at Oenophyta, which left them in control of much of central Greece (Thuc. 1.108.2),[14] and soon afterward of the Saronic Gulf as well, through their reduction of Aegina (Thuc. 1.108.4).

This aggressive policy of expansion culminating in the Athenian acquisition of a land empire could not be sustained indefinitely, particularly with the ongoing large-scale naval campaigns against the Persians in the eastern Mediterranean, and the Athenians soon suffered a series of reverses (Thuc. 1.109–111), prompting them to sign a five-year truce with the Peloponnesians (Thuc. 1.112.1).[15] During the period of the truce, the only recorded conflict between the Spartans and the Athenians occurred indirectly over control of Delphi.[16] Upon its conclusion, the Boeotians lost no time in exploiting Athenian vulnerability by revolting, and an Athenian army was defeated at Coronea, forcing the Athenians to withdraw from all of Boeotia (Thuc. 1.113). The uprising in Boeotia was soon followed by the revolts of both Euboea and Megara, and the Spartan invasion of Attica (Thuc. 1.114). The Athenians had no choice but to give up any formal claims to a land empire (not only central Greece, but all of their land outposts in the Peloponnese) in the Thirty Years' Peace they signed with Sparta in 446/445 (Thuc. 1.115.1),[17] the terms of which cemented the division of the Greek world into two great power blocs, Athens exercising hegemony by sea and Sparta by land.[18]

---

11 The original impetus for the outbreak of hostilities was rivalry between Athens and Corinth, rather than Spartan hostility toward Athens; so Holladay 1977 and Lewis 1997 *contra* De Ste. Croix 1972, esp. 187–190.

12 Hornblower (2011, 28–29, 32) is almost certainly correct that the object of the Spartan intervention was to restore their influence at Delphi.

13 So Diodorus 11.81.2–3, supported by Justin 3.6.10; see Badian 1993, 213 and Hornblower 2011, 32–33.

14 Probably in 457; on the date, see Rhodes 2010, 50.

15 On the problems associated with the dating of this truce, see Green 2006, 165 n. 351.

16 The so-called Second Sacred War; cf. Pownall 1998, 36–38 and Hornblower 2011, 28–29, 35.

17 The date can be calculated from Thucydides 2.2.1, 2.21.1; cf. Rhodes 2010, 57.

18 Thuc. 1.35.2, 1.40.2, 1.44.2; cf. Hornblower 2011, 36.

The Peloponnesian War (431–404)

Despite the mutual recognition of separate spheres of influence, the Thirty Years' Peace did not dissolve the tension that existed between Sparta and Athens. Although the terms of the peace allowed Athens a free hand in their brutal suppression of the Samian revolt in 440/439,[19] by the mid-430s a series of disputes broke out—Thucydides' "openly acknowledged grievances" (1.23.6), especially between Athens and Corinth, Sparta's most powerful and aggressive ally. Open warfare between Corinth and its disaffected colony of Corcyra led to the Corcyraeans approaching the Athenians with a request for a defensive alliance. As a neutral state, they were permitted by the terms of the Thirty Years' Peace to join either the Spartan or the Athenian bloc (Thuc. 1.35.2, 1.40.2), and the Athenians had their eyes upon their large navy, especially as they were well aware that war was in the air.[20] Inevitably, however, the Athenian acceptance of this defensive alliance resulted in open warfare between Athenian and Corinthian ships in the Battle of Sybota in 433 (Thuc. 1.47–54). In the aftermath of the battle, the Athenians became concerned about the continued loyalty of Potidaea, a city in the Chalcidice, an area crucial for Athenian economic and military interests. Although Potidaea was a tribute-paying member of the Delian League, it was also a Corinthian colony that maintained a very close relationship with its mother city. When the Athenians demanded that they tear down a portion of their fortification walls, send hostages, and sever their official ties with Corinth (Thuc. 1.56.2), the Potidaeans appealed successfully to Corinth, resulting in the Athenian siege of the city (Thuc. 1.58–65). Soon afterward, the Corinthians, supported by the Megarians and the Aeginetans (Thuc. 1.67),[21] put pressure on the Spartans to take action against the Athenians, and the Spartans issued an ultimatum, which the Athenians on Pericles' advice rejected (Thuc. 1.145).

The outbreak of the Peloponnesian War occurred in the spring of 431, when the Thebans attempted to invade the recalcitrant Boeotian city of Plataea, a loyal ally of Athens. This act of aggression officially broke the terms of the Thirty Years' Peace (Thuc. 2.7.1). Soon afterward, the Spartan King Archidamus led the Peloponnesian army into Attica, where he began the first of what was to be a series of annual invasions; hence the first phase of the Peloponnesian War is conventionally known as the Archidamian War (431–421). The Spartan strategy was to ravage Attic agricultural land in order to draw out the Athenians into a hoplite battle against the superior Peloponnesian land forces. Following Periclean strategy, however, the Athenians remained behind their walls, relying on their strong navy to import what they needed and to conduct reprisals on the Peloponnese by sea.[22]

In spite of the devastating plague that struck Athens in 430 (carrying off Pericles himself in the following year), at first the Athenian strategy was successful, culminating in the fortification in 425 of an outpost at Pylos in Messenia (ancestral homeland of the Spartan helots), which resulted in the shocking capture by the Athenian commanders Cleon and Demosthenes

19 On the Samian Revolt, see Thuc. 1.115.2–117; Diod. Sic. 12.27–29; Plut. *Per.* 24.1–2, 25–28; cf. Shipley 1987, 113–122 and Hornblower 2011, 36–37. For the war atrocities that the Athenians were alleged to have committed on their prisoners, see Duris of Samos, *FGrH* 76 FF 66 and 67. As Hornblower (2011, 37) observes, Thucydides ends his *Pentekontaetia* here, because "Samos was the last big violation of autonomy which Athens was to get away with".

20 Thuc. 1.44.2; cf. the careful financial preparations laid out in the Callias decrees, ML no. 58.

21 On Thucydides' downplaying of the grievances of Aegina and Megara, see Hornblower 2011, 112–114.

22 On the war strategy on both sides, see Rhodes 2010, 101–106 and Hornblower 2011, 157–160.

of 120 Spartan hoplites on the nearby island of Sphacteria.[23] Overconfident as a result of this unexpected success, the Athenians (induced by Cleon) continued to reject Spartan negotiations for peace (Thuc. 4.41.3–4), and made a serious miscalculation the following year in engaging in a pitched hoplite battle at Delium in an attempt to democratize Boeotia and sway it away from the Peloponnesian side.[24] The bloody defeat that the Boeotian army inflicted on the Athenian forces at Delium proved conclusively that the latter were no match on land.[25] After the disaster at Delium, matters went rapidly downhill for the Athenians, when the Spartan commander Brasidas succeeded in winning away from Athens a number of strategically important cities in the north (Thuc. 4.102–116). Nevertheless, it was only when both Cleon and Brasidas (the "pestles" of war in Aristophanes' *Peace*) died in the course of the unsuccessful Athenian attempt to recover Amphipolis that both sides finally agreed to negotiate for peace (Thuc. 5.16.1).

The resulting Peace of Nicias, named after the leading Athenian negotiator, was signed in 421 and essentially restored both Athens and Sparta to the status quo, as both sides were required (with a few exceptions) to restore any territorial gains they had made during the war (Thuc. 5.18–19). In other words, it addressed none of the issues that caused Sparta's allies to push for war, and it revealed the hollowness of the Spartan claim to be the "liberators" of Greece (Thuc. 2.8.4, 3.32.2, 3.59.4).[26] Thus, the Greek world returned to the state of unstable polarization with which the war had begun, and the peace that was supposed to last for a period of 50 years endured only for 6 (Thuc. 5.25.3). A coalition of the Spartans' disgruntled allies began to jeopardize their control of the Peloponnesian League, a situation exploited by the charismatic young Athenian aristocrat Alcibiades, who negotiated a defensive alliance between Athens and Sparta's traditional rival of Argos, as well as some other Peloponnesian cities (Thuc. 5.43–48). These anti-Spartan movements within the Peloponnese culminated in the Battle of Mantinea in 418, identified by Thucydides (5.74.) as the largest hoplite battle to have occurred for a considerable time, where at one fell swoop the Spartans reasserted their hegemony of the Peloponnese (Thuc. 5.75.3). As the uneasy peace continued, both sides began to seek out potential resources beyond mainland Greece and to search for new ways of achieving total defeat of their enemies, including the massacre of civilians (the Spartans at Hysiae in Argos, and the Athenians upon the reduction of the Dorian island of Melos; Thuc. 5.83, 5.85–111).

This new mindset left the Athenians vulnerable to the grandiose ambitions of the unscrupulous Alcibiades, who persuaded his fellow citizens (contrary to the caution urged by the seasoned general Nicias) to mount a large expedition to Sicily, ostensibly in support of their Italian allies, but in reality to gain an upper hand against their enemies in Greece (Thuc. 6.1, 9–24). Despite the unprecedented magnitude of the Sicilian expedition (415–413), it went wrong for the Athenians right from the start. The night before the ships were to set sail, busts of the god Hermes (the patron god of travelers) were mutilated, and in the course of their investigations the authorities uncovered evidence of the profanation of the Eleusinian Mysteries; public opinion considered these religious scandals to be part of a plot to overthrow the democracy in which Alcibiades himself was implicated.[27] Thus, the expedition

23 For the reaction of the other Greeks to the surrender of the Spartan hoplites at Sphacteria, see Thuc. 4.40.

24 Thuc. 4.75–76; cf. Buck 1994, 16–18.

25 Thuc. 4.89–101.2 with Lendon 2005, 78–90.

26 So Lewis 1992b, 432 and Hornblower 2011, 164–165.

27 Thuc. 6.27–29; on the atmosphere in Athens, see Rubel 2000, 178–232 and Rhodes 2010, 166–168.

began under a cloud and the situation did not improve once the Athenian armada reached Sicily. The Athenians discovered that their so-called allies were unwilling to support the effort against Syracuse (the most powerful and aggressive city in Sicily), and they failed to follow up their initial successes.[28] Worse yet, Alcibiades was recalled to stand trial for the profanation of the Mysteries (Thuc. 6.61), but escaped en route and made his way to Sparta (Thuc. 6.88.9), where he gave the Spartans two ultimately devastating pieces of advice: to intervene in Sicily and to fortify Decelea in Attica (6.91–92), which would cut the Athenians off from accessing supplies by land (cf. Thuc. 7.27–28). In the meantime, the Athenians successfully besieged Syracuse by sea, but failed to complete the circuit walls that would have blockaded the city completely. Athenian morale began to plummet, especially in light of the arrival of Peloponnesian reinforcements under the Spartan commander, Gylippus, and the chronic illness of the sole remaining Athenian general, Nicias, who requested to be recalled. Neverthless, the Athenians at home believed victory was still possible and sent a large fleet of reinforcements, commanded by the capable Demosthenes. Correctly surmising that the Athenians' only chance lay in striking immediately (Thuc. 7.42.3–5), he attempted a bold night attack, but was unsuccessful because many of the new recruits were unfamiliar with the terrain. At this point, Demosthenes advised cutting Athenian losses and withdrawing from Sicily, but was overridden by Nicias (Thuc. 7.47–49). Nicias only agreed to retreat after the Athenian position became untenable, but was further delayed by the seers' advice after an eclipse of the moon (Thuc. 7.50.3–4). Following a final desperate attempt to break out of the Great Harbor in Syracuse, the Athenians were forced to abandon their remaining ships and retreat by land, where they were slaughtered by the pursuing Syracusans and forced to surrender. Nicias and Demosthenes were executed, and the surviving Athenians were imprisoned in horrific conditions in the quarries in Syracuse (Thuc. 7.86–87.4).

Following the massive Athenian defeat in the ill-fated Sicilian expedition, the theater shifted to Ionia, as the Spartans resolved to make total war on the Athenians and become the undisputed leaders of Greece (Thuc. 8.2.4) in the final phase of the Peloponnesian War (413–404). To this end, the Spartans began to build up their fleet (Thuc. 8.3.1), foster revolts among the Athenian naval allies (Thuc. 8.5–6), and pursue Persian support in both aims. Although influential at first, Alcibiades soon lost the trust of both the Spartans and the Persians, and began agitating for a return to Athens, promising Persian help in return for the overthrow of the democracy. Mistrusted also by the leaders of the successful but short-lived oligarchic *coup d'état*,[29] ironically Alcibiades was recalled by the Athenian democracy, now embodied in the fleet at Samos, which immediately elected him general (Thuc. 8.81–82.1). Under Alcibiades' leadership, the Athenian fleet handily won a number of sea battles, as described by Xenophon in the first book of his *Hellenica* (Thucydides' narrative ends with the Athenian victory at Cynossema in 411), and the success of the navy provoked the restoration of the Athenian democracy. Alcibiades' string of successes ran out, however, with the defeat of his second-in-command at Notium in 406 (Xen. *Hell.* 1.5.10–14), and he failed to be re-elected general (Xen. *Hell.* 1.5.16–17). Even without Alcibiades, the Athenians were victorious over the Spartan fleet at a battle in the Arginusae Islands between the island of Lesbos and the coast of Asia Minor (Xen. *Hell.* 1.6.24–34), but the aftermath of the victory

28 For a useful summary of the Athenian weaknesses that ultimately crippled the Sicilian expedition, see Hornblower 2011, 174–178.

29 On the oligarchic revolution of 411, see Rhodes 2010, 168–175 and Hornblower 2011, 178–187.

left a tragic stain on the reputation of the Athenian democracy, for the commanders were put on trial in Athens for failing to recover the shipwrecked sailors (having been prevented by a storm) and were condemned to death (Xen. *Hell.* 1.7.1–35). The Spartan *nauarch* Lysander, whose friendship with the Persian prince Cyrus enabled him to strengthen the Spartan fleet with Persian subsidies, attacked the Athenian fleet at Aegospotami unprepared (in spite of the advice of Alcibiades, in his final appearance before his mysterious assassination) and was easily victorious in the last battle of the Peloponnesian War in 405 (Xen. *Hell.* 2.1.22–32). Thereupon Lysander blockaded the Piraeus and starved the Athenians into surrender in 404. The Athenians had no choice but to tear down their Long Walls and the Piraeus fortification walls, surrender all of their ships except 12, relinquish their overseas possessions, accept back their (mostly oligarchic) exiles, and acknowledge Spartan hegemony.[30] So ended the Peloponnesian War, with the Athenian exiles demolishing the walls of Athens to the music of pipe girls, "believing that this day was the beginning of freedom for Greece" (Xen. *Hell.* 2.2.23).

The Corinthian War (395–387/386)

The generation-long Peloponnesian War resulted in the collapse of the polarization between Athens and Sparta and left a lasting legacy on Greece.[31] The optimism that the defeat of Athens would bring freedom to the Greeks was demonstrated almost immediately to be illusory, as the insular Spartans proved utterly incapable of managing an empire beyond the Peloponnese and the brutality of their rule soon made them even more unpopular than the Athenians.[32] Furthermore, the Spartans' failure to destroy Athens at the end of the Peloponnesian War left their allies restless and unsatisfied (Xen. *Hell.* 2.2.19–20). Thebes, Megara, Argos, and Corinth immediately defied the Spartans by receiving exiles from the Spartan-backed government of the Thirty in Athens,[33] and the Boeotians and Corinthians refused to participate in the Spartan attempt to dislodge the Athenian democratic forces from the Piraeus (Xen. *Hell.* 2.4.30; cf. 3.5.5). Shared animosity to Sparta soon resulted in a *rapprochement* between Thebes, Corinth, and Athens, whose individual goals (respectively the hegemony of Boeotia, greater independence in the Peloponnese, and restoration of their sea empire) would all be furthered by challenging Spartan imperialism.[34] It was not until the Persians intervened directly by offering financial support for this anti-Spartan coalition (now joined by Argos) to foment war against Sparta in Greece (thereby forcing the Spartans to withdraw their forces from Asia Minor, where they were trying to liberate the Greek cities),[35] that the outbreak of the Corinthian War occurred.

The war began in 395 with a dispute engineered by the Thebans between Phocis and Locris, both of whom appealed to their allies. Lysander advanced to Haliartus in Boeotia, hoping that victory there would bring a speedy end to the conflict, but was defeated and

30 Xen. *Hell.* 2.2.19; Andoc. 1.80, 3.11–12; Diod. Sic. 13.107.4; Plut. *Lys.* 14.8.

31 On the effects of the Peloponnesian War, see Hornblower 2011, 190–216; cf. Welwei 2006, 535–537.

32 On the challenges facing Sparta in 404, see Buckler 2003, 1–3.

33 Xen. *Hell.* 2.4.1; Aeschin. 2.147–148; Diod. Sic. 14.6.2–3; cf. Buckler 2003, 5–8.

34 Corsi and Bernardini 2008, 33.

35 Xen. *Hell.* 3.5.2–12 (who patriotically denies that the Athenians accepted Persian gold); *Hell. Oxy.* 10.2; Paus. 3.9.8; Polyaenus *Strat.* 1.48.3.

killed before Pausanias could arrive with reinforcements.[36] The following year saw two large-scale hoplite battles at the Nemea River near Corinth (before the return of Agesilaus from Asia Minor) and Coronea in Boeotia (after his return).[37] Sparta's failure to gain a decisive victory resulted in a stalemate marked by ongoing conflict around Corinth for control of the isthmus. The Spartans were frustrated also in their attempts to end the war through diplomacy,[38] and it was only when the Persians became alarmed by Athens' growing naval resurgence and renewed control of the Aegean that peace was signed in 387/386.

Spartan Imperialism (387/386–371)

The King's Peace was negotiated with the Persians by the Spartan *nauarch* Antalcidas. In exchange for signing over the Greeks of Asia Minor to Persia, the Spartans were given a free hand within Greece (Isoc. 4.127), and the autonomy clause at one stroke dashed the hegemonial aspirations of Thebes over Boeotia, Athens' hopes to establish a new maritime empire, and the political union of Corinth and Argos, forcing Corinth to return to the fold of the Peloponnesian League.[39] The Spartans used the authority conferred upon them to enforce the terms of the treaty to punish the Arcadian city of Mantinea for evincing insufficient loyalty during the Corinthian War (Xen. *Hell.* 5.2.1–7), and then to suppress the increasingly powerful Chalcidian League, led by the city of Olynthus (Xen. *Hell.* 5.2.11–24, 37–43, 3.1–9, 18–20, 26). Even more egregiously, in 382 the Spartans occupied the Theban acropolis, garrisoned the city, and imposed a pro Spartan government (Xen. *Hell.* 5.2.25–36), a blatant violation of the autonomy clause.

This act of *hybris* (cf. Xen. *Hell.* 5.4.1) marked the climax of Spartan power and matters rapidly went downhill for them afterward. The first nail in their coffin was the recapture of the Theban acropolis by a group of exiles who had taken refuge in Athens; they assassinated the pro-Spartan rulers and forced the Spartan garrison to withdraw.[40] Successive Spartan military incursions into Boeotia were limited and unsuccessful, and the attempted raid on Attica by the Spartan *harmost* Sphodrias succeeded only in exacerbating tensions with Athens.[41] The Thebans re-established the Boeotian League and the Athenians founded a second Athenian naval confederacy, of which the Thebans were founding members. Mutual antipathy for Sparta did not make easy bedfellows, however, and the Athenians became increasingly nervous of the rapid growth of Theban power, particularly after Pelopidas' reorganization of the Theban army around a core unit of elite warriors, the Sacred Band, led to the unexpected defeat of a larger force of Spartans at the Battle of Tegyra in 375 (Diod. Sic. 15.37; cf. 15.81.2; Plut. *Pel.* 16). Tensions culminated in renewed peace negotiations in 375, but the peace was short-lived thanks to continuing conflict between Sparta and Athens, as well as aggressive Theban expansion, including the destruction of Athens' loyal ally Plataea

36 Xen. *Hell.* 3.5.17–25; cf. Buckler 2003, 79–83.

37 Xen. *Hell.* 4.2.18–23 and 4.3.15–20; cf. Buckler 2003, 85–95.

38 On the failed peace agreements of the late 390s, see Buckler 2003, 139–152 and Rhodes 2010, 227–228.

39 On the King's Peace, see Xen. *Hell.* 5.1.31 and 36; Diod. Sic. 14.110.2–4; cf. Badian 1991 and Buckler 2003, 166–180.

40 Xen. *Hell.* 5.4.1–12; Diod. Sic. 15.25; Buck 1994, 71–87.

41 On this puzzling episode, see Xen. *Hell.* 5.4.20–24, with Buckler 2003, 220–225 and Buckler and Beck 2008, 79–84.

(Isoc. 13; Paus. 9.1.8; cf. Xen. *Hell.* 6.3.1). In 371, both Sparta and Athens agreed to put aside their differences, and hammered out a peace agreement, which the Thebans refused to sign when the Spartans would not recognize their hegemony over Boeotia (Xen. *Hell.* 6.3.18–20; Diod. Sic. 15.50.4). Resolving to force the Thebans to comply with the terms of the peace, the Spartans sent an army into Boeotia, which engaged with the deepened Theban phalanx at Leuctra. Unable to cope with novel military strategy, the Spartans were severely defeated and the battle ended Sparta's traditional hegemony of Greece.[42]

The Theban Hegemony (371–362)

After their decisive victory at Leuctra, the Thebans immediately struck at Spartan control of the Peloponnese with a series of successful invasions, in the course of which the Spartans witnessed their own territory devastated for the first time (Xen. *Hell.* 6.5.24–32; Diod. Sic.15.64–65). Worse yet, the Thebans permanently liberated Messenia (Diod. Sic. 15. 66.1; Paus. 4.27), not only putting an end to centuries of Spartan domination but also crippling Spartan infrastructure. Although the Thebans also intervened in both central and northern Greece, they did not impose direct control on their allies outside of Boeotia,[43] which led to constant shifting of alliances and jockeying for position among previously subordinate states, particularly in the Peloponnese. This Peloponnesian infighting culminated at the Battle of Mantinea in 362, when despite the expectation that the battle would determine the leadership of Greece, the death of the Theban leader Epameinondas led to a stalemate, and "there was even more confusion and disorder in Greece than there had been before," as Xenophon states in the pessimistic conclusion to his *Hellenica* (7.5.27).

The Road to Chaeronea (362–338)

The indecisive result of the Battle of Mantinea left a power vacuum in Greece, exploited by a new player on the scene, Philip II of Macedon, hailing from a region that was traditionally considered a backwater on the periphery of the Greek world. Upon his sudden accession to the throne in 360/359, Philip energetically began to transform Macedonia into a major power. After centralizing his kingdom and reorganizing the Macedonian army, Philip began to expand, both in the northern Aegean, where the Athenians soon became concerned about maintaining control of their grain supply, and in Thessaly, where he was invited to settle the perennial rivalry between powerful elite families. This Thessalian connection offered Philip an opportunity to intervene in central Greece, in the so-called Third Sacred War (356–346), instigated by the Thebans and the Amphictyonic League that controlled Delphi against the Phocians, who seized the sanctuary and used its resources to hire a mercenary army. A string of victories by the Phocians and their mercenaries resulted in the Thessalian summoning of Philip, who experienced a rare defeat in a first engagement in 353. The following year,

42 Xen. *Hell.* 6.4.1–15; Diod. Sic. 15; cf. Buckler 2003, 289–293.

43 Buckler 2003, 298–299 and LaForse 2006, 544–546.

however, Philip returned to Thessaly with a larger army and was victorious against the Phocian army in the Battle of the Crocus Field, which established both his control of Thessaly and his reputation as a pious defender of Apollo.[44] Nevertheless, the war dragged on for another six years while Philip consolidated his gains in the north, leaving the Athenians little choice but to reluctantly sign a peace with him in 346. The same year, Philip was invited by the Thebans to march south and settle the Sacred War; the Phocians abruptly surrendered (Buckler 2003, 443–452).

The uneasy peace between Philip and Athens did not last long. The orator Demosthenes inflamed emotions in Athens against Philip,[45] who responded by laying siege to two of Athens' allies in the Chersonese and seizing an Athenian grain fleet; the Athenians thereupon renounced the peace.[46] When Philip used the outbreak of the Fourth Sacred War to advance to central Greece,[47] Demosthenes persuaded the panicked Athenians to forge an alliance with the Thebans.[48] The combined armies of the Athenians and the Thebans attempted to bar Philip's passage south in 338 at Chaeronea in Boeotia. The result was a resounding victory for Philip, whose 18-year-old son Alexander annihilated the Theban Sacred Band (Diod. Sic. 16.85.5–86; Justin 9.3.9–11; Polyaen. 4.2.7). The Battle of Chaeronea left Philip in control of the Greek city-states, ending their long-cherished autonomy.

Conclusion

Their victory in the Persian Wars ushered in a series of internal wars that plagued the Greek city-states for the next century and a half, rendering them unstable and ultimately unable to mount an effective defense against a new external invader. For much of the fifth century, the division of the Greek world into two power blocs, with Sparta the acknowledged military leader by land and Athens by sea, provided a *modus vivendi*. Eventually, however, Athens' openly imperialistic behavior and territorial aspirations began to cause alarm among Sparta's allies, leading inexorably to the outbreak of the Peloponnesian War. The 27-year length of the war not only left both Athens and Sparta in considerably weaker positions but permanently altered the balance of power in the Greek world. In the fourth century, no individual state had the resources to achieve long-lasting hegemony, and Greece became increasingly fragmented and subject to short-lived alliances of convenience. An ongoing series of inconclusive internal wars ensued, fostered by the Persians (whose goal was to keep the Greeks disunited and therefore unable to recover the Greek cities of Asia Minor), which left the Greeks vulnerable to conquest in 338 by the capable and ambitious Philip of Macedon. Ironically, it was Philip who united the fractious Greek city-states under his leadership, forcing them to come full circle in his planned campaign against Persia (eventually carried out by his son Alexander the Great), where they would be revenged for their suffering during the Persian wars.

44 On Philip's exploitation of his role in the Sacred War, see Buckler 1989 and Pownall 1998.

45 Buckler 2003, 467–477.

46 Dem. 18.87–94; Theopompus *FGrH* 115 F 292; Philochorus *FGrH* 328 F 162; Diod. Sic. 15.74.2–77.3; Plut. *Phoc.* 14.

47 On the so-called Fourth Sacred War, see Pownall 1998, 53–54.

48 Dem. 18.169–179; Aeschin. 3.14041; Philochorus *FGrH* 328 F 56a; Diod. Sic. 16.84; Plut. *Dem.* 18.

CHAPTER 4

Hellenistic Land Warfare

Edward M. Anson

Usually seen as beginning with the conquests of Alexander the Great, Hellenistic warfare in terms of armament and tactics actually began with the innovations of the conqueror's father, Philip II. He brought changes to land warfare that were to dominate until supplanted by the Roman legions in the second century. Siegecraft, which prior to Philip barely existed in the Hellenic world, also became a serious component of warfare during the reign of that monarch and especially in that of his son. Hellenistic naval warfare might be seen as beginning with the Carthaginians and the creation of the quadrireme, as claimed by Aristotle (Pliny *NH* 7. 207), or perhaps with Alexander in his siege of Tyre (Curt. 4.3.14), bringing to a close the centuries of domination of the trireme. But it is in infantry warfare that Philip's and Alexander's contributions are most noted. Philip introduced a whole new heavy infantryman equipped with a pike instead of the stabbing spear that had dominated land warfare in the Western world for more than two centuries. Along with his son Alexander, he enhanced what is today commonly called "Combined Arms," the integration of differently equipped military units, into a complementary force. Where, prior to these commanders, it was the heavy infantry that principally fought with light-armed troops most often engaging one another as skirmishers and cavalry either to defend the vulnerable flanks of these heavy infantry units or to hunt down fleeing infantrymen. The two Macedonian kings and especially the latter coordinated many different arms representing the flower of land forces both from the Greek world and, with Alexander, that of Asia. After Alexander's death, changes emerged over time that diminished the skillful use of combined arms. Much of this was due not so much to deliberate changes in tactics but to the political changes that saw the breakup of Alexander's empire and the emergence of a number of competing states.

A Companion to Greek Warfare, First Edition. Edited by Waldemar Heckel, F.S. Naiden, E. Edward Garvin, and John Vanderspoel.

Philip II

To understand the dramatic change in warfare, and in civilization itself, it is necessary to begin with those reforms initiated by Philip II. He transformed warfare by making the most of the resources available to him and saving Macedon from what might have been its extinction as a political entity. His accession to the throne in 359 came on the death of his brother, the former king, on a battlefield along with 4,000 Macedonian troops. Not the most auspicious of beginnings. The Illyrian victors from the northwest part of the Greek peninsula were now encamped in Macedon with much of the northern and western areas of the country either occupied or allied with the invaders. In a little more than two decades, Philip brought Macedon from seeming ruin to the most powerful state in the eastern Mediterranean. Much of this transformation resulted from his revolutionizing the instruments of war.

Prior to Philip, Macedon had been a land ruled ostensibly by an autocrat whose theoretical powers included control of foreign policy, the military, the state religion, and natural and human resources. In reality these powers were much curtailed. Most of the highland regions—Upper Macedonia—were under the control of local aristocratic families who paid only occasional lip service to the authority of the lowland king, and at Philip's accession to the throne, these were either allied with or subjects of the Illyrians. Even in the lowland plain, powerful aristocrats controlled the localities. The king had no true bureaucracy, but relied on these aristocrats to perform many of the functions of government. They were his military commanders and his administrators. They also made up the Macedonian cavalry, which was the best in the Western world. These aristocrats were the king's *Hetairoi*, his companions. With these individuals the king enjoyed a close personal relationship and shared his leisure activities. This relationship with the aristocracy was one that reminds many modern historians of the society described in Homer's *Iliad*.[1] While the Macedonian cavalry was a force to be reckoned with, unfortunately, the same could not be said of the Macedonian infantry, which was lightly armed, ill trained, and mostly under the control of their aristocratic officers, who were also their civilian overlords. The vast majority of the population were either tenant farmers or dependent pastoralists bound to some local aristocrat.[2] This situation was quite different from that of the Greek communities to the east and south whose infantries were based on long-established traditions of middle-class heavy infantry, the so-called hoplites who dominated the Classical Age. These soldiers wore bronze breastplates and greaves, carried a circular, three-foot in diameter, shield, and a roughly seven-foot long thrusting spear. They fought typically in a compact unit, the phalanx. Cavalry and light-armed troops played a secondary role, protecting the flanks and rear of these formations, and often pursuing the defeated after the battle was won. Macedonia, then, at the start of Philip's reign, was disunited with much of the interior dominated by a powerful, land-holding aristocracy and the coast by independent Greek cities, and, at the time of his accession, suffering foreign occupation.

To drive out the Illyrians, Philip created the famous Macedonian phalanx, which would dominate Western warfare for almost two centuries. Faced with a population that could not

1 Anson 2013, 24–25. See also Carlier 2000.
2 Anson 2008b, 17–30.

afford the hoplite panoply and insufficient royal revenues, Philip was forced to change the very nature of the infantry panoply and tactics. He also had to deal with a Macedonian army that had been virtually annihilated in the recent battle. Whereas the hoplites were previously the offensive key to victory, Philip made his new infantry primarily a defensive force. In his new army, the infantryman was equipped with a 15-foot pike (the *sarissa*),[3] a 2-foot diameter shield hung from the neck and shoulder, and little else in the way of defensive armor. This new panoply was cheaper and the soldier required far less training. Over time, the pikes lengthened so that in the later Hellenistic period they might be as long as 24 feet (Polyb. 18.29.2). Such formations of tightly packed pikes could easily defend against frontal assaults by either hoplites or cavalry. These troops resembled the later Swiss pikemen of the early modern era, a similarity noted by Nicolo Machiavelli.[4] The inspiration for this innovation by the new Macedonian king is unclear. Modern historians cite earlier experiments with longer spears by other commanders, or their occasional use by neighboring tribal peoples, but it is also possible that Philip simply adopted and adapted the long spears traditionally used in hunting.[5] Whatever the inspiration, it was Philip who realized the potential and brought it to fruition.

While others may have seen the potential, they lacked the means to make a pike phalanx effective.[6] To a great extent invulnerable to frontal attacks on level ground, a pike phalanx was incredibly vulnerable on the flanks and rear, and found it difficult to maintain its impenetrable front on broken ground (Polyb. 18.30.11, 31.5–6). In order to compensate for the weaknesses in this new infantry formation the Macedonian king maintained a force of infantry guards trained both as pike bearers and as hoplites, initially called the *Pezhetairoi* and later the *Hypaspists.*[7] These troops represented a standing, national, professional force, and typically occupied the right wing. When used as hoplites, they gave more flexibility on the wing, most often engaging the enemy infantry first. Beginning as a royal guard, they likely numbered only a few hundred. Their eventual strength was 3000. More mobile than the "typical" Macedonian phalangite, they were regularly equipped for hand-to-hand warfare. The most significant change for these elite troops was that the hoplite's heavy metal breastplate had been replaced with one of linen or leather, reinforced with iron plates. As Alexander marched deeper into Asia, more units were equipped in the manner of the Hypaspists. This was especially the case after the dismissal of the allied Greek heavy infantry of hoplites (Arr. *Anab.* 3.19.6; cf. Diod. Sic. 17.17.3). It has also been recently argued that as Alexander continued into Asia he required more of these modified hoplite units, and many of his *sarissa* battalions were converted, and acquired the name of *asthetairoi.*[8]

---

3 Theophrastus (*Hist. Pl.* 3.12.2) gives the length of the longest Macedonian *sarissa* as 18 feet and Asclepiodotus (*Tact.* 5.1) confirms by stating that the shortest Macedonian pike was 15 feet and that the longest was not more than 18 feet. These different lengths may be the result of using longer *sarassai* in later ranks so as to extend the number of spear points that protruded before the front rank (Asclepiod. *Tact.* 5.2). Arrian (*Tact.* 12.7) gives the length of the *sarissa* as 16 feet.

4 Machiavelli 1965, 47.

5 Anson 2010b.

6 Anson 2010b, 65–66.

7 Markle 1978, 491–492; Anson 1985a, 247–248; Anson 2010a, 63 and n. 88; Anson 2013, 50.

8 Anson 2010a, 81–90.

But the key to Philip's new army was that he possessed a great advantage over those who may have experimented with the pike earlier—the native cavalry of Macedonia. Not only would the cavalry protect the vulnerable phalanx but, in Philip's new model army, it would become the principal attacking force. With the Macedonian infantry occupying the full attention of the enemy, the cavalry, in what has been described as the "hammer and anvil" tactic, would probe for weaknesses that could then be exploited for victory. The infantry became the anvil on which the hammer of the cavalry would smash the enemy. With this new army, Philip subdued his enemies and his son Alexander conquered the Persian Empire. Macedonian cavalry was so expert that, in a wedge formation, it could punch through a weakened or gapped infantry line.[9]

This use of what is currently referred to as "Combined Arms" was augmented further by Philip and especially by his son Alexander with the incorporation of light-armed infantry and cavalry. In addition to the heavy infantry units, Philip's new model army included archers, slingers, and significant numbers of light-armed infantry, the peltasts.[10] Light infantry typically wore little armor except a helmet and fought as javelin men, slingers, or archers. Their formations were open as opposed to the dense, compact nature of the phalanx. The elite light infantry were the *Agrianes*, javelin men from a dependent ally centered in what is today Bulgaria. These were likely incorporated into Philip's army and were used extensively by Alexander. They are often included in that commander's flying columns where speed was paramount. Typically, battles began with attacks of slingers, archers, and javelin men, whose attacks from range were designed to disrupt the cohesiveness of the enemy infantry line. But they were also often used to shield the flanks and defend against enemy cavalry, and, in Asia in particular, against chariots and/or elephants. Alexander's army that crossed to Asia included 7,000 additional light-armed troops. To these, as he penetrated deeper into Asia, were added additional light-armed auxiliaries from those areas brought under his domination. In 329, Alexander included a unit of 1,000 Iranian horse archers (*hippotoxotai*) in his grand army. These were used to great effect at the Battle on the Hydaspes against the chariots of the Indian prince Porus. Additionally, in this battle, Bactrian, Sogdianian, and Scythian cavalry were employed by Alexander (Arr. *Anab.* 5.12.2–3). Earlier, Persian scythed chariots at Gaugamela also proved ineffective. These chariots, outfitted with blades that extended from the wheel axle three feet on each side (Liv. 37.41.7–9), were met by Alexander's Agrianians and other javelin men who disabled most of their drivers with their javelins. Those chariots that did make it through the barrage of javelins charged the Macedonian phalanx, which simply opened up avenues in their ranks through which the chariots passed harmlessly (Arr. *Anab.* 3.13.5–6).[11] Later, in Babylon, Alexander incorporated both Iranian cavalry in traditional equipment and Persian infantry units. In 323, the Macedonian king received 20,000 Persian infantry and additionally units of Cossaeans and Tapurians (Arr. *Anab.* 7.23.1).

9 Worley 1994, 157.

10 Later during the Antigonid dynasty of Macedonian rulers, peltast came to refer to the heavy infantry, pike bearers, of the phalanx. The change in definition in Macedonia came from the *pelta*, the small round shields carried by the members of the phalanx (Juhel and Sekunda 2009, 107).

11 See, however, Heckel, Willekes, and Wrightson 2010.

Chaeronea and Combined Arms

The battle that gave Philip hegemony in Greece, fought near the city of Chaeronea in 338, demonstrates the effectiveness of his new model army.[12] Here his army numbered roughly 30,000 mostly Macedonian heavy infantrymen, 2,000 heavy Macedonian cavalry, known as the Companions, and an unknown number of light infantry and cavalry. Heavy cavalry typically wore breastplates, shoulder guards, and helmets; they also carried lances, often shortened versions of the infantry *sarissa*, and secondarily swords. Light cavalry were usually unarmored and equipped with bows or javelins. Philip faced a similarly sized combined Greek force consisting primarily of Athenian and Theban hoplites. The Athenians took up a position on the left wing opposite Philip and the Hypaspists, the Thebans the right, with other allies in the center. The Greek army was arranged in a defensive position, its left flank anchored by the rising foothills of Mount Thurium, the right resting against the Cephisus River. The foothills and the river bed were to be covered by light-armed infantry and light cavalry by both sides. The Greek line, about two miles in length, was consequently relatively secure on both flanks, leaving little room for the unopposed operation of enemy cavalry or light-armed troops, and thus seemingly negating Philip's distinct advantage in cavalry. Facing this strong Greek defensive position, the Macedonian king decided on a complicated series of maneuvers to take advantage especially of the inexperienced Athenians. At first, he advanced his forces in echelon, with his right, as usual, under his personal command and containing the elite Hypaspists, here armed as hoplites[13] and perhaps already 3,000 strong, to engage the Athenians before committing the Macedonian left against the more experienced Thebans. Given the nature of the battlefield and the size of the respective armies, the battle would be fought at least initially with full frontal infantry assaults. However, during the actual battle the integrity of the long infantry line was difficult to maintain. Terrain was uneven, injuries or deaths could cause gaps, simply someone tripping and falling could create problems. In the particular case of the Greek allies, the different contingents had not previously fought together and were assembled according to nationality. Forces so arranged would tend under pressure to gravitate toward their compatriots, which could cause gaps in the line. Given that the flanks were well guarded, Philip's goal was to enable his son Alexander, who commanded on the left with the Companion Cavalry, to penetrate the Greek line.[14] Whatever opposition cavalry appeared would be easily overcome by the superior forces under Alexander's command. To accomplish the goal of penetrating the Greek line, Philip planned, after initial contact with the Athenians, for the Macedonian right, at Philip's command, to begin a slow withdrawal up some rising ground near the foothills of the nearby mountain. This would stretch the Greek line, as the Athenians attempted to overwhelm their opponents (Polyaen. 4.2.2, 7).

The Athenians finding the enemies before them giving ground began what was apparently an over-enthusiastic and disordered advance. With the Athenians attacking wildly to their left and the Boeotian right maintaining its position along the riverbed, as Philip had planned, the allied center began to separate as these troops found themselves pulled between the Athenians

12 On the battle see Diod. Sic. 16.86; Polyaen. 4.2.2, 7; Frontin. *Strat.* 2.1.9; cf. Plut. *Alex.* 9.2; Arr. *Tact.* 16.6–7.

13 Markle 1977, 323, 329; Heckel and Ryan 2006, 63–64; Anson 2010a, 82.

14 Markle 1978, 490–491; Anson 2017.

moving to the left and the Boeotians standing firm on the right. For a time the battle was a struggle of competing infantries all along the line, but under the continuous probing of the Macedonian pikes, a gap or thinning in the line appeared through which Alexander and the cavalry charged, turning to attack the exposed Theban flank.[15] Without stirrups, heavy cavalry could only attack infantry under special circumstances—when they fled, when gaps or thinning occurred in their line, or in assaults on the vulnerable flanks or rear. Here, direct assault would be possible, especially when charging in a wedge formation. Under these circumstances the momentum of the group served both to keep the pack in formation and to hit the line with tremendous force. The dangers in such an operation were great. If one horse in the front went down, it could lead to a cascading event. Also, if the infantry line was not broken by the initial charge, the cavalry was likely to be surrounded and overwhelmed by the opposing infantry.

After Alexander's breakthrough, the Theban right wing, including the famous Sacred Band, found themselves fully engaged frontally by their Macedonian heavy infantry counterparts, and attacked simultaneously on their left and rear by the Macedonian Companion Cavalry. On the Macedonian right, Philip, having withdrawn to higher ground, stopped his feigned retreat and charged, breaking the spirit and coordination of the Athenian phalanx. With the Athenians in full flight, Philip's right pivoted and attacked the allied line, which now broke and fled. The result was a crushing defeat for Philip's enemies. A thousand Athenians lay dead, with 2,000 captured (Diod. Sic. 16.86.5–6). While numbers are not given for the Theban, Boeotian, and other allied dead, the totals were probably equally high, and the famous Sacred Band had been destroyed (Plut. *Pel.* 18.5).

While the core of the Macedonian infantry was composed of those bearing pikes (*sarissai*), as noted, modified hoplite armament was an important component of both Philip's and Alexander's Macedonian infantry. The troops were more flexible than the traditional *sarissa* bearers, even though, under the tutelage of Philip and Alexander, *sarissa*-bearing units were capable of moving quickly into varying formations, square, wedge, column, etc. but they required level and clear ground with no obstacles (Polyb. 18.31.5–6) and were ineffective in small units or as single soldiers (Polyb.18.32.9; cf. 11.15.7–16.3).[16] *Sarissai* were weapons designed exclusively for the battlefield and in close array. As with broken ground, anything that could create gaps in the formation could spell disaster. Crossing broken terrain could also destroy a phalanx's cohesion. At the Battle of Mantinea, a small ditch was sufficient to result in the defeat of the Spartan pikemen (Polyb. 11.15.7–16.2), and at Pydna in 168, while the Roman commander described how the Macedonian phalanx filled him with fear and amazement, that general drew it onto broken ground where it was annihilated (Plut. *Aem.* 19.1–3). In instances where roads were difficult or where speed was essential, "lighter-armed" heavy infantry would be used. Most importantly the *sarissa* phalanx was not very flexible. As Nicholas Hammond states, "[the *sarissa*] was unsuitable for skirmishing, besieging, street-fighting, ambushing, [and] mountaineering."[17] Despite Aristotle's belief that "hoplites are useless without orderly formation" (Arist. *Pol.* 4.1297B), and even though they were most effective in, and designed for, warfare in the compact phalanx, hoplites were

15 See Worley 1994, 160; Willekes 2013, 300 with n. 496, 319.

16 In the contest in 326/325 between the Athenian wrestler armed with only a club and wearing no body armor and the Macedonian phalangite fully equipped, the latter proved no match for the former (Diod. Sic. 17.100.2–101.2, Curt. 9.7.16–23).

17 Hammond, 1980a, 53.

capable of individual combat especially against opposing troops in disarray when the cohesion of the enemy infantry collapsed, or in general against Asiatic heavy infantry forces that tended to wear less defensive armor. In all likelihood, Philip adopted the *sarissa* because this form of warfare was inexpensive. As noted, Macedon had no tradition of heavy infantry and the state, especially at the time of Philip's accession, was not in a position to afford the equipment or pay for the training of a large force. Moreover, their training would have required more time than was typically available to the non-professional soldier. The use of hoplites was a different proposition entirely.

This use of combined forces is clearly seen in the Battle of Issus in 333 between Alexander the Great's forces and those of Darius III (Arr. *Anab.* 2.8.5–11.7; Diod. Sic. 17.33–34). This clash took place in what is today southeastern Turkey very near the Gulf of İskederun. Here the Macedonian king faced an army of far greater numbers; even though a total of 500,000–600,000 is mentioned in our sources for the total size of the Persian force (Diod. Sic. 17.33.4; Arr. *Anab.* 2.8.8), 100,000 infantry and 20,000 cavalry is a commonly suggested estimate.[18] The Macedonian infantry was anchored on the right by the Hypaspists; the center was composed of the *sarissa* bearers, and the left wing covered by Greek allied hoplites. Alexander ordered his left wing to remain in contact with the sea in order to protect that flank from being turned by the larger Persian army. Alexander's Companion Cavalry held the right, with the Thessalian cavalry and other Greek cavalry on the left. Those troops along the sea were to prevent the larger enemy forces from endangering his phalanx (Arr. *Anab.* 2.9.1).

In front of the cavalry on the right, Alexander posted the *prodromoi* (these were cavalry who carried longer pikes than the *xysta* used by the Companions), the Paeonian light cavalry armed with javelins and swords, archers, and the Agrianians. Other cavalry and archers screened the center. On Alexander's left the light infantry comprised the Cretan archers and the light-armed Thracians. Other light-armed troops occupied the center in front of the phalanx. Numbers of Greek hoplites were kept in reserve. As the battle progressed, Alexander's left was almost overwhelmed by the superior numbers of the Persians, but the Persian line became over-extended and a thinning occurred near its right flank. Into this weakness in the line charged the Companion Cavalry, in a wedge formation, followed by the Hypaspists. This charge, similar to the one at Chaeronea also led by Alexander, brought victory.[19] The Greek mercenaries in Persian employ had held their own against the Macedonian phalanx and Alexander's left was almost overwhelmed (Arr. *Anab.* 2.10.4, 11.2).

After Alexander's death, in the great contest between Eumenes of Cardia and Antigonus Monophthalmus, the best generals among Alexander's Successors, cavalry was again the decisive force in all three of their battles. In 319, Eumenes lost the engagement at Orcynia through the treachery of the Cardian's cavalry commander who deserted in the the course of the battle (Diod. Sic. 18.40.5–8; Plut. *Eum.* 9.3). At Paraetacene three years later, it was Antigonus' cavalry that salvaged a draw from what could have been an overwhelming defeat by exploiting a gap in Eumenes' infantry line (Diod. Sic. 19.30.9–10) and roughly two months later won a victory by capturing Eumenes' camp, which led to the latter's surrender by his own forces.[20]

---

18 While this is the estimate found in most modern accounts, Bosworth (1988, 57) is correct that "it is impossible to give even an educated guess."

19 Arr. *Anab.* 2.10.3; Diod. Sic. 17.33.5, 34.2.

20 Diod. Sic. 19.42.3, 43.8–9; Polyaen. 4.6.13; Plut. *Eum.* 16.10–11, 17.2–4; Just. 14.3.3, 12.

Hellenistic Innovations

While Philip's and Alexander's innovations changed the nature of warfare by introducing new weapons, new formations, and a greater use of combined arms, these were all, as one might say, variations on established themes. What was truly new to Hellenic warfare was the introduction of the elephant.[21] Sometimes mistakenly compared to the tanks of modern warfare these impressively sized animals, however, seldom seriously affected the outcome of battle, and in some of those cases where they did, it was often to the detriment of their own army. Slingers, archers, javelin men, and caltrops were all effective against these animals, the last used effectively by Ptolemy in the Battle of Gaza in 312 (Diod. Sic. 19.84.4). Alexander the Great in his Battle on the Hydaspes River used his light infantry to kill the animals' drivers and to surround the elephants, which were now injured by the javelins (Arr. *Anab.* 5.17.3). In 316, Antigonus Monophthalmus' cavalry archers in the Battle of Paraetacene rode out of range and inflicted severe injuries on Eumenes of Cardia's elephants (Diod. Sic. 19.30.2). Also, by maneuvering formations to let them pass harmlessly between the ranks and then attacking their legs and sides with javelins and swords experienced infantry would effectively negate them. Wounded or frightened elephants, unlike mechanized vehicles, had minds of their own and would often turn and injure their own forces. This was especially the case if their driver or the lead elephant was killed. Alexander faced elephants at least twice, at Gaugamela and on the Hydaspes. At Gaugamela, 15 were stationed in front of the Persian infantry (Arr. *Anab.* 3.11.6), but it has been argued that this is in error since they are only mentioned by Arrian and are never listed as engaging in the battle.[22] They were, however, fully engaged on the Hydaspes. In 190, in the Battle of Magnesia, the Seleucid King Antiochus' phalanx under attack on its flanks and rear formed a square with elephants in its center. The Roman ally, the Pergamene King Eumenes had his archers and slingers concentrated their fire on the elephants, causing them to panic and run through the Seleucid infantry destroying the cohesion of the square (App. *Syr.* 6.35).

Those actions where elephants proved effective were most often against forces unfamiliar with them. This was especially the case with horses who were commonly terrified of these animals unless they had previously been acclimated to them.[23] In 275, in a battle between an invading force of Gauls and an army hastily assembled by Antiochus I, the Seleucid king, the elephants proved to be the source of the Seleucid victory. Antiochus was caught by surprise by the Gallic invasion. The Gauls had a sizable force of heavy infantry, 20,000 cavalry, and 240 war chariots. Against this formidable force the Seleucid king could muster only a small force mostly of skirmishers and light-armed troops, but also 16 elephants. The latter were arranged on the flanks and in the center to meet the chariots. Apparently the Gauls had never before encountered elephants. Even before they were within arrow range, they turned and fled the field (Lucian, *Zeuxis* 8–11). These battles where elephants proved effective would also include Pyrrhus' first victory over the Romans at the Battle of Heraclea in 280. Here, as the battle reached a stalemate, Pyrrhus brought forth his elephants, which to this point had been held in reserve. The Roman cavalry had no experience with these creatures and fled before them, resulting in the rout of the Roman forces (Plut. *Pyrrh.* 17.3). Subsequently, the

21 See also Chapter 15 in this volume.
22 Recently Charles 2008, 9.
23 Kistler 2007, 20.

Romans in their contests against Pyrrhus negated their effectiveness through the methods already described.

There are few cases where elephants were used effectively against forces familiar with their presence. The most notable of these is the Battle at Ipsos in 301 where they were partially responsible for the victory of Lysimachus and Seleucus over their rival Antigonus Monophthalmus. The elephants blocked the return of Antigonus' cavalry, which, while victorious over that of the opposition, had over-pursued the fleeing enemy, leaving one flank of the phalanx unprotected. The result was the defeat and death of Antigonus (Plut. *Demetr.* 28–30). Elephants were also one key in the Roman victory over the Macedonians at the Battle of Pydna, where they faced forces familiar with them, but despite the usual counter measures, nonetheless proved effective (see the section on Later Battles: Raphia, Magnesia, and Pydna).

In the years that followed the great conqueror's death, cavalry continued to retain their importance, but seldom again became the "Hammer" envisioned by Philip and executed by Alexander. The numbers of heavy cavalry steadily decreased over time. This was not the case, however, during what is called the Second War of the Successors, fought from 320 to late 316.[24] In what was probably the best documented battle of the entire Hellenistic era, heavy cavalry proved the salvation of what might have been a devastating loss for one of the parties. In late October or early November of 316, the armies of Eumenes and Antigonus drew up for battle in the Median district of Paraetacene, located somewhere on the fringes of the Dasht-i Kavir, likely southwest of the modern city of Isfahan. Antigonus' army contained 28,000 heavy infantry, 15,000 light infantry,[25] 10,600 cavalry of whom approximately 6,900 were light cavalry, and 65 elephants; while Eumenes' forces numbered 35,000 infantry,[26] of whom 17,000 were heavily armed, approximately 2,000 heavy cavalry, 4,300 light, and 125 elephants. Both armies were of Asian majority. While Antigonus' phalanx was likely mostly Macedonian, much of the rest of his forces, especially the light-armed infantry and cavalry, were recruited in Asia, and Eumenes' heavy infantry was largely Greek mercenaries and Asian pikemen.[27] Hellenistic armies in Asia came to be overwhelmingly Asian.[28]

Eumenes arranged his army with his best troops stationed on the right. Here were the Argyraspids, Alexander's former Hypaspists, who had decorated their shields with silver and hence the new name, and Eumenes' own personal infantry guards, themselves numbering more than 3000.[29] Given the numerical advantage that Antigonus possessed in heavy infantry, especially those equipped with the *sarissa*, the Argyraspids were on this occasion equipped with pikes,[30] and Eumenes' Hypaspists with hoplite spears. This flank was protected by 2,900 elite heavy cavalrymen, personally selected by Eumenes himself, including his 900

24 For a discussion of the controversial chronology, see Anson 2014, 116–121.
25 See Anson 2015, 191 n. 73.
26 See Anson 2015, 191 n. 74.
27 Anson 2015, 191.
28 Olbrycht 2013, 159–178.
29 Anson 1988, 131–133.
30 In the later Battle of Gabene, the Argyraspids overwhelmed Antigonus' phalanx and, when subsequently exposed to cavalry attack, were able to form a square and successfully retreat (Diod. Sic. 19.43.1, 4–5). Unlike the Spartan experience at the Battle of Lechaeum in 391 where an unsupported hoplite regiment was attacked by light-armed troops and 250 of them were killed (Xen. *Hell.* 4.5.13–18), a pike-bearing phalanx during the Battle of Magnesia in 190 faced with similar circumstances formed a square and the long pikes deflected the projectiles and kept the phalangites reasonably safe (App. *Syr.* 6.35).

"companions," his personal cavalry bodyguard, and all under the Cardian's command. The entire phalanx was positioned behind a screen of elephants and light-armed troops, with the right cavalry wing screened by an additional corps of 40 elephants. On his left, Eumenes stationed 3,400 mostly light cavalry, anchored on the rising foothills, making any flanking maneuver by Antigonus' forces difficult. In the infantry center, moving from left to right, were stationed 6,000 mercenaries armed as hoplites. Next came 5,000 "men of many races" armed with the *sarissa*. Following along the line were the Argyraspids, and then Eumenes' infantry guards (Diod. Sic. 19. 27–29).

Antigonus, in response to Eumenes' deployment, stationed approximately 7,000 light cavalry on his left flank under the command of Pithon. These included 1,000 mounted lancers and archers from Media and Parthia and 2,200 Tarentines,[31] drawn up in open order and instructed to avoid any frontal attacks. They were to impede the progress of Eumenes' best cavalry and elephants. Next to them were 12,000 hoplites, mercenaries, and Asian allies; and to their right, 16,000 *sarissa*-bearing infantry. The right flank was guarded by a few hundred light cavalry and 3,300 heavy cavalry, which, with the exception of 300 directly associated with Antigonus, were under the command of Antigonus' 20-year-old son, Demetrius. Like Eumenes, Antigonus placed his phalanx behind a screen of elephants and kept another corps of the animals with him and his cavalry on his right. After some initial skirmishing, the screening elephants and the light-armed troops on both sides retreated through the ranks of their respective infantries, which then advanced to battle. Only the 40 elephants protecting Eumenes' right flank apparently saw action and then only against the enemy's light cavalry.

With their armies thus disposed, both forces advanced in an oblique fashion, with their right wings leading and their left held back, hoping that their respective right wings would crush the enemy's left before their own left was destroyed. Consequently, Antigonus had told Pithon to attack Eumenes' opposing cavalry immediately to slow its advance and give Antigonus' right wing more time to win the battle. While Pithon's troops initially caused great confusion on Eumenes' right and inhibited this wing's advance, Eumenes reinforced the harassed units with additional light cavalry from his left, with the result that Pithon and his forces were driven back to the foothills. Elsewhere Eumenes' phalanx, even though inferior in numbers, proved superior in ability. Antigonus' infantry was forced back to the hills as well. Antigonus now faced defeat, if not annihilation. But, by ignoring those who urged him to retreat, Antigonus saved himself and his army. As Eumenes' phalanx pursued their beaten opponents, a gap opened in the line between the phalanx and the cavalry on Eumenes' left. Antigonus with his heavy cavalry detachment charged through the break. With his left threatened, Eumenes recalled his victorious right. Antigonus now assembled his defeated forces and formed them up in a line along the foothills. Although it was already growing quite dark, both generals rallied their forces and prepared to renew the struggle. By midnight the two armies, after many attempts to outflank one another, formed into line about 3.5 miles from the site of the previous battle. By now, however, both armies were too exhausted to continue. In the battle Antigonus' forces suffered the loss of 3,700 infantry and 54 cavalry, with 4,000 wounded. Eumenes' troops lost 540 infantry and very few cavalry; his wounded amounted to 900. (These may be heavy infantry losses only, Diod. Sic. 19.30–31.)

31 These were perhaps originally associated with the city of Tarentum, but over time the term came to be applied generally to any javelin-armed cavalrymen.

While Antigonus had lost more troops, the battle had ended in a stalemate. Subsequently, in the Battle of Gabene, Antigonus was able to effect the capture of Eumenes and thus end the Second War of the Successors. This victory was the result of a cavalry victory of sorts. During the actual battle part of Antigonus' light cavalry had captured Eumenes' camp, putting them in possession of the wives and possessions of the Argyraspids in particular. These troops, even though the result of the battle had been remarkably similar to that at Paraetacene, surrendered themselves and Eumenes to Antigonus (Diod. Sic. 19.43.8–9).

In time, cavalry diminished in importance and phalanxes increasingly became almost exclusively *sarissa* bearers, with pikes reaching as much as 24 feet in length (Polyaen. *Strat.* 2.29.2). These phalanxes were protected on the flanks by light-armed infantry and/or cavalry, whom they supplanted as the decisive component of the army. The lengthening of the pikes made this unit more invulnerable to a frontal attack, but even less maneuverable. Polybius describes the opposing phalanxes at Sellasia in 222 (2.65.1–7) and again at Mantinea in 207 (11.15.6, 16.1) as primarily armed with pikes. At Sellasia there were 14,000 *sarissa* carriers of whom 10,000 were Macedonian (Polyb. 2.65).[32] In these battles many of the light infantry were armed, however, more heavily than the usual light-armed troops. These carried large oval shields and weaponry that varied from long thrusting spears to javelins. Troops so armed were known as *Thureophoroi*. If these troops were more heavily armored still, with a breastplate, they were called *Thorakitai*. While distinguishable from other light-armed soldiers, they were none the less to be included in their number. While Alexander the Great's Hypaspists often accompanied light-armed troops, they also served regularly in the phalanx. The *Thureophoroi* and the *Thorakitai* appear not to be associated with the latter. They seem to have been a lighter version of traditional hoplites. In his description of the army of Antiochus III in 209, Polybius includes both the *Thureophoroi* and the *Thorakitai* with the light-armed troops (Polyb. 10.29.4–6). This distinction is also made in a battle near the Arcadian city of Caphyae in 219 at the beginning of the Social War. In this battle these troops were initially stationed on the wings, perhaps serving here as Philip's and Alexander's hoplite units did to protect the flanks of the phalanx (Polyb. 4.12.3, 12). However, unlike the Hypaspists, they clearly do not move in any type of formation (Plut. *Phil.* 9.1–2; Polyb. 4.12.7; Plut. *Cras.* 25.7).

Nowhere are these changes from the legacy of Philip and Alexander more apparent than in the decline in the quality of cavalry. This change in Hellenistic warfare was due to a combination of factors. One such factor was an effect of Philip's initial reforms, which changed the dynamic of Macedonian society. His creation of a landed peasant class who peopled his infantry and his centralizing of power in royal authority both diminished the *esprit de corps* of the aristocratic class, not to mention their role as highly skilled mounted warriors. Alexander's conquests only accelerated this process. While Alexander crossed to Asia with a ratio of infantry to cavalry of 6:1, the later Hellenistic period seldom saw a ratio of 10:1, and often even higher. In one of the few battles in the third century where cavalry played a significant role, Sellasia, this ratio in the victorious Antigonid army was still 23:1 (Polyb. 2.65.5; Liv. 34.28). Moreover, this was mostly light cavalry. The reduction in the numbers of heavy cavalry was even more dramatic in the Hellenistic kingdoms of Asia and northeast Africa. In Asia there was little tradition of heavy cavalry and in Egypt not much tradition of

32 The 3,000 Achaeans are described as "picked." Moreover, Polybius (2.65.2) records the presence of peltasts, Agrianians, and Gauls, which suggests that Polybius is being precise in his differentiation.

cavalry at all, apart from chariots. It apparently was not until the campaigns in central Asia of the Seleucid King Antiochus III that the Parthian "cataphracts" were introduced into Hellenistic warfare, but generally only served in the eastern Hellenistic armies. Whereas Philip and Alexander's heavy cavalry, the Companions, wore armor and carried lances, with cataphracts, horse and rider were virtually encased in armor similar to that of Medieval knights (Liv. 35.48.3).[33] However, even with these troops, cavalry in the Hellenistic era primarily skirmished with other cavalry. They were seldom employed against infantry. As noted, heavy cavalry was critical in Philip's and Alexander's battle plans. Cavalry now increasingly took on a secondary role, most units being light in weaponry and armament. After the deaths of Alexander's immediate Successors, cavalry seldom won battles, which increasingly became confrontations between opposing pike phalanxes.

Later Battles: Raphia, Magnesia, and Pydna

Many of these changes are seen in the Battle of Raphia in 217 during the Fourth Syrian War between Ptolemy IV Philopator of the Kingdom of Egypt and Antiochus III the Great of the Seleucid Empire. Antiochus' army comprised the 10,000 elite, pike-carrying Silver Shields (Argyraspids) and 20,000 so-called Macedonians, named for their also being armed with *sarissai*, for a total of 30,000 heavy infantry. The grand army also included 32,000 light-armed troops, including at least 4,500 archers and slingers[34] 6,000 cavalry and 102 Indian elephants completed the Seleucid force (Polyb. 5.79.3–12). Ptolemy's forces are more difficult to calculate. Polybius mentions 70,000 infantry, 5,000 cavalry, and 73 elephants (Polyb. 5.79.1). This roughly equates to his description of the creation of the army as consisting of 3,000 royal guards, 2,000 "peltasts,"[35] another elite unit of the phalanx, 8,000 Greek mercenaries, an additional 25,000 phalangites, 3,000 Libyans, and 20,000 Egyptians, all equipped in the Macedonian style (Polyb. 5.65.3–5, 8–9). These Egyptians were newly trained to fight alongside the Macedonians. Of the other troops there are listed specifically 700 "house" cavalry, likely heavy, 3,000 native Egyptian and Libyan horse cavalry, 3,000 Cretan archers, 4,000 Thracians and Gauls, and 2,000 others not clearly identified (Polyb. 5.65.5–10). Immediately before the actual battle, Ptolemy's line, moving from left to right, is described as follows: the royal guard, the peltasts, 3,000 Libyans, the 20,000 Egyptians, and 8,000 Greek mercenaries (Polyb. 5.82.3–6). The 25,000 phalangites noted earlier have apparently disappeared from the narrative. While various solutions to the problems of the Ptolemaic numbers have been suggested,[36] the solution may simply be that they stood in the

33 On cataphracts and their armor, see James 2004, 110–113, 129–131, and Chapter 14 in this volume.

34 Some light infantry units are not clearly differentiated.

35 These Ptolemaic peltasts are not to be confused with peltasts of the Classical Age who were light-armed troops, most often armed with javelins. In the Antigonid army, the peltasts were the elite of the phalanx (Walbank 1957–1979, 1.274, 390).

36 The difficulty is that if the 25,000 figure represents an additional force of *sarissai* bearers, they are not specifically noted in Polybius' delineation of the Ptolemaic forces at the actual battle. If the 25,000 phalangites include the 20,000 Egyptians and represent a total of these and the Libyans and perhaps another 2,000 others, the math works, but does not coincide with Polybius' overall total of 70,000 infantry. The 70,000 figure is likely accurate. In the actual battle it is the charge of the phalanx troops under the command Andromachus and Sosibius that carries the day. The unspecified 25,000 are earlier listed under the command of Andromachus (Polyb. 5.65.3) and the 20,000 Egyptians under that of Sosibius (Polyb. 5.65.10).

center of the phalanx between the Libyan and the Egyptian phalangites and are not specifically noted. Ptolemy's force was accompanied by 73 elephants of the African stock (Polyb. 5.84.9). According to Polybius, African elephants could not bear the smell, sound, and sight of their Indian counterparts and consequently were of little use (Polyb. 5. 84.5–6).[37]

After five days of skirmishing, the two kings arrayed their troops for battle. Both placed their phalangites in the center. The two pike phalanxes would then battle each other (Polyb. 5.82.2) with their flanks covered by light-armed infantry. In front of these light-armed infantry on the wings of both armies were their respective elephant corps and, even further on the wings, their cavalries. At the beginning of the battle, the elephant contingents moved to engage their opposite. Ptolemy's force on the left retreated in panic and crashed into their own light infantry. Taking advantage of the chaos on the Ptolemaic left, Antiochus defeated the Ptolemaic cavalry and light-armed troops, now in disarray. On the Ptolemaic right the reverse took place with the Ptolemaic commander avoiding the elephants and effectively engaging the enemy's cavalry and light infantry. Virtually in a vacuum, the two phalanxes engaged one another. The fighting continued until a charge by the *sarissa*-wielding Ptolemiac phalanx defeated its opponent (Polyb. 5.84–85). Neither cavalry was engaged in the primary struggle between the phalanxes.[38]

The other change begun by Philip was the developing professionalism of armies. Philip used traditional mercenaries especially for garrison duties, but he also created a professional Macedonian infantry guard, the Hypaspists. During Alexander's decade-long campaigns the entire Macedonian force became increasingly mercenary in its characteristics and professional in its abilities. The camp had become these people's home and the army their profession. In the years following the conqueror's death, armies became mercenary forces fighting for those who were successful and who paid well. In part, in an attempt to cement these soldiers' loyalty to a particular commander, lands were given out by the Hellenistic kings in return for military service. This was, indeed, similar to the practice instituted by Philip II in his creation of the original *pezhetairoi*. This practice had cemented the loyalty of Philip's new infantry and had similar, though diminished, effects on these troops in the new Hellenistic states. Armies were no longer citizen forces called to duty in time of need. The kingdoms of the Hellenistic world maintained professional forces. Initially in the armies of Alexander's Successors, this change was most noticeable in the proclivity of defeated armies to join the forces of their victors, as seen earlier in the case of Eumenes of Cardia. This practice was ameliorated later since so many of these troops held land back in the territory of their immediate commander where their families resided.

The last great battles of the Hellenistic age were fought against the armies of the Roman Republic at Magnesia in Asia in 190, and at Pydna in Macedonia in 168. In the former, the total Seleucid force was 70,000 strong and represented many different nationalities and fighting styles. This number included the *sarissa*-bearing phalanx of 16,000 troops (Liv. 37. 40.2; App. *Syr.* 6.32). These phalangites anchored the center and were divided into ten sections of 1,600 men *Syr.* 6.32). Interspersed between each section were two elephants (Liv.

37 Polybius also states that the Indian elephants were intimidating because of their greater size and strength. It is now generally conceded that this was a myth and that the African elephants likely used by Ptolemy were from Eritrea. These are large "savanna" elephants and consequently larger than their Indian counterparts (Brandt 2014, 82–90).

38 For a modern detailed account of this battle, see Bar-Kochva 1976, 132–137.

37.40.2).[39] Covering the entire front of the phalanx were skirmishers (App. *Syr.* 6.33, 35; cf. Liv. 37.42.3). On the right side of the phalanx, King Antiochus placed in sequence 1,500 Galatian infantry, 3,000 cataphracts, the king's *agema* or royal cavalry guard of approximately 1,000 hand-picked Median horsemen equipped with slightly lighter armor than the cataphracts, and an additional mixture of light cavalry drawn from many other nations, supported by 16 elephants.[40] After these come the royal infantry guard, the Argyraspides,[41] and 1,200 Dahae horse bowmen;[42] then, 3,000 Cretan and Trallian light infantry, 2,500 Mysian archers, 4,000 mixed Cyrtaean[43] slingers, and Elymaean[44] archers covered the flank of this wing. On the left stood another 1,500 Gallicians and 2,000 Cappadocians, 2,700 additional light-armed troops, 3,000 cataphracts, and 1,000 horsemen of the "royal squadron," traditional heavy cavalry. In front of this body of cavalry were the scythed chariots and a detachment of Arab camel archers. These were followed by Tarentines, 2,500 Gallician cavalry, and 7,000 archers of mixed nationality; then, 4,000 mixed Cyrtaean and Elymaean slingers; and finally 16 elephants (Liv. 37.40.6–14). The ultimate length of this Seleucid front was so large that those in the center could not make out the end of each flank (Liv. 37.41.4).

Unlike their counterparts the Roman army of roughly 30,000 heavy infantry (App. *Syr.* 6.31) was practically uniform with respect to both men and equipment.[45] They were armed with javelins, "short swords," and protected by helmets, breastplates, greaves, and full-body shields. There were two Roman legions and two legions of Latins and allies, each containing 5,000 men. The Romans occupied the center of the formation with the Latins, their Italian allies, on the wings.[46] To the right of the heavy infantry were 3,000 light-armed troops,[47] many of these supplied by King Eumenes of Pergamum; then, 3,000 cavalry, 800 of which were furnished by Eumenes, the rest being Romans. Anchoring this wing were 1,000 mercenary light infantry and cavalry.[48] On the left were stationed four squadrons of cavalry anchored against a river that was protected by steep banks. Some 2,000 Macedonian and Thracian volunteers guarded the Roman camp.[49] The 16 elephants were placed in reserve behind the Roman legions.[50] Skirmishers likely covered the fronts of both heavy infantry forces, but are not noted by our sources with respect to the Roman forces.

The battle began with a charge by the Seleucid flanks. On the right their cavalry, commanded by the king, broke through the opposing infantry and cavalry, and made an

39 Appian (*Syr.* 6.32) relates that there were 22 elephants on the flanks of each section.

40 Livy states that Antiochus had 54 elephants (Liv. 39.37.13), with both wings having 16 and 18 interspersed in the phalanx, there are two missing. These may be on either end of the phalanx.

41 Appian (*Syr.* 6.32) calls these troops cavalry. While their positioning far on the right wing might appear peculiar, Livy (37.40.7) is clear that these are infantry (Briscoe 1981, 349).

42 *Hippotoxotai*. Appian (*Syr.* 6.32) records 200 Dahae.

43 Cytaea was a town on the Caspian Sea.

44 Elymais was a region located near the Persian Gulf.

45 Appian records 30,000, Livy gives no total, the units identified in Livy add up to 29,000.

46 Appian (*Syr.* 6.31) says that the Romans occupied the right with the Latins on their left, but Livy's account of the battle makes this impossible.

47 These were apparently peltasts in the traditional sense. The unit included the Achaean *Caetrati* who were armed with "slings, javelins, and other light ordinance" (Liv. 31.36.1, 35.27.5).

48 See Briscoe 1981, 347–348.

49 Liv. 39.37; App. *Syr.* 6.31.

50 These were apparently the much-maligned African elephants (Liv. 39.37.13).

unsuccessful attack on the Roman camp (Liv. 37.42).[51] On the left, the scythed chariots attacked the forces of Eumenes, but that commander ordered the slingers, archers, and other light-armed troops to attack the horses. The battle also began in the midst of a thunderstorm, which further frightened the horses. These, when injured, became unmanageable, and many turned and charged their own ranks, disrupting the cavalry forces on the left, where the scythes were especially effective against the Antiochean cataphracts whose heavy armor made them less maneuverable.[52] Eumenes followed this success with a cavalry charge that chased the opposing Seleucid cavalry from the field, thus exposing their phalanx to attack (App. *Syr.* 34–35; Liv. 37.41.9–12). Antiochus' attack on the Roman left also ultimately failed through the counterattack by Romans and another part of the Pergamene cavalry under the command of Eumenes' brother Attalus (Liv. 37.43).

While Eumenes' assault on the Seleucid right was still in progress, the Romans and their allies charged the Seleucid phalanx, which quickly formed itself into a square to protect its exposed flanks. At first the Romans were pushed back by the *sarissai* of their opponents, but under attack from the legion's javelins and the light-armed infantry and cavalry gaps began to appear and soon the battle was over. As gaps appeared in the square, the Roman legionnaires charged in. Panic now seized the Seleucid forces and most of the survivors of the battle surrendered. The battle ended in a complete Roman victory (App. *Syr.* 6.34–35; Liv. 37.42).

At Pydna, the Macedonian army met the Romans in the climactic engagement that saw the true end of Macedonian independence.[53] In this conflict the Romans had fewer than 43,000 foot (cf. Plut. *Aem.* 16.4; Liv. 44.38.5), of whom 26,000 were heavy infantry (Liv. 44.21); the Macedonians had 39,000 infantry (Liv. 42.51.10), approximately 20,000 weilding pikes.[54] The cavalry forces were roughly equal, about 4,000 each (Liv. 42. 51.10). The site was a plain very suitable for the phalanx, with nearby hills that gave the skirmishers and light-armed troops opportunity for retreat and flank attacks (Plut. *Aem.* 16.5). The Romans occupied the center of their infantry line with the allied Latin, Italian, and Greek infantry on their flanks. The cavalry was placed on the wings, with the Roman right being supplemented by 22 elephants. The pikemen took up the center of the Macedonian line, with the elite 3,000-strong "peltasts" forming the left of the phalanx. Lighter infantry guarded both flanks, where the Macedonian cavalry were also arrayed (Frontin. *Str.* 3.20). The stronger contingent was on the Macedonian right, where Perseus commanded the heavy cavalry. The battle began with the advance of the Macedonian phalanx, which initially pushed back first the Roman skirmishers and then the legions. As the phalanx advanced it passed onto broken ground and gaps began to appear in the once solid pike line.[55] Into these, the Romans rushed. On the Roman right flank, the elephants proved

51 Harl (2008, 260) states that Antiochus' charge came after the extensive Roman attack on the phalanx, but Livy's account divides the action into three theaters: front, left, and right. Therefore, even though Antiochus' charge on the right wing is placed after the Roman infantry attack, it likely occurred at the same time as that of the chariots and light cavalry on the left. Justin (31.8.6) has the Antiochean force routing the Roman legion on the right wing. This is accepted by Bar-Kochva (1976, 170), but rejected by Harl.

52 App. *Syr.* 6.33; Liv. 37.41.10–11.

53 In 149 a pretender to the Macedonian throne, claiming to be the son of the last king, raised a brief revolt against Roman authority. He was defeated in 148 in a second Battle of Pydna.

54 Liv. 42.51.3; cf. Plut. *Aem.* 13.3.

55 According to Frontinus this was the result of the deliberate strategy of the Roman commander to lure the Macedonians onto broken ground.

most effective, crushing the Macedonian left. The Macedonian elephants proved of no use whatever. The Roman contingents of light-armed troops and cavalry followed up the charge of the elephants and drove off the Macedonian left wing. Once the compact nature of the phalanx was broken and attacked on its flank and rear, it became virtually useless. *Sarissai* were never designed for close-in combat and the pikemen themselves wore insufficient armor. The 3000-man royal infantry guard fought to the death, and the Macedonians are reported to have suffered more than 20,000 dead or captured. The high casualties were the result of the flight of the Macedonian cavalry, most of the victims falling during the pursuit (Liv. 44.41; Plut. *Aem.* 18–19).

Polybius, the second-century historian, compared the phalanx to the Roman legion, and while he declared that "it is impossible to resist the charge of the phalanx, so long as it retains its proper formation and strength," he still conceded the superiority of the legion (18.29–32). He points to the inability of the phalanx to operate effectively on broken ground, but more importantly the legion's greater flexibility and the individual legionnaire's ability to operate singly or in a small group. While these are valid criticisms, it may not be fair to speak of Hellenistic weaknesses, when, in truth, the defeat of these Hellenistic armies was to a great degree the result of other Roman strengths. The phalanx was not initially created to win battles, but to create conditions so that the superior Macedonian heavy cavalry could probe an enemy's infantry for weaknesses that could then be exploited to victory. This superior Macedonian cavalry did not exist in the second century. Cavalry was still effective in this period, but not in the same way as it had been, especially in the time of Alexander. In these last two battles, cavalry was indeed very effective, but unfortunately for the Macedonians and the Seleucids it was effectively used primarily by their opponents. Furthermore, while Roman forces represented the combined resources of a united Italy, what had been Alexander's great empire had fragmented into three main kingdoms with a number of lesser states in Asia and various federations of cities in Greece. The original resources and strengths of Philip's and Alexander's armies were now dissipated throughout these various states and federations. Moreover, these forces never united against the threat coming from the West. Most of the time the Romans, as at Magnesia and Pydna, were allied with other Hellenistic entities against some Hellenistic state. In the case of Magnesia, it was the king of Pergamum who was critical to the Roman victory. Roman military strength was certainly supplemented by clever diplomacy. The Romans regularly intervened in Hellenic affairs, claiming to be "freeing the Greeks" from the grasp of this or that Hellenistic state. In the final analysis, it could be said that the Romans freed the Greeks right into the Roman Empire.

CHAPTER 5

Greek Warfare in Sicily

Melanie Jonasch

The historiography of Sicily between the eighth and third centuries recounts numerous violent conflicts over territorial control, political power, and social hierarchy. The expansion of Greek settlements on an inhabited island naturally led to war, just as city-state formation and consolidation led to competitive struggles and civil unrest. Because of the favorable position of Sicily and its natural resources, Sicilian Greeks confronted Mediterranean rivals, including Carthage and Athens, and later Rome.

During the initial stages of colonization, the struggle for land and resources took the form of widespread raids. Women and other non-combatants were affected only when seized as booty by the winners. At the end of the sixth century, when rival ambitions for political domination arose, battles came to involve larger numbers, specialized forces, elaborate equipment, and greater devastation and casualties. Such conflicts could be calamitous for the populace, who might be killed or sold into slavery after their cities were besieged and devastated. Often, however, conquered cities were disenfranchised and made tributary, and the civilian population did not materially suffer under the change of authority. By the fifth century, warfare in Sicily depended mainly on mercenaries and allies from non-Greek Sicilian settlements, mainland Greece, and Italy. Although mercenaries featured in Classical and Hellenistic warfare throughout the Mediterranean, Sicilian cities in particular relied on them. This policy guaranteed troop allegiance and prevented military coups by citizen armies. Yet reliance on mercenaries had far-reaching consequences when they attempted to integrate themselves locally after their military service.

Indeed, the average Greek Sicilian was likely affected less by combat than by preventive and preparatory military measures and by the social unrest caused by foreign troops. Warfare was costly, especially when mercenaries were involved. Even in times of peace, leaders resettled entire city populations to meet land or manpower requirements, and new settlers were

A Companion to Greek Warfare, First Edition. Edited by Waldemar Heckel, F.S. Naiden, E. Edward Garvin, and John Vanderspoel.

imported to reinforce populations. Numerous domestic conflicts over political hegemony, property, and constitutional forms resulted from this social instability. While town defenses displayed the strength of the *polis* to outsiders and protected the citizenry, they also reflected social organization and stratification. Out of town, strategic requirements shaped the landscape through the construction of new towns, military outposts, and border sanctuaries in key positions.

Warfare changed significantly between the eighth and third centuries. From at least the fifth century, cavalry, archers, and slingers commonly fought alongside hoplites. Because Sicily is an island with many harbors, naval warfare played as important a role as ground combat. Siege equipment such as ramps, towers, and artillery were used by the Carthaginians and Greeks by the late fifth century, and complex siege walls were built, as at Syracuse during the Athenian Expedition in 413. Dionysius the Elder later initiated a series of improvements such as increasing the size and power of warships. The late fifth and fourth centuries also witnessed a major evolution in fortification, of which Hermocrates' fortress at Selinus and the Euryalus fortress of Syracuse are prime examples. When the Romans besieged Syracuse in 214, they faced not only formidable fortifications but also the military inventions of the Syracusan engineer Archimedes.

The following account of violent conflicts in Sicily focuses on various Greek parties and their inter- and intracultural conflicts on the island, but includes the previously mentioned regional powers (Carthage, Athens, and Rome). The concluding section deals with social unrest and domestic conflicts. This artificial arrangement should not obscure the fact that several kinds of parties might participate in a given struggle. For example, Selinus and Carthage allied with one another against Syracuse and Acragas in 480 (Diod. Sic. 13.55.1), and in the mid-fifth century Hybla considered and then rejected an alliance with Sicels led by Ducetius (Diod. Sic. 11.88.6). Appeals to origin and ethnicity could justify military action or the creation of alliances and military support.

The main sources for this reconstruction are Thucydides, Herodotus, Diodorus, Polyaenus, Polybius, Plutarch's *Lives*, and various fragments of the Greek historians. Archaeological evidence is considered whenever possible.[1]

Greeks and Indigenous Peoples

Upon arrival

Scholars long depicted Greek occupation of Sicily as primarily offensive and unilateral, whereas recent research has emphasized the role of native communities. It remains that Greek colonization was accompanied by violence, forced evictions, and territorial appropriation. Thucydides (6.3) recounts the expulsion of local inhabitants during the foundations of Naxos (734), Syracuse (733), and Leontini (728), and excavations at these sites reveal Iron-Age structures summarily removed to create space for townscapes. Violent encounters likely occurred after the foundation of Gela in 688. Although Gela was not constructed upon an earlier settlement, the earliest Geloans apparently sacked the town of Omphace in the foothills of the Geloan *chora* (Paus. 8.46.2).

1 Further reading: Shipley 1993; De Vido 2016.

The non-Greek communities concurrently developing in the hinterland were soon drawn into the sphere of influence of their new neighbors, thanks to trade, treaties, and shared rituals. For example, Thucydides (6.4) describes an agreement between Greek settlers and a local potentate during the foundation of Megara Hyblaea in 729. A group of men from Megara had suffered several misfortunes while searching for a permanent settlement area, when the Sicel leader Hyblon offered them a site on the eastern shore of Sicily, where they founded their city with the approval of the local people.

Although this story points to varied relations between Greek settlers and their neighbors, it also demonstrates that Greek enterprises in Sicily were not automatically successful. In 580, the expedition of Pentathlus of Cnidus to Lilybaeum utterly failed after the settlers became involved in a local power struggle (Diod. Sic. 5.9). At the end of the century, when the Carthaginians had already staked their claim in western Sicily, the Spartan Dorieus and his men also met fierce resistance while settling this area.[2] Colonization was risky as well as complex. Nevertheless, many of these early conflicts can be interpreted as raids or skirmishes—false starts with few casualties.[3]

Colonial expansion and neighborhood conflicts

During the seventh and sixth centuries, Greeks seemingly spread along the northern and southern coasts of Sicily without major violence against indigenous populations. Yet the foundation of Acrae (663), Casmenae (644), and Camarina (599) by Syracuse curtailed the autonomy of non-Greek neighbors, as did the foundation of Acragas 50 miles west of Gela in 580. The tyranny of Phalaris of Acragas in the first half of the sixth century featured aggressive expansion. Diodorus (19.108) details the construction of several military outposts by the tyrant, and Polyaenus (5.1.4) refers to the submission of the unknown local settlement of Vessa to Phalaris. In the last decades of the sixth century, Acragas expanded across the hinterland of Sicily toward the northern shore.

During the sixth century, increasing territorial competition inspired new or rebuilt town defenses and numerous new hilltop strongholds in many locales. Then, in the early fifth century, many hilltop sites in the hinterland were abandoned, perhaps because of two political developments: the centralization of political power in Greek Sicily and the territorial hegemony exercised by the rising tyrannies.

In western Sicily, territorial competition had far-reaching consequences. Selinus, founded as an *apoikia* of Megara Hyblaea in 628, experienced uneasy relations with her Elymian neighbor Egesta from the early sixth century onward. Diodorus (5.9) mentions strife between the two cities beginning with the failed colonization attempt of Pentathlus. Since Selinus undoubtedly had expanding territorial interests, these affrays likely occurred over border issues. Despite the tension, thriving trade appears in the archaeological record, and Thucydides (6.6) names Selinus and Egesta as parties to the the only known agreement concerning intermarriage between a Greek and an Elymian settlement. In the last quarter of the fifth century, relations deteriorated so much that Egesta sought aid from more powerful

---

2 Hdt. 5.46; Antiochus *FGrH* 555 F1.

3 Further reading: Graham 1982; Albanese Procelli 1996; Leighton 2000; Lentini 2012; De Angelis 2016; Tréziny 2016.

cities.[4] While the resulting Athenian Expedition had a major impact on Sicily, intervention by Carthage in 409 led to the destruction of Selinus and Himera and to a significant shift in the balance of power (see the subsection on Carthage).

During the colonial period, then, relations between Greek *poleis* and their non-Greek neighbors were in flux. No evidence exists for a united front among the local populations. Instead, politically independent communities continually acted as both allies and adversaries of the Greeks. During wars with Athens and Carthage, Sicel communities are repeatedly listed as participants on either side. Meanwhile, recognizable native elements were vanishing from the material culture by the mid-fifth century.[5]

Indigenous ambitions for power

A significant moment in indigenous history was the uprising of Ducetius, an autocrat of native origin, in the mid-fifth century. Conflicts with Catane led to his alliance with Syracuse and to a joint victory over Catane (Diod. Sic. 11.76.3). Encouraged by this success, Ducetius turned against the city of Morgantina and organized a federation of independent Sicel communities challenging Greek supremacy. Next to a sanctuary for the indigenous Palici, he founded the city of Paliké as capital of the federation (Diod. Sic. 11.88.6). After various minor successes, the tide turned when he attacked the so far unidentified hilltop site of Motyon, manned by a garrison from Acragas (Diod. Sic. 11.91). After the combined forces of Syracuse and Acragas destabilized the federation, the Sicels turned against Ducetius, who fled to Syracuse and surrendered all land under his rule (Diod. Sic. 11.92).

While the rhetoric attributed to Ducetius regarding the sanctuary of the Palici suggests a certain ethnic self-awareness and "pan-Sicel" aspirations, his strategy resembled those carried out by the Syracusan tyrants, the expansionist Deinomenids. Ducetius aimed to create a stable power in the hinterland, as proved by his early moves against Morgantina and Motyon, and sought allies among independent communities for this purpose. The *synteleia* of communities, however, did not act cohesively under one leader, as some communities remained independent and most Sicel allies disagreed with his decisions in the wake of his assault on Motyon. Ducetius found refuge and forgiveness, not in his Sicel hometown, but in Syracuse, which exiled him to Corinth.

Upon his return to Sicily in 446, Ducetius founded the city of Kale Akte, on the northern shore of Sicily, with the consent of Corinth and Syracuse (Diod. Sic. 12.8), but died soon after in 440. Syracuse consequently regained control over all insurgent communities and perhaps destroyed the Sicel center of Trinacria, sometimes identified with the Ducetian foundation Paliké (Diod. Sic. 12.29). In sum, Ducetius was not a freedom fighter for the indigenous Sicilian cause, but an opportunist who became a major figure in Greek historiography. His attempt to gain political power is nevertheless the sole instance when non-Greeks cooperated in order to claim hegemony in Sicily.[6]

4 Thuc. 6.6; Diod. Sic. 12.82.3–7, 13.43.

5 Further reading: Graham 1982; Tréziny 2006; De Vido 2009; Adornato 2011; De Angelis 2016.

6 Further reading: Asheri 1992; Jackman 2006; Evans 2016.

Colonial Rivalries and Regional Power Struggles

Just as no common spirit existed among native populations, no ethnic solidarity appeared among Greek settlers vying for the two prizes of territory and power. The following subsections will analyze these motives for violent conflicts among Greeks.

Territorial conflicts

Although we know of only one early armed border conflict, one between Leontini and Megara Hyblaea in the sixth century (Polyaen. 5.47), violent disputes over boundaries certainly occurred regularly during the foundation of colonies. For example, the war between Syracuse and Camarina in the mid-sixth century was a border dispute in a broad sense, as the *causa belli* was the lack of loyalty by Camarina, founded by Syracuse as a buffer against the potential ambitions of neighboring Gela. Camarina revolted only a few decades after her foundation, with the support of Gela and her local non-Greek neighbors (Philistus, *FGrH* 556 Fr5). After a coalition of Syracuse, Megara Hyblaea, and Enna quelled the revolt, all inhabitants were apparently deported to Syracuse, as noted by Thucydides (6.5.3). Since archaeological evidence proves continuous settlement through the sixth and early fifth centuries, we must assume that Syracuse merely asserted its hegemony. Because of her position, Camarina remained a plaything of outside powers throughout her history. In the early fifth century, she was given as payment to Hippocrates of Gela after he defeated Syracuse in the battle of Helorus (Hdt. 7.154.3). When Gelon became tyrant of Syracuse in 485, he destroyed Camarina and transferred her inhabitants to Syracuse (Hdt. 7.156). The city was later repopulated by the Geloans and continued to mark the border between the Greek and Carthaginian spheres of influence from the fourth century onward.

Heraclea Minoa, built in the sixth century to secure the border of Selinus (Hdt. 5.46.2), shares this story. Clashes with the neighboring *polis* of Acragas likely had occurred in the late sixth century, after which Acragas donated spoils from Minoa to Athena Lindia in Rhodes (*FGrH* 240 F16). Furthermore, Herodotus recounts that the remainder of the unlucky Spartan expedition to western Sicily turned their attention to Heraclea and freed the city from Selinuntine dominion (Hdt. 5.46.2). By 500, however, it had been seized by Acragas and was subsequently occupied by unemployed mercenaries soon after the fall of the Deinomenids (*FGrH* 577 F1). Like Camarina, the city of Heraclea came under the authority of Carthage in the fourth century.

The ambitious *apoikia* of Acragas previously had expanded both west and east, to the detriment of her mother city, Gela. Although no evidence for violent encounters between the two cities remains, the initial expansion of Acragas likely incorporated Sicilian hilltop sites formerly within the sphere of influence of Gela. As proven by literary sources and archaeological evidence, Acragas' area of control stretched to Himera on the northern shore of Sicily by the first quarter of the fifth century, when Theron from the Emmenid family was tyrant in Acragas and Terillus was ruler in Himera.[7] The surrender of Himera to Theron of Acragas in 483 certainly prompted the issue of a series of coins struck in Himera with the Acragantine crab on the reverse.

7 Hdt. 7.165. Further reading: Luraghi 1994; Hansen and Nielsen 2004; De Luna 2009; Adornato 2011.

Tyrannies and imperialist aspirations

The conquest of Himera formed part of Acragantine strategy to gain regional dominion by expanding along the prominent river valleys toward the northern coast of the island. Gela conversely turned to the east, as the tyrant Hippocrates conquered Zancle, Naxos, and Leontini and then defeated Syracuse in the early fifth century (Hdt. 7.154). No evidence of major destruction remains from these cities, although the main gate of the fortified walls at Leontini was likely rendered unusable during this period. Hippocrates installed client-kings, as suggested by a reference in Pausanias (5.22.7) to a tyrant Aenesidemus of Leontini, who previously was in Hippocrates' bodyguard (Hdt. 7.154.1). The Geloan tyrant also installed Scythes of Cos as king in Zancle. Scythes' subordinate position became evident when Anaxilas, the tyrant of Rhegium, captured Zancle around 490 (Hdt. 6.23.1–3). Hippocrates, summoned for help by Scythes, did not assist his ally, but instead punished the king for losing his city and put him in chains. Hippocrates did nothing to help Zancle (Hdt. 6.23.4–5).

Hippocrates' rise to power in eastern Sicily required a military consisting of far more than the citizen soldiers of Gela, and so the tyrant increased his troop strength and his budget by spending the funds in the coffers of every captured city. The presence of Sicel mercenaries in his army is attested by Polyaenus (5.6.), but the relationship between Gela and her military personnel of native origin remains unclear. Very likely, alliances between Greeks and non-Greek locals allowed the former to pursue power while the latter profited by cooperating.

After the death of Hippocrates in 491, his former cavalry commander, Gelon, continued the expansionist program of Gela by seizing power in Syracuse. Having taken advantage of the Syracusan stasis that led to the expulsion of the *Gamoroi* (Hdt. 7.155.2), Gelon reinstated the exiles in Syracuse and installed himself as a ruler in 485. These measures did not involve much bloodshed, for the sheer military might of Gela forestalled any Syracusan resistance. As part of his takeover, Gelon transferred the inhabitants of Camarina and part of Gela to Syracuse (Hdt. 7.156.2) and granted citizenship to 10,000 mercenaries of unknown origin (Diod. Sic. 11.72.3). The rapid growth of Syracuse worried the neighboring city of Megara Hyblaea, which tried to oppose Syracusan expansion (Hdt. 7.156.2). Gelon therefore besieged Megara, transported all its elite citizens to Syracuse, and sold the *demos* into slavery (Thuc. 6.94.1). Archaeological evidence confirms the temporary abandonment of the city in the first half of the fifth century.

Through these massive population transfers, Gelon reshaped Syracusan society and its environment and economy. Elsewhere in Sicily, he and his army played a decisive role in the crucial battle at Himera against the Carthaginians in 480 (see the subsection on Carthage). After returning to Syracuse, he advertised the political and financial implications of this victory by building a series of commemorative temples (Diod. Sic. 11.25).

After Gelon moved his center of power to Syracuse in 485, he left his hometown in the hands of his brother Hieron, who succeeded Gelon as tyrant of Syracuse after his death in 478 (Diod. Sic. 11.38.2). During his reign of nearly 12 years, Hieron consolidated his power by forging alliances with Polyzalus, tyrant of Gela, and Theron, tyrant of Acragas. He, too, reorganized parts of the population, evicting the inhabitants of Catane in 476 and then refounding the settlement (which he gave the new name of Aetna) with settlers from the Peloponnese and Syracuse. The inhabitants of Naxos and Catane were transferred to Leontini (Diod. Sic. 11.49.1). In 472, he subdued Thrasydaeus, son of Theron of Acragas, who had raised an army to challenge Syracusan hegemony (Diod. Sic. 11.53.3–5). After Thrasydaeus was sent into exile, the *demos* of Acragas took control, and their city became the first in the region to abandon one-man rule, followed by Gela, Himera, and finally Syracuse in 466.

The period after the fall of tyranny in eastern Sicily did not witness large-scale territorial expansion, since the first years were devoted to restoring the exiled populations to their hometowns and to establishing peaceful coexistence. Sicily's mercenaries, who had challenged the formation of these young democracies, were expelled. This time of detente was disturbed only by Ducetius and his attempt to fill the power vacuum left in the hinterland by the former tyrants. The rivalry between Syracuse and Acragas reemerged when Acragas blamed Syracuse for lax supervision of the exiled leader and declared war (Diod. Sic. 12.8). The subsequent battle at the river Himera, won by Syracuse and her allies, resulted in a peace treaty.

In the last third of the fifth century, the power struggles between eastern Sicilian Greek *poleis* and Syracuse regained momentum. This time, Leontini led a rebellion against Syracusan hegemony with the support of Camarina, Rhegium, and Athens (Thuc. 3.86), a conflict that quickly involved many neighboring cities of Greek and native origin. In 424, a general Sicilian congress at Gela brought temporary peace through the efforts of Gela and Camarina, two traditional allies who happened to be on different sides in this conflict (Thuc. 4.58, 4.65).

The last quarter of the fifth century and the fourth century featured external threats from Athens and Carthage. Nevertheless, Sicily did not unite. Syracuse continued to strive for control and to secure neighboring agricultural lands for funding her mercenaries. Even after the catastrophic triumph of Carthage in the years between 409 and 406, and the subsequent peace treaty of 405, the Syracusan tyrant Dionysius the Elder tried to reclaim his authority in eastern Sicily. He regained control of Etna, raided the territory of Leontini, and transferred their inhabitants once again to Syracuse (Diod. Sic. 14.14–17). The populations of Catane and Naxos were sold into slavery and their territories distributed among his Sicel and Campanian allies. Dionysius also established mercenary garrisons at Adranon on the slope of Mount Etna (Diod. Sic. 14.37.5) and at Monte Turcisi on the edge of the plain of Catania. After returning from his campaign to Motya in 396, he gave the territory of Leontini to disgruntled mercenaries, transferred new settlers from southern Italy to Messana, and installed Greeks from the Peloponnese in Tyndaris, a new foundation on the northern shore of Sicily (Diod. Sic. 14.68).

In subsequent decades, Sicily fell prey to adventurers of various stripes. Hicetas, a former general of the Syracusan army, conquered Leontini in 341 and used it as a base to wage war against Dionysius the Younger, the successor to the Elder (Plut. *Tim.* 2.3). He then allied himself with the Carthaginians and besieged Dionysius on the island of Ortygia. The statesman Timoleon from Corinth, summoned by the *demos* of Syracuse, drove both Hicetas and Dionysius from the city and seized control of their territories (Diod. Sic. 16.68). He then transferred the inhabitants of Leontini to Syracuse and invited new settlers from mainland Greece, who immigrated in considerable numbers (Diod. Sic. 16.82).

In the final decades of the fourth century, the politician and commander Agathocles seized power in Syracuse by a coup. Several thousand aristocratic opponents who were not killed on the spot had to flee, and sought refuge in Messana and Acragas (Diod. Sic. 19.8). This banishment of political opponents spurred two revolutionary movements under the leadership of Dinocrates of Syracuse and Xenodicus of Acragas. Dinocrates, an exiled aristocrat from Syracuse, recruited his own army and aligned with the Carthaginians against Syracuse (Diod. Sic. 19.103). Xenodicus, a general elected by the Acragantines, proclaimed the freedom of Sicilian cities from Syracusan and Carthaginian dominion while secretly promoting his own interests (Diod. Sic. 20.31). Since Xenodicus quickly gained support in Gela, Enna, Herbessus, Leontini, Camarina, and Heraclea Minoa, Agathocles was forced to return from his African campaign against Carthage in 308 (see the subsection on Carthage). He defeated Xenodicus and his allies and regained numerous cities in western and northern Sicily, before returning to Africa (Diod. Sic. 20.56). Dinocrates still held several hilltop sites and recruited an ever-increasing number of soldiers. Only after the peace treaty with Carthage in 306 was Agathocles able to address this threat. Although superior in numbers, Dinocrates and his allies were defeated, and then surrendered on terms, Dinocrates being appointed general in Syracuse (Diod. Sic. 20.89–90).

The death of Agathocles in 289 and the quarrel over his succession led to the emergence of autonomous local powers that included new tyrannical regimes in Syracuse, Acragas, Leontini, and Tauromenium (Diod. Sic. 22.2, 22.8). Some violent conflicts were initiated by Acragas and Syracuse, who attempted in vain to replicate past success. More significant was the occupation of Messana by a group of Campanian mercenaries left unemployed after the death of Agathocles. A constant source of unrest and concern for Syracuse and her neighbors, these mercenaries were essentially responsible for the first intervention of Rome in 264 (see the subsection on Rome).[8]

Foreign Wars

Sicily and the Greek world outside Italy

Culturally, the Sicilian *poleis* formed part of the Greek world, as illustrated by their regular participation in the Olympic Games and their votive dedications at Panhellenic sanctuaries on the Greek mainland. Alliances and collaboration with mainland Greek powers were nevertheless rare, even as late as the early fifth century, when envoys of a Greek confederation asked Gelon for help in the war against the Persian Empire. Gelon offered generous support only on the condition that he become the leading commander of the Greek forces. After the Spartans and Athenians categorically denied his request, he remained uninvolved, but sent to Delphi a delegation with gifts for Xerxes, in case of a Persian victory (Hdt. 7.158–163).

During the Peloponnesian War, a Syracusan fleet under Hermocrates fought as a Spartan ally against the Athenians in Asia Minor (Thuc. 8.26), in recompense for Spartan aid under the general Gylippus during the Athenian siege of Syracuse several years earlier (see the subsection on Athens). Since the fleet participated in several battles on the Hellespont with limited success, then was destroyed in the Battle of Cyzicus in 410, Hermocrates was relieved of his command and sent into exile (Xen. *Hell.* 1.1.18).

More than a century later, Agathocles assisted the inhabitants of Corcyra against the Macedonian King Cassander (Diod. Sic. 21.2). The ulterior motive of his intervention, however, was the offer of the island as dowry to King Pyrrhus of Epirus, who married Agathocles' daughter, Lanassa (Plut. *Pyrrh.* 9).

---

8 Further reading: Meister 1984; Caven 1990; Asheri 1992; Lewis 1994; Luraghi 1994; Westlake 1994; Frasca 2009; De Vido 2013; De Angelis 2016; Evans 2016.

Southern Italy

More important than mainland Greece was southern Italy. The territories across the Strait of Messina naturally played a decisive role in Sicilian territorial expansion, especially the Chalcidian foundation of Rhegium, frequently annexed to the Syracusan sphere of influence. Such interaction began in the early fifth century, when Anaxilas, tyrant of Rhegium, seized power in Zancle, and renamed it Messana (Thuc. 6.4.6). As Anaxilas controlled both sides of the strait from the 490s, he gave his daughter as wife to Hiero I to prevent confrontation with Syracuse. He himself married a daughter of Terillus, tyrant of Himera, and therefore took the side of Carthage in the battle of 480 (Hdt. 7.165).

The other major *polis* on the Bruttian peninsula, Locri Epizephyrii, was a stable ally of Syracuse and an enemy of Rhegium since her foundation. During the Athenian advance on Sicily in 427, Locri allied herself to Syracuse, while Rhegium hosted the Athenian fleet (Thuc. 3.86). As Dionysius the Elder obviously would not leave this strategic tip of the Italian peninsula in hostile hands, he attacked Rhegium several times, but unsuccessfully, and also rebuilt Messana as a Syracusan outpost (Diod. Sic. 14.100). He also allied himself with local Lucanians, who were fighting Greek cities of Southern Italy. These *poleis* had united as the Italiote League (Polyb. 2.39.1–7). Together with the Lucanians, Dionysius gained a decisive victory over the League in the battle at the Eleporus River (Diod. Sic. 14.104). After the conquest of Rhegium, which at last fell to him in 386, and the destruction of Caulonia and Hipponium, Dionysius controlled the entire Bruttian peninsula. Alliances with the Lucanians and the defeated Italiotes extended his hegemony beyond Italy into the Adriatic Sea.

Long after the death of Dionysius the Elder Agathocles reestablished Syracusan hegemony over the Bruttians. In 295, after besieging and conquering Croton, he secured substantial (but not absolute) Syracusan supremacy in the region (Diod. Sic. 21.4).

Another Italian power, the Etruscans, traded with Sicily from the eighth century onward. Although the Etruscans and the Greeks never competed over the colonization of Sicily, several violent clashes occurred, the most famous of which was the triumph of Hieron's fleet over the Etruscans in the naval battle at Cumae in 474 (Diod. Sic. 11.51). Illustrating the importance of this victory is the Etruscan helmet dedicated at the Panhellenic sanctuary of Olympia with an inscription of Hiero I as a votive from the spoils of Cumae (*CIG* 16). In the mid-fifth century, a Syracusan expedition to Elba and Corsica resulted in a victory over the resident Etruscans (Diod. Sic. 11.88.4–5). In 384, Dionysius the Elder attacked the Etruscan city of Pyrgi and looted a sanctuary to procure funds for his war against Carthage (Diod. Sic. 15.14.3). Apart from these conflicts, the Etruscans appear in Sicilian history as mercenaries. They joined the Athenians in their expedition against Syracuse and served the Carthaginians against Agathocles, although a fleet from Etruria assisted Agathocles in defeating Carthaginian naval forces in 307.[9]

Athens

Because of its fertile soils and its position as a trading hub in the western Mediterranean, Sicily was of interest for the Athenians also. Alliances with Sicilian cities, including the

9 Athens against Syracuse: Thuc. 6.103.2). Carthage against Agathocles: Diod. Sic. 19.106.2. Etrurian aid: Diod. Sic. 20.61.6. Further reading: Purcell 1994; Torelli 1996; Musti 2005; Funke 2006.

Elymian town of Egesta and the *polis* of Leontini (Thuc. 6.6.2), date back to 454. Athens became a significant force in southern Italy with the Panhellenic foundation of Thurii in 443 under its Athenian *oikistes* Lampon, who also aided Catane against an expanding Syracuse during the late 430s (Justin 4.3). These early interventions on Sicilian soil were undertaken to secure a position on the island and to discourage an alliance between Syracuse and Sparta. Several years later, Athens was again summoned by her ally Leontini (Thuc. 3.86). The expedition of 427 followed. The Athenian fleet established a base in Rhegium and conducted several campaigns to the Aeolian Islands and Mylae (Thuc. 3.90). No battles against Syracuse are attested, and the peace of Gela in 424 prevented further interference in Sicilian politics (Thuc. 4.65).

A new opportunity presented itself in 415, when Egesta and Leontini asked for assistance against the expansionist policies of Selinus and Syracuse (Thuc. 6.6). Despite cautionary voices, the Athenians agreed on a military campaign against Syracuse and assembled an army of considerable size under the leadership of Nicias, Alcibiades, and Lamachus. The enterprise was ill fated, as the Athenians could not find allies in Magna Graecia and Egesta defaulted on promised money (Thuc. 6.46). After skirmishes against secondary settlements on the coastline, the Athenians set up base in Catane and prepared for battle. The first military encounter, near the extra-urban Olympieum, proved victorious for the Athenians, who were superior in tactics and military skill (Thuc. 6.65–72). During the following winter, however, Hermocrates reformed the military leadership in Syracuse, reinforced the fortifications, and sent envoys for help to Corinth and Sparta (Thuc. 6.72–73, 6.75).

In the summer of 414, the Athenians attacked again and occupied the high plateau of the Epipolae to the north of Syracuse (Thuc. 6.97). While the Athenians quickly built a semicircular wall from the Epipolae to the Great Harbor to cut off the supply routes of Syracuse, the besieged *polis* erected a counter wall to the north of the high plateau and prevented the Athenians from reaching the coastline. The Syracusans thus secured vital supplies. During the following winter, Syracuse supplemented and improved their fleet with support from Corinth and nearly all Sicilian cities (Thuc. 7.36). While Athens also received reinforcements, they suffered setbacks and disease due to the unfavorable position of their camp near marshland (Thuc. 7.46–47). They failed to retreat to Catane and reassemble their forces, since the Syracusans blocked the harbor and any escape by sea. A desperate attempt to break free failed, and a retreat overland succumbed to an attack by the Syracusan coalition. The Sicilian Expedition ended in a crushing defeat for the Athenians, most of whom were killed or sold into slavery (Thuc. 7.75–86). Athens made no further Athenian attempt to influence Sicilian affairs.[10]

Carthage

The Phoenicians frequented Sicily as early as the ninth century, and from the eighth century on they maintained settlements at Panormos, Motya, and Soloeis in cooperation with their Elymian neighbors. The first violent conflict attested by the sources dates to 510, during the futile attempt of the Spartan Dorieus to settle the western coast of Sicily. Not long afterward, in the first quarter of the fifth century, Carthage became more interested in political influence and territorial holdings on the island. When Terillus, the former tyrant of Himera who had

10 Further reading: Andrewes 1992; Brice 2013.

been defeated and exiled by Theron of Acragas, asked Carthage for help, Carthage tried to restore Terillus as tyrant by forming an alliance with Anaxilas of Rhegium and Greeks from Selinus. Hamilcar sailed to Sicily in 480, with many citizen soldiers and with mercenaries from all parts of the Mediterranean. Diodorus (11.20–21) provides a detailed description of the Carthaginian forces. The first attack on Himera caused Theron, who led the resistance in the city, to call for reinforcements from Syracuse. Gelon willingly accepted the mission and came with a considerable number of soldiers to Himera. After capturing a messenger carrying a letter from Hamilcar, Gelon was able to attack Hamilcar by surprise, kill him, and wreck his naval base. A peace agreement was then negotiated on comparatively favorable terms for the Carthaginians. The victory at Himera was nevertheless of vital importance for the stability of Greek Sicily, for the Carthaginians confined themselves to the Western Sicilian towns of Panormus, Motya, and Soloeis for the next 70 years.

Partly because Himera occurred around the same time as the Battle of Salamis, it became an important topos in the self-representation of the Deinomenids and a prominent event in the Greek collective memory.[11] Like the Persians, the Carrthaginians played the role of a defeated foreign threat.

A longer and more complex series of confrontations began in 410, when Egesta called again for aid against the threats of neighboring Selinus (Diod. Sic. 13.43.4–5). With a considerable army of citizen soldiers, mercenaries, and allies, Hannibal Mago, Hamilcar's grandson, besieged Selinus. After several days of fierce fighting, the city fell. The reinforcements sent by Syracuse and Acragas did not arrive in time (Diod. Sic. 13.55–56). While the delayed allies collected the surviving inhabitants, Hannibal raided Himera, where the Himerans were supported by 4,000 soldiers from Syracuse under the command of Diocles (Diod. Sic. 13.59.9). After a battle outside the city walls, Diocles evacuated Himera. He had heard rumors that Hannibal was deploying ships from Motya in an attack on Syracuse. In the course of the evacuation of Himera, the Carthaginians attacked and razed the city to the ground, and tortured and killed many of the fleeing inhabitants (Diod. Sic. 13.62). Archaeological evidence confirms the destruction of Himera and its subsequent abandonment in the late fifth century. Hannibal sailed back to Africa without further operations on Sicily.

Only some years later, in 406, did the Carthaginians vote for a new military expedition to Sicily (Diod. Sic. 13.80.1). Hannibal and Himilco, another member of the Magonid family, recruited an army of allies and mercenaries from across the Mediterranean. The Sicilian Greeks, having prepared for this attack, defeated a Carthaginian contingent of warships near the Elymian city of Eryx. The Carthaginians sent reinforcements and continued their expedition with an eight-month siege of Acragas, where Hannibal perished from illness (Diod. Sic. 13.85). A Geloan army, mercenaries, and a Syracusan force supported the Acragantines. After several battles, mostly favorable for the Greeks, Himilco sent warships from Motya and cut off the food supply of Acragas by sea. Facing starvation, the Greek generals evacuated the inhabitants of Acragas to Gela and left the city to be looted by Himilco and his troops (Diod. Sic. 13.88.6–89). Afterward, the Carthaginians established their winter camp within the city walls. These events distressed the Syracusans, who elected new generals including Dionysius the Elder, who became *strategos autokrator*, or sole commander, in the following turmoil.

11 Pind. *Pyth.* 1.72–80; Hdt. 7.166; Diod. Sic. 11.24.1.

In the summer of 405, Himilco departed Acragas to wage war against the neighboring city of Gela and her territory (Diod. Sic. 13.108.2–3). Despite a highly motivated Geloan army and generous reinforcements from Syracuse, the counterattack on the Carthaginians failed and Dionysius abandoned both Gela and Camarina. Thwarted by the outbreak of an epidemic, Himilco proposed a peace treaty that Dionysius willingly accepted, despite the harsh terms it imposed on the Sicilian Greeks. The Carthaginians gained dominion over all of western Sicily, and collected tribute from Selinus, Acragas, Gela, Camarina, and Thermae, where the survivors of the battle at Himera in 409 had relocated. Leontini, Messene, and the Sicels remained autonomous, with Syracuse being left to Dionysius.

Several years later, having established his rule and prepared Syracuse and eastern Sicily for war, Dionysius challenged the Carthaginians by attacking Motya, their most important trading base. Many Greek cities, including Selinus and Thermae, joined forces with his large army of Syracusan citizens, allies, and mercenaries. They quickly took Eryx, and besieged Motya using new weapons such as catapaults and quinqueremes financed by Dionysius (Diod. Sic. 14.47.4–48, 14.42.1–2, 14.41.3–6). Both proved very effective after Himilco arrived with his warships and counterattacked (Diod. Sic. 14.50). After a fierce battle, the Greek attackers forced their way into Motya, looted it, and took the inhabitants captive. Despite this success, Dionysius could not secure the western part of Sicily, and Himilco managed to regain Motya and Eryx during the winter (Diod. Sic. 14.55.4). For lack of reinforcements from inland communities, Dionysius had to retreat, and the Carthaginians advanced once more into eastern Sicily, taking Messana and establishing alliances with the neighboring Sicels. Dionysius meanwhile fortified inland settlements and enlarged his army with freed slaves and mercenaries (Diod. Sic. 14.57).

The ensuing battle, at sea near Catane, saw the Carthaginians overcome the Syracusan fleet despite the technical superiority of the Greek ships (Diod. Sic. 14.60). Dionysius retreated inside the walls of Syracuse, while Himilco invested the city by land and sea. During the siege, the Carthaginians again suffered from diseases that weakened their forces. Dionysius took this opportunity to attack their camps and burn their ships in the harbor (Diod. Sic. 14.72–74). Himilco fled with his citizen soldiers to Africa, while the Sicels returned to their home towns; the remainder of the Carthaginian allies were sold into slavery (Diod. Sic. 14.75). Not warmly welcomed at home, Himilco and his soldiers later faced a major uprising against the hegemony of Carthage in Africa.

Only in 393 did Carthage attempt once more to strengthen its position in Sicily, when Mago fought a series of battles with Syracuse (Diod. Sic. 14.90). In 392, a new peace treaty confirmed the conditions of 405, with the difference that the Sicels were now subject to Dionysius (Diod. Sic. 14.96).

According to Diodorus (15.15), hostilities between Carthaginians and Greeks resumed in the 370s, when Dionysius convinced some cities paying tribute to Carthage to defect. Mago then entered a pact with Greeks from Southern Italy and challenged Syracuse from both sides of the strait. A battle at an unknown site called Cabala resulted in a striking victory for Dionysius, who rejected a peace request and demanded the Carthaginians evacuate Sicily. Carthage replied by defeating the Syracusan army at another unknown site called Cronium (Diod. Sic. 15.16–17). After a peace treaty reestablished Carthaginian hegemony over western Sicily, the death of Dionysius in 367 temporarily ended hostilities between the two major powers on the island.

In 345, the Carthaginians involved themselves in Syracusan politics by supporting Hicetas, the tyrant of Leontini, in his efforts to gain control of Syracuse. During his campaign, Hicetas invited the Carthaginians to take over Syracuse, although Plutarch (*Tim.* 17) notes that they preferred to avoid battle with Timoleon, who had been sent from Corinth to help the *demos* end the ongoing political infighting. Timoleon, however, needing money to pay his mercenaries, invaded Carthaginian territory and convinced many Greek and native communities in western Sicily to change sides (Diod. Sic. 16.73). The Carthaginians then recruited a large army and crossed over to Sicily (Diod. Sic. 16.77.4). They confronted Timoleon and his troops in the territory of Acragas, near the river Crimisus. Despite the numerical superiority of the Carthaginians, Timoleon gained another victory after severe weather hampered his opponents (Diod. Sic. 16.80). Afraid of further military operations, the Carthaginians requested a new peace treaty that established the river Halycus as the border between the Syracusan and the Carthaginian spheres of influence (Diod. Sic. 16.82.3). This agreement led to nearly three decades without major conflicts between the two parties.

Hostilities resumed only under the rule of the Syracusan tyrant Agathocles. He attempted to gain control of eastern Sicily by subduing both autonomous settlements and Greek cities granted freedom in the treaty of 338, and upset the balance of power on the island. Diplomacy having failed, Carthage made war on behalf of the Syracusan exiles who had fled to Acragas during Agathocles' military coup in 316 (Diod. Sic. 19.103.1). Agathocles had renewed the peace with Carthage two years after that, in 314, but responded by besieging Acragas, a startling act of war. The Carthaginian fleet thereupon sailed into the harbor of Syracuse and set up camp on Mount Ecnomus, on the western banks of the river Himera (Diod. Sic. 19.108.1). In a counter-maneuver, Agathocles occupied Gela and pitched camp in a stronghold on the opposite side of the river. He attacked the Carthaginian camp and inflicted severe losses, but was forced to retreat to Gela after Carthaginian reinforcements arrived from Africa (Diod. Sic. 19.109). Hamilcar menaced Syracuse and convinced many cities, including Camarina, Leontini, and Messana, to come over to his side. In desperation, Agathocles sailed to Africa and attacked Carthage, a first for the Greeks, while Hamilcar and his troops were besieging Syracuse (Diod. Sic. 20.13.4).

With the element of surprise on his side, Agathocles raided the African countryside and camped near Carthage. Internal conflict and the absence of Hamilcar, enfeebled the city's defenses. Agathocles thus conquered numerous cities within the Carthaginian sphere of influence, or convinced them to change sides. Hamilcar, still besieging Syracuse, died in an ambush in 309.

In the second year of his campaign, Agathocles returned to Sicily to fight his political enemies (see the subsection on Tyrannies and imperialist aspirations) and left his son Archagathus in charge of the African campaign (Diod. Sic. 20.55.5). The Carthaginians exploited this shift and launched an attack on the inexperienced general, who suffered a disastrous defeat. Agathocles returned to Africa in 307, but failed to retrieve the situation and fled, abandoning most of his troops and even his sons (Diod. Sic. 20.69). Upon his return to Sicily, he concluded a peace mostly reinstating the terms of the 314 treaty. When he was planning to violate this treaty in 288 with a new attack on Carthage, he fell ill and died at the age of 72.

Fighting between Syracuse and Carthage continued, as the Carthaginians haunted the Sicilian coast and even raided Syracusan territory a decade later. The Sicilian Greeks appealed to the Epirot King Pyrrhus, who at that time was waging war against the Romans in southern Italy.[12] During his short intervention in Sicily, he won a victory at Messana and nearly succeeded in driving the Carthaginians out of Sicily. His refusal to negotiate peace, however, and his injurious behavior toward the Sicilian Greeks, quickly alienated them (Plut. *Pyrrh.*

23.5). Pyrrhus gladly left the island in 275, and the status quo ante supervened. The Carthaginians continued to roam the island and even collaborated with King Hiero II of Syracuse when Rome came to the aid of the Mamertines (Polyb. 1.11). The resulting First Punic War between Rome and Carthage was fought nearly exclusively on Sicily. In the treaty of 241, Carthage agreed to evacuate Sicily and to refrain from further action against Syracuse and its allies.[13] Like Athens before it, Carthage had dropped out of Sicilian politics.

Rome

Rome came to Sicily in 264 when the Mamertines asked for help in fending off the attacks of Syracuse and Carthage against Messana, which they controlled. After liberating Messana, the commander Appius Claudius Caudex planned to besiege Syracuse, but King Hiero II agreed to pay a fine and to supply the Roman army during their following confrontation with Carthage. Although no literary source speaks of Sicilian participation during this, the First Punic War, allied locals certainly provided Rome manpower, logistics, and technical support. In western Sicily, many Greek foundations under Carthaginian hegemony were sacked and destroyed by the Romans, including Acragas, Selinus, and Heraclea Minoa. The treaty ending the First Punic War made them Roman subjects and left Syracuse autonomous (Polyb. 1.62.8). In 227, Roman praetor Gaius Flaminius was entrusted with the administration of Sicily.

A pro-Carthaginian faction in Syracuse, however, convinced the new king, Hieronymus, to ally himself with Carthage, at war with Rome since 218 (Polyb. 7.3.9). Although King Hieronymus and his allies fell victim to a palace intrigue (Diod. Sic. 26.15), the Romans prepared to conquer this last independent piece of Sicily. In 214, they attacked by sea and land. The powerful fortifications of Syracuse withstood this onslaught and the new battle equipment developed by Archimedes bolstered the defenders. The siege lasted two years before the Romans gained the upper hand (Polyb. 8.3–7). After they received news concerning a Syracusan festival to Artemis, they climbed the outer walls during the celebrations and thus improved their position.[14] Seizing the rest of the city took several months more, and ended in the partial destruction of its buildings, the murder or enslavement of its citizens, and the transfer of its treasures to Rome (Polyb. 9.10). Under praetor Lucius Cincius Alimentus, Syracuse was incorporated into the Roman province of Sicilia.[15]

Internecine Conflicts

Our sources attest to much domestic upheaval and civil unrest in Sicily from the Archaic period onward. Syracuse alone suffered at least 27 episodes of social strife between the mid-seventh and the mid-third century. Several sources mention the expulsion of citizen groups, without giving the reasons that led to their removal. Thucydides, for example, says the Syracusan clan of the Myletidae took part in the foundation of Himera, without noting the

12 Plut. *Pyrrh.* 22; Diod. Sic. 22.8.

13 Polyb. 1.62.8. Further reading: Meister 1984; Asheri 1988; Bondi 2006.

14 Polyb. 8.37; Liv. 25.24.

15 Further reading: Lomas 1993; Hoyos 2015.

context of their departure (6.5.1). Tyranny began on Sicily with the coup of the general Panaetius against the aristocrats of Leontini in 608 (Polyaen. 5.47). The rise to power of the infamous Phalaris of Acragas three decades later involved the killing of many citizens and the abduction of women and children (Polyaen. 5.1). These and other struggles for or against one-man rule were a leitmotif of the pre-Roman, Greek period in Sicily. Sieges and mass-killings were frequent. Agathocles, with his private army of allies and mercenaries and his massacre of leading aristocrats, was the last of this breed (Polyaen. 5.3).

Factions and civil groups conducted violent actions, too. The *demos* expelled the land-owning elite, or *Gamoroi*, from Syracuse around 490 (Hdt. 7.155). They achieved this goal only by mobilizing the landowners' slaves in revolt against their masters (*FGrH* 566 F8a), so the struggle for a democratic regime became linked to economic and social emancipation. Economic motives often caused discontent and upheaval, especially rivalries over land. The dispute between the *demos* and aristocracy of Leontini concerning land distribution after the peace of Gela in 424 can serve as an example (Thuc. 5.4).

Civil unrest also resulted from diverging loyalties, as evidenced by Acragas, where pro-Syracusan factions were expelled both in 413, during the Athenian expedition (Thuc. 7.50), and in 394, after a defeat of Dionysius at Tauromenium (Diod. Sic. 14.88.5). Policies of displacement also led to conflicts, as evident after the fall of the Deinomenids in the mid-fifth century, when old and new citizens of Syracuse argued over the right to hold public office (Diod. Sic. 11.72).

Foreign mercenaries, especially those who helped tyrants win and retain power, played a major role in these conflicts. In Syracuse, mercenaries to whom Gelon had granted civil rights battled Syracusan citizens. After a series of violent confrontations in Syracuse and the hinterlands of Gela and Acragas, the Greek cities allowed the mercenaries to retain their possessions and to settle in the city of Messana (Diod. Sic. 11.76.5–6). Bands of mercenaries also occupied sites of native origin, including the cities of Omphace and Cacyrum in the hinterland of Gela (*FGrH* 577 F1).

Nearly two centuries after mercenaries settled in Messana, this city again played a decisive role in the conflict between unemployed mercenaries and Greek Sicilians. Campanian mercenaries demanded political rights in Syracuse after the death of Agathocles in 289, but were rebuffed by the Syracusans. After leaving Syracuse, the mercenaries seized the city of Messana, renamed it Mamertina after the war-god Mamers, and prospered by raiding the surrounding countryside (Diod. Sic. 21.18). As already noted, the attempts of Hiero II and the Carthaginians to rid themselves of their troublesome neighbors resulted in the First Punic War.[16]

The role of mercenaries, along with the frequent displacement of populations, made the history of Greek Sicily peculiarly turbulent. So did colonization and the role played by outside powers. Rome put an end to this era of wars and upheavals, but introduced a new era, marked by revolts and slave rebellions.

16 Further reading: Lintott 1982; Berger 1992; Tagliamonte 1994; Hansen and Nielsen 2004, 124–129; Trundle 2004.

PART II

Military Operations

CHAPTER 6

The Nature of Hoplite Warfare

Fernando Echeverría

The study of warfare in the Archaic period relies on a handful of epic narratives and lyric fragments, on an immense but problematic stock of iconographic material drawn from paintings and sculptures, and on a considerable yet still expanding collection of artifacts, mainly arms and armor.[1] Lacking, however, are detailed historical narratives such as we find during the Classical period, notably Herodotus, Thucydides, and Xenophon. Here is where the trouble begins.[2] The scattered, fragmentary, unsystematic, and sometimes even contradictory evidence for early warfare has tempted scholars to believe that there was a broad gap between the warfare and society depicted in the Homeric epics and those described by the Classical sources. To fill this gap, radical changes must be hypothesized.

In the late nineteenth century came the concept of "hoplite warfare," meaning that the Classical period possessed its own style of combat governed by a complex as well as distinctive set of rules and conditions. In the next century came the "hoplite ideology," meaning that heavy infantry forming a new socioeconomic class sought to transform Greek politics and society. In the course of these scholarly developments, many other subjects were enfolded into the study of hoplites–colonization, tyrannies, mercenary service, archaic legislation, ritualized warfare, property levels, citizenship. The need to bridge an alleged divide between Homer and Herodotus thus led to a reconstruction of the political and social history of the

1 Most noteworthy are a few lyric fragments by Archilochus and Tyrtaeus and a few vase paintings from the mid-seventh century thought to represent closed formations (the Chigi Vase). See Echeverría 2008, 74–85. For recent interpretations of the Chigi Vase, see van Wees 2000a, 2004, 170–172; Hurwit 2002; Viggiano and van Wees 2013b, 67–68.

2 Nature and types of sources for Greek warfare: Hornblower 2007; Millett 2013.

A Companion to Greek Warfare, First Edition. Edited by Waldemar Heckel, F.S. Naiden, E. Edward Garvin, and John Vanderspoel.

Archaic period around the hoplite as a military, political, and socioeconomic figure, the so-called "hoplite narrative" of recent decades.

The narrative is a coinage of Victor Davis Hanson's, meant to subsume older concepts.[3] It evidently embraces "hoplite warfare," a way of taking a Greek type of heavily armed infantryman, one well attested in archaeological as well as literary sources, and making him the origin of a new kind of combat centering on heavy infantry to the exclusion of light troops, missile troops, or cavalry. The new type of combat, in turn, led to changes in the nature of the political organization of Archaic Greece—changes due to military innovation. The emerging Greek political communities of the Archaic period reflected these new patterns and practices in Greek warfare.

The next intellectual step, seeing these changes as revolutionary, came from viewing the heavy infantryman from a socioeconomic perspective—not just as a particular type of Greek soldier equipped with a specific panoply of weapons and fighting in a certain kind of formation, the phalanx, but also a citizen with landed property and political rights. Hoplites dominated and transformed government as well as warfare, rendering both less aristocratic and more egalitarian. This blend of military and political history became known in the English-speaking world as the "orthodoxy."[4]

Missing was any connection between these changes and changes in Archaic attitudes and culture. The hoplite, already turned into an historical force as well as a citizen soldier, needed an ideology, and Hanson and earlier writers have supplied it, in the form of agrarian notions of economy, sacrifice, and hard work as well as civic notions of equality, homogeneity, moderation, and participation. The hoplite soldier thus became a multifaceted, all-encompassing figure who synthesized political, social, and ideological changes in the Archaic period. Elites had given way to a self-conscious and autonomous middle class, which legitimized itself through the practice of an exclusive and highly normative (or ritualized) style of warfare, one taken to be the most adequate, efficient, and economic way of war for the *polis.* Military change, social and political revolution, agrarian ideology, ritualized combat, democracy and egalitarianism, and class struggle coalesced in a single, coherent story.

The following pages will sketch the origin and evolution of the scholarly discussion on hoplite warfare, trace the ensuing revolution and narrative, and elicit implications for the reconstruction and understanding of the Archaic period in Greece. They will chiefly address the difficulties that the culminating hoplite narrative presents as a problematic, overstretched modern construct.[5]

The Cornerstone: Hoplite Equipment

Archaic Greek military equipment has received considerable attention, particularly since the publication of Snodgrass' seminal work.[6] A series of modern studies traces the origins and evolution of the main types of weapons and infers that a particular group of them

3 Hanson 2013.

4 Morris 1987, 196; Rawlings 2000, 234; Cartledge 2001, 155; Matthew 2009, 395; Krentz 2010; 45; Kagan and Viggiano 2013b. See recently Konijnendijk 2016, 1–2, 2018, 22; Lloyd 2017, 234–235.

5 The "hoplite revolution" as a "modern construct": Raaflaub 1993, 80. More on these theoretical difficulties: Lazenby and Whitehead 1996; Echeverría 2012.

6 Snodgrass 1964; see also 1965a, 1965b, 1967, 1971, 1980.

constituted the typical "pack" of the Greek heavily armed infantryman, identified as "hoplite equipment" or the "hoplite panoply."[7] This set consisted of six items: an enclosed metal helmet topped by a crest; body armor of different types (fundamentally bronze, but later also leather) but at first predominantly a heavy, double-plated bronze cuirass (the "bell cuirass"); a pair of bronze greaves shaped to fit the lower leg; a two-meter-long spear with bronze ends, a spear point, and a butt spike (the *sauroter*), which was the main offensive weapon; a short sword (40–60 cm) with straight or curved blade that is usually regarded as a secondary weapon; and a wide (90–100 cm diameter), bowl-shaped round shield (the *aspis* or Argive shield), with a wooden core and a protruding rim, occasionally covered with a bronze sheet and provided with two handles, a central armband (*porpax*), and a lateral handgrip (*antilabe*).

According to the most recent scholarship, the elements of the panoply appeared in Greece from the middle of the eighth century onward and spread piecemeal during the following decades. They appear in considerable quantities as offerings in sanctuaries and figure prominently in the iconography of the period.[8] The impact of foreign influences in Greek military technology of this period remains debated, but whatever this impact may have been, by the beginning of the sixth century the panoply seems to have been widespread and became the standard equipment of Greek heavy infantry forces during the Classical period.[9]

The characteristics of the different items of the panoply have generated little controversy, except for the weight of the Argive shield and the accumulated weight of the panoply: in the past, both shield and panoply were regarded as heavy and cumbersome, but recent estimates reduce their weight by half, with important consequences for the general interpretation of the capabilities of the Greek hoplite.[10]

Discussion has focused on the panoply in action.[11] It all starts with the Argive shield: its double grip and its width provide better handling and greater protection, respectively; these qualities are improved by its concave shape, which allows the inner side of the rim to rest on the shoulder and deflect blows. These very same qualities, however, together with its considerable weight, make the hoplite slow and heavy, while the double grip means that the shield protects only one side of the fighter. This shield thus exposes him to attacks from the unprotected side or the rear. The bronze cuirass offers protection at the cost of more weight, and the helmet makes it hard to see and hear. Although a heavily equipped and highly protected fighter, the hoplite is also slow, clumsy, and easily disoriented target.

7 Main types: Bottini 1988; Bol 1989; Jarva 1995, 2013; Baitinger 2001. Treatments and descriptions of the "hoplite panoply": Snodgrass 1964, 1967; Anderson 1970, 13–42; Cartledge 1977, 12–15; Anderson 1991; Hanson 1991b; Hunt 2007, 112–117; Krentz 2007a, 67–72; Jarva 2013, 397–412; Lee 2013, 147–149; Viggiano and van Wees 2013b.

8 Iconography of the hoplite panoply: Lorimer 1947; Greenhalgh 1973; Salmon 1977; van Wees 2000a; Viggiano and van Wees 2013b.

9 Particularly Assyrian and Persian influences. See Lee 2013; Raaflaub 2013a.

10 An estimated combined weight of 30–32 kg: Lorimer 1947; Cartledge 1977, 20, 1996; Hanson 1989, 55–88, 1991b; Dawson 1996, 48; Mitchell 1996, 89; Schwartz 2002; Jarva 2013, 398. The idea goes back to Rüstow and Köchly 1852, 44 and Delbrück 1887, 56. See Krentz 2010, 45–50 for discussion.

11 Lorimer 1947; Adcock 1957, 3–4; Greenhalgh 1973, 70–74; Cartledge 1977, 12–13, 1996; Salmon 1977; Hanson 1989, 1991a, 1991b; Bryant 1990; Mitchell 1996, 89–91; Schwartz 2002, 2009, 2013; Viggiano 2013. For a recent analysis of the discussion, see Krentz 2013a, 137–140, who situates the origin of the idea in Helbig 1909, 1911.

Given these drawbacks, the hoplite needed to be part of a closed formation, the phalanx. In isolation, he was well-nigh useless.[12] This conclusion led to another: phalanx equipment must be standardized, to make it easier for hoplites to serve together at close quarters.

This scholarly discussion depended on technological determinism: the peculiarities of the weapons determine the need for a particular method of fighting, and that method only. Archaeological evidence for use of the equipment thus implies certain military realities.[13] Conversely, written evidence for the existence of the phalanx implies the presence of hoplites. Determinism comes full circle.

Recently, scholars have questioned this sequential reasoning. It has been argued, for example, that the panoply was rarely complete (something attested by the iconography) and that the different items could be used in endless combinations, even including controversial types such as the Boeotian shield; as a result, military equipment was not standardized, but could depend to some extent on particular tastes and local traditions.[14] If there was no such thing as a standard panoply, conclusions about the role of hoplites in the phalanx are questionable, as are larger claims about hoplite warfare, to which we now turn.

Hoplite Warfare and Hoplite Ideology

Notions of "hoplite warfare" date back to the 1850s.[15] The English Classical scholar (and Classical Liberal) George Grote then posited that the kind of warfare practiced during the historical period in Greece differed from the heroic warfare of the epics because of introduction of the phalanx. Individual prowess became secondary to the discipline of the group.[16] In the following decades, a group of German scholars, some of them Prussian officers, built on Grote's ideas and established the principle that Greek war centered on pitched battles.[17] According to these writers, Greek wars were decided by short clashes of two closed

12 Lorimer 1947; Andrewes 1956, 32; Detienne 1968, 139 n. 108; Cartledge 1977, 13, 1996, 712; Bryant 1990, 498; Mitchell 1996, 89; Hanson 2000, 206; Schwartz 2002, 40.

13 This quality is used, for example, for chronological purposes: identifying the "hoplite panoply" in painted scenes allows us to date the appearance of the hoplite and, consequently, of the phalanx. This is the premise of most iconographical studies: Helbig 1911; Nilsson 1929; Lorimer 1947; Snodgrass 1965b, 1993; Greenhalgh 1973; Salmon 1977; Anderson 1991. See van Wees 2000a for a comprehensive discussion.

14 For discussion with further bibliography, see Jarva 1995; van Wees 2000a, 2004, 47–52; Lee 2013; Echeverría 2015.

15 General descriptions of "hoplite warfare": Adcock 1957, 2–11; Anderson 1970, 1–9; Ridley 1979; Hanson 1983, 1989, 1991a, 1991b; Bryant 1990; Wheeler 1991a, 2007a; Bowden 1993; Dawson 1996, 47–52; Mitchell 1996; Santosuosso 1997; Morgan 2001; Trundle 2001; Lendon 2005, 39–57; Krentz 2007a, 72–77; de Souza 2008; Hunt 2009, 229–30; Lee 2013, 153–157. Theoretical discussions on the nature of "hoplite warfare": Snodgrass 1965a, 1993; Detienne 1968; Cartledge 1977, 1996; Hanson 1983, 1989, 1991a, 1995, 2000; Krentz 1985a, 2007a, 2007b, 2013a; Connor 1988; Cawkwell 1989; Lazenby 1991; Wheeler 1991a, 2007a; Foxhall 1993; Mitchell 1996; Storch 1998; van Wees 2000a, 2004, 2013; Schwartz 2002; Rawlings 2007, 2013; Echeverría 2011; Lloyd 2017; Konijnendijk 2018.

16 Grote 1846, 106–107. The theme also appears in Fustel de Coulanges 1864. Kagan and Viggiano 2013b, 3, describe this background as "proto-orthodoxy".

17 Whom Konijnendijk 2018, 7–12 aptly calls the "Prussians": Rüstow and Köchly 1852; Droysen 1889; Bauer 1893; Beloch 1897; Lammert 1899; Delbrück 1900; Kromayer and Veith 1903, 1928.

formations of heavy infantry arranged in neat ranks, with little room for tactics, maneuvers, or inventiveness; battles were planned in advance, and troops other than heavy infantry were regarded as auxiliary. It has been argued recently that the Prussians elaborated this model of combat in order to produce a contrast between the Archaic and Classical periods, on the one hand, and, on the other, fourth-century developments that interested them thanks to the emergence of great generals such as Alexander the Great, envisioned as models for Germany's own military.[18] Whatever their motives, they were authoritative; for example, the dictum that the standard depth of the phalanx was eight ranks derives from them, not from the ancient sources.[19]

In the early twentieth century, Anglo-American scholars adopted the German model without reservations.[20] An orthodoxy had emerged: the simple battles of phalanxes, the exclusive presence of hoplites, the deterministic correlation between hoplite and phalanx, and the homogeneity of equipment. English-language writers made several contributions, first, formulating an important paradox of Greek warfare, this being the prevalence of the heavy, rigid phalanx in the rugged landscape of Greece; second, explaining the common practice of devastating enemy fields and orchards as a bait to lure the enemy to give battle; and third, focusing on hand-to-hand combat in closed formation (giving rise to the debate on the *othismos*, or "push," and the analogy of the rugby scrum).[21] These ideas have shaped the study of Greek warfare up to the present.

The elaboration of the ideological framework of hoplite combat came during the 1960s and 1970s. Since Grote, the hoplite soldier had been an anti-aristocratic, if not democratic, figure, but now, thanks to a group of French scholars, equality and ritualism became aspects of a new ideology.[22] The conditions of the panoply and the phalanx (homogenous equipment, fixed positions in the formation, restricted movements, and rigidity) were said to enforce strict solidarity and equality on fighters, and aristocratic individualism thus yielded to collective values (keeping formation, cohesion, rejection of individualistic behavior, and sacrifice).[23] The hoplite was now expected to collaborate in the mass of fighters for the common good, as in the fragments of Tyrtaeus; the egalitarian principle of the *polis* resulted from transferring these military values to the political and social spheres. Egalitarianism, in turn, reinforced protocols of the phalanx and hoplite combat, encouraging solidarity, anonymity, and simplicity.

The German scholars of the nineteenth century had already described hoplite warfare as simple and unimaginative because of the limitations of the phalanx.[24] From the 1960s onward, however, this simplicity was reinterpreted as a choice made by the hoplites so as to materialize their ideas, principles, and interests (and thus solve a problem noted by Grundy

18 I follow the treatment by Kagan and Viggiano 2013b as well as Konijnendijk.

19 Rüstow and Köchly's axiom 1852, 118–120, noted by Konijnendijk 2018, 7–10.

20 Grundy 1911; Gomme 1945–1981; Adcock 1957; Anderson 1970; Pritchett 1971–1991.

21 Grundy 1911; Fraser 1942; Pritchett 1971–1991 IV; Krentz 1985a, 1994, 2013a; Cawkwell 1989; Luginbill 1994; Goldsworthy 1997; Schwartz 2002; van Wees 2004, 188–191; Rawlings 2007, 93–97; Matthew 2009. For a recent analysis of the discussion, see Krentz 2013a, 143–148.

22 Grote 1846, 106–107; Rüstow and Köchly 1852; Meyer 1884–1902, v.2 (1893); Grundy 1911; Weber 1922; Nilsson 1929.

23 Detienne 1968; Mossé 1968; Vidal-Naquet 1968; Garlan 1972; Ducrey 1985.

24 Rüstow and Köchly 1852, 144–145; Droysen 1889, 93–94; Lammert 1899, 21.

and Gomme, which was that phalanx fighting was not feasible unless armies were thoroughly and profoundly unified).[25] A strict code of unwritten norms and protocols, one that emanated from the material conditions and ideological principles of the "hoplite class," would supply this degree of unity. The "simple battle" described by the Prussians thus became a "ritualized battle."

Ritualism drew on studies of Greek athletic competitions to conceive battle as a regulated contest or *agon*: a setting (the battlefield), a board of judges (the gods), two participants (the rival armies), and a prize (military victory). The rules for the competition were distilled by Ober in a list of 12 norms that included a formal declaration of war, sparing of "civilians," and the ransom of prisoners.[26] Ritualism seemed to fit the hoplite agenda of economy of effort, the social background of yeomen soldiers, uniform reliance on a heavy and cumbersome panoply, and the tidy reduction of war to quickly resolved battles. By the 1990s, it had become the prevailing view on the mechanics of Greek warfare.

The expanding construct of "hoplite warfare" met with criticism from revisionists writing after the 1960s.[27] They addressed the problems posed by the fractious evidence for hoplite combat and also embarked on a reconsideration of Greek warfare that has led to the gradual articulation of an alternative view of the Archaic period and of the origins of the *polis*. In the first place, Snodgrass questioned the notion of a uniform panoply, which he termed a "motley assemblage" of weapons and armor.[28] It spread "piecemeal" through Greece.[29] Snodgrass suggested that the adoption of the equipment and the introduction of the phalanx were not successive steps but instead were separated by a sizable lapse of time. He presented middle-class farmers as "reluctant hoplites,"[30] undermining the notion of a hoplite mindset. Snodgrass accepted other popular ideas, such as the military superiority of the hoplite and the phalanx, and hoplite military service as an important feature of social and political life, but with nuances. He inaugurated a gradualist line followed by some scholars in the ensuing decades.[31]

Another revisionist, the Homeric scholar Latacz, attacked the construct in a more profound fashion.[32] His analysis of the Homeric poems concluded that heroic combat was only a phase in Homeric battles and that an anonymous mass, arranged in a closed formation that he recognized as a forerunner of the Classical phalanx, did the most important fighting.[33] The impact of this discovery cannot be underestimated: Latacz had questioned the gap between Homeric and historical combat that Grote had established more than a century before and that had led to hoplite warfare being regarded as an important military innovation—an innovation for which Ritualism could provide a cultural context. Latacz's

25 Detienne 1968; Pritchett 1971–1991, v.2; Cartledge 1977; Connor 1988; Hanson 1989, 1991a, 2000; Bryant 1990; Ober 1991, 1994; Dawson 1996, 49–52; Mitchell 1996.

26 Ober 1994.

27 Snodgrass 1965a, 1993, 2013; Krentz 1985a, 2000, 2002; Foxhall 1993, 2013; Rawlings 2000, 2007, 2013; van Wees 2000a, 2001, 2004, 2013. Terming these writers "revisionist": Wheeler 2011, 79–104.

28 Snodgrass 1964, 1965a, 1965b, 1967, 1971, 1980.

29 Snodgrass 1964, 83–84, 89–90, 136–139, 193–204, 1965a, 110.

30 Snodgrass 1965a, 115, 1980, 101–102, 106–107.

31 Finley 1970; Salmon 1977; Murray 1980.

32 Latacz 1977.

33 Latacz 1977, 45–67, 224–245.

work thus struck a blow against a century of scholarship.[34] Most later writers, however, chose to embrace Latacz's idea of a proto-phalanx without questioning the larger notion of hoplite warfare and ideology.[35]

Hoplite Orthodoxy (cont'd)

During these decades, the orthodox Anglo-American view of hoplite warfare continued to develop. In light of the importance of both hoplite warfare and the related ideology, scholars argued that the nature of the Greek *polis* was essentially military, and thus reinstated the Aristotelian principle of the "community of warriors": military values and practices determined social change, political participation, and institutional organization.[36] The possession of weapons was a crucial political asset, as proved by successive revolutions and coups d'état in Archaic and Classical Greece. To be sure, changes occurred gradually, as a middle class of farmers (the traditional *demos*) undid aristocratic influences amid intense political struggles. As a result, new timocratic systems emerged.

The archaeologist Lorimer expanded on the principle of technological determinism.[37] She argued that tactical changes in the hoplite era resulted from "a single structural alteration," the double-grip system of the Argive shield.[38] The reason to account for the unprecedented spread of the shield was its (presumed) superiority, and the reason to justify the adoption of the phalanx was the peculiar characteristics of the panoply; its subsequent spread throughout Greece was attributed to its tactical superiority and to the dynamics of "peer polity interactions" and an "arms race."[39] A small change in equipment led to a chain reaction that put an end to "heroic" and cavalry combat.

Andrewes explored the political consequences of the rise of the hoplite.[40] He embraced the Aristotelian definition of Greek tyranny and Grote's and Weber's idea of the end of traditional oligarchies being overthrown because of collaboration between the new middle classes (once again, the traditional *demos*) and ambitious demagogues. Internecine struggle, or *stasis*, resulted from the clash between oligarchs and their foes, and often led to the establishment of tyrannies rather than (or as well as) some measure of reform. The rise of the hoplites need not lead straight to some new, stable political system such as Athenian democracy or Spartan monarchy cum oligarchy. To speak of the hoplites as political revolutionaries would be a complex proposition. Nevertheless, military change led to social transformation

34 As argued by Raaflaub (2005, 2008, 2013a, 2013b), who accepts the existence of a "proto-phalanx" and thus rejects the larger thesis.

35 Pritchett 1971–1991 v.4, 1–44; Hanson 1995, 2000. French scholars added a different political color to this scholarship: a hoplite "republic of Equals" in which "political and military systems were perfectly homologous" (Detienne 1968, 140; my translation). See also Greenhalgh 1973; Cartledge 1977, 1996.

36 Meyer 1893; Helbig 1911; Weber 1922; Glotz 1928; Nilsson 1928, 1929; Ehrenberg 1932, 1937. Recent critique of the link between Aristotle and hoplite history: Echeverría 2008, 93–103; van Wees 2013.

37 Lorimer 1947.

38 Lorimer 1947, 76.

39 An analysis of technological determinism and hoplites: Echeverría 2008, 193–248, 2010. "Arms race" applied to Greece and hoplite warfare: Lorimer 1947, 108; Andrewes 1956, 38; Cartledge 1977, 18; Salmon 1977, 96; Holladay 1982, 99–100. "Peer-polity interactions": Snodgrass 1986, 51–52.

40 Andrewes 1956.

that, in turn, prompted political upheaval.[41] Other scholars stuck to the more cautious language of a hoplite reform that ratified military and social changes.[42]

Another archaeologist, de Polignac, somewhat altered the hoplite orthodoxy. His study of rural sanctuaries during the Archaic period presented them as nodes in the ideological definition of the community as well as markers in the political definition of the landscape.[43] Because these sanctuaries had to be protected, they led to a new kind of warfare in which the community as a whole could be involved. This was "hoplite warfare" as de Polignac understood it. He had overthrown determinism as a reason for the introduction of the phalanx and shifted emphasis away from social and political struggles to territorial conflicts between *poleis*. Yet he did not reject the development of hoplite ritualism or ideology, or some kind of hoplite military reform.

A third archaeologist, Morris, endorsed this line.[44] He expressed skepticism about orthodox views, but subscribed to notions such as vertical social conflict between the *agathoi* and the *kakoi*, the emergence of the egalitarian principle of the *polis*, and what he called the "middling tradition."[45]

These writers prompted a period of exploration that produced revised or updated accounts of hoplite warfare, all attempting to insert new discoveries into the framework of a hoplite revolution or reform.[46] Most reinstated determinism and retained some sort of social struggle, while allowing for more complex social and political causality. The leitmotif of these accounts was the ever-broadening use of the term *hoplite*. As befitted their backgrounds, the nineteenth century Prussians used both "hoplite" and "phalanx" in a narrow, military sense, even if they used them tendentiously where evidence was lacking. Next came Classicists and historians speaking of the "hoplite panoply," "hoplite equipment," "hoplite shield," "hoplite forces/army," "hoplite tactics," and "hoplite warfare/battle."[47] By the late 1960s, scholars of almost all persuasions were speaking of a "hoplite assembly," "hoplite landowner," "hoplite farm," "hoplite ritual," "hoplite ideology," "hoplite ethos," "hoplite discipline," "hoplite *agon*/agonalism," "hoplite *polis*," "hoplite republic," "hoplite democracy," "hoplite status," "hoplite class," "hoplite system," "hoplite state," and so on.[48]

Fifty years ago, Snodgrass warned that "we should be hesitant in our application of the term [*hoplites*] to the earlier stages of Greek armament."[49] The conceptual caution he was urging went unnoticed, even though the need for it only increased, as the constellation of themes provided in the preceding paragraph were engulfed in the most ambitious concept yet, a "hoplite narrative."

---

41 Drews 1972 modified this view, arguing that the "hoplites" on which the would-be tyrants relied were not lower-class members of the community but mercenaries, and thus skipped the social transformation envisioned by others.

42 "Hoplite reform": Snodgrass 1965a, 1965b; Detienne 1968; Greenhalgh 1973. Previous scholars had naturally conceived the idea of a military and political reform or revolution: Lorimer 1947, 92; Andrewes 1956, 38. I think the phrase "hoplite revolution" more appropriate, considering the nature of the social, political, and military changes ascribed to it.

43 Polignac 1984.

44 Morris 1987, 196–202.

45 Morris 1987, 19–42, 175–183, 1996.

46 Bryant 1990; Bowden 1993; Mitchell 1996; Osborne 1996; Storch 1998. Even Snodgrass (1993) reconsidered his old position in the light of the recent developments.

47 Nilsson 1928, 1929; Lorimer 1947; Snodgrass 1964, 1965a, 1965b.

48 Snodgrass 1965a, 1965b; Detienne 1968; Cartledge 1977, 2013a; Salmon 1977; Bryant 1990; Lazenby 1991; Hanson 1995; Kagan and Viggiano 2013a. Nilsson spoke of a "hoplite-state" already in 1929 (p. 9), so the potential for the spread of the term was already present.

49 Snodgrass 1964, 204. For a recent consideration of the terms *hoplite* and *phalanx*, see Echeverría 2012.

The Hoplite Narrative

The final step toward the consolidation of the model of "hoplite warfare" came with the work of Hanson, perhaps the most influential Greek military historian of recent decades.[50] Hanson's work during the 1980s and 1990s, which was a reaction to the challenges made by the revisionists, became the main reason for the model's survival. Hanson reinstated the traditional elements of the model: the peculiarities of the hoplite panoply, the deterministic identification between hoplite and phalanx, Greek warfare as phalanx battle, the simplicity and ritualism of phalanx warfare, and the hoplite class as an anti-aristocratic and egalitarian social group.

In so doing, he took for granted a century's worth of assumptions. A cultural gap separated the "Homeric world" from the "historical world" of the *polis*, necessitating a period of changes and adjustments. Common to the two periods was some community of warriors, so that the Greek political community, the *polis*, was isomorphic with the army.[51] Technological determinism, that bane of military history, meant that superior new weapons immediately rendered old types obsolete, and therefore replaced them; new weapons entailed new tactics, and new types of fighters. As in Lorimer, ancient Greece experienced an unacknowledged arms race driven by the principle of superiority in design, performance, and effectiveness. A sort of class struggle—but one owing more to Aristotle or even Livy than to Marx—was axiomatic. Aristocrats battled the lower orders in the context of political and social unrest connected to the rise of tyrannies. The whiff of Marx notwithstanding, the *polis* was intrinsically egalitarian and democratic. The hoplite values of solidarity and equality impregnated the social and political institutions, such as assemblies and courts, that emerged from several centuries of unrest in the Archaic and early Classical periods.[52]

Hanson nonetheless made four distinct and far-reaching contributions of his own. First, he described an economic revolution at the beginning of the eighth century in Greece that allowed the exploitation of new farmlands and thus explained the rise of the "hoplite class," a group of non-elite landowners presented as middle-class independent farmers or yeomen. Second, he connected the ideology of egalitarianism and ritualism to the agrarian nature of the "hoplite class," and imported the term *agrarianism* to define the hoplite worldview and hoplite agenda, thus linking warfare to the economy. Third, he adapted John Keegan's ground-level approach to battle to ancient Greece and thus explored the "hoplite experience" of combat from a battlefield point of view, describing the actions, emotions, and concerns of the hoplites standing in the ranks of the phalanx.[53] Fourth, he presented a "hoplite system" as the origin and foundation of a distinctive, Western "world system," in which modern warfare and democracy are both the descendants and the unmediated counterparts of hoplite warfare and democracy.[54] Armed with these tools, Hanson broadened, deepened,

---

50 Hanson 1983, 1989, 1991a, 1991b, 1995, 1996, 2000, 2013.

51 For a recent discussion on the *polis* as a "community of warriors," see Berent 2000, 273–285. The fundamental passages are Arist. *Pol.* 1265b 28–29, 1268a 21–23, 1279b 2–4, 1297b 1–2, 1329a 11–12.

52 For a synthesis and critique of these assumptions, see Echevverría 2008, 29–71, with Kagan and Viggiano 2013a, and also the syntheses in Bryant 1990, 494–500; Wheeler 1991a, Bowden 1993, 47–49; Cartledge 1996, 685–686; Raaflaub 1997, 49–50; Krentz 2007a, 61–65; Hunt 2009, 231–234; Viggiano 2013.

53 Keegan 1976.

54 The problem of the disappearance and reappearance of the hoplite mode of fighting: Hanson 1989, 1995, 1996; cf. Snodgrass on 2013, 92–93 on the end of the hoplite.

and reinforced the orthodoxy of the previous generation, and turned it into a comprehensive, holistic "narrative"—his term. The impact of Hanson's scheme on Greek military and political history has been massive.[55]

Critics soon challenged his methods and preferences—his American-centric perspective, his tendency to anachronism and presentism, his cavalier treatment of the literary sources, and his omission of new archaeological data.[56] The initial, fundamental challenge concerned the mechanics of hoplite combat. Hanson and earlier proponents of the orthodoxy envisioned a massive collision of troops culminating in a shoving match, or *othismos*. Critics such as Latacz and Snodgrass had not rebutted this view, which was nonetheless controversial. Krentz entered the debate on the *othismos* and argued that hoplite battle consisted of a multiplicity of individual combats; this strengthened the gradualist position at first, but later invited connections with van Wees' new approach to Homeric combat. Van Wees reinterpreted the dynamics of Homeric warfare as an open and fluid battlefield in which the individual leaders moved freely followed by their "hosts," and engaged intermittently in the fighting only to retreat to the safety of the rearguard.[57] This marked the greatest shift in the interpretation of Homeric warfare since Latacz, with the consequences still to be explored. Van Wees argued that his open battlefield could be detected in the fragments of Tyrtaeus and Callinus, thus allowing the phalanx to operate in a looser, open fashion during the Archaic period. Krentz projected this kind of phalanx further, into the Classical era.[58] He differentiated between an "inclusive" and an "exclusive" phalanx, depending on the presence (or in the second case, the absence) of troops other than hoplites in the formation.[59]

Krentz also targeted an originally French contribution to the hoplite orthodoxy—the ritualistic character of warfare, in particular Ober's dozen rules of early Greek war. These rules resulted from misreading the literary sources, which either described religious duties, not rules, or reported rules without indicating how often they were followed, or how much weight violations would carry. Most telling was the *terminus post quem*, which was usually sometime in the fifth century.[60] This crucial contribution was especially damaging for Hanson's hoplite narrative, with its idealized view of Greek warfare as a bloody but straightforward and honorable encounter setting a standard for all of Western civilization. In gruesome truth, the literary sources report frequent Greek military brutality, not only toward foreigners, but also toward fellow Greeks, including both soldiers and civilians, and Krentz had helped explained why: the rules of war were few, tardy, and weak.

Van Wees contributed to the new trend less in regard to Homeric combat than in connection with several assumptions made by Hanson and older scholarship. Van Wees first attacked the 150-year-old assumption of a gap between the Homeric and the "historical" worlds. Then he attacked the centerpiece of both Hanson's views and those of the preceding

55 Schwartz 2002, 2009, 2013; Viggiano 2013.

56 Fundamentally by Foxhall 1993, 2013; van Wees 2000a, 2001, 2013. Further criticism on Hanson's political and ideological bias in González and López-Barja 2012.

57 Van Wees 1986, 1988, 1992, 1994a, 1994b, 1996, 1997.

58 Krentz reassessed the weight of the panoply and found it much lighter than previously thought. This finding, too, made it easier to envision an open battlefield. See Krentz 2007a, 70–71, 2010, 50, 2013b, 135–137. Rawlings also questions the burden of the hoplite panoply (2000, 246–249).

59 Krentz 2010, 2013a, 2013b.

60 Krentz 2000, 2002, 2007a, 76–79.

orthodoxy—the emergence of a hoplite middle class.[61] He argued that both analysis of the literary evidence and quantitative assessment of the economic conditions of the Archaic period revealed a polarized society. The sharp divide between rich and poor left no room for middle classes. The elite were the group mainly associated with the various phenomena regarded as "middling," such as the moralizing passages in Hesiod's *Works and Days*, the position of the Athenian *zeugitai*, the Greek norm of service in the militia, and the slogans of *isonomia* and the like. This critique endangered another assumption made by Hanson and the orthodox—that of a vertical, if not Marxist, class struggle. It focused attention elsewhere, on horizontal, not vertical, competition and conflict within the elite.

Other scholars entered the fray. Foxhall argued for unexploited and comparatively vacant landscapes during the Archaic period, thus questioning Hanson's pattern of "family farms" that supported the rise of a yeomen class.[62] Rawlings presented the wide range of military tasks and contexts that the Classical hoplite was expected to cope with, undermining the traditional connection between the hoplite and the phalanx.[63] Raaflaub, finally, explored the nature of ancient Greek democracy and the rise of the *polis* and connected them to the activities of the elite and the gradual definition of a "public" sphere. This political process interacted with the military transformations of the period, but slowly. It took centuries for closed combat and the Classical phalanx to result.[64]

This 20-year wave of scholarship affected hoplite warfare, scholarly assumptions, and Hanson's narrative. The cornerstone, the panoply, was not what scholars had thought, and it no longer could serve as part of the foundation of the large historical construct of hoplite warfare and Archaic society. Neither could the phalanx. This formation now had a problematic history, and the hoplite had a problematic relation to the phalanx. As for a middle class of hoplites during the early Archaic period, no such class existed. It followed that there could be no social reform to name after hoplites, still less a social revolution resulting from struggles between upper and lower classes. Ritual played an ancillary role in Greek warfare, and the same was true of rules obeyed in a ritualistic spirit. Democracy had as little to do with these military developments as Marx or Livy did, for throughout the period in question oligarchies dominated Greece. Hanson's agrarianism was misplaced, as was his concept of Archaic military history as an epochal starting point. Even Keegan's face of battle had a new mien, now that the *othismos* had yielded to more maneuver, yet also to more confusion.

More broadly, van Wees proposed an alternative narrative for the Archaic period, one presenting multiple social, political, and economic processes that prompted different responses at different times in different Greek communities. Vanished was Hanson's—and Grote's—sequence of causal steps taken from a single point of origin.[65] This possible reconfiguration of the past—as yet to be examined and evaluated—brings us to the topic of the present intellectual situation in the field of Archaic and Classical military history.

61 Van Wees 2001, 2002, 2006, 2013.

62 Foxhall 1993, 1997, 2013.

63 Rawlings 2000.

64 Raaflaub 1993, 1996, 1997. Other critiques of the hoplite narrative, aside from those noted in this section: van Wees 1995; Raaflaub 2013a.

65 van Wees 2013. Other views of the problems of narrative and causation: Echeverría 2008, 2010, 2011, 2012, 2015; Konijnendijk 2016, 2018; Lloyd 2017.

Future Prospects

In their Introduction to *Men of Bronze*, the most recent and influential synthesis on the subject, Kagan and Viggiano declare that they had intended that the 2008 Yale Conference inspiring the book should "bring together the leading scholars from both the orthodox and the revisionist schools of thought to examine the current state of the field."[66] Eventually, they lament, "Each side sharpened its position in response to the latest research ... instead of working toward a consensus."[67] Considering the history of hoplite warfare and its congeners, that outcome might have been anticipated. Since the appearance of the first criticisms of hoplite orthodoxy in the 1960s, the debate has veered between reactions and counter-reactions involving the "orthodox" and their foes. Consensus has rarely resulted.[68] When, for example, doubts about the "hoplite revolution" started to grow in the 1970s and 1980s, the orthodox reaction took the form of Hanson's "Western way of war" and hoplite narrative. The situation has been, and remains, polarized.

This, too, is no surprise. Each of the two positions entails implications that go well beyond soldiers and their weapons—implications for conceptions of historical change and causality, for the relations among political and social structures, and for the nature of the *polis*. On the one hand, the orthodox thinks of the *polis* as an advanced form of ancient state that was the product of rapid but logical social and military changes, and that, by inaugurating democracy and the rule of law, set important precedents. On the other hand, revisionists envisage the *polis* as "little more than a stand-off between the members of the elite who ran them," that is, a less orderly, less formal political community that did not result from a sequence of military and other changes and that led to democracy only in certain circumstances—circumstances in which the rule of law was a rallying cry rather than an institutional reality providing a precedent for modern societies.[69]

Despite the revisionists, the hoplite narrative rules the intellectual field, particularly outside Academia, where its capacity to produce simple, all-encompassing historical schemes is much applauded. Within the scholarly world, meanwhile, the narrative has lost ground, but the older notion of "hoplite warfare" remains the chief model for reconstructing early Greek military dynamics. Orthodoxy holds that "hoplite warfare" should remain the chief model until revisionists achieve "a coherent theory that even begins to replace the orthodox model."[70] The burden of proof thus rests with those who wish to overturn received views.[71]

This flawed claim merits rejection. "Hoplite warfare" is a general concept missing from ancient Greek sources, and impossible, indeed, to render into ancient Greek. The burden of

66 Kagan and Viggiano 2013a, xvii.

67 Kagan and Viggiano 2013a, loc. cit.

68 "The silence of the record is so impressive that one should doubt the presuppositions of the 'hoplite' theory" (Sealey 1976, 57); "The 'hoplite revolution' must be listed among the great non-events of history" (Frost 1984, 293); "There is no evidence whatsoever to support the theory that there was a hoplite reform" (Morris 1987, 198); "We should begin by expelling from the closets of our textbooks two skeletons that have lingered there far too long. One is the theory that tyranny and hoplite phalanx were directly connected [...]. The other skeleton is the theory of the 'hoplite revolution'" (Raaflaub 1997, 53). See also Nafissi 2009, 129.

69 Foxhall 1997, 119. Military law remains a neglected field. There is no chapter on the subject in handbooks such as Lipsius 1905–1915, or Harrison 1968–1971, and recent essays are few; see, however, Naiden 2018, 103–104 with refs.

70 Viggiano 2013, 126.

71 Viggiano 2013, 119.

proof rests with those who wish to give this concept the authority of some heading for an omnibus Greek military inscription that does not and cannot exist. This concept has, admittedly, garnered approval, but only because it seems to bring clarity to complicated and obscure aspects of ancient Greek warfare and politics. Alas, this clarity is illusory, as scholars since Snodgrass and Latacz have shown. Clarity will more likely result from analyses that allow for a greater degree of inconsistency and singularity among the hundreds of different Greek *poleis* of the Archaic period, as in the work of van Wees.[72]

The hoplite in "hoplite warfare" resembles the apex of an inverted pyramid. Identified as a reactionary revolutionary, a tyrant-cheering democrat, and a rigidly disciplined military amateur, he has carried too much historical weight for too long. We should put him to rest and pay due attention to other kinds of Greek military personnel described in this volume, from amphibious forces to cavalry.

72 Van Wees 2013. Krentz also attempted an alternative interpretation of "hoplite warfare" (2002, 35–37), while Lloyd (2017, 239–246) specifically addresses the diversity of the Greek world in his own account. Another alternative is Morgan 2001.

CHAPTER 7

Cavalry Battle in Greece and the Hellenistic East

Carolyn Willekes

Chariotry was present in the Greek world from the Bronze Age onward, but the question of when cavalry appeared on the scene is a matter of some debate. Cavalry are found in the Near East by the late ninth century; a tantalizing mention of riding in the *Iliad* does not refer to mounted warfare but simply indicates that the concept of riding existed.[1] Armed riders on Archaic vases, depicted in full hoplite panoply, appear to be mounted infantrymen; once they reached the battlefield they dismounted and fought on foot. The earliest "true" cavalry most likely comes from Thessaly where cavalry units of considerable size existed by the sixth century.[2] As Thessaly was one of the few regions in the Greek Peninsula ideally suited to *hippotrophia*, we should not find this surprising. The number of cavalry fielded in Thessaly varied, likely depending on the political situation; in 375, Jason of Pherae apparently had 6,000 cavalry at his disposal.[3] Macedon also had a relatively early cavalry tradition. Horses are found on Macedonian royal coinage from the beginning of the fifth century. Many depict riders who could either be hunters or light cavalry, but some portray armored horsemen.[4] Macedonia contained large plains, and another *hippotrophia*-friendly region, Boeotia, is where we find an established cavalry corps by the early fifth century. Boeotian cavalry fought on the Persian side at Plataea in 479 and their proficiency on that occasion suggests that this

1 *Il.* 10.498–514: Odysseus and Diomedes steal the horses of Rhesus and seem to be riding them back to camp. In Homeric Greek, *eph' hippon* commonly means "in a chariot".

2 Spence 1993, 23; Arist. *Ath. Pol.* 19.5 states the Thessalians sent 1,000 cavalry to help the Athenians stop the Spartan invasion of Attica, c. 511.

3 Xen. *Hell.* 6.1.8.

4 Thuc. 2.100.5, describing them as "Armed with breastplates, and excellent horsemen …"; Spence 1993, 26.

A Companion to Greek Warfare, First Edition. Edited by Waldemar Heckel, F.S. Naiden, E. Edward Garvin, and John Vanderspoel.

was not a newly formed unit.[5] Boeotia was regularly able to field a cavalry of at least 1,100, but this number could increase in times of crisis.[6] In Athens there may have been a cavalry of sorts during the sixth century, possibly as early as Solon's reforms, the so-called naukraric cavalry.[7] Each of the 48 naukraries in Archaic Athens were required to provide two horsemen as well as a ship, creating a "cavalry" of 96. These may have served as a coastal patrol, given their association with the naukraries and the fact that during the Classical period the Athenian cavalry frequently performed the role of "mounted scouts/guards."[8] Hence the naukraric horsemen may represent the earliest firm evidence for cavalry in Greece.[9] An anecdote in Plutarch suggests that Athens had a functional cavalry force by 480.[10] There were several reforms of the Athenian cavalry throughout the fifth century, as numbers increased first to 300, and then to 1000. By 431, the Athenian cavalry regularly numbered 1,000 with an additional 200 *hippotoxotai*.[11] During the fourth century the *hippotoxotai* were replaced by the *prodromoi*.[12] In the Peloponnese we do not find much of a cavalry tradition prior to the late fifth/early fourth century, a fact largely due to terrain.[13]

After the Persian Wars there was a growing awareness of cavalry's usefulness in battle and a gradual build-up of cavalry corps throughout mainland Greece. This is particularly so during and after the Peloponnesian War. The fourth century saw increased cavalry activity on the battlefield, both on their own and in conjunction with infantry.[14] Horsemen were central to the campaigns of Alexander the Great, who regularly deployed the allied Thessalian cavalry on the left wing, while he himself initiated the battle with a charge from the right wing with the Companions.[15] By the age of the Successors, cavalry appeared in numbers rarely seen before in the Greek world.[16] Thereafter, these numbers declined in mainland Greece, but horsemen remained central to warfare in the Hellenistic East, where the horse had been a primary feature of Near Eastern warfare since the Bronze Age.[17] The eastern Greeks also began to experiment with other forms of cavalry, in particular cataphracts.[18]

5 Hdt. 9.68.1; 9.69.2; Spence 1993, 19.

6 Each of the 11 districts in the Boeotian League contributed 100 cavalry, which would create a standing force of 1,100, but Diod. Sic. 15.26.1 tell us that in 396 the Boeotians sent 2,000 cavalrymen into battle; see Donaghy 2014, 90–91.

7 Spence 1993, 11, Poll. *Onom.* 8.106.

8 Bugh 1988/2016, 4–5; Anderson 1960, 130.

9 At the same time, we also need to consider the fact that Athens was asking the Thessalian cavalry for help. Anderson 1960, 130.

10 Plut. *Cim.* 5.2, where Cimon gives his bridle over as a dedication, stating that Athens now needed seamen, not horsemen.

11 Bugh 1988/2016, 39; Spence 1993, 15–17. See Thuc. 2.18.8: reference to Pericles' 1,000 cavalry and 200 *hippotoxotai*; Arist. *Ath. Pol.* 24.3 simply refers to 1,200 horsemen, likewise Aesch. 2.174 and Andoc. 3.7.

12 This change must have happened by 360 as Xenophon does not mention the *hippotoxotai* in his *Eq. Mag.*; Bugh 1988/2016, 221; Xen. *Eq. Mag.* 1.25.

13 For a detailed analysis of cavalry development in Greece, see Spence 1993, 1–33.

14 Cavalry at Cynoscephalae in 364: Plut. *Pel.* 32.2–4. Combined arms attack by the Phliasians under the command of Chares against the Sicyonians: Xen. *Hell.* 7.2.20–23: Epaminondas showed the possibilities of a combined arms attack at Tegea, unfortunately he fell in battle: Xen. *Hell.* 7.4.23–25.

15 Granicus: Arr. *Anab.* 1.13.1–17.2; Issus: Arr. *Anab.* 2.7.1–2.11.10; Gaugamela: Arr. *Anab.* 3.11.3–15.5.

16 During the Lamian War the allied Greeks fielded more than 3,500 cavalry (Diod. Sic. 18.15.2–3), while later on thanks to reinforcements from Asia Minor the Macedonian side under Antipater amassed 5,000 cavalry (Diod. Sic. 18.16.5); Eumenes has 6,300 cavalry in Asia Minor (Plut. *Eum.* 4.2–3; Diod. Sic. 18.29.3–30.1).

17 Antiochus III, at Raphia in 217, fields 6,000 cavalry and personally leads the charge against the enemy cavalry (Polyb. 5.79–86).

The Greek Horse

Any consideration of cavalry must begin with the horse itself. The animal was very much a product of its environment, both in European Greece and in the Near East. The Greek world by and large was not particularly suitable for horse raising, or *hippotrophia*, since there were few large, well-watered plains on the Greek peninsula, and this restriction made the horse an expensive commodity, closely tied to aristocrats and other wealthy persons.[19] Aristotle goes so far as to suggest that regions well suited to *hippotrophia* were more likely to develop oligarchic governments.[20] In the Greek world, horses were rarely used for day-to-day labor, being reserved for sports, spectacles, and warfare.[21] Even after cavalry became a regular part of Greek warfare there was often an underlying hesitancy on the part of the horsemen to commit fully to battle. This is not to say that they did not fight, but rather that they did not push themselves quite a much as they could have to ensure a more complete victory.[22]

The rationale behind this was quite simply the fact that horses were expensive and cavalry service imposed a significant financial burden.[23] Lead tablets found in the Agora and Ceramicus record the values of cavalry mounts as part of the *timesis* evaluation.[24] The bulk of the Agora tablets date to the second century and values range from 100 to 1,200 drachmas, with the average cost of a warhorse coming to approximately 700 drachmas.[25] This does not mean that the maximum price of a warhorse was 1,200 drachmas—Alexander the Great's Bucephalas famously cost 13 talents—but rather, it suggests that 1,200 drachmas was the maximum replacement value assigned by the state to a cavalry mount; those who were so inclined could certainly spend considerably larger sums on flashier or fancier mounts.[26] Purchasing the horse was only the initial financial outlay; there was also the cost of day-to-day maintenance, training, and equipment.[27]

18 Seleucid cataphracts open the battle at Panion between Antiochus III and Scopas in 200 (Polyb. 16.18). On cataphracts see Chapter 14.

19 Pl. *Leg.* 8.834b (no need for horses on Crete, hence, little *hippotrophia*); *Od.* 4.600–608 (Ithaca not suitable for horses).

20 Arist. *Pol.* 4.1289a 33–40; 6.1321a; Gaebel 2002, 56.

21 Isager and Skydsgaard 1992, 86.

22 Gaebel 2002, 134.

23 Bugh 1988/2016, 53; Hyp. *Lyc.* 1.16: Lycophron states that his involvement in *hippotrophia* for cavalry on behalf of the state has overtaxed his strength and resources. Xen. *Eq. Mag.* 1.10–12 mentions men who try to avoid cavalry service (presumably because of the expense) and ways to encourage new recruits to join and their parents to bankroll them. For the general cost of cavalry service in Athens see Spence 1993, 272–826.

24 The *timesis* evaluation in Athens was a current "market value" assessment of each cavalry mount. These records were essential as it seems that the state would replace a horse lost or injured beyond recovery in active service, but it would only provide funds to cover the current value of the horse lost. Bugh 1988/2016, 58; Kroll 1977, 97–100. For the Ceramicus finds, see Braun 1970.

25 Kroll 1977, 89.

26 The maximum 1,200 drachmas value also corresponds with the *katastasis* loan, another Athenian creation. The *katastasis* loan was given to Athenian cavalrymen starting out in service who, although well to do, may not have had the financial equity to hand to purchase a warhorse: the loan was meant to help with this. A cavalryman could only receive the loan once in his career and it had to be paid back before he could retire, at which point the funds went to a replacement recruit. (Bugh 1988/2016, 53–56).

27 Xenophon (*Eq.* 2.1–2) recommends sending the young horse to a professional trainer to be "started".

Greek horses were not large animals: by modern equine standards, they would be considered ponies.[28] We have a reasonably good idea of what the average Greek horse looked like thanks to its frequent appearance in the artistic record, which is remarkably consistent in its depiction of certain features.[29] The head presents a clean, refined profile with broad forehead, large, well-spaced eyes. The neck is of moderate length and, in the case of cavalry mounts, often thick and muscular, an indication of the Greek preference for riding stallions to war. A nicely sloped shoulder allows for a smooth, free-flowing gait and adjustability. The chest tends to be narrow, but always muscular: this is typical of mountain horses. The body is lean and compact, with moderately well-sprung ribs, suggesting an animal with a fair degree of endurance. The back is muscular, calling to mind Xenophon's reference to a "double-back" on his ideal horse.[30] The hindquarters are not particularly large, but well shaped and proportionate to the rest of the body. Artistic representations often emphasize these muscles by showing the horse shifting its weight back onto its hind end. The legs are slender and well formed with no excess fleshiness or hairiness, and they are long in proportion to the rest of the body. The overall impression given by an image of a Greek warhorse is one of a compact, muscular, lightly built animal. Such an animal would be suitable for an array of tactics on the battlefield, yet capable of carrying ample weight.[31]

The Near Eastern Horse

The horse was prominent in the cultures of the Near East from the mid to late third millennium onward.[32] *Hippotrophia* became a key component of the Near Eastern economy, particularly among the aristocracy who built up substantial herds for both civilian and military use. These horses were bred on stud farms that ranged from relatively small, private facilities to the large-scale, state-run operations seen on the Second Millennium Alalakh Tablets, the earliest record of state sponsored *hippotrophia*.[33] Herodotus mentions privately owned stud farms, while Strabo describes the royal stud at Apamea in Syria.[34] Although *hippotrophia* took place across the Near Eastern world, two regions were renowned for their horses,

28 The unit of measurement used for horses is known as a "hand." One hand equals 4 inches, the horse is measured standing on level ground from the base of its hoof to the withers. In modern usage, a pony is any equine that measures 14.2 hands and under; a horse is over 14.2 hands. These measurement designations did not exist in antiquity. Weight-carrying ability is not connected to height, but to conformation. Many horses in the Mediterranean and Near Eastern world would have ranged between 13 and 15 hands in height (Donaghy 2014, 95–97). See also Azzaroli 1985, 136–142 for ancient horse heights in general.

29 The emphasis on these features is consistent from the Bronze Age onward—seen in both the "abstract" geometric portrayals of equines and the far more realistic representations of the Classical and Hellenistic periods (Willekes 2016, 8–12). Xenophon (*Eq.* 1.1–17) provides us with a lengthy description of the ideal horse.

30 This term likely refers to the correct muscling of the *m. longissimus* muscle alongside the spine, which is reflective of particular training methods. Xen *Eq.* 1.11.

31 See Willekes 2016, 131–134 for a more detailed description of this horse type.

32 Equids begin to appear in Near Eastern art from c. 2400 onward, while they begin popping up in Near Eastern Literature during the Ur III period (c. 2100–2000). Cool-Root 2002, 204–206; Caubert 2002, 215–220; Gordon 1958, 18–19.

33 Tablets 293–331, which date to the mid second millennium. Wiseman 1953, 94–95.

34 Hdt. 1.192.2–3; Strab. 16.2.10.

Media and Armenia. The latter had been known as a key horse-breeding region from the second millennium onward. Hittite and Assyrian rulers were concerned with maintaining control over Armenia and its environs to ensure a steady supply of horses.[35] Media likewise provided horses for the Assyrians and was home to the Nesaean horse, an animal of unusual size that came to be regarded as the royal horse of the Persian Empire.[36] This was entirely due to the "Medic grass" growing there. Also known as lucerne/alfalfa, this type of grass has a protein content of close to 20% while most other premium grasses/hays reach only 7–10%. The higher protein content gave the Nesaean horse greater growth potential, resulting in a horse that was considerably "larger" than any of its counterparts in the ancient world.[37]

The Nesaean, indeed, was unique in the ancient world, easily distinguished by a powerful, heavyset body. The head is its most recognizable feature with a convex ram's head shape, a broad forehead, and well-spaced ears. The thick neck has a muscular crest. The chest, broad and muscular with a sloped shoulder, allows the animal to have a good length of stride, but the angles make it more suited to collective pulling or carrying, rather than galloping.[38] The back is short and broad: an excellent example of a "close-coupled" horse, which again indicates its suitability for collective effort. The hindquarters are relatively large but proportionate: they suggest a powerhouse of an animal capable of sudden starts and stops. The Nesaean horse does not have particularly long legs, but they are well formed with good, dense bone, making them ideal for carrying weight. The legs give the animal a sturdy, sure-footed appearance. The overall impression is one of dynamic power. A strong, compact body made this animal an excellent weight carrier. Although not likely to be fast over long distances, it was capable of short, explosive bursts of speed. Well balanced and maneuverable, this animal was happiest moving at a controlled pace. Its weight-carrying capacity made it suitable for heavily armed cavalry, and so the Nesaean type must have made up the ranks of the cataphracts.[39]

A second Near Eastern type was the lighter and leaner Turanian horse. The head is long with a flat, narrow profile and a delicate muzzle. The long and slender neck, set quite high on the shoulders, gives the animal an elevated head carriage. The horse would have easily been able to stretch the neck and head forward, which, combined with a long sloping shoulder, allowed the animal to lengthen its stride and cover more ground with a flowing, efficient gait. The chest is narrow but typically well muscled. Well-sprung ribs provide plenty of room for a robust respiratory and cardiovascular system indicative of an animal built for speed and endurance.[40] A long back ends in moderately sized hindquarters, suggesting that

35 See, for example, *Royal Correspondence of the Assyrian Empire*, Vol. II 757. The reference to mares suggests this was a breeding facility. Goodall 1977, 86–87 on the Lake Van region (Armenia) and horse breeding.

36 Hdt. 7.40.2–3; Strab. 11.13.7; Arr. *Anab.* 7.13.1.

37 Hyland 2003, 30. Large here does not refer to height but to breadth of body.

38 Collection refers to a way of moving in which the horse shifts its weight onto its hindquarters, lightening its forehand to carry itself in a balanced, upright manner. This allows for a greater degree of athleticism and flexibility while also putting less strain on the joints of the horse. Willekes 2015, 51–52.

39 See Willekes 2016, 119–120; Afshar and Lerner 1979, 44; Gabrielli 2006, 22–28; Azzarolli 1985, 146–149. Examples of the Nesaean horse can be seen on the Apadana Staircase at Persepolis, on gold plaques in the Oxus Treasure (BM 123947, BM 123946).

40 In the famous race between Persian and Thessalian horses recounted by Herodotus it must have been this lighter type of horse that Xerxes used, not a Nesaean (Hdt. 7.196).

this type was not one for sudden turns and bursts of speed. The legs are long and slender, but correctly shaped. This Turanian horse, adapted to life in an arid environment, could travel long distances with a comfortable, sweeping stride. A hardy, versatile animal that required very little feed, it was particularly well suited to light cavalry tactics, *hippotoxotai*, skirmishing, and scouting.[41]

Creating a Warhorse

Of all the uses to which horses have been put, from racing to plowing, warfare is the most remarkable, because of two equine traits. First, the horse is a prey animal, and second, it is also a herd animal. As potential prey, horses are controlled by a flight-or-fight instinct, and they will almost always choose the former response.[42] Horses evolved over millions of years to be able to outrun predators. They are naturally cautious and jumpy, ready to bolt at the first hint of danger, especially in unfamiliar surroundings.[43] Given this flight instinct, putting an equine into combat ought to have been impossible. Here the second trait comes into play: the power of the herd. Horses are social animals: left to their own devices, they will do anything they can to avoid being alone. They naturally form herds that are strictly hierarchical.[44] Body language plays an essential role in maintaining order in these herds, for although horses produce varied sounds, wild herds are not very vocal. They communicate silently by moving the ears, tail, and mouth, and by posturing. Both body language and hierarchy affected the training and use of warhorses. The importance of regular training is evident throughout Xenophon's two texts on horses and cavalry.

Yet even with thorough training the horse remained unpredictable, and unfamiliar situations could trigger its flight instinct, as numerous examples in the literary record show. Several of these instances involved cavalry being faced by unfamiliar animals such as camels and elephants. Despite the fact that these were also prey species, the sight, smell, and sounds they produced were enough to render the opposing cavalry useless.[45] However, regular exposure to such animals removed this instinctual fear, and horses

41 Willekes 2016, 120–123. These horses appear frequently in Assyrian art: see, for example, BM 124579, BM 124532, BM 124796, BM 124795.

42 Horses admittedly fight or become aggressive, typically when flight is restricted. Even two rival stallions will try to avoid a physical contest by posturing and using other body language, and resort to fighting only as a last resort. This is because fighting is inherently risky, for even the victor in a scuffle may be injured and thus find himself vulnerable. One of the most famous accounts of such a case is the "taming" of Bucephalus by a young Alexander the Great. Rather than "mean" or "dangerous," he is scared of his own shadows, but he lashes out in a fear response because he has nowhere to go. Once Alexander leaves him free to move, he flees, then realizes that he is no longer being restrained and relaxes, losing the "aggressiveness" he had previously displayed. Plut. *Alex.* 6.1–5.

43 For a discussion of the effect of sights, sounds, and smells of battle on the horse see Cantrell 2011, 11–24.

44 In the wild, two types of herds are found: first, the harem band, which is made up of a stallion, mares, and their young offspring; and second, the bachelor band comprising stallions who have not yet successfully challenged for and acquired a harem band of their own, and older stallions who have lost their harem band to a younger male.

45 Croesus' cavalry spooked by Persian camels: Hdt.1.80.3–5. Scythian cavalry thrown off by Persian donkeys: Hdt. 4.129. Alexander trying to cross the Hydaspes River with Porus' elephants stationed on the opposite side: Arr. *Anab.* 5.10.2, 5.11.4.

became capable of working alongside them. The armies of the Diadochic and Hellenistic periods often included both horses and elephants.

Greek Arms and Armor

There was no set cavalry "uniform" in the Greek world, the selection of arms and armor varying from *polis* to *polis*. For example, Thucydides notes that the Macedonian cavalry of the fifth century wore breastplates and were better armed than their opponents.[46] Certain weapons were synonymous with particular types of cavalry, such as the bow for the *hippotoxotai* or the *kontos*/lance for the cataphracts. Once again, Xenophon provides a detailed list of what he deemed the most suitable arms and armor for a standard cavalryman.[47] With regard to armor, he seems to believe that "more is better." He felt that a well-fitting breastplate with long *pteryges* was essential for comfort, protection, and ease of riding.[48] Next he recommends the use of neck protection and what he describes seems to be more than a simple gorget, as it should also sweep up to protect the lower part of the face much like a medieval bevor. For protection of the left arm, Xenophon recommends what appears to be an armored sleeve ending in a gauntlet—as the left hand held the reins, this was the steering arm and serious injury could result in loss of control—and for the right (fighting) arm, a vambrace to protect the forearm while still allowing full mobility, but there is no mention of hand protection. The *pteryges* of the breastplate covered the thighs/upper leg, while tall boots of durable leather protected the shins and feet.[49] On the topic of leg protection, he also mentions sidepieces, which swept over the top of the rider's thighs.[50] Xenophon's head protection of choice was the open-face Boeotian helmet, which allowed for a wide range of visibility and mobility.

Recognizing that protecting horses was essential, Xenophon also provides suggestions for horse armor. The side pieces were meant to protect not just the rider but the horse as well. In addition to these he mentions face armor (chamfron) and a chest piece (peytral). Horse armor, however, rarely appears in the Greek iconographic record, so what inspired Xenophon's recommendations? The answer is the East. Horse armor can be found on Near Eastern warhorses from a very early period.[51]

46 Thuc. 2.100.5.

47 Xen. *Eq.* 12.1–14.

48 Such a breastplate can be seen on the Hellenistic funerary stela of Nicanor. See Hatzopoulos and Juhel 2009, 430.

49 Several examples can be seen on the Parthenon Frieze, for example, BM 1816,0610.71.

50 There are no surviving examples of this type of armor from the Greek world, a fact suggesting that if it did exist it was likely made of a perishable material like thick quilt padding. An example may appear on the tomb of the Lycian Payava at Xanthos (Anderson 1961, plate 13A; also Anderson 1960, 7–8). It is not a far-fetched idea, as these sorts of armored side pieces do appear in the medieval record both as independent pieces of armor and as part of a saddle. Many thanks to Arne Koets for providing me with some examples of these side pieces. See also Fallows 2010, 109–113.

51 Spence 1993, 64–65. For example, relief panels on the bronze gates of Shalmaneser III show cavalry horses that appear to be wearing peytrals in the ninth century. Anderson 1961, plate 3A, an ivory plaque from Nimrud (ninth-eighth century), depicts a chariot horse that wears a protective blanket that is either quilted or armored. Curtis and Tallis 2012, catalog entry 41, is possibly a *phalera* depicting an Achaemenid period hunting scene (fifth-fourth century) including three horses, two of which seem to be wearing chamfrons see Curtis and Tallis 2012, entry 103. The Çan sarcophagus (fourth century) shows a horse wearing a chamfron while the rider wears what seems to be the type of protective neck armor described by Xenophon (Sevinç et al 2001, 396–397).

Xenophon spent time there and would have seen first-hand the additional protection offered by this equipment. To what degree his recommendations were adopted back in the Greek peninsula is unclear.

For weaponry Xenophon strongly recommends the use of javelins (*palton* or *xyston*) made from cornel wood as these were more durable than a longer spear (*dory*).[52] Measuring about 6 feet long, the javelin was also a versatile weapon, and for this reason he suggests that each cavalryman should carry two javelins, one to be thrown from a distance while charging the enemy and the other to be used in close combat.[53] Javelin throwing was a regular part of cavalry training and a mounted javelin event was part of the Panathenaic Games.[54] The use of the double javelins is regularly attested in the iconographic record.[55] The other option, a single, longer spear, measured 8–9 feet, including a spearhead and butt spike. This may have been the preferred weapon of the Macedonian cavalryman. It could not be used as a missile, but rather was intended for the charge and hand-to-hand combat.[56] As a secondary weapon, Xenophon prefers the short, curved, single-edged *kopis* or *machaira* to the long, straight, double edged *xiphos*. Ideally designed for use on horseback, the *kopis* could deliver devastatingly effective blows. It was with such a sword that Clitus the Black severed the arm of Spithridates at the shoulder.[57]

The absence of saddles and stirrups in antiquity is often commented upon in discussions about the effectiveness of cavalry, but the belief that cavalry without saddles and stirrups were less effective than their later counterparts, particularly as shock troops, is fallacious. Riding bareback or on a saddlecloth, vs. saddle and stirrups, requires a distinct but no less viable technique. An individual who has only ridden bareback will have a very secure seat and stable balance on the horse, whereas stirrups often give the rider a false sense of security.[58] In any event, Greek bareback cavalry were not riding against opponents with different gear, but against other bareback cavalry, so one side was as secure on its animals as the other. With regard to mobility and effectiveness, the absence of saddles and stirrups was a moot point.

---

52 Xenophon recounts a skirmish between Persian and Greek cavalry in which the Greek spears broke but the Persian cornel wood javelins did not (*Hell.* 3.4.13–14).

53 Throwing the javelin while the horse was in motion at a canter or gallop would add to the strength of the throw if the rider timed the release just right. Spence 1993, 50–51 also mentions the use of a throwing loop to add distance.

54 Xen. *Eq. Mag.* 1.21; see Spence 1993, plate 8 for an iconographic example of the mounted javelin contest.

55 For example, see Anderson 1961, plate 25, and Spence 1993, plate 7.

56 Arrian reports Alexander's spear shattering in battle at the Granicus (*Anab.*1.15.6); see also the Alexander Mosaic for the use of the longer spear.

57 Arr. *Anab.* 1.15.8 (**ὑποφθάσας δὲ αὐτὸν Κλεῖτος ὁ Δρωπίδου παίει κατὰ τοῦ ὤμου καὶ ἀποκόπτει τὸν ὦμον τοῦ Σπιθριδάτου ξὺν τῇ κοπίδι**). For discussions on weaponry, see Sidnell 2006, 31–34, and Spence 1993, 49–56.

58 In what for scholars is the infamous remark at *Anab.* 3.2.18–19, Xenophon tells his men that the opposing cavalry were "precariously balanced" on their horses. I agree wholeheartedly with Gaebel 2002, 113 that Xenophon was downplaying the effectiveness of cavalry to boost the courage of his men, who had no cavalry of their own and were at a severe disadvantage. As an experienced horseman Xenophon was fully aware of the potential of the Persian cavalry and knew, moreover, that the men on horseback were highly skilled riders and fighters.

Use in Combat

Cavalry charges often formed part of the opening engagement in a pitched battle. These speedy attacks could play a role in determining the course of the battle, and in some cases played a leading role, as at Olynthus.[59] Following the opening charge, horsemen would continue to engage with each other, and light infantry would often join them. Some light infantry specialized in this style of fighting, particularly the *hamippoi*.[60] At this stage of the battle, cavalry protected the phalanx against attacks from enemy horsemen. They could break up charges against the phalanx by riding against the opposing cavalry. This not only kept the flanks and rear of the phalanx safe, but also gave the infantry a chance to pause and regroup before pressing forward again.

Cavalry might also attack infantry formations. The cavalry would make repeated, rapid attacks and withdrawals against the flanks of the infantry, forcing them to break or loosen their ranks, at which point the cavalry could ride through them. Sources show a hoplite phalanx that did not have cavalry to protect it from flank attack was considered vulnerable.[61] It was also not unheard of for cavalry to attack a phalanx head on, as I believe occurred at Chaeronea.[62] To be sure, horses will not willingly charge what appears to be a solid object, and with its overlapped shields a hoplite phalanx would certainly give that impression. Attacking the flanks or rear of a phalanx presented an easier task: the spaces between the lines of hoplites provide gaps for the horse to push through. These gaps do not have to be large. If the horse can fit its head and neck through an opening, it will willingly push through, bumping and shoving the hoplites out of the way. At Chaeronea, Macedonian tactics created gaps in the line.

A tactic that begins to appear regularly in the fourth century is combined arms, or cavalry and infantry working in unison. The *hamippoi* were the most obvious choice for this, but other light-armed troops like peltasts were suitable as well. Light infantry were ideal for fighting among the cavalry, as they fought in looser formations allowing them to move easily between the horsemen. In many cases, the cavalry charge opened the engagement, then relaxed their own formation a bit, so infantry could join them.[63]

The role of the cavalryman did not end when enemy lines collapsed. In many cases, horses were key in the success or failure of any pursuit.[64] If either side lacked cavalry, pursuit was typically a short, simple affair. If the defeated side were able to flee the battlefield on horseback and the victors did not have horsemen to catch them, pursuit was futile, and if the victors were mounted and the defeated were on foot, pursuit would be brief and effective.[65] When both sides were mounted, the pursuit became problematic. The victors needed to avoid being carried away as they tried to run down the defeated, for both horses and men would be hurt and tired after fighting a battle, yet their adrenaline would be pumping, masking the

59 Xen. *Hell.* 5.2.41–42.

60 Aris. *Ath. Pol.* 49.1; Thuc. 5.57 and Xen. *Hell.* 7.5.23 suggest that the *hamippoi* were a Boeotian creation. Xen. *Eq. Mag.* 9.7 refers to infantry who fight alongside the cavalry, but he does not specifically call them *hamippoi*. Louvre 744 (relief 32) shows what appears to be a *hamippos* and cavalryman (Spence 1993, 58–60, plate 10).

61 Xen. *An.* 2.4; Thuc. 6.70.2.

62 Sears and Willekes 2016.

63 Xen. *Hell.* 3.4.23–23.

64 Xenophon alludes to the tactics of a pursuit at *Eq. Mag.* 5.1.

65 Xen. *An.* 3.1.2; 3.3.8–9.

warning signs of injury and exhaustion. These factors made horses more likely to stumble, fall, or collapse as their bodies shut down from overexertion.[66] If the defeated side could manage to outdistance its pursuers, horses and men would have been pushed beyond their limits to no purpose. Pursuing cavalry also might be ambushed, turning a victory into a defeat. The desire to pursue the enemy after a victorious battle needed to be weighed against these risks.[67]

Battles aside, horsemen fulfilled a variety of roles on campaign. They were ideal for reconnaissance, acting as mounted scouts, sometimes referred to as ***prodromoi***.[68] A mounted scout had several advantages over his counterparts on foot: from his elevated seat on the back of a horse he could see more, he could cover more ground, and he had a better chance of escaping if spotted by the enemy.[69] That scouts were armed is clear from accounts of impromptu skirmishes breaking out between opposing scouts, or even between scouts and armies on the march.[70] The scout's horse was a particular sort of animal: trained in combat, adaptable to different terrain, quick, and level headed. It did not become agitated when separated from its "herd" and ridden out on its own into unfamiliar territory. When on reconnaissance both horse and rider had to remain calm in what could easily be a tense situation, where a nervous, high-strung animal would be a liability.

Cavalry were useful when it came to maintaining and protecting supply lines, as well as moving livestock ("food on the hoof").[71] They protected unmounted foraging parties from attack by enemy infantry, but especially by cavalry, the usual troops used against plunderers.[72] Cavalry also provided a rapid and immediate response to raids. Whether bursting unexpectedly from the city gates or a secluded area, they could fall upon the raiders and cut them

66 Arr. *Anab.* 3.15.3–4 (Alexander has to give his horses a rest and loses contact with Darius); 3.20.1 (horses collapse from exhaustion); Plut. *Cleom.* 19.2 (Corinthian horses "ruined" from exhaustion); also Thuc. 4.96.6–8 (Boeotians successfully pursue whereas Athenians halt at nightfall). Xen. *Hell.* 4.5.16 reports a pursuit that was badly mismanaged. See also Spence 1993, 36–39 about the endurance of horses in general, and Diod. Sic. 19.80.2 for horses worn to the point of collapse on campaign. The main limit placed on a horse's endurance is its respiratory system, which dictates how fast a galloping horse can breathe. At a gallop the horse takes one breath per stride, inhaling during the suspension phase and exhaling during the stance phase, so the respiratory rate depends on stride frequency. When reaching top speed a horse lengthens its stride to cover more ground, but the breath frequency does not change even though the animal is working harder. Thus, the respiratory system becomes less efficient, causing the horse to alter the rhythm of its gallop, which becomes erratic. The animals try to force more air into their lungs by extending the suspension phase and shortening the stance phase and they begin making jerky head movements to compensate for the loss of rhythm/balance, all of which combines to put them at greater risk of lower limb injury as well. See Budiansky 1997, 225–228.

67 Alexander's pursuit of Darius III's fleeing troops at Gaugamela cost him the lives of a hundred riders and over a thousand mounts (Arr. 3.15.6; cf. Curt. 4.16.24).

68 This term can refer specifically to scouts, but during the fourth century it also describes advance troops who often opened battle, suggesting that the ***prodromoi*** were lightly armed cavalry who served a multitude of roles.

69 Xenophon (*Eq. Mag.* 4.4–5) describes the use of mounted scouts; during the disastrous Sicilian campaign the Athenians faced repeated harassment from mounted Syracusan scouts (Thuc. 6.63.3).

70 Xen. *Eq. Mag.* 4.4–5 mentions the importance of an advance guard acting as scouts. These mounted scouts/***prodromoi*** would have been very skilled riders, as they had to take their horses through varied unfamiliar terrain: mountains, forest, canyons, etc. In the event that they had to move at speed, the scout relied on the nimbleness and surefootedness of his horse while maintaining a secure seat.

71 Xen. *Hell.* 7.5.14: Epaminondas sends the cavalry to acquire and presumably move the Mantinean cattle; Homer (*Il.* 11.676–680) has Nestor recount an epic Elean raid that involved the movement of many livestock; one has to imagine that horsemen were somehow involved.

72 This was a key component of Pericles' defense strategy for Attica, and it was used by the Thirty in the aftermath of the Peloponnesian War; see Xen. *Hell.* 2.4.4.

down before wheeling and riding away to safety.[73] However, using horses to defend swathes of the countryside for any length of time would put a fair amount of wear and tear on the animals, especially in difficult terrain, as Thucydides shows in the case of the Athenian cavalry patrolling the land around the Spartan garrison at Decelea.[74]

Finally, cavalry played a defensive role by protecting columns when an army was on the move. As in reconnaissance, the horsemen had a better vantage point. They also maintained a looser formation than the bulk of the infantry, allowing individuals or small detachments to break away without throwing the unit into disarray. Of course, if cavalry were well suited to defending their own army on the march, they were equally proficient at attacking and harassing enemy troops who were on the move.[75]

Conclusion

The example set by Xenophon, who wrote extensively about both cavalry and infantry, has not inspired modern historians to put these two arms on anything like a par. As explained in Chapter 6 in this volume, scholars have concentrated their attention on heavy infantry, regarded not only as the chief military arm but also as the chief military influence on Greek politics and society. These scholars put cavalry in the unenviable position of being associated with foreigners, such as the Medes, and with aristocrats, such as Xenophon himself. Only regarding Macedon do scholars concede an important political and social role to cavalry. Even in Macedon, cavalry are few in number, and those numbers decline after Alexander, as shown in Chapter 9.[76]

Cavalry nevertheless played a bigger role in Greek warfare than numbers would suggest. Every large Greek army had some, if only because every large opposing army did. By fielding armies including cavalry and eventually artillery as well as infantry, the Greeks set a precedent that most armies would follow for over 2,000 years. This chapter has described the first stage in the development of this tradition.

73 Xen. *Hell.* 5.3.1–2.

74 Thuc. 7.27.5.

75 Xen. *Hell.* 7.1.20–21; Diod. Sic. 15.71.5–6.

76 Compared to horses, elephants have unrivaled value as a novelty, and artillery have unique value as an innovation; see Chapters 15 and 23, respectively.

CHAPTER 8

Siege Warfare

David Whitehead

A Definition and (Philonian) Basics

Although most readers of this chapter, like its author, will lack any personal experience of siege warfare, what we know of instances ancient or modern lets us formulate a definition of the phenomenon. My own (Whitehead 2008, 141) runs as follows:

> In a **battle** scenario, opposing forces meet on open ground and, to that extent, on the same terms. Then there is **siege** (or **blockade**), where the opposing forces can be characterized as attackers and defenders. The latter are inside a walled town or something similar, and hoping to be safe there; the others try to see to it that those hopes, by one means or another, are dashed. Of course, this over-simplifies, on several counts. Real life is less tidy than systems, so the two could overlap; some events are anyway not easily classifiable under either head—ambush, for instance. Nevertheless as a rough dichotomy it will serve [...] to encompass the majority of military confrontations on record.

Battles (and wars) being amply covered elsewhere in the present volume, my concern here will be with siege/blockade, which I shall sometimes refer to by its Greek-derived name: poliorcetics.

A survey of siege warfare in the ancient Greek world is best begun at (or near) its end, a viewpoint from which hindsight can add illumination and rigor to the fragmentary record of origins and subsequent development.

A Companion to Greek Warfare, First Edition. Edited by Waldemar Heckel, F.S. Naiden, E. Edward Garvin, and John Vanderspoel.

At some indeterminable time in the second half of the third century[1] an author called Philo of Byzantium wrote a multi-subject work entitled *Mechanike Syntaxis*: that is, *Engineering Compendium*. The majority of its (probably) eight or nine parts are lost, but the extant three all concern aspects of *military* "mechanics"—as, in all likelihood, was true of the whole. In any event, the *Belopoiika* is a technical manual on constructing various kinds of siege artillery;[2] and, central to our purposes here, the interrelated *Paraskeuastika* (*preparatory things*) and *Poliorketika* (*siege things*) provide—even in the abbreviated, epitomized form in which they have been textually transmitted—an overview of how sieges might be successfully conducted.[3]

What constituted such a "success" obviously differed for the two combatants. In the Diels-Schramm division adopted by most scholars, Philo's parts A–B (= the *Paraskeuastika*) together with C (= the first half of the *Poliorketika*) offer advice to the defending side in a siege, a facet to which we must return later. Part D reverses this perspective, looking instead through the eyes of the assailants, and it is here that Philo distills the essence of how sieges typically play out:

> D2–4: (2) He who intends to capture cities must, for preference, make the attack during a festival which they are celebrating outside the gates; otherwise, at grain-harvest <or> vintage time; (3) for by intercepting most people outside the city you might most easily capture the town; (4) otherwise, at night, when there is a storm or when the enemy are drunk at some public festival, approach the wall secretly with ladders ready and seize some of the towers.

> D72: If in your siege you cannot take the city by force because of its being strong (when approached) from every side, an attempt should be made to take it either by <stealth or> treachery or starvation. (The ensuing chapters then elaborate on these three strategies: "by stealth [*klope*],[4] at night" in D73–75; "by treachery [*prodosia*]" in D76–83; and "by starvation [*limos*]" in D84–85.)

Combining these two passages enables us to say that, in Philo's view, the paramount necessity for the defenders in a siege—meaning anyone who might become so—is to have walls, gates, and other defenses (e.g. ditches and palisades) strong enough to give a potential attacker pause before launching a frontal assault, and ideally deter him from doing so. Instead, he must rely either on the element of surprise (the preference in D2–4), or bring about collusion with some of the defenders, or blockade the community into famine and ultimate surrender.

---

1 All ancient dates are BCE unless otherwise indicated.

2 Marsden 1971, 105–184 (full text with facing English translation and notes). See also Rihll 2007, 148–160 and *passim*.

3 For full detail on Philo's *Paraskeuastika and Poliorketika*— is *On Sieges*, taken together—see Whitehead 2016: Introduction, Greek text with facing English translation, full Commentary, Appendices. (Earlier work on Philo which can still be consulted with profit: Diels and Schramm 1919 [1920]; Garlan 1973, paving the way for Garlan 1974, 278–404; Lawrence 1979, 69–107; Murray 2012, 283–301 (on the naval aspects of some sieges).

4 As a translation of *klope*, "stealth" is better than e.g. "secrecy" because it retains something of the Greek noun's sense: theft. In warfare, of any kind, *klope* could be regarded as underhand: stealing from the opponent his right to parity of combat. On the other hand, it might display admirable ingenuity. In short, a gray area of morality. See generally Whitehead 1988; Wheeler 1988, 32–33. Pritchett 1971–1991 collects and discusses instances of "surprise attacks" (2.156–76) and "ambuscades" (2.177–189).

Beginnings

It will be clear already that a complete account of siege warfare in the Greek world would need to start from the very earliest appearance of fortified settlements in the territories that circled the Aegean. But no such account can be given today, because the ancients themselves did not give one, or even adequate materials for reconstructing one. Archaeological remains are relevant, quite obviously, but they can answer only a limited range of questions.[5]

Instead we must begin with Herodotus. He registers, for the period c. 560–479, 24 sieges in which one side or the other (or both) were Greeks.[6] Most of these mentions are brief. Usually he confines himself to specifying the participants—sometimes including the name of the attacking general and (less often) the defending one—and the outcome. In 14 instances the attackers prevail, but explanations of how they did so are seldom given. A celebrated exception is Barca, a Greek outpost in Libya; captured in c. 512 by the Persians. After a protracted, evidently unsuccessful blockade, which included the equally ineffective resort to digging tunnels under the walls,[7] the assailants eventually achieved their goal by a ruse (see Hdt. 4.200).

The duration of the sieges in Herodotus varies considerably, but Barca's nine months is at the extreme upper end of the range. Naxos in 500 (Hdt. 5.33–34) and Soli in 498 (Hdt. 5.115) each took the Persians four months to overcome—the latter again by undermining; so did Potidaea in 488 (Hdt. 8.127–129). Otherwise we read of periods naturally expressed in days: the Spartans at Samos in 523 (40 days: Hdt. 3.54–56), where the attack failed, and the Athenians likewise at Paros in 490 (26 days: Hdt.6.132–136).[8]

Mention of the Spartans and the Athenians calls to mind what Thucydides says about them in the context of the southern Peloponnese in 463 or 462 (Thuc. 1.102.1–2):

> For the Spartans, the war against the men on Ithome [i.e. the Helot rebels who had entrenched themselves on that Messenian mountain] was dragging on, so they appealed for help to their allies, including the Athenians, who arrived with a considerable force under the command of Cimon. The main reason why the Athenians were asked was that they were thought to have expertise in siege-operations—something which, from the length of the siege already, was evidently a skill the Spartans themselves lacked; otherwise they would already have taken the place by force.

This characterization of the Athenians echoes, even down to the verb *teichomachein*, what Herodotus had said about them in 9.70.1–2; and that earlier context is noteworthy because there (Plataea in 479) the Athenians bring to bear their expertise not on a city wall, as the *teicho-* part of the verb might require if taken literally, but on a palisaded camp. (For another

---

5 Walls and other fortifications lie beyond my remit in this chapter. For general accounts see Winter 1971; Lawrence 1979; Adam 1982; and other contributors to the present volume.

6 To these can be added Heraclea, on the southern-facing coast of Sicily, successfully sacked by Carthaginian forces in c. 510 (Diod. Sic. 10.18.5).

7 An approach absent, as we have seen, from Philo's prescriptions for attackers in D2–4 and D72–84, but in fact quite copiously covered in D30–33 (and already foreshadowed in A1: those who build walls and towers should guard against it).

8 Athenian vs. Spartan sieges in this era were very much shorter, focussing on Athens' acropolis and the Spartan king Cleomenes. See Hdt. 5.64–65 (he takes it in 510) and Hdt. 5.72 (he loses it in 508). The Persians under Xerxes were to overcome it equally quickly in 480 (Hdt. 8.51–54).

such instance, Mycale in the same year, see Hdt. 9.102.2–4.) The Ithome Helots I would visualize as protected by more than a mere palisade but less than a fully engineered wall. Be that as it may, two inferences may be drawn from all this. One is that the Spartans, battlefield warriors *par excellence* (and, famously, inhabitants of an unwalled city themselves), had neither skill nor relish in any form of siege-and-blockade work. And the other is that the Athenians, though latterly thwarted at Thasos for three years (465–462: Thuc. 1.101), had had previous successes—of unstated duration, but presumably shorter; some much shorter—at Paros (as mentioned earlier), Andros in 480, Sestos in 479, Eion c. 476, Scyros c. 475 11, Naxos in 467, Aegina in 464, and Tanagra in the same year.[9]

When a walled city did succumb to siege within a matter of days, the explanation may sometimes be that Philo's "treachery" method had been at work. But the aforementioned case of Potidaea is a reminder that *prodosia* offered no guarantee of success. Before Artabazus' efforts there ultimately ended in calamity and defeat, as Herodotus recounts, he had suffered an earlier failure to take the town by collusion with its Scionian garrison commander. (Their means of communication with each other, via arrows that carried messages, became famous enough to be taken over from the historian by Aen. Tact. 31.25–27.)

Early Technology

In the simplest form of siege (not only in the period we have examined so far but also for ever afterward) the assailants, if presented with no opportunities for collusion or for tunneling and unwilling to set up a lengthy starvation blockade, would attack their opponents in situ. If possible, they would do so in one of the circumstances illustrated by Philo, where the latter would be caught unawares. With or without this advantage of surprise, their hopes of success would be pinned on impetus, over the gates and walls and into the town. Unless these structures were unfeasibly low (or weak for other reasons), the troops needed scaling ladders (*klimakes*) in order to proceed. Thucydides 5.56.5 gives a routine instance from early 418: "the Argives took ladders and went against Epidaurus, expecting to find it undefended because of the war and to capture it by force; but they returned unsuccessful."

But were more sophisticated means than mere *klimakes* available, by that time, to would-be besiegers? That question will be answered in the affirmative if credence can be placed in post-Classical accounts of two sieges that had taken place in the mid-fifth century:

> Pausanias 4.25.1–2 (date 455): Knowing that the Acarnanians of Oeniadae possessed a good land and had a long-term hostility towards the Athenians, (the Messenians of Naupactus) marched against them. They ... shut them inside the wall and began a siege. What ensued found the Messenians lacking in no human invention for siegecraft. They tried to enter the city *over* the walls, with ladders, and *under* them, with tunnels; and they were constantly destroying things by deploying whatever machines (*mechanemata*) could be quickly constructed. The inhabitants, fearing that if the city was taken they themselves would be killed and their wives and children enslaved, chose to withdraw under truce.

9 Andros, Hdt. 8.111; Sestos, Hdt 9.114–118, Thuc. 1.89.2; Eion, Thuc. 1.98.1; Scyros, Diod. Sic. 11.60.2; Naxos, Thuc. 1.98.4; Aegina, Thuc. 1.105.2–4, 108.4, Diod. Sic. 11.70.2–3; Tanagra, Diod. Sic. 11.82.5.

Diod. Sic. 12.28.2–3 (date 440): Pericles … renewed the siege (of Samos) both by land and by sea, making continuous assaults. He also did what nobody before him had done and constructed machines (*mechanai*), including what were named rams and tortoises—Artemon of Clazomenae being their actual builder. By besieging the city with vigor and bringing down the walls with the machines Pericles became master of Samos.[10]

The first of these passages is easily disposed of. No serious scholar accepts that Pausanias, for all his other merits,[11] is transmitting here anything other than a generic, rhetorically driven inflation of the Messenians' siege of Oeniadae.[12] But Samos may be another matter. Thucydides (in 1.117) says nothing of Artemon and his "machines," and for A.W. Gomme and others this silence is decisive.[13] I am not sure that it is, given two considerations: how comparatively briefly Thucydides covers pre-431 events, and the fact that in his (full) narrative of the Spartan/Peloponnesian siege of Plataea between 429 and 427 he does mention rams and other machinery.[14] Indeed, as early as 431 King Archidamus is found deploying "machines" (*mechanai*) against the walls of Oenoe (Thuc. 2.18.1).

Ephorus/Diodorus, quoted earlier on Samos, glosses the term "machines" (also in Plutarch) with a noteworthy phrase: "including what were named rams and tortoises." These were descriptive names drawn from the animal world. A farmer's ram butted, sometimes destructively, and so did a mechanical wooden *krios* when faced with a city gate or wall.[15] As to a "tortoise" (*chelone*), the distinctive feature of the animal of that name was its protective shell. Modern readers will be familiar with the Roman army's *testudo*, and will probably visualize the tactic where a body of troops combined their shields into a tight carapace (e.g. Liv. 34.39.6, 44.9.6–7). In fact, military *chelonai/testudines* are both attested in a variety of forms and functions, especially in sieges. Mobile, purpose-built structures (either carried or wheeled) could provide cover for troops as they performed a variety of tasks: digging, ditch-filling, ramming, and tunneling.

A larger question, but again one without a clear-cut answer, is this: To whom were the Greeks indebted for the idea of using such siege technology—beyond ladders—at all? If the colorful figure of Artemon of Clazomenae (carried along in his hammock) was worth clinging onto, this might look like Ionian applied science, borrowed straight from the Persians or others in Asia. On the other hand, the later "textbook" account of the origins of the battering *krios*, extant in the parallel accounts of Athenaeus Mechanicus and Vitruvius written during the 20s BCE, traces a more roundabout route, peopled by shadowy individuals called Pephrasmenos of Tyre and Geras of Carthage, and centering on a Carthaginian siege of Gadeira (present-day Cadiz) apparently in the late sixth or early fifth century.[16] When stripped

10 For more detail see Plut. *Per.* 27–28. The ultimate source of both writers is Ephorus: *FGrH* 70 F194.

11 Habicht 1985; Hutton 2005; Pretzler 2007.

12 So e.g. at length Garlan 1974, 131–134; in brief Kern 1999, 95, and Campbell 2005, 23 (again Campbell 2006, 42).

13 Gomme, Andrewes and Dover (1945–1981) 1.354. See also e.g. Garlan, Kern, and Campbell, cited in the preceding note.

14 This position is taken by Stadter 1989, 253 (commenting also that Ephorus was particularly interested in inventions). Note that he cites Potidaea by mistake for Plataea.

15 The latter might be vulnerable if they were stone-built. Against brick walls, the drill (*trupanon*) mentioned in the poliorcetic writers would have been more effective.

16 On Athenaeus and Vitruvius, see the editions of Whitehead and Blyth 2004 and (more sober) Gatto 2010. Also e.g. Callebat and Fleury 1986.

of its fanciful anecdotal detail, this has the advantage of suggesting influences upon Aegean Greece from the West, where (as we shall see in the section on The West) Dionysius I of Syracuse later became prominent.

The Peloponnesian War

The 20 years (431–411) of the great Peloponnesian War are covered in detail by Thucydides and, after mid-411, by the continuation of his narrative in Xenophon's *Hellenica*. From these two writers, supplemented (with all due caution) by Diodorus and Plutarch, a total of 60 sieges[17] are on record; and it is no surprise to find them depicting siege warfare in its full Philonian variety.

Recorded instances of communities brought to their knees by starvation (*limos*), as Polyaenus 2.33.1 maintains Thasos had been in 462, begin with Potidaea in 430/429, Mytilene in 427, Plataea the same year, and Acragas, by the Carthaginians in 406, and of course the sequence culminates with Athens itself, blockaded by sea and land into despair and capitulation.[18] The Athenian case stands out for several reasons, not the least of which is the fact that *limos* was the strategy employed by Lysander and the Spartans from the start, once they had gained control of the Hellespont. Elsewhere it was often the means resorted to after other approaches had failed. At Mytilene, for example, the Athenians' initial response had been to mount a surprise attack—Philo's *klope*—while the Mytileneans were outside the walls celebrating a festival of Apollo. (They were forewarned and it was cancelled, as at Thuc. 3.3.3–5.)

Another "hybrid" episode is the 431 Theban move against Plataea in spring 431, one that combines *klope* (a nocturnal attack by an advance force) with treacherous *prodosia* in the form of collusion with pro-Theban citizens who open the gates.[19] More straightforward instances of *prodosia* at work include Megara by the Athenians in 424, Torone by the Spartans the same year, Byzantium and Selymbria by the Athenians in 409, and Methymna by the Spartans in 406.[20]

Wherever serious fighting was expected to take place outside as well as inside the fortifications, both of the hegemonic powers by now had (as we have seen) "machines" at their disposal. Some scholars doubt whether, as a rule, Thucydides meant anything more by this term than scaling ladders.[21] I am inclined to believe that he did, chiefly because he seems quite capable of designating such objects *klimakes* when that is what they are.[22] Also, in his description of two particular sieges from the 420s quite an array of poliorcetic sophistication is on show.

---

17 This figure excludes Magna Graecia, for which see the subsection on The West.

18 Potidaea, Thuc. 2.70.1; Mytilene, Thuc. 3.27.1; Plataea, Thuc. 3.52.1, etc.; Acragas, Xen. *Hell.*1.5.21; Athens, Xen. *Hell.*2.2.1–23.

19 Thuc. 2.2.6; also Aen. Tact. 2.3–6.

20 Megara, Thuc. 4.66–74; Torone, Thuc. 4.110–114; Byzantium and Selymbria, Diod. Sic. 13.66, Plut. *Alc.* 30–31; Methymna, Diod. Sic. 13.76.5.

21 So e.g. Gomme in Gomme, Andrewes, and Dover (1945–1981) 2.334 (specifically on Thuc. 3.51.3); Campbell 2005, 17, 23 (again Campbell 2006, 33, 42).

22 Cf. already Whitehead 1990/2001, 193–194.

The first of these, mentioned in passing already, is the Spartans/Peloponnesians vs. Plataea in 429–427.[23] Though the small Boeotian city's eventual surrender was achieved by *limos*, that bare fact conveys little of the elaborate military moves and countermoves that preceded it: notably a double mud-brick *periteichismos*, encirclement, to seal the blockade[24] and, before that, an (experimental?) embankment or ramp against the city wall itself. The Plataeans, for their part, had engaged in tunneling (to destabilize the embankment) and had begun a crescent-shaped wall as an emergency line of defense in case the city wall itself was compromised. And "machines" had been brought into play by both sides: the attackers deploying rams (this is guaranteed by Thucydides' use of the term *embole* alongside *mechane*) and the defenders managing to neutralize them—some lassooed away, others crushed by heavy timbers.

Three years later came the Boeotians' assault on Delium. This site of a semi-derelict sanctuary of Apollo was within their own territory, near the eastern coast, but had now been seized and fortified by the Athenians. At the heart of Thucydides' narrative (4.76.4–5, 4.89–101) is a summary account of how the attackers, in 17 days, overcame the Athenians' timber defenses (4.100.1–4). Among other methods, all unspecified, they built and used a flamethrower.[25] This was a long timber pipe, mainly iron-clad, with a cauldron hung by chains at the assault end. The cauldron was filled with hot coals, sulfur, and pitch; the pipe was brought up on carts to the wooden wall; and bellows blew air down the pipe and into the cauldron, producing blasts of flame sufficient to set the wall afire and scatter the defenders. (What was presumably a similar device is mentioned in the course of a later event during the same year: the Spartan capture of Lekythos in Chalcidice, where the Athenian-built defenses consisted of a low-grade wall and some houses with parapets. But it never came into play: instead, the defenders' counter-construction, a rooftop tower from which stones and water could be dropped, collapsed, and the ensuing disarray allowed Brasidas' troops to storm in (Thuc. 4.115–116).)

With these exceptions, it would be hard to justify any claim that the character of siege warfare was transformed during the Peloponnesian War.[26] In particular the advent of first generation "machines" did not of itself make overcoming sturdy stone walls an easy matter. For that and other reasons, the record shows almost as many unsuccessful sieges as successful ones, from Archidamus at Oenoe in 431 to the Athenians' failure(s) in Sicily.[27] Human fallibilities exploited by surprise attacks and divided loyalties were often still the key.

---

23 Thuc. 2.71–78, 3.20–4, 3.51–68. Kern 1999, 97–112 is a sound modern account. See also in brief Campbell 2005, 20–22 (again Campbell 2006, 37–40).

24 Apart from this almost absurdly elaborate Spartan instance of *periteichismos*, the strategy was adopted primarily, in this era, by the Athenians, e.g. at Mytilene, Leucas, and Minoa. (They calamitously failed, of course, at Syracuse.) See generally, Campbell 2005, 17–19 (again Campbell 2006, 33–37). Later cases include Alexander the Great at Cyropolis in Sogdiana (329) and Leosthenes at Lamia (322).

25 Campbell 2005, 23 has a drawing. See also Garlan 1974, 140–141.

26 That said, some change can arguably be discerned at the margins. See for instance Kern 1999, 106–107 on what may be an increased resort to archers and/or slingers in support of front-line troops.

27 Oenoe: Thuc. 2.18. Athenian failures: before Syracuse itself, the Athenian failure at Messina (Thuc. 6.74), and indeed Inessa during the invasion of the mid 420s (Thuc. 3.103).

The West

We saw at the end of the section on Early Technology that a plausible route for the arrival of technological innovation into the Greek Mediterranean was via the Carthaginians and, before them, their kinsmen the Phoenicians. Whatever the truth of that, Diodorus' account of the Carthaginians' military return to Sicily (after an absence of seven decades) in the years 409–405 depicts them as equipped in ways that not only brought them immediate success but also provided an object lesson for the Sicilian Greeks in general.

What the Carthaginians achieved, under their commanders Hannibal and Hamilcar (Himilco), was to take four major cities: Selinus and Himera in 409, Acragas in 406, and Gela the year after that.[28] Hannibal had shipped across the Libyan Sea "machines for the sieges, and missiles, and all the other equipment," and already at Selinus some of this is deployed: "six [*sc.* mobile] towers, of exceptional height, and the same number of iron-clad rams." At Himera the "machines" that shake and eventually breach the walls are evidently rams. (Tunneling was also undertaken, but not to decisive effect.) At Acragas the Carthaginians attacked with "two enormous towers" but the defenders burned them, and ultimately it was lack of food that brought the eight-month siege to an end. At Gela, though "rams" were again a feature, at night the defenders repeatedly made good any breaches, and it was not until they abandoned the city that the Carthaginians could take possession of it.

The technological aspect of all this needs to be assessed with caution. Above all else, Diodorus' mention of "missiles" (*bele*) is problematic because it is ambiguous. In some contexts a *belos* is a bulky projectile—typically a large arrow-like bolt—that requires a mechanism to fire it, in other words, a catapult, the traditional term from the Greek *katapaltes/katapeltes*. But it is highly unlikely that such artillery was in existence in the late fifth century (as will be discussed further), so if the phrase "and *bele*" in Diodorus 13.54.2 is not simply an interpolation into Diodorus' text there, we are entitled to assume that what he (or his source) meant by it is ammunition for archers and slingers.[29]

On the other hand, the iron-clad rams present no difficulty; and of especial interest is the repeated mention of mobile towers (*purgoi*). Together with rams, these had been a feature of Near Eastern siege warfare since well before Classical times,[30] but this is their first attested appearance in the Mediterranean.

Diodorus' trustworthiness on this point is corroborated by the wholesale adoption (and then extension) of Carthaginian siege technology by someone who had been on their receiving end at Acragas and Gela. The power of Dionysius of Syracuse in eastern Sicily during the closing years of the fifth century was initially shaky, but, as soon as he had a secure grip on it, new military hardware—the lack of which had thwarted his efforts to take Leontini in 403 (Diod. Sic. 14.14.3)—became a priority. Diodorus (14.41.3–43.4) describes, under the year 399, this ferment of creativity and manufacture: a veritable arms-race, principally

---

28 Selinous, Diod. Sic. 13.54–57; Himera, Diod. Sic. 13.59.6–62.6; Acragas, Diod. Sic. 13.85–90, 13.108.2, Xen. *Hell.* 1.5.22; Gela, Diod. Sic. 13.108.2–111.2.

29 For the flexibility of the term *bele* in Diodorus cf. 13.55.4 (during the siege of Selinus women and girls keep the defenders supplied with them). A *belos* could also designate the artillery piece itself: so in e.g. Philo's *Belopoiika*.

30 See in brief Kern 1999, 17.

intended to catch and if possible surpass the Carthaginians.[31] Then, between 397 and 388, a sequence of sieges ensued against Carthaginian-held settlements and others within Sicily and across the straits to southwestern Italy (Rhegium).[32] The great set-piece in Diodorus' narrative is his account of Dionysius' 397 siege and capture of Motya, a prime Carthaginian supply base located a lagoon-protected offshore stronghold at the westernmost tip of Sicily (Diod. Sic. 14.48–53).

Besides deploying rams and "wheeled towers, six storeys high"[33] against Motya, Dionysius reportedly unveiled a wholly offensive weapon: "sharp-missiled [i.e. bolt-firing] catapults." The passage that includes the phrase (Diod. Sic. 14.50.4) states that these caused dismay among the defenders of Motya by virtue of their novelty. Diodorus has already asserted that "the catapult was invented at this time [i.e. 399] in Syracuse" (14.42.1).

While most present-day opinion accepts Diodorus' claim, as the alternatives look fragile, one aspect of it needs to be filleted-out and digested.[34] How did a catapault function? Diodorus himself is no help here, but the term used by Biton for Zopyros' designs (and also, impersonally, in a later artillery manual attributed to Heron) is *gastraphetes;* literally, belly discharger. Bizarre at first sight, the word picturesquely applies to a crossbow-like weapon held steady against the user's abdomen (and propped up, somehow, at the front).[35] Though some experts dissent, orthodoxy holds that the "catapults" used at Motya—and for half a century afterward—were *gastraphetai*. They shot pointed bolts, as Diodorus says, and they did so with a force (and, where necessary, a range) unattainable by archers, by dint of a more powerful bow, drawn back mechanically, than an archer's arm could manage. The bow itself would be composite, i.e. constructed out of three laminated horn–wood–sinew layers, like archers' bows. The propulsive power of the *gastraphetes*, similarly, was the tension of its drawn string. (A power source superior to tension, torsion, still lay ahead.)

How soon the use of this first generation of "catapults" spread beyond Sicily cannot be exactly traced. Dionysius will have done his utmost to deny his innovations to his enemies in the West. In Aegean Greece, however, there is evidence to suggest that by c. 370 the Athenians were in possession of the weapon—possibly as a gift from Dionysius himself—and a Spartan prince, the future King Archidamus III, had been impressed and dismayed by a demonstration of it.[36]

31 Aspects of the passage, such as rhetoric flourishes, invite comparison with Xen. *Hell.*3.4.17–18, where the Spartan King Agesilaus turns Ephesus into a "workshop of war" in 395, but the weaponry and training is meant for campaigning in the field.

32 Good summary in Campbell 2005, 26–28 (again Campbell 2006, 47–54).

33 Diod. Sic. 14.51.1 They were equipped with gangways (*epibathrai*), allowing troops to cross into the city: 14.51.7, 14.52.4.

34 See e.g. Rihll 2007, 26–45. Two alternatives have been canvassed: invention (*a*) in the Near East, in or before the eighth century, or (*b*) in southern Italy, not long before Dionysius. Option *a* fails for the lack of evidence across three centuries and more. Option *b* focuses on Zopyrus of Taras/Tarentum, whom a post-Classical treatise on artillery by Biton (Marsden 1971, 61–103) mentions twice as a designer of catapults, used at Miletos and at "Kyme in Italy" (a.k.a, Cumae, on the Bay of Naples). However, Zopyrus cannot be dated with any certainty: first half of the fourth century (e.g. Marsden 1971, 47, 98); perhaps late fifth century (Campbell 2003b, 3, elaborated at 5; again Campbell 2006, 49, 50).

35 See Campbell 2003b, 3–5 (cf. Campbell 2006, 49–52).

36 Marsden 1969, 65–66.

Aeneas

This chapter began with Philo's *Mechanike Syntaxis*, which reflects poliorcetic theory and practice in the second half of the third century, and I will return to him in my concluding section. But Philo had a predecessor, and at this point in our survey, the mid-fourth century, he makes his appearance.

His name was Aeneas. There is an attractive possibility (it can be no more) that he should be identified as Aeneas of Stymphalus, an Arcadian general who briefly features in an episode from c. 366 recounted in Xenophon's *Hellenica* (7.3.1). Be that as it may, a combination of *termini ante* and *termini post quem* do place him in the middle of the fourth century; and his writings show him to have been—unlike Philo—a man of first-hand military experience.[37]

Lamentably, only a work on siegecraft is extant. Polybius (10.44) quotes Aeneas on another subject, fire-signaling, but this passage appears to have been excerpted from a treatise of Aeneas' called *Preparations-Book* (*Paraskeuastike biblos*), one that the work on siegecraft cites four times, while also referring to a treatise called *Procurement-Book* (*Poristike biblos*) and a forthcoming *Encampment-Book* (*Stratopedeutike biblos*). Aeneas' coverage of siege-related matters makes him a significant precursor of Philo's.

The title of the siegecraft treatise is lost, but, given the form of his other titles, it likely was *Siegecraft-Book* (*Poliorketike biblos*). Nevertheless, modern readers may be forgiven for preferring the gloss given in the principal ancient manuscript: *(Tactical Treatise on) How to Survive under Siege*. Like Philo, Aeneas recognizes the perspectives of both attackers and defenders, yet almost entirely confines itself to the needs of the latter.

The concluding 9 of the 40 short chapters or paragraphs into which the work is conventionally divided help us visualize the inhabitants of a typical Greek *polis* of the era "under siege" in the most immediate and literal sense. Aeneas tells them how best to protect their walls and gates against attack, both human and mechanical; how to deal with incendiary devices; how to detect and thwart the operations of sappers who are seeking to tunnel under the fortifications; how to create the illusion that the defending forces are larger than they truly are; and in general how to deploy available manpower to best advantage.[38]

This material (chs. 32–40), occupying some 15% of the treatise as a whole, would be valuable enough in itself. So would ch. 31, a long, showpiece devoted to what was plainly a major enthusiasm of the author's: steganography and cryptography, the transmission and receipt of secret messages.[39] It is replete with historical instances, drawn from Herodotus and elsewhere, and at its heart is an elaborate theoretical scheme—quite obviously one of Aeneas' own devising—for winding thread through 24 holes bored in in an astragal, a knucklebone-shaped die. Even so, the true revelations of the *Siegecraft* come in chs. 1–30.

Though lacking overall coherence, chs. 15–30 (or chs. 15–31) deal with operations against an enemy who, from the city's standpoint, is already known to be on his way, and even within sight, whereas chs. 1–14 mostly cover preparatory measures prior to that—and indeed in

---

37 The modern era of work on Aeneas (with copious citation of older work) was initiated by the simultaneous publication of Bettalli 1990 and Whitehead 1990/2001; but Illinois Greek Club 1923 (a.k.a. the Loeb edition) may still be used with profit. Recent perspectives: Pretzler and Barley 2017/18.

38 In one notorious instance, womanpower (women create a static, *trompe l'oeil* tableau of armed men: 40.4–5).

39 The difference between steganography and cryptography proper is that in the former the very existence of the message is concealed; in the latter it is (or can be) overt but undecipherable.

some instances prior to ***any particular*** hostile force being in the offing. In 11.3–6, for example, Aeneas recounts an otherwise unattested episode from the history of Chios. An attempt to capture the place through the treachery—Philo's ***prodosia***—of one of its officials revolves around his persuading his colleagues that, "as it was peacetime," the state's defenses, physical and human, could be dismantled with impunity.

Two crucial, related inferences flow from this treatment. First, Aeneas was contemplating potential sieges at least as much as actual ones. He was envisaging any and every ***threat*** of siege or invasion—a threat that was chronic as well as occasionally acute, and was the backdrop of daily life for typical small-to-medium communities. Ch. 10 thus proposes a striking series of peacetime measures that comprise the proclamation of martial law, with curfews and other restrictions on civil liberties and freedom of movement. Second, Aeneas adopts the position of the individual(s) whose responsibility it is to maintain the ***polis'*** security. Danger can arise at least as readily from conspiracy, disaffection and subversion as from warfare and siege; from the enemy within—apt to engage in Philonian ***prodosia***—as well as well as the enemy without. No one is, or should be, above suspicion: soldiers (especially any mercenaries) and civilians alike; foreigners, whether resident or visiting, as well as citizens. Gatekeepers must be not only intelligent and alert, but also loyal and immune to financial or other temptations (ch. 5). And the community must be of one mind—in Greek, displaying ***homonoia*** (ch. 14)—if it is to maximize its chances of survival in the uncertain world of the mid-fourth century.

Establishing this date for Aeneas and his treatise is a matter of balancing considerations that push it later against those that keep it earlier. Aeneas makes lavish use of illustrative episodes that, where they can themselves be dated, show a thin scatter from the sixth and fifth centuries followed by a heavy concentration in the first four decades of the fourth, with nothing demonstrably beyond c.360–355. The Macedonians are conspicuous by their absence; and, within a discussion of countermeasures against the enemy's "machines," which might include towers and rams (ch. 32), catapults receive only a passing mention (32.8). These two facts are probably connected. In any event we must address them in following section on Macedonian Siegcraft.

Macedonian Siegecraft

In siegecraft as in warfare as a whole the Macedonians have only a modest claim to attention until the reigns of Philip II (359–336) and his son Alexander III (336–323). Each of them undertook a sequence of notable sieges, and Alexander inherited from his father and maintained in his own right a corps of poliorcetic experts whom in some instances we know by name. Under Philip the chief expert is a Thessalian called Polyidus, an author (of a lost *On Machines*) as well as a technician.[40] Some of Polyidus' equally versatile pupils, such as Diades, accompanied Alexander into Asia.[41]

What Diades and his colleagues provided for Alexander is relatively well-documented, and (unsurprisingly) addressed the needs of an itinerant siege train operating in unfamiliar terrain very far from home. In particular, Diades devised towers of modular construction; that is,

40 On Polyidus see in brief e.g. Garlan 1974, 208 and index s.v.; Whitehead and Blyth 2004, 84; Whitehead 2016, 185.

41 Whitehead 2015.

ones that could be carried along in their component parts, assembled for use, and taken apart again for a march to another fortification. What, in the previous generation, had been invented for Philip by Polyidus and others is more uncertain; but an attractive hypothesis put forward by E. W. Marsden and still influential today puts torsion catapults at the heart of it.[42]

Torsion catapults—which, without rendering the older *gastraphetes* instantly obsolete (especially as a handheld weapon), did represent a significant advance—introduced two main novelties. One, as the modern name suggests, was a different and more potent principle for discharging the projectile. Instead of simple tension, the requisite kinetic energy came from two hanks of hair or sinew, each gripped tight in a timber frame so as to create a spring mechanism. When they were twisted back by a pair of wooden arms, thrust was transmitted to the missile by a stout "bow"-string. Both greater power and larger sizes became possible via improvements in the component parts. The second advantage of the torsion catapult came with the realization (whether immediate or subsequent) that its design could be engineered, as required, to fire stones (*lithoi, petrai*) rather than the familiar arrow-like bolts.[43] Thus, in the historians and other Greek sources, both *lithoboloi* and *petroboloi* join the relevant vocabulary alongside *oxubeleis.*[44]

The earliest case of mechanical stone throwers in the field is difficult to determine. Some consider it to be an episode from 353, where the Macedonians are at the receiving end: the military anecdotalist Polyaenus (second century CE) describes Onomarchus and his Phocians routing Philip's forces in a blind valley with "stones and stone-throwers".[45] No noun for "machines" is reliably part of the text, which might simply be a tale of stones hurled by hand. If it is, Alexander at Halicarnassus in 334 and Tyre in 332/331 seem to be the earliest episodes.[46] Philip used bolt throwers (sc. torsion *oxybeleis*), as at Perinthus and Byzantium, which he failed to take in 340. Any reputation he gained as an exponent of the poliorcetic arts will have rested on earlier successes obtained by various means, such as Amphipolis (357), Pydna (356), Potidaea (356), Methone (355), Olynthus (348), and Halos (342). And not even Alexander, oddly, is recorded as having used mechanized stone throwers again. Their next attested deployment comes in 307, when one of his Successors, Demetrius, captured the Munychia fortifications of Piraeus by means of "catapults and *petroboloi*" (Diod. Sic. 20.45.7). His siege of Cypriot Salamis featured *petroboloi* capable of discharging heavy stones at the walls.[47]

42 Marsden 1969, 18–24, 56–64 (esp. 60); again Marsden 1977. See also e.g. Campbell 2003b, 8–9.

43 "It seems that the first torsion catapults were designed to fire only arrows" (Campbell 2003b, 15). Marsden 1969, 15–16 and *passim* alludes to non-torsion stone throwers, but not within the fourth century (except for the dubious Polyaenus episode discussed at n. 45).

44 For *oxubeleis* see already under section 5 (on Dionysius at Motya). Strictly speaking all three words are adjectives implying the noun catapult, but they were often used as substantives in their own right.

45 Polyaen. 2.38.2. Machines are postulated here by e.g. Marsden 1969, 59; Hammond and Griffith 1979, 269; Krentz and Wheeler 1994, 219; Rihll 2007, 60–61 (comparing the rout Polyaenus describes to the Charge of the Light Brigade). For scepticism see e.g. Campbell 2003b, 15–16 (cf. Campbell 2006, 93).

46 Halicarnassus: Arr. *Anab.* 1.22.2: "large stones were discharged by the *mechanai* from the towers" Tyre: Diod. Sic. 17.42–45.

47 Diod. Sic. 20.48.3. This passage is unusual among historical narratives in purporting to specify the machines' caliber (for want of a better word): three talents, i.e. c.180 lb. This is information we would wish to have had across the board; but it is given—for *oxybeleis* and *petroboloi* alike—mainly in the poliorcetic writers. Together with the evidence of surviving ammunition stores, one from Cypriot Salamis itself (and securely datable to 311: summary in Campbell 2003b, 19, with 20–22 on other caches), this offers no basis for rejecting Diodorus' three talents outright (cf. Whitehead 2016, 279–280) but does place it at the extreme upper limit of a range that more usually came to span c. 10 to c. 60 lbs.

Reasons why a city might survive a siege were of course many and various, but one consideration that we should not overlook is the enhancement of walls, gateways, and other fortifications in response to the advance of mechanized siege technology, beginning with rams. Archaeological remains sometimes provide a degree of corroboration,[48] but often they are difficult to date; and literary testimony on the topic (besides Aeneas) is all but non-existent. With more and better evidence it might be possible to trace over time the balance of advantage between attackers and defenders, both as regards particular cities or regions and across the Greek world as a whole. Without it, we can progress little beyond the obvious: even a detached observer such as Aristotle grasped the importance and value of strong fortifications (whereas his unworldly teacher Plato had deplored them as injurious to moral health).[49]

The Early Hellenistic Period

After the relatively rich documentation of the reign of Alexander, our narrative sources about sieges and siege warfare mainly narrow down to Diodorus and Plutarch.[50] They can be supplemented with material from the post-Aenean poliorcetic writers, but alongside benefits this can sometimes bring (or exacerbate) problems of credibility. One of the features of the period in the military sphere as a whole is gigantism and excess in construction. At sea this manifests itself in ever-more grandiose warships.[51] The terrestrial counterpart is larger and more elaborate siege engines.[52] Their existence is not in doubt, but believing some of the actual dimensions claimed is another matter.

One striking, though elusive, term here is *helepolis;* literally, "city-taker." Originating as a personal epithet (of Helen of Troy and other figures from mythology), in and after the late fourth century it came to apply to siege supertowers associated with engineers and others. The earliest known of these persons is a Macedonian called Posidonius, whose *helepolis* for Alexander's siege of Tyre (332–321) is noted by Biton.[53] Frustratingly, the specifications concentrate on the basic structure of this wheeled tower. Beyond several mentions of assault bridges, there is no clear statement of what it was intended to do in the field, and indeed how it differed from the *purgoi* already in standard use.[54]

48 See e.g. Marsden 1969, 126–139; Campbell 2003b, 7 (again Campbell 2006, 55); Rihll 2007, 50–54.

49 Plat. *Lg.* 6.778D–779A (cf. 760A–761A for the defense of the surrounding territory); response in Aristot. *Pol.*7.1330b32–1331a18. The two passages are quoted and juxtaposed in Whitehead 2016, 133.

50 Bosworth 1988, 295–300.

51 Murray 2012.

52 "Siege Warfare and Naval Warfare" (Tarn 1930/1984, 101–152), though outdated on some points and arguable on others, remains well worth reading.

53 See Marsden 1971, 70–73 with 84–90 (illustrations at 87); Lendle 1983, 38–58. In calling Posidonius the earliest-known builder of a tower described as a *helepolis*, I discount the dubious testimony, in a papyrus list of notable engineers, that Polyidus had built a *helopolis* (sic) for Philip's failed siege of Byzantium in 340: see Whitehead and Blyth 2004, 84 n.13. (This accepted by Campbell 2003b, 4, 6; cf. Campbell 2006, 59–60.).

54 Biton writes that its height was 50 cubits, i.e. some 75 feet. That might seem large until we note that, according to Arrian (*Anab.* 2.21.4), Tyre's walls where Alexander attacked them were twice as high, at 150 feet. Even if that figure is an exaggeration (so Bosworth 1980–1995, 1.247–248), they were surely higher than Biton's *helepolis* could reach. In fact our various sources for this siege mention other *purgoi*, such as the two on the mole (depicted in Campbell 2005, fig. D = Campbell 2006, 66–67), and there is no foolproof way of differentiating between them all.

Somewhat more helpful in this regard are the descriptions of a *helepolis* built by an Athenian engineer called Epimachus for the second stage of the assault on Rhodes in 304 by Demetrius, whose association with siege operations gave him the epithet *Poliorketes*. Its overall height was almost 150 feet, and its nine tapering stories housed catapults as well as assault troops.[55] Demetrius had already deployed a *helepolis* of the same dimensions, but not associated with a named engineer, at Salamis in 307, and a *helepolis* would feature again in two of his sieges on the Greek mainland: Argos in 295 and Thebes in 291.[56]

Nor were *helepoleis* and other *purgoi* the only siege engines to manifest this early Hellenistic taste for outlandish size. The Athenaeus-plus-Vitruvius summary cited earlier gives prominence to a certain undatable Hegetor of Byzantium and his colossal *krios*-topped *chelone*. Tellingly, we are never told that this monstrosity—"an absurdly ponderous one-piece battering-mantlet, with a lavishly spiked front (an attempt at a prototype of the Tank) which the men inside could not move with enough impetus to cause damage"—ever entered active service.[57] Had it done so, its movable roof-mounted ram, perhaps 80 feet long, would have taken very little time to buckle and snap off.

This tendency to gigantism and excess was real, but practical limits usually constrained it. Thus, what might be termed the routines of many sieges unfolded much as they had always done, with success from the assailants' point of view being no more likely an outcome than failure. Under the latter head, we hear more of cases where the attackers might have succeeded but for the intervention of a third party favorable to the defenders.[58] After the death of Demetrius *Poliorketes* in 283, grand set-piece sieges pass unrecorded—even if a few might have merited that description—until Syracuse in 213–211. There, despite the applied ingenuity of Archimedes, the Romans under M. Claudius Marcellus succeeded where Athenians and Carthaginians alike had failed.[59]

Earlier, in the 280s and 270s, another ambitious Hellenistic monarch, Pyrrhus of Epirus, gained his proverbially backhanded reputation for victory in battles rather than sieges, so from our perspective his importance lies in the fact that his Thessalian henchman Cineas produced an epitome of Aeneas for him.[60] It is also beyond doubt that Philo of Byzantium was familiar with the writings of Aeneas; and to Philo we can now return.

55 Marsden 1971, 84; Whitehead and Blyth 2004, 134–138 with 190.

56 Salamis: Diod. Sic. 20.48. Argos: Ath. 10.415A. Thebes: Plut. *Demetr.* 40.

57 Lawrence 1979, 50. For comparable scepticism see also Lendle 1981, 343 (distancing himself from Lendle 1975, 45–86); Whitehead and Blyth 2004, 187–189 with figs. 12–15. The contraption is given more credence by Campbell 2003b, 19–24 (again Campbell 2006, 87–92).

58 For examples see Macedonian forces at Lemnos in 314 (Diod. Sic. 19.68.3–4); Cassander at Corcyra in 299, foiled by the arrival of Agathocles (Diod. Sic. 21.2.1–3); and further, *vis-à-vis* Philo, in the section on Philo.

59 The primary accounts are Polyb. 8.3–7 and Plut. *Marcell.* 14–20. Good general treatment in Kern 1999, 262–267. Walbank 1957–1979, 2.69–78 has denser detail.

60 Whitehead 1990/2001, 6 with n. 14; Chaniotis 2005, 58–9, 218.

Philo

Like the work of Aeneas, Philo's *On Sieges* has not survived in the full form in which he wrote it. The extant epitome lacks the epistolary preface that begins his *Belopoiika*, and probably also lacks some historical illustrations of the kind that so enliven Aeneas.[61] But no matter: the majority of Philo's statements about and recommendations for his subject have been transmitted intelligibly enough to allow us to see in considerable detail the nature of siege warfare at the time. (He writes implicitly for the part of the world he lives in, the Seleucid Empire, which should caution us not to presume that everything he addresses and anything he recommends is applicable throughout the third-century Greek world.) Here, brief synopses must suffice.[62]

The *Preparatory Matters* (*Paraskeuastika*), parts A and B, entirely concern the defensive measures that a city expecting siege should implement. First and most basic comes consideration of its foundations, towers, gateways, and types and configurations of walls, together with extramural structures such as trenches and palisades.[63] The layout and sequence of the constituent topics sometimes lacks coherence, but by A30 a theme has emerged that is the remainder of part A until A87's concluding mention of (lost) drawings. Philo stresses the paramount need to tailor wall circuits to the nature and contours of the terrain. Six generic ways of doing so are briskly described: an "old" one and, by implication, five new(er) ones, including a "meandering" line and configurations of semicircles and of serrations.[64]

Archaeology has had only patchy success in matching up Philo's typology with surviving remains. That, in conjunction with scattered, impenetrable details in his specifications means that the obvious question we would wish to ask—how far was he describing what already existed, how far suggesting novelties—has no simple answer.

In part B, Philo's remit of *Paraskeuastika* turns to provisioning. If an enemy might blockade the city, and indeed do so effectively (as was not always the case), how could it be ensured that the inhabitants have enough to eat?[65] The answer Philo gives is twofold. (*i*) There must be advance planning to stockpile and store a large amount of grain and other foodstuffs; and (*ii*) this food must be used intelligently and sparingly. (Almost as an afterthought, in B54, advice is given on purifying any foul water.) Under *i*, Philo offers designs for sunken grain pits and raised granaries, and (in B30) he counsels cities to keep one year's

---

61 Faint traces can be seen in the allusions to Rhodes in A17 and A59, Ake (?) and Babylon in B53. Another toponym may or may not be lost in A44, concerning the engineer Polyidus.

62 For comprehensive discussion and documentation of all matters arising see Whitehead 2016.

63 Such outworks appear to have been routine, standing aspects of a city's defenses by the time of Philo, and had in some instances been installed much earlier (so e.g. Sidon in 351/0, if Diod. Sic. 16.44.5 is to be believed). See generally, Winter 1971, 273 (expanded at 329–331), 283–286. The most celebrated instance, from the third quarter of the third century, was the Euryalos fortress (third stage) at Syracuse: Winter 1971, 280–283; Lawrence 1979, 288–299.

64 The "old" one apparently as at Rhodes (A59, and already A17), a city with which Philo was personally familiar, as he says in the *Belopoiika*. Alexandria was another such city, but not mentioned here.

65 While Philo's idea of prudent provisioning focuses on food, B49–53 remembers to cover relevant non-edible items of equipment and material too; it also deals with technical personnel.

supply of grain in storage, on a rolling basis; other evidence seems to indicate that this amount, while not unfeasible, was an aspiration rather than a norm. Under *ii*, both private gardens and appropriate public spaces are earmarked for a policy of horticultural productivity reminiscent of the DIG FOR VICTORY! campaigns in both World Wars of the twentieth century. Featured there and throughout the section is a botanical encyclopedia's worth of ancient Mediterranean and Near Eastern fruits and vegetables to be eaten either fresh or preserved; and particular attention is given to systems of rationing and to appetite-suppressant recipes.[66]

Part C suffers just badly as part A from a disjointed order of topics but does manage to address a third-century version of what Aeneas had covered in his chs. 32–40: how best to counter artillery fire, human and mechanical frontal assaults, and tunneling; also guard duty, patrolling, and passwords. Of especial interest to modern readers are C45–48 on sustaining morale (with particular reference to mercenaries), and C72–73 on care for the wounded.

Finally, part D, as we have already seen, switches point of view to that of the attackers. Their best hope of success, in Philo's opinion, is a surprise attack (D2–4), but the remainder of the lengthy discussion (D5–111) assumes that this has not proved possible. Thus, before consideration of the use of stealth (D73–75), treachery (D76–83) and a starvation blockade (D84–85), D5–71 concern themselves with what to do both before (D5–23) and during (D24–71) an assault that the defenders are fully anticipating. Psychological warfare plays a prominent role here. D12–16 give instances of proclamations for psychological advantage, D59–60 recommend wrong-footing the defenders' expectations, and D65 speaks for itself: "Do not skimp on money, whether in bribery or in other expenses; once you have taken the city you may recoup many times as much". Following the treatment of *limos*, twin additional heads address the matter of how to respond to the arrival of enemy reinforcements by land (D86–100) or by sea (D101–110).

Overview

Philo's part D ends with a single-sentence conclusion: "Anyone besieging cities in this way is especially likely to capture them while suffering nothing ruinous himself" (D111). Considered as precisely that, a conclusion to part D, the claim is reasonable enough. Surprise attack; resisted attack; collusion with the defenders; starvation blockade—the commanders of an attacking force seem to enjoy options galore. Yet the *Paraskeuastika* and *Poliorketika* taken in their entirety assert that many of these strategies could be thwarted; and the historical record of sieges across the 350-year period surveyed here confirms that, very often, they were. The advent and development of a range of mechanized weaponry, we may be tempted to think, maintained a balance of advantage for assailants; yet artillery in particular could be and was deployed by defenders also; and fortifications themselves, broadly speaking, were more formidable at the end of the third century than they had been in earlier times.[67]

66 Part B concludes with specifications for a long-distance signalling system: a pair of large water-filled vessels into which, by careful synchronization, calibrated rods descend (B55–57). Here Philo shows most clearly his debt to Aeneas; cf. Polyb. 10.44.

67 Whitehead 2016, 417–424 proffers (with all due caution) the walls of Thessalian Scotoussa—described in an inscription from the early second century—as "typical" of that era.

In any case, siege narratives themselves and the synoptic treatments by both Aeneas and Philo demonstrate that the human side of poliorcetics remained much the same across our period.[68] (So too did the likelihood of cruel treatment of defeated communities, crueler than of armies on the losing side in a battle.) [69] Treachery, whether motivated by avarice or conviction, worked just as well at Chalcis in 357/356? as at Seleucia Pieria in 219.[70] Examples of a *limos* blockade pursued to the point where the besieged are alleged to have had resort to cannibalism come alike from Potidaea in winter 430/29 and from Tauromenium in 132.[71] Even short of that extremity, the record of hunger delivering cities to their besiegers embraces Herodotus on the Athenians vs. Sestos in 479 (where the inhabitants were reduced to "boiling and eating their bedstraps") all the way to Plutarch's *Demetrius*, on the great besieger vs. Athens in 295 (where, the story went, a father and son were seen fighting over a dead mouse that had dropped from their ceiling).[72] The enduring consolation for possible victims was that success by any method was never a foregone conclusion.

68 Aeneas and Philo are compared and contrasted, under ten heads, in Whitehead 2016, 33–59.

69 Sources on the treatment of captured cities (Kern 1999, 135–162, 190–193, 227–236) make for grim reading. A regular punishment was death for men, enslavement (*andrapodismos*) for women and children, but further atrocities could be inflicted at the victors' will. Clemency, as shown by e.g. Cimon (in Cyprus in 450: Diod. Sic. 12.3.3) or Callicratidas (Diod. Sic. 13.76.2) was rare enough to be noteworthy.

70 Chalcis, Aen. Tact. 4.1–4; Seleucia, Polyb. 5.60–61.

71 Potidaea, Thuc. 2.70.1; Tauromenium, Diod. Sic. 34/35.2.20.

72 Hdt. 8.114–118, Plut. Dem. 33–34.

PART III

Military Personnel

CHAPTER 9

The Organization of Greek Armies

F.S. Naiden

The organization of Greek and Macedonian armies is a subject with three parts: first, the number of men enrolled and engaged; second, the number, size, and kinds of units; and third, the number, ranks, and duties of officers, especially those in supreme command. All three parts point to both practical and cultural explanations for the characteristics of Greek military organization. Comparisons with later armies are illuminating, for they show how long some of the characteristics of Greek military organization persisted.

Modern parlance distinguishes among conscripted, professional, and mercenary soldiers, and none of these categories has a Greek equivalent. In Homer, where we must gingerly begin, these categories are irrelevant. In the Greece of the city-states, the universal obligation of citizen military service makes the term "conscripts" misleading, while the lack of salaries and military schooling makes "professional" somewhat misleading as well, even for the Spartans to whom scholars often apply it.[1] Greeks were the most famous mercenaries of antiquity, but these troops were commonly called "strangers," "wage earners," or simply "soldiers."[2] This trio of terms implies that mercenaries did not form a distinct type, but were interchangeable with other soldiers, save for their origin and pay. Greek sources mainly distinguish soldiers according to their equipment, such as the heavy shield that gave hoplites their name, and according to their being mounted or on foot.

As for ranks among common soldiers, the Greek sources never mention them. Officers have ranks, but the terms for these ranks mostly designate command of combat units, such

1 Spartan "professionals": Hunt 2007, 130. A critique: van Wees 2004, 84–86. The only general treatment of the bibliography for professionalism among soldiers: Teitler 1977, ch. 1.2. Philip's officers as the culmination of previous advances: Naiden 2007.
2 These terms: Foulon 1995.

A Companion to Greek Warfare, First Edition. Edited by Waldemar Heckel, F.S. Naiden, E. Edward Garvin, and John Vanderspoel.

as *lochos*, a band or company, yielding the agent noun *lochagos*. The modern equivalent, "captain," has wider meaning. By the same token, Greek lacks categorical terms such as "company officers" and "general officers," and instead merely distinguishes between command of small and large units. Terms for supreme commander reveal political factors at work in Greek military organization. Just as their subordinates are merely agents of some command function, top commanders are commonly magistrates, and thus agents of the *polis*. Some are kings.

The Macedonian or Macedonian-led armies of the Hellenistic period inherit much of this structure, and add two complications, artillery and "Companions."

Besides using modern terms cautiously, any treatment of Greek military organization must reckon with the biases of the sources, which are mainly Greek historians beginning with Herodotus and Thucydides. These writers have much more to say about some *poleis* than others, more to say about hoplites than other troops, and less to say about officers than about supreme commanders and the mass of soldiers. Compared to these weaknesses, ancient carelessness or ignorance about numbers is a glaring fault far easier to detect, even if it is not easier to repair.[3]

The secondary literature on this subject effectively begins with late nineteenth-century Germany, where a number of military officers, notably Hans Delbrück, produced surveys and handbooks that remain essential.[4] These writers treated organization thoroughly, but relied on Common-Era military manuals and readily compared ancient units and ranks to those of the nineteenth century. This chapter will rely on sources written before Christ, and will distinguish between ancient and modern kinds of rank. A Spartan *mora* or Macedonian *taxis* will still be regarded a unit that an officer such as Delbrück could have marched across a parade ground, but it will not be regarded as a unit of soldiers with military pensions, seniority in rank, service decorations, or the other features that separate the modern military from the ancient, and also from civilian life. Compared to modern soldiers, Greeks and Macedonians were very much citizens, hirelings, or companions, and were organized accordingly.

Numbers

Almost all reports of numbers of Greek and Macedonian soldiers (and of their allies and opponents) are estimates that end in zeroes, and sometimes we cannot tell how many zeroes there should be. At times we cannot tell whether estimates are for effectives or only for available manpower. To check many estimates, we need to know how many citizen age classes were called to service, but we almost never do. Estimating Greek numbers by comparing enemy numbers is absurd. Delbrück proved some estimates of the Greeks' enemies to be worthless, and his arguments cast doubts on many other estimates. Worse still, the ancient figures for Greek soldiers often do not distinguish among kinds of troops. In these cases,

3 Thus Delbrück 1900, 1.1, citing Livy and Polybius. An example not discussed by Delbrück: Parke 1933, 114, n. 4, observing that high figures for troops in Sicily in the Classical period are for manpower only. Criticism of some of Delbrück's conclusions: Kromayer and Veith 1903–1907.

4 Besides Delbrück 1900, see Rüstow and Köchly 1852, Bauer 1893, and Kromayer and Veith 1928. The one recent work of this type: Pritchett 1971–1991. The only monograph on the best-known Greek army: Lazenby 1985. For Macedon, see Milns 1976, with references beginning with Droysen 1877, and also Heckel 2016a, 260–281. For the fourth century, which is the best known, see Anderson 1970.

scholars commonly assume that the troops were mostly heavy infantry, the best reported type of soldiers, and presumably the most numerous.

Aside from heavy infantry, the most numerous category was perhaps soldiers' servants, all of them slaves. This was a large group about which we can only say that in some armies every heavy infantryman had a slave, and in others many probably did. Even this deduction depends on an argument from silence, which is that reports of other ratios are rare and sometimes dubious. The highest report is seven helots per Spartan at Plataea, although not seven per soldier serving with the *perioikoi*, Sparta's Lacedaemonian "neighbors." The lowest report is one slave for every ten infantrymen in the army of Philip II, perhaps a ratio not applicable to the army of Alexander once it acquired a large baggage train.[5]

Another large group was light infantry, or *psiloi*, some of them citizens, the same as many or most hoplites, but often recruited from abroad or from outlying, tribal regions of Greece. Often the sources fail to report any number for them, and when they do, they are not directly or indirectly basing their report on a muster roll as accurate as those kept for citizens. For tribal troops, there may have been no muster roll. Many trireme crews saw military action, and we often have accurate reports of the number of ships involved, but we never have reports of the number of rowers converted into light troops during an engagement. Homer illustrates this gap in our information. His fighting men double as rowers at 50 per ship. For nearly 1,000 ships, the very doubtful total is 50,000 soldier-sailors.

As for artillery, we know that missile-throwing machines had crews of three to seven, and stone-throwing machines had crews up to ten.[6] That ought to spell a siege train of hundreds, and, under Alexander and Demetrius Poliorcetes, a train of several thousands, but no ancient source tells us. Nor do they tell us how many essential non-combatants served in leading armies—doctors and nurses, for example, or surveyors such as traveled with Alexander, or even grooms for horses and elephants.

For cavalry we know much more, especially for *polis* cavalry and the cavalry of Alexander. These estimates are probably accurate to within some hundreds. The proportion of cavalry to total combatants, however, is very small: To cite leading Classical armies other than Thebes and Thessaly, the ratio of cavalry to other arms is one to ten in Athens, and even less in the army of the Peloponnesian League. Later, the proportion rose, and in the armies of the Successors the ratio was one to five, six, or seven in some major battles. The highest ratio for a large force was one to two, attained by Alexander. By the third century, the role of the most important Macedonian cavalry, the Companions, diminished.[7]

For the Archaic period, all these issues vanish for lack of evidence. Greek sources report army size for only one Archaic belligerent—Eretria, which fielded 3,000 infantry, 600 cavalry, and 60 chariots in the Lelantine War.[8] Assuming that the 3,000 were hoplites, we can trust our next piece of evidence, the total of 38,700 hoplites reported for Plataea by Herodotus.[9] The two dozen *poleis* fighting there ought to have been able to field about a dozen times as many hoplites as a leading mercantile city of two centuries before. Only one figure in Herodotus' list is an apparent round number, the 10,000 Lacedaemonians. Since the two

5 Seven: Hdt. 9.28. One: Frontin. *Str.* 4.1.6.

6 Kromayer and Veith 1928, 233.

7 Delbrück 1900, 1.7. Last mention of the companions: Gaza, 312 BC (Diod. Sic. 19.82.3). Later Hellenistic period: Hatzopoulos 2001, 33–35.

8 Assuming we accept Str. 10.1.10. Ten thousand infantry for all of Thessaly in the late 500s: Ael. *Tact.* 18.1.3.

9 Hdt. 9.28–29. Rejected by Delbrück 1900, 1.9, preferring 20,000 hoplites.

dozen states at Plataea were from southern, not northern, Greece, they were so weak in cavalry that they did not field any. They could have used the 8,000 cavalry promised, but not delivered, by Gelon of Syracuse (Hdt. 1.158).

The largest permanent allied army of the fifth century was that of the Peloponnesian League, possessing some tens of thousands of hoplites. Against them, the Athenian Empire fielded 13,000 hoplites at most, plus a comparable number of garrison troops.[10] Yet the Peloponnesian League is not sure to have fielded more than 13,500 in any one engagement.[11] The largest single Athenian force during the Peloponnesian War reached only 13,000. Even at Chaeronea, the most important Greek battle of the fourth century, the Athenians fielded only 10,000.[12] Boeotia did better than either Athens or Sparta, but only in the years of Epaminondas' ascendancy, when it recruited enough allies to field 30,000 hoplites at the battle of Second Mantinea (Diod. Sic. 15.84.4).

The cardinal feature of Greek military manpower in the Classical period is mercenary infantry's becoming more numerous than citizen infantry, first abroad and then in Greece and Magna Graecia. Herodotus and Xenophon report tens of thousands of Greek mercenaries serving in the Near East in the Archaic as well as Classical periods. At least one large army of these mercenaries was under Greek command—the Ten Thousand of Clearchus and then Chirisophus and Xenophon.[13] In Greece, 6,000 of the infantry of the tyrant Jason of Pherae were mercenaries, and 8,000 served under the Athenian commander Timotheus.[14] These two fourth-century forces were bigger than any total reported for the Peloponnesian War, in which the largest reported force is 1,500, recruited by Brasidas (Thuc. 4.80.5).

Magna Graecia presents a different picture from the beginning of the Classical period onward. As Nicias told the Athenians, Sicilian governments relied largely on mercenaries, not citizen soldiers, and relied on them even for heavy infantry, and so mercenary totals ran into the tens of thousands as early as the battle of Himera in 480.[15]

Compared to mercenaries, other alternatives to citizen soldiers were small. Sparta never armed more than 6,000 helots at a time. The highest figure for Athenian metic hoplites is half that, and these were garrison troops.[16] A third, apparently neglected, choice would have been to rearm light troops as hoplites, but there is no report of this, only of the opposite, which was Iphicrates' supposedly turning hoplites into peltasts.[17] In contrast, we do know that Hellenistic changes in infantry equipment led to new terms for heavy and also lighter infantry.[18]

---

10 Thuc. 2.31.2, 2.13.6. Seven years later, only 7,000 fight at Delium (4.96).

11 Thirteen thousand five hundred: Nemea Xen. *Hell.* 4.2.16–17. Beloch 1906, 108, reckons 27,000 Peloponnesians invaded Attica yearly during the Archidamian War, and gives estimates of other Classical forces.

12 Delbrück 1900, 1.1 estimated total Athenian forces at the start of the Peloponnesian War as 45–50,000.

13 Hdt. 2.63.1; Xen. *An.* 1.1.10–12. Cf. the 10,000 that serve Artaxerxes (Diod. Sic. 16.43).

14 Jason: Xen. *Hell.*6.4.1. Timotheus: Isoc. 15.111.

15 Diod. Sic. 11.21.3, 11.62.3.

16 Xen. *Hell.*6.5.28–29; Thuc. 3.13.6. Rüstow and Köchly 1852, 90–92, hold that some Spartan *lochoi* were mostly for garrison service, others for field service. The oldest and youngest would have served in these garrison *lochoi*. Similarly, they hold at p. 96 that Athenians without a *kleros* for field service as at Lys. 16.15–16, were assigned to home duty. There is, however, only one report of older soldiers being organized separately, Thuc. 5.72.3–4, about Argos.

17 Diod. Sic. 15.44.2–4, Nep. *Iph.* 11.1, an idea debunked by Best 1969, 102–110.

18 *Thureophoroi*, heavier *thorakides* (Polyb. 4.12.3), and *chrysaspides* for heavy infantry; *argyrasipides* for the medium or flexible infantry called hypaspists; *machimoi* for sundry Egyptian infantry. So also *kataphraktoi* for mailed cavalry.

By the time of Alexander's Successors, larger armies resulted from two new trends, large states or alliances deploying big forces and the use of Asian troops. Many reports are round numbers from Diodorus, but the 154,000 men reported at Ipsus and the 143,000 at Raphia, the two most important battles of this period, indicate armies of a different order of magnitude than any previous one besides Alexander's in India, a force the king broke up as soon as he reached the Indus delta.[19]

With totals like these, the Greek states and the Kingdom of Macedon could not compete, but 48,000 Macedonians reportedly fought at Crannon—a figure comparable to the Macedonian forces of Alexander and Antipater combined—and the armies of the biggest Greek federation, the Achaeans, who were Macedon's biggest Greek enemy, must have been comparable. To attack Cassander, Demetrius hired 15,000 mercenaries, double his number of Macedonians (Diod. Sic. 20.110.4).

In the Archaic period, Greek *poleis* raised troops largely by tribe. This implied groups of unequal size and quality. Sparta was perhaps the first to organize men by locality, and Thessaly and Macedon were among those who followed this example.[20] Now groups were likely to be of standard size, and in all three of these cases they were likely to be of standard quality, thanks to some drill.[21] Alexander's Successors followed the Persian example and raised troops partly by province. That meant more troops but less standardization. To compensate for uneven quality, armies of the late Classical and Hellenistic period included picked troops, the most famous Greek unit being the Theban Sacred Band, and the most famous Macedonian unit being the Silver Shields.[22] Under the Successors, important satraps had elite troops. Size and selectivity progressed together.[23]

The shift from Homeric crews of spearmen and a few archers to tripartite masses of infantry, cavalry, and artillery did not alter a leading feature of ancient military manpower. Arms, dress, and martial qualities depended on the economy and mores of some locality or region.[24] Sparta produced no slingers, and Rhodes, which did, produced nothing like the *enomotiai*, or sworn bands, that all Spartan infantry joined. Ancient manpower was not only more diverse than the manpower of contemporary Western states, but more diverse than the military manpower of the British Isles hundreds of years ago. Turning the war bands of a Greek valley or an Aegean island into armies was correspondingly difficult. This transition depended partly on soldiers forming units, the next topic in this chapter.

19 Ipsus, Raphia, and Crannon, below: as at Delbrück 1900, 1.7.

20 Villages: Rüstow and Köchly 1852, 37, 92. Criticism: Bauer 1893, 306; Anderson 1970, 237–239.

21 Although the Thessalian cavalry, not infantry, were famous for their command and control (Asclepiod. 7.2, Ael. *Tact.* 18.2).

22 Perhaps the largest force of select troops: 3,000: Kromayer and Veith 1928, 132, unless Polyb. 5.65.5 refers to a permanent Greco–Macedonian unit of 3,700.

23 Eumenes' elite force, and others': Diod. Sic. 19.27.2–28.3. The tactical and operational effect of the successors' following the Near-Eastern practice of granting lands by military tenure is little studied, and perhaps unknowable; for Egypt, see Fischer-Bovet 2014a.

24 Thus it was rare for soldiers from one tribe or region to use equipment typical of another tribe or region; for example, the "Macedonian-style" Africans and others at Polyb. 2.65.3, 5.65.8, 5.79.4, 5.82.2–10 with Bar-Kochva 1976, 68–69.

Units

For units, as for manpower, we know most about hoplites, less about cavalry, and least about light troops and artillery. For the Archaic period, of course, we know little, and too much of what we do know comes from Homer, who reports only two kinds of combat groups, both difficult to visualize. The Homeric *stichoi*, or lines, lack commanders, and the Homeric *pyrgoi*, or towers, lack any plausible shape. Only one group, the Myrmidons, have fixed subdivisions.[25]

The Classical period presents a different problem—when and how types of units evolved. When exactly did the Spartan *hippeis*, or cavalry, stop riding?[26] When did the Macedonian foot become the heavy infantry called *pezetairoi*, or foot companions? When did the Macedonian hypaspists cease to be "shield-bearers" and become a crack force?[27] The most important such question concerns the evolution of the term *lochos*, which originally meant "band." When and how did it come to be a unit, and then a subunit with subunits of its own? In the face of these questions, this section will confine itself to military units at their most elaborate, which was in the fourth and late fifth centuries.

The smallest unit of heavy infantry was the file, and the file could not become an operational unit, as opposed to a way of counting and marching men, until it performed maneuvers. The first attested maneuver is the mock retreat, in which either the first or the last man in the file withdrew, and the rest followed and reformed behind the leader. The Spartans performed such a retreat at Plataea.[28] Polyaenus says that Macedonians performed mock retreats, and that Thebans did under Epaminondas' contemporary, Pammenes.[29] The Lacedaemonians and the Boeotians were both able to rally files in disarray and order them to counterattack.[30] The length of the file might vary; eight was common, but not universal. (Only one source, Xenophon's *Hellenica*, reports a fixed size for a smaller infantry unit, and this is a garrison unit of five.) [31]

Some number of files—once again there was no single number—formed a squad or platoon. The best known were the Spartan *enomotiai* of some 25 to 40, depending on the number of age classes called into service. If an army lacked a unit of this size, it might have a unit called a "Fifty," which did not necessarily have that many men.[32] In Sparta, the Fifty had two *enomotiai;* the same was true among Xenophon's mostly Peloponnesian Ten Thousand.

25 The contrary view: Bauer 1893, 302.

26 A related question: after becoming infantry, did the *hippeis* first serve in the front line, and only later become specialized, as happened in the case of the Sacred Band at Thebes? See Figueira 2006, 72–73. The Sacred Band, as it happens, were originally ἡνίοχοι και παραβάται (Diod. Sic. 12.70.1), as at Kromayer and Veith 1928, 64 and Pritchett 1971–1991, 2.222, both dating the switch to infantry to some time well before Delium.

27 A shift that Milns 1976, 93, found too difficult to imagine, and thus rejected. A summary treatment of this force: Heckel 2013. Hellenistic developments: Spendel 1915, 45–53.

28 Naiden 2017a, 63–65, giving Classical sources.

29 Polyaen. *Strat.* 4.2.4 by Philip II at Chaeronea, 4.12.13 by Craterus, 5.16.2 Pammenes.

30 Spartans leading others: Thuc. 4.16.6: Thebes: Ar. *An.* 1.8.4. Additional orders given during battle: Konijnendijk 2018, 149–50. Boeotian command and control: Cawkwell 1972, 260 n. 12.

31 Phlios: Xen. *Hell.*7.2.6. Leaders of five and ten: Xen. *Cyr.* 2.3.21, the only Greek source reporting this Near-Eastern practice.

32 The highest number reported: 64 at Thuc. 5.68.3.

Or the unit of this size might be called a *lochos*, usually translated as "company," of 50 or 70, or at most 100, as sometimes among the Ten Thousand.[33]

Whatever the size or the term, this unit comprised a group that an officer could command with a shout or even a nod, provided that the man at the head of each file could, like a sergeant, control the men behind him, perhaps assisted by another officer at the rear of the file.[34] At Sparta, the enomoty introduced an additional layer of control. Without *enomotiai* or "Fifties," control would diminish, as apparently at Athens.[35]

A small, fixed number of *lochoi* formed a larger unit that the Spartans came to call a *mora*, the Athenians and Macedonians a *taxis*. The *mora* reportedly ranged from 500 to 900 men, again depending on the number of age classes called into service. Six hundred was a common size, and 572 is the size reported by Thucydides for the comparable unit not called a *mora* at the first battle of Mantinea (Thuc. 5.67.1). This figure of 572 is, in fact, the only one for any of these units that is an exact number for a single occasion, and thus warns us against mistaking round numbers or estimates for full information, yet it also shows that large infantry units tended to have at least 500. The largest of these units was the Macedonian *taxis* of a nominal 1,536 and the second largest was the comparable unit at Argos, numbering 1,000. The Athenian *taxis* might be the same size, but only if most or all age classes were serving.

Mercenary armies are harder to describe because of vagaries in organization. *Lochoi* were commonly recruited by officers called *xenologoi*, who doubtless often became combat commanders of *lochoi*. *Strategoi* or taxiarchs commanded a variable number of *lochoi*. Among the Ten Thousand, the most important generals commanded up to almost a dozen *lochoi*, yet others commanded only several.[36] No one general commanded more than some 2,000 men, a figure comparable to the Macedonian *taxis* and also to the infantry regiment of the nineteenth and the infantry battalion of the twentieth century CE.

The reason for this regimental size was the same as for the size of the *lochos*—battlefield convenience. The leader needed to communicate with a small, tidy group of some half a dozen subordinates—to spot them in column or in line and give them orders through gestures, trumpets, or messengers. In better armies, such as those of Lacedaemonians and mercenaries, the commander would have to order his subordinates to wheel as well as advance or retreat, and even in worse armies, such as that of Athens, he could order them to advance, retreat, and reform.[37] Half a dozen officers could all be known by name, a practical and psychological advantage in the field and in council.[38]

The same reasoning explains how many *morai* or *taxeis* comprised the main body of heavy infantry in many Classical and some Hellenistic armies. Sparta had six *morai*, and earlier they had seven comparable units. The Macedonian army of Alexander had six or briefly seven *taxeis* of *pezetairoi*, and the force left with Antipater was about the same size.[39] The Argives

33 Mercenary *lochoi* of 50: Xen. *An.* 1.2.25. Of 400: *Xen.* An. 6.2.12, 6.2.16, 6.3.2. Aristides commanding 300: Plut. *Arist.* 14.5. For the Theban Sacred band, see immediately following.

34 An *ouragos* at the rear: Wheeler 2007a, 207.

35 Naiden 2017a, 79, 83. The Athenians could form a square (Thuc. 6.67; 7.78.1).

36 Sizes: Xen. *An.* 1.1.11, 1.2.3. The number of *lochoi* under the generals serving Agesilaos no doubt varied also (Xen. *Hell.* 3.4.20). *Xenologoi* as commanders: Nussbaum 1959.

37 Messengers: see n. 73 below. Trumpets: Diod. Sic. 14.52.1–5 with Krentz 1991. Gestures: Curt. 3.2.13.

38 The practical advantages: Xen. *Cyr.* 5.36–38.

39 Sparta: Lazenby 1985, ch. 1. A 7th *taxis*: Milns 1966.

had six units of a thousand. Like a Union brigadier at Gettysburg, the commander of this body of infantry had half a dozen "colonels" under him. Too many colonels would have been unwieldy, but controlling thousands of men without colonels would have been impossible. If the commander was fighting in the front line, like a common soldier (or a low-ranking Spartan officer), this would not matter, once combat started, but commanders did not always fight in the front.[40] Sometimes they stood or rode alongside, and surveyed the fighting.

No source says that larger permanent units existed. The Spartans, for example, never had a unit larger than the *mora*, and the Macedonians never had one larger than the *taxis*. *Morai* and *taxeis* often formed a phalanx, but a phalanx was a formation, not a unit. Among the Macedonians, one general might control the left end of the phalanx, and another general the right.[41]

There is little evidence for unit structure in the armies of the Successors, but the Macedonians and Greeks were presumably organized the same way, while the larger number of native infantry formed looser groups.[42] In Egypt, for example, the native infantry consisted of units of 50, 100, 500, and 1,000, all multiples of five and ten as opposed to the diverse multiples found among Macedonians and Greeks.[43] The phalanx might number 16,000, as compared to the nine or ten thousand of Alexander.[44]

The emphasis that this chapter gives to tactical requirements does not mean that infantry units originated entirely or mainly on some tactical basis. Large units were often organized by tribe, and in Athens (although not Sparta or Argos) they remained that way, even after Clisthenes changed the number and composition of the tribes.[45]

For both the *taxis* and the Homeric band, a local origin helped with recruiting and *esprit de corps*. Yet the best Greek infantry of the Classical period, the Lacedaemonians, were recruited through collective oath swearing.[46] The glue holding men together was civic and religious, not local or familial. For mercenaries, of course, the glue was a contract, not an oath. It was professional and pecuniary, and once again not local or familial.[47]

Perhaps because of the need for horsemanship and drill, as well as because of greater cost, cavalry units were smaller than infantry. In Athens, which is best documented among the Greeks, the largest unit, the hipparchy, comprised 500 divided into five tribal contingents. Smaller groups were apparently temporary.[48] The 1,800 Macedonian companion cavalry of

40 Naiden 2017a, 73; Wheeler 1991a, 148–150.

41 Arr. *Anab.* 2.8.4, 3.11.10, where Craterus controls the left flank of the heavy infantry and Parmenio controls it all, and thus controls the right in person.

42 What evidence exists is conveniently collected in Spendel 1915.

43 Lesquier 1911, 69, but failing to note that these unit sizes do not reflect the Greek and earlier Macedonian practice of creating units reflecting tactical requirements of phalanx fighting. See also Kromayer and Veith 1928, 122–126.

44 Magnesia: App. *Syr.* 32. Cynocephalae: Liv. 33.4.4–5.

45 Arguing that the Spartan army was never recruited locally, i.e., by *obai*: Wade-Gery 1944, 116–121, with refs. The opposite view: Lazenby 1985, 51–52. Locally elected polemarchs in charge of Boeotian districts: Xen. *Hell.* 5.2.25, 30. Four districts in Thessaly: Arist. fr. 497.

46 *Pace* Rüstow and Köchly 1852, 92.

47 The earliest extant examples of the text, or likely text, of such a contract: Xen. *An.* 1.3.1, 7.2–3. A survey of documentary evidence: Griffith 1935, ch. 10.

48 Xen. *Hipp.* 1–2, also reporting reports files of 10 and squares of 100. There were surely officers leading temporarily formed small groups, as at Thuc. 5.84.2. Rüstow and Köchly 1852, 95, hold only 600 cavalry were for field service. Further references: Bugh 1988/2016, chs. 3, 5.

Alexander eventually consisted of 100-man *lochoi*, 200-man *ilai*, at two *lochoi* per *ile*, and hipparchies of 400 to 800, or two to four *ilai* per hipparchy. (Earlier, it consisted of one hipparchy of eight *ilai*, one double strength).[49] Like the Spartan infantry, the Macedonian cavalry had more layers of control than its opponents. We have little information about the units of the two biggest Greek cavalry forces, those of Thessaly and Boeotia, but if Jason of Pherae devised the diamond-shape formation, the Thessalians must have had a hierarchy of units more or less like the Macedonians.[50]

Like cavalry, picked troops had small units: three for the 300-man Spartan *hippeis* (Xen. *Lac.* 4.3), and some larger but unknown number for the 1,000 picked troops at Argos. In contrast, light troops are seldom said to have any units, the two exceptions being the archers at well-documented Athens and the light troops serving with the Ten Thousand.[51] Xenophon knows of *lochoi* of 60 for peltasts, and *taxeis* of probably one or two hundred for other light troops. Unlike the *lochoi* of the heavy infantry, these may have been temporary units. Iphicrates surely divided his peltasts into units, too, as shown by a report that he used them to perform a mock retreat—the only such retreat by light troops.[52] About artillery units nothing is known.

Just as infantry did not form permanent units larger than the *mora* and *taxis*, forces of combined arms did not form permanent units. Even in the best-organized large army, that of Alexander, there was no unit to designate the infantry, cavalry, and light troops commanded by Parmenio at the Granicus River, Issus, and Gaugamela. Instead these troops comprised a wing assembled for the purpose of fighting the battle. About the armies of the Successors we know less, and so it is at least possible that these armies had wings or other permanent elements comparable to something bigger than a regiment, but no early source says so, and it is very unlikely such a large unit existed among the Macedonians, who were not more numerous in these armies than in the armies of Alexander and Antipater. At most, we can say that under the Successors the phalanx might be larger than under Alexander—no more than 25,000 phalangites at Raphia.[53]

Unsurprisingly, the structure of units influenced rank structure. Roughly speaking, Greek armies had few ranks of general officers and no ranks for ordinary soldiers, but relatively numerous ranks for company and regimental officers. Much needs to be said about this neglected topic.

---

49 Here as at n. 36, see Errington 1990, 242–243. I assume that the tetrarchy was a temporary forerunner to the hipparchies, as at Bosworth 1980–1995 ad Arr. *Anab.* 3.18.4. Additional unknown subunits: Sekunda 2007, 331. *Lochoi* only after Gaugamela: Spendel 1915, 8. See also Brunt 1963.

50 I.e. no information other than the handbooks of the Common Era, Ael. *Tact.* 18.1–3, mentioning Jason, and Asclepiod. 7.2–10, according to which each diamond of 40 cavalry had four officers, and four diamonds formed an *ile*. Asclepiod. 7.11 supplies artificial multiples of these *ilai*. An important Hellenistic cavalry force: the Aetolians (Polyb. 18.22.5), but no organizational details.

51 Xen. *An.* 1.10.7, 6.5.11.

52 Polyaen. *Strat.* 3.9.49. Discussions of Iphicrates' reforms by Best 1969, 102–108, and Lippelt 1910, 65–67, do not deal with this issue. A half dozen soldiers serving under Iphicrates who may or may not have been *lochagoi*: Sekunda 2017, 69–80.

53 Polyb. 5.65.8–9, but with difficulties discussed by Anson, Ch. 4 in this volume. Sketchy evidence for an Antigonid and Lagid subdivision of the phalanx differing from the earlier subdivision: Rüstow and Köchly 1852, 136–137; Kromayor and Veith 1928, 132.

Rank

Rank has several aspects: first, rank itself; second, seniority allowing one officer to command another of the same grade; and third, choice of assignment, which makes one officer of a given rank more important than another. These three factors would seem to be universal. How, after all, can there be a chain of command without ranks and without seniority to distinguish among those of the same rank, and how could assignments not differ from one situation to the next? Yet this trio of factors did not exist in ancient Greek armies. Seniority did not divide those in the same rank, and for most ranks the range of assignments was small.[54] Once an officer lost his command, his seniority and his rank went with it. When Epaminondas ceased to be Boeotarch, for example, he became a common soldier (Nep. *Ep.* 7).

Even more than with regard to units, information about ranks clusters around the armies of Sparta and Macedon, with Athens providing a sketchy contrast and the Ten Thousand a mercenary variant. About the Successors we know far less than about the Macedonians of Alexander.[55]

For Sparta, the best information is Xenophon's for the early fourth century. In the heavy infantry the lowest ranking officer, or *archon*, was the file leader (or perhaps the file closer, or *ouragos*, if he, too, was an *archon*). Next came the enomotarch, the *pentekoster* or "Fiftier," and the *lochagos*. They all served under the polemarch commanding the *mora*, and the polemarch commonly served under the king, who commanded the army, save for smaller deployments a polemarch would lead. Reckoning a *mora* as 600 men and the infantry as 3,600, this body of men included 96 enomotarchs, 48 Fiftiers, 24 *lochagoi*, 6 polemarchs, and a king, or 175 higher officers with an officer-to-soldier ratio of 1 to 20. Add four file leaders per band, and the ratio falls to 1 to 6; add four file closers, and it falls to 1 to 3. The military hierarchy may also be described per unit: commonly eight file officers per enomotarch, then two enomotarchs per *pentekoster*, two *pentekosteres* per *lochagos*, four *lochagoi* per polemarch, and six polemarchs per commander in chief, or ratios of 8 to 2, 2 to 4, and 6 to 1. This army thus had two low ratios, first, officers to men, and second, subordinates to superiors. In the second ratio, there was no bulge due to any large number.[56] The chain of command was six links long.

Slightly different was the 4,400 heavy infantry at First Mantinea, described by Thucydides. It contained 112 enomotarchs, 28 *pentekosteres*, 7 *lochagoi* acting as polemarchs would later, and a king, with an officer-soldier ratio of 1 to 30. The file leaders and file closers were presumably the same as later. The hierarchy per unit was eight file officers per enomotarch, four enomotarchs per *pentekoster*, four *pentekosteres* per *lochagos*, seven *lochagoi* per commander in chief, or ratios of 8 to 4 and 7 to 1.[57] The chain of command was five links long. This structure was nearly the same as Xenophon's.

54 The only chapter-length account, there being no monograph on the subject: Anderson 1970, ch. 4. See also Wheeler 1991a.

55 The chief piece of evidence: Amphipolis inscription, c. 200, referring to *hegemones*, as given at Hatzopoulos 2001, 76–81. This word may refer to a *lochagoi* or file-leaders. Hipparchies for sundry Egyptian cavalry: Fischer-Bovet 2014a, 127.

56 Two full accounts with references: Lazenby 1985, ch. 1, but supposing 32 enomoties per *mora*, and Anderson 1970, 225–251.

57 Lazenby 1985, ch. 2; Anderson, loc. cit.; and van Wees 2004, Appendix 2, with bibliography on this long-disputed issue, especially whether Thucydides omitted a link in the chain.

Much more different was the army of Alexander. The lowest ranking officer was once again the file closer and next came the file leader, called a dekadarch, even though the file commonly had 16. Third was the *lochagos* and then the pentakosiarch, or commander of 500, and the taxiarch. Among the hypaspists, the lower ranks are unknown, but the higher ranks were pentakosiarch, chiliarch, and the supreme commander of the 3,000 hypaspists. Down through Gaugamela, the supreme commander for each of these forces happened to be a member of the family of Parmenio—the old man himself and his son Nicanor. Reckoning a *taxis* as 1,500 men, and the number of *taxeis* as 6, the heavy infantry contained 36 *lochagoi*, perhaps 18 pentekosiarchs, 6 taxiarchs, and a general; or 253 high-ranking officers with an officer-to-soldier ratio was 1 to 36. Add 96 *ouragoi* and 96 decadarchs for the files, or 192 lower-ranking officers, and the ratio falls to 1 to 18. The ratio for the hypaspists was perhaps somewhat lower.

As with Sparta, the military hierarchy may be described per unit: one file closer for each file leader, 16 file leaders per *lochagos*, 2 *lochagoi* per pentekosiarch, 3 pentakosiarchs per taxiarch, and 6 taxiarchs per phalanx commander, or 1 to 16, 2 to 3, and 6 to 1. For the hypaspists, we know of two pentakosiarchs per chiliarch, and three chiliarchs per supreme commander. The low ratio of subordinates to superiors resembles Sparta, and the chain of command had six links, as at Sparta. The chief difference between the two armies was the bulge at the second Macedonian link, a bulge reduced if we supply four *ektaktoi*, or supernumerary officers, to each Macedonian *lochagos*.[58]

Democratic Athens fielded up to 13,000 infantry divided into ten *taxeis*, each led by a *strategos*, a taxiarch, and some number of *lochagoi*. There is no evidence whatever for Athenian file leaders and file closers. One piece of evidence implies there cannot have been any such officers: This is the report that the Athenian soldier did not have an assigned place in line, and merely had to stay wherever he was.[59] Given that Sparta had four *lochoi* per *mora*, and Macedon had six per *taxis*, the number of Athenian *lochoi* per *taxis* cannot have been more than half a dozen, and likely was three or four.[60] The 13,000 infantrymen would thus have had at most 100 officers—60 or 80 *lochagoi*, 10 taxiarchs, and ten *strategoi*, their being no post of commander in chief. The ratio of officers to soldiers was 1 to 130 or more. Cut the mobilization in half, and it was still 1 to 65 or more. If the hierarchy is described per unit, it ran from 6 or 8 *lochagoi* per taxiarch to one taxiarch per *strategos*, a sequence of 8/6:1:1. The bulge is at the bottom, unlike Macedon. The chain of command had only three links. The Athenian army lacked both higher officers and subordinates, both levels of command and coteries of leaders.[61]

Compared to Sparta and Macedon, Athens was so underdeveloped as to seem abnormal. The truth was perhaps the reverse: Sparta and Macedon were hyperdeveloped. Most *polis*

58 As at Errington 1990, 242–243. Sekunda 2007, 331, doubles the size of the *lochos* and reduces the chain of command by one link.

59 Thuc. 5.66.2; all Greek versions of the ephebic oath (Rhodes and Osborne 2003, no. 88i; Lycurg. *Leocr.* 76; and later sources, viz. Plut. *Alc.* 15.7, Philostr. *VA* 4.21, Poll. 8.105, Stob. *Flor.* 43.48).

60 Units smaller than *lochoi* doubted by Rhodes 1981, 685. Accepting such units: Kromayer and Veith 1928, 49 n. 3, suggesting 4 *lochagoi* per *taxis*.

61 Rüstow and Köchly 1852, 96, suggest that the taxiarchs often commanded garrisons, thus depriving the Athenians of this link in the chain of command.

armies were in between, but closer to Athens.[62] The most important other army, that of Boeotia, may have had *lochoi* of 300, at least to judge from the size of the picked *lochos*, the "Sacred Band," and so when the Federation mobilized 6,000 men, as it did at Leuktra, it may have consisted of 20 *lochoi* somehow distributed among the ten Boeotarchs who were not serving as commander in chief, a post reserved for the eleventh, Epaminondas. The infantry would thus contain 31 high-ranking officers, for a ratio of officers to soldiers that was about the same as at Athens.[63]

Did the Successors follow the Macedonian model? In regard to their Macedonian troops, they largely did.[64] Their Greek mercenary troops surely followed the model Xenophon set forth for the Ten Thousand. A *lochagos* usually led 100 men, and sometimes he or another officer led a *taxis* of 200 assigned some special task. All *lochoi* and *taxeis* in a given detachment reported to a lieutenant general, or *hypostrategos*, and all served under the general, or *strategos*, who led the force. The mostly Peloponnesian Ten Thousand also had Fiftiers and enomotarchs, and if these officers' units were the same size as their Lacedaemonian counterparts, there were two enomotarchs for each Fiftier, and two Fiftiers for each *lochagos*.[65] The ratio of officers to men is impossible to calculate, but the ratio of subordinates to middle-ranking officers is six subordinates for each *lochagos*, and some unknown number of *lochagoi* per general and lieutenant general, a sequence comparable to Sparta. The number of links in the chain of command equaled Sparta.

So much for heavy infantry; for cavalry there is less to say. Only Xenophon provides a contemporary report, and he provides it only for Athens: 3 officers per file of 10, 1 commander per company of 100, and 1 phylarch for 500, or a ratio of a little more than 1 to 3. Per unit, two lower-ranking officers served under each file leader, five file leaders served each company commander, or phylarch, and five phylarchs under the hipparch, or 2 to 5 to 5 to 1, a chain with four links for only 500 men. Macedon and Thessaly, if not Boeotia, were surely comparable.[66]

Information about officers of light troops is almost entirely missing, the same as information for light units. Although 1,300 peltasts served at Kounaxa, a battle for which we know the names of nine Greek mercenary generals, Xenophon names only two peltast officers, Pasion and Menon, anywhere in the *Anabasis*.[67]

62 In Thessaly, evidence for *pezarchoi* under polemarchs: *IG* ii2 175. In Corinth, *lochagoi* under *strategoi*: Thuc. 4.44.1–3, 5.57.2.

63 Boeotian Fifties etc.: Lammert 1921, 446. Six or perhaps seven thousand at Leuktra: Anderson 1970, 196–198. Unknown number of Boeotian *lochoi*: Xen. *Hell.* 5.2.30, Polyaen. *Strat.* 2.3.1.

64 For Egypt, Fischer-Bovet 2014a, ch. 4, reports phalanx infantry chiliarchies of 1,024 under two pentachosiarchs with 512, each with two syntagmarchs of 256, each with 2 taxiarchs of 128, to which 16 file leaders apiece may be added, or 1/2/2/2/8, or a ratio of 1 to 13. For calvalry cleruchs, the hipparchy of 400–500 includes 2 ilarchs, each with 2 *lochagoi*, each with 8–10 *dekanoi*, or 1/2/2/8–10, or a ratio of about 1 to 10. For Seleucid difficulties, but no estimates, see Bar-Kochva 1976, 75, 92–93.

65 *Taxeis* of 200: Xen. *An.* 1.10.7, 6.5.11. *Lochoi* of various sizes: 50 (Xen. *An.* 1.2.25), 70 (Xen. *An.* 4.7.8), but mainly 100 (Xen. *An.* 3.4.21). *Hypostrategos*: 5.6.25. Lacedaemonian subunits: Xen. *An.* 4.2.3. Another rank, *hypolochagos*, appears only in siege operations (Xen. *Hell.* 5.2.13). In late writers, *lochagos* may mean "file leader," as at Polyaen. *Strat.* 3.9.10.

66 For Athens, see n. 59. For Macedon we lack information about lower-ranking officers; for Thessaly, information from any source that is BCE. For Boeotia, we know only that four *tele* comprised 1,000 cavalry at Delion (Thuc. 4.96.5).

67 Xen. *An.* 1.2.3, 1.2.6, with Lippelt 1910, 46, 72.

The one position outside the tripartite division of hoplites, cavalry, and other troops, that of the supreme commander, is the most complicated, even though the number of officers is very small. From 506 onward, the Spartans established unity of command by keeping one of their two kings home while the other went on campaign, or by putting one polemarch in charge. The Boeotian Federation known to us through Thucydides and Xenophon also established command unity, as ten Boeotarchs yielded up authority to one of their number, first Pagondas and later Epaminondas. They also established unity of command for lesser efforts such as some led by Pelopidas.[68] The Macedonians, like the Spartans, followed their king into battle, or, if he was absent, some commander the king had appointed.

In Athens, however, unity of command did not prevail among the *strategoi*. In one case, Marathon, command rotated from one general to another. In many cases, several generals shared command, the most senior or influential often dominating his colleagues, as Nicias did during the Sicilian expedition. Only for smaller efforts was one general commonly in charge; to clarify his powers, he might be termed an *autocrator*. Athens was perhaps less typical in this respect than in regard to numbers of officers. Any Greek *polis* with a single polemarch had a commander-in-chief, and Athens herself had one during the Archaic period.[69]

Save for some subordinate Macedonian officers, we have not yet noticed officers who were outside the command pyramid. Only at Sparta and among the Macedonians are some titles and duties of the supernumerary officers preserved. Every Spartan polemarch, or regimental commander, had his *symphoreis*, or aides-de-camp. The Spartan king had aides among those who are described as sharing the king's quarters, one that significantly was called "the public tent." The king's bodyguard of 300 *hippeis* must have included individuals serving as aides.[70]

For the army of Alexander, we know more than about the Spartans. Like earlier Macedonian kings, but especially like his father Philip, Alexander maintained a circle of companions, or *hetairoi*, who included many individuals with military responsibilities but without command of combat units. Some were famous men: the secretary Eumenes, the administrator Hephaestion, the engineer Aristobulus, the artillerist Demades, the armorer Gorgus. Under Philip, the *hetairoi* numbered 800, and under Alexander they surely grew in numbers.[71] The king's bodyguard, another elite, furnished commanders, too, among them Ptolemy, the future pharaoh of Egypt.

The term "companion" had a Homeric flavor, one especially appealing to Alexander, and important for understanding the *esprit de corps* that contributed to the success of Macedonian arms under these two kings. Just as the officers of the early modern period were "officers and gentlemen," the officers of Philip and Alexander were "officers and companions." Not even Sparta conferred this sort of identity on officers. To have done so would have kept officers and men from all being *homoioi*, or "peers." By the same token,

68 Pagondas: Thuc. 4.91.2. Epaminondas: Xen *Hell.* 7.5.4–14, 18–24. The collegial vote at Diod. Sic. 15.53 is either anomalous or erroneous. Pelopidas: Diod. Sic. 15.67.3, 15.80.1.

69 Aside from ten *strategoi*, two Athenian hipparchs (Arist. *Ath.* 61.4). Achaean dual command, one head of infantry, and one of cavalry: Kromayer and Veith 1928, 131–132.

70 For personnel in the "tent," see Anderson 1970, 68–70. Perhaps *hippeis* were the messengers at Xen. *Lac.* 13.5, *Hell.* 5.4.7.

71 Theopompus *FGH* 115 F 225, an estimate that may or may not include the highest-ranking army officers.

the Successors called their top soldiers and advisers *philoi*, a term comparable to Roman *amici*, but not to Homeric or Macedonian companions.[72]

Although Alexander killed a number of companions, he apparently did not degrade or expel any, and so the status of companion was permanent. The same stability may have characterized the officers in charge of the Spartan *hippeis* and the personnel of "the public tent." In contrast, Athenian and Boeotian generals held office for one-year terms. For the crime of serving longer than one year, Epaminondas stood trial.[73] Besides having more officers, and more kinds of officers, the Spartans and Macedonians thus had a more stable complement of officers. The one Athenian general to hold his post in the same theater of war for many years was Iphicrates, and he was commanding peltasts on the Persian payroll.[74]

As bears repeating, almost all ranks of officers had titles relating to the command of units in combat—leader of 50, 500, or 1,000, or leader of a band. One of the exceptions, enomotarch, referred to the swearing of an oath by squad members, and so this title, too, involved a unit. Only the terms for top commanders, such as *polemarch* and *strategos*, lacked this specificity. In modern terms, no Spartan colonel did anything but lead a regiment, and no lieutenant did anything but lead a platoon. Ancient Greek command was strictly functional, and the functions acknowledged were few. There was no general staff, nor any staff for units such as *taxeis*. Staff work was done, to be sure, but never specialized.[75]

The ancient Greek chain of command was modern enough that we have repeatedly used the terms "company" and "regiment," but not so modern as to allow "brigade," "division," or "corps." We have no occasion to use "fire team" or "battery." Greek units and their commanders mostly evoke the modern period as far as the mid-nineteenth century, a stretch of time that, unlike the Middle Ages, possessed organized infantry, and, like antiquity, relied on lines of men armed with spears or pikes, or, if gunmen replaced some or all of the pikes, a time that still let gunmen fix bayonets and fight like Homeric warriors. As with weapons, so with transport. The Greek trio of servants, mules, and horses dominated armies down to the advent of railroads, but not beyond. So also with the Greek trio of music, messages, and flags, communications basics prior to the telegraph and balloon.[76] And so also with the trio of infantry, cavalry, and artillery, which persisted until armor and airborne modified it in the twentieth century.

Having described numbers, units, and ranks, this chapter turns to explanations for why Greek armies were organized as they were, and why organization changed in the Hellenistic period.

72 Spendel 1915, 17–18.

73 Buckler 1980, 142–146.

74 Anomalous Iphicrates: Pritchett 1971–1991, 2.122.

75 Xen. *Lac.* 13 accounts for staff work now designated G 6, 7, and 9, or communications, training, and civil affairs, but not G 1–5 personnel, intelligence, operations, logistics, and planning, and G 8 finance. From 334 BCE onwards, Athens had G 7 through the training of ephebes, but had it in the fifth century only if the *peripolarchos* (Thuc. 8.92.2) trained as well as mustered the youngest soldiers. See Chankowski 2010, 139–142 with references. *Ephebeia* in fourth-century Thebes: Oehle 1905. In the Hellenistic period this institution spread yet became less important.

76 Flags as at Polyb. 2.66.11, but not standards, which were common in Near Eastern armies, as at Xen. *An.* 1.10.12.

Practical and Cultural Aspects of Organization

Some years ago, when Everett Wheeler divided ancient military historians into "Mad Hatters" and "March Hares," one group being practically and operationally inclined, and the other inclined theoretically and culturally, he used language that suggests two influences on Greek military organization and how it developed, one practical and one cultural.[77] The practical influence involves every physical factor that affected warfare—weapons, fortifications, terrain, climate, food and water supplies, animal transport, manpower, fitness, and training. These numerous factors fall into three groups, technical, environmental, and human. The chief technical factor was the prevalence of spear, javelin, and sword rather than longer-range weapons. This trio was easier to mass produce, and required less training, which is a human factor. By the same token, footmen were easier to train than riders, and horses—an environmental factor—were few. The result was a preponderance of infantry, not simply hoplites. Heavy and light infantry both were organized in a distinctive fashion, one with a long chain of command, and the other with a short one. The second most important technical factor was the impregnability of stone walls, a factor that changed only with the development of artillery in the fourth century. The siege train emerged as an important part of armies under Philip II, and it doubtless was organized in a distinctive fashion we cannot guess at.

The two characteristic items of hoplite equipment, the heavy shield and the metal helmet, have been an overemphasized technical factor. On the one hand, both were used in the Near East; on the other hand, neither was used by the Macedonian *pezetairoi*. These two items affected tactics, but they did not affect organization from top to bottom.

Terrain was a quintessentially Greek factor that reinforced the importance of both hoplites and light infantry. Hoplites could control Greece's narrow roads, fields, and orchards, while light troops could control rugged terrain. Save in the north, cavalry lacked room to operate, and so elaborate cavalry organization is reported only in Macedon, and may have existed only there and in Thessaly.

Together with the rugged terrain, the climate made Greek warfare largely seasonal. Philip and Alexander would break this pattern, but Alexander broke it most in the Near East, where the climate had made year-round fighting normal.[78] Seasonal fighting meant annual or frequent call-ups, and so units might readily be enlarged or reduced, another characteristic feature of Greek armies. Once again the Macedonians broke the pattern, but select units throughout Greece were maintained at full strength, the largest of these units being the Argive select *lochos* of 1,000.

Limited food and water supplies were one reason Greek armies were not large enough to have divisions and corps as opposed to regiments and companies, and they were also one reason that large armies stuck to roads leading to wells and markets. Armies would have been larger and moved more freely if the Greeks had developed quartermasters, signaling, surveying, cartography, and other logistics, but the lack of units and officers for these kinds of work shows that the non-combatant aspects of war, all too complex today, were rudimentary in Classical antiquity.

Another reason for small size was that Greek and Macedonian officers communicated with the troops by word of mouth and by primitive signals visible or audible at a distance of

---

77 Wheeler 2011.

78 The only full winter breaks occurred in Egypt, 332/331, and Central Asia, 329/328, 328/327.

hundreds of yards at most. Officers wore transverse crests on their helmets, or extra plumes, two distinguishing marks visible from only dozens of yards away.[79] These handicaps reinforced the limits imposed by terrain, climate, and reliance on citizen manpower.

Yet another factor limiting the size of armies was poor fitness. Gymnasia kept many Greek citizens fit, but hardly kept most adult males fit. Many or most free adult males of military age were isolated peasants, poor city dwellers, or others suitable for garrison duty. Lack of drill also reduced fitness. To be sure, Spartans, mercenaries, picked troops, and cavalry all drilled from the late Archaic period onward; Macedonians and Thebans later joined them, and so did peltasts under Iphicrates, Chabrias, and the leaders of the Ten Thousand. Most soldiers drilled very little. The Athenians did not introduce artillery drill until 332.[80] Deploying these ordinary men against enemies like the Spartans was risky—one reason that pitched battles against the Spartan *morai* were rare.[81]

Drill and other training cost money, and so these military options competed with the allure of mercenaries. In the Classical and Hellenistic periods, when Greek cities grew richer, they often preferred to spend their increasing revenues on mercenaries rather than on their own men, or even on their own artillery. This choice to shed the blood of hirelings rather than citizens leads us to the second, cultural explanation for Greek military organization. This explanation chiefly involves political factors, but also religious ones. The former varied greatly according to type of government, but the latter did not.

Political influence on military organization began with recruitment, which more or less responded to constitutional developments. At first, units were tribal, like society, and were little more than bands. Next, units were recruited according to newly organized civic districts, as seems to be the case in Sparta and was surely the case in Boeotia. In the case of mercenaries, the catchment area for recruits was larger than a tribe or *polis*. In the Hellenistic period, it was the whole Greek world, a change corresponding to the emergence of the Successor kingdoms, the kingdom of Sicily, and the large federations of Achaeans and Aetolians.

Politics also affected the selection of officers. The *poleis* elected not only all their top officers but also some if not all their lower-ranking officers. By the same token, the *poleis* subjected officers to trial in civilian courts. As Nicias said, disobeying the Athenian Assembly would lead to a trial before a civilian court with little understanding of the circumstances.[82] The likelihood that a defeated or merely unsuccessful general would be put on trial, and even charged with treason, encouraged many to go into exile, weakening Athens' military (and naval) leadership. Besides having too few officers, Athens had too many who feared for their lives.[83]

In Macedon, however, the king appointed all officers, not just top commanders such as Parmenio, and these officers faced military and not civilian courts.[84] In Sparta, the *gerousia*,

79 Wheeler 1991a, 140–141, to which add Alexander's extra plume (Plut. *Alex.* 16.7). The purple cloak worn by Macedonian generals (Plut. *Eum.* 8.12) would have been visible from farther away, especially if they were mounted. Eumenes gave purple garments to more numerous officers than had Alexander and Philip.

80 Marsden 1969, 70–71.

81 Another possible reason: a reluctance on the part of the *perioikoi* to fight far from home. No *mora* went to Asia Minor, or apparently went farther north than Chalcidice (Xen. *Hell.* 5.2.3; 5.3.9 are volunteers). A negative view of athletics: Alexander's, as at Plut. *Alex.* 4.5, *Reg. apophth.* 179d, *Fort. Alex.* 331b.

82 Thuc. 7.48.4. For similar passages, see Hamel 2015, 67–68.

83 As shown by the catalogues of Pritchett 1971–1991, 2.6–10, and Hamel 1998, 41–74.

84 The military court being the king's council of war, for which see Naiden 2018, Appendix 3, nos. 4, 23, 27, 28.

a court dominated by commanders and ex-commanders, tried officers in a more sympathetic spirit, and unwritten Spartan law made it easier to charge officers with offenses less severe than treason.[85]

Fundamental to Athens' political and legal difficulties was the commingling of diplomatic, military, and naval authority in the office of general. The college of ten generals was the most important part of the executive branch of government, and so the rival legislative branch regarded it jealously. The Athenians thought they had reason for this jealousy: The tyrant Pisistratus had begun his career as polemarch, or commander in chief, and so had other tyrants. To prevent another tyranny, the Athenians sacrificed unity of command. In Boeotia, similar fears confined the commander in chief to a one-year term.[86]

To compensate for weak leadership, Greek cities might hire mercenary commanders. From the fourth century onward, some of the best commanders were bandied about like modern professional athletes, working now in one city, now in another.[87] To some degree, these men were military professionals—the products of a military ethos and training to be distinguished from the adolescent training, or *agoge*, of Sparta. By the time of the Successors, such men could pursue an informal military curriculum and commanders could read letters and memoirs, not to mention histories of a largely military character. This literature had no contemporary parallel anywhere except in ancient China.[88]

Throughout the Greek world, religion was intertwined with supreme command. In Macedon and Sparta, the commander in chief of the army was also the chief priest of the nation, and could use sacrifice and other rituals to justify his decisions and boost his prestige. As noted in Chapter 24 in this volume, Spartan kings manipulated sacrifices, and Alexander did more of the same, and did not confine himself to rituals. By having himself declared the son of Zeus-Amon, he manipulated his own identity, turning himself from national priest into international god. This use of religion failed to achieve its objective. Instead, discontent among the officers led to a conspiracy against the king, to the execution of Philotas, and then to the murder of Clitus, who was practically, if not officially, Philotas' replacement, not to mention the related execution of Parmenio, long the commander of the heavy infantry.[89]

In contrast to both Spartan and Macedonian kings, *polis* generals were not chief priests. At most they could conduct hepatoscopy to help them make some operational decisions, as described by Xenophon, who conducted this ritual frequently once he emerged as the leader of the Ten Thousand. Xenophon took pains to have his generals present during the ritual and

85 The *gerousia;* identifiable because of the sentence of exile imposed on some commanders, one only this court could impose. See MacDowell 1986, 148, citing Xen. *Lac.* 10.2, Plut. *Lyc.* 26.2. A charge other than treason: Thuc. 5.72.1.

86 The effect of a past tyranny on command structure elsewhere is hard to assess. Corinth, for example, had multiple *strategoi*, but their relations with the legislature are little known, even in the case of Timoleon (Plut. *Tim.* 3.2, Diod. Sic. 16.65.8–9, 16.66.1). A different view of this political factor: Harris 2015, stressing Athens but neglecting polemarchs elsewhere.

87 E.g. Agesilaus at Plut. *Ages.* 36.3, Diod. Sic. 15.92.2; the Theban Lacrates at Diod. Sic. 16.44.2; and the Theban Pammenes at Diod. Sic. 16.34.1.

88 Including not just Sun Tzu but six other works translated by Sawyer 1993. For India at this period there is only the heterogeneous *Arthashastra*. For Assyria, remarkably, there is no evidence of military literature written for or even about officers other than kings. The origins of the Roman genre of *commentarii* are unknown, but the first securely known military memoir is that of M. Aemilius Scaurus, as at Cornell 2013, 3.18 F 3, 5, 7.

89 Misuse of religion being essential to the "orientalism" criticized by Droysen 1833, 145–148. Later views of the subject, including Badian's own critique of "orientalism": Badian 1996.

thus avoid religious disputes. This way of handling ritual strengthened the chain of command rather than disrupting it.[90]

If practical factors tended to accentuate the resemblance of Greek armies to later ones, these cultural factors tended to accentuate differences. Alexander's or Iphicrates' early modern counterparts, such as Frederick the Great and Napoleon, were not national priests, military magistrates, or mercenary captains. By the mid-nineteenth century, only one major army elected any of its officers, the volunteer forces of the United States. The experience of the Civil War convinced the Americans to abandon this important political link between modern army organization and the organizations of Athens and the other Greek cities that inaugurated democratic warfare. The future belonged to officers resembling the quasi-professionals of Macedon, Sparta, and the Hellenistic monarchies.

Concluding Remarks

Organization has a bearing on various topics in Greek military history. A man in a Macedonian phalanx, or in a Thessalian cavalry diamond, could not be brave in the same way as his counterpart in a Homeric band. Except in Macedon, ideals of Homeric companionship gave way to ideals of solidarity and uniformity that corresponded to standard units. Organization even affected religion, for military leaders could either serve as priests or consult them. Priests such as the king of Sparta enjoyed a psychological advantage, but bore a corresponding responsibility.

The nineteenth-century writers who pioneered this subject took organization for granted, and proceeded to analyze it. For recent writers, organization is a social marker as well as an operational essential. The Greek aspect of Greek military organization is as important as the military aspect.

90 Naiden 2013, 175–181. For other aspects of religion and warfare, especially the effect of religion on the experience of combat, see Chapter 24 in this volume; for religion and propaganda, see Chapter 31 in this volume.

CHAPTER 10

Generalship

Joseph Roisman

This chapter focuses on the expectations and demands of Greek generalship and how they evolved over time. In spite of changes, military leadership retained important basic characteristics that went back at least to the age of Homer. Space constraints and the range of the topic have forced us to exclude from the discussion (unless relevant) commanders ranking below generals, and naval generalship, which resembled command on land in many respects.

Generalship in Homer's *Iliad*

In full awareness of the contested historicity of what is known as the Homeric world, I join scholars who view it as largely historical and who date many of its aspects to the Dark and early Archaic Ages. I also agree with those who grant the masses of soldiers in Homeric battle scenes a greater role than suggested by the poet's focus on individual heroism.[1] This is not to deny the military contribution of prominent individuals. In his illuminating publications on Homeric warfare, H. van Wees has argued that most of the fighting was done by men of high social rank and military prowess, accompanied by relatively small bands of companions, attendants (***therapontes***), and followers who came from their area. They were distinguished for being "front fighters" (***promachoi***) in combat of a fluid nature. They moved about the battlefield freely, coming, going, and joining others as they saw fit.[2] But these warriors also

1 Latacz 1977, and in part van Wees 1997. Still relevant is Albracht 2005 (German original: 1886).

2 Van Wees 1986, 1988, 1992, esp. 61–258, 345–405, 1997. Homeric battlefield: e.g. Cuillandre 1944; Clay 2011.

A Companion to Greek Warfare, First Edition. Edited by Waldemar Heckel, F.S. Naiden, E. Edward Garvin, and John Vanderspoel.

led groups larger than their immediate fellowship. Moreover, scholars have given less attention to the significant similarities between these and later generals.

The Homeric commander, or king, was a military, social, and political leader whose authority was based on his preferred position and status within his kinship and friendship groups, his local power, and his personal prowess as a warrior and/or wise counselor. He often served as the chief identifier of his followers, and his performance and reputation reflected on them. He was rewarded accordingly with greater honors and material possessions.[3] Agamemnon, who may be described as the supreme commander of the Greeks, had all these attributes, but also acted under considerable constraints. To exercise power in his coalition army he needed the cooperation of other kings, some recognized as his betters in fighting (e.g. Achilles) or in wisdom (e.g. Nestor). His competition with, but also dependence on, other leaders explains why he often seeks their approval before taking action (2.369–374) and why he is not always the initiator of a military plan (e.g. 7.323–344). When he differs from other members of the military elite, he cannot overrule them (e.g. 9.1–87; cf. 14.81–134). Hector, the chief commander of the Trojans and their allies, also practices leadership by advice, although he seems to grow more authoritarian as the fighting progresses.[4] Athenian generals and even some Spartan kings of the Classical Age would have understood this kind of contested-but-collegial generalship.

While persuasion seems to have characterized leadership within the elite, the relationship between commander and his troops was more asymmetrical. In one episode, Odysseus disciplines kings and other individuals of high status using persuasion and "soft words," but when he came across "one of the people," he resorts to verbal abuse and even blows. He uses the same means to silence one Thersites, described as a troublemaker who questions Agamemnon's right to riches at the expense of the troops' labor. Thersites even suggests that the Achaeans return home and let the king find out how useless Agamemnon is without his army (2.1–269). Similar challenges to the general's authority and privileges were voiced later in democratic Athens, but Homer and his elitist audience considered such complaints contemptible. In the epic, a united military elite could keep legitimate authority synonymous with high status, military record, and social and economic privileges.

Homeric commanders met additional challenges to defending their authority. First among their concerns was ensuring that the troops did not shirk fighting. Fear of death or injury, the desire to strip the enemy dead of their valuable weapons, and even the wish to go home moved many to let others fight at the forefront. Therefore, we find various leaders stirring up the troops' enthusiasm to fight, praising good performance on the battlefield, but also rebuking or threatening with death those who would not engage the enemy, before and during the fighting.[5] At times, the appearance and the participation of a commander in the fighting at the front boosts the morale of his troops (e.g. 7.4–7). Generals of later periods also exhorted their troops before battle, often with speeches full of patriotic themes. Like Homeric commanders, they invited displays of courage and appealed to the soldiers' sense of honor and fear of disgrace. It is less common to find them stirring up courage in mid-battle

3 Homer *Il.* 6.5–12; 12.310–323. Henceforth, all references in parenthesis in this section are to Homer, *Iliad*.

4 12.49–109, 210–250; 18.243–314.

5 E.g. 2.365–392, 439–440; 4.230–254, 268, 289; 5.519–532; 6.66–71 (booty); 8.273–299 (praise); 11.210–213, 285–290; 12.256–276; 15.346–351; see also van Wees 1988, 13.

or trying to prevent soldiers from keeping to the rear, perhaps because they left such duties to lower-ranked officers and fellow soldiers.

Homer says little about preparations for battle. We are not told who placed a leader and his company in the battle lines, and even if the choice fell to the leader in question, he probably consulted the supreme commander. Nevertheless, there is enough in the epic to show that Homeric commanders shared with later generals the responsibility for planning and organizing the fighting. They divided their force into often pre-existing units and appointed commanders to them in a way that facilitated control and execution of military tasks.[6] The Cretans, for example, had one leader at the front and another at the rear. Such commanders did not always stick to their positions, but the initial arrangement ensured unity of purpose and probably cohesiveness (4.250–254). In an admittedly rare case, commanders divided guard duty among seven leaders, gave them equal-size units, and later inspected to see if they did their job properly. It suggests a chain of command and follow-up supervision.[7]

Like later generals, Homeric commanders sought military intelligence about the enemy's intentions and dispositions before acting (10.250–411). Nestor arranges his army before battle by placing chariots and horsemen in front, infantry at their back, and cowards in the middle so they would be forced to fight (4.291–308). On another occasion, he proposes defensive measures for the Greek camp that allow both good protection and movements of forces in and out (7.323–344). Hector makes plans for his forces besieging the Greek camp, and for his city at a distance, that ensure rest and food for the army and the horses, and prevents a Greek flight or surprise attack (8.489–543; cf. 19.154–183). Rudimentary as these plans are, they belie the disorganized and spontaneous style of warfare ascribed to Homeric armies and generals. Military historians have argued that Classical Age commanders evolved from warrior generals, who merely set the battle lines and then joined the fighting, into battle directors, who made tactical and other changes after the fighting began.[8] But Homer suggests early examples of the latter art of command, mentioning leaders whom men look to (4.322), and who go where help is needed (11.147–152, 521–542) and are expected to reverse defeat and lead or withstand a counterattack, especially by raising courage and morale (5.460–609). Together these actions suggest that Homeric generalship had more in common with later military leadership than is sometimes supposed.[9]

Generalship in the Classical Age

The Archaic Age, c. 800/700–500, formed a significant chapter in Greek military history, witnessing the evolution of the Greek *polis* and its army into their Classical form. The wars of the mainland Archaic Greeks often amounted to raids and border disputes. Military leaders tended to be aristocrats, elected to office as in Athens, born into it like the kings of Sparta and Macedon, or seizing power as tyrants. We shall not discuss, however, the Archaic art of command because the ancient sources on it are uninformative.

6 2.362–368, 474–477; 4.422–429; 9.50–87; 12.49–109; 16.168–173; cf. 3.1.

7 10.50–99, 180–193; cf. 12.86–104.

8 E.g. Wheeler 1991a; 2007.

9 See, however, Moore 2013, 457, who denies that generalship existed "in any real senses" in the Homeric epic.

Nevertheless, the sources' relative silence suggests a decline in individual contribution in favor of collective efforts and values.[10]

There was no "typical" general in the Classical Age, if only because personality played an important role in military leadership. Yet Classical generals shared a number of attributes and practices. They often came from privileged backgrounds, enjoyed public esteem, and, when successful, were granted honors and material rewards that distinguished them from lower-ranked warriors. Conversely, when unsuccessful or at fault, they could be held accountable for their performance and conduct, either through legal procedures as in Athens, Boeotia, and Sparta, or in the form of riots and discontent as in Syracuse under the tyrant Dionysius I and in Macedon under Philip II. It was desirable, though not necessary, for generals to have prior military experience. In states where generalship was collegial (e.g. Athens or the Boeotian Confederacy), and even where it was not (e.g. Sparta or Macedon), a military leader often consulted colleagues and other officers before making an operational decision. The ancient sources emphasize the commander's need to prepare for the campaign and be involved in the details of its execution. Generals mobilized and trained the forces at least of their native city. They were supposed to ensure that their troops were well disciplined, well supplied, and properly camped. It was equally important that the soldiers' morale be high and that the commander enjoy their good will (fortified by rewards). Generals were also expected to study the terrain, obtain good intelligence about the enemy, establish effective communication with their subordinates, and plan surprise attacks and other stratagems when needed.[11]

Before battle, generals used divination in search of good omens to alleviate anxieties about the outcome, and were expected not to test fate when the omens were bad. Like Homeric leaders, they exhorted the troops before setting up the lines of battle (often placing themselves on the right wing with their best forces) and joining the fighting. Although the Classical art of command expanded to allow the general greater influence over, and better reaction to, the dynamics of the battle, this kind of "general-soldier" never disappeared, and neither did the values associated with his role. Like the Homeric hero, he was supposed to be courageous, ambitious for honor for himself and his city, bellicose, and competitive, all attributes that inspired the troops and strengthened their loyalty to him. Also like Agamemnon and Hector, many generals led coalition armies and employed troops of different military proficiencies, over whom they had to establish their authority. Unlike the Homeric heroes, however, the Classical general was not under pressure to seek out and excel in personal encounters with the enemy, and he was much more involved in coordinating movements of different forces. In modern parlance, he was a multitasker. He fought at the head of large armies and small units, on land and sea, in the open field and in urban settings. On campaign and in battle, he was supposed to adapt his plans and actions to changing challenges, and be alert and opportunistic. The last traits indicate the importance the ancients ascribed to the role of personality in generalship, as opposed to the modern preference for structural and institutional perspectives.

These observations apply to military leaders on both the strategic and the tactical level. Indeed, some generals combined the role of tactician and strategist by shaping their cities'

10 Archaic warfare: Singor 2009; Schwartz 2009; and Chapter 1 in this volume.

11 Ancient sources on generalship, e.g. Xen. *Mem.* 3.1.7, 2.1–4, 4.1–12, 5.22.23; *Cyr.*, e.g. 1.6.9–46, 2.1, 3.36–31; Onosander *The General;* and see Polybius. Modern treatments: e.g. Anderson 1970, esp. 67–83; Wheeler 1991a; Moore 2013, 458–463; and Roisman 2017, which forms the basis for parts of this section.

policies and war plans. It will be useful, then, to illustrate how prominent Classical generals dealt with tactical and strategic challenges.

Pericles of Athens combined the role of commander of land and maritime operations with that of a strategist who prepared his city for the Peloponnesian war. He could play both parts because, in fifth-century democratic Athens, generals could secure their hold on office through reelection and were often leading politicians. Over a long career, he expanded and defended the Athenian empire by adding allies, founding colonies, and subjugating cities deemed a threat to Athenian hegemony, such as Samos. In the Samian campaign, facing a fortified city with a strong navy and allies, he had to split his force to meet a Phoenician fleet that was rumored to be coming to Samos' aid (but in fact never did). Pericles overcame these challenges by exploiting his larger fleet and deeper pockets as well as a new siege technology, which he employed to breach Samos' walls.

As a strategist, Pericles relied on what he believed were Athens' undisputed advantages: its mighty navy, its wealth, its strong fortifications, and an empire that provided revenues and allies. Acknowledging the Peloponnesians' superiority on land, he refused to meet them in battle when they invaded Attica. Although he tried to contain their devastation of the countryside with his cavalry, and sent his navy to raid the Peloponnesus from the sea, he essentially waited the invaders out. Lest Athens get entangled in more conflicts, he advised against expanding the empire, hoping the Spartans would sue for peace when they realized that they could not win. Although Pericles prevailed, he had to modify his strategy significantly, allocating more capital and defending the city more robustly than he had planned. He also overestimated Athens' financial resources and the Spartans' desire to end the war. Pericles' risk-averse personality affected both his generalship and his strategy. As a general, he preferred to use overwhelming force, which was not always affordable, and his plan of curtailing Athens' military involvement in new conflicts was not always feasible.[12] The strategy was largely unsuccessful, partly a victim of opposition from Athenians who wished to confront the invaders.

Unlike Pericles', the careers of most generals in the Classical Age were limited to tactics. Demosthenes, son of Alcisthenes (not the famous orator), was one of Athens' leading generals in the Peloponnesian War.[13] It has been argued that he advocated a new strategy of bringing the war into the territory of Sparta and her major allies. But Demosthenes was more a tactician than a strategist (he was also far less involved in politics than Pericles) and his leadership demonstrated the adaptability that was required of an active campaigner in that period. He suffered a heavy defeat in his first campaign in Aetolia (426) because he was ill prepared for an encounter with skillful javelin throwers, had poor military intelligence, and fought on unfamiliar terrain. He also received limited local support because he failed to reconcile his allies' conflicting agendas. His next engagements in the region (426), against joint forces of western Greeks and Peloponnesians, showed that he had learned his lesson. He won two consecutive victories by making good use of the terrain to ambush numerically superior forces. He coordinated the movements of allied heavy and light infantry and, thanks to a good military intelligence, was able to ambush, predict, and counter the enemy's actions.

Demosthenes faced a different set of problems at Pylos in Messenia (425). After an Athenian victory over a Peloponnesian fleet in a nearby bay, the Athenians put the small

---

12 Pericles: e.g. Kagan 1991; Lehmann 2008; Roisman 2017, 101–143.

13 Demosthenes: Roisman 1993; Wylie 1993; Cawkwell 1997; Roisman 2017, 145–185.

island of Sphacteria under a maritime blockade. There was a Spartan force of about 420 hoplites and an unknown number of helots on the island. After a fire exposed these troops to view, Demosthenes and his colleague, Cleon, formed a plan of attack that took into consideration the enemy's size and positions, its manner of fighting, and the island's topography. It also exploited the Athenian numerical superiority in hoplites and especially light infantry. First, Demosthenes surprised the enemy by landing on the island at several places under cover of darkness. Refusing to engage the Spartans in a hoplite battle, he skillfully employed his light-armed troops, who shot at the Spartans from the surrounding hills and used hit-and-run attacks. The surrounded Spartans suffered heavy losses and exhaustion. After short negotiations, the 120 surviving Spartans surrendered, to the astonishment of many Greeks, who wondered why they did not fight to the death like the famous Leonidas and his Spartans at Thermopylae. Their capture damaged Sparta's reputation and gave Athens an important asset: the Spartans stopped invading Attica out of fear for the fate of their prisoners. As for Demosthenes, his subsequent career was far less successful, and he met his death in Athens' failed Sicilian expedition (413).

The career of Dionysius I, tyrant of Syracuse in 406–376, illustrates a different but common challenge that faced Classical generals: siege warfare.[14] At war with Carthage in 397, Dionysius besieged its ally Motya in western Sicily. Motya was a well-fortified city on a small island in the middle of a lagoon, and Dionysius brought a large army and navy against it. Shortly after his arrival, a Carthaginian fleet coming to Motya's aid surprised him, destroying his merchant ships on its way there and then blocking the exit from the lagoon. Using his engineering knowhow and the many hands available to him, Dionysius built a wooden path on his side of the lagoon that enabled him to haul 80 triremes to the sea. He then coordinated an attack on the enemy from his ships and from the nearby land, by means of projectiles, including catapults, which Dionysius introduced to the world. The Carthaginians fled for home and Dionysius was free to focus on Motya. He constructed a mole to the island from the Sicilian shore and used it to roll rams and siege towers that dismantled the walls and rained arrows and missiles on the defenders. The Motyans used inventive countermeasures and were able to oppose him even after their walls were breached. They also made death traps of the bridges that Dionysius built for his infantry to attack the defenders. The tyrant then employed a stratagem that misled the enemy into thinking he had temporarily halted fighting while, in fact, sending a unit to seize a bridgehead from which he was able to capture and plunder the city. The odds against the Motyans were uneven to begin with, but Dionysius proved himself one of the best generals of the Classical Age. Undeterred by setbacks, he used his considerable organizational skills, superiority in numbers, units of different military proficiency, and military technology to overcome both resourceful enemies and obstacles natural and artificial.

Dionysius was a member of a small group of tactical innovators, while the majority of Classical generals remained conventional in their manner of command and warfare. A few notable exceptions deserve mention. In 390, in the battle of Lechaeum, the Athenian general Iphicrates improved upon Demosthenes' aforementioned tactics at Sphacteria by using light-armed peltasts in hit-and-run attacks against 600 Spartan hoplites and their cavalry companions. Fully exploiting the traditional Spartan hoplites' vulnerability to attacks on their unprotected side, their limited persistence in chasing the enemy, and their disorderly

14 Dionysius: Caven 1990; Roisman 2017, 227–273.

regrouping, Iphicrates sent against them his nimble, elusive peltasts. These troops attacked with their javelins and disengaged when the enemy got too close, thereby nearly annihilating the Spartans. Iphicrates is also credited with the creation of a phalanx-like infantry in arms and tactics (370s), later perfected by Philip II and Alexander.[15]

The Theban Epaminondas, too, had the limitations of the traditional Spartan style of fighting in mind when he faced them and their allies in the battle of Leuctra (371).[16] Generals commonly concentrated their hoplite force (supported by light infantry) into a stronger right wing that faced the enemy's weaker left. When possible, the phalanx moved to the right to outflank the opposing left wing and attack its unprotected flank or rear. Epaminondas went against these conventions. In spite of his numerical inferiority, he placed many more troops in his left wing than the Spartans had in their strong right, putting his best unit, the elite Sacred Band, at their head. He then led his left wing at an angle that encouraged the enemy to stretch its lines and try to outflank him. The Spartans' movement created a gap between them and their allies, exacerbated by confusion from an earlier defeat of the Spartan cavalry and their retreat into the phalanx. Seizing the moment, Epaminondas went on a frontal and outflanking attack that inflicted on the Spartans the worst defeat in their history, leading eventually to their loss of hegemony in Greece.

Arguably towering over all these generals was the Macedonian king, Philip II (reign: 359–336).[17] Like Pericles, Philip was both a strategist and tactician. In broad terms, his strategy centered on defending his country from predatory neighbors and other powers, securing it by expanding his control over neighboring territories, and creating a Macedonian hegemony in Greece. His means were military force, diplomacy, and personal ties with local leaders. His tactical measures were more original. He turned a weak Macedonian army that had just suffered a devastating defeat into an efficient and much bigger fighting force. He created a well-trained infantry phalanx equipped with relatively light armor and a *sarissa*, a pike that was longer and hence deadlier than the hoplite spear. The phalanx moved in compact formations that were sufficiently flexible to protect it from attacks and to penetrate the opposition. Philip also increased the size of his cavalry and trained them to attack not just their counterparts, as was commonly done, but also infantry lines by using a wedge formation. Furthermore, he utilized specialized units such as archers, javelin throwers, and siege experts. The result was an army much more professional than its contemporaries, thanks to its discipline, training, mobility, and, above all, experience gained in many campaigns. Philip too learned primarily from his own experience to become a versatile, effective military leader.[18]

In addition to tactical changes, two other major developments affected the character and expectations of Classical generals. Both had to do with the nexus of military leadership, politics, and finance. In Greece, the correlation between military and political leadership was as old as Homeric kingship, and for much of the fifth century, generals in democratic Athens continued to double in politics and the military. From the late fifth century, however, these domains diverged, and many generals' careers began to center chiefly or exclusively on the

15 Sekunda and Burliga 2014.

16 Epaminondas in Leuctra: e.g. Tuplin 1987; Hansen 1988 (questioning Epaminondas' tactical contribution); Buckler 2003, 289–295; Roisman 2017, 286–298.

17 I favor the traditional ascription of military reforms to Philip, and see Müller 2010, 168–169. For the following see Griffith 1980; Gabriel 2010, 54–96; Sekunda 2010, 449–452.

18 See Chapter 4 in this volume.

military.[19] Some found employment with foreign rulers, often with the blessing or the tolerance of their native state. For example, Iphicrates of Athens and Charidemus of Oreus commanded mercenary forces for Thracian rulers and Persian officials, and the Athenian Chabrias fought for rebel Egyptian kings. Even in exile, mercenary generals tried to avoid conflicts of loyalty or interest with their home states, while trying to gain their followers' personal loyalty. This was, for example, the case with some of the Greek commanders who served in Cyrus the Younger's unsuccessful expedition against his brother, the king of Persia.[20]

Mercenary generals had a paymaster, but for other generals, financing a campaign or war was a major concern. Shortage of funds could prevent the launching of a campaign or abort it, and affected the discipline and loyalty of citizens and mercenaries alike. The state might allocate monies to military actions, and the Athenian general was legally accountable for how he spent them. Yet insufficient or unavailable funding often forced generals to fend for themselves. Some excelled as fundraisers. The Athenian Alcibiades spent much of his time in office persuading or compelling allies to contribute to the Athenian war chest against Sparta. Across the lines, the Spartan Lysander was able to rebuild and maintain the very large navy with which he won the war, thanks to financial support from the Persian authorities, with whom he maintained an excellent relationship. Some generals raided foes and even friends for income, and so did their troops, showing that not much had changed since Homer. We hear of Athenian fleet commanders charging cities and vessels "protection" money, and the resourceful Iphicrates sold captives and hired out his sailors as field workers in Corcyra. Generals might line their own pockets, but they were also expected to bring loot or money home.[21]

In sum, generals of the Classical Age were expected to prepare their armies materially, physically, and mentally for conflict. Military intelligence and experience told them what to expect of the enemy and how to use tactics that might include an element of surprise or similar stratagems. When possible, they shaped the course of battle by responding on the spot to emerging challenges while providing an example of martial valor.

Hellenistic Generalship

In his account of the Second Punic War, the Hellenistic Greek historian Polybius digresses to offer reflections on what makes a successful general. In his analysis, meticulous planning and preparation is a general's most important skill. The commander must keep his plan and other information to himself, and know how day, night, and the different seasons affect the movements of troops. He is required to inspect the terrain for the right place and time for action, to arrange for secure means of communication, and to study applicable sciences such as astronomy, geometry, and mathematics (Polyb. 9.12–21). There was little in Polybius' practical approach that other military authors, or Greeks of earlier times, would have found new or objectionable. Indeed, Hellenistic generalship largely resembled its predecessors.[22] In

19 Isoc. 8.54–55; Plut. *Phoc.* 7; Lengauer 1979; Hamel 1995, but see also Tritle 1993; Roisman 2017, 9–14.

20 Hired generals: Pritchett 1971–1991, 2:59–116; Trundle 2004, 75–79, 134–135, 148–159.

21 Alcibiades: Rhodes 2011, 55–94; Lysander: Roisman 2017, 189–198. "Protection" payments: Demosthenes 2.28; 8.24–29. Iphicrates: Diod. Sic. 15.47.7; Xenophon *Hell.* 6.2.37; Polyaenus 3.9.30. Profits: e.g. Demosthenes 3.31; 20.77, 80. See also n. 20.

22 See also e.g. Baker 2003 and Chaniotis 2005, 31–36, as opposed to Roth 2007.

Hellenistic sources, as in earlier ones, the good leader is self-controlled, diligent, and a problem solver. He manages the campaign in a way that exploits his advantage, uses stratagems when at a disadvantage, and shows an ability to recover from setbacks. Like the Homeric hero, he is brave and ambitious to excel on the front line. Also like some of his predecessors, success at war could result in and justified the general's attainment of supreme power (as a king), and encouraged the prosecution of an aggressive, expansionist policy.[23]

Hellenistic generalship encompassed traditional elements with an openness to new means of war. These included war elephants often placed with light infantry in the front line against opposing elephants, infantry, and (more effectively) cavalry.[24] Generals also used war chariots to break enemy formations at the beginning of a battle, sometimes unsuccessfully, as at Magnesia. They used catapults with greatly improved fire power in battles and especially in sieges. Demetrius Poliorcetes' famous siege of Rhodes included unprecedented huge towers. Less dramatic was the lightening of the infantry's panoply. None of these developments, however, revolutionized warfare and its basic rules of engagement, especially on the city state level. More significant changes occurred in the big Hellenistic kingdoms, but even these had more to do with size than substance. The kings' vast territories and their frequent wars led to the creation of standing armies and spread-out garrisons. They were also forced to delegate command to subordinate commanders, especially when engaged in simultaneous campaigns. Distant wars involved logistical challenges such as long supply lines and the need to carry provisions, because living on the land was not always sufficient. Kings also needed to generate revenues for themselves and their subjects, sometimes resorting to raids and the collection of war booty, in which, according to one scholarly view, success or failure could make or break a king. Yet disloyalty and indiscipline among troops and officers could equally originate in political, personal, and other noneconomic grievances.[25]

The diverse nature of Hellenistic military leadership in action can be shown through the two ends of its continuum. On one side is Alexander the Great, who ushered in the Hellenistic age, and whose battle against the Indian ruler Porus mixed old and new styles of fighting and command. On the other is the third-century Athenian general Epichares, whose achievements were less than negligible in comparison, but important enough for his contemporaries to shower him with honors.

In 326 Alexander clashed with Porus on the River Hydaspes.[26] The Macedonian king enjoyed a numerical and especially a qualitative advantage in troops, of which, as we shall see, he took full advantage. With an army of about 30,000 infantry, 4,000 cavalry, 300 chariots, and between 85 and 200 elephants, Porus waited for Alexander to cross the river. To meet the double challenge of natural obstacles and military opposition, Alexander made meticulous preparations that would have made Polybius proud. He kept his plans secret, and built a main, visible camp that led Porus to expect him to cross from there. Like Lysander at Aegospotami or Dionysius at Motya, he established a routine suggesting that

23 The general's qualities: Beston 2000. From generals to kings: Chaniotis 2005, 57–62. For the following see Sabin 2007, esp. 402–416, 430–433.

24 See Chapters 7 and 15 in this volume.

25 Raider kings: Austin 1986; Roth 2007, 371–372. *Pace* Roth 2007, the "indirect approach" strategy was not a Hellenistic or even fourth-century innovation, but can be recognized already in Pericles' and Demosthenes' generalships.

26 On the battle of the Hydaspes, sources, and bibliography see e.g. Bosworth 1996, 9–21; Worthington 2014, 243–250; Howe 2016. For the figures of the opposing armies, see Heckel 2008, 160–162.

he had no immediate plans of attack. The crossing took place from a hidden, distant place and at an unexpected time. In an impressive logistical feat, Alexander had transferred dissembled boats over a distance of about 160 miles. He used them and other transportation to cross the river under cover of night and bad weather, first to a mid-river island and then to the other bank. With him came 6,000 heavy infantrymen, about 1,000 lighter infantry, and 5,000 Macedonian and Asian cavalry. Because these forces were "the strongest part of his army," as the historian Arrian puts it (5.15.3–4), he could afford to leave behind many more troops, who were instructed to join the battle only later. So far, Alexander's generalship showed clever use of topography and weather, logistic prowess, tactical deception, and the taking of a calculated risk.

He initially encountered Porus' son, who led an advanced force of cavalry and chariots, each carrying five archers and a driver. Alexander attacked first with Asian mounted archers, who harassed and put the Indians on the defensive, allowing Alexander time to organize. After ascertaining the size and composition of the enemy and its separation from the main Indian force, he formed a plan of attack on the spot, using his larger cavalry to overwhelm their opponents. The defeat was complete, partially because the Indian chariots got stuck in the mud and could offer no help. Both in this engagement and in the subsequent battle, Alexander showed no tactical innovation, only good use of familiar maneuvers. It was the Indians who encountered something new in this encounter, unused as they were to cavalry attacking in squadrons.

After learning of his son's defeat, Porus moved his army to less muddy ground. He placed his elephants at intervals as the first line of offense or defense, stationing the infantry behind and among them, and the cavalry on his wings fronted by chariots. Except for the presence of elephants and chariots, Porus' arrangement resembled traditional Greek battle lines, making the challenge Alexander faced less than formidable, especially as Porus left him the initiative in charging. After giving his infantry a rest from their quick march (another good score on the Polybian scale of generalship), Alexander again used his superior cavalry to decide the battle, and in a way that accurately predicted the enemy's reaction. He attacked Porus' cavalry on the right and sent a cavalry force to the left to pursue expected reinforcements to the attacked wing, instructing his infantry commanders to attack the Indian elephants and infantry only when they saw them in disorder as a result of his charge.

At first, Alexander harassed and confused the cavalry on the enemy's left with his Asian mounted archers. He then forced them to fight at the front against his cavalry and at the rear against the units chasing the Indian reinforcements from Porus' right. Opportunistic as ever, he hit them when they were disorganized and rearranging their lines. The Indian cavalry retreated into their infantry and elephants, which were now in disorder and sandwiched between Alexander's cavalry and infantry. The elephants turned around to fight Alexander's cavalry and were attacked with perfect timing by his Macedonian phalanx, who used their sarissas against the mahouts and their animals (the Macedonians already had some experience in fighting elephants). Alexander moved in for the kill, confining the enemy with his cavalry on three sides and his solid phalanx at their front. The result was mass slaughter. Porus surrendered in a dignified manner, and Alexander kept him on the throne and even expanded his realm. His generous treatment of Porus was intended to encourage neighboring rulers to choose surrender over war, while his imperial strategy relied on local leaders who were dependent on him for power and territory. Tactically, the battle on Hydaspes shows that, remarkable as Alexander's generalship and battle management was, he owed much of his success to an army that operated like a well-oiled machine—and to an enemy that was its mirror opposite.

About two generations later, the Athenian commander Epichares served in what is known as the Chremonidean War (260s) between the Macedonian King Antigonus Gonatas and an alliance made up of Athens, several other Greek states, and the Egyptian King Ptolemy II. We know of Epichares' career from a partly preserved decree that the frontier deme (town) of Rhamnus proposed in his honor.[27] It appears that Epichares' main virtues as a military leader amounted to defending the essential food resources of the city and the countryside, and to using his personal wealth to help citizens and others. At the time of his generalship (268), Athens' situation was difficult: a Macedonian garrison controlled its harbor of the Piraeus, and Antigonus was besieging it and raiding its territory. Epichares, whose military experience included commanding the Athenian cavalry, was in charge of the coastal region that included Rhamnus. At the outbreak of hostilities, he took several measures to protect the local food supply, moving grain and other produce to a safe place in Rhamnus, and buying 500 *medimnoi* (25,500 kilos) each of wheat and barley, which he sold at the official price to citizens and soldiers. His purchase alleviated anxiety about the grain supply, a constant worry at Athens since at least the Classical era. Epichares also protected local production of food by stationing lookouts to warn harvesters of raiders and to summon help. He built a stoa where the guards could rest and be readily available, added two watch towers, and increased the number of watchdogs and provided them with food, all at his own expense. His defense of the region also included punishing residents who guided incursions by raiders or pirates. Epichares then negotiated the ransoming of citizens and slaves, thereby helping to restore local productivity. Lastly, he provided shelter and provisions for troops sent by Ptolemy to help Athens, sparing citizens the need to quarter them in their houses.

It is noteworthy that, according to the extant decree, the closest Epichares came to any danger was his presence among the lookouts in the country. Yet even though he was unlike the Homeric generals and Alexander in personal bravery, he enjoyed his peers' recognition and honors. For his fellow Athenians, civic-mindedness and public benefaction were as important as courage and battle prowess in a military leader. Epichares, who could not boast the latter virtues, was still an admired general. It did not mean that the definition of good Greek generalship had completely changed, but that it was flexible enough to incorporate new characteristics.

27 Epichares' inscription in translation: https://www.atticinscriptions.com/inscription/IRhamn/3. See also Habicht 1997, 144–145; Oliver 2007, 128, 130, 139–140.

CHAPTER 11

Greek Camps and Camp Followers

Mauricio G. Álvarez

To my daughter Sonali, who has fulfilled Alexander the Great's dream in my family

In contrast to the numerous publications dealing with Roman *castra*, there is a near-total absence of any similar works on their Greek counterparts. For a long time, researchers have mostly either avoided the question or claimed only that Greek armies were organized in such a rudimentary fashion that nothing of interest can be said about them.[1] The explanation for this unusual situation is to be found in Polybius 6.42, where the historian states that ancient Greeks, unlike Romans, did not have a single unvarying type of camp, preferring to adapt their camps to the natural advantages of the ground on which they were sited. This section, it should be noted, follows his detailed description of the way in which Roman camps are laid out (6.27–41), a text that has no parallel in the case of the Greek world.

Because of the technological revolution in warfare since the beginning of the twentieth century, we have lost sight of the important role played by army encampments in ancient times.[2] Militarily speaking, it is inconceivable that an army (whatever its nature, and however limited its military knowledge may be) should lack a minimum degree of organization in this sphere. Only thus is it possible to guarantee that the troops can recover their energies by being able to rest in a secure and suitable location at the end of each day of campaigning. For these reasons we can safely say that the way in which a camp is laid out always constitutes a visual expression of the nature of an army, including its level of organization and professionalism. Hence the military machine of the Romans accompanied a corresponding perfect organization of its camps, the changes undergone

1 Bibliography in Álvarez, 2013, 23–34, the first monograph on this issue; solely on Alexander's army, Karunanithy 2013, 186–206; also see Alonso and Álvarez 2017.

2 Lochée 1778, 67 ("On the choice of camps and posts, frequently depends the success of a campaign, and sometimes even of a war"); also, Ferraz 1801, 1; Veg. 1.1.

A Companion to Greek Warfare, First Edition. Edited by Waldemar Heckel, F.S. Naiden, E. Edward Garvin, and John Vanderspoel.

over time by the former being reflected in the latter.[3] From this perspective, it is therefore unacceptable to think that such professional armies as those of the Greeks (and those of Macedon or Sparta in particular) would not have introduced some degree of order into their camps.[4]

Moreover, we know that in the fourth century Aeneas Tacticus wrote a book on the art of encampment, *Stratopedeutike Biblos*, which has not survived, this being the first recorded military manual devoted exclusively to the subject (Aen. *Tact.* 21.2). It enables us to suppose that by the late Classical period Greek armies had developed methods of organizing their camps to a sufficient extent as to justify a treatise.[5] The purpose of this chapter is therefore to show to what extent, after collating and analyzing the available sources, it is possible to describe the Greek art of castrametation.

Setting Up and Laying Out a Camp

The criteria followed by Greek generals when setting up a military camp—*stratopedon* being the Classical Greek word—were simple.[6] From a logistical point of view, they looked for a nearby spring, pond, or river that would ensure sufficient supplies of water for all the troops, located wherever possible close to towns or villages where food could be acquired or requisitioned from the local population.[7] Tactically, they preferred to site the camp on a hillside or hilltop, to make it more easily defended.[8] If the camp was located on level ground, then it was always better to improve its defensibility by using available ditches, gullies, watercourses, or walls, and even the *temene* of temples.[9] Greek generals also took into consideration the sanitary conditions of the location, which they reckoned to be of prime importance.[10] All these concerns were magnified in the case of a fleet looking for a place to anchor, ships in a harbor being more vulnerable than an army in the field.[11] Ships being beached were the most vulnerable of all, as shown at the battle of Arginusae.

3 Dobson 2008, 56–65; 71–72; 110–119.

4 See, e.g. Droysen 1889, 88–90; Delbrück 1900, IV, 3, 287–288; Adcock 1957, 62, n. 47; Saglio 1962, 940 941; Krentz 2007b, 162 167.

5 Xen. *Lac.* 12, and *Cyr.* 8.5.2–16 could be considered as earlier approaches, despite their singular nature.

6 For an in-depth analysis of the Greek castrensian evolution, from Homer to Alexander, see Álvarez 2013, 35–50, who also deals with its possible oriental origin.

7 On water supplies (a very important concern, as noted by Engels 1978, 18), see e.g. Hdt. 1.75.3, 9.25.1–3, 9.57.1; Xen. *An.* 1.4.1, 1.4.9, 1.4.10, 2.5.1, etc.; river waters were usually not enough to supply the Persian armies: Hdt. 7.21.1, 7.43.1, 7.58.3, 7.108.1; Diod. Sic. 9.22.4. Food requisitioned, see Hdt. 1.62.1, 1.76.1, 1.76.3, 5.63.3, 6.77.1, 9.114.1; Thuc. 3.107.1–3, 6–66.1, 7.78.4; also see Álvarez, 2013, 51–9.

8 Hdt. 9.19.2; Thuc. 2.20.1–3, 3.97.3–2, 3.11.1–2, 4.44.1, 4.46.1, 5.58.1–2, 7.82.3; a hilltop represented a better place to encamp than within the city walls: Thuc. 4.57.1–2. Xen. *Hell.* 6.4.4, 14; Xen. *Lac.* 12.1. However, this option was not without its problems: Xen. *Hell.* 4.6.7; Xen. *An.* 6.3.6–9; Polyaenus, *Strat.* 1.38.2; Diod. Sic. 11.22.4. Also see Álvarez 2013, 59–62.

9 Defenses: Hdt. 7.176.5; D.S. 11.30.5; *temene*: Hdt. 1.75.3, 6.108, 19.57.1; Diod. Sic. 10.28; Thuc. 3.96.1, 4.90.1–2; Plut. *Ages.* 14.2. The fairness of such behavior was debated: Thuc. 4.97.2–4, 4.98.1–6. Temples usually had sources of water, very convenient for an army on campaign (Guettel 1988, 161–165); also, Xen. *Hell.* 1.3.7, 4.1.41, 2.2.8, 5.2.25, 6.5.27. Álvarez 2013, 66–8.

10 Hdt. 9.41.2; Thuc. 7.47.1–2; Polyaenus, *Strat.* 2.30.3 (considered as the first example of bacteriological warfare in history). Xenophon was also familiar with such problems: Xen. *Cyr.* 1.6.15, 6.1.23. Arist. *Pol.* 1327a. It was thus essential to move camp frequently: Xen. *Lac.* 12.5; Veg. *Mil.* 3.2; Ferraz 1801, 141, 242–248; Gomme 1970, 424; Anderson 1970, 61, 161; Álvarez 2013, 62–64.

11 Hdt. 3.136.1; Diod. Sic. 9.12.5. Álvarez 2013, 64–66; Casson 1971, 85–92, 175–82; Morrison and Williams 1968, 289–313.

Although there is no general statement on this matter in our Greek sources, its importance was generally accepted and debated. Thus, Darius' ghost explains the Persian defeat as being due to poor judgment in this regard (Aesch. *Pers.* 790–794), while Thucydides explains the lengthy duration of the Trojan siege as a consequence of the erroneous decision made by the Greeks as to where to pitch camp (Thuc. 1.11). Furthermore, Herodotus shows a strong interest in all these issues, while Xenophon's *Anabasis* reveals their importance for the Ten Thousand.[12] Finally, Alexander's army had mastered this science, despite receiving a number of shocks and surprises due to the unfamiliar country they were traversing (Arr. *Anab.* 6.25). In the end, the decision, right or wrong, about where to establish the military camp defined many of the operations during the Peloponnesian War, and can be found behind momentous episodes in Classical history.[13] It is therefore evident that such criteria existed, and it was generally recognized that a good commander should take them into account.[14] The most striking thing of all, however, is that the rules we have just referred to are very similar to those that would be applied by the Roman legions a few centuries later, because the tactical or logistical needs of a regular army in antiquity were identical.[15]

The commander-in-chief was responsible for deciding where to set up camp.[16] Reasonably, he would be assisted by, or delegate his decision to, another officer who would be responsible for organizing the camp, with an advance guard being sent ahead of the main column of troops to scout locations.[17] No record of such an officer has come down to us, but clues indicate that it was the baggage train personnel, with its commander, who assumed this responsibility.[18] We can envisage them setting up camp in the rear during the Battle of Cunaxa, while the rest of the army is making ready to fight, so that the Greek soldiers, finally defeated, were able to return safely to their tents for dinner (Xen.

12 The importance given by Herodotus to logistical matters was also appreciated on the Persian side: Hdt. 3.9, 3.25.1–2, 3.25.4–7, 3.136.1, 4.203.2–3, 7.29.1, 7.58.3, 7.118.1, 8.22.1, 8.97, 8.115.2–3, 9.15, 9.41.2, 9.96–97. The relevance of choosing adequate locations is expressed by Artabanus in Hdt. 7.49.1, who foresaw the disaster at Magnesia: Hdt. 7.188.1–3.

13 Marathon: Hdt. 6.102.1, 107.2–3; Thermopylae: Hdt. 7.176.5, 7.207–209; very precisely described in the case of the Battle of Plataea: Hdt. 9.25–51; Diod. Sic. 11.30.4–5; Sicilian campaign during the Peloponnesian War: Thuc. 6.22, 6.42.1.

14 Xenophon sums them up in *Cyr.* 1.6.10, 5.4.40; mistakes made by Greek commanders in this regard were denounced: Thuc. 7.4.

15 On Roman armies: Ps.-Hyginus, 50; Veg. *Mil.* 1.22, 2.2, 8; Diod. Sic. 12.24; Le Bohec 2004, 180, 218; These criteria persisted until modern times: Lochée 1778, 16, 57, 65, etc.; Ferraz 1801, 129–133; Greeks supplying an army on campaign: Pritchett 1971–1991, 1.30–52; Engels 1978, 22; Hdt. 7.187.1; relying on private providers and personal initiative: Ar. *Ach.* 198–200; Xen. *An.* 1.5.5–7; the army became an effective weapon of destruction on friendly or enemy ground: Hdt. 7.118.1; Xen. *Lac.* 12.5; *Cyr.* 3.3.23; Polyaenus, *Strat.* 2.1.21; the baggage train, which accompanied the army, was a necessity and a major concern: Hdt. 9.39; Thuc. 4.128; Xen. *An.* 3.2–3.

16 Xen. *Lac.* 13.10; Arist. *Eth. Eud.* 1227a; implied in Hdt. 9.26.1.

17 The role played by scouts is clearly shown in Xen. *An.* 1.8.1, 2.2.4, 7.3.6; an ideal model is Gadatas: Xen. *Cyr.* 5.4.40; it was essential for fleets to know beforehand the most suitable places to anchor: Hdt. 3.136.1; so Hippias guided Xerxes to Marathon: Hdt. 6.102.1; also, merchants have always been a rich source of geographical knowledge for armies: Mayer 1995, 436.

18 Xen. *Lac.* 13.4; Probably, also, Lycurgus (*Leocr.* 59); Pritchett (1971–1991, 1.30–52) considers this officer among those who regularly gathered with the Spartan king before starting the march; among Alexander's troops there was an officer called *skoidos*, responsible for the military camp (Hesychius *s.v. skoidos*; Pollux 10.16 calls it *skeuophylax* and *skeuoros*); after Alexander, Plut. *Demetr.* 23.4, there appears a *stathmodotes*, playing the same role; an official called *stromatophylax* oversaw the process of pitching Alexander's tent: Plut. *Alex.* 57.4.

An. 1.10).[19] These troops at the rear were—as was also the case with the Romans—older soldiers who were organized in *taxeis*.[20] They had to measure out the camp, prepare the ground, and distribute the various areas.[21] Quite another thing was breaking camp, a task in which everyone had to collaborate.[22] Greek soldiers always remembered this duty (and curiously enough not that of pitching camp) as one of the most unpleasant of their life in campaign (Xen. *An.* 5.1.2). Accordingly, armies kept an up-to-date record of the manpower that had to be accommodated, this being the only way to calculate the surface area to be assigned to the whole army and to each unit. The manner of doing so was to perform regular troop reviews, accompanied by an ever-increasing amount of bureaucracy.[23]

Once at the chosen spot, armies of different origins would each occupy their own camp, keeping a certain distance from each other. In military expeditions in which Greeks and barbarians joined forces, the distance from one camp to another was directly proportional to the degree of mutual trust, there never being any thought of sharing a camping ground. Such distance was a better gauge of relationships than any verbal account, as is shown by the evolution of the relation between the Ten Thousand and the rest of Cyrus' forces. They were always some ways apart, as evinced by the fact that Cyrus' concubine had to run away and find refuge among the Greeks when her master was defeated and his camp was plundered by Artaxerxes' forces (Xen. *An.* 1.5.6). After this battle, relations between Greeks and barbarians deteriorated and consequently the distance from one camp to another increased, mirroring greater mistrust (Xen. *An.* 2.4).[24] A clear separation was also observed when an expeditionary force consisted of allied Greek armies from different *poleis*, as at the city of Charmande, again vividly described by Xenophon (Xen. *An.* 1.5.10–14). In this camp, the passageways dividing the Greek troops were broad enough to allow the deployment of Menon's forces against Clearchus's soldiers. A third group of soldiers, under Proxenus' command, had the room to take a stand in between the two phalanxes.

A clear camp layout was also essential because it defined the jurisdiction of each military leader. Each commander controlled the soldiers encamped within the boundaries assigned to him, and no others, so if soldiers changed location they might fall under the jurisdiction of a new commander. This was the case of the 2,000 mercenaries who deserted their own chiefs, Xenias and Pasion, to camp among Clearchus' forces (Xen. *An.* 1.3.7).[25]

Within each campground, space was distributed by *taxis* (or, alternatively, the corresponding *lochos* or *mora*), with the commander of the baggage train of each *taxis* going ahead to organize his plot of land. This created a grid of thoroughfares, such as the one along which the cart bearing Masistius' dead body was paraded in front of the ranks of the encamped Greeks

19 Note also Xen. *An.* 1.5.12, 2.2.16–17.

20 Xen. *Anab.* 1.3.1, 5.3.1, 4.1.12–14, 6.4.1; Thuc. 5.72.3. In Rome, veterans were free from all service except for defending the camp: Goldsworthy 2003, 49.

21 Perhaps Polyaenus, *Strat.* 2.1.21; Pl. *Resp.* 7.526d; in Alexander's time, Karunanithy 2013, 190–191 (although the instances refer to city-planning and not to military camps).

22 Probably, some arrangements would have got under way the night before the march, as shown in Xen. *Cyr.* 5.35.40.

23 Xen. *An.* 1.2.7–9, 1.2.14–17, 1.7.10, 5.1.16, 7.1.7; in Alexander's army: Arr. *Anab.* 7.5.3; Diod. Sic. 17.17.3, 17.65.2–4, 17.76.2, 17.89.3, 17.110.3.

24 Other examples: Hdt. 9.17.2; Xen. *An.* 1.2.17. Also, Xen. *Cyr.* 4.2.20.

25 Proximity to the commander was probably the decisive element when camping in archaic Greece: Hom. *Il.* 10.150–156; also: Hdt. 2.154.1–5, 9.48.1, 9.78.1; Xen. *An.* 4.3.10–14. Among Roman forces: App. *Hisp.* 7.34.

before the battle of Plataea.[26] No sources show any subdivisions of the grid according to small units, even though such subdivisions were universal in Roman camps.[27]

The plan of a Greek military emplacement included certain essential common areas, the first and foremost of which was the tent of the commander-in-chief. He did not enjoy a fixed position within the layout of a camp; rather, the decision as to where to put him was made on two criteria: to find a position that would ensure safety while at the same time affording a commanding position that would facilitate decision-making and relaying orders. The ideal spot would be the center.[28] This, however, was not always the case, at least if we go by Arrian's report about Alexander's army in Tyre (2.20.9–10).[29]

The commander's quarters were, accordingly, better than his soldiers' tents, and conveyed his prestige. Innovations were occasionally introduced by ambitious individual commanders but proved unsuccessful on the whole. In this regard, the impact of oriental practices became evident from the Persian Wars onward.[30] In the case of Alexander the Great, the commander's quarters served as the royal court—the heart of an empire on the move—and so it comprised a small city of tents continuously evolving with regard to its structure and symbolic elements.[31] Curtius Rufus called this complex the *regia* or *praetorium* in order to distinguish it from Alexander's private abode, a *tabernaculum*.

In any case, a broad, open space lay in front of the commander's quarters. Altars and statues of the gods stood here (and perhaps the holy fire that accompanied Spartan armies). This piece of ground was large enough to enable many if not all troops to attend sacrifices or gatherings. It fulfilled the function of an *agora*, constituting the core of religious and political life in the community of citizen-soldiers.[32]

Additionally, every camp had areas set aside for plunder, the army's war chest, prisoners, the elements of the baggage train (oxen, mules, and carts), and the flocks of sheep or goats that commonly traveled with the army.[33] No information regarding the place that each of these spaces occupied within the camp has come down to us: we know only that the fruits of plunder were kept under permanent close guard in an inner area reserved for this purpose.[34]

In every campaign, soldiers were followed, and their needs supplied, by a train of merchants, women, and children. The camp location and business conditions for the merchants were agreed before departure in negotiations with the commander, who could change them

---

26 Thoroughfares: Polyaenus *Strat.* 2.3.11, 3.13.1; Xen. *Lac.* 12.5; Hom. *Il.*10.65–66. For a possible Hippodamian grid applied to Alexander's camp, see Karunanithy 2013, 188. Masistius: Hdt. 9.25.1.

27 While in camp, only the *taxiarch* had a say (and accordingly the responsibility) for anything that happened inside his place assigned to him and he was thus the authority to be called before a court over issues among encamped soldiers: Lys. 13.79; Isae. 4.18; Dem. 54.3–4; Lys. 16.15–16; also, Arr. *Anab.* 1.21.1. No mention is ever made of any other lower-ranking officer in any connection with camp life.

28 Xen. *Cyr.* 8.5.8; Aen. Tact. 1.8–9; Hom. *Il.* 10.150–156; Polyaenus *Strat.* 4.8.2; also, in Rome, Polyb. 6.27.1.

29 Also, Arr. *Anab.* 6.25.4–5.

30 Hdt. 6.56.1 for Spartan kings, probably an overstatement; Xen. *Hell.* 1.1.30, *Cyr.* 4.5.2, 39.

31 Alonso and Álvarez 2017; Spawforth 2007, 82–120; Karunanithy 2013, 191.

32 This is probably the place which Euripides used as the setting for his *Iphigenia at Aulis*; also, Xen. *An.* 6.4.14; *Lac.* 13.3–5; Paus. 3.26.7, 4.27.1–3; Thuc. 8.83; Hdt. 5.75.2, 8.6.2, 8.64.

33 War booty: Xen. *An.* 4.7.27, 5.1.2; *Lac.* 13.3–5; Hdt. 1.89.3, 9.80.1; Diod. Sic. 11.33.1. Animals: Hdt. 8.19.1; Polyaenus *Strat.* 3.11.15, 4.4.1. Prisoners: Diod. Sic. 9.21.1; Thuc. 1.54–55. Even the Greeks who defended the pass at Thermopylae took animals with them for their camp sacrifices: Hdt. 7.219.1.

34 Probably, latrine areas were assigned at some distance from the main camp (Xen. *Lac.* 12.4), as was the case in the nineteenth century (Ferraz 1801, 96–97).

according to circumstances at any time. The considerable profits made by these sutlers and their fear of being plundered naturally led them to accept any orders given by the military authorities.[35] Some Greek generals made military use of them, as shown by the stratagem of the Corinthian Ariston in Syracuse. The Syracusans and the besieging Athenians had been fighting off and on for hours, when Ariston persuaded the Syracusan admirals to send a message to the city, and have the market moved to the seaside, opposite the ships. That, Thucydides says, "enabled the admirals to land the crews and dine at once close to the ships, and shortly afterwards, to attack the Athenians again when the enemy were not expecting it" (Thuc. 7.39). Seeing the Syracusans eating, the Athenians disembarked and set about getting their own dinner. Then the Syracusans suddenly manned their ships, caught the Athenians unawares, and bested them in the ensuing fight.

This episode shows that ancient armies and navies did one another the customary courtesy of taking breaks for meals. The service provided by the sutlers was thus subject to an unwritten rule of war. In some cases, including this one, one side relied on merchants sent from a nearby town, whereas the other side had to rely on merchants who accompanied them on campaign. Managing the food supply was a perpetual, ever-changing challenge to commanders.

The women and children who accompanied armies formed what Thucydides called *ho ochlos* (Thuc. 7.78.2), all of them sharing the fate of the army and turning the military camp into a "city on the move."[36] Protecting this virtual city was a vital concern to commanders, too, if only because it was vital for soldiers, as shown by the willingness of some of the soldiers of the Diadoch Eumenes to desert him once their slaves and mistresses were captured by Eumenes' opponent, Antigonos the One-Eyed.[37]

The application of the Hippodamian model or of Pythagorean mathematical methods to camp planning probably became common practice during the Hellenistic period.[38] Even earlier, Xenophon was aware of the advantage of regular models when he advocated establishing a fixed, unvarying layout for the distribution of troops in his description of Cyrus' camp in the *Cyropaedia*.[39] Needless to say, Greek political polycentrism made it impossible to impose a universally accepted castrensian canon. In any event, we lack sources showing how regular models were applied in any particular case.

Guards, Security Measures, and Entrenchments

Except on very few occasions, Greeks did not think of their military camps as permanent bases from which to control territory, but rather as temporary bases of operations. Consequently, a camp was not considered to be a prime objective, and field defenses were thus generally established only to prevent a surprise assault (usually at dawn). Camp security rested on two fundamental pillars, the first being to make the best possible use of territorial features and the second to set guards and lookouts, while at the same time seeking to reduce

35 Their allocated space may have been delimited by posts, as in cities: Ar. *Ach.* 719–725. The market was under the constant watch and supervision of officers entrusted with this responsibility: Xen. *An.* 1.5.12. See n. 15.

36 Rostovtzeff 1941, 1.146.

37 Diod. Sic. 19.37–44; PLut. *Eum.* 16–19.

38 Álvarez 2013, 75–120; Karunanithy 2013, 186–189.

39 Polyb. 6.42; Xen. *Cyr.* 8.5.2–16.

as far as possible the labors of citizen-soldiers on campaign.[40] A first defensive line of guards (*phylakes*) was located on the perimeter of the camp itself, its main duty being to protect entry points and prevent unauthorized access to the camp zone, especially at night, when bonfires pinpointed army limits. It was these soldiers who prevented Alexander of Macedon, for instance, from entering the Greek camp at Plataea (Hdt. 9.44). A second line, consisting of advance guards (*prophylakes*), was located a little further away from the camp. They were distributed in small groups, probably on horseback, for the early detection of approaching enemies. Their roles were somewhere between those of a guard and those of a watch, probably working on a 24-hour shift basis, with the relief taking over in the morning.[41] Finally, we should mention the lookouts (*skopoi*), who were located some distance away on high ground at strategic points around the camp. They constituted a third line of defense, by serving as the army's eyes.[42]

During the fourth century, guards and lookouts acquired increasing recognition, to the point that Xenophon as well as Aeneas Tacticus regarded them as essential to the study of military camps,[43] yet their use was not systematic.[44] Armies of citizen-soldiers were not eager to perform the duties of life on campaign, as Leon of Thurii clearly expressed to his comrades: "Well, I, for my part, gentlemen," he said, "am tired by this time of packing up and walking and running and carrying my arms and being in line and standing guard and fighting, and what I long for now is to be rid of these toils" (Xen. *An.* 5.1.2).[45] Greek military camps, in short, did not reply on a permanent, static defense, such as walls, entrenchments, and ditches, but rather on a temporary, active defense consisting mainly of guards and sentries. The reason for this choice was the nature of Greek warfare, which centered on defeating the enemy in combat, not on holding positions. This, however, by no means excluded the possibility of armies erecting fortifications to defend themselves; they were built, but only in certain circumstances.

The most typical—and increasingly frequent—case was that of the camps set up after landing in enemy territory by sea, when soldiers commonly erected a wooden palisade (*eryma* or *stauroma*) to protect their beached ships. The most famous case was that of the Achaean forces in Ilium, who built a wall around their camp and vessels.[46] Fourth-century generals often defended their ships with an entrenchment, as was done by Demosthenes at Pylos, the Athenians at Syracuse, and Mnasippus at Corcyra.[47] The

40 The best example of its ideal application would be the Spartan system as described by Xen. *Lac.* 12: Álvarez 2013, 172–173.

41 Some examples in Xen. *An.* 2.1.7–8, 2.3.1–3; *Eq. Mag.* 4.10; *Cyr.* 6.3.9; also, common around cities: Diod. Sic. 13.72.4.

42 Xen. *Cyr.* 6.1.46; *An.* 5.1.9; *Hell.* 6.2.29; *Eq. Mag.* 4.10.

43 Xen. *Cyr.* 1.6.43; Aen. Tact. 21.2.

44 Xen. *Oec.* 20.8; *An.* 5.1.9.

45 English translation by Carleton L. Brownson. Guard duties on campaign were such a general experience among Greek citizens that Aristophanes made good use of them to wrest laughter from his audience; for instance, Ar. *Ach.* 1145–1150.

46 The case in point needs further explanation, which is beyond the scope of this chapter (see Alvarez 2010, 151–186); in fact, the Achaean *teichos* is described by Thuc. 1.11.1 as *eryma*, understanding the case from its fourth-century point of view.

47 Thuc. 4.9.1, 6.66; Xen. *Hell.* 6.2.23; also, in Thapsus, in Thuc. 6.97; Spartan forces were more reluctant to use it, as in Speiraeum: Thuc. 8.10; the opposite case was Iphicrates: Polyaenus, *Strat.* 3.9.17.

reason behind such a decision is easy to understand: the troops were few and had usually landed in enemy, or at least potentially dangerous, territory. Great care thus had to be taken to prevent the enemy from damaging the army's ships by a sudden attack and leaving its members stranded. The army naturally continued also to deploy guards and lookouts, both seaward and landward.[48]

Furthermore, troops operating inland on foreign soil would set up walls and small fortifications to secure their position, while at the same time making use of topographical features. For example, the last wall built by the Athenians at Syracuse served to fortify the harbor that they were using as an anchorage, but without employing men needed for other purposes (Thuc. 7.60.1–2). These, however, were exceptional and very dangerous situations, in which the army was making every effort to survive.[49]

Another occasion for the construction of fortifications was troops wintering in enemy territory with a view to renewing hostilities in the spring. There are very few examples of this practice, since such undertakings ran counter to Greek practice in the Classical period.[50] It happened chiefly during the Athenian expedition to Sicily, when the distance of the return journey to the metropolis was so great that the construction of winter camps was imperative. Such also was the case with the Greek military camps at Catane (Thuc. 6.64.3, 6.65), Naxos (Thuc. 6.74), and Chios (Thuc. 8.40). The Syracusan chiefs already foresaw this possibility when getting ready for an Athenian landing on their territory (Thuc. 6.37.2).

A further possibility arose in the case of siege.[51] In the face of a large besieging force, it was essential to erect fortifications around the threatened city and fortify the camps in order to protect the outnumbered defenders and prevent them from being driven out of the positions they were holding. This was a highly dangerous and costly measure. The best description of it appears in Thucydides' account of the Athenian siege of Mytilene (Thuc. 3.5–30), although important evidence of such kind of fortifications is to be found in many other episodes such as the sieges of Potidaea (Thuc. 1.63–64), Melos (Thuc. 5.114), Plataea (Thuc. 2.78), and Epidaurus (Thuc. 5.75), as well as of the forts on the Epipolae (Thuc. 6.98.2, 7.5, 7.23–24).

If we ask the general question of whether the Greeks fortified their camps, there can be no categorical answer. We can affirm that during the Classical period the Greeks did not fortify their camps unless obliged to do so by exceptional circumstances. They generally considered the presence of a ditch and palisade surrounding a camp as more of an obstacle than a help, since the enclosed area could become a veritable death trap for the hoplites. Palisades or ditches hindered the swift formation and movements of a phalanx, and Greek hoplites outside their formation, and thus deprived of the cohesion and security provided by their fellow warriors, were at the mercy of their opponents.

According to a widespread, traditional view, the best defensive wall was that formed by soldiers' shields arrayed in a phalanx on open ground. This truism gained credence because of defeats suffered by the Persians at the hands of the Greek phalanx during the Persian Wars and also during the expedition of the Ten Thousand into the Persian Empire in the service

48 About the military camps of landed troops, see Álvarez 2013, 121–128.

49 Other examples are the Greek wall in the Thermopylae pass (Hdt. 7.176.5) or the Ten Thousand's fortification in Calpe (Xen. *An.* 6.4–5).

50 Probably this is the case in Hammond 1987, 331–347.

51 On siege camps, see Álvarez 2013, 109–120.

of the pretender Cyrus the Younger.[52] In Sparta, this truism inspired another, which was that Sparta did not need city walls, since it had infantry to guard it. It remains true that in the fourth century some Greeks dissented. Iphicrates, one of the leading generals in this period, is said to have entrenched his camp even in friendly country (Polyaenus, *Strat.* 3.9.7).[53] In Alexander's time the *vallum* was probably a much more common feature surrounding military camps.[54]

Campaign Life and Military Solidarity

An important part of a Greek citizen's life took place inside a military camp amid countless hardships, as Aeschylus' herald recalls in the play *Agamemnon* (555–565). Any regular day started at first light or even earlier, for troops on the move had to reach the planned destination before noon in order to avoid walking during the hottest hours of the day. The leader of the army, joined by his officers, would attend dawn sacrifices outside his tent, while his men said their prayers, each according to his custom, as Plato recalled of Socrates during the siege of Potidaea (Pl. *Symp.* 220). Once this ritual had been completed, the order for breakfast was given by heralds, the camp gates were opened, and the camp and forward guards were relieved.[55] During the rest of the day troops busied themselves fetching firewood and water, foraging, performing guard duties, looking after the animals, and visiting the market that traveled with the army. They might rest in their tents, or they may go plundering or sight-seeing, or only shopping, in nearby villages and towns. The hustle and bustle was never-ending, as busy or idle soldiers mingled with animals, women, children, traders, and servants.[56]

Sometimes idleness would lead to excesses due to the combination of wine and gambling. The officers might remedy the situation, or might find themselves overmatched. In most armies—Sparta especially being excepted—the troops were not obliged to drill frequently, so little or no time was set aside for such a purpose.[57] The officers had to devise ways of keeping soldiers busy. Otherwise the lack of activity might reduce combat capacity or even spark conflict within the camp itself (Hdt. 1.63.1).

52 A representative case is the Persian defeat at Plataea (Hdt. 9.70), where hundreds of Persian soldiers were slaughtered inside their fort; other examples are the Persian forts at Mycale, which also failed to ensure the safety of the Persian troops (Hdt. 9.96–97), and the battle of Cyrus' mercenaries against the Drilae (Xen. *An.* 5.2.5–21) and at the town of Tinos (Xen. *An.* 7.4.14–19). Elsewhere in Xenophon, fortifications allowed hoplites to be killed like sheep inside their pen (Xen. *HG.* 3.2.3–4). All this experience is summarized by Xen. *Cyr.* 3.3.28, echoing the traditional Greek ideas about war, for which see Pl. *Leg.* 778d; Alc. fr. 112 (10 LP); Aesch. *Pers.* 349; Lycurg. *Leoc.* 44, 47; and Plut. *Lyc.* 19.12.

53 Likewise, Epaminondas seems to follow the same principles: Xen. *Hell.* 6.5.30, 7.5.8; also, Arist. *Pol.* 1330b; fortifications seem to be common among Sicilian Greeks: Diod. Sic. 11.20.3–5, 11.21–22.

54 Karunanithy (2013, 187–188, 192–194) considers fortification a common feature of Greek camps since Philip II; it is important to note that Alexander transformed his soldiers' camp at the Tanais into a city by walling the space: Curt. 7.6.25 (*muro circumdedit*); only once does Curtius (9.6.1) speak about a *vallum*, probably in the case of *stativa* camps (Curt. 8.1.10) or strongholds; also, Dem. 6.23, 18.87; Polyb. 5.2.5; Engels 1978, 17, n. 19.

55 Hdt. 9.92.2; X. *Hell.* 6.5.17, 7.1.16; *An.* 3.2.1, 4.3.9, 6.4.15.

56 If the famous *graffiti* left by Greek mercenaries at Abu Simbel are anything to go by; anecdotes and adventures in camp were commonplace in people's memories: Ar. *Vesp.* 237 ff., 355 ff.

57 Spartan training: Xen. *Lac.* 11.8–10. Spartan athletics: *Lac.* 12.5. Thrasyllus' soldiers performed physical exercises (Plut. *Alc.* 29.2).

One solution was to organize games and competitions with prizes for the winners of races, wrestling bouts, and other contests. This would become a regular practice in Alexander's army. In the navy, however, it was vital to carry out frequent exercises to ensure perfect coordination between the crew and their officers in performing maneuvers, hence, all fleets performed drills from early times, perhaps from the introduction of the trireme. All aboard ship, from rowers to admirals, realized that skill, stamina, coordination, and practice could determine the outcome of an engagement, as Nicias reminds the Athenians in a letter in which he told the Assembly of his concerns.[58] Yet even sailors did not accept an excess of zeal in this matter, as the Phocaean Dionysius experienced at Lade (Hdt. 6.12.1–4)

If troops had to march, but did not expect to fight, they would busy themselves with varied tasks once they reached the designated camp site. Besides putting up tents and huts, they dug latrines, fetched water, and gathered or chopped wood. It is a common misconception that because of the mild, dry, and sunny Mediterranean summer climate a simple bivouac (*aulizomai*) would suffice. In fact, the fierce sun in the middle of the day made it necessary for the troops to find shelter in a tent or hut, the former being the most common choice.[59] Greek tents (*skenai*) were relatively easy to put up and take down again, as they were constructed from wooden posts and tanned hides, sewn together, and probably secured with guys and metal pegs, as are their modern-day counterparts.[60]

At the end of the day, the call for the evening meal (*deipnon*) was sounded before night fell. Each group of soldiers lit a fire by their tent and gathered to prepare and eat the last meal of the day. The officers, too, met with the commanding officer in his quarters to have supper together, converse, discuss plans, receive orders, and finally, make libations to the gods. This was one of the most enjoyable moments of a soldier's day, as we discover when we read Xenophon's account of the march of the Ten Thousand. When the meeting concluded, each officer marched off to his quarters to transmit orders, give the watchword, and station the guards, after which the call for silence was given and the camp entrances were closed, no one being allowed to leave or enter thereafter.[61]

Camp life was regimented.[62] Aside from the signals at dawn and dinner time, heralds and horns announced soldiers' duties throughout the day. Plutarch (*Nic.* 9.5) tells how during the Peace of Nicias the Athenians gladly called to mind the saying "In peace the sleeper is awoken not by the trumpet, but by the cock." Since the sequence of signals and duties was very much the same in all Greek armies, it was easy to hear what the enemy was doing at any given time, so long as he was within earshot, and easy to guess even when he was not. Astute

58 Nicias: Thuc. 7.14.1. Fleet exercises: Xen. *Hell.* 1.1.16, 6.2.27–31. Iphicrates used the time his men spent aboard ship to train them and keep them fit: Xen. *Hell.* 6.2.27–31.

59 Veg. *Mil.* 3.2, Álvarez 2013, 129–133.

60 Arr. *Anab.* 4.19.1. The excellent quality of these tanned hides and the skill with which they were worked enabled them to be unstitched and used as rafts (*keleks*) for ferrying troops across the great rivers in Asia, after which they could be sewn together again for return to use as the soldiers' tent roofing: Xen. *An.* 1.5.10, 3.2.7; Arr. *Anab.* 1.3.6, 3.29.4; Curt. 7.5.17–18; Álvarez 2013, 133–142; Heckel and Jones 2006, 27; Karunanithy 2013, 195–196.

61 At Plataea, Alexander of Macedon was unable to enter the Greek camp at night and had to wait at its limits: Hdt. 9.44; also, *Military Decree of Amphipolis*, 3.28–30 (*SEG* XL, 524).

62 The best account of camp routine: Xen. *Lac.* 12.5–7, most of which seems to apply to other armies.

commanders could use this information to implement stratagems that would enable them to surprise the enemy (Hdt. 6.77–78).

Greek campaigns involved a common deep and powerful experience of life. This shared experience led men who used the same tent, or *syskenoi*, to participate in a kind of small-group fellowship known as *syskenia*.[63] Members of the group ate together and formed a logistical unit, providing mutual assistance and collaboration for the duration of a campaign (Xen. *An*. 5.8.5–7). Yet the bonds of *syskenia* transcended the sphere of mere material organization.[64] This makes Thucydides' description of the Athenian withdrawal from Sicily all the more painful for Greek eyes:

> [Soldiers] fell to entreating and bewailing until their friends knew not what to do, begging them to take them and loudly calling to each individual comrade or relative whom they could see, hanging upon the necks of their tent-fellows in the act of departure, and following as far as they could, and when their bodily strength failed them, calling again and again upon heaven and shrieking aloud as they were left behind. (Thuc. 7.75.5)

We know little about the mechanics of this important institution but it seems to be the result of a free decision by the members to join forces for the duration of a campaign. In many cases, it would have developed from earlier ties of family or friendship.[65] Within the framework of civil life, tent-fellows are called on to help decide lawsuits in which events occurring during a military campaign are of relevance, a context in which the close fellowship between tent-sharing soldiers makes them particularly important witnesses (Isaeus 4.18, 9.4). In Classical times, the public aspects of this kind of fellowship do not seem to go any farther, but in the Hellenistic period, *syskenoi* maintained their ties in civil life for religious and funerary purposes.[66]

Under Alexander, camp solidarity developed in a unique fashion, as observed by Bradford Welles. Pointing to the transformation of the Macedonian camp from a castrensian entity into an enlarged migratory group, Welles remarked that Alexander encouraged this for reasons other than the troops' happiness. It made them forget Macedon and make the camp their home, thereby enabling them to bring up a new generation loyal only to themselves.[67]

Conclusions

Greek army camps evolved from the archaic forms described in the *Iliad* to the large and complex camps of Alexander, a shift that invites us to revisit the opinion of Polybius given at the start of this chapter. The Hellenistic historian was seeking to emphasize

63 Unsurprisingly, Hesychius equals *syskenia* to *syssition* in his dictionary (s.v.); Pl. *Leg*. 625e.

64 Plut. *Alc*. 4.4; Pl. *Symp*. 219e–220b; it could mean social rejection: Lys. 13.79.

65 Leading the group was the *presbytatos apo skenes*, who was probably the oldest in age and/or in military experience; he was also known as *skenarchos* (*SEG* II, 60). See Xen. *An*. 2.1.10, 3.1.34, 6.5.13; *Cyr*. 4.2.34; *Lac*. 13.7; in Rome, the leader was called *caput contubernii*. Family ties and friendship: Plut. *Alc*. 4.4, 7.3; Pl. *Symp*. 219e–220; D. 54.3–4; Álvarez 2013, 136–142.

66 Although Hdt. 1.65.5; in Hellenistic times, *SEG* II.60; IG. XII.2.640; Launey 1950, 1001–1036.

67 Welles, 1970, 34. Also see Alonso and Álvarez 2017.

the Greeks' lack of a system—of a fixed layout for distributing their units in camp. Xenophon anticipated him in the *Cyropaedia*. Yet the Greeks did have settled camping practices that gradually became more elaborate and more formal. Aeneas Tacticus surely described them in his lost book on encampment. This chapter has attempted to summarize them.

CHAPTER 12

Greeks in Foreign Service: The Case of the Achaemenid Empire

Jeffrey Rop

The conventional narrative of the ancient encounters between Greece and Persia is book-ended by the Greek victories in the Greco–Persian Wars and the Macedonian conquest of the Achaemenid Empire. A prominent feature of this narrative is the unique and superior Greek heavy infantry hoplite. The hoplite's peculiar equipment and densely packed phalanx fighting formation is often said to be one reason that the Greeks defeated the Persians in the fifth century,[1] and is often used to explain why many Greeks were later recruited for Persian armies in the fourth.[2]

This chapter offers a different perspective on Greco–Persian interactions by surveying their military collaborations from the sixth to the fourth centuries. It identifies two aspects of Greek military service for the Achaemenid Empire that are overlooked by or challenge the conventional narrative. The first is that Greeks fought alongside Persians more frequently than is usually recognized, and that many battles portrayed as Greek against Persian in fact included thousands of Greeks on the Persian side. The second is that the Greeks who performed military service for the Persians did so in a wide variety of roles, not only or even predominantly as traditional heavy infantry hoplites. This material is presented in Table 12.1.

1 Recently Cawkwell 2005, 115; Rahe 2015, 320–321; Konijnendijk 2012, 1, n. 4 catalogs further references.
2 Rop 2019a, 2–9.

A Companion to Greek Warfare, First Edition. Edited by Waldemar Heckel, F.S. Naiden, E. Edward Garvin, and John Vanderspoel.

Table 12.1 Greek Military Service in the Persian Empire

Date	*Campaign/event*	*Greeks*	*Role(s)*	*Source(s)*
525–522	Conquest of Egypt	Samos	Navy	Hdt. 3.44
		Phanes of Halicarnassus	Intelligence	Hdt. 3.4
		Aeolia, Ionia	Navy*	Hdt. 3.1
		Military colonists	Garrison	Hdt. 3.25
513	Scythian Expedition, Conquest of Thrace*	Aeolia, Ionia, Hellespont	Navy	Hdt. 4.88–89, 97–98, 141–144
499	Invasion of Naxos	Aeolia, Ionia	Siege, navy	Hdt. 5.32–34
492	Invasion of Greece	Ionia*	Naval	Hdt. 6.42–43
490	Invasion of Greece	Aeolia, Ionia	Naval, infantry*	Hdt. 6.98–116
480–479	Invasion of Greece	Eastern and islander Greeks	Navy, infantry*	Hdt. 7.89–95, 8.66, 85–89, 9.99–104
		Thessaly	Guides, ambassadors, cavalry	Hdt. 7.196–197, 233, 8.27–35, 9.31
		Thebes	Guides, infantry, cavalry	Hdt. 8.50, 9.31, 38, 67–69
		Phocis, Doris, Locris	Infantry	Hdt. 8.66
Ca. 449	Service for Megabyzus	Athens*	Unknown	Ctesias F14.36–40
441	Attack on Samos	Unknown*	Unknown	Thuc. 1.115
430	Attack on Notium	Arcadia	Unknown	Thuc. 3.34
415	Betrayal of Pissuthnes	Athens	Unknown	Ctesias F15.53
412	Capture of Amorges	Peloponnese	Unknown	Thuc. 8.28
412–404	Peloponnesian War	Sparta and allies	Navy	Thuc. 8.5, 29, 37, 45, 57–58; Xen. *Hell.* 1.5, 1.6.18, 2.1.11–15
401	Rebellion of Cyrus	Sparta, Boeotia, Thessaly, et al.	Heavy and light infantry	Xen. *Anab.* 1.7.10
		Phalinus	Military advisor	Xen. *Anab.* 2.1.7
397–393	War against Sparta	Athens (Conon)	Navy	Diod. Sic. 14.39, 79–85; Xen. *Hell.* 4.8.1–12
388–387	The King's Peace	Sparta	Navy	Xen. *Hell.* 4.8.16, 5.1.25–36
373–372	Invasion of Egypt	Athens	Navy, marine	Diod. Sic. 15.29, 41–43
364	Tour of the Aegean	Thebes	Navy	Diod. Sic. 15.79
356–355	Service for Artabazus	Athens	Navy, infantry	Diod. Sic. 16.22; Dem. Schol. 4.19; FGrH 105 F4

(*Continued*)

Table 12.1 (Continued)

Date	*Campaign/event*	*Greeks*	*Role(s)*	*Source(s)*
354–353	Service for Artabazus	Thebes	Heavy* infantry, cavalry	Diod. Sic. 16.34; Frontinus 2.3.3
349	Revolt of Cyprus	Athens	Siege, navy*	Diod. Sic. 16.42.6–9
343–342	Conquest of Egypt	Thebes	Marine* hoplites, siege	Diod. Sic. 16.44.2, 46.8–9, 49.1–3
		Argos	Navy, marine	Diod. Sic. 16.44.2–3, 47.3, 48.3–5
		Aeolia, Ionia	Navy*	Diod. Sic. 16.44.4
		Mentor	Heavy* infantry, siege	Diod. Sic. 16.47.4, 49.7–50.8
340	Defense of Perinthus and Byzantium	Athens, Rhodes*, Cos*	Navy, garrison	Diod. Sic. 16.75.1–2, 77.2; Pausanias 1.29.10
336–335	Macedonian invasion	Memnon	Infantry	Diod. Sic. 17.7
334–331	Macedonian conquest	Athens and other Greeks (Memnon)	Heavy infantry (Granicus), garrison, navy	Arrian 1.13–16, 20–23; Diod. Sic. 17.19–21, 22–27, 29
		Samos	Naval support	
		Unidentified Greeks	Anatolian garrisons	Arrian 1.17.9, 19.1–6, 24.4, 26.5, 29.1–2
		Pherae, Acarnania, Phocis, Aetolia	Heavy infantry (Issus, Gaugamela)	Arrian 2.13.2–3, 16.2; Curtius 5.11–12
		Sparta and allies	Heavy infantry (Megalopolis)	Diod. Sic. 17.62–63; Curtius 4.1.39–40, 6.1

* denotes speculated identity or role

From Pelusium to Marathon

The first appearance of Greeks in the Persian military took place during Cambyses' invasion of Egypt in 525. Polycrates, the tyrant of Samos and a Persian client, provided 40 triremes for this campaign (Herodotus 3.44). Cambyses also made use of intelligence provided by Phanes, a Greek defector from Egypt (4.4). The Ionian and Aeolian Greeks of Anatolia also participated, but Herodotus does not describe the nature of their service (3.1). It is plausible that they contributed hoplites, but equally so that, like Polycrates, they provided ships to the Persian fleet. The Greeks who resisted the Persians in Egypt did so as hoplites and are said to have fought well in defeat at the Battle of Pelusium (3.11). Cambyses briefly used them as garrison troops in Egypt during their subsequent invasion of Ethiopia but dismissed them from service upon his return (3.25). In a notable contrast, the Persians retained the Jewish military garrison at Elephantine.[3]

3 Fischer-Bovet 2014a, 23–24, 36–37.

The next major Persian military campaign involving Greek soldiers was Darius' Scythian expedition around 513. The majority of the Persian force during this campaign were infantry (4.136); the Ionians, Aeolians, and Hellespontine Greeks provided the fleet. Greek ships were used as a bridge for the Persian infantry to cross the Bosporus, and the Ionians built another bridge for the army at the Ister River, defending it until the Persians returned from their invasion inland (4.88–89, 97–98, 141). Darius departed for Asia, leaving his general Megabazus to conquer the Hellespont and coastal Thrace (4.143–144). Herodotus does not mention Greeks on this operation, but part of the fleet may have remained with him given its proximity to the sea.[4]

The Persians once again made use of Greek naval expertise in an attack on Naxos in 499 (5.32–34). The strike force was composed of Persian, Greek, and other subject troops. It was supported by 200 triremes; many if not most Greek. Not only had Ionians and Aeolians provided the naval support for the Scythian expedition, but Herodotus mentions that one of the ships was from the Greek city of Myndus. Moreover, the Ionian Revolt that immediately followed this campaign's failure was probably possible in large part to the appropriation of ships built with the King's money.[5] This fleet defeated the Phoenicians near Cyprus earlier in the war (5.112), and numbered 353 triremes by the end of the war in 494 (6.8).

In 492, the Persians dispatched a force against mainland Greece to punish Athens and Eritria for supporting the Ionian rebels. It is possible that this invasion included Greeks who had defected to Persia at the Battle of Lade two years earlier. Their recent disloyalty was not necessarily disqualifying, since Cypriote ships fought for the king at Lade despite their own earlier involvement in the rebellion (6.6). After mustering in Cilicia (6.43), the Persian fleet spent time in Ionia while sailing to meet the army at the Hellespont (6.42). They then proceeded together along the coast of Thrace in the northern Aegean.

The campaign failed when its fleet was wrecked in a storm off the coast of northern Greece (6.44), but another expedition was launched in 490. It, too, mustered in Cilicia before sailing to Ionia (6.95), where this time Herodotus reports that Ionian and Aeolian Greeks joined in an unspecified role (6.98). Sailing directly across the Aegean, the Persians captured Naxos, Carystos, and Eretria before suffering defeat at the Battle of Marathon in northern Attica (6.99–116). Given the route and its launch from Ionia, Greek ships were certainly part of the fleet, and Greek sailors and marines may have even fought on the Persian wings as at Marathon.[6]

Throughout this early period, Greek naval personnel could serve as sailors, marines, engineers, or rowers put ashore to fight as light infantry, this last being possible at Naxos and during Cambyses' invasion of Egypt. The status of Greek personnel varied, too. Those sent by Polycrates were allied troops; those serving Darius were subject troops; and some individuals, especially ship captains and helmsmen, may have been paid for their services.

4 Cawkwell 2005, 50–51.

5 Briant 2002, 153.

6 Van Wees 2004, 221–223 notes that a trireme's rowers could be repurposed to serve as lightly armed infantry. The Persian center in the battle was manned by Persians and Sacae, but Herodotus omits the composition of the wings (6.113). Krentz 2010, 140–142, 159–160 discusses mention of Ionians at Marathon in the Suda and the possibility that the Persian force included hoplites.

From Thermopylae to Aegospotami

In the modern imagination, the Greco–Persian Wars are a clash of civilizations pitting the Greek West against the Persian East. Yet Greek forces participated on the Persian campaign to Marathon, as we have seen, and even more served in the invasion of Xerxes in 480–479. The Greeks of Asia provided ships, while thousands from the mainland served as hoplites and cavalry as well as guides and negotiators. The reality is that this war pitted eastern, northern, and central Greeks aligned with the Persians against a coalition of southern Greeks led by Sparta and Athens.[7]

According to Herodotus, Greeks provided nearly 40% of the 1,207 ships in the Persian fleet: Cyprus contributed 150, the Dorians of Asia 30, the Ionians 100, the Greek islanders 17, the Aeolians 60, and the Greeks from Pontus 100. Some of the Cypriote ships were from Phoenician Salamis, and an unidentified number of Greeks were tasked with defending the bridge over the Hellespont during the campaign (7.89–95). Although Herodotus undoubtedly exaggerated the size of the Persian navy, it is notable that it was remembered still in Herodotus' day that Greeks had provided many ships to King Xerxes.

Upon entering Greece, the Persian army relied upon logistical support from northern Greeks and acquired Greek ground troops (7.32). Most prominent to join were the Thessalians (7.196). While their cavalry was not suited for the terrain at the Battle of Thermopylae, the Thessalians had fought against the Phocians there roughly 20 years earlier (7.176). They were surely the guides that Herodotus mentions assisting Xerxes on the approach to Thermopylae (7.197), and probably used their connections in the area to recruit locals who led Persian forces along with the Anopaea path, much as in their earlier war against Phocis (7.215). One of these was Ephialtes of Malis, who not coincidentally later went into exile in Thessaly (7.213). The Thessalians negotiated the surrender of the Thebans at Thermopylae (7.233), then protected Persian sympathizers in Doris (8.29–35). They guided the Persians through the Cephisus Valley (8.27) and may have secured the loyalty of the 1,000 Phocians on the Anopaea path, probably the same who fought for the Persians at Plataea in 479.[8]

The three-day battle at Thermopylae coincided with a series of engagements on the sea at Artemisium. Herodotus claims that the Ionians participated unwillingly on the first day, yet only a single Greek ship deserted the Persians (8.10–11). Greeks were probably among the otherwise unidentified detachment of 200 triremes destroyed in the attempted circumnavigation of Euboea (8.7). Islander and eastern Greeks were the most likely to have pilots familiar with the route and, notably, many of the Greek divisions enumerated at the outset of the campaign are never again mentioned by Herodotus. The third day of battle at Artemisium saw heavy fighting with significant losses on both sides. Herodotus refers to the entire Persian force simply as *barbaroi*, but his comment that the size of the King's fleet made it unwieldy is an indication that the Greeks were involved (8.15–17).

The Greek coalition fleet withdrew to Salamis upon learning of the results at Thermopylae (8.21). As Persian forces advanced into Attica, Greeks from Doris, Locris, Melos, Carystus, Andros, Tenos, and Boeotia joined them with ships and troops. Herodotus is not specific

---

7 In-depth studies on Greek collaboration with Persian during the invasion of Xerxes include Westlake 1936; Gillis 1979; Graf 1979; Waters 2014.

8 Rop 2019b, 422–421.

which group supplied which or in what number other than speculating that they roughly equaled the Persian losses at Thermopylae and Artemisium (8.66). Herodotus reports that the Thebans led the Persians through their own region, encouraging the destruction of both Thespiae and Plataea (8.50). They also would have shown the way into neighboring Attica, and may have helped raze the countryside if not the city of Athens itself (8.50).

Meanwhile, the coalition inflicted a defeat on the Persian fleet after luring them into the straights between Salamis and the coast of Attica. The king's line was manned by Phoenicians on the right (western) wing opposite the Athenians and Ionians on the left (eastern) wing against the Spartans (8.85). The anecdote of a Samothracian ship destroying an Athenian and then capturing an Aegenitan vessel complicates this scheme (8.90) but may have taken place after the Persian lines were disastrously entangled (8.89). Herodotus is silent on how the rest of the king's Greeks fared, aside from highlighting the exploits of two Samian captains (8.85).

Xerxes withdrew from Greece after the defeat at Salamis, leaving part of his army in Thessaly for the winter (8.133). This force took position in Theban territory in the spring and confronted a coalition army not far from Plataea (9.15). In the days prior to the battle, Persian cavalry, perhaps joined by mounted Greek allies (9.67–68), harassed coalition forces and, guided by the Boeotians, attacked their supply lines (9.38).

Greek forces made up around half of the contingents in the Persian array at the Battle of Plataea. The Persians took the left wing, with the Medes, Bactrians, Indians, and Sacae next in line followed by the Boeotians, Locrians, Malians, Thessalians, and Phocians, as well as allies from Macedon (9.31). Herodotus guesses that the Greek allies amounted to no more than 50,000 (9.32), but his narrative of the battle focuses on the Persians and their Theban allies against, respectively, the Spartans/Tegeans and Athenians. It is unclear what the rest of the Persian force did, although Herodotus claims that the non-Theban Greeks fought badly on purpose (9.67). The most notable Greek contribution for the Persians was the annihilation of 600 Megarians and Phliasians by Theban cavalry during their retreat (9.69).

The final major engagement of 479 took place in Anatolia at Mycale, where coalition forces defeated a Persian army off the Ionian coast. Herodotus writes that many Greeks deserted the Persians during the battle, but, other than stating that the Milesians were detached to guard local mountain passes, he does not provide details that would allow us to know their tactical role or armament (9.99–104).

Following the establishment of the Athenian-led Delian League in 478, there is little evidence that Greeks served in Persian armed forces for the next 30 years. This changed with the thawing of Persian–Athenian relations following the suppression of the revolt of Inarus in Egypt in 454 (Thuc. 1.109–110) and the subsequent Peace of Callias (Diod. Sic. 12.4).[9] Not coincidentally, Megabyzus, the Persian official most responsible for both achievements, took 6,000 Greeks—probably Athenians—into his service, and even brought some with him to court around 449 (Ctesias *FGrH* 688 F14.36–40).

Over the following decades there were no military campaigns requiring significant mobilization in the western Persian periphery, and the king did not acquire Greeks for conflicts elsewhere in the Empire. Still, Greeks were probably among the 700 soldiers sent by Pissuthnes the satrap of Lydia to restore exiles at Samos in 441 (Thuc. 1.115), and Arcadians helped him seize control of Notium in 430 (3.34). Pissuthnes later rebelled against the king and was betrayed by Greeks under the command of Lycon of Athens around 415 (Ctesias

9 Hyland 2018, 18–34.

F15.53). Likewise, his son Amorges was employing a division of Peloponnesian Greeks when he was captured in 412 by the Spartans (Thuc. 8.28).

These Spartans had themselves recently come into the service of the king, thanks to the alliance negotiated in 412.[10] Persian subsidies helped build three separate Spartan fleets over the second half of the Peloponnesian War, with the first two destroyed at Cyzicus in 410 and at Arginusae in 406. The third, which annihilated the Athenians at Aegospotami in 405 and essentially ended the Peloponnesian War, was constructed almost entirely with Persian funds provided by the Persian prince Cyrus, in charge of the western satrapies. Thus by the end of the War in 404, the Empire ruled western Anatolia once again and was the naval paymaster of the most powerful Greek state on the mainland.

From Cunaxa to Perinthus

The most famous episode of Greek military service for Persia undoubtedly took place during the rebellion of Prince Cyrus in 401. Mustering an army that included 10,600 Greek hoplites and 2,500 Greek light infantry,[11] Cyrus confronted and was defeated by King Artaxerxes II near the village of Cunaxa outside of Babylon. His Greeks included an official Spartan delegation, and the rest were recruited through the relationships that Cyrus had established thanks to his connections with Sparta during the Peloponnesian War.[12] Xenophon mentions that Phalinus, an individual in the entourage of the satrap Tissaphernes fought for the King at Cunaxa (Xen. *An.* 2.1.7).

The performance of Cyrus' Greek hoplites at Cunaxa is often said to have demonstrated a Persian deficiency in heavy infantry, which in turn led the king and his satraps to depend upon Greek mercenary hoplites throughout the fourth century until the Macedonian conquests.[13] Yet Greeks served the king in other roles more frequently during this period, and soldiers who were probably traditional hoplites would not again appear in a Persian army for nearly 50 years.

The next Greek to fight for the king was the Athenian Conon, who joined the Persian navy in 397 during a time of war between Sparta and Persia. Conon coordinated the recruitment of Athenian sailors and ships for the king, and helped destroy the Spartan navy at Cnidus in 394. The alliance that Conon negotiated between Athens and Persia did not last, however, and by 387 the Spartans had come to terms with Persia and used the king's support to end the war in Greece. One reason their Greek rivals were willing to submit to King's Peace, which heavily favored the Spartans, was the destruction of an Athenian fleet by the Spartan diplomat and admiral Antalcidas, whose own ships had been funded by Persia.[14]

The next substantial Greek force to serve the King served under the Athenian general Iphicrates, who participated in a failed invasion of Egypt in the 373–372. Rather than a traditional hoplite phalanx, Iphicrates deployed a mix of naval and marine units that helped

10 On Persian support of Sparta during the Peloponnesian War, see Roberts 2017, 222–227, 255–284; Hyland 2018, 53–121.

11 Lee 2007, 44–48.

12 Rop 2019a, 77–87.

13 For example, Trundle 2004, 6–7, 73–74.

14 Asmonti 2015, 116–178.

carry out an amphibious assault against a fortress at the Mendesian mouth of the Nile.[15] The king shifted his patronage from Sparta to Thebes in 367, and three years later a Theban fleet funded by Persia toured the Aegean.[16] In 356, the Athenian Chares secured a brief alliance with the satrap of Hellespontine Phrygia, Artabazus, who gave shelter to the Athenian fleet and allowed it to supply itself with plunder from the territory of his nearby rival, Tithraustes.[17]

Finally, in 354 the Theban general Pammenes joined Artabazus with an army of 5,000 soldiers that probably included hoplites in the traditional sense. Although our sources do not identify them as such, circumstantial evidence supports this view. Pammenes had defeated a Phocian army at the Battle of Neon in Greece just prior to marching to Anatolia, where he led them to victory in a land battle against an unidentified Persian rival of Artabazus. Our sources describe his army as including cavalry and foot soldiers; if the latter were indeed hoplites, they would be the first to fight in a battle for the Persians since Cunaxa in 401.

Greeks continued to serve the king in primarily naval and marine capacities throughout the 340s. In 349, the Athenians sent Phocion to join Evagoras II, a longtime friend of the city and of Phocion's mentor Chabrias, in an attack against Salamis on Cyprus. Their force had 40 triremes and 8,000 men and was under the authority of Idrieus of Caria, whom the king assigned the task of suppressing a rebellion on the island (Diod. Sic. 16.42.6–9). Next, Greeks from Thebes and Argos joined the Persian conquest of Egypt in 343–342. Diodorus describes the Theban force as 1,000 hoplites, but their deployment in the vanguard leading an initial assault across a narrow and deep canal would suggest they were not of the traditional type. Similarly, 3,000 Argive troops were deployed along with 5,000 elite Persians and 80 triremes as part of an amphibious assault force that seized important canal routes located behind the Egyptian front line.[18]

The Persian invasion force also included 6,000 Greeks from the king's territories in Asia Minor and 4,000 more led by Mentor of Rhodes. The function of the former is not described, but it is most likely that they served in a naval capacity. The desert terrain that land armies were forced to cross to reach Egypt made seaborne supplies a necessity,[19] and the recent rebellions on Cyprus and in Phoenicia may have meant that the king looked to other subjects like the Ionians and Cilicians to provide his fleet. Mentor's Greeks are the most likely to have been hoplites, since they had fought a land battle against a satrapal army while serving the rebels of Sidon before betraying the city to the king. They then entered his service.[20] During the invasion of Egypt they besieged Bubastis, and Mentor personally negotiated the surrender of a number of the Pharaoh's Greek mercenaries (16.49–50).

In the aftermath of the invasion of Egypt, Mentor carried out several minor campaigns on behalf of the king in the region until his death c. 340, at which point he was replaced by his brother Memnon. Around that same time the king ordered his satraps to cooperate with Athens to prevent Philip of Macedon from conquering the coastal cities of Byzantium and Perinthus.[21] Our sources provide few details about these campaigns, but, since the

15 Ruzicka 2012, 114–121.

16 Buckler and Beck 2008b, 199–210.

17 For the service of Chares and Pammenes for Artabazus, see Rop 2019a, 119–147.

18 Ruzicka 2012, 182–198; Rop 2019a, 158–164.

19 On the challenges faced when invading Egypt, see Kahn and Tammuz 2008; Ruzicka 2012, 14–16; Fischer-Bovet 2014b.

20 Rop 2019a, 154–158.

21 Rop 2019a, 179–180.

Macedonians were laying siege against both by land, they must have involved seaborne resupply and reinforcement. Once again, then, the king made use of Greek naval resources to conduct military operations.

In this period, Greek mercenaries come to the fore, but the Persians continue to use subject troops in the service of Anatolian satraps as opposed to the Great King himself, and naval forces with military capability remain significant. Iphicrates' men rowed as well as fought, and perhaps the troops that invaded Cyprus and Egypt did, too. Amphibious warfare was widespread, and the long, inland march of the Ten Thousand into the center of the Persian Empire was atypical.

The Macedonian Conquest

Many Greeks entered into Persian service as heavy infantry during the Macedonian invasions of 336–335 and 334–331. In the first campaign, Memnon of Rhodes led a small ground force that presumably included Greek hoplites. In spite of his reputation for military brilliance, however, Memnon's accomplishments during this period were modest. Most notably, he appears to have been absent from the major Persian victory over the Macedonian vanguard near Rhoetium.[22]

Memnon and between 4,000 and 7,500 Greek hoplites were present at the Battle of the Granicus River in 334, the first engagement of Alexander's invasion.[23] Defeated by the larger Macedonian army, he and 2,000 Greeks, including the Athenians Ephialtes and Thrasybulus, eventually gathered to resist a Macedonian siege at Halicarnassus. After the city fell, Memnon received command of Persian maritime forces in the Aegean, where he took Chios and most of Lesbos before dying during the siege of Mytilene. Although many of the Greeks from this campaign were summoned to join the king's force gathering in Syria, the Persian counteroffensive in the region continued for some time. Notably, one of its leaders was the Athenian general Chares, who was captured defending Mytilene in 332.[24]

Athenians participated heavily in the first years of the Persian resistance in Anatolia. Another general, Charidemus, reportedly joined Darius in Babylon before being executed for insubordination. Many of the soldiers captured at the Granicus and later Halicarnassus were from Athens, as the city repeatedly sent embassies to Alexander requesting their release from captivity. Greeks beyond Athens also joined the Persian resistance to Alexander throughout his conquests. The Thebans had attempted to rally the rest of Greece against him with promises of support from the king in 335, the Samians provided safe harbor to the king's fleet during the siege of Miletus in 334, and Greeks were among the garrison troops at a number of Anatolian cities that the Macedonians captured during their advance along the coast.[25]

Most notably, Darius counted as many as 15,000 Greek hoplites in his army at the Battle of Issus in 333. Their commanders included Aristomedes of Pherae, Bianor of Acarnania, Patron of Phocis, and Glaucus of Aetolia. Around 8,000 who survived Issus joined the

22 Rop 2019a, 182–185. See also the discussion in Heckel 2020, 41–57.

23 Devine 1986, 270–272.

24 Briant 2002, 823–828.

25 On the Greeks who fought for Persia against Alexander, see Rop 2019a, 201–206, 228–230.

Spartan army, funded in part with Persian money, that the Macedonians crushed at Megalopolis in 331, while between 2,000 and 4,000 others fought for the Persians at Gaugamela that same year. The Greek survivors of this battle fled with Darius, and turned themselves over to Alexander following the king's assassination by Bessus. Their surrender marked the end of more than two centuries of Greek military service for the Achaemenid Empire. In this, the last period surveyed in this chapter, Greeks figure as Persian allies, as was true of the Athenians serving in Anatolia, and as mercenaries, as was true at the Granicus and Issus battles, and perhaps as subject forces, again in Anatolia. For the first time, Greeks in Persian service travel to the Greek mainland, and serve there, fighting in the interests of both Persia and some Greek states.

Conclusion

The modern fixation on the distinctiveness of the hoplite and on the relatively infrequent wars between Greeks and Persians obscures their long and varied history of military cooperation. Ionian and Aeolian Greeks made important contributions to the Persian navy from the sixth century onward, while the Spartan navy effectively functioned as the Persian Aegean fleet in the late fifth century. Throughout the fourth century, Athenians and other Greeks repeatedly served the Persians as sailors and marines. Greek rowers also served as light infantry, a phenomenon that the sources may underreport, and they evidently served as naval officers as well as below decks. Greeks were frequently tasked with garrison and guard duty in the western regions of the Empire, and during Xerxes' invasion played important roles as guides and negotiators. By contrast, the Persians deployed the traditional Greek hoplite phalanx on only a few occasions, most of them defeats: at Plataea in 479, Cunaxa in 401, in Hellespontine Phrygia in 354, perhaps in Egypt in 343–342, and finally against Alexander at Granicus, Issus, and Gaugamela. Hoplites, and other troops, varied in status sometimes subject troops, sometimes allies, and increasingly mercenaries. The Great King could be their sovereign, their ally, or their paymaster. As this brief survey reveals, Greek military service for the Achaemenid Persian Empire was more frequent and varied than is often recognized.

PART IV

War with Non-Greeks

CHAPTER 13

The Royal Elite of the Achaemenid Army

Michael B. Charles

This chapter focuses on the core of the Achaemenid army, in particular what might be described as those forces most closely associated with the king and that seem to have been on permanent duty within the Achaemenid homeland—to use a late Roman term, the "palatine" army.[1] This standing army on permanent duty differed from local garrisons and from the forces of the satraps, who seem to have had their own forces on which to rely, both in times of peace and in war. The palatine army evolved from one drawn from Iranian peoples, such as Persians, Medes, and possibly even Elamites, to one that drew heavily on Greek mercenaries.

First, however, it is important to stress how much we rely on the writings of Greco-Roman authors such as Arrian, Diodorus and Curtius Rufus, writing long after the fall of Achaemenid Persia. Some sources wrote closer to the era that they were describing, such as Herodotus and Xenophon, but their accounts contain manifold difficulties, such as a reliance on sources now lost to us, or an interest in subject matter that differs from our own. Sources from the Ancient Near East do of course exist, but, in general, they tell us more about military formations emanating from conquered peoples rather than those most closely associated with the Great King. Happily, some few literary and visual sources help us to understand the more permanent Iranian forces under the king, and this chapter will draw upon these to supplement sources dealing with Achaemenid Persia from a Greek or Roman perspective. Regardless,

1 The reader will find that some of the points or conjectures made herein are simplified versions of elaborate arguments that I have made elsewhere. I would particularly like to acknowledge the work of Pierre Briant, Nick Sekunda, and Christopher Tuplin. Even if I have not always agreed with them, I have found the thoughts of these scholars invaluable, and trust that the interested reader will seek out their work. Translations in general follow the relevant Loeb edition, modified where deemed appropriate.

A Companion to Greek Warfare, First Edition. Edited by Waldemar Heckel, F.S. Naiden, E. Edward Garvin, and John Vanderspoel.

it is always important to bear in mind that Greco-Roman sources for Persia, even those who had actually visited the empire, such as Clearchus of Soli, often supply information relating to the Achaemenid military for literary effect, rather than to give an accurate account of the forces of interest to us in this chapter.

"Palatine" Infantry

There is a strong literary tradition of an ample standing infantry force being associated with the Great King. Darius I, in his famous *Behistun Inscription* written after his victory over those opposing his royal designs, writes that "The Persian and Median army, which was under [the control of] me, that was a small thing" (2 line 18).[2] Small perhaps in the context of all the troops at the king's disposal, but a tradition in the Greco-Roman sources reports an Iranian infantry force of 10,000 strong. This myriad would appear to be what Herodotus refers to as the "Immortals" (ἀθάνατοι), surely the best known of all Achaemenid formations. The name itself is likely a mistaken translation of a Persian term, such as *anušiyā* or "followers [of the Great King]," a line of reasoning that accords with the terms employed on the *Behistun Inscription.*[3] Similarly, Isocrates writes of "the army which wanders about with the king" (4.145), which I presume refers, in the main, to what Herodotus calls the Immortals.

Despite this evidence, the evocative term Immortals—which suggests some kind of almost supernatural military prowess—has stuck, and Herodotus (7.83.1) provides an explanation of what the name referred to. He says that they were called thus because, if a member of the myriad died or fell sick, he would be immediately replaced, so that there "were never more or fewer than 10,000." Perhaps this is unremarkable to us in modern times, where standing armies can run into the hundreds of thousands, but we should recall that standing Greek armies of the time were extremely small, or even close to non-existent in some *poleis*, with armies only being assembled in times of need. Although his *Cyropaedia* is hardly the most reliable of sources, Xenophon (*Cyr.* 7.5.68) also refers to Cyrus, the father of the Achaemenid Empire, setting up a force of 10,000 Persian "Spear Bearers" (δορυφόροι) to "keep guard about the palace day and night, whenever he was in residence; but whenever he went away anywhere, they went along drawn up in order on either side of him."

This reference is curious, however, given that Xenophon calls the group "Spear Bearers," rather than use the epithet, Immortals. Even more curious is that probably the most famous description of a Persian army, penned by Herodotus (7.40.1–41.2), also fails to use the term. In his description of the army of Xerxes about to cross a bridge over the Hellespont on its way to attempt to subjugate Greece, this author presents a detailed description of not only the various allies, but also those Persian troops who marched around their king. I have

2 Translation of Schmitt 1991, 57; cf. that of Kent 1953, 123: "The Persian and Median army which was with me, this was a small (force)." Note that, at the same *locus*, a certain Hydarnes was made their leader; cf. Hdt. 7.83.1, where the "Immortals" are commanded by "Hydarnes son of Hydarnes." The *locus* also corroborates, to some degree, the view that the name ἀθάνατοι was a mistranslation of the Persian *anušiyā* or "followers [of the Great King]"; on this, see n. 3. This also perhaps affirms the suggestion (discussed at n. 82) that the Immortals were composed of Persian *and* Median troops.

3 Herodotus (7.211.1) interestingly says that it was the king who called these troops Immortals.

written on this description in detail elsewhere,[4] and arrived at the following breakdown of the Persian infantry being described:

- 1,000 with golden-pomegranate-bearing spears reversed marching before the king (referred to as αἰχμοφόροι or "spearmen");
- 1,000 with golden apples on their spears stationed closest to (and directly behind) the king (referred to as αἰχμοφόροι or "spearmen");
- 10,000 composed of another 1,000 bearing spears with golden pomegranates, and 9,000 carrying spears with those of silver (all are referred to simply as πεζοί or "foot soldiers").[5]

Once again, a tradition appears of a 10,000-strong force of Persian infantry associated with the king. The other two groups of 1,000 infantry represent something of a problem, however, that we shall examine in the second section of this chapter. Suffice it to say, for the moment, that this *chiliad* (or two!) represents the infantry bodyguard of the Great King, and not the regular standing infantry. Furthermore, the notion of 10,000 Persian infantry reappears at 7.55.2 of Herodotus' narrative, while the Immortals, now described thus, remain with the senior military commander Mardonius at 8.113.2 when Xerxes returns to Persia.

Numbers aside, there is a reasonably strong tradition of some sort of Persian infantry force that, even if it is not a royal bodyguard, serves as a core infantry force. What became of this force, which Herodotus *sometimes* refers to as the Immortals, is not clear after Xerxes' reign. Pausanias (6.5.7) preserves a story of three Immortals being involved in mortal combat with a famous Greek pankratiast at the court of Darius II, who became king in 423. Then these infantrymen go missing until an appearance in the Roman writer Curtius Rufus' description (3.3.13) of the troops of Darius III at some point before the battle of Issus against Alexander the Great in 333.[6] I have written elsewhere that Curtius' description of these Immortals is fanciful, and that it probably reflects the penchant of Greco-Roman authors for describing Achaemenid military processions.[7] This is especially likely given that neither Arrian nor Diodorus refer to such a force in their descriptions of Alexander's battles against Darius III.

If the so-called Immortals were eventually disbanded, we are, of course, prompted to ask why this occurred. Perhaps the answer has to do with the fighting role of these troops. Regardless of questions of armor, which we will look at in the penultimate section of this chapter, soldiers described as "spear bearers" were meant to assume a role in the center of a battle. Perhaps we might call them, somewhat anachronistically, "heavy infantry," that is, troops intended to fight at close quarters in massed formations. We encounter such troops on two occasions in Herodotus' narrative on Thermopylae (480 BCE). In the first instance (7.211.1), after the unsuccessful attempts of other units under Persian command, including the Medes, to break the lines of the Spartan King Leonidas and his men, the Immortals are thrown into the fray. These, too, are rebuffed and fail to penetrate the narrow pass, where the Persians' greater numbers proved to be of no advantage. Later, when Xerxes is informed of a path that will enable the Persians to get to the Spartan rear, the Immortals are again deployed (7.215).

4 See Charles 2011, 116–124; see also id. 2015a, 17–18, where I persisted with my previous thinking on the matter.
5 Charles 2011, 119.
6 Note that 10,000 spearmen (*decem milia hastatorum*) are also located behind the king's chariot at 3.3.20 of Curtius' narrative; cf. Hdt. 7.41.1. In effect, Curtius might be describing two lots of Herodotus' Immortals!
7 Charles 2011, 25, with id. 2015a, 22–23.

Yet in Persian eyes their function might have been different, given the common use of larger defensive wicker shields (γέρρα).[8] If that type of shield was indeed used by the Immortals, holding the line may be a more important task than making assaults.[9] In any event, battlefield tactics changed after the Persian wars described by Herodotus. General-purpose troops able to offer greater flexibility came to the fore, such as the so-called Iphicratean πελτασταί, who were not peltasts in the original Thracian sense but were rather a kind of "ersatz" hoplite.[10] To some degree at least, these supplanted the line-of-battle hoplite infantry that had once been the mainstay of Greek forces. Since the Persians understood that a significant military threat was emanating from across the Hellespont, they may have developed a general-purpose infantry arm of their own. This force, I think, was the *Kardakes.*

At the battle of Issus (333), as the curtain was about to close on the Achaemenid Empire, Arrian (*Anab.* 2.8.6) describes a force of 60,000 "κάρδακες" arrayed around a central force of 30,000 Greek mercenaries. Arrian describes these *Kardakes* as "heavy infantry" or ὁπλῖται. As I have argued elsewhere, Arrian was not referring to the familiar hoplites of the Peloponnesian Wars, or indeed to peltast-like infantry, but was rather referring to general-purpose infantry.[11] The real heavy infantry at Issus were surely the Greek mercenaries, so placing peltast-like troops on either side of them does not quite ring true. Indeed, the fact that the Greeks held the center of the line suggests that they were serving as replacements for the missing Immortals.

At this stage in the Achaemenid Empire, there does not seem to be great importance attached to preserving core units of ethnic Iranians. In such a multicultural empire, preserving such distinctions may have come to be been regarded as an anachronism. Perhaps a large, standing force of Iranians was feared as a possible tool to be used by dissenting Persian nobles or close royal relatives against the king. And so Greek mercenaries may have come to be regarded, in time, as not only excellent soldiers, but also the most trustworthy ones, particularly given that their commanders were also Greek.

Much uncertainty, however, remains about the *Kardakes.* I would dismiss Strabo's contention at 15.3.18 that "*Kardakes*" refers to Persian recruits, and give preference to Briant's argument that they were a multiethnic force *commanded* by Persians, with Persians also being part of their rank and file.[12] Numbers, of course, are a distraction, for we have no way of being sure that Arrian, in his description of Issus, was not following the sort of sources that aimed to augment Greek success against Near Eastern foes by exaggerating the numbers of the enemy. Whatever the case, the *Kardakes,* so named, do not reappear at Gaugamela (331 BCE). We must presume that they had been wiped out at Issus, or that Arrian was using a different source for his account of Gaugamela; that is, one that did not identify any of the Persian infantry as *Kardakes.* Perhaps some of these were among the mixed Persian cavalry and infantry on Darius' left wing (Arr. *Anab.* 3.11.3). At Gaugamela, a depleted force of Greeks—the remnants of the 30,000 at Issus—and the Great King's bodyguard, the so-called "Apple Bearers"

8 Note that the phrase "defensive shield" is not as tautological as it might initially appear. Indeed, the large, curved rectangular *scutum* of the Roman legionaries of the Principate not only protected the soldier carrying it, but also its large metal boss was intended to be punched against opposing soldiers, a disarming move that would enable the double-edged *gladius* to be thrust into the enemy's lower torso.

9 Note that there seems to have been different sorts of shields used by the Persians.

10 See Sekunda 2007, 328, 339, with Diod. Sic. 15.44.2–3.

11 Charles 2012b, *passim.*

12 Briant 1996, 121. For other views, see Sekunda 1988a, 42.

(μηλοφόροι), presumably the descendants of the elite spearmen described by Herodotus in the context of Xerxes' campaign against Greece, held the center, as Arrian (3.11.5) indicates.[13] It is this infantry bodyguard that we will look at next.

The Infantry Guard

As we have seen, Herodotus alludes to an elite group of infantry who carried spears with golden apples, presumably at the butts of their spears. Although he does not call these troops Apple Bearers, they seem to be the same men, for guard infantry are so described in later times. Witness, for example, Arrian's account of the armies of Darius, where no numbers are given, and the account of Diodorus, where they are merely described as "brave and numerous."[14]

A conundrum, of course, is presented by Herodotus' description of *two* groups of elite infantry in Book 7, one with golden pomegranates on their spears, and another with golden apples. I support Sekunda's proposition that what is being described *is two halves of the same unit*, one being placed in front of the king, and the other behind.[15] This is likely why the former troops carry their spears reversed, in deference to the king behind them, who might not have appreciated the spears of his own bodyguard being pointed in his direction. Perhaps Herodotus was relying on two sources, one that associated apples with the unit, and the other pomegranates. Herodotus, being unaware that the sources were talking about one and the same unit, simply doubled up. It could well be that, even if their total number was 1,000, and not the 2,000 that Herodotus seems to imply, 500 men were posted in front of the king, and 500 behind—if we assume that Herodotus is describing a marching order that actually occurred, something of which we cannot be certain.

Of interest, too, is that, in his description of the battle of Plataea, Herodotus writes that the supreme commander in Greece, Mardonius, was accompanied by a thousand "picked men," described as "the best (τούς ἀρίστους) of the Persians" (9.63.1). While these have often been regarded as a cavalry unit, I see no reason to reject my previous assertion, prompted by Flower and Marincola, that these men were infantry.[16] Thus, they were likely the infantry guard, also described as "cuirass wearers" or θωρηκοφόροι earlier in the *Histories* (8.113.2).[17] That Mardonius was mounted at Plataea is beside the point, for ancient military commanders were not accustomed to directing battles on foot. A further clue is that Herodotus had previously used the term "the best (οἱ ἄριστοι) and most noble of the Persians" with clear reference to an infantry unit (7.41.1). This adds further weight to the view that Mardonius' guard was what later ancient writers called the Apple Bearers, although

13 Arrian (3.11.7) writes that the Greek mercenaries, together with the Persian guard contingents, were stationed opposite the Macedonian phalanx as they were regarded as the only troops with the ability to offset this powerful force.

14 Arr. *Anab.* 3.11.5, 3.13.1, 3.16.1; Diod. 17.59.3. As I have written previously (Charles 2011, 127), Diodorus' description "does not particularly sit well with a unit size of 1,000," but, then again, Diodorus might not have had any information at hand regarding their numerical strength, and so this is simply his embellishment.

15 Sekunda 1992, 6–7.

16 Charles 2011, 123, with id. 2015a, 19, following Flower and Marincola 2002, 218.

17 Masaracchia 1990, 215, observes that the troops in this passage seem to be fighting in an infantry and not cavalry encounter.

why these men should be protecting Mardonius, rather than Xerxes—who had long departed for home—is difficult to explain.

Another reason why the Apple Bearers were 1,000 strong is the term used for their commander, the Chiliarch (χιλίαρχος), which we shall capitalize in order to distinguish him from other commanders of chiliads. Questions remain regarding the function of this senior officer. Suffice it to say that the Chiliarch, a Greek translation of the Persian *hazārapatiš*, often appears to have wielded enormous influence, so much so that, on some occasions, he was even regarded as the second most important person in the Achaemenid Empire.[18] As commander of the king's infantry guard (although only one Chiliarch is attested to have led them), he was closely involved in the protection of the king, and, by extension, could influence who was granted a royal audience.[19] Perhaps the influence of the Chiliarch depended on who held the office, and on his closeness to the king.[20]

To sum up, the infantry guard were 1,000 strong, under a Chiliarch.[21] I reiterate my previous contention that the main tradition preserved by the Greco-Roman sources is correct in reporting both a Persian palatine infantry of 10,000, often known as Immortals, and a unit of 1,000 crack infantry serving as a royal bodyguard often called Apple Bearers.

There remains the troublesome tradition that the 1,000 Apple Bearers were not separate from the Immortals but instead were drawn from their ranks, as in Athenaeus (12.514b). This report, borrowed from Heraclides of Cumae, has satisfied many modern scholars.[22] It contradicts Herodotus' presentation of elite spear-bearing infantry as separate unit(s) *outside* the 10,000 Persian infantry. Herodotus does say that one unit of the 10,000 had spears with golden pomegranates, which seems to suggest that the 10,000 were composed of ten chiliads. This, at least, conforms to our understanding of the decimal system supposedly employed by the Achaemenid Persians. My view is that Herodotus, in producing his description of the Persian army at 7.40.1–41.2, is likely drawing upon one main source and supplementing it with others, which has led to possible chiliads of golden pomegranate bearers as well as to two units of elite "Spear Bearers."

Whether there was a pathway from the Immortals to the Apple Bearers is uncertain, even if I once suggested such a thing.[23] It may be that the Apple Bearers drew upon the sons of the elite of Persian society, in much the same way as the "Kinsmen Cavalry" (συγγενεῖς ἱππεῖς), which we will discuss in the next section of this chapter, appear to have been composed of members of a certain exalted class of Persians.

18 Nep. *Conon* 3.2, perhaps exaggerating for literary effect.

19 For Artabanus, see Charles 2015b, 287–288. Even then, Diodorus (11.69.1) calls Artabanus' men "Spear Bearers" or δορυφόροι. That said, the term δορυφόροι is often found in the context of guards at the palace, so I have no trouble equating these Spear Bearers with the pple Bearers," as other ancient authors call them.

20 Charles 2015b, 299–300.

21 It follows that in Book 7 Herodotus, or his underlying source(s), erred in describing the Spear Bearers as two groups of 1,000 men. In 8.113.2 and 9.63.1, however, the historian rightly reports an elite Persian infantry unit of 1,000.

22 Herakleides, *FGrH* 689 F 1. Modern views: Charles 2011, 120–125.

23 Charles 2011, 132.

The Cavalry Guard, or Kinsmen Cavalry

If the Apple Bearers were 1,000 strong, one might reasonably expect that the Kinsmen Cavalry, as their mounted counterpart, would be of the same strength. There are two main pieces of evidence. First, Herodotus' describes a select group of 1,000 cavalry at 7.40.2 ("a thousand horsemen, selected from all the Persians"; ἱππόται χίλιοι ἐκ Περσέων πάντων ἀπολελεγμένοι) and 8.113.2 ("the thousand horse"). Second, Diodorus asserts at 17.59.2, in the context of the battle of Gaugamela, that an elite unit designated the Kinsmen Cavalry, numbered 1,000 horsemen. Curtius' testimony at 3.9.4 that Darius III's "customary bodyguard" (*assueta corporis custodia*) was 3,000 strong does not seem to align with other accounts of the Great King's horse guard. Then again, Curtius presents much perplexing information about the Persian army. As we have seen, he even countenances the continued existence of the Immortals before Issus. Let us put his puzzling figure of 3,000 to one side.

Of course, a problem is that Herodotus mentions *another* group of 1,000 Persian cavalry at 7.41.1. The language used to describe it, however, is remarkably similar to that used to describe the group at 7.40.2—witness ἵππος ἄλλη χιλίη ἐκ Περσέων ἀπολελεγμένη ("a thousand horsemen picked out of the Persians"). But only a single unit of picked cavalry (described simply as "the 1,000 horse") appears at subsequent descriptions of the Persian army at 7.55.3, when the Persians cross the Hellespont, and at 8.113.2, when Mardonius chooses which men will remain with him after Xerxes' return to Persia. As with infantry guard, it may once again be that Herodotus' sources have got the better of him, and that there was only one unit of genuine guard cavalry, which, once again, had been split into two groups of 500 men.[24]

Note that, in this description of the Persian army on the march in Book 7, we also find reference to a group of 10,000 cavalry, a cavalry match for the infantry Immortals. This is also likely the group referred to at 7.55.2, where they are simply described as "the horsemen" (ἱππόται) in another assembly of Persian troops, although no number is given. This notion of a cavalry equivalent to a myriad of Persian infantry is exclusive to Herodotus. For example, while Xenophon (7.5.68) refers to Cyrus establishing a force of 10,000 spearmen, which in a way corroborates Herodotus' Immortals, he does not mention a cavalry equivalent. A 10,000-strong cavalry force was present at the Granicus (334 BCE), according to Diodorus (17.19.4), but these were satrapal troops of various ethnicities, as the battle took place at some distance from the Persian heartland.[25] What Herodotus, or rather his source/s, was referring to here with the 10,000 cavalry is uncertain. My best guess, if Herodotus and his source/s were indeed describing *something* that existed, is that they were alluding to a large body of Iranian cavalry, perhaps a permanent force, but more likely not. One does not usually find a cavalry corps being the same size as the infantry, aside from exceptional cases such as Hunnic or Mongol armies, which relied almost exclusively on mounted troops to do the bulk of the fighting.

Of course, Achaemenid armies had recourse to a large cavalry component. But a good many of these mounted troops, it would appear, came from *outside* the Iranian homeland of the Achaemenids, or were recruited for specific campaigns. Arrian's description (3.11.3) of

24 Charles 2015a, 20.

25 Arrian (*Anab.* 1.14.4) gives a much larger figure in any case. I cannot see these *loci* as supporting the idea of a Herodotean standing cavalry myriad associated with the Great King.

Persian cavalry and infantry being mixed on the left wing at Gaugamela, together with groups of other ethnicities, probably refers to such troops. In Herodotus' view of the world, cavalry of this nature would not have been grouped together with the properly Iranian permanent troops marching around the Great King in Book 7.

According to Xenophon (*An.* 1.7.11), a group of 6,000 cavalry of unstated origin under the command of the noble Artagerses were stationed in front of Artaxerxes II at the battle of Cunaxa (401 BCE), where Cyrus the Younger, claimant to the throne, sought to defeat his relative. These troops, however, were supposedly easily dispersed by a force of 600 cavalry under the royal pretender (*An.* 1.8.24), so they do not appear to have been of high quality. As a result, one hesitates to connect them with the Herodotean myriad of horse.[26] So, we must either believe Herodotus, or rather his source/s, or contend, in the face of a lack of corroborating evidence, that there was no standing cavalry myriad. Let us leave the matter here, as further conjecture merely muddies turbid waters.

To return to the Kinsmen Cavalry, as they are generally described in the post-Herodotean literature, this unit appears to have existed from at least the time of Xerxes' attempted subjugation of Greece right up until the death of Darius III. In that sense, there is a parallel to the Apple Bearers, which are similarly present—differences in nomenclature aside—in accounts of the early years of the empire through to the very end. Indeed, Darius, hiding from Alexander, who sought to capture him in the wake of the disaster at Gaugamela, might even have been accompanied by members of this unit, together with remnants of the Apple Bearers, right up until his death. This is especially so if my previous (admittedly unprovable) suggestion that Pompeius Trogus' *cognati*, preserved in Justin's epitome (11.15.1) as those accompanying the king, could refer to the "Kinsmen [Cavalry]" rather than merely a certain class of (aristocratic) Persian.[27] Certainly, Arrian (3.16.1) tells us that "many" (πολλοί) of the Apple Bearers, together with the Bactrian cavalry and the Kinsmen Cavalry accompanied Darius in his flight from Gaugamela (331 BC).[28]

While the commander of the Apple Bearers is usually described as *the* Chiliarch, or the Chiliarch *par excellence*, as I have previously described the role,[29] it is uncertain whether the commander of the Kinsmen Cavalry normally enjoyed a similarly high status. Perhaps it might be argued that infantry, being without horses, were a more appropriate "intimate" bodyguard of the king, and so their commander enjoyed a more privileged position on account of his usual proximity to the ruler.[30] Then again, it is not improbable that, in some cases, the commander of the king's cavalry guard enjoyed a more prominent standing,

26 See Charles 2015a, 22.

27 Charles 2016b, 61–62. Curtius (3.3.14) refers to 15,000 "kinsmen of the king" (*cognati regis*), but Curtius is not a particularly reliable source for information about Achaemenid institutions. When ancient authors refer to "Kinsmen," it is difficult to work out whether they are referring to a social group, or the cavalry guard. A good example is Diodorus' description (17.20.2) of the battle of the Granicus (334), where 40 "Kinsmen" (συγγενεῖς) were present. I am now not very sure of my view (see Charles 2015a, 23–24) that these were a detachment of an elite cavalry unit.

28 My thinking on this was influenced by the presence of Nabarzanes, who seems to have been connected in some way with the "Kinsmen Cavalry," either as their commander, or perhaps as commander of *all* the guard troops of the king, at the betrayal of Darius. See Charles 2016b; see also id. 2016a, 395, with extensive references to earlier scholarship.

29 See Charles 2016b, *passim*.

30 That, however, did not stop the mounted *Germani* protecting the Caesars in their palaces, these being the emperor's most intimate guard in the early Principate.

depending, perhaps, on his personal relation to the king. Scholars have recently argued about such matters in a number of articles.[31] Suffice it to say that the reality, like many aspects of our understanding of Achaemenid Persia, is not fully recoverable.

Why properly Persian cavalry do not appear more frequently in the sources, despite Xenophon's emphasis on the social status of Persian cavalrymen in his *Cyropaedia*, is a quandary. Tuplin responds that, while the use of cavalry increased over the Achaemenid era, their status vis-à-vis the infantry is unlikely to have changed, their relatively infrequent appearance in iconography being indicative of this.[32] An important remark is that, "[t]here is remarkably little sign that the Greeks found Persian cavalry inherently awesome or exotic, and there is no perceptible demonizing of the Iranian horseman."[33] The famous report of Herodotus that Persians were taught to ride, to shoot, and to tell the truth (1.136) reflects this attitude, no matter how little it has to do with Persian education, or with Persian comparisons of infantry and cavalry. If Tuplin's view holds, it follows that the relative silence in the sources regarding truly Iranian cavalry is not surprising after all.

Arms and Armor

It has often been thought that the vast majority of Achaemenid troops went into battle without body armor. This, it would appear, is an exaggeration based on Herodotus' statement at 9.62.3 regarding Persian troops lacking the protective armor of Greeks. At Plataea (479), the Persian infantry under Mardonius' command were, according to Herodotus, at a disadvantage compared to the Greek hoplites opposing them as they were "unarmored" (ἄνοπλοι). As some form of corroboration, Pausanias (1.40.2–3) also writes of armored Megarian troops making short work of some unarmored troops under Persian command, although we are not told their ethnicity—perhaps they were not genuinely Iranian troops in any case, and so are not of interest to this study. The testimony of Herodotus and Pausanias is augmented by visual representations of Persian troops without any kind of defensive arms.

Visual depictions, however, can be misleading, as Persian iconography, such as illustrations from the royal palaces at Susa and Persepolis, tends to depict soldiers in court settings, where armor would have proved largely unnecessary, and certainly burdensome. The Principate provides similar depictions of Praetorians in scenes around the Roman emperor, such as the Cancelleria Relief from Rome itself. While weapons are still carried, the troops often limit themselves to wearing a helmet, and dispense with the cuirasses we know that they wore when on campaign. To return to Persia, overt displays of arms and armor within the palace and its surrounds, in any case, could have been interpreted as weakness on the part of the Great King. Instead, troops in light order, though trained as line-of-battle infantry, contributed to the portrayal of the king as being personally secure and thus emphasized the "civil" nature of his rule. For the same reason, Roman troops did not regularly wear armor within the city in the early Principate, that is, when the fiction of the emperor being less of a warlord and more of a father figure was still in play. Then again, depictions of Persian infantry on

31 See Meeus 2009, 287–310; Collins 2012, 159–167, building on id. 2001, 259–283.
32 Tuplin 2010, 179–182.
33 Tuplin 2010, 180.

Greek pottery, as well as the so-called Alexander Sarcophagus, generally show them without body armor.

In addition, those Persian troops conscripted for certain campaigns might *not* have been heavily armored, particularly if they were intended to act mainly as missile troops, the Achaemenid penchant for barrages of arrows being well attested.[34] Yet those troops who were intended to act as line-of-battle infantry, i.e. the sort that we have dealt with in this chapter, quite likely wore some sort of armor in the field. Despite his earlier statement about unarmored Persians, Herodotus (7.61.1) writes that the Persians under Xerxes wore what I have interpreted elsewhere as a form of scale armor—although the passage is garbled.[35] This *locus* would appear to refer to infantry, as we read of the Persians carrying a large wicker shield. Although they wore a cuirass, it appears that they were without helmets, instead being equipped with cloth headwear called πῖλοι. Herodotus refers to similar body protection for cavalry at 9.22.2, where the mounted Persian noble Masistius is described as wearing a purple tunic over a cuirass (θώραξ) made from golden scales. At 7.84, we read that the Persian cavalry are equipped like the infantry, which I take it means that they also normally wore a cuirass, but wore a helmet of hammered bronze or iron, rather than the cloth headgear of the infantry.[36]

The Medes (7.62.1) were equipped like the Persians. Indeed, according to Herodotus (7.62.1), the Persians wore a type of armor adopted from the Medes. Clearchus (49 Wehrli = *FHG* II 304), preserved in Athenaeus' *Deipnosophistae* (12.514d), confirms a tradition of Persians borrowing Median military practices, such as the palace guards adopting the apple-butted spears of the Median guards (hence the term Apple Bearers).[37] So, Herodotus is at least consistent with other sources on this point. Yet Herodotus (1.135), again presumably mixing up his sources, elsewhere writes that, although Persian soldiers might wear Median dress, because of its beauty, they used "the Egyptian cuirass in war." Although the Egyptian soldiers are described in Herodotus as wearing cuirasses at 7.89.3, the Assyrians (7.63) are described as being equipped in similar fashion to the Egyptians, and wore "linen cuirasses" (λινέους θώρακας). My guess, after reflecting on the argument of Stylianou, together with references elsewhere in Herodotus to ornate Egyptian linen cuirasses for votive purposes (2.182.1, 3.47.1–3), is that these Egyptian-style cuirasses of Book 7, called "shoulder-piece corselets" by some scholars, were indeed likely to be made of linen.[38]

Confirming that even Persian troops could wear linen, Plutarch (*Alex.* 32.5) writes that Alexander wore a cuirass of "two-ply linen" at Gaugamela, this cuirass having been captured at Issus. While this cuirass was likely the armor of a noble, and therefore probably a cavalryman, it at least suggests that such cuirasses were used by the Persians up to the very end of the Achaemenid period. Before we finish with Herodotus, also consider his aforementioned reference to "cuirass wearers" (θωρηκοφόροι) at 8.113.2. My conjecture—impossible to prove—is that these men are guard infantry. Some translators of Herodotus have seen these

34 It is always amusing to recall the Laconic quip of the Spartan Dieneces, who, upon being informed that the enemy was so numerous that their arrows, when they shot them, blocked out the sun, responded happily that, "if the Medes hide the sun, we shall fight them in the shade and not in the sunshine" (see Hdt. 7.226.2).

35 Charles 2012a, 260.

36 On Achaemenid cavalry equipment, see Bittner 1985, 180–225, 274–288; Nefedkin 2006, 5–18.

37 See Charles and Anagnostou-Laoutides 2016, 149–164.

38 See Stylianou 1999, 345. I have listed a number of other references to linen cuirasses from the Near East at Charles 2012a, 262 n. 21.

men as cavalry, yet the context refers to these troops *and* "the 1,000 horse" (and, of course, the Immortals) as those men chosen by Mardonius to remain with him in Greece, together with an array of other troops. Here, then, could be evidence that the elite Persian infantry would normally wear body armor.

Other sources even less reliable than Herodotus refer to armored Persians. Pockmarked as it is by anachronisms, Xenophon's *Cyropaedia* is always of dubious utility with respect to the Persian military, but it frequently refers to Persians wearing body armor. In *Cyropaedia*, at 2.1.16, for example, we read that Cyrus determined that the Persian commoners, instead of being archers and javelin-throwers, should be fitted with a cuirass, and should carry the large wicker shield. The word "cuirass" (θώραξ) is also associated with Persian soldiers, and possibly infantry, at 1.2.13, 2.1.21, 2.2.9, and 2.3.17–18.[39] Likewise, Strabo (15.3.18), in his troublesome passage describing Persian military recruits as *Kardakes*, writes that they wore a cuirass of iron scales. Unfortunately, it is difficult to believe these sources, since Strabo in particular wrote long after events. On the matter of body armor, let us conclude that *some* of the Persian troops, and quite likely the palatine or guard forces, did wear cuirasses, and that the Persian cavalry, particularly those of noble background, also wore such protection. While the notion of unarmored Persians probably refers more to non-ethnic Persians, or Persians who had cast aside such armor for some reason (perhaps a lack of serviceability late in the campaign), it might also have resulted from some Persians wearing a garment *over* their armor—as would make sense given their Middle Eastern origin.[40]

A few words on offensive, not defensive, equipment. While soldiers likely could be equipped with bow and arrow, as we see on visual depictions of Persian infantry from the Iranian heartland, their main weapon in the line of battle, was a thrusting spear, a shield of some sort being held in the other hand. I think it more likely that the well-attested large wicker shield was the province of the infantry recruited for particular campaigns, rather than an accoutrement of the standing, palatine forces. Yet the short length of the Persian spear seems to have been a disadvantage against Greek hoplites, whom we might imagine as classic heavy infantry, that is, troops intended strictly for the line-of-battle, especially at the very center of an engagement. Herodotus (7.211.2) tells us that, at Thermopylae, the Immortals fought against the Lacedaemonians with spears that were shorter than those of the Greeks, with their reach not being as great as a result. One would presume that the length of the Persian spear changed after that, although we cannot be sure.

Concluding Remarks

Although the Great King could call on a vast array of subject peoples to provide infantry and cavalry, the properly Iranian force under the command of the king was comparatively small. Although the 10,000-strong Immortals, which seem to have constituted the core of this palatine force, disappeared at some point after Xerxes' reign, Greek mercenaries were called on to provide the main heavy infantry component of the standing army, while a mysterious force known as the *Kardakes* likely provided a general-purpose infantry. The armies defeated by Alexander the Great, which had these highly trained Greek mercenaries positioned in the

39 For further detail on these *loci*, see Charles 2012a, 265.

40 Charles 2012a, 260 n. 14.

center, and perhaps numbered 30,000, were thus superior, in terms of line-of-battle infantry, to the armies that confronted the Greeks at Marathon and Plataea a century and a half earlier. The Persians had improved, just as the forces of Alexander may be said to have improved on the Classical Greeks (except for the well-drilled Spartans).

With respect to permanent Iranian cavalry, the only formation that appears to have existed throughout the Achaemenid dynasty is the Kinsmen Cavalry, which appears to have been only 1,000 strong, and thus were mounted counterpart of the infantry Apple Bearers. A myriad of horse only appears once or possibly twice in *The Histories*.[41] Yet no other reference to such a grouping is detectable.

Of course, an overview such as this one merely attempts to make sense of a tight array of literary sources, many of which contradict each other—many of which, indeed, are internally contradictory, notably Herodotus.[42] The other important point to grasp is that, as Briant and Meeus have pointed out, we should not expect to find continuity throughout the Achaemenid era.[43] Rather, we must contemplate a woefully incomplete picture of the Persian military, and take solace that we are able to put together a few pieces of the puzzle.

41 Twice if the horsemen at 7.55.2 are the same as the earlier-mentioned 10,000 Persian cavalry.

42 I have previously referred to the "extreme limitations of the documentary sources available to us" (Charles 2015b, 300), and find my doubts have only increased whenever I have returned to them.

43 Briant 1994, 292–293; Meeus 2009, 304.

CHAPTER 14

Parthian Warfare Under the Early Arsacids[1]

Marek Jan Olbrycht

The Arsacid Parthian state (248 BCE–CE 226) emerged in the Trans-Caspian steppes of western Central Asia and in northeastern Iran in the mid-third century. By 140, it turned into an empire stretching from the Euphrates to Afghanistan, including sedentary Iranians (Parthians, Medes, Hyrcanians, etc.), nomadic tribes (including the Dahae), Greeks, Babylonians, Jews, and other peoples. The Parthians enjoyed renown as a people with particular military skills. In his *Philippic Histories* epitomized by Justin, the Roman historian Trogus (flourished under Augustus), praises the Parthians as strong rivals of Rome who were beset by the Romans "in three wars, during which they faced the greatest commanders of Rome when she was at the height of her power; and of all the people worldwide they alone emerged not just on equal terms but as victors," although they were "plagued by serious conflicts with the Scythians and other neighbors" (Justin 41.1.7–8).[2]

The evidence for Parthian warfare comes from varied sources—literary, epigraphic, numismatic, archaeological, and iconographic.[3] This chapter covers the rule of the Early (or Older) Arsacids (c. 248 BCE–CE 11), roughly corresponding to the Hellenistic period. While scrutinizing Parthian warfare of that age, some developments will be discussed that belong to the reign of the so-called Later Arsacid dynasty (CE 11–226).

1 The present chapter was written thanks to funding from the Institute for Advanced Study at Princeton (USA) and the Gerda Henkel Stiftung (Germany).

2 On military aspects of Parthian history, see Koshelenko 1980; Olbrycht 1998, 2003.

3 Only a few studies have focused on selected aspects of Parthian warfare so far; see Koshelenko 1963, 1980; Pugachenkova 1966; Khazanov 1968, 1971; Coulston 1986; Wilcox 1986; Nikonorov 1987, 1994, 1998, 2013a; von Gall 1990; Mielczarek 1993; Olbrycht 1993, 2003, 2010a, 2010b, 2012, 2015, 2016; James 2004; Hauser 2006; Pilipko 2006; Wójcikowski 2014.

A Companion to Greek Warfare, First Edition. Edited by Waldemar Heckel, F.S. Naiden, E. Edward Garvin, and John Vanderspoel.

Army Organization

The Arsacids created a state with a significant number of fighting men to draw upon, and with a strong, efficient army. They employed an inclusive approach to subjugated peoples and kingdoms that allowed for the strengthening and broadening of the Parthian military power. Justin's account of Arsaces I (248–211/210), in relating the military triumph of a leader from the steppes in the northeast, focuses on two aspects: that Arsaces raised a large army (41.4.8) and that he built fortresses and strengthened the cities (41.5.1). Thus, military affairs were crucial from the very beginning of the Arsacid dynasty.

The armed forces of the Parthian kings may be termed a composite, as they comprised 1) a permanent army, including the royal guard units and royal garrisons in the major cities, fortresses and governors' troops in the provinces; 2) armed forces mobilized in the royal domain in the event of war, including both royal troops and private armies from the clan estates such as the Sūrēn and the Kārin; 3) the armed forces of dependent kingdoms (such as Adiabene, Media Atropatene, Characene, and Elymais); 4) mercenaries (e.g. of Greek, Sacan, or Sarmatian origin); 5) the troops of allies, often not distinguishable in the sources from mercenaries.[4] Every Parthian nobleman who was a member of the *āzādān* (free men) class was obliged to do military service and, depending on the value of his estate, to supply a unit of soldiers (Justin 41.2.5–6).[5]

The numerical strength of Arsacid armies varied. Phraates II (133/132–127) was able to muster an army of 120,000 soldiers against the forces of Antiochus VII Sidetes in 129/128.[6] The population of Parthia proper, Hyrcania, Areia, most parts of Media (not occupied by Antiochus' forces), and the tribal levies of the Dahae furnished this force.[7] According to Justin (41.2.6), the entire Parthian army that appeared on the battlefield under Phraates IV against Mark Antony in 36 consisted of 50,000 horsemen. At the same time, another army totaling at least 40,000 infantry defended Phraaspa and other strongholds in Media Atropatene (Strab. 11.13.1–3). Surenas' army at Carrhae (53) numbered at least 10,000 cavalry including 1,000 cataphracts (Plut. *Crass.* 21.6). More than likely there were a few thousand more horsemen, since Plutarch did not include the forces of Sillakes, governor of Mesopotamia.

Parthian kings often led their armies in person, e.g. Phraates II around 127 against the nomads (discussed further later), Phraates III in 66, Orodes II in 53, and the latter two against Armenia in 36, and Phraates IV against Mark Antony the same year.[8] Minor local operations were conducted by provincial governors or appointed commanders, such as Himerus "the Hyrcanian" against rebels in Babylonia (120s) and Sillaces and Sūrēn in Mesopotamia against Crassus in 54–53.[9] Those were usually members of the major Parthian clans including the Sūrēn and the Kārin.[10]

4 For details, see Olbrycht 2016.

5 On the interrelations between the social division and military organisation in Parthia, see Koshelenko 1980; Olbrycht 2003.

6 Porphyrios *FrGrHist* 260F 32 (19).

7 Olbrycht 2016, 311.

8 Phraates III, Dio 36.51; App. *Mithr.* 104; Orodes II, Plut. *Crass.* 21.5; 22, the latter plus Phraates IV, Dio 49.25.2.

9 Babylonia: Diod. Sic. 34.19.21; Justin 42.1.3; Poseidonios, *FGrHist* 87 F13. Sillaces: Plut. *Crass.* 21.4, 33.3f.; Dio 40.12.2; Flor. 1.46.8; Oros. 6.13.3; Ruf. Fest. 17.2.

10 Plut. *Crass.* 21.6; Tac. *ann.* 12.12.3–12.14.2.

Weapons

Horse riding epitomized the Parthians and their cavalrymen used a set of weapons that was a novelty in Iranian warfare. This set included a bow (with a gorytos and arrows), a sword, a dagger, and different types of javelins and spears. Some soldiers were equipped with lassos, hatchets, and slings. There was a regional diversity in arms and armor.[11]

In the coin iconography of the early Arsacids, the key element is the bow as an attribute of the royal power that became the most common symbol of the Parthians. Another motif, often used by the Arsacids on their coins and in minor arts, is the gorytos (Fig. 14.1). The best-known Parthian sculpture, the *Prince from Shami* (first century), demonstrates typical Parthian costume and weapons. The figure features a v-necked jacket, trousers, and leggings. The prince has two daggers attached at the right and left hip (Fig. 14.2).[12]

Perhaps there has never been a people more renowned for their archery skills than the Parthians. Originally the Parthians used a composite bow of the Scythian type, often asymmetric.[13] Longbows appeared in Parthian coinages in the reign of Mithradates II (122–87).[14] They may have been introduced in the course of the second-century Parthian wars against the nomads in Bactria and eastern Iran.

The Parthian bow distinguished itself by the efficiency due to its reinforced construction. Bone laths for bows (second century BCE to first century CE) are known from nomadic tombs at Gyaur 4 and Tarym-kaya 1 in Chorasmia (Uzbekistan[15]) that can be ascribed to tribes related to the Dahae. Some Parthian terracotta reliefs show riders with bows that apparently are provided with bone laths.[16] The famous Yrzi bow, found at Baghouz near Dura Europus and dated between the first century BCE and third century CE, was reinforced by ear bone laths.[17] No bone laths have been discovered on Achaemenid sites, like Persepolis, so such reinforcements were conceivably introduced in Iran by the Dahae and the archers of the Arsacids.[18] The Parthians used arrows with iron (and bronze) heads. Arrows with three-sided or bullet-shaped heads were the most efficient for piercing armor.[19] Plutarch (*Crass.* 24.4, 25.6) underscores that the Parthians could pierce enemy breastplates and armor, as at Carrhae (Plut. *Crass.* 25.6, 18.3). There were various kinds of gorytos depending on the type of bow.[20]

At some point, the Parthians began to use the so-called Sasanian bow.[21] Sasanian bows were doubly convex composite weapons (up to 1.5 m), with long, straight, or slightly curved ends.[22] The handle and ears were stiffened with bone or horn laths. An early depiction of the

---

11 On Parthian weapons, see Nikonorov 1987; von Gall 1990, passim; Mielczarek 1993, 61–62; Olbrycht 2015, 358–369.

12 Olbrycht 2015, 360–361.

13 Litvinskii 1966, 57.

14 Litvinskii 1966, 58.

15 Yablonskii 1999, pl. 35.17–27; 55.6–14.

16 Coulston 1985, 240.

17 Brown 1937; Coulston 1985, 239–240.

18 Coulston 1985, 240.

19 On Parthian arrowheads found in south Turkmenistan, see Pilipko 2015, 32, 66, 78, 97, 173, 183.

20 Winkelmann 2006, 141–143.

21 Bows of this type often appear later on Sasanian vessels and this is why some scholars use the term "Sasanian bows" (Maenchen-Helfen 1973, figs. 7–9, 228–232).

22 Nikonorov 1997, II, fig. 44m-0.

Fig. 14.1 An Arsacid Gorytos. Terracotta slab from Old Nisa (second century BCE). National Museum, Ashgabat. Turkmenistan.
Photograph by Marek Olbrycht

Sasanian bow appears in the coinage of the Indo-Greek ruler Apollodotus II (c. 80–65).[23] Very large Partho-Sasanian bows are known from the ivory panels found at Takht-e Sangin in Bactria (dated between mid-first century BCE and mid-first century CE), and from plaques discovered at Orlat in Sogdiana (first to second century CE).[24] A graffito in a private house at Hatra (probably second century CE) shows a gazelle hunter with the top part of his bow and bowstring visible. The stave ends in a narrower, straight section, which is probably a straight ear typical of the Partho-Sasanian bows.[25] Images from Dura Europus depicting bows feature a variety of types.[26]

The long sword (80–130 cm) with a double-edged blade was an essential offensive weapon of the Parthian cavalrymen. Such swords, worn in wooden scabbards, were intended for thrusting and slashing from horseback.[27] A Parthian sword appears on the Parthian stele found at Ashur (early first century CE).[28] Long swords were also weapons widely used by

23 See Nikonorov 1997, II, fig. 22d; 1997 I, 47, 53.
24 Tacht-e Sangin plaques: Olbrycht 2015, fig. 3; Orlat plates: Olbrycht 2015, fig. 4.
25 Venco Ricciardi 1998, fig. 7. Cf. figs. 4, 5.
26 James 2004, fig. 117.
27 Olbrycht 2015, 358–360.
28 Olbrycht 2015, fig. 20.

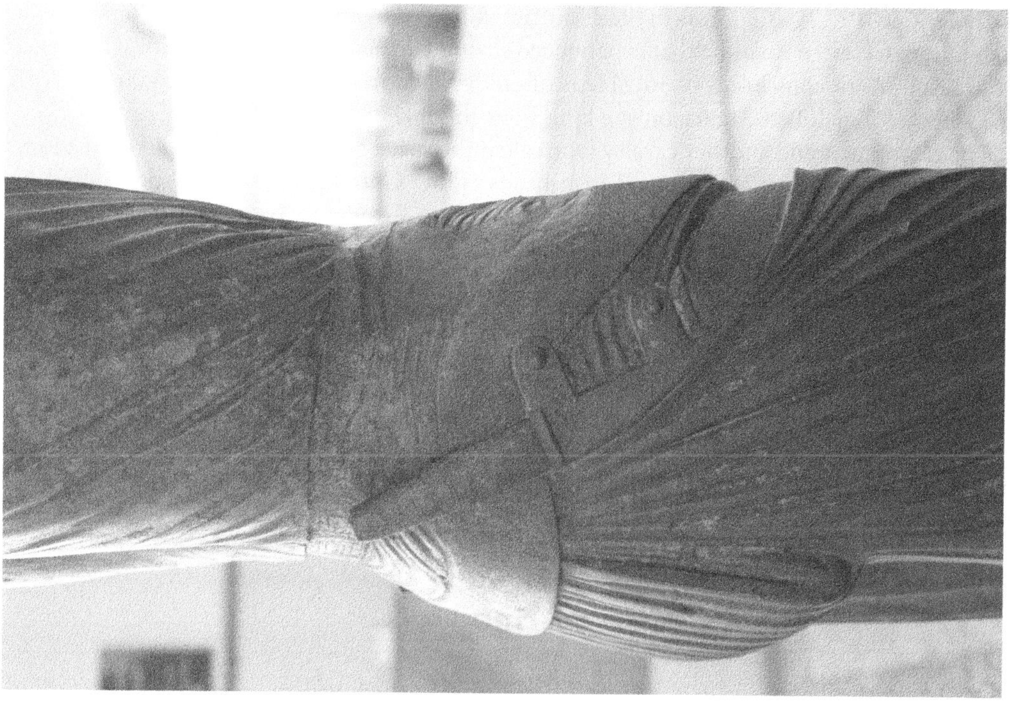

Fig. 14.2 The Prince of Shami (first century BCE). Bronze sculpture, fragment. National Museum, Tehran. Iran.
Photograph by Marek Olbrycht

nomads of Central Asia in the Achaemenid age and have been discovered on post-Achaemenid nomadic necropolises in Bactria, Sogdiana, and Chorasmia.[29] Unlike Parthian-period, armies, however, Achaemenid armies rarely used them. Parthians themselves may have used shorter swords in light cavalry units (Dio 39.29.3–4).

The sword and dagger scabbards worn by the Parthians were often suspended from the waist belt by means of a special device made of jade, called the scabbard slide. These were introduced in Iran in the Arsacid period. A scabbard slide is represented in the Parthian relief from Ashur (early first century CE).[30]

The most common Parthian dagger type seems to be the ring-pommel dagger worn in an elaborate scabbard. Of special significance were the scabbards with four-sided applications or projections, depicted in art monuments and in coins and known from original finds. This sort of dagger originated in the Altai region and China, then it began to occur in Bactria and Sogdiana (second century BCE), and was adopted by the Parthians.[31]

Under Orodes II (57–37), daggers began to be represented in Arsacid coinages as a special royal attribute, the same as the bow. At the beginning, the ring-pommel daggers appeared, with scabbards usually provided with four rounded applications at the sides and one at the chape. This type can be seen on the coins of Orodes II to Artabanus II, i.e. in the period c.

29 Nikonorov 1997, II, fig. 33a,b, 34m.
30 Olbrycht 2015, fig. 20.
31 Olbrycht 2012, 505–509, 2015, 360–369.

57 BCE–CE 40.[32] It also occurs in the countries impacted by Parthia, including Commagene, a kingdom in eastern Anatolia that experienced strong Parthian cultural and political influence under King Antiochus I (c. 70–36). He is often depicted in an ornamented royal garb, with an elaborate four-lobed dagger on the king's left thigh (Fig. 14.3).[33]

The cataphracts and another cavalry type called the spear-bearers (*kontophoroi*) used a long spear called the κοντός.[34] These had iron heads (Plut. *Crass.* 27.2). *Kontoi*-spears used by Parthian cataphracts could transfix "two men at one go," evidently because of their length.[35]

Weapons of note were found in the Parthian fortress of Old Nisa (Turkmenistan), mainly in the Big Square House, dated to the second to first century.[36] However, most of the arms discovered there were likely captured from defeated enemies and deposited in the Arsacid Nisa treasury.[37] One of them is a battle hatchet made of gilt silver. This weapon was likely a trophy from a victorious campaign by one of the Parthian kings, possibly by Mithradates II (c. 122–87) against the Sacae. Among the Sacae, the hatchet was an attribute of power.[38] In all likelihood, similar hatchets were used by the Parthians but the evidence is not conclusive. There are representations of the hatchets at Dura Europus, Tang-e Sarvak, and on Indo-Sacan coins of Spalirises and Azes I.[39]

Armor

The cataphracts were usually clad in full body armor, which was of several types, including scale, lamellar, and hybrid armor. Plutarch (*Crass.* 24.1, 25.7) uses the general term θώραξ for Parthian body armor. Justin speaks of *lorica plumata* for the armor that covers both man and horse entirely.[40] Sallust describes Parthian cataphracts fighting at Tigranocerta in 69 as wearing complete iron armor and their horses as being provided with textile trappers reinforced with fastened iron plates resembling scales.[41]

The heavy armor of the cataphracts is depicted on Indo-Sacan coins of the first century BCE. It consists of four main elements: a corselet with a high collar, armguards, thigh-guards (*pteryges*), and a helmet.[42] The Indo-Sacae used the same type of armor as the Parthians. The cataphract corps of Sūrēn at Carrhae, regarded as a typical Parthian cavalry force, comprised units from eastern regions of the empire, likely including Sacan contingents from Sacastan and Arachosia.[43]

32 Olbrycht 2012. Depicted are also other variants of daggers, see Winkelmann 2006, 143–144.
33 Antiochus I's daughter married Parthian king Orodes II (*SEG* 33.1215).
34 Plut. *Luc.* 28.3; *Crass.* 24.3; 25.8; Dio 40.24.1; Heliod. 9.15.1.
35 Plut. *Crass.* 27.2; Heliod. 9.15.
36 Pilipko 2006.
37 See Pilipko 2006, 259.
38 For details, see Nikonorov 2013a, *passim.*
39 Nikonorov 2013a, *passim.*
40 41.1.10: *munimentum ipsis equisque loricae plumatae sunt, quae utrumque toto corpore tegunt.*
41 *Hist.* 4.64.4, ed. Maurenbrecher: *equites catafracti, ferrea omni specie.* 4.64.6: *Equis paria operimenta erant namque linteo ferreas laminas in modum plumae adnexuerant.* For Parthian trappers made of metal, see Plut. *Crass.* 24.1; Justin 41.2.10; Arr. *Parth.* Fr. 20 = Suda s.v. θώραξ.
42 Bopearachchi and Sachs 2001, 328, fig. 4, fig. 8A–G.
43 Olbrycht 1998, 115–117.

Fig. 14.3 Antiochus I of Commagene (c. 70–36 BCE), king of Commagene, and his dagger. Relief at Antiocheia on the Nymphaios. Turkey.
Photograph by Marek Olbrycht.

The armor of Parthian riders was made of iron and untanned leather (Plut. *Crass.* 25.7). The use of leather could be twofold: either it was worn as underlay for metal armor elements, or as independent protection.

Scale armor seems to have been the most common type in Parthia. Fragments of scale armor sets were discovered at Old Nisa in the Big Square House (second century).[44] A painting of scales was found in the Tower Building at Old Nisa (second century), and there are several representations of scale armor from western Parthia dated to the first to second century CE.[45]

An armor set discovered in the Arsenal of Ai Khanoum in Bactria (first half of the second century) consisted of a scale corselet, a leg-guard, a pair of shoulder-guards, and a piece of horse armor (plastron or flank protection).[46] Numerous pieces of scale armor were found in the Iberian palace of Dedopolis Gora (Georgia), dated to the first century BCE to first century CE.[47] Lamellar armor appears in the Parthian relief at Tang-e Sarvak (second century CE) as protection for the horse (trapper) and rider.[48]

---

44 Pugachenkova 1966, 33–34; Pilipko 2006, 264–265.

45 Painting: Pilipko 2006, 267–268. Other representations: Mielczarek 1993, fig. 13 (terracotta from Babylonia); Wójcikowski 2014 (graffito from Dura).

46 Nikonorov 1997, II, fig. 13b–e, p. 7.

47 Gagoshidze 2008, 129–132. Besides, a set of mail (*lorica hamata*) was discovered at this site (ibid., 132–133).

48 von Gall 1990, 64.

An important component of cataphract armor was the armguards. A pair of shoulder-guards made of iron scales/lamellae and leather was discovered at Bactrian Ai Khanoum (first half of the second century).[49] Two sets of armguards consisting of overlapping rings of graduated sizes (made of iron or steel) come from the Indo-Sacan site of Sirkap/Taxila (first century CE).[50]

At Tigranocerta the legs of some cataphracts were unprotected and the Gallic soldiers are said to have attacked the riders' limbs effectively, but usually the legs of Parthian riders were covered by leather guards or metal plates.[51] Such armor pieces appear on Indo-Sacan coins (broad flank protection of metal plates) and in the relief at Tang-e Sarvak (Fig. 14.4).

The Parthians used high collars to protect the neck. Such a collar is documented on coins of Tanlismaidates, a ruler from western Bactria in the first century who was subject to the Arsacids. Similar collars are known from the sculpture of Khalchayan and from the Orlat plaques.[52]

Military helmets of Parthian cataphracts were made of metal (bronze or iron) and were usually low, oval or semi-spherical in shape. Tanlismaidates, for one, wears a low oval helmet, suited to the head, with a neckguard and plume. The Parthian rider from a British Museum terracotta (first to second century CE) wears a semi-spherical helmet with a band at its base, and with a visor. The helmet protects the ears.[53]

A typical cataphract helmet appears atop a sculpted figure from Khalchayan depicting a Sacan warrior. The Khalchayan helmet features an oval shape with a protruding visor. The bowl is fitted to the head and has a neckguard, and under the helmet the warrior wears a cap made of felt to protect against heat and absorb enemy blows. Similar helmets worn by Indo-Sacan rulers are represented on their coinages. A Sacan iron helmet of oval shape, discovered at Taxila, features cheek pieces and is reinforced by horizontal bands attached to it, and by a boss on the summit.[54] Additionally, Parthians made use of leather helmets.[55] In some neighboring regions, such as Iberia and Albania, helmets of hide were commonly used (Strab. 11.4.5). Under the Later Arsacids, conical helmets often occur.[56]

At Old Nisa, a helmeted sculpted head was discovered in the Square Hall Building (second century). The helmet is of the Attic type and features cheek pieces.[57] Such helmets were used by Greek or Macedonian soldiers enlisted in contingents of Parthian armies or garrisons. It is likely that the helmeted head belonged to a composition depicting figures of Seleucid princes, as enemies or relatives, stressing the glory of the triumphant Arsacid house.

Some cavalry and infantry carried shields, but their horses received comparable, but more complex, protection from armored trappers.

---

49 Nikonorov 1997 II, fig. 13b–e.

50 Marshall 1951, 550.

51 Plut. *Luc.* 28.4 with Mielczarek 1993, 60.

52 Abdullaev 2012, figs. 5, 6.

53 Mielczarek 1993, fig. 13.

54 Nikonorov 1997, II, 13, fig. 26a–d (Indo-Sacan coins), fig. 30a (helmet from Khalchayan), fig. 35c (helmet from Taxila).

55 Tac. *ann.* 6.35 (*galea*) with von Gall 1990, 70.

56 Nikonorov 1987, 13–14; Mielczarek 1993, 57.

57 Nikonorov 1997, II, 8, fig. 17.

Fig. 14.4 Parthian Leg Armor. Rock III (first to second century CE). Iran. Photograph by Marek Olbrycht.

The Cataphracts: Formative Period

Arsacid cavalry tactics, based chiefly on the combined use of the cataphracts (or *kontophoroi*) with their long spears, and of horse-archers with powerful bows, proved extremely efficient, and enabled the Arsacids' victories over the Seleucids, the Roman Empire and other enemies. To avoid confusion one needs to define what is meant here by western Asian "cataphracts" (from the Greek κατάφρακτος), since not every type of heavy armored cavalry may be legitimately denoted by this term.[58] One may describe the western Asian "cataphracts" as a type of cavalry with the riders wearing heavy armor and armed with a long spear (called *kontos/contus* in sources) as their principal weapon, and with a sword, dagger, and frequently a bow and arrows for offensive warfare. Moreover, cataphract mounts were usually clad in metal armor or in trappers of leather.[59] Another salient point is cataphract tactics: they attacked in close battle array, and were effective only if deployed in coordination with light horsemen, best of all with mounted archers as a screening and supporting combat force.

Apart from the cataphracts, there was also the *kontophoroi* (spear-bearers), in which riders (often armored) used long spears, but the horses wore no armor. Cassius Dio (40.15.2) describes the Parthian armed cavalry forces as consisting of mounted archers and

58 On the cataphracts, see Pugachenkova 1966; Khazanov 1968, 1971, 71–80; Nikonorov 1987, 1994, 1998; Mielczarek 1993, 51–67; Olbrycht 2010a, 2010b, 2015, 339–374; Abdullaev 2012.

59 Khazanov 1971, 71–74; von Gall 1990, 70.

spear-bearers, "most of them being cataphracts" (ἱπποτοξόται δὰ καὲ κοντοφόροι, τὰ πολλὰ κατάφρακτοι, στρατεύονται).

Cataphracts emerged from the development of cavalry in the Achaemenid Empire. The military traditions of the Central Asian Massagetae provided Achaemenid commanders with a model for heavy cavalry using armor and long swords.[60] A major advance came with the reforms implemented by Darius III (336–330) and his commanders following the Persian defeat at Issus (333). Darius III's new heavy cavalry combined armor for horse and rider with the use of long spears imitating Macedonian *xysta* and *sarissai* (Diod. Sic. 17.55.1. Cf. Curt. 4.9.3). The use of long spears by heavily armored riders and attempts at combining heavy cavalry with mounted archers allow us to speak of proto-cataphracts in the Achaemenid army at Gaugamela (331).[61] The proto-cataphracts thus resulted from Central Asian nomadic cavalry tradition, Achaemenid military innovation, and the Macedonian challenge.

After the death of Darius III, Spitamenes put up bitter resistance to Alexander's army in Sogdiana and Bactria (329–328).[62] The methods of offensive warfare Spitamenes used, involving a combination of mounted archers and lancers like cataphracts, foreshadowed the Parthian art of war under the Arsacids. The final stage of development took place in the Trans-Caspian area in the late fourth and third centuries, when cataphracts began to carry a long spear modeled on the Macedonian long spear (*xyston*) and pike (*sarissa*).

The oldest depiction of a heavy armored horseman who may be called the cataphract comes from Chorasmia, an area traditionally connected with the steppe peoples of the Massagetae and the Dahae: a relief from Khumbuz-tepe shows an armored horse and rider with a spear (end of the fourth to the beginning of the third century). The peoples of Chorasmia and the Trans-Caspian steppes undoubtedly saw the need to develop cataphracts as a force that could fight the phalanx, plus cavalry of the Macedonian type as well as cavalry from Iran and from the steppes. At the same time they developed units of *kontophoroi*. A depiction of an Iranian rider with the long spear (*kontophoros*) comes from Koi-krylgan-kala in Chorasmia. Another depiction on a ceramic fragment recently found at Kalaly-gyr 2 in Chorasmia shows a rider with a spear held with both hands (late fourth to third century).[63] Original fragments of a scaled corselet and laminated leg-guard were discovered at Chirik-Rabat in the Syrdarya basin.[64]

In Western Asia, the cataphract cavalry are first mentioned under this name as part of the army of Antiochus III. Since Antiochus III marched against the Parthians and Greco-Bactria in 209–206, this is where the Seleucid drew his inspiration from this encounter.[65] Convinced of the effectiveness of the Parthian cataphracts, Antiochus III made them the main striking force at Panion in 200 (Polyb. 16.18–19). At Magnesia (188), the Seleucid cataphracts were able to crush both cavalry and infantry units of the Roman army, but weak command on the part of Antiochus caused his eventual defeat.[66]

---

60 Hdt. 1.215; Strab. 11.8.6 with Olbrycht 2004, 144–147.

61 Olbrycht 2010a, 2010b.

62 Olbrycht 2004, 128–150.

63 See Olbrycht 2015, fig. 25a,b,c: reliefs from Khumbuz-tepe, Koi-krylgan-kala, and Kalaly-gyr 2.

64 von Gall 1990, 68.

65 Antiochus III in Parthia and Hyrcania: Polyb. 10. 27–31; Justin 41.5.7.

66 Liv. 37.40; App. *Syr.* 32. Antiochus fielded 6,000 cataphracts and 1,200 mounted archers. For Seleucid cataphracts, see Mielczarek 1993, 67–72. The first scholar to draw attention to the Parthian genesis of Seleucid cataphracts was W.W. Tarn, who asserted that the "Parthian re-organisation of Iran" in terms of new warfare practices was accomplished by the first century (1930/1984, 72).

The arms and armor of the Seleucid heavy cavalrymen may be reconstructed thanks to the frieze of the stoa surrounding the temple of Athena at Pergamon (second century). The reliefs display the arms looted from the defeated enemies of Pergamon, including the Seleucids. One of the reliefs depicts a cuirass with shoulders and *pteryges*. A segmented arm-guard is similar to that from Ai Khanoum. The frieze also depicts segmented head-guards and breast-guards for horses.[67]

Mention should be made of Parthian horse equipment. The Parthians used four-horned saddles. An early representation of such a saddle comes from a graffito (first century) in the Square Hall Building of Old Nisa.[68] The horned saddle, fastened to the horse's body by straps, was made of a wooden framework and leather.[69] Apparently, horned saddles developed to provide the heavily armored rider with stability, as the Parthians did not use the stirrups. Such saddles are depicted in the scenes from Kosika (first half of the first century, discussed later) and in Parthian minor arts.[70] The horned saddle likely was an element of cataphract equipment from the beginning of the Arsacid era. The use of the horned saddle may be linked with the "the introduction of heavy cavalry, that is by the beginning of the Parthian period."[71] Other types of rigid saddles with high front and rear arches were also used.

The Parthians used the most famous horses of the ancient world, the "best and the largest" Nisaeans, plus similar mounts from Parthia proper and Armenia. Strabo (11.3.7) underscores that the Parthian horses "had a shape of their own." Greater Media in western Iran was renowned as a center for horse raising.[72]

The Arsacid Cataphracts and Other Cavalry Types

The combat record of the Arsacid cataphracts (Greek: κατάφρακτοι, Latin: *cataphracti* or *catafracti*), *kontophoroi*, and horse archers (*hippotoxotai*) reaches us mainly from accounts of wars between the Arsacid Empire and Rome. Additional information derives from coinages, minor arts, rock reliefs, and from original arms found at archaeological sites.

Arsaces I and his successors gained surprising victories in Iran and were able to defeat the Seleucids. During his invasion of Parthia in 209, Antiochus III feared the Parthian cavalry most, but Polybius's description breaks off, so we do not know about this war in detail. Parthian cataphracts were evidently unrivaled, and thus were imitated by Antiochus III. Similarly, cataphract cavalry are attested in a number of kingdoms of Western Asia that were influenced or directly subjugated by the Parthians—Adiabene, Armenia, and the Caucasus kingdoms of Iberia and Albania. Cataphracts developed in Bactria of the pre-Kushan und Kushan period.[73]

67 Mielczarek 1993, 70–72.
68 Nikonorov 2002, 21–27.
69 Details in: Herrmann 1989, 763–769; Nikonorov 2002.
70 Parthian objects depicting saddles: Herrmann 1989, 766, pl. Va–b.
71 Herrmann 1989, 764.
72 Polyb. 5.44.1; Strab. 11.13.7; Arr. *Anab.* 7.13.1. Cf. Hdt. 3.106; 7.40. See Tarn 1930/1984, 76–83.
73 Nikonorov 1997; Olbrycht 2015, 373, *passim*.

The first detailed reports of Parthian cataphracts come from the battle of Tigranocerta in 69, against a Roman army under Lucullus.[74] Lucullus was particularly afraid of these horsemen. Parthian cataphracts from Media Atropatene and Adiabene became part of the army of King Tigranes II of Armenia, who considered cataphracts the most important fighting force in his army (Plut. *Luc.* 28.2). This cavalry force thereafter flourished in Armenia. Artavasdes II of Armenia commanded a force of 6,000 cataphracts in 36.[75]

Plutarch in his account of the battle of Carrhae (53) offers the most detailed description of the Arsacid cataphracts and mounted bowmen in action.[76] At Carrhae, the Parthians demonstrated the effectiveness of the tactics of combined cavalry types. The Parthian force led by Sūrēn comprised about 10,000 soldiers, including 1,000 cataphracts. One should add a few thousand soldiers more for Sillaces' contingent, but we lack exact figures. The Roman force led by Crassus numbered about 42,000 soldiers. Parthian archers poured a hail of arrows into the Roman units, and cataphracts charged at them, disrupting their ranks. Thousands of Romans were killed or wounded, and the vanquished Romans retreated to Carrhae. During the ensuing negotiations between the Parthians and the Romans, Crassus was detained and put to death. An important point to note is Sūrēn's very well-organized logistics. He had a squad of 1,000 camels to carry his supply of arrows. Moreover, he also had a highly efficient intelligence service, which collected information from the rulers of Parthia's western marches (such as Edessa).

Valuable information on western Asian cavalry types may be gained from iconographic evidence. In the Parthian stronghold of Old Nisa, fragments of a late second-century frieze discovered in the Tower Building may depict episodes in the wars between the Parthians and the peoples of Central Asia conducted under Mithradates I (165–133/132), Phraates II (133/132–127), and Mithradates II (122–87).[77] One of the victorious riders can be identified as a spearman, another is a fugitive.[78] Both riders are without armor, their weapons are bows and spears. Some fragments depict figures with scale armor.[79] A classical Parthian cataphract is depicted at Tang-e Sarvak (Rock III) in southwestern Iran, roughly dated to the first to second century CE (Fig. 14.4).[80] He wears tube-like, segmented armguards. His legs and neck are protected by lamellar armor, while his breast seems to be additionally guarded by plate armor or by a textile gambeson. The horse is protected by a lamellar trapper.[81] The warrior holds his spear across the horse's' neck.

Indo-Sacan coinages of the first century show that the Sacae and the Parthians used the same types of cavalry and armament on account of military and cultural ties. The Sacan kings Sparilises and Azes I are shown on their coins as fully armored cataphract riders.[82] On some coins Sparilises is dismounted, grasping a battle-axe (hatchet) and a Scythian bow.[83] Judging

74 Plut. *Luc.* 26–28; Frontinus 2.2.4; App. *Mithr.* 85. Eutropius (6.9) and Festus (15) refer to Tigranes' heavy cavalry as *clibanarii*, using a late Roman term.

75 Plut. *Crass.* 18.3, 19.1; Plut. *Ant.* 50.5; Strab. 11.4.4, 11.14.9.

76 Plut. *Crass.* 23.8–27.2. See also Cassius Dio 40.21.1–24.1. Cf. Nikonorov 1995.

77 Pilipko 2006, 267.

78 Pilipko 2006, figs. 24, 25.

79 Pilipko 2006, 268.

80 Olbrycht 2015, fig. 28a,b.

81 See von Gall 1990, 64.

82 Nikonorov 1997, II, fig. 26c–d; Bopearachchi and Sachs 2001, fig. 8.

83 Nikonorov 1997, II, fig. 26a–b.

from the coin imagery (dominated by cataphracts but also displaying kings as archers) it is clear that the Sacae employed a Parthian combination of horse-archers and armored riders, or cataphracts.[84]

A Sarmatian tomb at Kosika on the lower Volga yielded an early first century silver vessel depicting mounted archers and *kontophoroi* armed with huge spears but lacking any body armor.[85] The vessel bears a Greek inscription reading, "Ampsalacus made this in the reign of King Artavasdes," who must have been king of Media Atropatene or Armenia, vassal kingdoms of Arsacid Parthia. Both kingdoms were engaged in far distant trade with the Upper Aorsi on the lower Volga.[86] The *kontophoroi* were not only Sarmatians but also Iberians, Iberia and Albania being closely connected to the military developments in the Sarmatian world.[87]

Another type of cavalry, the Dahan horse-archers, distinguished themselves in the wars against Alexander and later, in the Hellenistic period.[88] A little-noticed entry in the *Suda* s.v. Ἀγαθός implies that the martial skills of the Dahae became proverbial, and they earned the epithet "the good" because they were excellent warriors, that is, "good javelin men, and good too at hand-to-hand fighting."[89] The unidentifiable source very likely derives from Seleucid historiography. Effective fighting in these circumstances required some defensive armor (helmet and lamellar or scale armor for the rider) and suitable weapons including swords and daggers, so we can conclude that the Dahae were not only lightly armed missile troops.

Nomads such as the Dahae made another important contribution to Arsacid cavalry warfare, the kettle drum. The Parthians used the drum to give signals in battle, eschewing the trumpet used by the Romans (Justin 41.2.8). Plutarch confirms that the Parthians roused their men to action not by horns or trumpets, but by percussion instruments (*Crass.* 23.9, 26.4). Kettle drums would exert strong psychological impact upon enemies in battle.[90]

Parthian Infantry

Cassius Dio (40.15.2) informs us that Parthian infantry was small, "made up of weaker men" who were archers. However, during the campaign of Antiochus III in Parthia and Hyrcania, light Parthian infantry, including javelinmen and archers, played an important part in the war. Similarly, Parthian infantry bravely defended Syrinx in Hyrcania. Parthia proper, Hyrcania, Susiana, Elymais, and Media were able to furnish tens of thousands of infantry, including archers, slingers, and javelinmen, mostly without body armor.[91] In particular, Media Atropatene could provide as many as 40,000 foot soldiers (Strab. 11.13.2). These infantrymen proved their combat value when they defeated the army of Mark Antony at Phraaspa in 36.

84 Rosenfield 1967, 125.

85 Olbrycht 2015, 373.

86 Βασιλέως Ἀρθεουάζου Ἀμψάλακος ἐποίησεν, as at Olbrycht 2015, fig. 26. The Upper Aoprsi: Strab. 11.5.8.

87 Plut. *Luc.* 31, Strab. 11.4.4–5; 11.14.9.

88 Olbrycht 2004, 158–170.

89 Ἀγαθός: [...] τοὺς δὲ Δάους ἀγαθοὺς μὲν ἀκοντιστάς, ἀγαθοὺς δὲ καὶ ἐν χερσὶ ποιήσασθαι μάχην.

90 Details in Nikonorov 2000.

91 See Olbrycht 2016, *passim*.

Arsacid infantry included some heavy foot soldiers furnished by Greek-Macedonian cities like Seleuceia-on-the-Tigris, Seleuceia-on-the-Eulaeus/Susa, Babylon, and Merv (Antiocheia-in-Margiana). Local garrisons, such as that at Susa, consisted of Greeks equipped with helmets, body armor, and shields. The arms and armor of a Greek soldier in the Parthian age may be reconstructed on the basis of a bronze figure from Mesopotamia depicting a late Hellenistic soldier (first century), now in the Louvre. The soldier wears a helmet, cuirass with armored sleeves, and a leather belt.[92] A terracotta slab found at Bactrian Kampyrtepe (Uzbekistan) depicts a soldier in Greek-Macedonian armor dating to the first half of the second century. The soldier has a helmet of the Montefortino type, a cuirass with thigh-guards (*pteryges*), greaves (*knemides*), and a shield. His offensive weapon is the *xiphos*-sword.[93] Such soldiers may have served in eastern Parthian garrisons, including Merv (Antiocheia-in-Margiana), a city with a Greek-Macedonian population. Thousands of Greek soldiers were captured after the defeat of Antiochus Sidetes in 129/128, and they were forced to join the Parthian army of Phraates II. At some point they rebelled against the Parthian king and slew the Parthians in battle, supporting nomadic invaders (ca. 127: Justin 42.1.4–5).

Parthian Fortifications

The earliest Arsacids attached great importance to the erection of fortifications and strongholds.[94] Fortified centers formed a crucial element of Parthian defenses against the aggressive neighbors, including the Seleucid state, Greco-Bactria, some nomadic tribes of Central Asia, and Rome.

One of important foundations by Arsaces I was the city of Dara, "on Mount Apaortenon" (i.e. Apavortene), the first Arsacid residence mentioned in the sources.[95] Dara, a major, early Arsacid fortified center, was located in the mountains between the southern fringes of the Kaahka region in Turkmenistan and Kalat-e Naderi in Iran.[96] Ancient sources speak of an early Parthian residence at Asaak, where a holy Arsacid fire burned (Isid. *Stathm.* 11). It was located in Astauene, likely in the Bojnurd region in Iranian Chorasan. The stronghold of Old Nisa, located near Bagir (Turkmenistan), was called Mihrdātkirt in Parthian times. We may suppose that some constructions in Old Nisa belong to the early Arsacid period before Mithradates I (165–133/132). During the second century, Old Nisa became a dynastic cult center.[97]

Syrinx, in Hyrcania, is important for the study of Parthian fortification thanks to Polybius' account of the attack made by Antiochus III in 209. The city was protected by "three moats, each not less than thirty cubits broad and fifteen deep, and each defended at its edge by a double row of palisades, and behind all these was a strong wall" (10.31.8). In spite of

92 Rostovtzeff 1935, 234, fig. 46. He assumes that the figure represents a Parthian governor or vassal.

93 For a detailed study, see Nikonorov 2013b.

94 Justin 41.4.8, 41.5.1; Amm. 23.6.4, with Koshelenko 1963; Olbrycht 1993, 131–151.

95 Justin 41.5.2-4; cf. Plin. *HN* 46.

96 See Olbrycht 2014.

97 Throughout the early period, the capital was Hecatompylus, often identified with Shahr-e Qomesh near Damghan. It remained an Arsacid residence from Arsaces I on. Other Parthian royal residences were located in Hyrcania, in Ecbatana after 148, and in Ctesiphon, beginning in the first century. A survey of early Parthian residences: Chaumont 1973, 197–222.

defenders' sallies, Seleucid soldiers filled up the moats and breached the wall, thus capturing the city. From Polybius' account, it appears beyond doubt that the fortifications of Sirynx were built according to the requirements of Hellenistic fortification theory (three rings of moats with palisades and a *proteichisma*). Greek engineers in Parthian service probably did this work.

Under the Arsacid rule (from c. 160), Merv became a powerful stronghold on Parthia's frontier with Central Asia. In place of the destroyed Alexandreia/Seleuceia-in-Margiana, Antiochus I rebuilt the city, henceforth called Antiocheia. Merv's walls, preserved to a height of 7 m, were up to 6.6 m thick at the base. They were made of dried brick on a foundation of compacted clay blocks. Possibly, Erk-kala was also fortified under the Seleucids.[98] At an early stage (Phases 1 and 2), the Seleucid wall had a shooting gallery, a *proteichisma*, and a defensive platform. Merv's defenses were then greatly expanded under the Parthians (Phases 3 and 4; Fig. 14.5).[99]

Parthian poliorcetics is rarely described in sources, but in the second century the Parthians did not find it insuperably difficult to capture Ecbatana in Media, Babylon and Seleuceia-on-the-Tigris in central Mesopotamia, and Edessa in northern Mesopotamia. In about 54, Babylon was taken by Orodes II and Sūrēn as a result of famine caused by a long siege, but Seleuceia-on-the-Tigris was apparently stormed.[100]

Parthian Strategy

The Arsacid strategy changed over time. Expansionist conceptions predominated under Arsaces I, Mithradates I, Mithradates II, and Orodes II (c. 58/57–37). All these kings engaged in large military campaigns for conquest. Other kings, such as Phriapatius (c. 185–170), emphasized domestic development and avoided major confrontations. Finally, there were times, as under Arsaces I, Arsaces II, Mithradates I, Phraates II (133/132–127), and Phrates IV (c. 37–33), when the Arsacids were forced to pursue defensive strategies against their neighbors' designs—especially those of the Seleucid Empire, Greco-Bactria, and Rome.

In the early period, the foremost adversary was the Seleucid state; on occasion, Greco-Bactria posed a threat. Both these states were finally defeated under Mithradates I and Phraates II by the 120s. Beginning in the 90s, Arsacid political perspectives began to encompass Rome, which was to be Parthia's most powerful rival. Yet Parthia faced other formidable enemies in Bactria and Transoxania, and around 140–120 Greco-Bactria gave way to the strong nomadic tribes of the Sacae and Yuezhi (Tocharians). Some Sacan tribes moved southward into Drangiana/Sacastan and Arachosia.

In response to these various onslaughts, the early Arsacids strove to create something of a sanctuary in the Trans-Caspian steppes. Combining agricultural and urbanized areas in Parthia proper (Chorasan and southern Turkmenistan) with steppes (on the Uzboi) afforded the Arsacids freedom of movement and let them avoid all-out confrontations and then mount counterattacks. Such was the case of Seleucus II's invasion in 230 BCE, when Arsaces I took

98 Tashkhodzhaev 1963, 95–118; Filanovich 1974, 38.
99 Zavialov 2007.
100 Plut. *Crass.* 21.7; Justin 42.4.2-4.

Fig. 14.5 The Defenses of Merv (Gyaur-kala). Section of the city walls (southwest corner, Hellenistic and Parthian periods). Turkmenistan. Excavated by V. Zavialov.
Photograph by Marek Olbrycht

shelter with the Apasiacae (part of the Massagetae) and then launched a victorious counterstrike. To some extent, Arsaces II repeated this strategy against Antiochus III.[101]

From the reign of Mithradates I, the Parthian Empire also embraced all of Media and Babylonia, so the Arsacids adopted a new, imperial strategy. Northern Iran, with a center of gravity in Parthia proper (Chorasan) and Greater Media, became a bastion of Arsacid rule. In the northeast, the Arsacids made Merv and Margiana their chief bastion. Babylonia, dotted with many cities, including Seleucia on the Tigris, became the empire's economic center of gravity in the first century. Starting from 54 to 53, when Roman armies under Crassus invaded Mesopotamia and threatened the Babylonian cities, the Arsacids faced a challenge to their strategy, and in response Mithradates II (around 92) and Orodes II (in 41/40–38) tried to establish a buffer against the Romans by subjugating areas to the west of the Euphrates, including northern Syria and Commagene.[102]

101 Olbrycht 1998, 65–66.

102 Rostovtzeff 1935, 163 rightly stresses that for the defense of the Mesopotamian plains and for occasional attacks on the Roman provinces, the Parthian army "proved an excellent instrument, if in skillful hands".

The Arsacids also devoted much energy to controlling Armenia, beginning with Mithradates II, who conquered the region (Justin 42.2.6). It remained a bone of contention between Parthia and Rome for centuries; Caucasian peoples, including the Iberians and Albanians, eventually joined in the tussle.

To summarize, nomadic influences on Arsacid Parthia proved enduring in regard to armament, combat forces, and tactics.[103] These influences mingled with the Achaemenid-Hellenistic traditions of Parthia proper. Justin thus speaks of the "native" arms of the Parthians yet also of their "Scythian" arms: *armorum patrius ac Scythicum mos* (41.2.4). We should, however, amend this phrase to acknowledge the Macedonian-Greek influence manifest in Macedonian spears as the prototype for Parthian riders' shafted weapons and in Greek-Macedonian engineering.[104] The resulting synthesis, epitomized by cataphract cavalry, dominated warfare throughout Western Asia.[105]

103 A "steppe shift" in the warfare of Arsacid Parthia is stressed by several experts: Rostovtzeff 1935, 162–163; Nikonorov 1994, 1998, 2005; Coulston 1986; Olbrycht 2003, 2010a, 2010b.

104 Another view: Hauser 2006, denying the nomadic factor in Parthian warfare, "the idea of the Arsacid army as a basically traditional nomadic force has to be challenged for various reasons" (Hauser 2006, 298). He fails to engage fully with the evidence and insights in Olbrycht 2003, which demonstrates a nomadic contribution to the development of Arsacid cavalry tactics and armaments alongside Iranian and Macedonian-Greek factors. See also n. 103.

105 Tarn 1930/1984, 74 is right in his assertion: "When the Parni from the steppes originated the Parthian monarchy, their aristocracy must have brought this mode of fighting [sc. the heavily armored cavalry] with them".

CHAPTER 15

Elephants in Hellenistic Warfare

Christopher Epplett

Virtually no aspect of Greek society remained untouched by Alexander's conquests, including the military. The century prior to Alexander's reign had, of course, already witnessed important developments in this sphere, including the evolution of siege equipment and tactics,[1] but the Macedonian conquest of Persia and beyond was to add one more important facet to the Hellenistic armies that followed: war elephants. Initially, they had been one of the more unpleasant novelties encountered by Alexander's army, but they were nonetheless adopted in short order as an important instrument of war in the Greek world. Alexander himself did not have time before his death to integrate elephants properly into his own army, but his former generals, the so-called *Diadochoi*, were not slow to recognize the potential of war elephants and incorporate them into their own forces. From this point on, these animals became a staple of Hellenistic warfare, whose use spread to the armies of Carthage and Rome. Ultimately, however, the mixed combat record of war elephants, as well as the enormous expense required to maintain and train them, led to their demise as a military force.

Alexander's most famous encounter with war elephants is undoubtedly the Battle of the Hydaspes. But the Macedonian army had first encountered these animals some years earlier. According to Arrian, the massive army assembled by Darius III at Gaugamela in 331 included a contingent of elephants deployed in advance of the king's own squadron: the only specific number of elephants Arrian gives is the 15 supplied by, as he puts it, "the Indians on this [i.e. the western] side of the Indus" (Arr. *Anab.* 3.8.6, 11.6). These war elephants appear to have been part of the forces that the Persians levied from the tributary states on their eastern frontier in the period leading up to the final showdown with Alexander.[2]

1 See Chapter 9 in this volume.
2 Briant 2002, 757; cf. Arrian. *Anab.* 3.8.1-3.

A Companion to Greek Warfare, First Edition. Edited by Waldemar Heckel, F.S. Naiden,
E. Edward Garvin, and John Vanderspoel.

The animals played no significant role in the battle. Diodorus and Curtius do not even mention them in their accounts of Gaugamela, and Arrian says only that the elephants were captured in the Persian camp in the aftermath of victory.[3] The cumulative evidence suggests that, at the time of the battle, war elephants were a relatively new addition to Darius' army, and that they had not yet been effectively integrated with the other units. Indeed, there is no record of the Persians deploying war elephants in their previous battles with Alexander: as Charles suggests, Darius' last-ditch attempt to destroy the Macedonian army at Gaugamela may have prompted him to use all available weapons in his arsenal, even those that the Persians had not yet tested in battle.[4] The apparent ineffectiveness of Darius' war elephants was perhaps due to the Persian cavalry not yet being accustomed to the elephants' odor, which prevented the latter from being deployed near or advancing in concert with the Persian horsemen once the battle started.[5] As we shall see, a horse's natural disdain for the smell of elephants was an important consideration whenever the latter were added to an army, and played a role in at least one other major battle of the Hellenistic era.[6]

Despite the less than awe-inspiring performance of Darius' war elephants, Alexander appears to have been intrigued by them and their military potential. For the remainder of his campaign, he was happy to augment the number of elephants he captured at Gaugamela by whatever means presented themselves, a fact not lost upon his new subjects. Not long after Alexander's victory at Gaugamela, the satrap of Susa presented him with numerous gifts including 12 elephants said to have been imported from India by Darius, as a means of ingratiating himself with his new master (Curt. 5.2.9–10).

In all likelihood, the *paradeisos*, or royal game preserve, at Susa was one of the sites where the Persians maintained their fledgling force of elephants, which would explain how the satrap had a number of these animals at hand to offer to Alexander upon his arrival at the city. The Persians also appear to have kept elephants in the *paradeisos* at Babylon: the early fourth-century writer Ctesias of Cnidus, cited by Aelian, recorded seeing elephants in Babylon uproot date palms at the behest of their Indian trainers, evidently a training exercise for attacking structures and fortifications.[7] The evidence for such elephant facilities under the Persian Empire, like the evidence for their deployment in the Persian army prior to its defeat by Alexander, is limited; in all likelihood, however, the infrastructure for war elephants that had been created by the Persians continued to be used by both Alexander and the Seleucids.[8]

When Alexander and his army crossed the Hindu Kush and descended into the Indus valley, they encountered a new situation respecting the military use of elephants. While the Persians, as we have seen, had only recently begun to experiment with them, war elephants had been a staple of Indian armies for centuries prior to Alexander's arrival.

---

3 Arr. *Anab.* 3.15.4; cf. Fuller 1960/1998, 166; Charles 2009, 29–31.

4 Charles 2008, 12–13.

5 Scullard 1974, 64. The discrepancy among the ancient sources as to what, if any, role Persian war elephants played at Gaugamela has led to a number of different theories being proposed by modern scholars, including, for example, the claim that the elephants were only part of the Persian baggage train at Gaugamela, and therefore did not participate in the actual battle. For a full discussion of such theories, see Charles 2008, 14–20.

6 Cf. Glover 1948, 4–5.

7 Ael. *NA.* 17.29. Ctesias, in fact, is the first known Greek whose testimony on elephants was based upon personal observation rather than hearsay; see Wellman 1905, 2249.

8 Epplett 2007, 210–211. Regardless of whatever elephant facilities the Persians may have possessed, it is highly unlikely, as Charles (2008, 10–12) concludes, that any sort of captive breeding program was in place.

Indian War Elephants

Various Sanskrit texts attest to the elephant's presence on Indian battlefields from at least the early first millennium.[9] War elephants play a pivotal role in the early Indian epic, the *Mahabharata*. Not only are the rival forces of the Pandavas and Kauravas both said to include tens of thousands of war elephants, but one, in particular, is featured in a famous episode of the epic. The Pandava commanders, faced with the unconquerable martial prowess of the Kaurava leader, Drona, are forced to resort to a dishonorable lie in order to compel his surrender. If they can convince Drona that his son, Aswhatthaman, has been killed in the fighting, he will lay down his arms. The problem with this scheme is that Drona will only believe the report if told by one of the Pandava leaders, Yudhishthira, renowned for his refusal to ever tell a lie. The Pandavas ultimately circumvent this difficulty by killing a war elephant, which happens to share its name with Drona's son. Hence, when Yudhishthira subsequently tells Drona that Aswhatthaman has been killed, thereby compelling his surrender, he is literally not lying, although the morality of this scheme leaves something to be desired![10] By the time of the *Mahabharata*'s composition, then, war elephants were a commonplace of Indian warfare—creatures as highly esteemed as blooded horses in other armies.

Greek sources, in particular those derived from the writings of Ctesias on Persia and India, also attest to the ubiquity of war elephants in India centuries before Alexander the Great. First of all, Diodorus, in describing the reign of the semi-mythical queen Semiramis of Assyria, recounts the massive defeat she suffered at the hands of the Indian king, Stratobates, largely as a result of his overwhelming force of war elephants. According to Ctesias, the queen had attempted to offset Stratobates' force in an unorthodox fashion: she prepared 300,000 elephant dummies made of ox-hide, each with a camel and driver inside; the intention was that her enemies would be terrified when they saw that Semiramis could apparently field as many war elephants as the Indian king. Unfortunately for the queen, however, Stabrobates and his soldiers discovered Semiramis' stratagem in short order, and in the ensuing battle, her army was overwhelmed by the Indian war elephants.[11]

One aspect of Ctesias' description may relate to actual practice, and that is the use of dummies to acclimatize cavalry to the presence of elephants: Semiramis is said to have employed her mock-ups in such a fashion prior to the onset of her campaign against Stabrobates, and we know of one other instance when a Hellenistic ruler used such facsimiles in an attempt to make his horses comfortable with the appearance and smell of elephants.

Ctesias also described Cyrus the Great's final campaign against King Amoraeus of the Derbices, southeast of the Caspian Sea, a campaign that cost the Persian king his life. According to Ctesias, the war elephants that fought on the side of Amoraios played a pivotal role in Cyrus' defeat and consequent death. During the battle in question, the Derbices ambushed the Persian cavalry with elephants that had previously been kept in reserve, and then, in the confused struggle that followed, Cyrus was mortally wounded by a javelin

9 Scullard 1974, 28.

10 *Mahabharata* 7. 191; cf. Kistler 2007, 8–9. I thank my colleague, Dr. Hillary Rodrigues, for this reference.

11 Diod. Sic. 2.16.8–10; 2.18.6–19.6; Kistler 2007, 14–18.

thrown by one of the mahouts.[12] Unfortunately, we do not possess a Persian account of the end of Cyrus' reign with which to corroborate the information given by Ctesias.[13]

For our purposes, the most interesting aspect of Ctesias' account is the mention of elephants in the army that defeated the Persian king; these were supplied by Indian allies of Amoraeus. In addition, Ctesias states that an Indian threw the fatal javelin at the Persian king. This portion of Ctesias' account, if it has any basis in historical reality, suggests that the Indian practice of supplying elephants to their allies had a long pedigree prior to the Battle of Gaugamela; in addition, the Indians supplied not only the animals, but also mahouts to guide them in battle. When it came to states such as Persia, inexperienced in elephant warfare, the Indians would also have had to supply the attendants, trainers, and other support staff who kept the animals in good health and fighting trim between appearances on the battlefield.

Elsewhere in his work, Ctesias (as cited by Aelian) provides a general summary of the use of elephants in Indian warfare:

> When the Indian king goes to battle against his enemies a hundred thousand elephants of war form the vanguard. And I learn that another three thousand of the largest and strongest bring up the rear, and these have been trained to overturn the enemies' walls by attacking them when the king gives the order; and they overturn them by the weight of their chest. Such is the account given by Ctesias, who writes that this is hearsay.[14]

The alleged number of elephants employed by the Indians is dubious; by contrast, the use of war elephants both on the battlefield and in siege warfare is amply attested in ancient sources.

More and better information comes from a contemporary Indian source, the *Arthashastra*, ascribed to Kautilya, chief minister of the late fourth-century Mauryan emperor Chandragupta.[15] Book X outlines in detail the tasks and functions of war elephants:

> Their functions are: marching in front; making new roads, camping grounds and fords; protecting the flanks; helping to cross water and climb or descend from mountains; entering difficult or crowded places; starting fires or extinguishing them; using elephants alone for victory; reuniting [one's own] broken ranks and breaking up the [enemy's] unbroken ranks; protecting against dangers; trampling the [enemy's] army; frightening [by mere presence]; causing terror [in the enemy's army by action]; adding magnificence; capturing and liberating [battle positions]; destroying ramparts, gates and towers; and bringing [one's own] and carrying away [captured] wealth.[16]

---

12 Ctesias. fr. 9; Lenfant 2004, 112. Ctesias' account of Cyrus' death differs somewhat from that of other Greek writers: Herodotus, for example, relates that Cyrus was killed in battle against Queen Tomyris of the Massagetae, with no mention of any elephants in her army (Hdt. 1. 201–214). It appears, however, that Greek writers like Herodotus and Ctesias present the same basic account of the end of Cyrus' reign, insofar as the Persian king died in battle fighting against a Scythian tribe on the northeastern frontier of his empire.

13 Lenfant 2004, LXIII–LXIV; cf. Briant 2002, 49–50; Kistler 2007, 19.

14 Ael. *NA* 17.29; Ctesias *FGrH* 688F 45b; Lenfant 2004, 187. As Lenfant [CXL–CXLII] notes, Ctesias' use of the term "king of the Indians" is problematic, since there was certainly no single ruler of the Indian subcontinent at the time of Ctesias' writing. Perhaps the man being referred to was one of the local princes of the Indus basin, who may have had an equivalent status to the semi-autonomous dynasts of the western Persian Empire.

15 Scullard 1974, 242; Rangarajan 1992, 6–8; Sick 2002, 131–132, 136.

16 *Arthashastra* 10.4.14; Rangarajan 1992, 658.

Beyond their ability to rout the enemy through fear alone, the *Arthashastra* mentions the requisite "magnificence" of elephants, which could overawe and intimidate a potential enemy before a battle had even begun. As we shall see, the psychological effects of war elephants, particularly upon enemy troops unfamiliar with them, was a serious matter, widely credited with playing a role in Alexander's ultimate decision to halt his march to the East.

The passage just cited mentions elephants fording bodies of water, but elsewhere, the text goes into more detail about their role in amphibious warfare. Not only could they cross most rivers with relative ease and secure a bridgehead, they could also, on occasion, serve as bridges themselves. Book VII of the *Arthashastra* describes an enemy fort surrounded by water on all sides being attacked by troops crossing on the backs of elephants made to stand in a row between the banks of a river.[17] Presumably, some sort of planking or level surface would be placed on the backs of the animals once they were in position.

Their frequent mention in the *Arthashastra* attests to the importance of elephants not only on the battlefield, but also in Indian society, as do the penalties for stealing or harming elephants. Not only was the killing of such animals generally punishable by death, but the penalty for merely stealing or injuring an elephant from the royal herd was also death.[18]

The *Arthashastra* also discusses elephants in diplomacy and statecraft, noting that an independent ruler could be won over by an invitation to a prestigious event—a festival, a wedding, or an elephant hunt. A ruler subservient to another should organize an important undertaking such as an elephant hunt only after obtaining his superior's permission.[19]

Upon his arrival in India, then, Alexander confronted an enemy with long experience of, and a high regard for, war elephants. Seemingly impressed with their military potential, Alexander intensified his efforts to augment his own elephant corps. It is no doubt significant that the Indian leaders who met with the Macedonian king shortly after he crossed the Hindu Kush promised to bring him 25 elephants. Arrian is not explicit on this point, but it is reasonable to assume that Alexander had previously demanded the animals as a token of submission (Arr. *Anab.* 4.22.6). Not all of the local rulers, however, surrendered themselves or their animals to the Macedonian king without a fight, and Alexander shortly thereafter captured additional elephants in battle.[20]

At this early stage in his Indian campaign, Alexander had evidently come to recognize not only the military potential of elephants, but their diplomatic value as well. As Charles notes, such animals appear to have been frequently used as a bargaining chip in negotiations between the Macedonian king and various native rulers.[21] Like his Indian counterparts, Alexander began to stage elephant hunts soon after his arrival in the Indus basin, both as a means of augmenting his supply of war elephants and of demonstrating his own royal status to local rulers.

> Here he seized a few barbarians and learnt from them that the Indians of the district had fled ... but that they had left their elephants grazing there near the river Indus. He commanded them to guide him on the way to the elephants. Many Indians are hunters of elephants, and Alexander

17 *Arthashastra* 7.10.3–37; Rangarajan 1992, 580.
18 *Arthashastra* 2.2.8, 4.11.7; Rangajaran 1992, 458, 651.
19 *Arthashastra* 5.6.10–13, 7.15.22–29; Rangajaran 1992, 146, 627.
20 Scullard 1974, 65–66; Charles 2010, 332–333.
21 Charles 2010, 345–346.

> took pains to have them among his attendants, and at this time had their help in elephant hunts. Two of the beasts threw themselves over cliffs in the pursuit and perished; the rest were captured, permitted riders to mount them, and were [marshaled with] the army. (Arr. *Anab.* 4.30.7–8)

The Macedonian king not only captured elephants for his own army, but also enlisted experienced hunters from the local population to help him do so. Arrian's reference to the captured animals being incorporated into the Macedonian army is unfortunately vague, but it appears unlikely that they were already organized into formal units or the like at this early juncture.[22]

Lack of sufficient trained, coordinated elephants may have prompted the king's decision to avoid using elephants against the local Indian ruler, Porus, who had thus far refused to submit to the invaders. Alexander had evidently not had the time to mold his animals into an effective fighting force, and an apparent lack of the requisite trained personnel may even have made it difficult to keep them in good health. It is perhaps significant that when the Macedonians captured an additional 30 elephants shortly before the battle against Porus, Alexander turned them over to his ally, Taxiles, without even attempting to add them to his own elephant corps.[23] Success in the upcoming battle depended upon the superior mobility and maneuverability of the Macedonian army, which would be ill served by a lumbering force of war elephants (Charles 2010, 334–335).[24]

Another factor in the Macedonian king's hesitancy to employ his elephants in battle was the potentially disastrous effect they might have upon the rest of his army, in particular the cavalry. The latter had played a decisive role in previous victories, and was to do so again in defeating Porus' cavalry at the Hydaspes.[25] If, as seems likely, Alexander had not yet had time to accustom his horses to elephants, he was certainly not going to risk disrupting the battle-tested cavalry for the sake of deploying unproven war elephants.[26]

For the same reason, Alexander did not intend to attack Porus' sizable force of elephants with his cavalry.[27] How, then, would the Macedonian infantry deal with them? Alexander could rely not only upon the tactical versatility of his light troops, but also on the advice of his Indian allies. The Macedonian light infantry, evidently acting upon such advice, was first able to harry Porus' elephants from a distance with missiles such as javelins, and then drive the animals from the battlefield by advancing in open order and hacking at their legs and trunks, causing at least some of the injured elephants to turn upon Porus' infantry. This grisly task was accomplished with axes and, according to Curtius, curved cleavers that he calls

22 Charles 2010, 333.

23 Curt. 8.13.3–5; Scullard 1974, 66.

24 As Scullard (1974, 265, n. 34) notes, the transport of Alexander's elephants across the Hydaspes River in order to use them against Porus likely would have ended up being a logistical nightmare in and of itself; cf. Kistler 2007, 39–40.

25 Fuller 1960/1998, 193–197.

26 Tarn 1930/1984, 93; cf. Charles 2010, 340–342.

27 Even prior to the battle proper, Alexander avoided crossing his cavalry across the Hydaspes while Porus' elephants still guarded the opposite bank: see Fuller 1960/1998, 185–187; Scullard 1974, 67–68; Kistler 2007, 32.

kopides. Advice and expertise offered to Alexander and his *Diadochoi* by their Indian allies and trading partners could be just as valuable as the animals and mahouts they supplied.[28]

Despite the fact that war elephants did not play a decisive role at the Hydaspes, they did play an important role in the Macedonian propaganda that followed the battle, yet another indication of the impression they had made upon Alexander and his troops. One of the more famous examples is a series of medallions minted to commemorate Alexander's triumph. The obverse of one set of these medallions depicts a mounted warrior, armed with a spear, pursuing two other soldiers mounted on the back of an elephant, while on the reverse a helmeted warrior, holding a spear or scepter in one hand and a lightning bolt in another, is being crowned with a wreath by the goddess Nike. The subject matter of this medallion type, as well as its message, has been a source of contention among scholars since its discovery; most likely, the obverse depicts Alexander himself pursuing the defeated Porus on his war elephant, while the reverse drives home the theme of victory, portraying the Macedonian king, in all his martial glory, being crowned by the goddess of Victory.[29]

Several years later, on the eve of his death, Alexander had at least begun to take some steps to incorporate formally war elephants into the Macedonian military, measures that were subsequently expanded upon by the *Diadochoi*. One piece of evidence comes from Diodorus' description of the ornate funerary cart built to carry Alexander's body back to Macedonia.[30] Among the cart's lavish decorations were painted panels depicting various divisions of the Macedonian military, including a group of war elephants, each with an Indian mahout and a Macedonian soldier on its back (Diod. Sic. 18.27.1). In his description of the myriad of troops who surrounded Alexander's royal tent in Babylon shortly before his death, Athenaeus calls the elephant force an *agema*, a Greek term that normally denoted an elite formation. The use of this particular term indicates that elephants had begun to be organized as a formal part of the Macedonian military.[31]

The Successors

In the age of the Successors, elephants became an established feature of Hellenistic warfare and, indeed, a prized commodity, as the various claimants to power sought to strengthen their forces. The regal connotations of the elephant were not lost on the *Diadochoi*, just as they had previously been obvious to Alexander.[32]

The relatively rapid emergence of war elephants in Hellenistic warfare was due to their versatility. Most commonly, elephants were interspersed with bodies of light infantry on the battlefield, a combination often likened by ancient writers to a wall buttressed by towers

28 Arr. *Anab.* 5.17, Diod. Sic. 17.88, Curt. 8.14.24–30 with Tarn 1930/1984, 97–98; Fuller 1960/1998, 197–198; Bosworth 1996, 18–19; Epplett 2007, 214; Kistler 2007, 35–36; Charles 2010, 339, 341–342.

29 Fuller 1960/1998, 188; Holt 2003, 118–30, 151–153; Charles 2010, 348.

30 It was diverted to Egypt by Ptolemy, in defiance of the wishes of Perdiccas (Arr. *Succ.* 1.25).

31 Ath. 12.539 with Scullard 1974, 75; Epplett 2007, 215–216; Alonso Troncoso 2013, 255–256; cf. Charles 2010, 344–345.

32 As Tarn (1930/1984, 94) and Adcock (1957, 54) both suggest, one reason for Seleucus' evident enthusiasm for war elephants is that he had commanded part of the Macedonian infantry which fought against Porus' elephants at the Battle of the Hydaspes, and had therefore had an excellent, close-up opportunity to observe their fearsome nature and fighting potential.

along its length. The animals could fight either on their own or with soldiers, armed with long spears or bows, mounted on their backs. Support infantry protected the elephants from attack in the flank or rear, as happened with Porus' elephants at the Hydaspes. In the Seleucid army at least, a complement of 50 light infantry appears to have been assigned to each animal. One problem with such an arrangement, however, was that the elephants, unlike the infantry supporting them, could not easily retreat through their own lines: in such an instance, retreating war elephants would often throw their own armies into disorder.[33]

War elephants not only could attack and repel enemy infantry and elephants, but also prove effective against cavalry, especially those with little or no experience of them. Indeed, war elephants found their most effective application on the Hellenistic battlefield in their use as a cavalry screen. Hence, at the Battle of Ipsus in 301, Seleucus used his massive force of war elephants to prevent Demetrius' cavalry, after its initially successful charge, from returning to aid the rest of his father's army. Elephants thus sealed the doom of Antigonus the One-Eyed.[34]

Hellenistic generals also employed their elephants in tasks reminiscent of those described in the *Arthashastra*—no doubt tasks suggested by the Indian personnel acquired along with the animals. The elephants of Perdikkas, for example, played a pivotal role in his attempts to force a crossing of the Nile against Ptolemy in 321. They formed the vanguard of the force that Perdikkas sent across the river to attack the so-called Fort of Camels, and advanced to the base of its walls, allowing the troops mounted on their backs to begin tearing down the fortifications. This attack ultimately failed, however, and Perdikkas subsequently made a second unsuccessful attempt to cross the Nile at another point, at which time his elephants were posted upstream of the rest of the army so as to act as a breakwater and lessen the force of the current.[35]

The psychological effect of elephants appears once again during Antiochus I's so-called Elephant Victory over the Gauls in the early third century. According to Lucian, the mere sight of Antiochus' 16 war elephants was enough to cause the much larger Gallic army arrayed against him to flee the battlefield.[36] Tarn suggests they were meant to prevent the Gallic cavalry from participating in the battle.[37] Similarly, Diodorus

33 Glover 1944, 259; Adcock 1957, 54; Bar-Kochva 1976, 82; Epplett 2007, 227; Sabin 2007, 419–421. War elephants appear to have been deployed with heavy infantry only on the rare occasions when special circumstances demanded it: Bar-Kochva 1976, 82–83, for example, suggests that the Seleucids deployed their elephants in conjunction with heavy infantry at the battle of Beth-Zechariah (162) both because of the nature of the terrain and the large number of Jewish light infantry opposing them. For elephants at the Battle of Magnesia (188), see Bar-Kochva 1976, 171; Sabin 2007, 420–421.

34 Plut. *Demetr.* 29 with Tarn 1930/1984, 96; Adcock 1957, 53–54; Epplett 2007, 220–221; Kistler 2007, 66–67. According to Strabo (15.2.9), Seleucus had been given 500 war elephants in 303 by the neighboring king Chandragupta in return for the ceding of considerable territory on his eastern frontier, and it was these elephants which later played a decisive role at Ipsus. The figure of "500," however, has proven to be a source of controversy among many scholars, some of whom, like Scullard (1974, 97–98), view it as too high. Others, however, see the number as plausible: Kistler 2007, 64–65; Engels 2017, 112–113. As Alonso Troncoso (2013, 260–262) notes, the prominence of the elephant motif on Seleucid coinage following the battle reflects not only their importance in the battle, but also Seleucus' enhanced prestige among the remaining *Diadochoi* thereafter.

35 Diod. Sic. 18.33–35; Kistler 2007, 45–46 .Ptolemy celebrated his victory over Perdiccas, as well as his subsequent triumph over Demetrius and his elephants at Gaza in 312, by minting tetradrachms depicting Alexander wearing an elephant scalp and the horns of Ammon, likening his own mastery over elephants to that of the deceased Macedonian king: see Alonso Troncoso 2013, 256–258.

36 Luc. *Zeux.* 9 with Kistler 2007, 76–77.

37 Tarn 1930/1984, 96–97; cf. Sabin 2007, 420.

records the anxiety of the defenders of Megalopolis upon the approach of Polyperchon's army in 318, an anxiety prompted in part by the force of elephants under the latter's command.

> the whole city was deep in activity, owing both to the spirit of the population and to the danger that was foreseen. Indeed, word had spread abroad concerning the magnitude of the royal army and the multitude of the accompanying elephants, which were reputed to possess a fighting spirit and a momentum of body that were irresistible. (Diod. Sic. 18.70.2–3)

Coskun describes similar episodes in other battles.[38]

Alexander's Successors pursued every available method and spared no expense in procuring as many elephants as possible for their respective armies. Seleucus, most famously, in preparation for the upcoming showdown with Antigonus, ceded his easternmost territories to the Indian ruler Chandragupta in return for a substantial force of war elephants. These, as noted earlier, played a decisive role in Antigonus' defeat and death at Ipsus.[39] Subsequent Seleucid rulers, in the face of a growing rivalry with the Ptolemaic Empire, attempted to cut off all exports of Indian elephants to Egypt and thereby monopolize the supply of these valuable animals. In response, the Ptolemies sent hunting expeditions south, into modern-day Eritrea, Ethiopia, and Sudan, to capture the elephants native to those regions for their own army.[40]

The consensus of modern scholars holds that the animals captured by these Ptolemaic hunters were forest elephants, as opposed to the larger bush species of equatorial Africa. This would explain Polybius's oft-cited comment (5.84) that the Indian elephants of Antiochus III at the Battle of Raphia were larger and stronger than their Ptolemaic counterparts, a statement that would not be true if Ptolemy IV had bush elephants at his disposal. A recent study of modern-day elephant DNA in Eritrea, however, concludes that the Eritrean population are bush, not forest, elephants. The elephants captured by the Ptolemies in the region (including those that fought at Raphia) were thus the former rather than the latter variety.[41] Polybius would be proved incorrect. Yet it is entirely possible that the modern elephant population in Eritrea differs in its genetic makeup from that which inhabited the region in antiquity.[42]

The Ptolemies and Seleucids sought to import Indian personnel in addition to the elephants themselves, particularly in the early Hellenistic period. Accounts of various battles between the *Diadochoi* often refer to the Indian mahouts driving the war elephants.[43] Later, however, Hellenistic rulers appear to have used their own personnel to train and look after

38 Coskun 2012, 61–68. Pyrrhus' elephants had a similar psychological effect upon the opposing Roman troops in his victory at Heraclea in 280: Wellman 1905, 2253; Adcock 1957, 56; Kistler 2007, 84–86.

39 Adcock 1957, 54; Epplett 2007, 222–223. Eudamus, a Macedonian officer who had been assigned to remain behind with Taxiles after Alexander's departure from India, murdered the king's one-time adversary, Porus, after Alexander's death in order to obtain his current herd of 120 war elephants. Subsequently, he was rewarded by Eumenes with the princely sum of two hundred talents for the animals (Diod. Sic. 19.14.8–15.5 with Epplett 2007, 221–222).

40 Casson 1993, *passim*; Grainger 2010, 100–101.

41 Brandt et al. 2014, *passim*.

42 Charles 2016c, 61–64.

43 Cf. Sick 2002, 134–135, 139.

their elephants, rather than importing specialists. Our evidence is limited, but it appears that Ptolemy II, for example, faced with the previously mentioned Seleucid blockade, took steps to have his elephants trained by his own personnel. Aelian relates that the king, at one point during his reign, was presented with an elephant that could understand commands given in Greek. Apparently, the animal in question was something of a trendsetter, as the popular belief up to this point, according to Aelian, was that elephants could only understand the Indian language.[44]

The necessary infrastructure for war elephants in the Hellenistic period is ill attested. The only royal elephant enclosure known for certain to have existed in the Seleucid Empire is that at Apamea, set up by Seleucus and subsequently maintained by his successors (Strabo 16.2.10). Given the extent of Seleucid territory, however, it is almost inconceivable that other such enclosures did not exist elsewhere: the two likeliest locations are Bactra, an important way station for trade between India and the Seleucid Empire, and Babylon.[45] As for the Ptolemies, the limited evidence available indicates that their main elephant enclosure was outside Memphis, relatively close to the eastern border with the Seleucid Empire.[46]

One important question is whether or not the Hellenistic states attempted to breed elephants in such enclosures. Only a few Hellenistic rulers, like Antigonus the One-Eyed and Pyrrhus, are known to have possessed elephant cows, although it appears unlikely that the Seleucids and Ptolemies would not have attempted to procure breeding stock as well. Even a secure supply of elephant cows, however, would not have guaranteed a sufficient number of elephants for any given army. Like many other animals, elephants are much less successful breeding in captivity than in the wild, and given the long gestation period of elephant young, a cow could only give birth every three years at the most. Any breeding program put in place by Hellenistic rulers, therefore, would not have replaced animals lost over time to warfare, illness, and natural causes. In addition, the considerable expense involved in raising even a single elephant from infancy to maturity discouraged breeding these animals in captivity, as opposed to importing or capturing adult specimens. As a result, Hellenistic states remained dependent upon mature elephants imported from regions like India and the Horn of Africa.[47]

The reliance of the Hellenistic states upon their Indian suppliers should not lead us to conclude that they themselves were incapable of innovation. The most notable development in this regard appears to have been the placement of towers or turrets on the backs of elephants, in particular for archers. The evidence is scattered and, at first sight, contradictory, but it nonetheless suggests that a Hellenistic ruler, Pyrrhus of Epirus, introduced this tactical innovation.[48] At this same time, an effective, albeit unusual counter-measure was devised against war elephants. According to Aelian, the Romans were able during their war against Pyrrhos to drive off his elephants by releasing squealing pigs against them; the elephants, having a natural aversion to the sound of these creatures, swiftly broke ranks and fled. Similarly, when

44 Ael. *NA.* 11.25; Epplett 2007, 219.

45 Epplett 2007, 224–225.

46 Casson 1993, 258–259; Grainger 2010, 101.

47 Bar-Kochva 1976, 79; Charles 2016c, 57–58.

48 Scullard 1974, 240–242; Charles 2009, 33–34. Kistler 2007, 23, 85, maintains that Indian war elephants were equipped with such towers as early as the sixth century. In his opinion, the real innovation of Pyrrhus was stationing archers in these turrets, which turned them into "mobile artillery platforms." As Charles 2009, 34–35, notes, however, with the sole exception of a questionable reference at Diod. Sic. 17.88.6 there is no report in the extant accounts of Alexander's campaign that the war elephants he encountered in India bore turrets on their backs.

Antigonus Gonatas attacked Megara only a few decades later, the Megarians are said by Aelian to have driven off the Macedonian king's elephants by lighting pigs on fire and releasing the screeching animals against them.[49]

Another tactic was the use of spikes hidden in the ground. Damis, the leader of the Megapolitan forces resisting Polyperchon, turned the assault of the latter's elephants into a fiasco by concealing spiked planks in the ground along their line of advance; many of the animals, maddened by pain, ultimately turned upon Polyperchon's own troops.[50]

The Demise of the War Elephant

The mixed combat record of war elephants was an important factor in their eventual disappearance from Hellenistic armies. When deployed effectively, as at Ipsus, these animals could play a decisive role. But all too often, when elephants faced stiff resistance or countermeasures such as those devised by Damis, they would end up killing almost as many of their own soldiers as those of the enemy in their confusion. Under such circumstances, the troops assigned to guard the flanks of the elephants would often be among the first to be trampled to death.[51] The disappointing performance of their African war elephants at the Battle of Raphia in 217 spelled an end to the Ptolemies' interest in this particular branch of their army.[52] By the later Hellenistic period, many rulers had reached the same conclusion as the Ptolemies.

The Seleucids maintained their force of war elephants longer than their rivals, but by the mid-second century, they too had abandoned them. Issues of cost and supply were exacerbated by the growing Roman military presence in the eastern Mediterranean. The secession of Bactria from Seleucid rule in the 240s appears to have seriously hampered efforts to import Indian war elephants and personnel. Subsequently, the Roman expropriation of the Seleucid war elephants at Apamea after their victory over Antiochus III at Magnesia was another severe blow to the Seleucid elephant corps. Later rulers did attempt to rebuild it, but the effort was short-lived. The last campaign involving war elephants was launched against Judaea in 162, and shortly thereafter, a Roman embassy ordered the entire Seleucid elephant herd be destroyed, on the grounds that it violated the earlier treaty with Antiochus III.[53]

By this juncture, the Romans could employ their own war elephants against Hellenistic states that had long since abandoned their use. The Romans had their own period of experimentation with these animals following their victory in the Second Punic War, and it was during this time that they included war elephants in the army that ultimately defeated the last king of Macedon, Perseus, at Pydna in 168. Faced with this now unfamiliar threat, the king

49 Ael. *NA*. 1.38; 16.36 with Scullard 1974, 113–116.

50 Cf. Tarn 1930/1984, 95–96; Kistler 2007, 54–56; Sabin 2007, 420–421. Similarly, Ptolemy was able to halt the advance of Demetrius' elephants at the Battle of Gaza by positioning a long chain with iron spikes in front of his troops posted opposite to them (Diod. Sic. 19.83–84 with Adcock 1957, 55; Kistler 2007, 59–61).

51 Glover 1948, 6. It should also be noted that Pyrrhus' war elephants, by blocking the gate out of the city in their confusion, directly contributed to the king's death during his abortive attack upon Argos in 272 (Epplett 2007, 220; Kistler 2007, 94–95).

52 Scullard 1974, 137–144; Burstein 1989, 10–11.

53 Bar-Kochva 1976, 77–82; Epplett 2007, 225–226.

attempted a variety of unsuccessful measures to counter the Roman elephants. First, he had a special contingent of *elephantomachae* trained to combat them, equipped with spiked shields and helmets so as to prevent the elephants from grabbing the soldiers with their trunks.[54] He also used wooden elephant dummies to acclimatize his cavalry to the sound and smell of elephants prior to the battle, a tactic very similar to that allegedly employed by Semiramis against the Indian king Stabrobates centuries earlier.[55] We cannot be certain, of course, but it is tempting to speculate that Perseus, in his desperation, may have hit upon this particular idea while browsing through the pages of Ctesias concerning Semiramis. Regardless, the performance of the anti-elephant corps against the Romans proved to be a profound disappointment to the Macedonian king.

Although the heyday of war elephants in the Greek world lasted for less than 200 years, their appearance on the battlefield nonetheless forms an interesting chapter in Hellenistic warfare. Alexander was clearly fascinated with their military potential as well as their propaganda value. He did not have time to make extensive use of the war elephants before his sudden and premature death, so it was left to his officers, men such as Seleucus and Ptolemy, to exploit their potential. Over the next century and a half, war elephants became one of the most visible and intimidating expressions of royal power in the Hellenistic world. Ultimately, because of their decidedly mixed combat record, as well as the prohibitive cost and effort required to maintain them, they disappeared from the Hellenistic battlefield: not coincidentally, their decline mirrored that of the states that had used them as a symbol of royal power. A fitting final verdict upon the military effectiveness of Hellenistic war elephants can perhaps be given by the Roman writer Quintus Curtius Rufus who, writing centuries after the fact, and with the benefit of hindsight, inserted the following statement into the speech given by Alexander to his men on the eve of the Battle of the Hydaspes:

> It [the power of elephants] is a doubtful kind of strength, and rages more violently against its own men; for it is driven against the enemy by command, against its own men by fear. (Curt. 8.14.16)

54 Zonar. 9.22, Livy 44.41.3–5 with Charles 2014, 196.

55 Polyaenus, *Strat.* 4.21, Diod. Sic. 2.17.3 with Epplett 2007, 230–123; Charles 2014, 196 n. 30.

CHAPTER 16

Thracian Warfare

Emil Nankov

Thracian warfare has been a subject of interest for several branches of scholarship studying Classical antiquity. Perhaps one of the most enduring narratives has been produced by historians and philologists.[1] This is not a surprise granted the available body of literary sources left by Greek and Roman authors.[2] Despite the inherent bias and limited chronological scope of the written records, they continue to exert profound influence on modern scholarship dealing with Thracian military organization and tactics.[3] This chapter begins by making cautious use of these records to provide a summary account of Thracian warfare—native and foreign armies, their weapons, their military operations, and the political results. This account will raise questions to be addressed in the second part of the chapter, which describes alternative sources, especially archeological ones, and shows how they contribute to a better understanding of Thracian sieges, battlefields, weapon hoards, and militaria found in graves and tombs.

Historical Précis

Thracians were described by the Greek historiography as *ethne*, traditionally and somewhat inaccurately translated nowadays as "tribes," while the *polis* structure never took root on Thracian soil. These are fundamental premises no scholar dealing with ancient Thrace can

1 Best 1969; Mihailov 2015, 267–305.
2 Katsarov 1913, 31–36; Fol 1969, 103–126; Tacheva-Hitova 1971; Kolev 1975; Launey 1987, 366–398, 404–406; Webber 2011, xx–xxiii.
3 Webber 2011, 136–142; Sears 2013, 234–289; Stoyanov 2015, 436–437.

A Companion to Greek Warfare, First Edition. Edited by Waldemar Heckel, F.S. Naiden, E. Edward Garvin, and John Vanderspoel.

escape. By default the overall picture that has come down to us is piecemeal and, very often, unavoidably generalized. Despite these drawbacks scholarly narratives strive to produce a consistent trajectory of the historical development of Thracian warfare relying heavily on the Greek and Roman authors.[4]

The story begins with Homer who introduced the Cicones and Abantes as fighting in close quarters using spears and swords. The Paeonians, on the other hand, emerged as fighting with spears and "crooked-bowed" as well (Hom. *Il.* 2.547, 848, 21.155). The Thracian nobility also practiced chariot fighting (Hom. *Il.* 14.287, 20.48–487, 21.205). These sketchy snapshots present a composite image of the Thracian warrior from the Greek heroic past positioned within a wide temporal dimension, from the Late Bronze Age until the Late Archaic period (late sixth century). From the later historical tradition, most notably Herodotus, Thucydides, and Xenophon, one learns about the existence of different troop types and Thracian military tactics. The most common are the light-armed peltasts carrying two javelins, a crescent-shaped shield (*pelte*), and short swords (*encheiridia*). They wear fox fur hats, chitons covered with long cloaks called *zeirai*, and long boots of deerskin (Hdt. 7.75.1). This iconographic type of Thracian warrior circulated, predominantly on red-figure pottery, in the Greek world during the Classical and Hellenistic periods. Heavy infantry, on the other hand, or the *hoplites* derived from the Greek *poleis*, remained a foreign concept for the Thracians. The most significant role, reserved traditionally for Thracian nobles, was played by the cavalry, armed with spears, javelins, swords, and bows. The Odrysians and the Getae were especially notable as horse breeders. The figure of 50,000 horsemen, for instance, mustered by Sitalces during his march against the Macedonians in 429 is staggeringly high (Thuc. 2.98.3–4), a testimony to the vast potential of an economy based on horse breeding.

In the multicolored mosaic of Thracian *ethne*, several qualities commonly expressed in military contexts and traditionally ascribed to Thracians immediately stand out. Thracians had the reputation of being good and courageous fighters, often depicted as fierce and warlike (Hdt. 7.111.1). They preferred marching and fighting under the cover of night.[5] Upon giving battle they issued vociferous war cries, singing songs and performing war dances, while loudly clashing their weapons (Tacit. *Ann.* 4.47). The lack of heavy infantry allowed for greater flexibility and a mobile style of military conduct; very often the element of surprise assumed priority. The existence of the wedge formation, which the northern Thracians allegedly learned from the Scythians has been doubted, but on one occasion in 335, Arrian says, the Tribalians positioned themselves against Alexander, divided into a center and two flanks.[6] On account of their maneuverability, Thracian *peltasts* became sought-after mercenaries after the march of Xerxes in Greece in 480, and especially during the Peloponnesian War and the war between Cyrus the Younger and Artaxerxes in 401.[7]

In the Hellenistic period, Thracian troops and techniques rulers had a profound impact of Greek warfare. For instance, the Athenian general Iphicrates, who married a daughter of Cotys (383–359), introduced reforms to the cumbersome panoply of Greek hoplites that

---

4 See n. 2.

5 Hdt. 6.45.1; Xen. *Anab.* 7.3.37–41.

6 Arr. *Tact.* 16.6 has been questioned by Stoyanov 2015, 436, with lit.; Mihailov 2015, 279–280. Cf. Arr. *Anab.* 1.2.5–6.

7 Hdt. 7.75.1; Thuc. 4.129.2, 5.6.2, 7.27.1; Xen. *Anab.* 1.2.9, 2.2.7.

showcased the effectiveness of Thracian peltasts during the Corinthian War (395–387).[8] After the Macedonian conquest of Thrace by Philip II and Alexander III, Thracian cavalry and light-armed troops (*psiloi*) began to appear under the supervision of Macedonian and Thracian commanders.[9] This practice continued during the Macedonian Wars, when Thracians fought in the armies of Philip V and Perseus against the Romans in the battles of Cynoscephalae (197), Callinicus (171), and Pydna (168).[10] Thracians who served as mercenaries in the Ptolemaic armies eventually settled in Egypt as military colonists.[11]

Alternative Sources for Thracian Warfare

The two sources needed to improve this sketchy description of Thracian warfare are iconography and archaeology. Neither is new, but archaeology in Thrace has been advancing recently, creating opportunities to re-examine typologies in Thracian warfare through synthetic studies and reconceptualization.[12] First, iconography, for which diverse architectural, sculptural, and visual media date to the Classical and Hellenistic periods. Sometimes iconographic subjects appear as stand-alone studies; more often the emphasis falls upon their subsidiary role as a visual prop for historical narratives.[13] Obvious cases in point are the studies on Thracian warriors depicted on Athenian red-figure pottery Persian reliefs, Thracian toreutics, as well as on the figural compositions painted in Thracian tombs of the Early Hellenistic period, e.g. Kazanlak, Alexandrovo, Sveshtari, and Ostrusha.[14]

Second, and more important, archaeology, presenting diverse but daunting material.[15] Traditionally, the largest number of weapons and armor come from burials spanning the entire first millennium, thereby constituting the backbone for the creation of artifact typologies or horizons of "rich burials."[16] We are dealing mostly with panoply deposited in sumptuous graves or tombs associated with the aristocracy rather than the common soldiery.[17] Another direction of research centers on weapons and armor thought to be Greek imports. Lagging behind are interpretations of excavated settlements, even though artifact distribution at urban sites provides new ways of understanding ancient warfare.[18] Recent discoveries in the Thracian fortified residences in Kozi Gramadi and Sotirya have shown convincingly that inland Thrace has much to offer in battlefield studies. Although deposition of weapons

8 Best 1969, 102–110; cf. Mihailov 2015, 293–295.

9 Heckel 1992, 332–334, with references.

10 Liv. 33.3–10, 31.39.10–11; Plut. *Flam.* 7–8; *Aem.* 18.3.

11 Velkov and Fol 1977, 22–96; Bingen 2007, 83–93.

12 Archibald 1998 197–208, 247–258; Webber 2011; Stoyanov 2015; Teleaga 2015, 109–121, 152–156.

13 Valeva 2005; Rabadzhiev 2014.

14 Pottery: Zimmermann 1980, 441–446; Lezzi-Hafter 1997, 359–366; Hermary 2010, 485–486. Persian reliefs: Vassileva 2015, 322. Thracian toreutics: Marazov 2005, *passim*; Webber 2011. Tomb compositions: Valeva 1992, 2005, 83–84, 93–96, 97–102, 103–104; Nankov 2007, 41, 43, 2010, 44, n. 14, 50; Stoyanov 2008a.

15 For the difficulties, see Buykliev 1998. The wide spectrum of archaeological contexts within which various classes of objects are unequally distributed: Stoyanov 2015, 426–435.

16 Domaradzki 1998, 44–64; Teleaga 2008; cf. Stoyanova 2015.

17 Archibald 1998, 197–208; Georgieva 2016.

18 Imports: Teleaga 2008, 323–327. Settlements: Stoyanov 2008b. Mapping artifact distribution: Lee 2001; Hagerman 2014.

is better attested on some mountainous sanctuaries in the Rhodopes, future work will reveal whether this reflects a Thracian perception particular only to that region.[19]

For weapons, bullets, and other artifacts, work on-site is less important than the publication of holdings in private collections.[20] The issue with this growing multitude of axes, swords, spears, helmets, cuirasses, etc. is that they come from disturbed complexes. Such compendia are nonetheless used by scholars because of their interpretative potential in understanding Thracian myth and royal ideology.[21] In addition, some artifacts, such as the sling bullets from collection "Vatevi," point to identifiable battlefields or military movements.[22]

Third, archaeometric studies on metalwork from Bulgaria have proved promising, as with the handgrip of a Greek hoplite shield retrieved from a burial on the Vishegrad peak near Topolovgrad. ICP-AES analyses showed an unusually high percentage of brass in the alloy, which was taken as proof for the non-Thracian origin of the shield.[23] Finally, isotope and chemical analyses on several sling bullets from Dobrudzha and ***emporion*** Pistiros demonstrated that the lead used to manufacture them originated in the Chalcidice and Laurium.[24]

In response to this range of evidence, scholars have concentrated on artifact typologies. Morphological features of weapons are deemed instrumental for putting forward typological sequences.[25] Best served are artifacts accompanied by stratigraphic data as well as by a reliable chronology extracted from coins or amphora stamps. Distribution maps serve to illustrate development and direction of influences from one region to another. Most attention has gone to helmets on account of their considerable quantity in Thrace and to the alternatives of local manufacture and Greek imports.[26] For instance, the manufacture of bronze arrowheads is most securely documented through the discovery of several molds originating from northeastern Bulgaria.[27] In the case of the ***rhomphaia***, single-edged blades attached to poles, the evidence for Thracian production, albeit quite logical, remains largely circumstantial.[28] Conversely, equipment like elastic greaves made of bronze—the hallmark of hoplites—are often considered Greek imports, while sumptuous silver ones with plastic decoration, sometimes described as "ceremonial," form a homogenous group (Agighiol, Vratsa, Zlatinitsa, and Golyama Kosmatka), perhaps produced specially for Thracian markets.[29] The anachronistic use of the "bell-shaped" cuirasses in Thrace provide an intriguing example, where the armor would receive a Thracian "touch" in order to become functional in the local milieu.[30] Claims that itinerant Greek craftsmen from the Black Sea coast (or even Thasos)

19 Residences: Nankov 2015a. Rhodopes: Bayrakov 2015, 224–225 n. 1175.

20 Ivanov 2000; Marazov et al. 2005, 2012; Marazov 2011a; Torbov 2016.

21 Marazov 2011b, 165–181. Often these items complement the splendor of Thracian gold and silver as part of traveling exhibitions abroad (Zhurvalev and Firsov 2013, cat. nos. 7, 11, 15–18, 21, 57–62).

22 Manov and Torbov 2016.

23 Vasileva 2015, 149–150, figs. 6–7, and 154–155.

24 Kuleff et al. 2006, 246, 251, 254, table 1; Avram et al. 2013, 240 n. 45.

25 Helmets: Dimitrov 2005a, 2007; Ognenova-Marinova and Stoyanov 2005; Stoyanov 2015, 430–431. Curved knives: Torbov 2018; ***akinakes***: Topal 2017. Arrowheads: Buykliev 1991. Spears: Leshtakov 2015.

26 For technological issues of local manufacture, see Vasilev 1979, 1980.

27 Arrowheads of Type G1 after Robinson 1941, nos. 2097–2100, reported by Minchev 2013, figs. 1, 4–7.

28 Mikov 2010, 223–224, fig. VI.

29 Imports: Teleaga 2008, 249–251. "Ceremonial": Webber 2011, 49; Stoyanov 2015, 431. Thracian markets: Agre 2011, 45–72; Dimitrova 2015, 173–178, figs. 141–145.

30 Dimitrov 2010, 219–20; Parvin 2010, 227.

manufactured Thracian armor remain speculative.[31] Epigraphic documents reveal ethnic diversity among armorers, but not the location of workshops.[32]

Ways Forward

Rather than produce an inventory of Thracian weapons, this chapter will now seek to capture the dynamic nature of warfare by tracing the biography of certain weapons and focusing on the various archaeological contexts within which they appear. What follows is a set of examples, not an attempt to draw general conclusions from expanding evidence.

Emphasis falls on the Macedonian expansion in Thrace because this expansion ranks among the most richly documented and archaeologically visible episodes in Thracian history. Across the Thracian countryside, it scattered traces of evidence for settlement organization, trade, burials, and not least military tactics and equipment.[33] We begin with sieges, and proceed to battlefield sites, weapon hoards, and burials. One challenge is how to distinguish Thracians borrowing new technology from foreign troops using it in Thrace. Another is the incidental burying of military paraphernalia, which should be considered not in isolation, but in conjunction with mortuary data.

The Greek city of Olynthus was besieged by the troops of Philip II of Macedon in 348. Sling bullets inscribed with his name, along with the names of several of his military commanders, have long been known from the town. On the basis of a close analysis of the distribution pattern of weapons, most notably sling bullets and arrowheads, Lee argued that the town was in fact subdued after fierce fighting inside the city walls, in the streets, even within house courtyards. In contrast with the peaceful act of treachery mentioned in the historical accounts. Lee introduced the notion of "urban combat" by laying emphasis on the *physicality* (my italics) of ancient warfare.[34]

Illustrating this notion is the site at Adjiiska vodenitsa, commonly identified with *emporion* Pistiros, which has yielded evidence of repeated destruction during the fourth and third centuries. The chronology of these events has changed quite frequently, but published artifacts show that Pistiros suffered because of numerous sieges. Projectiles such as arrowheads, catapult bolts, sling bullets, and a catapult stone ball predominate in the debris in houses, pits, streets, and fortifications. Weapons used in close combat, like spears and swords, are few.[35] The bronze triangular socketed arrowheads suggest the use of composite bows, while iron

31 Ognenova 1961, 526; Vasilev 1979, 70–71; Ognenova-Marinova and Stoyanov 2005, 525, but cf. Dimitrov 2007, 58–60; Stoyanov 2015, 431.

32 Note the pair of greaves stamped with the name of the smith Agathanor found in a grave near Asenovgrad (Ivanov 1948b, 105–106), and a scabbard of a late first century date, inscribed with the name of certain Seuthes from the region of Stara Zagora (Dana et al. 2016, figs. 1–3, 64).

33 Archibald 1994, 465–475; Archibald 1998, 231–316; Delev 2015.

34 Robinson 1941, 418–443; Lee 2001, 16, fig. 4; cf. Cahill 2002, 45–49, fig. 9.

35 Destruction: Nankov 2015b, 403–404, with lit. cited. Chronology: Bouzek et al. 2013, 227; Bouzek 2013, 288. Projectiles: Archibald 2007, 80; Nankov 2008, 42 n. 32. Weapons: Archibald 2002, 121, pl. 35: 10/11; Bouzek 2007, 249, pl. 56; so also the fragmentary cheek-piece of a Chalcidian helmet (Ivanov 2017, 490–491, fig. 5.4).

bolts testify to arrow-shooting artillery.[36] The picture painted by the lead sling bullets will come into full view after the publication of a constantly growing collection. One bears the name of a Macedonian commander (Tyrvakos), associated with Philip II, Alexander III, or Zopyrion.[37]

At the Macedonian colony of Kabyle, founded by Philip II in 341, evidence of military encounters comes from the fortified area and the immediate hinterland.[38] An unpublished lead sling bullet found in the agora, one with the name of Anaxandros—a Macedonian general of Philip II—signals Macedonian activity during the 340s, while confirming the presence of slingers.[39] Three bronze triangular arrowheads (Robinson's Type G1) have been found along the fortification walls, and the discovery of a curved iron knife and two laurel-shaped spears may be related to campaigns of Philip V in 183, as in the literary sources.[40] In all, the remains resemble those at Pistiros.

The largest number of military artifacts comes from the fortress near the village of Dragoevo, Shumen district, a site sporadically excavated through test trenches in the 1980s. The weapons were found inside the fortress, as well as along the slopes surrounding the fortified area, and published in part in several articles and a small exhibition catalog.[41] The breakdown according to type is the following: 3 swords of the *xiphos* type, 2 bronze battle-axes (dated to Early Iron Age), 150 spears (of which 29 are published as Roman), 7 butt-spikes, 79 spurs, 13 bodkin arrowheads with tang (Robinson's Type E), 14 bronze triangular arrowheads (Robinson's Type G1), 13 bronze and iron Cretan arrowheads (Robinson's Type D1), 10 sling bullets, none of which are inscribed, 1 iron bridle, 9 buckles, and 7 distributors for horse harnesses.[42] The copious offensive weaponry and cavalry equipment has been taken to indicate Macedonian professional soldiers, perhaps mercenaries.[43] The two sarissae and butt-spikes suggest a Macedonian siege of Dragoevo, and the Cretan arrowheads suggest the Cretan archers who were frequently hired by Alexander the Great.[44]

The Thracian town of Seuthopolis, founded c. 310 by Seuthes III, met its demise as a result of siege operations undertaken most likely by the Seleucid king Antiochus II Theos in 252.[45] Since it was never occupied on the same scale during later periods, military artifacts may be treated as a snapshot of what happened to the town as a result of the 252 siege. In this Seuthopolis closely resembles Olynthus besieged by Philip II in 348. Armor is entirely

36 Socketed arrowheads: Type G1 after Robinson 1941, nos. 2097–2100. Bows: Domaradzki 1995, 66, nos. 1–2; Bouzek and Musil 2002, 47, fig. 3.18, 2, 51; Bouzek and Domaradzki 2010, 223–225, fig. 11.1, 6; Bouzek et al. 2013, 227–228. Bolts with bodkin heads: Hagerman 2014, 85. Bodkin with tang: Type E after Robinson 1941, nos. 1972–2026; Domaradzki 1995, 66, nos. 3, 4, 9; Katincarova 2007, 40, fig. 22, pl. 9; Nankov 2008, 40; Bouzek et al. 2013, 227–228.

37 Bouzek et al. 2013, 227. Tyrvakos: Avram 2011, 198–189; Nankov 2016, 284 nn. 9–12, map 1.

38 The site: Nankov 2015b, 402–403. New fortification, most likely after the Macedonian takeover: Nankov 2008, 36; Delev 2014, 193. The finds: Domaradzki 1991, 59–60.

39 Domaradzki 1991, 63 no. 51; Nankov 2008, 40; Nankov forthcoming.

40 Tancheva-Vasielva 2007, 37, fig. I. Philip V: Domaradzki 1991, 59–60; Tancheva-Vasileva 2007, 38, 41, figs. II.7, 8, V.4; Torbov 2018, 14, 28, 112.

41 Atansanov 1995, 35–91; Atansanov et al. 2006, 67–68.

42 Atanasov and Todorov 1999, 62; Atanasov 2003, 2005, 2006; Atanasov et al. 2006, 68.

43 Atanasov 1995, 29–30.

44 Ashley 2004, 47–48. The bodkin arrowheads are of the same kind as those found at Olynthus (Lee 2001, 13; Hagerman 2014, 87–88).

45 The town: Archibald 1998, 312–315; Nankov 2015b, 404–405 with lit. Antiochus: Nankov 2008, 41–42.

absent, except for a bronze cheek-piece from a Chalcidian helmet.[46] The offensive weapons recovered are various: 5 spears, 3 javelins, 7 bronze triangular arrowheads (Robinson's Type G 1), 2 of them inside the charred remains of a quiver, and 15 curved knives. Related items include an ivory scabbard chape for *xiphos*, 20 pyramidal bodkin heads with socket, 19 stone balls of different calibers, and 4 sling stones. Noteworthy are two bronze appliqués to a helmet bearing an image of the goddess Athena.[47]

These items come in clusters located in the agora, streets, houses and the citadel. The highest concentration is inside the citadel, where the cheek-piece, a curved knife, and six arrowheads, including the quiver, were found. Perhaps it is not a coincidence that 7 of 17 stone balls were uncovered there as well. The reason may lie not only in the importance of the citadel as a *basileia*, but also in its function as a refuge. It would seem that, after an artillery bombardment attested by the stone balls and aided by the fire of archers and slingers, the town saw hand-to-hand fighting along the streets, in the agora, and in houses. The ivory chape may have been stripped from the *xiphos* scabbard during the turmoil inside House 2, while the helmet decorations may have fallen off during street fighting.[48] Intense urban warfare took place.

Battlefields

Among the military remains found on battlefields, inscribed sling bullets are perhaps the most significant, helping to generate narratives that supplement written records. Like coins, they are both a material objects and epigraphic records, but unlike coins, they are removed from circulation once employed in battle.[49] The vast majority record the names of commanders of slingers.[50] On the one hand, these projectiles fill gaps in the itineraries of known historical figures; on the other hand they can contribute new names.

Battlefield archaeology in Bulgaria, still in its infancy, revolves around the study of sling bullets. Since the first publication of sling bullets by I. Venedikov in 1952, short reports from museum collections or publication of private collections, very often with no provenience, emerged with increased frequency.[51] During the last decade came a surge of newly excavated items, thanks particularly to excavation and surface scatter finds from *emporion* Pistiros, the fortified Thracian residence at Kozi Gramadi, Plovdiv district, and most recently from the fortified site Kolopenishteto, Sliven district. The discovery of sling bullets and catapult arrowheads from these sites brought new perspectives regarding the scale of the Macedonian campaigns in inland Thrace, south and north of the Balkan range.[52]

The site at Kozi Gramadi has so far yielded the largest sample of sling bullets retrieved from archaeological excavations in Bulgaria. Of the total of 157 lead bullets, only 18 are

46 Stoyanov 2005, 649.

47 Weapons: Buykliev 1994, 160–161. Other items: Nankov 2007, fig. 1, 2008, 40, fig. 18. Helmet decoration: Ognenova-Marinova 1984, 288–289, fig. 12; cf. Marazov 2006, 50.

48 Nankov 2007, 44; Chichikova 2016, 45.

49 Ma 2010, 166; Scott and McFeaters 2011.

50 Ma 2010, 171; Avram et al. 2013, 259–293.

51 Venedikof: Avram 2011, 195–198. Later: Atanasov and Todorov 1999; Paunov and Dimitrov 2000; Manov and Torbov 2016.

52 Pistiros: Hristov and Manov 2011; Hristov 2012; Hristov 2014. Kolopenishteto. Nankov 2015a, 3–4, fig. 1a–d; Leshtakov et al. 2016. Inland: Nankov 2015a; Nankov 2016.

inscribed with the names of Anaxander and Cleobulus, which are famously known from the sample at Olynthus.[53] Among the other weapons from the site, noteworthy are nine pyramidal, bodkin arrowheads with socket, two double-bitten axes, and one shield umbo.[54] In my view, the artifact distribution points to a siege that ended quickly, with very little fighting inside the fortified area, the tower-gate, and the monumental building. The picture may change after proper publication of the entire sample.[55]

Ten lead sling bullets, of which five carry the names of Anaxander and Cleobulus, originate from the mountainous site of Kolopenishteto (1054 masl), located near the village of Sotirya, Sliven district. Nearby were found ten iron arrowheads of the pyramidal, bodkin type with socket and with tang (Robinson's Type E), and also mostly with bent tips.[56] More damage appears on some of the sling bullets, revealing a battle in which Thracians stood their ground against Macedonians. The mountainous topography of Kolopenishteto resembles closely that of Kozi Gramadi, which explains why Philip II operated not only in the plains along valleys of major rivers, where he preferred to found his colonies such as Kabyle, but also allocated light-armed troops to deal with the aristocratic seats of Odrysian overlords nestled high in the Sredna Gora and Stara Planina mountains.[57]

Weapon Hoards

The best-known example of a Bulgarian weapon hoard comes from the battle of Abritus in 251 CE, in Razgrad region. Numerous Roman militaria have cropped up miles away from the supposed battlefield. Perhaps this signals looting by locals who gathered weapons from the battlefield following the Roman defeat.[58] Another hoard consisting of weapons (***pilum***, shield ***umbo***, a frame of an arrow-shooting ***ballista***, and a tent peg), as well as digging tools (pickaxes, spades, mattocks, and a plowshare) may be connected to the events surrounding the battle of Abritus. It was discovered by chance near the village of Elenovo, Nova Zagora district.[59]

Recently, a Roman military camp dated to the Third Macedonian War was identified through rescue excavations in the Middle Strymon region. The character of the site was determined through a preliminary analysis of the spatial distribution of hobnails, fibulae, coins, arrowheads, torches, and tent pegs, resembling remotely the blueprint of Roman camps known from elsewhere.[60]

In contrast to this later evidence, the evidence from Classical and early Hellenistic Thrace is patchy. At Olonesti, Moldova, unrepaired body armor consisting of six helmets and six greaves has been retrieved; more such armor, consisting of eight Chalcidian helmets, was

---

53 Hristov and Manov 2011, 24–27; Hristov 2014, 124.

54 Arrowheads: Hagerman 2014, 80–85. Other items: Hristov 2011, 116, fig. 1; Hristov 2012, 81, fig. 3, 82, fig. 6; Hristov 2014, cf. Nankov 2015a, 8.

55 Nankov 2015a, 5–9; cf. Hristov 2012, 84; Tzvetkova 2018, 108–110.

56 Bullets: Leshtakov et al. 2016, 869–871, figs. 1–3; Nankov 2015a, 3–4, fig. 1a–d. Arrowheads: Leshtakov et al. 2016, 871, fig. 3, 4–6, reporting finds in nearby "Haramiata." Tang: Hagerman 2014, 80–85.

57 Nankov 2015b, 406–407; cf. Popov 2017, fig. 7; Tzvetkova 2018, 139.

58 Radoslavova et al. 2011, 25, 28–29, maps 2–4; Ball 2014, 98–99.

59 Kayumov and Minchev 2013.

60 The village of Polenitsa, Sandauski municipality, as in Alexandrova 2017, 279–281, fig. 2.

found near the village of Yudelnik, Ruse district.[61] Similar are chance finds collected from river beds, such as the Corinthian helmet from Chelopechene, Sofia district and the Chalcidian helmets from Silistra and Budesti, described as *Flussfunde*.[62]

The helmets from Olonesti clearly betray a series of violent encounters with, no doubt, lethal outcome for their owners. Among the weapons doing the damage are arrowheads (Olonesti), spears, and swords (Yudelnik).[63] Scholars have convincingly connected the Olonesti find with the disaster of Zopyrion's army against the Scythians (331), and the Yudelnik find with the Macedonian campaigns of Philip II (339) or of Lysimachus (313) against the Gaetae.[64] One scholar has linked the helmets of the so-called *konos* type recovered in the Veliko Turnovo district and also the only Boeotian helmet known from Bulgaria, discovered in the Ruse region, to the Macedonian invasion in Thrace.[65]

Militaria in Graves and Tombs

For most of their history the Thracian military elite found it necessary to supply the deceased with armor and weapons for their last journey. This custom contributes greatly to our knowledge about Thracian panoply. M. Domaradzki was the first to realize the potential of the material that he dubbed "Thracian rich burials." Rather than arrange weapons in familiar typologies, he sought to trace the Odrysian kingdom from its heyday until the Roman era through the way the military aristocracy portrayed themselves in death.[66] This in turn made it possible to see how Thracian warfare adapted itself to the emergence of new weapons and the invention of new fighting techniques. In this section, we cite cases involving cuirasses, spurs, swords, rhomphaia, and composite bows.

In Thrace, the bronze cuirasses of the so-called "bell-shaped" type found use not as hoplite armor but as armor for cavalry during the late fifth century, long after it had gone out of use in Greece.[67] By the time of Philip II and Alexander III, it seems to have been replaced by the scale-leather corslets, supplemented with collars or neckguards, or *peritrachelia*.[68] These corslets and *peritrachelia* stand out in the "rich equestrian burials" in upper Thrace along the courses of Tonzos and Hebros rivers, where, as already noted, many Odrysian dynasts chose to stand their ground against the invading Macedonians. These "second-horizon" burials, to use Domaradzki's phrase,[69] contain nearly identical sets of military equipment (helmets, greaves, appliqués for horse trappings, and swords).

61 .Sergeev 1966, 137–142, figs. 8–15; Ognenova-Marinova and Stoyanov 2005, cat. nos. 16–25, Table IV, 1–5; Lazar 2009, 26, no. 7.

62 Teleaga 2008, cat. nos. 931–932, Karte 42; Lazar 2009, 26, cat. nos. 4 and 19, figs. 2, 6.

63 Sergeev 1966, 137–140, figs. 8–11, showing that the damage went unrepaired. The weapons: Vasilev 1983, 6–9, figs. 1–10; Sergeev 1966, 138, fig. 10.

64 Sergeev 1966, 141–142; Archibald 1998, 252–253.

65 Dimitrov 2002, 187, 2006, 248–249, figs. 1, 3. The helmets come from near the village of Merdanya and from Kazanlak.

66 Domaradzki 1998, 44–64, see also Archibald 1998, 198–207; Georgieva 2016.

67 Ognenova 1961, 504–526; Dimitrov 2010, 219–220; Parvin 2010, 227.

68 Parvin 2008, cf. Ognenova 1961, 527–538; Tacheva-Hitova 1971, 55.

69 Domaradzki 1998, 44–46.

Among them are the Peichova tumulus, the site at Starosel, the Dalakova tumulus, and the sites at Sliven, Zlatinitsa, Koylmen, and Branichevo.[70]

The advent of spurs presents a complicated case. It was long thought that Celts brought them into Thrace during the second and first centuries. Hence the high concentration of spurs deposited in graves or chance finds situated exclusively in northwestern Bulgaria. Recent discoveries have shown, however, that they were introduced sooner, on the arrival of the Macedonians, in both northeastern Bulgaria, occupied by the Getae and in the upper Thrace valley.[71] In the fortress of Dragoevo, Shumen district, 71 spurs were singled out as "Macedonian" and likely associated with Macedonian infantry and cavalry in northeastern Thrace during the reigns of Philip II and Alexander III.[72] A few decades later, spurs appear in rich equestrian burials in the Golyama Kosmatka tumulus, Shipka, and near Dolna Koznitsa, Kyustendil municipality.[73] They further appear in the equestrian representation of a Getic dynast in the Sveshtari tomb dated to the first quarter of the third century.[74]

Swords are not readily identified with Thracians, in spite of the oft-cited passage in the *Iliad* (13.576, 807). That they were considered a weapon of last resort is valid for Thrace as well, where the supremacy of the spear and shorter javelins was never challenged (Marazov 2005, 35–40), as can be seen on Athenian red-figure pottery and in painted Thracian tombs (Hermary 2010, 484–486; Nankov 2010, 44, nn. 14, 50). The Greek *kopis*, often described generically as *machaira*, was recommended by Xenophon (*On Horsemanship* 12.11–12) to be used by cavalry, because curved one-edged weapons performed better when used from an elevated position. Their curved shape and off-center gravity point allow the user to strike with a greater force, very much like a battle-axe. Evidently, Thracian cavalrymen appreciated Xenophon's remark well, because the *kopis* accompanied them in graves during the Classical and the Hellenistic periods, e.g. Duvanli, Zlatinitsa, Kabyle, and Golyama Kosmatka tumulus.[75]

The iron *rhomphaia*, that peculiar cross between a spear and a sickle, is possibly a Thracian invention that evolved from a farming implement. Thracians in Macedonian service carried it during the First and Second Macedonian Wars, but apparently only in the mountainous west and central Rhodopes.[76] The slashing quality of the weapon made it effective against cavalry in the hands of light troops with small shields (*thureoi*).[77] Why did this weapon come into use at this time, and not earlier? We do not know. Attempts to identify the weapon on scenes in painted Thracian tombs, e.g. Alexandrovo and Sveshtari, have proved

70 Atanasov 1995, nos. 87–88; Parvin 2008; Agre 2011, 72–83.

71 Torbov 1998; Stoyanov 2003, 198, 200.

72 Spurs of Type II after Atanasov 2006, 27; list of other examples in Baitinger 2004, 368–369. Dating and provenience: Nankov 2016, 286–287; see also Baitinger 2004, 371.

73 Dimitrova 2015, 164, 305, 330, nos. 56, 91; Stoyanova 2015, 119, fig. 8/1.

74 Stoyanov 2008a, 64, figs. 2–3, 8, 2015, 432.

75 Agre 2011, 91; Stoyanov et al. 2013, 265, 267, fig. 23; Zhuravlev and Firsov 2013, 92–93, cat. no. 18; Dimitrova 2015, 325–328, no. 84; Torbov 2016, IX–XI.

76 The invention: Kolev 1975, 20–21; Buykliev 1994, 159; Webber 2011, 61–67; Torbov 2016, XII–XIV. Place of use: Mikov 2010, 209, table 2, reporting mortuary finds in this region.

77 Sekunda 1983, 284–285. His opinion that the weapon is an adaption of the Greek *drepanon*: ibid., 275–276.

unsatisfactory.[78] An acceptance of this hypothesis still needs to explain why Thracians began to deposit *rhomphaiai* in graves during and after the First and Second Macedonian Wars, but not earlier.

The Thracians, like the Scythians, made famous use of the composite bow, which appears in both grave assemblages and artistic representations.[79] Graves of the late fifth and early fourth centuries have thus produced large numbers of bronze arrowheads (Robinson's Type G1). The differing needs of hunters and warriors may explain various projectile sizes and separate quivers.[80] Although Robinson's Type G1 arrowheads continued to be used during the Hellenistic period, they crop up in settlements such as Pistiros, Kabyle, Dragoevo, and Seuthopolis, not in graves. They had proved ballistically inferior to Cretan arrowheads (Robinson's Type D1) and to catapult projectiles made of iron (Robinson's Type E), which arrived in Thrace with the Macedonian troops of Philip II, Alexander III and the Successors.[81]

Sling ammunition deposited in graves present another puzzle. Round stones or pebbles put in leather pouches appear conspicuously alongside other weapons in the aristocratic equestrian burials at Golemanite, Veliko Tarnovo, and Aghigiol, but scholars differ about whether they are ammunition. Perhaps they were thrown by hand rather than by a sling.[82] Odrysians are mentioned as skillful stone-throwers (*petroboloi*) by the early fourth century.[83] Lead sling bullets cast in matrices, on the other hand, arrived in Thrace only with the Macedonians.

---

78 Dimitrov 2005b, 290, fig. 7; Paunov 2005, 369; cf. Stoyanov 2008a, 62; Mikov 2010, 224; Nankov 2010, 36–37. Sekunda (1983, 283–284) suggests that when Thucydides described the Dioi in the Rhodopes as *machairophoroi* during the Peloponnesian War he meant *rhomphaiophoroi*. See Paunov 2005; Dimitrov 2006; Mikov 2010; cf. Marazov 2005, 48; Webber 2011, 61–67.

79 Buykliev 1991; Marazov 2005, 40–45; Georgieva 2016, 271.

80 E.g. the Golemanite site, the Svetitsa tumulus, and the Duvanli, Chernozem-Kaloyanovo, and Zlatinitsa sites, as at Agre 2011, 100; Stoyanov 2015, 427.

81 Hagerman 2014, 80–85.

82 The finds: Stoyanov 2015, 427. By hand, not sling: Dedyulkin 2015, 51; cf. Pritchett 1991, 65–67.

83 Xen. *Hell.* 2.4.12; Polyaen. *Strat.* 3.9.62; cf. Xen. *An.* 5.2.12. An example of doubtful evidence: A sling bullet found in a grave of a ten-year-old child in Apollonia Pontica (Venedikov 1948, 18, fig. 21 with Avram 2011, 198). The fact that no weapons accompany the deceased in the excavated necropoleis rules out the possibility that it was deposited as a grave gift. The child was likely a casualty of an unrecorded siege of the town by Philip II in 339, as were at least a hundred individuals, the cremated remains of which were buried in a *polyandreion* (Damyanov et al. in print a). Other sling bullets from Apollonia Pontica carry the names of Philip's commanders, e.g. Potalos (Robinson 1941, 434–435, nos. 2244–2248).

PART V

Technical and Economic Context of Greek Warfare

CHAPTER 17

Logistics

Stephen O'Connor

Introduction

The approach of this chapter is different in three important ways from previous treatments of the logistics of Classical Greek land warfare. Firstly, it sets the logistics of Classical Greek armies in a comparative historical framework.[1] Secondly, it considers the particular social, economic, institutional, and ecological contexts in which Greek armies supplied themselves. In so doing, the chapter identifies the fundamental logistical challenges facing Greek armies—and how those challenges, and the solutions Greek generals and their soldiers found (or did not find) to meet them, were often different from those facing armies in other times and places in pre-industrial Europe and western Asia.

Thirdly, this chapter also differs from previous treatments by paying closer attention to the military thought of Xenophon, the first Greek military thinker we know of to stress the importance of logistical requirements for armies (Wood 1964, 53). A man with an exceptionally wide experience of war and campaigning (Lee 2017), Xenophon developed an intense interest in military leadership: the question of how leaders, and especially military leaders, could do the most for their men to succeed infuses all of his writings (esp. Gray 2011). Strikingly, Xenophon emphasized throughout his discussions of military leadership how critical the right planning of logistics and, principally, food supply was for the success of generals and their armies. Placing Classical Greek military supply within a comparative framework and in its wider societal setting, then, this chapter will show why Xenophon was right to think that solving logistical problems was a task of the first importance for any Classical Greek general—and why, in particular, the food supply of his soldiers was the most urgent of those problems.

[1] Starting points for this work: Erdkamp's introductory chapter (1998, 11–26) is the best survey of the sources and methods of food supply used by pre-industrial European armies. More broadly, Leighton 2017 is the most useful general introduction to the logistical problems faced by armed forces.

A Companion to Greek Warfare, First Edition. Edited by Waldemar Heckel, F.S. Naiden, E. Edward Garvin, and John Vanderspoel.

Food Supply: Setting Out on Campaign

In Xenophon's most developed vision of military leadership, the fictional narrative of the *Cyropaedia*, the main character, the Persian prince Cyrus, is presented to a contemporary Greek audience as a paradigmatically ideal leader and general (e.g. Stadter 1991; Hutchinson, 2000, 180). Cyrus, however, only completes his development as an ideal general after a conversation with his father, Cambyses, just before his first military campaign, fills in the gaps in his understanding of the duties of a commander (Stadter 1991, 474–475, 484–485). Cambyses stresses repeatedly to Cyrus in this conversation the fundamental importance of logistical security, and especially security of food supply, for an army's success (Xen. *Cyr.* 1.6.7–11, 17–18, 26). Given the absolutely critical importance of provisioning in his advice to his son then, it must strike us as paradoxical when Cambyses tells Cyrus in the same dialogue that one of the major reasons a general must constantly attend to his men's provisioning is because Greek armies brought very little food out on campaign (Xen. *Cyr.* 1.6.17):

> But the worst burden of all is to support an army in idleness. For not only are the mouths in an army very numerous but the supplies they start with are exceedingly limited, and they use up extravagantly whatever they get, so that an army must never be left idle.[2]

But contemporary evidence supports Cambyses' statement that Classical Greek armies took very little food with them on campaign. Athenian soldiers departing for land campaigns in the fifth century were typically ordered to bring along three days' supply of food (Pritchett 1971–1991, vol. 1, 34). This amount is also the quantity of provisions most frequently evidenced for other Classical Greek military forces setting out on overland marches (Pritchett 1971–1991, vol. 1, 32–34). The number of days' supplies soldiers were ordered to bring with them at the outset of a campaign sometimes varied, however, according to the tactical requirements of the operations they were about to undertake, and the opportunities they would have during those operations to replenish their supplies (Erdkamp 1998, 77). Thus, for example, the Arcadian forces that occupied Oneion in 367—in order to prevent the Thebans from being harassed by enemy forces on their march through the Isthmus of Corinth—took seven days' provisions with them (Xen. *Hell.* 7.1.41) since they knew that they would have to remain immobile in uninhabited and hostile territory for some time in order to achieve their mission. But the circumstances of this operation were unusual (as we will see, the operations of Classical Greek armies were usually characterized by their mobility, and soldiers usually had the opportunity to replenish supplies during their operations); the figure found here of seven days' supplies is, accordingly, the largest attested figure for an amount of provisions carried by a Classical Greek land force in mainland Greece.

All of these provisions were either bought by individual soldiers from their own money in markets in their home cities (e.g. Ar. *Pax* 368, 1182) or provided out of the resources of soldiers' own households (e.g. Ar. *Ach.* 1095–1101). Their caloric base was *alphita*, roughly milled barley groats (Moritz 1949); the standard 2 l per day would have provided somewhere around 70 per cent of the 3–3,500 calories an ordinary Greek soldier, standing somewhere around 5'7" and weighing around 150 lbs, would have needed to retain fighting fitness (O'Connor 2013). Because barley grains were roasted in Classical Greece as part of the process of removing their glumes prior to milling (Amouretti 1986, 123), *alphita* had two

---

[2] This and all subsequent translations in this chapter are from the Loeb Classical Library.

particular advantages for men on the march. Firstly, it was long-lasting: kept properly, it could remain edible for two to three months. Secondly, and even more convenient for men on campaign, it did not need to be cooked. Instead, *alphita* was consumed in the form of *maza*, when the barley groats were mixed with some form of liquid (e.g. water, olive oil,[3] honey), without any cooking, before being kneaded into whatever form desired. *Maza* could therefore be pre-made, or prepared just before a meal (Amouretti 1986, 124–126). The soldier's *maza* was accompanied by *opsa*, that is, foods added to supplement the grain base of the diet. These were obviously a matter of personal taste, but Xenophon's Cyrus told his men to bring sharp, pungent, and salty foods to add flavor to their meals (Xen. *Cyr.* 6.2.31). We hear, accordingly, of soldiers bringing out on campaign cheese, salt fish, onions, olives, garlic, and salt in addition to their *alphita* (e.g. Ar. *Ach.* 1095–1101; Ar. *Pax* 368, 529, 1128–1129; Eupolis, fr. 275 (PCG); Xen. *An.* 7.1.37). All of this, *alphita* and *opsa*, was washed down with wine, probably somewhere between 0.5 l and 1 l per day. Wine provided important vitamins not present in grain (Bresson 2016, 122) and—even when, as customary, the wine was diluted with two or three parts water[4]—more calories; 1 l of wine contains around 700 calories (Roth 1999, 35).

Although private traders always accompanied Greek armies setting out on land campaigns, too, sometimes in very large numbers (Diod. 14.79.2), they did not add any substantial amount of food to the supplies brought out from home by soldiers (O'Connor 2015). This was because the high costs of carrying grain and other basic foods overland using pre-industrial technology left very little room for traders to profit from selling these foods to soldiers, especially as soldiers were able (as we will see) to resupply locally on campaign with cheaper (in *polis* markets) or free (through foraging) food. Rather, these traders followed Greek armies (just as they did other pre-industrial Eurasian armies (O'Connor 2015, 191–203)) on campaign primarily in order to purchase any captured plunder at low prices, with the aim of profiting from its resale on their return to friendly territory. Xenophon still thought it was a good idea for careful generals to encourage the presence of traders on campaigns as a supplementary source of provisions, inasmuch as the small amount of food they carried for sale could constitute a sort of minor risk-buffering mechanism for the provisioning of an army (Xen. *Cyr.* 6.2.38–39 with O'Connor 2015, 184–185). But private traders never played a structurally important role in the provisioning of any foreign campaign undertaken by a Greek army, and so never contributed to contradicting Cambyses' statement that Greek armies departed on overland campaigns with very little in the way of food supplies.

"Everything Else that an Army Might Need": Other Logistical Requirements

Logistics is, of course, not just about the provision of food supplies for armies. It is the science (or, probably better, art) of supplying armies with everything they need in order for them "to live, move… and fight" (Leighton 2017)—or, as Xenophon in his *Memorabilia* has Socrates tell an aspiring general, a successful military leader, before anything else, has to prepare food supplies *and* everything else that an army might need (Xen. *Mem.* 3.1.6). Just as

[3] Olive oil was used in cooking in Classical Greece mainly as a binder of grain products and for more elaborate types of cooking such as the glazing of meats (Amouretti 1986, 181–182; Foxhall 2007, 89–91). As such, apart from sometimes being used in the making of *maza*, it did not play any major role in soldiers' diets on campaign.

[4] See what follows for the sources of this water.

with the amount of food they brought with them, the number of other items soldiers took from home would have varied according to the needs and aims of the campaign they were departing for: men responding to a border raid would need much less materiel to survive and retain their full operational capacity than, say, the Peloponnesian and Boeotian army invading Attica for 40 days in 430 (Thuc. 2.57.2). That said, for almost all but the briefest operations, soldiers would have needed some basic items to subsist in the field and to remain an effective fighting force: weapons, armor, clothing, cooking utensils, drinking cups, tools of various types, some (rudimentary) medical supplies, packs and other containers for all of these (and their food), and straps to fasten everything together.[5]

These items, in total, were much greater in bulk and weight than the few kilograms of food and wine soldiers took from home. A hoplite's arms and armor could range in weight from 10 kg to over 30 kg of unwieldy objects to carry; even a light-armed soldier's weapons and armor might weigh 10 kg or so in total (Krentz 2007b, 150–151; Lee 2007, 111–117). Bedding, clothes, cooking equipment and everything else men thought necessary to bring on campaign could also add another 5–10 kg of awkward, bulky weight per soldier (Lee 2007, 126). We have no real idea of the size or even appearance of the hand-mills Greek soldiers brought with them to grind grain foraged in enemy territory but they, too, must have added quite a bit of bulk and weight to the carrying requirements of a typical Greek army.[6]

The supply of this equipment did not, however, present any major logistical challenge for generals of Greek armies. Firstly, each individual soldier was responsible for providing his own weapons and armor, clothing, cooking gear, bedding, shelter, and other articles for personal use.[7] Secondly, and even more crucially, (nearly) all of these items were durable and did not need constant resupply.[8] Rather, the rate of replacement for objects such as pots, drinking cups, and clothes on the brief land campaigns undertaken by Classical Greek armies must have been absolutely minimal (van Creveld 1989, 47–48; Leighton 2017). Since they were fighting a warfare dominated by edged weapons such as spears and swords, neither did Greek soldiers need resupply of ammunition.[9] Therefore, once soldiers had loaded up their arms and armor and other personal articles before setting out on campaign, they would no longer have a need or demand to replace them, and so generals did not have to concern themselves with providing opportunities for the resupply of equipment. This remains true even when we add to the items already mentioned as necessary for Greek soldiers to bring from home axes for cutting down wood for shelters (and for cutting down the fruit trees of enemies (e.g. Xen. *Hell.* 4.5.10, 6.5.22)), sickles for foraging in enemy territory, and wooden pegs (Ar. *Ach.* 761–63) for uprooting enemy crops.

---

[5] Good discussions at van Wees 2004, 104; Krentz 2007b, 150–154; Lee 2007, 109–127.

[6] Hand-mills: Xen. *Cyr.* 6.2.31; Frontin. *Str.* 4.1.6; Polyaenus *Strat.* 3.10.10. Foraging: see below.

[7] See e.g. Cruickshank 1954, 83; Anderson 1970, 39–40, 59; van Wees 2004, 104–105. Unusually at Sparta, due to that community's exceptional institutional focus on group cohesion and lack of market structures, there was public organization and provision of a uniform military equipment, at least for cavalry (Finley 1981, 29–30; Hodkinson 2000, 221–226). Elsewhere, private purchase of arms and armor in markets and from workshops was usual (e.g. Ar. *Pax* 1210–1264 (with Harris and Lewis 2016, 1–2); Aen. Tact. 30.1–2; Xen. *Hell.* 3.4.17).

[8] See Leighton 2017 for the crucial distinction between the supply requirements of expendable and non-expendable items for military forces.

[9] See Gomme 1945–1981, vol. 1, 114; van Creveld 1989, 47 for "edged weapons." Neither was resupply of projectiles for light-armed troops an issue for Greek generals: these were either retrieved after use (arrows, javelins) (Anderson 1970, 117) or simply collected in the field (stones).

"(Nearly) all of these items" were long-lasting; there were two minor exceptions to Greek soldiers' self-reliance in equipment. Firstly, although there are several references in contemporary sources to soldiers bringing bedclothes (*stromata*) on campaign (e.g. Ar. *Ach.* 1136), these bedclothes could be made more comfortable or replaced altogether by *stibades*, field beds made out of materials such as grass, leaves, and rushes found in the vicinity of camps.[10] This was one reason (the other being the Mediterranean climatic regime armies fought under) Xenophon's Cyrus could give his soldiers the advice to replace the weight of their *stromata* with an equal weight of food (Xen. *Cyr.* 6.2.30). Secondly, while some men clearly did use tents as shelters to sleep in, made as far as we can tell from leather skins,[11] it is also clear that at least as often, if not more so, men slept in shelters made from whatever materials (such as sticks and branches or woodwork from houses and farm buildings) were available around their camp (Anderson 1970, 61–62). The practice of improvising shelters was so common, in fact, that Aeneas Tacticus, in his advice to communities for how to make it difficult for invading armies to camp in their territory, recommended that all materials in the countryside which could be useful to the enemy for making shelters be removed or hidden away (8.3). The materials needed to improvise both camp beds and field shelters were so commonplace and so easily collected, then, that they did not make any major exception to the fact of Greek armies' self-containment in hardware (Cruickshank 1954, 70; Leighton 2017).

The transport of their soldiers' military equipment was also not a matter of concern for Greek generals since, outside of Sparta (Xen. *Lac.* 11.2, 13.4), Greek soldiers, rather than their states, were responsible for the organization and supply of the transport of their food and equipment (e.g. van Wees 2004, 104–105). It was usual practice for citizen hoplites and cavalrymen, drawn from the wealthier men of their cities (van Wees 2004, 55–59), to be accompanied on campaign by their own slaves employed as attendants.[12] This was a simple extension of the practice of well-off Classical Greek men to always bring a slave to attend to their needs when traveling (Whitehead 1982, 120). These slaves carried their masters' provisions as well as much of their equipment as they could. The provision of food to slaves was a matter for their masters: although some owners might be especially stingy (Theophr. *Char.* 30.7), most owners realized the value of maintaining their investments and so provided their slaves with roughly the same number of calories they consumed (O'Connor 2013, 349–350). Slave attendants could therefore be (relatively speaking) expensive to maintain; but they did have important advantages over other forms of transport in that they could negotiate more difficult and varied types of terrain, while needing less care, than baggage animals (Hammond 1983, 28; Lazenby 1994, 6; Leighton 2017).

Slave attendants were therefore a taken-for-granted presence in Greek armies, as comes out clearly from a Xenophontic description of a disorganized army (Xen. *Oec.* 8.4):

> [an army in disorder was] a confused mass, an easy prey to enemies, a disgusting sight to friends and utterly useless—donkey, hoplite, slave baggage-carrier, light-armed troops, horseman, cart, all huddled together. How are they to march in such a state, when they hamper one another, some walking while others run, some running while others halt, cart colliding with horseman, donkey with cart, slave baggage-carrier with hoplite?

[10] See e.g. Ar. *Pax* 348 with Olson 1998, 142; Xen. *Hell.* 7.2.23; Polyaenus *Strat.* 3.9.19.

[11] See Xen. *An.* 1.5.10, 3.2.27 with Lee 2007, 122; Krentz 2007b, 164 n.31; also e.g. Theophr. *Char.* 25.5.

[12] See e.g. Pritchett 1971–1991, vol. 1, 49–51; Spence 1993, 284–285; van Wees 2004, 68–69.

This description also shows that pack animals—donkeys—and carts were also presumed constituent parts of Classical Greek armies. Animals in Greek armies carried fewer types of items than in other pre-industrial armies since slave attendants performed the tasks of carrying most of the personal effects of hoplites and cavalrymen. Pack animals and carts were still needed, however, for heavier, more awkward items of personal equipment but also for the more elaborate baggage of richer soldiers; for tents for those who had them; and often, too, for weapons and armor.[13] In addition, it is also probable that soldiers below the richest tier of men often banded together to use one or a couple of pack animals to share the burden of carrying their supplies and equipment (van Wees 2004, 105; Lee 2007, 102–103).

The few specific references we have for the species of pack animal accompanying Greek armies are consistent: donkeys were the animals used for this task (Lazenby 1994, 6–7). This is unsurprising inasmuch as soldiers had to provide their own baggage animals, and donkeys, because of their low costs of reproduction and maintenance (Isager and Skydsgaard 1992, 87, 89), were the predominant pack animal in the agricultural and wider economy of Greek cities (Griffith 2006, 224–225). Donkeys as pack animals had several advantages over carts as a mode of transport: they did not need roads (very useful in Greek warfare since unconstructed tracks were common means of travel overland in Greece (e.g. McHugh 2019, 4, 7–8)); they could therefore travel more quickly and for more hours per day in areas of broken or mountainous terrain (very often the type of terrain Greek armies operated in); and, in general, they were less of a hindrance and occupied less room both while marching and in camp.[14]

Pack animals thus offered greater flexibility of movement than carts. But carts were still a feature of Greek land campaigns.[15] Why? Firstly, although there are no specific references identifying the animals drawing the carts following armies in the Classical Greek world, the ability of carts to match the marching pace of armies means that the main means of wheeled transport were almost certainly always the mule-drawn, two-wheeled carts used everywhere in the Classical Greek world for draft transport,[16] rather than the heavier four-wheeled wagons drawn by oxen used for especially heavy freight (Cruickshank 1954, 70–71; Lee 2007, 134). These mule-drawn carts were light, quick, and maneuverable[17] but could still transport much heavier loads than pack animals: up to 1 ton compared to the 100–200 kg a donkey could carry (Raepsaet 2002, 192, 2008, 588–589). In addition, items transported by donkeys needed to be divided into loads of even weight before being placed into packsaddles or wooden-framed panniers (Erdkamp 1998, 70; Lee 2007, 136–137); this restricted the use of donkeys and necessitated the use of carts for larger, more awkward items. So, especially for marches over short or medium distances to war zones accessible through carriage roads (such as, for instance, the Spartan campaign to Mantinea in 418 (Thuc. 5.72.3)), carts could and did make a significant contribution to the transport capacity of an army.[18]

The donkeys and mules which provided transport for armies needed to be fed. But their feeding was not a major problem for the leaders of Greek armies not only because the

---

[13] See Lazenby 1994, 4–5 for references to pack animals carrying arms and armor.

[14] Erdkamp 1998, 70–71; Roth 1999, 202–203; Mitchell 2018, 110.

[15] See Cruickshank 1954, 69, 71–72 for the structure and main ideas of this paragraph.

[16] Mules, although more expensive than donkeys because of their greater costs of reproduction (Isager and Skydsgaard 1992, 87; Raepsaet 2008, 587), are the usual animals shown on contemporary vases or reliefs as drawing carts in Classical Greece (e.g. Lorimer 1903, 140; Raepsaet 2002, 168–173; Kourinou 2007, 88).

[17] Cruickshank 1954, 69; Griffith 2006, 204; Raepsaet 2008, 591–592.

[18] See Cruickshank 1954, 71–72, and, generally, on the extent of and transport possibilities afforded by Greek roads, e.g. Bresson 2016, 81.

animals' care was the responsibility of their owners but, crucially, also because, comparatively speaking, the number of pack and draft animals with Greek armies was very small. This was because, as we have seen, slave attendants did much of the work performed by baggage animals in other armies—but also, and more importantly, because Classical Greek armies brought much less materiel and personal baggage on campaign than other pre-industrial military forces and therefore required much less transport capacity.

This important point can be illustrated by means of some explicit comparisons with the transport requirements of other armies operating in pre-industrial Europe and western Asia. So, for example, while generals and wealthier soldiers in Greek armies, just as any wealthy man in the Greek world setting out for a long-distance journey overland,[19] may have taken enough baggage on longer campaigns to need a pack animal or two (in addition to a slave attendant) to carry their belongings, Persian kings on campaign, Roman generals, and early modern European commanders, in stark contrast, needed hundreds, sometimes thousands, of pack animals and wagons to carry their household's and staff's personal effects.[20] Another contrast lay in Greek soldiers' consumption of *maza* rather than bread; because of this, there was no need for any bread-making facilities for Greek armies, and therefore no need for the hundreds or thousands of horse-drawn wagons necessary in early modern European warfare to transport brick ovens (together with the necessary flour) to armies (e.g. Lynn 1993a, 19–21).

In another difference with later European armies, too, Greek armies took with them very little or nothing in the way of construction materials. Greek camps were generally unfortified, relying instead on sentries for their defense (Anderson 1970, 59–66). Even in the rare circumstances where camps were fortified, as when campsites turned out to be in unexpectedly vulnerable positions (e.g. Xen. *Hell.* 4.4.9) or faced an unusually high threat of imminent attack (e.g. Xen. *Hell.* 6.5.30), the wood for palisades and other defensive structures was cut down from the surrounding countryside (van Wees 2004, 107). In contrast, Roman armies carried pre-made wooden stakes for palisades; these taken together with the specialized tools carried for camp-making added some 15 kg per soldier to the transport requirements of Roman armies (Roth 1999, 74).

The transport requirements of Classical Greek siege operations were also minimal compared to later European armies. Even in those cases when an assault on fortifications was conceived as the aim of a military operation, Greek armies did not bring much in the way of siege trains. Greek generals and their *poleis* would have needed for these operations to organize some carts and wagons (by hiring them from private individuals[21]) to transport ladders (e.g. Aen. Tact. 36), battering rams (Aen. Tact. 32.3), mining and tunneling tools and shelters (Aen. Tact. 32.11–12), flame-throwers (Thuc. 4.100.2–4), and a small number of wooden towers and (from the beginning of the fourth century) catapults (e.g. Aen. Tact. 32.1, 8). Nor were many animals needed to carry materials for siege works since, apart from some iron for building clamps (e.g. Thuc. 4.69.2), Greek armies improvised their circumvallation walls and siege-mounds from materials collected around the settlements they were besieging (e.g. Thuc. 2.75.1–78.1, 4.69.2). All this was a world away from the massive

[19] See e.g. Xen. *An.* 5.3.11–12, Howe 2008, 91, for men traveling overland to religious festivals being accompanied by baggage animals.

[20] Persian: Briant 2002, 256–257; Roman: Roth 1999, 89–90; early modern European: e.g. van Creveld 1977, 6.

[21] This statement can be made with confidence, although there is no specific contemporary evidence for how these carts or wagons were procured or organized, since Greek *poleis* contracted out to private owners of wagons for all other heavy public transport needs (e.g. Burford 1960).

increase in transport requirements imposed upon late medieval and early modern European armies by the introduction of gunpowder artillery (e.g. van Creveld 1977, 12; Ansani 2019). Already by the 1490s, the transport capacity needed to move the prerequisite mounted guns, shot, gunpowder, and associated tools and equipment could be so great as to reach 12,000 draft horses (the number Charles VIII of France was reported to have needed to transport his ordnance to Italy in 1494 (Ansani 2019, 359)).

Since, then, Classical Greek armies needed animals to transport much, much less than other armies in other times and places in pre-industrial Near Eastern and European history, they had very many fewer animals to feed. The second important reason why the feeding of the pack and draft animals of Greek armies did not present a significant logistical problem for Greek generals was that the dietary requirements of their comparatively few donkeys and mules were easily met. While we do have a few ancient references to armies carrying or preparing provisions—probably barley—for their transport animals (Xen. *Cyr.* 6.2.25, Hdt. 7.25.1, 7.187.2), mules and donkeys did not need hard grain fodder to remain effective as working animals. Donkeys especially can maintain working capacity through long stretches of thin feeding (Griffith 2006, 224); another "major military advantage in using donkeys is that they can feed on scarce pasturage and fodder of the poorest quality, and can survive on leaves, thorns, and thistles" (Roth 1999, 65). Although mules benefit from a couple of kilograms of hard grain fodder in their diets, they, too, are able to subsist entirely from grazing and "alternative fodder" such as brush and weeds for prolonged periods of time (Isager and Skydsgaard 1992, 106–107; Roth 1999, 65–66).

So, the feeding requirements for the animals of Greek armies were comparatively very small and could easily be met through a range of plant and tree matter widespread and easily available in every terrain Greek armies crossed. This is still true when we add to the animals that needed to be fed by Greek generals the dozens of sacrificial sheep and goats that followed Spartan and other Greek armies (Xen. *Lac.* 13.3 with Lipka 2002, 214)—these sheep and goats, too, could be fed by rough pasturage (Bresson 2016, 134). It even remains mostly true when we add the food required for cavalry horses. This was often non-existent since many (especially smaller) *poleis* in the Classical period had no cavalry: in his advice to cities making arrangements against invaders, for example, Aeneas Tacticus never takes it for granted that cities will have a cavalry force to defend themselves (6.6, 15.5, 26.4). Even in the regular cavalry units that were established by some cities, the number of horses was small, ranging from a few dozen to a few hundred mounts (Spence 1993, 1–33). Although modern field manuals presume that horses need hard grain fodder to remain operational, and Athens (at least) did provide a daily monetary allowance to cavalrymen to pay for most of the costs of the grain (barley) their horses ate (Spence 1993, 272), the few horses on Greek land campaigns would have been able to meet much or all of their dietary requirements through grazing, as they did in later European warfare (Perjés 1970, 14–15; Roth 1999, 64).

All of this helps to explain why fodder needs are never raised as a major logistical concern for any Greek army—in contrast to later European warfare, where fodder requirements for the much greater numbers of pack and draft animals and cavalry mounts in armies could determine the timing, character, and entire outcome of campaigns.[22] The paradigmatic Cyrus does, as part of his reconnaissance for locations for campsites in enemy territory, include availability of grazing among his considerations for choosing a site (Xen. *Cyr.* 5.4.40). But once a location with suitable grazing for the army's animals was found (in friendly or

[22] Perjés 1970; van Creveld 1989, 46–47; Lynn 1993b, 139–141.

enemy territory), they could be fed in and near the camp, especially as horses (Xen. *An.* 7.2.21) and other animals could graze at night. And since Greek land warfare was (comparatively speaking) highly mobile, with armies rarely spending more than a couple of nights in a campsite, generals and soldiers would be on the move well before their animals would have stripped the ground bare of grazing (Roth 1999, 128; Lee 2007, 178, 182–183).[23] These considerations all help to explain why, even in a Greek world that was characterized by varied climatic regimes and thus widely differential amounts of grazing (not much in Attica and Aegean Greece compared to lots in central Greece and the Peloponnese (Chandezon 2003)), we never hear in any source of a Greek army operating in their usual Mediterranean environment in logistical difficulties because of a lack of fodder, or having to strike camp to obtain fodder, or needing to mount complex, centralized operations to collect fodder—all in complete contrast to later European warfare (e.g. Erdkamp 1998, 23).[24]

Donkeys (20 l per day), mules (20 l), and horses (30 l) (Roth 1999, 66–67) needed water to do their work, too (although again very conveniently for Greek armies on the march, donkeys can survive for some time on meager amounts (Griffith 2006, 224)). Soldiers and slave attendants also needed water, of course; 2 l per day (mixed in with their wine) to maintain their fighting capacity (Roth 1999, 37, 119) in addition to what they needed for cooking and washing (Xen. *Hell.* 7.2.23 with Coulton 2002, 95). Water for both animals and men was usually obtained from the countryside the army passed through. In Cyrus' model preparations for his army's long-distance march to engage the Assyrians and Croesus, for example, he tells his soldiers to bring wine to drink, but only enough till they became used to drinking water (Xen. *Cyr.* 6.2.26): in other words, rather than telling his men to bring water, Cyrus assumes it will be available in the devastated country the men would be marching through. Accordingly, the few occasions we hear of armies carrying water on the march are in unusual circumstances, as when the Athenians on their doomed retreat in Sicily in 413 took water from a settlement to carry with them after they learned that there was very little water for many *stadia* to come in the direction of their march (Thuc. 7.78.4); or when Iphicrates, finding out that there was no water for the next two days of his march, ordered his men to carry water for one day and marched at night to save carrying any more (Polyaenus *Strat.* 3.9.47).

Greek armies needed campsites, then, with good accessibility to water sources where men could fill skins and bottles for their meals and washing, and where pack and draft animals could be led to drink (Xen. *Cyr.* 8.1.44). This was easily done in friendly or neutral territory where the provision of water was part of the set of conventions of how military forces were to be received (Thuc. 6.44.2) and where, accordingly, armies could simply use the sophisticated water systems of Greek cities (e.g. Crouch 1993). In friendly or neutral, as well as enemy, territory, Greek generals also liked to encamp in or near religious sanctuaries, which always had water sources for purification and butchering purposes (Krentz 2007b, 162–163). But, even outside of sanctuaries, gaining access to water in enemy territory was never a major concern for generals or their soldiers. As one scholar put it (Cruickshank 1954, 68), "in Greece itself there

---

23 It is notable that the two instances (one actual (Xen. *An.* 2.4.10), one fictional (Xen. *Cyr.* 6.3.5, 9)) Xenophon mentions of men collecting fodder at a distance from their camp concern large armies encamped in one place for unexpectedly lengthy amounts of time in Near Eastern, semi-desert environments.

24 Xenophon does once advise the Ten Thousand to burn their carts so that their yoke-animals would not become the army's generals (*An.* 3.2.27). But this advice was given in an unprecedented tactical situation for a Greek army, when the Persian forces under Tissaphernes were using their exclusive superiority in cavalry to constantly harry the Ten Thousand in the open country of northern Mesopotamia (e.g. Xen. *An.* 3.3.19–20). In this exceptional case, it made sense to remove anything, baggage train included, that could slow the mercenaries' march (Cruickshank 1954, 69; Erdkamp 1998, 74–75).

is usually water to be found, if one knows the countryside." So, Xenophon, in his advice on where to hunt hares, deer, and boar in the Greek countryside, mentions in passing pools and streams in cultivated territory as well as *hydata*, watery places.[25] Aeneas Tacticus, too, takes for granted rivers as a usual feature of cities' rural territory (1.2, 8.1).[26] Cisterns, wells, and other sources of standing water were also usually available for exploitation by armies as can be seen from Aeneas's advice to cities to make the still waters in their territories undrinkable for invaders (8.4). The general availability of water in Greece explains why Xenophon's model leader Cyrus took care—but also could be confident—before entering enemy territory that he could gain information on campsite locations with good access to water (Xen. *Cyr.* 5.4.40).

As with grazing and water, the firewood—or, more accurately in the Greek case, the brush, twigs, and leaves (Bresson 2016, 72)—armies needed for warmth, light, and cooking *opsa* ([Dem.] 54.4) was obtained in or around their camps (Krentz 2007b, 162; Lee 2007, 210–214, 2020, 44–45). The practice of gathering brush and other materials for fires in the vicinity of camps was so common that, in the winter of 426/5, members of a Mantinean-dominated Peloponnesian force were able to use the pretense of gathering firewood to leave their camp and desert their Ambraciot allies (Thuc. 3.111.1). Although choosing a halting site with enough material nearby to make campfires might seem to be one of the easier tasks of generals, it may not have been; in the absence of any other form of source of heat, demand for brush and branches was so high in the Greek world that firewood was often scarce (Bresson 2016, 72–75). That said, no Greek land force is ever attested as lacking in or being unable to obtain firewood in its immediate surroundings, or as having to carry firewood through areas bare of it.

So, Classical Greek generals led armies that were virtually self-contained in military equipment and hardware, and whose grazing, water, and firewood requirements could be easily supplied from the environment they passed through. If we expand our definition of logistics to its full extent and include communications and services (Leighton 2017), too, this situation did not change. Greek armies did not possess any special communications equipment (Anderson 1970, 67): using shields, fires, and smoke as their signaling devices, they did not need any external supplies to communicate beyond brushwood. Services in typical Greek armies were basic, too. Spartan armies had designated craftsmen (Xen. *Lac.* 11.2) as well as dedicated seers, doctors, and flute players (Xen. *Lac.* 13.7). But while seers were part of most other Greek armies (e.g. Krentz 2007b, 158), other Greek soldiers usually took care of their own simple technical needs and relied on friends for medical care in the field (Cruickshank 1954, 86–89). Greek armies, then, just as most other pre-industrial armies, were completely self-contained in services, too (Leighton 2017).

Food Resupply: In the Field

Given all of these considerations, it becomes clear why Xenophon's Cambyses, and his Socrates (Xen. *Mem.* 3.2.1, 3.4.1–2), focus their logistical advice on the food supply of armies. Cambyses and Socrates did not need to raise materiel or munitions as a concern to

[25] Xen. *Cyn.* 5.34, 9.11, 10.19 with Rackham 1990, 96.

[26] Unfortunately for Greek generals, however, these rivers were not regular, wide, or slow enough to be used for the transport of equipment and food supplies, in contrast to the rivers of central and northern Europe (e.g. Mallett 1974, 173; Parker 1988, 76).

their charges since there was no need for their resupply once on campaign. In contrast to, say, the *Historia Augusta*, which gives fodder as the first, most pressing logistical requirement for imperial Roman armies (Roth 1999, 61), Xenophon's Cambyses makes no mention of fodder in his teaching of military leadership to Cyrus, nor does Socrates in his advice on generalship, because food for the small number of transport animals and cavalry mounts of Greek armies could be sourced with ease through grazing. Food for the men was different. As Cambyses emphasized to Cyrus, because Greek soldiers set out on campaign with so few provisions, it was always an urgent matter for generals to ensure opportunities for their men to replenish their food supplies (Cruickshank 1954, 75; Erdkamp 1998, 22). This was all the more true since the ability of Greek soldiers to function effectively as fighting forces was intrinsically tied to their generals' ability to provide them with chances to obtain food—since without food, as Cyrus told his men (Xen. *Cyr.* 6.2.26), soldiers "should be unable either to fight or live."

The methods used by Classical Greek generals for the resupply of their men's food differed depending on whether they were operating in friendly or enemy territory. If an army needed to travel through the territory of friendly (and/or neutral) cities on its way to its theater of operations, generals sent ahead to request permission from these cities to use their market (*agora*).[27] Cities could refuse military forces the right to access their market (e.g. Xen. *An.* 5.5.6), but much more often granted access (e.g. Xen. *An.* 4.8.23, 6.2.8, 7.6.24). In so doing, cities were simply extending a common practice in the Greek world of organizing markets for groups of foreigners passing through their territories, even if markets for armies were more usually held outside city walls because of the understandable fear of allowing large numbers of armed men inside the gates of cities (Descat 1993, 151–155). Whether in its regular marketplace, or in a temporary *agora* established outside the city walls, the friendly city organized and operated the market. Thus, cities could also halt the operation of a market given to an army when they wished: so, having provided an *agora* to the Ten Thousand, the people of Heraclea Pontica, in response to an attempt at blackmail by the army, collected their property from outside the city walls, and "moved the market within the walls" (Xen. *An.* 6.2.8). But this was exceptional. Much more commonly, soldiers were able without problem to use their pay provided by the state to buy the *alphita*, *opsa*, and wine (and whatever food they thought necessary for their slave attendants) they needed to replace or supplement the supplies they had brought from home.

Once in enemy territory, however, the situation was completely different. Unless there were exceptional circumstances, Classical Greek armies usually acquired their provisions in hostile territory through foraging,[28] that is, through the collection of food supplies from the area around their march route or camp (Roth 1999, 118). Xenophon presumes in several generalizing discussions throughout his works (e.g. Xen. *Cyr.* 3.3.16, *Oec.* 5.13) that armies invading enemy territory would subsist there on the provisions they obtained through foraging. These presumptions were not limited to Xenophon. In a discussion of the military force of his ideal *politeia*, Plato's Socrates tells his interlocutor that his soldiers would not burn their enemies' houses or devastate their farmland but only take their annual harvest (Pl. *Resp.* 470D-E). Aeneas Tacticus, too, thought that invading land forces would forage for their provisions and recommended, accordingly, measures to prevent this (8.1, 8.4).

---

[27] See Xen. *Hell.* 3.4.11, 5.4.48 with e.g. Anderson 1970, 53.

[28] See e.g. Gomme 1945–1981, vol. 1, 16; Cruickshank 1954, 75; van Wees 2004, 105–106.

Greek generals needed to plan carefully in order to provide their soldiers with the opportunity to subsist through foraging. Firstly, they needed to carry out diligent reconnaissance: accordingly, before invading enemy territory, Cyrus obtained information on possible campsite locations that had accessibility to grazing, water, *and* provisions for his men (Xen. *Cyr.* 5.4.40). Secondly, generals leading military forces aiming to engage in foraging also needed to ensure their forces had tactical superiority (Erdkamp 1998, 123–130). This was because soldiers needed time to slowly disperse over the area around their camp or march route to forage effectively for food.[29] That said, most generals leading invasion armies could usually assume tactical superiority: the lack of permanent standing forces in Greek cities prevented quick responses to invasions, and so armies could forage safely until the invaded *polis* decided and organized itself (if it eventually did so) to send out a force to engage the invaders in battle. It was only in exceptional cases that Greek invading armies were not sufficiently tactically strong to be able to depend on foraging for their provisions. So, in 454/3, an Athenian force had to retire from Pharsalus in Thessaly because of lack of food supplies before they had achieved their objective; although they had controlled the land around their camp, they could not go beyond its immediate vicinity to forage because of the Thessalian cavalry, the strongest and most numerous cavalry force on the Greek mainland, operating against them (Thuc. 1.111.1 with Gomme 1945–1981, vol. 1, 324). Finally, Greek generals had to plan their invasions to arrive in enemy territory at a time when their soldiers could find ripe grain to forage (e.g. Rawlings 2007, 144–145). A failure to time an invasion properly could halt an entire campaign, as when Spartan-led Peloponnesian forces invaded Attica in 427 before the grain in the fields had ripened and consequently had to return home after 15 days because of lack of food (Thuc. 4.6.1–2). The reliance on foraging for food supplies meant, too, that winter campaigns could never be more than brief raids (Rawlings 2007, 145).

The right planning of foraging operations by Greek generals was crucial for the success of the armies they led since they usually could not count on any external food resupply from home (or any operational base). It was only in extraordinary situations in fact—when armies which had been expecting a quick battle had been unexpectedly forced to remain immobile in the field in the presence of nearby enemy forces (Anderson 1970, 7–8, 53–54)—that Classical Greek states sent out supply convoys to forces operating in the field. Supply convoys were sent to the Greek forces encamped at Plataea in 479 (Hdt. 9.39.2, 9.50–51), but this was to supply an extraordinarily large Greek force that had been forced to remain stationary in a position that enabled no foraging (because of Persian possession of the territory (Boeotia) in front of it and devastation of the area behind it (Attica)).[30] In addition, Plataea was the culminating battle of a Panhellenic war for survival, and the effort against the Persians would have mobilized resources (from a range of states) not normally available for Classical Greek wars (Cruickshank 1954, 63–64). In 457, the Athenians also sent out a supply convoy to a force that had been in the field an unusually long time, guarding the passes at Mount Geranea (to prevent Peloponnesian forces from returning from Phocis to their home *poleis*), before proceeding to Tanagra (where the Peloponnesians had diverted to on learning of the Athenian position at Geranea), and remaining encamped there (in the presence of the Peloponnesians) after a battle (Diod. Sic. 11.80.2–4).

[29] E.g. Harari 2000, 306–307. Greek examples of scattering to forage: Xen. *Eq. mag.* 7.9; Aen. Tact. 16.4–5.

[30] Gomme 1945–1981, vol. 1, 16; van Wees 2004, 280 n.11 with Hdt. 8.50, 8.142.3.

The fate of these supply convoys—both failed to get through to the armies which they had been equipped to provision because of attacks by cavalry (Persian and Thessalian, respectively)—points up a major reason why supply convoys were only exceptionally sent to Classical Greek forces in the field. Like medieval European armies, Greek armies "plunged into enemy territory like a stone into a pool of water" (Harari 2000, 325). In other words, Greek land armies moved rapidly on very narrow fronts and were unable to secure the invaded territory they passed over. Because Greek invading forces' fronts were proportionately very small to the distance any supply convoy would have to cover to reach them, and because Greek armies could not offer any protection to any convoys attempting to reach them, any supply convoys sent out to Greek armies were extremely vulnerable to enemy attacks.[31]

The (near) impossibility of provisioning through supply convoys required Greek armies, then, to be totally dependent on foraging for their food resupply in enemy territory. Reliance on foraging, in turn, meant that invading Greek armies were unable to maintain control or exert continuous strategic pressure over the territories they invaded (Erdkamp 1998, 22). But this problem of securing and controlling territory—which faced all pre-industrial armies—could be and was solved by other pre-industrial European states: in order to secure the possession of recently invaded territory, series of magazines could be constructed to provision and protect supply lines constantly connecting an invading force as it advanced with food supplies sent from its home territory or an operational base.[32] The Spartan general Dercylidas did once establish a supply dump at Atarneus in 398 (Xen. *Hell.* 3.2.11 with Anderson 1970, 53). But he made no attempt to use Atarneus as an operational base from which he could develop supply lines to his forces operating in the field. Rather, he filled the city full of supplies simply and only so he could use Atarneus as a convenient halting-place if he ever happened to be operating in its vicinity again. No Classical Greek state ever had the means (requisitioning, taxation-in-kind, private contractors operating on a sufficient scale) to acquire supplies to fill magazines for their land forces (Anderson 1970, 53; Erdkamp 1998, 14–18).

Conclusion

One of the most striking things about land warfare in Classical Greece is the complete lack of development of logistical structures. There were never any supply lines developed to connect armies to bases; the second and last ever attested supply convoy was sent in 457. Seasonality always continued to determine the limits of campaigning: just as the Spartans and their allies had had to leave Attica in 427, so the Thebans and their allies had to leave Laconia in 370 because all the grain there had been used up and winter was coming on (Xen. *Hell.* 6.5.50). Inability to forage always continued to mean the end of campaigns in enemy territory: just as the Athenians had had to leave Thessaly in the mid-450s because the Thessalian cavalry had prevented them from spreading out in search of food, so, too, the Thebans, the great land power of mainland Greece in the late 370s and 360s, had to end

[31] See Cruickshank 1954, 58, 71; Hanson 2005, 214 for discussion of the problem in a specifically Greek context.

[32] Definition of supply line: Roth 1999, 157; Harari 2000, 311 n.43. See Roth 1999, 284–285, Lynn 1993b, Howard 1976, 71 for examples (from the first, seventeenth, and eighteenth centuries C.E., respectively).

operations in Thessaly in 368 for lack of provisions because of attacks from local cavalry forces (Diod. Sic. 15.71.4–5).

All of these features of land warfare resulted from the fact that no Classical Greek state ever developed internal administrative structures, or harnessed the means of allied communities, to amass the means necessary to support land forces in the field with substantial state supplies.[33] But this was not inevitable. Italian city-state armies, too, relied on foraging for their provisioning in enemy territory in the fourteenth and early fifteenth centuries and encountered, accordingly, all the limits and obstacles Classical Greek armies faced. But strengthened state structures, including the establishment of permanent military administrations, and the increased use of requisitioning from subject communities in the fifteenth century enabled the most powerful Italian states to construct magazines, develop supply lines, and extend their foreign campaigns over whole seasons.[34]

But, in its logistical organization, at least, Classical Greek warfare, unlike its late medieval Italian counterpart, remained structurally sclerotic. The private, often unmethodical nature of Greek logistics—and its consequences—meant, for example, that Socrates, in his discussion of the training of soldiers for his ideal community, had to take into account "the many changes in their drinking water and food" they would face on campaign (Pl. *Resp.* 404b); and that Alcibiades, reminiscing at a drinking party about Socrates' endurance on a campaign to Potidaea, could tell his companions that the Athenians there had been forced to go without food "as often happens on campaigns" (Pl. *Symp.* 220a). It is in this context that we should return to Xenophon's ideas on military leadership. In contrast to contemporary military experts who had a narrow and misguided focus on tactics (Delatte 1933, 8–9, 23–24), Xenophon's wide military experience gave him a crucial insight—that satisfying the daily demands of provisioning on campaign was as crucial to the success of a general's men as the few minutes or hours they would spend fighting man-to-man against their enemies.[35] Xenophon never forgot that armies needed security of food supply to function effectively as operational military forces. Or as his Cyrus put it much more succinctly: without food or drink "it was not possible to fight a campaign or do anything else" (Xen. *Cyr.* 4.2.34).

---

[33] Cruickshank 1954, 73, 85; Anderson 1970, 53; van Wees 2004, 107.

[34] See e.g. Mallett 1974, 108–114, 123–129, 139–140, 188, 227–228; Mallett 1984, 140; Blastenbrei 1987, 265–267.

[35] Gray 2011, 47. See Perjés 1970, 26 for the everyday importance of provisioning versus the fleeting moments of fighting on campaign.

CHAPTER 18

Fortifications

Silke Müth

Fortifications were—and many today still are—an important part of the Greek landscape, having inspired numerous reports—and detailed drawings—by nineteenth-century travelers.[1] For a very long time, however, these fortifications were studied only as practical, functional buildings.[2] Only from the late 1950s were fortifications understood in historical, social, and regional terms.[3] In the 1970s, they at last found a place in the study of Greek architecture.[4] In the 1980s and 1990s, several conferences studied fortifications in the light of their socioeconomic and cultural aspects.[5] A boom of fortification research in the new millennium opened more fields of interpretation: geographic conditions and the building methods as influences on monumental forms, urban and symbolic alongside military functions, and a range of historical, economic, sociocultural, and regional contexts.[6] Besides being military necessities, fortifications are architectural monuments with multiple purposes and historical witnesses with much information to impart.

1 E.g. Leake 1830; Curtius 1851–1852; Bankel 1986.

2 Formal aspects: de Rochas d'Aiglun 1881; Droysen 1889. Practical aspects: Krischen 1922; von Gerkan 1935; Krischen 1941.

3 E.g. Maier 1961.

4 Winter 1971; Lawrence 1979.

5 Leriche and Tréziny 1986; van de Maele and Fossey 1992; Debord and Descat 1994.

6 Many of these projects were united in the network "Fokus Fortifikation"; see Frederiksen et al. 2016; Müth et al. 2016. A conference in Haifa in 2017 ("The Art of Siege Warfare and Military Architecture from the Classical World to the Middle Ages") gave another focus on the subject, as does Ballmer et al. 2018 on prehistoric fortifications.

A Companion to Greek Warfare, First Edition. Edited by Waldemar Heckel, F.S. Naiden, E. Edward Garvin, and John Vanderspoel.

Typological Development of Fortifications

Long after Mycenaean fortifications had attained impressive, elaborate forms,[7] fortifications re-emerged in the Early Iron Age.[8] Our few reliably dated examples, mostly in Ionia and on the Greek islands, stand on high ground close to the coast. Some of them enclosed settlements, whereas hilltop fortifications, as at Emporio on Chios, offered refuge to the settlement next to them.[9] The space within was rather small, except in the case of Paphos and Salamis on Cyprus, Paphos perhaps even representing an early form of a *Geländemauer* or great circuit.[10] Most walls were free standing; at Hypsele on Andros, however, houses were included in the lower defenses.[11] Walls at these times could be up to 10 m wide, as at Smyrna and Salamis on Cyprus; equal or greater height is to be expected. It is noteworthy that the widest walls enclosing the largest spaces are located closest to the Near Eastern military powers of Assyria and Lydia.[12]

Gates were axial or tangential, the former being more common. Paphos and Smyrna show long approaches guarded by towers or platforms.[13] Until the sixth century, it is difficult to distinguish these two kinds of structures, since they are not preserved to a decisive height. However, the use of the word *pyrgos* (tower) in sources as early as the seventh century may provide evidence that towers existed earlier. Flanking devices 3–8 m long are found at Paphos and Minoa on Amorgos, for instance, but should not be considered indubitable elements of Early Iron Age fortifications.[14]

In early Archaic times, reliably dated fortified sites appear in the northern parts of Greece and Sicily; in the middle and late Archaic period they spread to the Italian Peninsula, North Africa, and the Black Sea region.[15] From the seventh century onward, walled sites are found in plains, and the spaces being enclosed increase considerably.[16] The great circuit type emerges in places such as Miletus, Leontini on Sicily, Phocaea in Ionia, and Samos;[17] it is often difficult, however, to determine how much of the intramural space was built up in these times.[18] Once again, great circuit walls and the widest curtains (up to 18 m at Smyrna and 10.75 m at the acropolis of Idalion on Cyprus) are located closest to foreign empires. Yet wall width also depended on topographical factors and generally increased for fortifications built in the plains rather than on hilly ground. Where preserved or restorable with certainty,

7 Simpson and Hagel 2006; Gauss 2018.

8 For an overview of Early Iron Age and Archaic walls, see Hellmann 2010, 299–309; Hülden 2018.

9 Hansen and Nielsen 2004, 137; Hülden 2018, 94.

10 Maier 2008. Under this German term is commonly understood a wall circuit that not only includes the area of a settlement and its acropolis, but also distinct parts of unbuilt terrain, and the trace of which makes use of the natural formations of the terrain, cf. e.g. Winter 1971, 111–115.

11 Frederiksen 2011, 70–71, 150–151, tables 1–2, map 1.

12 Frederiksen 2011, 90, 94, table 9.

13 Nicholls 1958–1959; Maier 2008; Frederiksen 2011, 72–73.

14 Winter 1971, 153; Frederiksen 2011, 34–38, 73.

15 Frederiksen 2011, 77, 80–81, 85, map 1–4; Hülden 2018.

16 Frederiksen 2011, 77, 90, tables 6, 8, 10, 12.

17 The extent of the late seventh-century walls on Kalabak Tepe cannot be fixed with certainty: von Graeve 2006.

18 Frederiksen 2011, 52 n. 21, 90, 149–150, 159–160, 182–183; Hülden 2018, 98 with refs.; Kienast 1978, 99–102.

Archaic walls vary between 5 m and 8.5 m high, so the average might be around 6 m.[19] Usually, fortification walls were free standing, but in Vroulia on Rhodes one part of the wall constituted the back wall of a long line-up of houses.[20] Information concerning wall walks and battlements is scarce, as Archaic walls are generally not preserved up to this height. Visual art suggests that the battlements were usually crowned by square merlons.[21]

From the middle of the seventh century, a simple form of the "indented trace" appears at several places, so the efficacy of flanking devices has been recognized. Towers and platforms both serve as flanking elements, and the largest flanking devices reach lengths of over 13 m on Cyprus and Rhodes.[22] Most towers and platforms are square, as before, but in middle Archaic times circular towers appear at Halae (East Locris) and Leontini. In the late Archaic period, the series of round towers at Megara Hyblaea (Sicily) provides the first securely attested sequence of towers at roughly regular intervals along Greek fortification walls.[23] A series of six towers in the late Archaic fortifications of Thasos served as lookouts rather than as fortified barriers.[24]

With regard to gate types, the axial gate became predominant in Archaic times, although tangential gates persist, as is shown by the late Archaic West Gate of Megara Hyblaea.[25] In rare cases such as Eretria and Halae, gates of a mixed type appear. Towers or platforms reinforcing gates occur from the middle Archaic period onward.[26] Compared to the Early Iron Age, gates are slightly wider. Although dependent on topographical particularities, defensive needs, and the frequency and character of traffic, the design of a gate may also reflect aesthetic factors.

Outworks are only rarely found in early fortifications, the first example being a seventh-century ditch at Megara Hyblaea.[27] In late Archaic times, ditches become more common. Highly sophisticated outworks combining earthen ramparts or berms, ditches, and glacis make their first appearance at Alalie on Corsica and Paphos; they exceed 20–50 m in width. Again, such works faced hostile neighbors, a pattern for Greek fortifications before Classical times.[28] Isolated towers in rural locations are also common. A lone watchtower of the late sixth century on the island Thassopoula, north of Thasos, served as a Thasian outpost toward the continent.[29]

---

19 Frederiksen 2011, 94–95, tables 9, 11, 13. These numbers are much higher than the average of 4 m estimated by Winter 1971, 130.

20 Frederiksen 2011, 52, 197. Date: 700 to the mid-sixth century.

21 Winter 1971, 13; Lawrence 1979, 36; Childs 1991; Frederiksen 2011, 38–40. At Samos, fragments of covering tiles probably belonging to the fortification wall of the sixth century indicate the existence of square merlons: Kienast 1978, 46–47. The only Archaic merlons found as yet, from the mid-sixth century fortifications of Thasos, provide an astonishing contrast to this, but seem to have served mainly for decoration: Grandjean 2011, 408–411.

22 Frederiksen 2011, 79, 83–84, 88. 97, tables 9, 11, 13.

23 Frederiksen 2011, 79, 83–84, 88, 91–92, 97. Another example of a regularly spaced succession of towers of the late sixth century is the fortification of Eleusis: Frederiksen 2011, 135–137. The fortification of Larisa on Hermos, the dense sequence of towers of which is pointed out by Winter 1971, 153, cannot be securely dated: Frederiksen 2011, 67, 158.

24 Grandjean 2011, 418–429.

25 Frederiksen 2011, 95–96 (*contra* Lang 1996, 39). 163.

26 Frederiksen 2011, 83, 87–88, 96–97.

27 Frederiksen 2011, 79; *contra* Miller 1995, 45, 258.

28 Frederiksen 2011, 88–89, 97–98, 127–128, 177–179.

29 Grandjean 2011, 410 n. 102.

On Paphos, a rare example of temporary siege construction survives: ramps, mines, and passages from the time of the Persian attack, around 498, tell of hard battles and the means and ruses used by attackers and defenders.[30] Also rare are examples of forts that are largely intact, and thus easy to identify, such as the late Archaic fort of Phylla in central Euboea.[31]

Fortification techniques in Classical times did not develop much until the end of the fifth century, a remarkable delay in light of the Persian wars decades earlier. Greek military planning of this era evidently concentrated on field battles, not sieges.[32] Nevertheless, the Themistoclean city wall of Athens—about 6.4 km in length and reinforced by a moat—was an impressive project of the *Pentekontaetia*, although the remains of only two towers have been found.[33] In due course, long walls protected the connections to Phalerum as well as Piraeus, Athens' two ports, both of them fortified. Further examples of long walls are those from Corinth to Lechaeum and from Megara to Nisaea.[34] Around the middle of the fifth century, Pericles extended the earlier fortifications of the important sanctuary of Eleusis, which displayed a masonry quality unrivaled in fortification architecture of the time.[35] These big projects all took place at the epicenter of Greek war and commerce, the isthmus of Corinth and adjoining Attica.

Progress in the fifth century came mostly at the borders of the Greek world, not on the mainland. The earliest great circuit equipped with towers at roughly regular intervals seems to appear around the mid-fifth century at Stratus in Acarnania.[36] The Athenian colony of Amphipolis in Thrace, established in 437, possessed an impressive wall circuit around 7.5 km long; it shows finished, regular ashlar masonry and includes an elaborate drainage system. The circuit possessed square, round, and half-round towers without regular distribution.[37]

On the mainland, forts arose in order to protect supply routes and to establish military bases during the Peloponnesian War, such as the coastal forts of Oenoe, Rhamnus, Thoricus, and Sunium.[38] Only the fort at Sunium features a regular succession of towers flanking the curtains.

At the very end of the fifth century, a new kind of landscape fortification emerged in Sicily with the circumvallation of the huge, unsettled plateau of Epipolae north of Syracuse by Dionysius I, who aimed at preventing the city from being cut off from its hinterland. The circuit of roughly 18 km was completed only toward the middle of the fourth century.[39] Sicily was also the place where around 400 engineers of Dionysius I invented the catapult, which was to transform fortification technique.[40] Curtains, towers, and gates having to withstand the new weapon had to become stronger and thicker; most walls were now constructed entirely of stone, whereas earlier fortifications were usually built of mudbrick on a stone

30 Maier 2008, 32–39.

31 Sapouna-Sakellarakis et al. 2002; Hellmann 2010, 347–348; Frederiksen 2011, 14–15.

32 Hellmann 2010, 314.

33 Lindenlauf 2003; Theocharaki 2011; cf. Thuc. 1.89.2–93.8, 2.13.7.

34 Winter 1971, 111, 239.

35 Adam 1982, 197, fig. 112.

36 Ley 2009, 153, 157–159.

37 Lazaridis 1986, 2003, 21–43.

38 Mussche 1961; Winter 1971, 42–44, 161 n. 41, 303; Adam 1982, 209–211; Karlsson 1992; Munn 1993, 3–15; Petrakos 1999, 47–80; Goette 2000, 44–61; Hellmann 2010, 343–345.

39 Beste and Mertens 2015.

40 Marsden 1969; Garlan 1974, 155–200; Lawrence 1979, 43–44; on the development of other means of siegecraft, 49–52.

socle. Flanking devices large enough to house catapults had to appear at regular intervals. Surviving examples of this first generation of catapult towers mainly surround fortifications influenced by the Thebans (Messene, Siphae, and Eleutherae). They contain wide openings instead of narrow loopholes or proper windows protected by shutters.[41] In Asia Minor, Carian walls provide the most impressive examples of fourth-century fortifications reflecting these developments.[42] The course taken by curtain walls reacted to the new developments as well: from the second quarter of the fourth century onward, fortifications of Asia Minor often feature the systematic form of the indented trace, as at Erythrae and Colophon.[43]

As a rule, towers now guarded gates. Already in the fifth century, courtyard gates were guarded by towers on the field side, early examples being the Dipylon and the Sacred Gate at Athens.[44] This type gained popularity in the fourth century and could appear in two forms: as an axial gate with a round or rectangular courtyard such as the Arcadian and South Gates of Messene, or as a tangential gate, as at Stymphalus and Mantinea.[45] Posterns were rare until the later fourth century and were designed for pedestrian and agricultural traffic rather than for defense.[46]

Forts and guard towers were relatively common in the fourth century—the Boeotian forts of Siphae, Chorsiae, and Eleutherae, the Attic forts of Panacton and Phyle, Megaran Aegosthenae, and the many towers spread along the Attic–Boeotian border.[47] The Attic Dema wall, a land wall of more than 4 km on the pass between the Parnes and Aegaleus Mountains, may also belong to the fourth century, although its date is debated.[48] It consists of short stretches in saw-tooth pattern with small posterns between them, a type of curtain that otherwise first appears in Hellenistic walls.

By the end of the Classical period, fortification technique was quite advanced. Yet the invention by Macedonian engineers of the powerful torsion catapult, widely used in Alexander's campaign, made even more elaborate defenses necessary.[49] Now the theory of fortification developed, the most prominent example being Philo of Byzantium's ***Poliorketika*** (for which see Chapter 8 in this volume).[50] The new generation of catapult towers (late fourth/early third century) featured sides of more than 10 m to house larger catapults, plus stronger, two-faced walls and larger windows. Examples of such towers appear at the huge *Geländemauer* of Heraclea on Latmos, but interspaced with smaller towers. Further examples appear at Ephesus, at Iasos, and on Samos.[51] Experimenting with tower shapes became

41 Ober 1987.

42 McNicoll 1997, 15–45; Pedersen 2010.

43 Martin 1947; Adam 1982, 179; McNicoll 1997, 60–70. The same appears at Arcadian Gortys (Müth and Ruppe 2016).

44 Knigge 1988, 56–73.

45 Messene: Schwertheim 2010. Stymphalus and Mantinea: Winter 1971, 224, 230; Lawrence 1979, 333–334; Adam 1982, 176–178.

46 An exception are the posterns in the northern wall of the Epipolai: Beste and Mertens 2015, 60. On posterns and passive or active defense, cf. Winter 1971, 234–251; Schwandner 1977; Lawrence 1979, 335–342; Ober 1985, 160–163; McNicoll 1986, 309; Pimouguet-Pédarros 2000, 35–38, 59.

47 Adam 1982, 217–218; Camp 1991; Munn 1993, 3–15; Cooper 2000; Hellmann 2010, 350–351. Van de Maele 1992 dates Aegosthenae to the third century.

48 Ober 1985, 149–152; Munn 1993.

49 Marsden 1969, 1971; Garlan 1974, 201–270; Lawrence 1979, 44–49.

50 Marsden 1971; Lawrence 1979, 39–107; Hellmann 2010, 317–319.

51 Ober 1992.

common: we find the first pentagonal towers, as on Hellenistic Samos, at Lycian Oenoanda, and at Gadara, in Jordan, and even hexagonal ones, such as those of Argos and again Samos.[52] The battlements of the curtains also changed: toward the end of the fourth century at the latest come closed battlements with windows, the first certain example being Heraclea on Latmos.[53] Nevertheless, crenellated battlements continued in use.[54]

Unparalleled in strength for their time were the early third-century fortifications of Rhodes, particularly on their southern, landward side. Here, a new type of curtain with casemates provided shelter for troops and storage room.[55] The wall walk, more than 7.5 m wide, provided enough space to set up catapults. Immense towers with side lengths of 17 m and walls 3.5 m wide stood in short intervals of about 40 m. Closely spaced sally ports reveal an active system of defense.[56] Another wall features immense outworks including jogs pierced by sally ports and a ditch. In middle Hellenistic walls, two-story wall walks appear at Side and Perge.[57]

The indented trace now combined with towers and sally ports at the pointed angles, as at the south walls of Miletos and Apollonia in Cyrenaica.[58] The great circuit, however, lost popularity in Hellenistic times, as the disadvantages of it became obvious: against large, well-equipped attacking armies, long circuits with requisite manpower and means of defense became more and more costly and elaborate. Thus, Hellenistic fortifications from the late third century tended to shorter circuits (e.g. Dura-Europus), or *diateichismata* built to shorten an existing one (e.g. Heraclea on Latmos).[59]

The tendency toward representative gates, already visible in Classical examples such as the Arcadian Gate at Messene, grows stronger in the Hellenistic period, a particularly impressive instance being the ornamented facade of the Zeus and Hera gate at Thasos from around 300.[60] Gates have various plans, as illustrated by the South Gate at Pergamum with its angled entrance ways.[61] Some were protected by huge defensive installations, the most advanced battery systems in Sicily being located at fort Euryalus (fourth and third centuries), guarding the northwestern entrance to the Epipolae. Comparable defenses barred the northern entrance to the Acropolis of Selinus (end of the fourth century).[62] These elaborate and

52 Kienast 1978, 83–91; Lawrence 1979, 87; Adam 1982, 58–59; McNicoll 1997, 124–126; Hoffmann and Bührig 2000.

53 Krischen 1922, 49–51 (date of Herakleia: around 300); Maier 1961, 104–105; Winter 1971, 140–141; Kienast 1978, 95; Lawrence 1979, 367–370; Adam 1982, 37; Ober 1992, 166–167. Chalkis as the often mentioned earliest example as well as other pre-Hellenistic ones are neither certain nor reliably dated. For a reasonable dating of the walls of Herakleia in the last decade of the fourth century; cf. Hülden 2000.

54 Adam 1982, 37–39.

55 The Rhodian curtain type is described by Philo of Byzantium, *Poliorketika* 1.17–19.

56 Filimonos-Tsopotou 2004, 84–115.

57 McNicoll 1997, 144, 128. The date of these two fortifications may have to be lowered to early Imperial times, at least for most of the curtains of Perge; see Lohner-Urban and Scherrer 2016; Martini 2016.

58 von Gerkan 1935, pl. 16; Adam 1982, 66–68; Garlan 1992.

59 Hellmann 2010, 323; Caliò 2012.

60 Meesene: Giese 2010, 86–88; Müth 2010, 82; other examples mentioned by Lauter 1986, 74 are of Classical date. Hellenistic: Lauter 1986, 73–74. This theory, although challenged by Lohner-Urban and Scherrer 2016 because of probably lower chronologies for some monuments, nevertheless holds true for other examples. Thasos: Grandjean 2011, 265–267, 541–546.

61 Lorentzen 2010; for comparisons see Adam 1982, 90.

62 Mertens 2003, 128–177; Beste and Mertens 2015, 127–240; Beste 2016.

efficient systems remained unparalleled throughout Hellenistic times, even if large battery stands were also built at other places, such as Thessalian Goritsa (around 250).[63]

Hellenistic forts still survive in many locations. At Kydna in Lykia, a fort was built in the third century, obviously in order to protect the near city of Xanthos.[64] Many other territories were guarded by forts, such as those of Rhodes, Miletus, and Smyrna.[65] An example of a strong fort of highly representative value built by a regional sovereign appears on Mt. Carasis, high in the mountains of Cilicia.[66] A particular subtype are the *tetrapyrgia*, quadrangular forts with four towers at the angles, which mainly occur in Asia Minor, for instance at Theangela and in the region of Limyra.[67] A Hellenistic land wall runs on the mainland terrain adjoining the peninsula of Iasos in Caria.[68]

Single towers were widespread, an exception being Sicily.[69] From the Greek mainland this phenomenon spread to Asia Minor, where around Ephesus a series of Hellenistic towers was found.[70] A high number of towers appeared on the Cyclades to protect farmsteads and other property against pirates.[71]

After this outline of chronological tendencies in the development of fortifications must come some caveats. Scholars often try to date monuments according to typology, but economic or functional concerns may keep fortifications from following trends. Other fortifications might be ahead of the times. Moreover, regional particularities in development have to be respected. Altogether, dating on the basis of typological criteria is insecure, often providing only rough estimates to be supported by more reliable methods.

The Building Process: Materials, Techniques, Masonry Forms

Diodorus provides a vivid description of the enormous construction site, several kilometers long, for the north wall at Epipolae, and of the huge amount of building material and manpower involved.[72] Great amounts of stone had to be quarried, transported to the site, worked, and set up, or mudbricks had to be produced and laid in enormous numbers. These practical factors influenced the shape and appearance of the monument to a considerable degree, and the same is true of lesser projects.[73]

Until well into Classical times, fortifications would normally consist of a socle of two faces of stone masonry filled with rubble, topped by plastered mudbrick.[74] This technique

63 Bakhuizen 1986.

64 Adam 1982, 116–165.

65 Hellmann 2010, 351–353.

66 Radt 2011, 2016.

67 Marksteiner 1994; Pimouguet-Pédarros 2000, 117, 128–133, 267–270.

68 McNicoll 1997, 111–117.

69 Hellmann 2010, 353–357.

70 Jobst 1978. See also Heraclea on Latmos: Krischen 1922, 47; Hellmann 2010, 355.

71 Hellmann 2010, 355–357; for Amorgos cf. Marangou 2005; for Andros cf. Koutsoukou and Kanellopoulos 1990.

72 Diod. Sic. 14.18.1–8. The speediness of the process may here be exaggerated indeed, but the numbers of workmen and carts needed may be a realistic estimate (Beste and Mertens 2015, 256–257).

73 In this field, the research of J.-C. Bessac is fundamental, e.g. Bessac 1986; Bessac and Leriche 1992 for a methodological overview.

74 Bessac and Leriche 1992.

continued throughout the Late Classical and Hellenistic period, but far more rarely, except in the dry climate of the Hellenistic Near East.[75] All-stone fortifications, in two faces with inner rubble filling, appeared on Sicily from the late sixth century, in Thasos in the early fifth century, and in Acarnania from the middle of the fifth century at the latest, but in central Greece these structures became common only in the fourth century.[76] They were less prone to weathering than mudbrick, and entailed much less frequent maintenance. Inner compartment walls, already known in Archaic times for wall socles,[77] were first used in Acarnania and Sicily from the middle of the fifth century.[78] A closer bond between outer faces and filling resulted from the use of binders, this technique and the use of inner walls being termed *emplekton*.[79] Introduced to the Greek mainland and Asia Minor in the second quarter of the fourth century, *emplekton* quickly became standard for fortification construction. The alternative of solid stone construction was sumptuous and rare, a famous example of it being the Hellenistic wall of Rhodes.[80]

Stone and mudbrick were the main building materials for Greek fortifications. Mudbrick production was cheap and easy, requiring only untrained workers, with suitable earth available nearly everywhere.[81] Quarrying stones was more costly and laborious, for material often had to be transported some kilometers. Whenever possible, the builders would, however, quarry the rock in the immediate vicinity of the construction. The geological character of the deposit and the quality of the rock always influenced the masonry forms employed in the wall: soft materials are quarried more easily than hard ones and therefore generally produce more regular forms. Hard rock, in contrast, had to be quarried according to the layers and fractures of its natural deposits, so it might yield polygonal, trapezoidal, or rectangular blocks. For economic reasons, these shapes mostly determined the masonry forms of fortifications.[82]

For a long time, scholars defined masonry styles, ascribed them to chronological periods, and dated fortification walls accordingly, an often invalid method.[83] Builders of fortifications sometimes employed better material than what was at hand, constructed parts of their monuments in special forms, or treated the blocs in a decorative manner, all of which affected "style."[84] The availability of materials, resources, and time could also affect it, as might changes in construction techniques.[85] Another variable was personnel. Variations in stonework and building technique from one section of a monument to another might indicate

75 Leriche 1993; Hellmann 2010, 320.

76 Miller 1995, 42–44; Ley 2009, 238–239, partly *contra* Karlsson 1992, 75–76 and Cooper 2000; Grandjean 2011, 385–386.

77 Frederiksen 2011, 72, 78, 92–93.

78 Karlsson 1992, 67–79; Mertens 2003, 98–99, 111; Ley 2009, 240.

79 Karlsson 1992, 67–79.

80 Filimonos-Tsopotou 2004, 76–79.

81 Bessac and Leriche 1992, 79–81.

82 Bessac 1986; cf. also Ley 2009, 202.

83 The method: Wrede 1933; Scranton 1941. The issue of validity: Ley 2009, 178; Frederiksen 2011, 63–64; Brasse and Müth 2016; cf. Grandjean 2011, 392–400. Chronological successions of different masonry forms may, however, be established if they are confined to certain sites, e.g. the city wall of Samos (Kienast 1978, 40–46).

84 For examples, see Ley 2009, 208–209; Müth 2010, 67–69, 78–79; Frederiksen 2011, 72–73; Brasse and Müth 2016; Müth et al. 2016a.

85 An example for changing masonry techniques in face of the growing menace of the Parthians are the city walls of Dura-Europus (Bessac 1997, 2:75–101, 267).

different master-builders or teams of workmen, with varying skills and traditions, while typological variations might be a sign of different architects at work. Sometimes joints identify construction sections assigned to different workshops. Quarry or stone cutters' marks can help reconstruct the practical organization of the building process as well.[86]

Lastly, functional criteria influenced masonry forms and construction techniques. Forms reacted to particular topographical situations: wider and stronger curtains and flanking devices often appear in lower areas vulnerable to hostile attack, while on steeper ground the defenses were reduced in strength.[87] Representative and symbolic functions could also be responsible for the choice of certain masonry forms.[88]

Functions

Protection certainly constituted the foremost aim of ancient fortifications, but most if not all fortifications had other functions not to be overlooked. First, protective functions were not restricted to the military aspect: fortifications could ward off brigands, pirates, wild animals, and natural threats due to dangerous or dirty industries and professions.[89] Furthermore, a wall circuit often separated settlements from *necropoleis*.[90] It defined the inside and outside of a town and thus acted as a border, to which sacral functions could be attributed, or which affected trade and everyday life according to *polis* regulations.[91] The course of a town wall inevitably affected the organization of space. Gates and posterns opened up for both streets and extramural paths, as at Priene, and gates and even towers placed on axes in a street grid defined civic, public space, as at Dura-Europus.[92] The course of a wall might surround the town in a geometrical form.[93] *Diateichismata* might have defined neighborhoods.[94] An existing fortification would also affect further spatial development and with it civic planning.

Already before Classical times, a wall circuit was a constitutive element of a town.[95] Seen from afar, it identified the town, and in this capacity as a figurehead it conveyed symbolic messages.[96] We thus find early examples of embellished or ornamented city walls, and of outsized circuits, or elements of circuits, placed in the landscape in a scenographic way. Towers were concentrated where they were not necessary for defense, but appreciated best by passers-by. Fortifications thus displayed power, independence, strength, wealth, and artistry all at once. They could serve as symbols for ethnic pride, civilization, and urbanity. The defensive

86 Bessac 1986.

87 A good example is Messene, as at Müth 2010.

88 Brasse and Müth 2016, 98–99; Müth 2016b, 189–191; Müth et al. 2016a, 138–145.

89 Müth 2016a.

90 Hölscher 1998, 64; Frederiksen 2011, 60 n. 89.

91 Hölscher 1998, 64, 69–73. For the Greek *polis*, the collection of duties does not seem to have been common practice: Hansen 2006a, 104.

92 Leriche 1996, 206; Ruppe 2007, 291–296.

93 Hellmann 2010, 346–347; Müth 2016a.

94 Sokolicek 2009, 31–33.

95 Ducrey 1995; Cobet 1997, 249–253; Hansen and Nielsen 2004, 135–137; Frederiksen 2011, 103–120, *contra* Hülden 2018.

96 Details on this subject: Müth 2016b; Müth et al. 2016a; also Hellmann 2010, 294–299; Ruppe 2010, 142–148.

functions of Greek fortifications, however, never seem to be neglected. Only in the Roman world may we find examples with a negligible defensive use or none at all.[97]

Historical and Regional Contexts

Considering the larger picture of Greek military history, fortifications were important assets. Their most important effect was to deter enemies from any attempt at attacking a city. Even if a city wall was attacked, the defenders were in an advantageous situation, and their chances of holding out were high, the greatest danger coming from betrayal (for these circumstances, see Chapter 8 in this volume). For storming a wall, expensive siege machinery or artillery was needed, and the death toll was high, another advantage for the defenders. Aside from urban defense, fortifications could protect outlying settlements, agricultural terrain, roads, passes, watercourses, and anchorages.[98] It thus paid well for all kinds of settlements to undertake the high investment of building fortifications, wall circuits being foremost.[99]

And a very high investment it was: a huge amount of material and work was needed, often in difficult and remote terrain, and it inflicted an immense financial strain. In a thorough analysis of building the fortifications of Dura-Europus, J.-C. Bessac estimated 1,000 men worked for more than two and a half years to complete the 4 km long structure, using c. 486,000 modular blocks (95,000 m^3) of gypsum.[100] Even if the workers were mainly the inhabitants themselves, this meant interrupting their regular work. Workmen from abroad were expensive and had to be housed and fed. The socioeconomic implications of wall building are a neglected part of the history of fortifications and their impact.[101]

On a larger scale, regional defenses of connected fortified settlements, forts, watchtowers, and signal towers have been suggested for many parts of Greece, notably Attica.[102] Recent studies, however, show the need for caution in interpreting interconnected fortifications spread over a large area:[103] in the case of the territory of Eretria, rural fortified sites proved to be erected rather by the local population, not by the *polis*.[104]

In another respect, fortifications are indeed regional, for they reflect regional and local traditions in architecture, craftsmanship, and military technique, all of which influenced their construction and shape. Nevertheless, innovations in fortification technique spread quickly. Regional phenomena exist mostly on the level of architectural elements and construction details.[105]

97 Müth et al. 2016a.
98 Fachard 2016a.
99 Ducrey 1986.
100 Bessac 1997, 2:252–254. This is roughly equivalent to the relation of time, workforce and length in Diodorus' account of the construction of the north wall of the Epipolae, 14.18.1–8 (cf. Mertens 1999). The estimation of Hellmann 2010, 309–310 that the expenses for building of a fortification are roughly comparable with those for a sanctuary might be true only for the comparison of a very large and elaborate sanctuary to a rather short circuit. The investment for building an average fortification with its enormous length and many particulars such as platforms, towers, and gates must be estimated as being many times as much. See also Ducrey 1986, 134–135.
101 Bessac 1986.
102 Ober 1985; Munn 1993.
103 Fachard 2016a, 2016b.
104 Fachard 2012. For recent studies concerning other regions, see Balandier 2016.
105 For such regional phenomena cf. Müth and Ruppe 2016 with further literature and examples.

From the historian's perspective, to be sure, fortifications may serve as sources for regional and local history.[106] Their mere erection may bespeak wars or military threats, and their destruction may testify to attacks or natural catastrophes. Structures like ramps or mines indicate sieges; hurried repair suggests continuous conflict, while thorough rebuilding demonstrates a long period of security. Abandonment may signify the end of occupation of the site, whereas disrepair consorts with continued occupation mostly in peacetime, and reuse may attest to reoccupation or new dangers. Extensions or reductions can account for a growing or diminishing population, and changes like reinforcements or embellishments may attest to changes of function.

Fortifications were an important feature of Greek military history. Already in Archaic times, most major settlements were probably walled, and later a wall circuit became nearly indispensable, while rural fortifications guarded the landscape. Apart from their military use, these buildings played a considerable role in economic and social life, helping define Greek public spaces. They were far more than practical structures. Like war itself, they affected, and in turn were affected by, many aspects of Greek culture and society, from government budgets to engineering and visual art.

106 Cf. Leriche 1995, 9; Cobet 1997, 253; Laufer 2016. An example of the use of a fortification as a source for the history of a site is given for Thasos by Grandjean 2011, 367–383.

CHAPTER 19

Military Intelligence

Frank Russell

> Smoke rising from the plains and fire signals for friend and foe have disappeared, as if a breath had blown them away, beyond land and sea.
>
> —Aelius Aristides, *Roman Oration* 99.990

Users of military intelligence such as Nicias in Athens, Evagoras in Cyprus, and Alexander in Asia, operated in radically different contexts. Nicias, a *strategos* in the democracy of fifth-century Athens, was caught in a tension between the limited intelligence resources of his state, on the one hand, and his own informants and spies, on the other hand, and unsuccessfully faced the dilemma of influencing the decisions of the *ekklesia* at home and his colleagues abroad.[1] The kings of Cyprian Salamis in the fourth century enjoyed a sophisticated and organized intelligence apparatus, applicable to domestic security and to concerns about the great powers, especially Persia.[2] Alexander, later in the same century, was almost continually on the move, with pressing demands for geographical and logistical information in addition to the typical political and military pressures, but employed a rudimentary intelligence apparatus centered on his person to good effect.[3]

1 Russell 1999, 121, 131–132, 226–234. In democratic states, strategic decision making was collective, and there were no reporting bodies on which the *ekklesia* could rely, although there are intelligence aspects to some magistracies (*episkopoi* at Athens, Balcer 1977; *proxenoi* generally, Gerolymatos 1986). Information in the *ekklesia* is provided by advocates of policies (Aeschines and Demosthenes come readily to mind), *strategoi*, current and former envoys (*presbeis*), and the like. Intelligence tends to be persuasive rather than deliberative. See Russell 1999, 103–139 for a more detailed discussion of political intelligence.

2 Russell 1999, esp. 108–114. The kings of Salamis are extraordinary, combining Greek and Persian elements in their intelligence practice, and expanding on both.

3 Engels 1980; Russell 1999, esp. 114–115, 226–234.

A Companion to Greek Warfare, First Edition. Edited by Waldemar Heckel, F.S. Naiden, E. Edward Garvin, and John Vanderspoel.

If we turn from the horizon of strategic concerns to the middle ground of operational intelligence, we find that in most polities the acquisition and implementation of intelligence was in the hands of the commanders of troops, whether kings, *strategoi*, or *hipparchoi*, and that their assets and methods were consistent over space and time (although a degree of specialization develops in the fourth century), and differed principally by defensive or offensive context. It is worth noting, however, that spies, who today address strategic concerns, were for the most part used by military commanders as operational assets.[4]

For espionage in military contexts, the foundational passage is found in Xenophon's *Cavalry Commander* (4.7–8):

> It is also necessary to have given thought to spies (*kataskopoi*) before the outbreak of war, so that they may be from states friendly to both warring parties and also from merchants; for all states always receive those importing goods as men well-disposed to them. Fake deserters (*pseudautomoloi*) can be useful as well. Do not, however, neglect to be on guard (*phylakes amelein*) even if you have confidence in your spies; rather it is always essential to make preparations as if they have come reporting the enemy to be at hand. For even if your spies are entirely reliable, it is difficult to provide timely information, since many mishaps occur in war.

We will return to our *kataskopoi* later, and examine them in connection with reconnaissance; for now, let us examine what it means not to "neglect to be on guard."

"To Be on Guard": Operational Intelligence in Defensive Contexts

In the first place, guards served at night. These watchmen were called *phylakes*, *nyktophylakes*, or *prophylakes*—never *skopoi*, perhaps since their primary sense was hearing rather than sight. In Homer, we find the Greek commanders worried that sleep might overcome watchmen's bleary eyes.[5] Indeed, they typically ruined their night vision by sitting around fires, although a few commanders (e.g. Seuthes and Xenophon) were thoughtful enough to conceal *prophylakes* well outside the range of firelight, to ambush approaching enemies and deserting friends.[6] Nor did *phylakes* enhance their hearing when they carried bells as they patrolled the walls of a city, or when they amused themselves by singing "guard songs."[7] However, nightwatchers were assisted by the sharp ears of dogs, who are mentioned in an appropriate simile in Homer, and were actually assigned to garrisons and guard posts (*phylakeion*, *phourion*, *pyrgos*, etc.) by the third century or before.

By day, guards could look out over the countryside and the margins of the state—the *chora*. Traces of such a guard are still visible through material remains, especially of towers. Even as the beacon fires blaze in succession across the lines of Aeschylus' *Agamemnon*, so too the towers of classical Greece stretch back in a series all the way to the Bronze Age. Structures

4 Russell 1999; see also Richmond 1998.
5 Homer *Iliad* 10.180–193.
6 Russell 1999, 29–33.
7 Russell 1999, 33–34.

identified as guard posts and watchtowers are attested in the Early Minoan period, and arguments have been made for systems of towers in communication with each other.[8] A complex system of lookouts and *phryktoria* (fire signals) atop peaks and ridges has been reconstructed for the Pediada region of Crete c. 1900–1700; their connection with military surveillance is difficult to establish, but a fascinating possibility.[9] On the mainland, Mycenaean lookouts and watchtowers are attested both in isolation and in groups, with the capability of signaling to each other or to settlements.[10] Mycenaean era Pylos Linear B tablets record the watch on the coast of the western Peloponnesus and thus give a glimpse of a system of coastal surveillance.[11]

Even as the ruins of Mycenaean towers crumbled, new structures were constructed throughout the Greek world. While much attention has been devoted to the towers and forts between Attica and Boeotia, these are by no means singular. The island of Siphnos alone boasts the remains of at least 77 towers, most of which were constructed between the sixth century and the end of the fourth century.[12] While particularly well represented, Siphnos is by no means singular: more than a score have been identified on islands such as Thasos and Ceos, and there is no lack of examples in other islands and districts.

A collection of towers in an area does not necessarily equate to intense military surveillance. Agricultural or mining uses can replace, or be combined with, military functions.[13] Even in demonstrably military contexts, towers are rarely necessary to keep an area under surveillance—an extra few meters of height offers little advantage in areas that are hilly and not heavily wooded. Furthermore, their expense in labor and materials is considerable. It has nonetheless been demonstrated that towers were integral elements of surveillance systems. Bold reconstructions of the Attic–Boeotian frontier proposed by scholars such as Ober and Munn, while challenged in particulars, have recently been affirmed by the application of view-shed and landscape approaches to the study of other areas.[14] For instance, in southern Euboea, Gardner and Seifried have ably reconstructed a surprisingly dense and enduring concentration of towers in the peninsulas flanking the port of Geraestus.[15] While some of these have limited view-sheds, and may be connected to agriculture or domestic security, a

8 Minoan: Alusik 2008, 2016; Mycenaean: Castleden 2005, 35–36, 62; Field 1984; Loader 1995, 99–100.

9 Panagiotakis 2004, 180; Panagiotakis et al. 2013: "A *soros* is a large pile of pyramidal shape, with a flat top, of usually red baked earth mixed with roughly dressed stones on top of a hill … Sections of these *soroi* reveal their function, since layers of red baked soil alternate with irregular layers of ash to suggest fires … Such evidence recalls the fires lit on peak sanctuaries, but equipment to suggest cult is quite lacking. It is more plausible to see the *soroi* as a kind of a lookout and signal station".

10 Lord 1939; Field 1984; Loader 1995, esp. 99–100.

11 Ventris and Chadwick 1956, nos. 56–60; Russell 1999, 24–25 ad loc.

12 Young 1956.

13 See, for instance, Morris and Papadopoulos 2005. Cf. also Osborne 1986; Nankov 2009, 335–337; Seifried and Parkinson 2014.

14 Ober 1985; Camp 1991; Munn 1993. Reconstructions of tower systems with observation as a defining characteristic can be found across the Greek world, in such areas as: Locri Epizephyrii (Visonà 2016); Achaea Phthiotis (Chykerda et al. 2014a, 2014b); Opountian Lokris (Nankov 2009); Mantineia (Topouzi et al. 2002; Maher 2018); Eretria (Fachard 2012); southern Euboea (Gardner and Seifried 2016); Miletus (Ma 2000); and Sagalassus (Vandorpe and Waelkens 2007); one may possibly include Heraclea on Latmos (cf. De Staebler 2016a, 62, 2016b, 124). In contrast, some *poleis* may have watchtowers that are not demonstrably interconnected (e.g. Molossia, cf. Nakas 2016).

15 Gardner and Seifried 2016, 154–155. The Paximidi peninsula, at the tip of Euboea, has among the densest concentration of towers (18 towers datable from the classical era alone, at a density of 0.82 per km2).

substantial number afford superb views of the port, bay, and the vital shipping route skirting the coast.[16] Moreover, groups of contemporary towers are often inter-visible, and can be demonstrated to be connected by paths and roads. The Geraestus network was capable of relaying signals to Andros, Keos, and to the mainland.[17]

Towers were also useful for signaling with fire, smoke, or flags.[18] Their fixed location and height offer an advantage over observation posts—less chance of confusion or even deliberate disruption due to nearby fires.[19] While it is unlikely that complicated mechanisms for signaling, like those described by Aeneas Tacticus and Polybius, were anything but exceptional, towers could provide a facility to house water-clocks, and materials and platforms for fires and *semeia*. More importantly, and doubtless more typically, towers served the valuable function of providing a fixed and conspicuous point from which signals could be expected and seen. They might also serve to mark zones and boundaries: in Euboea, they represented a commitment on the part of the Athenians to the area during the fifth century; in Calabria, they defined borders and zones of control.[20] In some instances, the masonry of the towers is of high quality; some feature adornments and aesthetic elaboration. It follows, then, that towers are meant to be seen as well as to see. They attest to the influence of their builders, whether these are states or individuals.

Finally, towers provided protection for small, isolated, and vulnerable groups of men. We know from Xenophon and Aeneas Tacticus that *phylakes* and *hemeroskopoi* were liable to be killed or captured, and were in fact targeted. To be sure, scholars have rightly doubted the ability of small forts and towers to provide a preclusive, or even porous, defense of the frontier. Such was probably never the intent of their builders; rather, they provided secure bases for watches and patrols.[21]

Whatever their advantages, towers and forts (*phrouroi*) were neither mobile nor hidden, but we can infer from epigraphical evidence that they contributed to systems that included a mobile and concealed element, the *peripoloi*. These patrols of light-armed soldiers stationed in the *chora* not only watched for external threats but also local security (e.g. guarding against brigands, and controlling low-intensity insurgencies, as the *orophylakes* at Amyzon in Caria).[22] They even policed people and property (e.g. the capture of runaway slaves by the *paraphylakes* of Thessaly, Miletus, and Heraclea).[23] They used towers and forts as their bases of operations and as waystations. Our earliest allusion to *peripoloi* puts them in the seventh century BCE, as part of the story of the rise of the Orthagoras of Sicyon, from *peripolos*, to

16 Gardner and Seifried 2016, esp. 169.

17 Gardner and Seifried 2016, esp. 154; although the material arrangements are not established, it would be reasonable to conjecture some sort of similar arrangement for the Hellespont; see Krotscheck 2010; Sergidis 2012.

18 For more detailed discussion, see Russell 1999, 145–150.

19 Thuc. 3.22.7 and Russell 1999, 212 ad loc.; Polyaenus, *Strat.* 6.19.2.

20 Funari and Pollini 2005, esp. 338, 341–32; Visonà 2016.

21 And, while this may seem trivial to those who have never spent a night on a ridge in bad weather, they also afforded shelter.

22 Light-armed soldiers: Clouet 2015, esp. 148–149, 150, 176 re. *SEG* 26.1306. Local security: Dmitriev 2005, 212; Chaniotis 2008, 141–142.

23 Chaniotis 2008, 139–141. It is, perhaps, worth noting that these functions are analogous to those assigned to towers in the ongoing debates over their purposes—the combination of roles for the *peripoloi* is an implicit argument for an analogous combination for the towers.

commander of the patrol (*peripolarchos*), and then to tyrant.[24] In the Classical period, they were recruited from citizens or mercenaries or both.[25] Kent argued that, in this period, they were deployed on the Attic frontier only during time of war, but their numbers and postings may have varied according to the perceived threat.[26] Certainly, larger states like Athens had more resources to sustain a continual watch, but there is evidence that it was possible even for smaller states like Teos.

Not surprisingly, the clearest picture we have of the watch in the *khora* is Attic and Hellenistic. Couvenhes has admirably reconstructed the temporal and functional frameworks. Using the garrison at Rhamnus as a focal point, he has traced a transition from *peripoloi* (in the fifth century) to ephebes (in the fourth to third century) to a mix of citizens and mercenaries (including Macedonians in the period 261–229), demonstrating how the demographics change with the political, social, and international situation. An inscription honoring Epichares, *strategos* of the chora by the sea, is particularly helpful for understanding the role and nature of his role in maintaining social order and regional security.[27] Epichares was elected to the post by the demos, having already served in the capacity of cavalry commander (*hipparchos*). He is honored for having well and nobly (*kalos kai philotimos*) cared for the guard, and for having preserved the fort for the people in time of war.[28] His role is by no means limited to defending people and property: as a benefactor of sorts, he ransomed prisoners taken by raiders, distributed food, refurbished the sanctuary of Nemesis, and paid for a stoa. As the director of the watch over the *chora*, he built watchtowers, supplied extra dogs and food for the guard,[29] punished those who guided raiders, assigned some watchers to forts and towers, and posted others as concealed lookouts in vantage points (*skopiai*) on the hills.

This last distinction, between towers and *skopiai*, is highlighted in a number of places. Lookouts on hills are attested from the time of Homer, and later are called *hemeroskopoi* (watchers) and *hemeroskopiai* (lookout points).[30] The reference to hidden men (*kryptoi*) assigned to watch by Epichares is in keeping with instructions given by Xenophon and Aeneas

24 Forbes 1930, 76.

25 Kent 1941, 348.

26 Kent 1941, 346–347.

27 *SEG* 24.154. Command of the watch on the frontier became more specialized over time. Over the course of the fifth century, the Athenians transitioned from focusing on tribe to focusing on task. In the fourth century, the garrison forts fell under the command of the strategoi *epi ten choran*. The chora was subdivided in the early third century, and the eastern part (including the forts at Rhamnous and Sounion) was assigned to the *strategoi* of the coastal region (*epi ten choran ten paralian*), while the western (dominated by Eleusis), was given to those *epi ten gen ep' Eleusinos*.

28 Similar, almost formulaic, attributions of justice to *strategoi* in charge of Attic frontier districts are commonplace enough to make one wonder if they are an attestation to a real problem with irresponsible or abusive counterparts (cf. Teos). It is perhaps telling that Glaucon, ignorant though he may have been, was portrayed by Xenophon (*Mem.* 3.6.11) as holding the opinion that the *phrouroi* did more harm than good.

29 For dogs and the nightwatch, see Forster 1941; Russell 1999, 29–31; Oliver 2007, 139–140 and n. 6; Nankov 2009, 145; cf. also McIntire 1965 for an analysis of modern use.

30 For *hemeroskopeion* as a place-name, e.g. in Spain (Ellinger 2010, 473, associated with a sanctuary of Artemis and a pirate base, Strabo 3.4.6); so too *skopia*, e.g. Skopje, in the Northern Republic of Macedonia.

Tacticus to keep *hemeroskopoi* concealed.[31] A dedicatory inscription for Philotheus of Phrearrioe, *strategos epi ten paralian*, was commissioned by his *kryptoi* and others who served under him; in this instance, his *kryptoi* were members of a unit that was commanded by Athenodorus of Oa,[32] suggesting specialization. The *strategos* Archander recruited *kryptoi* from among his soldiers. He posted some of the younger (or "less experienced") in "the tower," and others on the heights (*skopiai*), which were further away and hence more dangerous. In contrast, he assigned the older (or "veteran" men) to the task of protecting the harvest.[33] These testimonies do not permit a clear definition of the *kryptoi*, and we would be wise to follow Couvenhes and Knoepfler, who recognize these troops as distinctive without seeking to find a professional unit.[34]

Although *peripoloi*, ephebes, *xenoi*, *kryptoi*, and *hypaithroi* ("those under the open sky") served in different times and roles on the Attic borderland, we must ask ourselves to what degree the provisions of a wealthy, powerful state like Athens are relevant to the Greek experience as a whole.[35] The evidence is skanty, but parallels in a decree from the small state of Teos give some hope that Attica is not unique.[36] Here we find a garrison commander (*phrourarchos*) elected for a term of four months, paid and carefully monitored by the state, and expected to behave in a manner not unlike Epichares (nobly and justly), despite the fact that he is posted to a recently annexed territory to protect against revolt as much as raids. The Tean *phrourarchos* enrolls at least 20 citizens, equipped more like *psiloi* than hoplites, and three dogs. In role and expectations, he resembles his counterpart in Athens at least as much as the Athenian resembles his ancestors in the *phrouria* of the fifth century.[37]

Operational Intelligence in Offensive Contexts

The measures described earlier enabled a defense to react effectively—to counter an attack, or to withdraw to safety. An attacker, on the other hand, needed to prevent an effective response while discovering and exploiting vulnerabilities. It was also in an attacker's interest

31 Xenophon notes that hidden outposts cause the enemy to feel insecure (*Hipparch.* 4.10–12, Russell 1999, 27); Aeneas Tacticus (6.6) recommends that *hemeroskopoi* be sent to their posts before first light to elude detection. I follow here the restoration and reading of Knoepfler 1993, 328–330, who rejects Vanderpool's restored "[pur]go<u>s" in view of a visible "ntos" and hence prefers an ending in *–ontos* or *–antos* (so also Garlan). Knoepfler reads "kruptous" in the text as a substantive: hidden men ("*homme camoufle*" or "*sentinelle a couvert*" or even "*espion*" citing Aristophanes *Thesm.* 600).

32 *SEG* 41.87 lines 24–25; Knoepfler 1993, 335: "les kryptoi formaient un corps de troupe bien distinct." For Philotheus as former *strategos epi tous xenous*, see *SEG* 41.87 and Oliver (2007, 183), who dates this to 234/233.

33 Couvenhes 2014, 51–52 and nn. 33–35 cites an unpublished decree shared with him by Knoepfler. His analysis is far more detailed and thoughtful than my aformentioned condensed statement.

34 Couvenhes 2014, 51: "Le processus de recrutement des kryptoi s'effectue par sélection à partir de l'ensemble des soldats (stratiotai), disponibles dans la forteresse; les paroikoi ne semblent pas concernés. Ce processus de sélection est dicté par les circonstances troublées du moment. L'activité des kryptoi apparaît comme une spécialisation mais il n'y a pas de différence de nature entre les kryptoi et les soldats, simplement une différence d'affectation".

35 Oliver 2007, 180. Oliver rightly notes that *kryptoi*, *hypaithroi*, and *paroikoi* are attested only at Eleusis and Rhamnus, but their functions are paralleled elsewhere.

36 *SEG XXX* 1376, esp. 9–10, 19–20.

37 Cf. Lambert 2018; *SEG* 26.18 (c. 424/423), concerning an Athenian phrourarchy in Halieis, provides a relevant counterpart to both inscriptions.

to minimize the ability of the defense to make informed decisions —and Greeks frequently sought to do so by preventing or delaying discovery of their intention and presence through concealment (by terrain or darkness) and speed, as well as by ruses, corruption, and deception.[38] In addition, an attacker needed to maximize his own capacity to deploy effectively, and thus sought out viable routes, potential obstacles and defenses, resources such as water, and the like. Failing personal familiarity with the terrain (certainly possible in inter-*polis* warfare, but less likely abroad), he would need willing or captive guides.[39] To avoid ambushes, and to bring his forces to bear at the right time and place, he needed to know the location and possible actions of his foe. To meet these requirements, Greek commanders had recourse to mobile assets.

Like their counterparts atop hills or in towers, the men employed in reconnaissance might be called *skopoi;* unlike them, they could also be called *kataskopoi*, *proskopoi*, or by the more generic label of *prodromoi* (forerunners).[40] Homer uses *episkopoi*, when telling a tale of the first successful reconnaissance, and the first abject failure, in the tenth book of the *Iliad*.[41] For all its epic context, the episode is a stylized but coherent version of a reconnaissance mission—a team is sent to sneak close enough to the enemy to see what they are doing, overhear what they are saying, and report back. If possible, they look to waylay a straggler to interrogate him to learn the enemy's intent. Our characters are a lone scout (Dolon) or a small team (Odysseus and Diomedes), disguised or lightly equipped, relying on stealth and mobility, and moving as far past the margins of safety toward the enemy as they dare. The historical example of Aratus' group of three men reconnoitering the defenses of Sicyon by night is hardly less colorful.[42]

While night reconnaissance was always conducted by a few men on foot, numbers and missions varied by day. In Homer, we find Odysseus repeatedly sending three men ("two men and a herald") into new lands to learn what manner of people live there.[43] In the Classical era, groups of three are typically associated with espionage, embassies, and official investigations.[44] Most reconnaissance was undertaken by platoon-sized groups, typically composed of *psiloi* among the smaller poleis and in mountainous regions, or of cavalry among the larger states, and those with open country (Boeotia, Thessaly, Macedonia, Syracuse, etc.).

The exact numbers of these teams can only be inferred. On their great expedition to Sicily, the Athenians initially sent 30 horses with riders, a force obviously incapable of facing the powerful Syracusan cavalry. Given that they intended to acquire more horses after their arrival, it is likely that they believed that the 30 would suffice as scouts until they could field an effective force. Among the Ten Thousand, 40 horsemen were cobbled together and put

---

38 Russell 1999, esp. 190–225.

39 Russell 1999, esp. 54–60. For superb analyses of the disorienting effect of terrain, see Stoll 2013; also Stoll 2016, esp. 87–88.

40 In the Homeric epics, men employed in either surveillance or reconnaissance could be called *episkopoi*, no doubt from the practice of being stationed upon (*epi*) a lookout (*skopia*). By the classical era, *episkopos* became associated more with the political realm (Balcer 1977), and finally religious office. The term *kataskopos* is a somewhat curious variation on *skopos*, with the *kata* bearing faint echoes of "down" (as opposed to on the heights?) or "against" (with respect to a function of uncovering hidden watchers or ambushes?), although it could merely be an intensifier; cf. *LSJ* s.v. κατά, esp. E.I, III and V.

41 *Il.* 10.206–563.

42 Plut. *Arat.* 7.4–8.3; Russell 1999, 30–31, 2013, 113.

43 *Od.* 9.88–90, 147–149, 10.100–102.

44 Espionage: Russell 1999, 136–137. Investigators, 87; envoys, 63–70.

under the command of Lycius, and later Timasion, who were repeatedly given the task of riding ahead to reconnoiter.[45] Spence reasons that the main role of the Athenian *prodromoi* in the fourth century was reconnaissance, and estimates their number at 50.[46] There is no compelling reason to believe that a cavalry unit had to keep together when operating.

In much of Greece, small groups of cavalry tended to follow paths rather than run cross country. One of their tasks was to find and explore routes ahead of a marching army. Light-armed troops operated on the flanks or in the hills with the primary purpose of locating the enemy and discovering ambushes.[47] They avoided combat, but often sought to question or capture those they encountered. Large forces, including cavalry in more open areas, could act as screening forces or vanguards, with some ability and inclination to join in combat, rather than purely for reconnaissance. They bought time for deployment or maneuver before battle.

As is typical, Xenophon's *Anabasis* is rich in examples. By the time they reached Thrace, the Ten Thousand had broken up, and Xenophon's contingent was passing through rough country. Xenophon received word from his cavalry, who had encountered and questioned some travelers, that a Greek army was being besieged by Thracians. Xenophon kept the travelers under strict guard (ἐφύλαττεν ἰσχυρῶς), so that he might use them as guides. He posted watchers (*skopoi*) and prepared his men. Timasion went ahead with the cavalry so that nothing would be missed (σκοπείτω τὰ ἔμπροσθεν, ὡς μηδὲν ἡμᾶς λάθῃ). Some of the light-armed troops escorted the army on the flanks and heights, so they could signal if they saw anything. Rather than trying to conceal his approach, Xenophon tried to give the impression of a larger army by lighting widespread fires. After halting for the night, his men camped on a hill, from which they could see the campfires of the Thracians about 40 stadia distant. Then, after stationing guards (*phylakes*), they slept. At daybreak they offered prayer to the gods, and set forth at the fastest possible pace, once again sending Timasion ahead with cavalry and captive guides. They found no battle, but learned from locals an outline of what had happened.[48]

We can see here the shift between offensive and defensive postures, the complementary use of cavalry and light-armed troops, the need to seize opportunities (and people), ruses to deceive the enemy, and ultimately the responsibility of the *strategos* to conduct a variety of instruments and players. Just as in the case of the men posted to the garrison at Rhamnus, we see a degree of specialization—certain commanders and troop types are used repeatedly for reconnaissance, both here by Xenophon, as well as among the Greek states (e.g. Sparta's use of the Skiritai, Alexander the Great's use of the Paeonian cavalry).

It would be remiss to leave the scene of Xenophon in Thrace without noting a detail indicating another dimension of operational intelligence that is lost to us in the present: prayer, sacrifice, and divination. Oracles and diviners (*manteis*) had a role in Greek warfare—indeed, one could almost claim that *manteis* were their only professional intelligence agents. While we may smile at the story of Croesus in Herodotus, we must acknowledge that states like Sparta and Athens appointed officials who bore the responsibility for consulting oracles and preserving their responses (the *pythioi* attached to the Spartan kings, for instance). Xenophon

---

45 Xen. *Anab.* 3.3.20, 4.3.22 and 25, 4.7.24, 6.3.14, 7.3.46.

46 Spence 1993, 134, 149–150. Cf. Bugh 1988; Russell 2013.

47 S. Barefield, who competes in "eventing" (cross-country horseback races), stated that before she rides a course in competition, she walks it on foot to reconnoiter it. Riding at speed over an unknown course is dangerous; doing so pursued by missile fire or enemy cavalry even more so.

48 Xen. *Anab.* 6.3.10–23.

took pains to justify the validity of divination, even though he and Silanus, an Ambraciot *mantis* who accompanied the Ten Thousand, could not trust each other. And, indeed, from the story of Agamemnon and Calchas onward, we find problems attendant on conflicts between religious and secular authorities.[49] While Nicias, Xenophon, and the Spartans represent the more religious elements in the Greek world, even the likes of Pericles could not afford to dismiss divine voices from the deliberations of *strategoi*. (For more on this subject, see Chapter 24 in this volume.)

Final Thoughts

While landscape archeology, view-sheds, and GIS applications have expanded our knowledge of surveillance in the last two decades, those who practiced intelligence in antiquity tend to remain obscure, unless glimpsed through a few inscriptions. There are no Greek counterparts to Roman scouts such as Tiberius Claudius Maximus, the *explorator* who became the subject of a very public and dramatic story when he brought the head of Decebalus to Trajan.[50] The closest would be a creature of fiction, Odysseus, whose covert missions—stealing into Troy dressed as a beggar and plotting against the suitors in the same disguise—always conform to heroic convention thanks to spectacular carnage capped by a glorious epiphany contrasting with his humble appearance. He bears little relation to the ordinary men who watched from tower or hilltop. Like shadows without a name or a face, the *skopoi* of the Greek world creep through scattered sources as through the brush.

49 Xen. *Anab*. 5.6.16, 27–30.

50 His story is portrayed on the column of Trajan, a celebrated monument then and now, and recounted with pride on his tombstone (see Speidel 1970 for discussion).

CHAPTER 20

The Economics of War

Johannes Heinrichs

Greek sources skip over economic details, unknown to the authors in most cases and thought to be stylistically unappealing to the reader. For pre-classical times we learn next to nothing about public or private expenditure for arms and armor, army equipment and warships, fortifications and naval installations. Information is scarce even for later periods. Maintenance for troops and animals[1] will have been organized for the most part by public authorities, but military "pay" was not introduced before the Persian Wars.[2]

Exceptions to the rule are the historians Thucydides and Xenophon, both socialized in democratic Athens where public finance was a permanent issue of debate open to all contemporary citizens; the same applies to Attic orators such as Demosthenes.[3] Yet in oligarchic *poleis* access to pertinent information was strictly limited. It may, however, turn up where we least expect it: clay tablets from late Mycenaean archives at Pylos and Knossos, from c. 1220–1200, contain detailed data on military organization, armament, and service.[4]

1 Daily supply rates for troops on campaigns: Engels 1978, 123–126 (personnel), 126–130 (animals). Costs depend on factors that at times can be extreme, cf. Xen. *An.* 1.5.6.

2 For an apparent exception, see van Wees 2010 (*IG* XII 9, 1273f., 10–16, cf. *SEG* 41 [1991], 725), to be dated to 525/00, cf. Cairns 1984; *pace* Walker 2004, 192ff. who however pleads for a late start for minting at Eretria.

3 See e.g. his detailed proposal to finance a task force against Philip (Dem. 4.28f.).

4 See Chapter 1 in this volume.

A Companion to Greek Warfare, First Edition. Edited by Waldemar Heckel, F.S. Naiden, E. Edward Garvin, and John Vanderspoel.

The "Trojan War": Logistics and Booty in the *Iliad*

So little is written about the economic background of the mythical war against Troy that Thucydides (1.11), taking fiction for history, attempts a rationalized reconstruction. His perspective is that of the Athenian *strategos* he had been in 424, well aware that the maintenance of troops was indispensable. Since the Greeks, fighting far from their homes, could neither bring along food sufficient for ten years nor organize a permanent line of supply, they had to grow grain in the region of Troy—or so the historian concludes. His problem lies outside the epic's focus. From the *Iliad* itself—the version we read is limited to some 40 days in the final year—all we get are highlights, such as booty from conquered towns in the Troad (Thebe and Chryse: Book 1), spoils, both normal (bronze, iron, leather, slaves: 7.473ff.) and prestigious (prizes in Achilles' funeral games for Patroclus: Book 23), and ransom for Trojan princes sold abroad (24.751ff.) and for Hector's body (24.275–276, 555–556, 594ff.). For the epic's purpose, this sufficed. By expecting more, we repeat Thucydides' mistake.[5]

Pre-classical Greece

In the Archaic *poleis*, those who could afford it had to pay for their own arms and armor, and so did those who owed the *polis* an obligation in return for a parcel of land granted by the state, such as the Athenian settlers on Salamis in the late sixth century.[6] Poor men served as light-armed troops and skirmishers.[7] Since it was impossible to maintain war funds in the pre-monetary periods, public authorities had to make impromptu decisions on who had to contribute what, both in kind and in effort.[8]

In the early sixth century, Peisistratus, who had not yet assumed the tyranny, gained popularity among his subordinates (Hdt. 1.59.4) as a commander in an Athenian campaign against Megara. We may guess that he had made generous grants from his private stock to poorer fellow soldiers. Generally, this was an aristocratic strategy in archaic Greece: social benefits in exchange for political support.[9] It also helped to finance wars.

Booty could at times compensate for the lack of income, but honors granted for valor soon took on the character of modern medals or crosses.[10] On the one hand, an inscription on a votive offering from Priene, dedicated probably in the second half of the seventh century, says that Pedon, an Ionian mercenary under the Egyptian king Psammetichus, had received

5 Donlan 1997a.

6 *IG* I2 1, cf. ML no. 14.8ff. (with comm.): "They shall provide their weapons themselves, for 30 drachmas each man. Afterwards, the *archon* shall control the weapons ..." At the end of the sixth century (the decree cannot be much older, since the obligation is defined in money), 30 drachmas seem rather moderate for arms and armor, particularly as the sum had to be invested only once, not regularly on an annual basis. In turn, compared with the Solonian quotas, it was 1/10 of the yearly income prescribed for a *zeugites*.

7 Cf. Tyrtaeus fr. 11.35ff. Gerber (Loeb, *Greek Elegiac Poetry*).

8 Trundle 2010.

9 As a democracy, Athens provided meals and wages, and charged wealthy citizens with liturgies, also of military nature, such as building and operating a war ship (*trierarchia*) or to maintain horses (*hippotrophia*); see Davies 1967. By this means, the potential for private benefits was reduced.

10 New Comedy poets could grossly exaggerate its amount, e.g. Menander (*Aspis* 34ff.; cf. 83, from the very end of the fourth century).

"a golden bangle, and a town,"[11] meaning its regular tribute. On the other hand, in 598 Antimenidas, the brother of the poet Alcaeus, received from the Babylonian king, King Nebuchadnezzar II, nothing more than a sword with a "handle of ivory bound with gold" for a decisive victory in a personal duel.[12]

Xerxes' Invasion of Greece, 480 and 479

In 480 and 479 Greek contingents were in the field or aboard their ships for long terms of service far away from their cities. Herodotus has little to say about how their maintenance was organized and financed, whereas his reports about Persian logistical preparations are detailed; from these reports we learn that Persian subjects had to bear a large proportion of the costs.[13] Even the Macedonian ruler Alexander I, who at that time certainly controlled land in the central Chalcidice, was required, in consequence, to contribute to the construction of the Athos canal—employing personnel whom he had to maintain at his own expense—with the exception, perhaps, of the grain that was delivered by the Persians (Hdt. 7.23.4). The Macedonians got no wages—subjects had to perform military service free of charge—and the canal was, after all, a military project.

Partly, at least, to pay these costs, the first "royal" Macedonian coins were minted at this time. Their denomination was suitable for buying daily rations obtained at markets near the canal (Hdt. 7.23.4).[14] Alexander thus availed himself of the economic infrastructure of Greek cities on the Chalcidice and avoided the costs of bringing supplies from Macedonia proper.[15]

The Greek preparations for defense followed the age-old practice that maintenance had to be borne by those who requested troops. Quite early, a majority of Thessalian cities had called for Greek assistance in defending the Tempe Valley at its eastern point (Hdt. 7.172). Hence, they were liable for food and fodder. The Greeks left their position soon, warned by Alexander, probably on information from then Aleuad Larisa: siding with Xerxes, it could cut Greek supplies from the rear when the Persian troops arrived in the region.

The next defensible point, Thermopylae on the Malian Gulf, could be supplied with Euboean grain from the south by ship, but had to be abandoned when the Greek fleet was prepared to withdraw from its forward position at the northern cape of Euboea (Artemisium). The third constriction was the Isthmus of Corinth, the gateway to the Peloponnese. Resistance had been planned here from the start, so supplies will have been stored in large amounts. Peloponnesian resources, however, could not feed extra contingents from Attica, the Megarid, Phocis, and elsewhere, and so the Greeks had negotiated in advance with Gelon of Syracuse for grain and ships (Hdt. 7.157ff.). Evidently, the tyrant sent both. Since he was an ally of the Corinthians but not a full participant in the war, we cannot know under which

11 *SEG* 37 (1987), 994, dated under Psammetichus I, 664–610.

12 Alcaeus fr. 350 (Campbell [Loeb, *Greek Lyric* I]); such prestigious swords occur already in the Mycenaean palace at Pylos (*PY Ta* 718 with Ventris and Chadwick 1956, 346–347).

13 Hdt. 7.21, more detailed for the Greek cities on the Thracian coast: 7.118–120.

14 Tetrobols on the Chalcidean standard compatible with the regional coins of the canal zone.

15 The "Confederate" coins of the Arcadians started in 480/479, once more equaling a soldier's daily maintenance (triobols on the Aeginetan standard). For the same rate, cf. Thuc. 5.47.6 (treaty of alliance between Athens and several Peloponnesian states in 420) and Xen. *Hell.* 5.2.21 (Peloponnesian campaign against Olynthus in 382). Kraay 1976, 97 with pl. 16 no. 288; cf. Heinrichs 2015, 72ff. (with photos).

terms he supported the Greeks. If he sold the Sicilian grain, the intermediary was the Corinthians, since the Isthmus was their responsibility. We may suppose that Sicilian grain played a major role also in the following year (479).

At Athens, poor families could not do without income during extended periods of military service. The situation was even worse than elsewhere since the Athenians had to maintain a fleet of 200 triremes, each with a crew of c. 200 men, totaling 40,000 per day. Though only poor families will have been supported, the city was financially overburdened.[16] Against this background, we must understand Themistocles' collecting of silver by force from some of the Cyclades (Hdt. 8.121ff.) after Salamis: these islands had sided with the Persians and now had to bear more than a fair share of the Athenian war costs. Since the city's dies for minting coins from the collected silver had certainly been plundered by the Persians along with the tools for making new ones, the iconography of the series in question is of utterly low quality.[17]

The *Pentekontaetia* Through the Early Fourth Century

Athens' armada would be needed in the future, all the more so as the Greeks expected an imminent renewal of Persian attacks. But the devastated city did not have the means to bear the costs, whereas many Greek cities on the coast of Asia Minor saw an opportunity to break away from the Persian Empire—in the event that the sea before their harbors came under Greek control—with the help of Athenian ships. The consequence was the foundation of the Delian League (478/477): the Athenians protected with their triremes an ever-growing number of Greek city states in Asia Minor and the Aegean, whereas the protected regularly contributed money. In the beginning, some contributed ships, but most contributions were converted, either voluntarily or by force, to monetary payments. The original assessment of annual contributions amounted to 460 talents (Thuc. 1.96.2), along with booty sufficient to keep under sail an appropriate federal fleet for some months of the year, and at best to build up reserves. The allies' individual contributions can be calculated from lists inscribed on steles, the so-called Athenian tribute (quota) lists (*ATL*), spanning the period from 454/453 to 406/405. By far the majority of them are heavily damaged, and therefore of limited value for assessing the development of the League's finances.[18] At the beginning of the Peloponnesian War, at least, an average of 600 talents per year was coming in, on top of a reserve of 6,000 talents (Thuc. 2.13.3).[19]

During the war, the reserves rapidly diminished. Athens reacted by raising rates substantially in 425/424 (ML 69); in 413 the allies were burdened temporarily with a duty of 5% on all imports and exports (Thuc. 7.28.4) instead of fixed dues, and in 412 the Periclean core reserve of 1,000 talents had to be touched (Thuc. 8.15.1). Probably in 406, to rescue Athens' vessels from blockade in the harbor of Mytilene, the gold and silver contained in Athenian temple dedications (Thuc. 2.13.4) was melted down for coinage: silver-plated bronze

16 A pertinent episode in Ps.-Aristotle (*Ath. Pol.* 23.1) was obviously invented during Ephialtes' conflict with the Areopagus in the late 460s.

17 Kraay 1976, 60, with pl. 10 nos. 185–186 (Seltman groups C and E).

18 Meritt et al. 1949, see also ML no. 39 (with comm.).

19 Feyel 1999; Gabrielsen 2008; Malitz 2008; Meister 2008.

denominations[20] and, for the first time in Athens' history, (full) gold coins.[21] These had been the very last resources, just sufficient to finance one last fleet. Along with it, the Athenians lost the war in 405 and had to capitulate in the following year.

When Athens' enemies, originally Sparta and her Peloponnesian allies, entered the war, they knew well that their financial resources were utterly insufficient to build and maintain a fleet. A stele on the Spartan *akropolis* with lists of war contributions of Sparta's friends all over the Aegean and the coast of Asia Minor, in money and kind, was therefore meant to be a demonstration of political support rather than of economic strength; the preserved total in money is risible in comparison with Athenian revenues.[22] Yet for warfare on land far less money sufficed: nearly all that was needed was carried by the attackers or taken from the invaded territory of Attica. The Peloponnesians thus restricted themselves to a strategy of limited invasions with their superior land troops—a strategy intended to destroy as much agricultural infrastructure as possible, in the hope of demoralizing and dividing the Athenians, while creating difficulties of supply. The strongly fortified Piraeus remained in hand, however, and so grain continued to come in from the Black Sea regions. In turn, the Peloponnesians got compensation for damage done by the Athenians to their own fields from Sicily. It was clear that the conflict would be decided by the side that would cut its opponent's grain supply.

In 414, Athens took the initiative with its Sicilian expedition, which aimed ultimately at the conquest of Syracuse. When the plan failed, a portion of the Athenian fleet fell into the hands of the Peloponnesians, who nevertheless commissioned 100 new ships (Thuc. 8.3.2). Hence, from 412 on, the radius of sea warfare could be extended to the Asian coast and eventually to the Hellespont, the gateway to the Black Sea. Yet the Peloponnesians now faced the same problem as Athens in 479/478: a fleet was at hand, but not the means to maintain it. Athens had solved the problem by imposing tribute payments on her numerous allies in the Aegean and on the Asian coast. As a first step, the Lacedaemonian leadership tried to stop the flow of tribute payments in order to weaken the Athenian fleet decisively before the Hellespont could be blocked.

At this point, Greek cities, Persian satraps, and eventually the Great King entered the game.[23] They promised to bear the operating costs of the Peloponnesian fleet, which in return would contribute to driving the Athenians out of the Greek cities of Asia Minor. The concept seemed simple but proved to be difficult, for various reasons. A decisive factor was the lack of coordination between the Great King and his deputies in the West, who at times were unable to pay their share of the immense costs.[24] Nor could the Great King enter the war with a fleet of his own. These problems were finally overcome by the Spartan (vice-)

20 Klöppel 2014 (with photos) is rightly cautious in identifying the Athenian series in question since many private forgeries existed, along with full standard imitations primarily minted by Persian authorities to pay the Peloponnesian crews. Specimens of the pertinent series (Weiser 1989, 267ff. with pl. 16–18) often occur with test cuts.

21 Kraay 1976, 68ff., with pl. 11 nos. 202–203. In the late 360s, gold denominations of 1.5 and 1 obol were struck from votive offerings at Pisa, the site of the Olympian sanctuary, to pay units of the Arcadian League fighting against the Eleans (Xen. *Hell.* 7.4.33): Hoover 2011, 113–114 (with photos).

22 Loomis 1992.

23 Thuc. 8.5.4ff., 8.37.4, 8.58.5.

24 Thuc. 8.29, 45.6.

admiral Lysander in cooperation with the Persian prince Cyrus:[25] The Peloponnesian fleet, severely damaged in 406 in the battle at the Arginusae (southeast of Lesbos), was repaired and expanded with much Persian money. Peloponnesian crews who had escaped to Chios had to work on farms temporarily for their living; when they lost their jobs in late autumn, their plans to plunder Chian farms and villages to secure their maintenance could only be prevented by paying them naval wages for a while, so they could leave and serve again in the new Spartan fleet. With it, Lysander lured the Athenian ships into a trap in 405 (at Aegospotami) that decided the war.[26]

It was over in 404, after nearly 30 years. During such a long period, many men had earned their living as rowers and mercenaries. Now they were commodities for anyone with political ambitions—and the money to hire them. In 401, the younger Cyrus was the first in a series. Xenophon who accompanied the Persian pretender's Greek troops to Babylonia gives detailed reports of the negotiations about pay rates and rewards promised in return for the final victory over the regular Great King.[27] By accepting that, the Greek mercenaries bore a risk, but they had no choice, even when payment was interrupted because of a temporary lack of means.[28]

Philip and Alexander

In Macedon, in 359, Philip succeeded his brother Perdiccas III, who had fallen in a decisive battle against the Illyrians to the west, along with a large proportion of his army. That was not all: in the north, the Paeonians tried to profit from the catastrophe by an invasion, as did several Thracian groups from the east and the Athenians in the center, both supporting Macedonian pretenders and hoping for political concessions from their candidates. And just beyond the Thermaic Gulf the Chalcidians under Olynthian hegemony were an ongoing threat.

Macedon was generally a poor country, and after Perdiccas' defeat not much of its army remained: men were killed, arms and equipment lost. Reforms were indispensable, and Philip secured them by means of "gifts" of clearly considerable value. We have no clue where the money for all that came from. The Upper Macedonian kingdoms in the west and north[29] certainly contributed a substantial share, if only because some of their cities had been lost, and others were utterly threatened. So Philip could well hope for money and fresh troops: after all, it was in his cousins' own interest to reconquer and effectively protect their territories. Winning victories in the west and the north with his reformed army, Philip acquired new resources from the booty, strengthening his authority in Macedonia and with his troops. His next steps also improved his long-term revenue from harbors (Pydna in 356, Methone in 354, then located a mile or so from the coast) and kept the Athenians out, as did the annexation of Amphipolis (in 357), a rich source not only of silver and timber but also tolls.

---

25 For Persians minting tetradrachmas of the Athenian owl-type to pay the Peloponnesian fleet see Weiser 1989, 267ff.

26 Xen. *Hell.* 2.1.1ff., 18ff.

27 Xen. *An.* 1.3.21, 1.4.13.

28 Xen. *An.* 1.2.11–12.

29 Heckel 2017b studies their influence in the whole of Macedonia under Philip.

As he gradually consolidated his financial position, those who had threatened him in 359—with the exception of the Thracians in the east—were also restrained.

Philip turned his attention to this problem in 356, answering calls for help by several places in "the mainland possessions of the Thasians (*Thasion epeiros*)," located on the Thracian coast between the Strymon and the Nestus. Among the most important was Crenides, situated in the hinterland east of Mt. Pangaeum. Thus, Amphipolis ceased to be an exposed Macedonian outpost, and many Thasian settlements on the mainland came under Philip's control and would contribute to his revenue.[30]

Moreover, Philip profited from the gold mines east of Crenides, as his contemporaries observed.[31] In the sixth and fifth centuries, the mines—under the name of *Skapte Hyle* ("Dug-up woodland," in the Lekani area)—had belonged to the Thasians who had exploited them intensively, extracting yearly revenues of gold worth 80 talents (of silver, then the equivalent of about 6 Euboean talents of gold, approximately 157 kg).[32] In 363, the mine district was lost to the Athenians who, after their defeat in 405/404, lacked the means to continue protecting and exploiting it. During what was probably an interval of Thracian supremacy in the region, mining activities decreased or even ceased, until around 360 the Thasians made a fresh attempt that quickly ran into trouble with neighboring Thracians—and caused the appeal to Philip. His intervention, combined with a new engagement in the mining district, not only brought him into permanent conflict with Thracian groups in the Crenides–Lekani area but also, according to Diodorus' source (Theopompus?), put an end to Philip's financial problems (Diod. Sic. 16.8.6–7):

> Afterwards he went to the *polis* Crenides. He enhanced it by many new citizens and changed its name after himself to Philippi. The gold mines on its territory (*chora*) which were totally insignificant and poor in their fittings he improved so much that they yielded a revenue of more than 1,000 talents (26,200 kg, on the Euboean standard). From these he rapidly amassed wealth, and due to the large amount of his money he permanently extended the supremacy of the Macedonian kingdom, for he minted a golden coin that was called after himself *Philippeios* (sc. *stater*). With it, he gathered a large force of mercenaries and induced many Greeks to become traitors of their cities.

Diodorus' source is eager to explain Philip's minting of large amounts of gold staters. Yet numismatic research proves the contrary: according to G. Le Rider, compiler of the authoritative *corpus* of Philip's coins,[33] the minting of gold staters only started more than a decade after the foundation of Philippi. Moreover, they were produced with a very small number of dies, so the output under Philip himself was small.[34] The very large majority of preserved *Philippeioi* are posthumous, minted from Persian gold that Alexander had seized in the Achaemenid residences; further quantities were minted by the Successors from Asiatic metal they collected as tribute. Concerning Philip's Macedonian gold, we may ask whether, after

30 Loukopoulou 2004.

31 Koukouli-Chrysanthaki 1990, with map 2 on p. 501 (based on surveys: gallery mouths indicated by crossed hammers, slagheaps by thin pyramids). The mining district is situated in the North and East of modern Kavala in the Lekani region, not in Mt. Pangaeum.

32 Cf. Hdt. 6.47. 2, referring to another mine on the island itself.

33 Le Rider 1977.

34 Le Rider 1977, 428ff., Le Rider 1999, followed by e.g. Kraay 1984, 14; Price 1991, 85ff.

many decades of Thasian and Athenian exploitation of *Skapte Hyle*, a substantial proportion came from the mine that he controlled in the Chalcidice, northwest of Stagira, after his victory over the Chalcidians in 348, at the latest. They had minted staters, presumably with gold from these mines, as the inhabitants of Amphipolis had done with gold from mines in the Strymon/Pangaeum area, and the people from Crenides with the first fruits of the Lekani mines, all of them in very moderate numbers.[35] Philip's wealth in gold cannot have substantially exceeded the previous amounts.

The costs of his military engagement in Thrace, for building forts and keeping them manned, as near modern Kalyva above the Nestus valley and thence to the west to block the corridor from the Nestus to the northern plain of Philippi (near modern Drama[36]) were financed by other means: by tribute, e.g. from the Chalcidian cities since 348/347, by selling booty and slaves, by exacting tolls and harbor taxes, and, of course, the traditional trade in timber.[37]

His activities in Thessaly and central Greece were presumably financed by those who summoned him. He himself had to bear the costs of the campaigns to the east, conducted on his own initiative. Expensive projects failed, such as the sieges of Perinthus and Byzantium in 340/339; had he been successful, he would have gained control of the Bosporus with its tolls. The project anticipated the only realistic solution for the problem caused by Philip's very substantial military costs: regular tribute paid by the Greek cities in Western Asia Minor. The First Athenian League provided the blueprint. Athens had lengthy experience in the region and special relations with many cities there, so Philip courted the *polis*—despite her most influential orator. Philip did not, however, succeed in cooperating with Athens, and Demosthenes continued to complain. The eventual war between Philip and Athens was in part the result of the failure of the two sides to solve this question of military costs.

Alexander[38] started with rather empty coffers, though we need not believe that he inherited from Philip only a few gold and silver cups and not more than 60 talents.[39] When he crossed the Hellespont in 334, he had but 70 talents at his disposition, according to Aristobulus, and borrowed another 200, according to Onesicritus.[40] Underestimated as these figures may be, they show that Alexander was short of money at the outset,[41] the more so if he had to pay back loans made to his father and himself. This might explain why he continued minting the old *Philippeioi*-staters along with his own new gold type (both from

35 Kraay 1976, pl. 26, 479 (Chalcideans); pl. 28, 509 (*Thasion epeiros*: Crenides) and 510 (*Philippon*: Philippi), cf. SNG ANS 7.1 no. 663f.; Gaebler 1935 Taf. 1, 1f. (Amphipolis).

36 Literature on Philip's forts is scanty and generally restricted to references in archaeological circulars; see Archibald 1994, 468–469, 1998, 234. In map 51 of the *Barr. Atlas* (D2) five bolt-like linear arrangements are indicated, each indicating three forts, across the Nestus valley and in the corridor between the Lekani mountains (Sapaike) and Rhodope. Yet since some of these forts were still manned in the Byzantine period, without further archaeological research it is impossible to know how many go back to Philip. To the southwest of modern Kalyva there was at least one in a Classical ashlar construction (autopsy in 2000, then deep in the forest and accessible only on a dirt track). The concentration illustrates the significance of the route to the west, especially in Philip's days when the future Via Egnatia near the coast had not yet solved the problems in the swamp-covered delta of the Nestus.

37 Bissa 2009, 31ff. (precious metals), 117ff. (timber).

38 An approach: Kremydi-Sicilianou 1999.

39 Arr. *Anab.* 7.9.6 (from the Opis speech!).

40 Plut. *Alex.* 15.2.

41 Heckel 2008, 41.

333/332 on, from Persian booty), on identical standards: possibly it had been stipulated that the debt be repaid only in *Philippeioi*.[42]

Hellenistic War Costs: Kings and Cities[43]

Alexander's Successors and then the kings of the Hellenistic dynasties in Asia and Egypt disposed of substantially higher sums than did the Greek states in Europe. Much of their revenue went to their armies and fleets,[44] partially for new and utterly expensive types of military equipment such as war elephants[45]—at Ipsus, Seleucus employed 480 from his total of 500.[46] Huge contingents of mercenaries[47] were hired in times of tension, and dismissed after decisive battles. The core troops were permanent, often granted parcels of land (*kleros*) in military settlements attached to civilian towns (*klerouchiai*).[48] There they spent the winters, and non-campaigning periods, with their families,[49] but could be mobilized quickly whenever necessary. Such places played a role not only in colonization but also in regional strategy and economy. During campaigns, troops were often billeted in towns and villages, at the expense of the civil population; limitations or even exemptions were a privilege that had to be sought for from the king himself.[50]

Cities, too, often waged war, at times even in regional conflicts. At the very end of the third century Philip V of Macedon, who demanded for himself large parts of the Aegean and Caria as the heritage of his family, had built up a strong fleet to conquer these regions. An alliance directed by Attalus of Pergamum and the Rhodians prepared for defense (Polyb. 16.2–12). Although it is not explicitly stated,[51] the city of Cos was probably among the allies. Since a war fund was either unavailable or insufficient, the city collected money from citizens and inhabitants. Three steles set up in different public places recorded exactly who had contributed how much, in the first section without destinations for the sums paid (a large proportion must have been spent on mercenaries), then for ration money. Best preserved is the first stele (*IG* XII 4,1, 75.7ff.):

> It has been resolved (by the Council): Those who have committed themselves (to pay a share) ... to instantly announce their names to the Assembly (*ekklesia*), and the People (*damos*) shall vote by a show of hands about the fair amount of the donations (*doreai*) which, when resolved, shall be (immediately) raised.

To each name the donated sum is added, from 8,000 drachmas (two men, one of them the petitioner) down to 30 drachmas (1 man; in most cases between 100 and 500 drachmas),

42 Price 1991, I 106.

44 For general considerations, see Austin 1986; Anson 2014, 189–195, for special data Champion 2014, 152–162, for a model calculation of revenues and war costs of Antigonus Monophthalmus, ib. 183–186.

45 Sabin 2007, 419–421; Alonso Troncoso 2013. See Chapter 15 in this volume.

46 Diod. Sic. 20.113.4, cf. Plut. *Demetr.* 28.6. Anson 2014, index s.v. "elephants"; Champion 2014, 157.

47 Griffith 1935; for discussion since then see Krasilnikoff 1992; Trundle 2004.

48 Billows 1995, 160–178.

49 Loman 2005.

50 Heinrichs 2018, 272–311.

51 Kleu 2015, 147.

one more example of the age-old principle that material engagement in a community pays off in political authority—and contributes to the stabilization of public finance. Lacunae in the text make it impossible to calculate the totals of men and money, substantial in either case.

The pay rates for soldiers/mercenaries (*misthophoroi*) are not specified, but in the third century we may assume 6–8 obols per day for the rank and file. As explicitly noted (l. 315ff.), ration money (*siteresion*[52]) was paid as well, being calculated per year on a basis of nine months of service. A common soldier received 3.36 obols per day, some commanders approximately double that amount (1 drachma), and one man nearly eightfold (4 drachmas). If we compare Seuthes' offer in 400 to the Ten Thousand (Xen. *An*. 7.36)—one Cyzicenian (gold coin) per soldier (*stratiotes*) per month, twice as much for a captain (*lochagos*), and the fourfold for a general (*strategos*)—we may conclude that at Cos the eightfold rate was paid to the commander-in-chief. Monthly pay rates of 30–40, 60–80, and 240–320 drachmas can thus be calculated. Ration pay was extra: (100 obols being) 16.66 drachmas for the rank and file, 30 for commanders, and 120 drachmas for the commander-in-chief, adding up to average monthly composite rates (*opsonia*) of 51 drachmas and 4 obols, 100 and 400 drachmas respectively. Both in Hellenistic kingdoms and cities war costs were major and at times even dominant economic factors—just as they had been already in the Mycenaean kingdoms.[53]

Concluding Remarks

Generally speaking, military costs rose in the course of Greek antiquity, and they also become more complicated, for wages and ration pay were added to the increasing cost of equipment, which itself became more complicated as equipment became more elaborate. These changes reflected, and contributed to, the gradual monetization of the Greek economy, and they also reflected the increasing size of Greek political entities, along with an increasing urban population and a rise in the number as well as the size of Greek cities. These demographic changes made larger armies possible, and thus increased the demands being made on Greek military budgets. Scanty though the sources for war economics are, they do tell a story of expansion, monetization, and diffusion. Economics became an ever greater part of war.

52 Burrer 2008; cf. Couvenhes 2006.

53 See Chapter 1 in this volume.

CHAPTER 21

War and Slavery in the Greek World

Peter Hunt

Introduction

Classical Greece was one of the few true slave societies known to history. Especially during the classical period, slaves constituted a large proportion (>20%) of the population and were crucial to the economy of city-states throughout the Greek world. The recent tendency to view the Helots as some type of slaves, if accepted, would make the role of slavery in ancient Greece even more pervasive. Hence, connections between slavery and war, another widespread and central institution, are *a priori* probable. And, indeed, slaves played various roles in war, and war influenced both the practice and ideology of slavery. Three connections stand out.

First, slaves fulfilled crucial economic roles and contributed to the wealth that allowed Greek states to fight more effectively and intensely. This role was particularly important in the Classical period as the role of money in warfare rose in importance, most strikingly at Athens where the state's wealth allowed it to pay for large navies to conduct lengthy and distant campaigns. Slaves also filled military support roles: most conspicuously one slave typically served as an attendant and porter for each hoplite on campaign. But a large slave population could also be a weakness. Belligerent states sometimes—perhaps regularly—encouraged slaves to desert when they invaded an enemy's territory. In some circumstances, armies could get help from or even foment uprisings among unfree populations, most conspicuously the discontented Helots.

Second, slaves constituted a large pool of potential military manpower: for example, male slaves at Athens outnumbered male citizens. Practical and ideological problems surrounded the recruitment of slaves as combatants, but the pressures of war often forced Greek states to overcome these hesitations and scruples. In particular, the Athenian navy regularly included

A Companion to Greek Warfare, First Edition. Edited by Waldemar Heckel, F.S. Naiden, E. Edward Garvin, and John Vanderspoel.

a significant proportion (>20%) of slave rowers and the crews in the navies of other cities could comprise an even higher proportion of slaves. The promise of freedom induced some Helots to serve as professional soldiers for Sparta, especially on long and distant expeditions, which the Spartans themselves tried to avoid. At some major battles Helots, presumably light-armed, were present in numbers greater than those of the Spartans, but their role in the fighting is obscure.

Third, capture in warfare was the most conspicuous way that people lost their freedom. Whether war was actually the most important source of slaves is a more difficult question, but there is no doubt that warfare not only contributed to the number of slaves but also influenced Greek attitudes toward slaves.

A Source of Power or an Achilles' Heel?

The crucial role of slaves in the ancient economy is most obvious in Athens, about which we have the best information. First of all, silver was by far Athens' biggest export and the vast majority of the laborers, both underground and in ore processing, were slaves. Silver production alone may have employed more than 20,000 slaves at its peak and, famously, paid for a large part of the fleet with which the Athenians won the battle of Salamis.[1] Second, although Athens had to import grain to feed itself, agriculture was still the largest single sector of its economy. The predominance of slave labor on the farms of the rich is undeniable and is manifest, for example, in the advice on slave management in Xenophon's *Oeconomicus* (e.g. 7.35, 11.14–16, 11.23, 12.3–12.20, 13.6–14.1). Mid-sized farms, mainly owned by men liable to hoplite service, also employed a few slaves each.[2] So the main characters in Aristophanes' plays are often middling farmers and hoplites. They distinguish themselves from the rich, but nevertheless they typically possess from a couple to ten or more slaves.[3] Third, the role of slaves in urban crafts is known both from lists of works on public building projects, such as the Erechtheion temple, where they constitute about a quarter of the work force of known status (Randall 1953, 201), and from references to the labor force in large workshops—some producing arms—such as the shield shop of Lysias and his brother, which employed up to 120 slaves, and the sofa and sword workshops of Demosthenes' father, whose workers must have constituted most of his 50 slaves (Lys. 12.19; Dem. 27.31). Due in large part to our poor evidence, historians differ widely about the proportion of the population of Attica who were slaves. I believe that they constituted between 25% and 45% of the population, but other scholars argued for estimates outside of that (already wide) range (Hunt 2018a, 66–69).

We have even less knowledge about the extent of slavery in other Greek cities, but our meager evidence suggests that Athens was not that exceptional. The approximately 1,000 captives from a Corcyraean naval defeat consisted of 80% slaves (Thuc. 1.55.1). If this percentage held for the entire Corcyraean fleet of 110 ships, then Corcyra would have had to possess more than 17,000 *male* slaves out of a population of 70,000 at the very highest (Hansen 2006b, 95–96). But Thucydides describes yet another island, Chios, as having more

1 Isager and Hansen 1975, 42–49; Lauffer 1979, 142; Rihll 2001. Hdt. 7.144.
2 Jameson 1992; Osborne 2010; *contra* Wood 1983.
3 Ehrenberg 1974, 165–168; Garlan 1988, 61. On slaves in Greek comedy see Akrigg and Tordoff 2013.

slaves—proportionally one assumes—than any Greek state other than Sparta (8.40.2). Although Athens was relatively wealthy and probably possessed a higher proportion of slaves than most other Greek cities, all of them seem to have considerable slave populations.

Thucydides' description of the number of Chian slaves implies that Sparta had even more slaves. Imported chattel slaves were rare or non-existent at Sparta, so Thucydides is following a common Greek habit of referring to the Helots as slaves, certainly the usage the Spartans preferred. I believe that the Helots are best classed as serfs rather than slaves, but this may be becoming a minority opinion.[4] In any case, their participation in war was more important than that of chattel slaves in most Greek city-states; this justifies the category of the "unfree," which Karl-Wilhelm Welwei employed in his magisterial treatment of the role both of slaves and Helots in ancient Greek warfare (Welwei 1974). And populations like the Helots existed in other Greek states (Van Wees 2003). Thessaly with its *penestai* is only the most conspicuous example.

The connection between the economic role of the Helots and Sparta's war-making capacity is particularly clear and direct. The Spartan elite were able to focus their lives on military training because Helots took care of the agricultural production on which the economy was based. The Helots were subjected to share-cropping: they had to contribute a large portion, perhaps one half, of their produce to their Spartan masters. Some supervision was no doubt necessary, but for most of the archaic and classical period this system enabled the Spartans to be the only professional soldiers in Greece. Their king, Agesilaus, reportedly once responded to his allies' complaints about the small number of Spartans on an expedition by saying, "we provide more *soldiers* than you do," and he had some justification for this tactless claim (Plut. *Ages.* 26).

The way that slave labor contributed to military strength in other city-states was less direct, but its impact was still great. War in the classical period became more and more a matter of wealth (see also Chapter 20 in this volume). Money was needed to build and man ships, to build walls, and to pay soldiers during long campaigns, especially sieges. From the archaic period on, dynamic and thriving economies allowed the intensification of warfare throughout Greece. Our evidence suggests that slavery played a major role in these dynamic economies. The presence of slaves also made it possible for production and farming to continue even when many free citizen males were away at war, but we should not oversimplify the various ways that this would work in practice.

Slaves contributed to the ability of Greek states to wage war. They also constituted a vulnerability for Greek states because they did not, in general, want to be slaves. The conditions of slaves in classical Greece varied greatly and some slaves were content enough with their positions and prospects so that they did not attempt to escape. On the whole, slaves' tendency to vote about slavery "with their feet" provides strong evidence for their widespread discontent: references to fugitive slaves as well as to the precautions and countermeasures of masters are sprinkled throughout every genre of ancient Greek literature as well as in our archaeological evidence: for example, ancient farms sometimes had stone towers that locked from the outside.[5] Slaves were particularly prone to run away in wartime or even on occasion to collaborate with the enemies of their city.

4 Luraghi 2002; Lewis 2018, 125–146; *contra* Hunt 2016.

5 Hunt 1998, 102–115; Morris and Papadopoulos 2005.

Greek city-states were relatively small, so slaves only had to cover a short distance to escape to another political entity: a slave from the city of Athens needed only to cover about 30 miles to get to Megara. But, in peacetime, escape to a nearby city would not guarantee a slave's freedom. We can assume a fair degree of cooperation between slaveholders, and in every Greek city they possessed power and had a stake in making flight a dangerous and unattractive option for slaves. Our best evidence for this cooperation is a reference to its breakdown: one of the grounds of complaint of the Athenians about Megara before the Peloponnesian War was that the Megarians were "harboring" runaway slaves (Thuc. 1.139.2). Presumably, the Athenians expected that the Megarians, were they not hostile to Athens, would not harbor runaway slaves but would assist their return.

Open war changed this expectation entirely. Masters could not even pursue their runaway slaves into a hostile state—as Megara was hostile to Athens during the Peloponnesian War. In one fourth-century case, a master pursuing runaway slaves by sea—presumably having obtained some information about their whereabouts—was himself captured by hostile privateers and enslaved (Dem. 53.6). In Aristophanes' *Clouds*, a character complains that he cannot even beat his slaves "on account of the war," that is, they would run away if they were treated harshly (Ar. *Nub.* 6–7). Most telling, in Aristophanes' *Peace* (421 BCE) the chorus lists slaves contemplating flight as one of the classes of people who oppose making peace (Ar. *Peace* 451). Escaping slavery was even easier during an enemy invasion or siege. Most striking, when the Spartans established a permanent fort at Decelea in the last third of the Peloponnesian War, Thucydides reports that over 20,000 slaves, most of them skilled, ran away (7.27.5). His emphasis on the slaves' skill highlights the economic damage caused by their flight.

Philo of Byzantium, a Hellenistic collector of military stratagems, even advises besieging armies to promise freedom to the slaves of the beleaguered city (*Poliorketika* 5.4.14–15). His stated goal is not the economic harm that slave desertion caused. Rather he explains that, after such a proclamation, the city will not be able to arm its slaves and will have to feed them better. Slaves were a considerable part of the population and, during a siege, the less they ate, the more the citizens had to eat. Philo supposes that the promise of freedom would prevent the citizens from starving their slaves for fear of their desertion. Philo also refers to the arming of slaves as if it were a regular practice during sieges—a remark we shall revisit later. Its success may have depended on the citizens convincing the slaves—or at least some groups of slaves—that the invaders were the enemies of all and that, in this situation at least, slaves and masters had a common interest in beating them off. Philo assumes that promising freedom would undermine the citizens' chance of persuading slaves that this was the case. The masters within the city could counter the announcement with their own promise of freedom for loyal slaves who fought for the city. The asymmetry remained that a slave's freedom was a costly proposition for his or her master, but not for an invading army.

Different circumstances could lead to different tactics. In 373, Mnasippus, a Spartan general, attacked Corcyra, gained control of the countryside, and invested the city (Xen. *Hell.* 6.2.5–26). As its population began to starve, many of them, including slaves, fled from the city. Mnasippus wanted to maximize the number of hungry mouths within (cf. Xen. *Hell.* 2.2.2) and so he made a proclamation that any deserters, regardless of status, would be sold as slaves. Even this did not prevent starving people from fleeing the city and so Mnasippus eventually drove them from his lines "with the whip." Even the money to be gained by selling people as slaves was less important than starving the city into submission. The Corcyraeans were equally brutal in their calculations. They did not allow the slaves among the deserters back within their walls and many died outside, presumably of starvation, thirst, or exposure.[6]

On occasion, a state could do better than just damaging an enemy's economy by encouraging desertion; they could incite revolt. The Athenians did this to Chios in the last period of the Peloponnesian War. Upon hearing of the revolt of Chios, the Athenians freed the slave rowers in the Chian naval squadron serving with them; they imprisoned the citizens (Thuc. 8.15.2). Although Thucydides does not note a connection, this unusually generous treatment of the slave rowers was probably a prelude to the tactic the Athenians adopted in Chios. Thucydides explains that, "the slaves of the Chians are more numerous than any other single city except for that of the Lacedaemonians. At the same time, on account of their number, they are punished harshly for their crimes" (8.40.2). So, when the Athenians tried to put down the revolt of Chios and set up a fort there, most of the slaves there not only deserted to the Athenians but themselves did the most damage to Chians' farms in the countryside (Thuc. 8.40.2). Indeed, 20 years later, the Athenian general Iphicrates reportedly convinced the Chians to comply with his demands by spreading the rumor that he was collecting weapons to arm their slaves against them (Polyaenus, *Strat.* 3.9.23).

Better known is the long history of outside states encouraging revolts among the Helots and Messenians subject to the Spartans. Sporadic signs of this appear in the archaic period, but our information gets better in the mid-fifth century. Then the Spartans called on their allies to help them put down a Helot revolt centered on Mt. Ithome. These allies included Athens, but the Spartans grew suspicious of the Athenians' sympathies and were afraid they might "do something radical," that is, help the Messenians (Thuc. 1.102.3). The Spartans sent the Athenian contingent and only the Athenian contingent home, an insult that represented a critical step toward hostile relations between Sparta and Athens. When the Messenian rebels finally left the Peloponnese under a safe conduct on condition of never returning, the Athenians settled them as citizens at Naupactus, a city on the Gulf of Corinth, which the Athenians had captured and later used as a naval base. During the Peloponnesian War, the Athenian general Demosthenes, who had spent more than a year at Naupactus and had served extensively with the Messenians there, pressed for and accomplished the fortification of Pylos—with crucial support from the Naupactine Messenians (Thuc. 3.94–98, 3.107–108, 3.112, 4.9.1). Pylos not only served as a magnet for Helot desertion but was also a center for raids and ravaging by Messenians from Naupactus, who provided a garrison (Thuc. 4.41.2–3). As a result, the Spartans feared a more general uprising. Thucydides lists worries about Pylos as one of several reasons that the Spartans were inclined to make peace with the Athenians, "the Peace of Nicias," in 421 (Thuc. 5.14.3). Thebes' liberation of Messenia in 369 and the fortification of the city of Messene represented the final and decisive success of the strategy of damaging Sparta by taking advantage of the discontent of its Helots. The great Theban general Epaminondas boasted of these accomplishments on his epitaph: "By my strategies Sparta was deprived of its glory, and holy Messene received back its children at last" (Paus. 9.15.6). Even after Messenia was free, Aristotle described the remaining Helots in Laconia itself as "waiting to take advantage of any Spartan defeat," a situation made more dangerous by the fact that Sparta's neighbors supported them (Arist. *Pol. II* 1269a37–b7; cf. 1272b18–20).

6 Xen. *Hell.* 6.2.15. Parallels include the Caesar's siege of Alesia (Caes. *B Gall.* 7.78).

Since all Greek states possessed substantial unfree populations, any incitement of slave desertion or discontent risked retaliation. This risk—that the tactic would lead to escalation rather than to any great advantage—must have curbed some states. Thus, we usually hear about the incitement of desertion when there was some asymmetry involved—against Sparta, for example, with its large population of discontented Helots. And historically the threat of future escalation has only sometimes prevented a state at war from seizing an immediate advantage. In some of the cases that we have discussed, we hear about a practice because it was noteworthy and unusual. But in other passages, we find no indication that anything radical or unprecedented is going on. Our limited evidence seems, on the whole, to suggest a normal pattern in Greek warfare rather than special measures.

Slaves in Land Warfare

The use of slaves as support personnel on campaign falls between the function of slavery as the economic basis of Greek cities' military capacity and the participation of slaves in the actual fighting. One slave attendant typically accompanied each hoplite on campaign (e.g. Thuc. 3.17.4). The attendant would help carry the hoplite's gear and particularly his armor, so heavy and uncomfortable that a hoplite would not don it until just before the actual fighting began (Hanson 1989, 60). These attendants were usually the hoplites' own slaves. In one law-court case, a defendant, a guardian accused of defrauding his wards, reportedly forced his young nephew to serve as his attendant on campaign. This is depicted as a typical example of his brutal conduct toward his family, since the nephew was performing a duty usually done by a slave (Isae. 5.11).

As the Athenian siege of Syracuse floundered, the slave attendants of both hoplites and cavalry began to desert to the enemy (Thuc. 7.75.5). In more normal conditions, slaves in these roles could help their masters fight in one way or another. One stratagem required grooms to join a cavalry formation holding up lances, presumably spares, or imitation lances to make the number of cavalry look larger to the enemy (Xen. *Eq. mag.* 5.6). Hoplite attendants could throw stones, dispatch injured enemies, and help devastate and plunder the enemy's countryside. By their very presence on campaign and at battles, such attendants incurred risks. Indeed, one official Athenian casualty list includes a slave attendant (*IG* I^3 1144 line 139–140: cf. Paus. 1.29.7). But they were not essentially combatants.

Before we turn to the topic of slaves as actual combatants, one preliminary consideration requires our attention. The high prestige of military service was at odds with how the Greeks viewed slaves. A strong strain of Greek thinking linked military prowess, bravery, and service or sacrifice for the community with the highest ideals of manhood and thus with the prerogatives of citizenship. Slave participation in warfare was awkward in terms of this ideology. For, although the notion that slavery was simply a matter of bad luck and implied no inferiority can be found, it was more common to look down upon slaves and in particular to despise their supposed lack of courage.[7] This stereotype drew on the fact, discussed later, that many slaves had lost their freedom in war. Some historians, aware of

7 This prejudice only applied to male slaves and so it was female slaves that Greek authors depicted when they wanted to show slavery as an undeserved disaster (Hunt 2011, 36–38).

these strains of thought, approach our evidence presupposing that slave participation cannot have been frequent: it was too much at odds with Greek thinking about slavery and military service. I have argued elsewhere that this was an awkward topic for our sources to dwell on and we should approach our evidence with the expectation that slave participation has been underreported as a result.[8]

These considerations do not pertain equally in all cases, since the participation of slaves in warfare was not a single unitary practice. It is often impossible to generalize, but rather we need to try to consider each case individually, and in particular to distinguish between the role of chattel slaves versus Helots, the use of slaves in sieges or armies or navies, the standard employment of slaves versus emergency measures, and the connected issue of whether the slaves were promised their freedom for their service or remained slaves. I shall briefly consider first the emergency role of slaves in land warfare, especially sieges, then Helot soldiers, and then slave rowers on warships.

One goal of proclaiming a promise of freedom for the slaves of a besieged city—a tactic we discussed above—was to prevent the city from arming its slaves. This implies that the military use of slaves in sieges was common. Such a tactic would have involved some risks—as a story about a slave revolt in Syracuse during the Athenian siege suggests (Polyaenus, *Strat.* 1.43.1.)—but a siege was almost always an extreme emergency with the massacre of the males and enslavement of the rest of the population a possible outcome. The danger of the situation may often have trumped the danger of arming slaves.

Rarely do we hear of slaves armed to fight on land and the details are sparse and sometimes suspect. Pausanias mentions the grave of the slaves who fought with the Athenians at the battle of Marathon and received public burial there.[9] Herodotus records that slaves—admitted to citizenship and presumably armed after a disastrous loss—took over Argos after the calamitous defeat by Sparta at Sepeia (490s) and were driven out years later.[10] On the one hand, fear of such an occurrence may have discouraged cities from arming their slaves. On the other hand, war is often an emergency and our sources are not so detailed that we would expect the arming of some slaves to always be reported.

Civil war, *stasis*, in the Greek city-states was usually bitter and hard fought, since the fate of the defeated was usually death or exile, with confiscation of all property in either case. Thucydides tells us that at an early stage of the Corcyraean civil war, both sides, the democratic and oligarchic factions, sent messengers into the countryside to promise freedom to those slaves who joined their side (Thuc. 3.73). Most of the slaves went with the democrats, perhaps out of hatred of their masters, mainly the wealthy owners of large farms. Thucydides presents his detailed description of Corcyra as a model and precursor of the *stasis* that infected many Greek cities (Thuc. 3.82.1), so we may suspect similar tactics in the many cases of *stasis* about which we have no detailed account. About a century later, the League of Corinth forbade the freeing of slaves for political purposes, so it was presumably still a worry and perhaps often a reality (Dem. 17.15, cf. 10–11).

8 Hunt 1998, 1–7; cf. Van Wees 1995, 160–161.

9 Paus. 1.32.3 with Notopoulos 1941; Lacey 2011, 129–130.

10 Herodotus 6.83 with Scott 2005, 579–588; Hornblower and Pelling 2017, 199. See also Arist. *Pol.* V 1303a6–8.

Helots

Sparta's Helots indirectly contributed immensely to its military power by freeing the ruling class of Spartiates from agricultural labor. They also took part directly in land warfare to a greater extent than did the chattel slaves of other states. This may at first seem puzzling, since the Helots were notoriously rebellious, as we have already explored. Part of the explanation for this apparent contradiction may lie in different groups among the Helots: for example, the distant, discontent, and rebellious Messenians as opposed to the Laconian Helots, more closely tied to their Spartan masters and hoping to improve their lots by cooperation rather than resistance. For various reasons this cannot be the whole story, but history provides other examples of discontented groups recruited into the military when a state decided that the advantages outweighed the risks, however real. That the recruitment of Helots was dangerous cannot be denied: for example, Xenophon writes that on campaign the Spartans guard their arms dumps "from friends not enemies" and later refers to the "friends" as "slaves," that is to say, Helots. He adds that the same motivation—not worries about the enemy attacking—is behind the rule that soldiers carry their spears with them at all times (Xen. *Lac.* 12.2, 12.4). But, despite the obvious risks, Helots played important roles in Sparta's military. These fall into three main categories.

First, Helots were present on every Spartan expedition as hoplite attendants, as various stories and practices attest: thus, a Helot led his blind Spartan master into battle at Thermopylae (Hdt. 7.229). Although we associate the battle of Thermopylae with the 300 Spartans, Herodotus reports 4,000 Greek casualties there and writes that "naturally some were Helots"—his numbers would imply about 2,000 Helot casualties.[11] On an earlier occasion, Cleomenes ordered the Helots on campaign to set some woods on fire, killing the Argive soldiers who had retreated there and completing Sparta's devastating victory at the battle of Sepia (Hdt. 6.80). The Helots may have been more numerous and played a more active role than the attendants in other armies. Or were some of them actually light-armed soldiers?

My second category includes a couple of cases where our sources report explicitly that the Helots were recruited *en masse* for a major battle, at Plataea (479) and at Mantinea (418).[12] At Mantinea, Thucydides simply reports that the Helots were mobilized in full force, *pandemei*, for the battle (Thuc. 5.64.2; cf. 5.57.1), but he does not describe what such a considerable force did during the battle. In the case of Plataea, Herodotus states that in attendance on each Spartiate were seven Helots, "every one armed for battle" albeit "light-armed."[13] Herodotus even includes the Helots with the Lacedaemonian and Tegean contingents, who alone fought the Persians after many of the Greeks had retreated.[14] Of course, Helots died in the battle. These received their own tomb; the rest were rewarded by being allowed to collect the spoils, some of which they embezzled (Hdt. 9.85.2, 9.80.1–3).

11 Hdt. 8.24.1, 8.25 with Van Wees 2004, 61, cf. 83–85. In fact, the Helots may have been present at Thermopylae in the same seven to one ratio as at Plataea (Hunt 1998, 31–32).

12 Taking so many potentially rebellious Helots on campaign or leaving them behind may have been equally dangerous for the Spartans.

13 Hdt. 9.29.1; see also 9.10.1, 9.28.2, 9.61.2.

14 Hdt. 9.61.2; the Athenians fought against the Persians' Greek allies.

Some historians emphasize that the Helots were lightly armed and argue that they did not play much of a role. The Helots could have skirmished before the battle, hung on the flanks of the hoplite formations during the battle, and participated in the pursuit. But it was the hoplites—Spartan, *perioikoi*, Tegean, and Athenian—that did the real fighting.[15] Or perhaps Herodotus was wrong to place the Helots so definitively on the battlefield and they were mainly guarding supply lines (Welwei 1974, 123). Our ancient sources often neglect the role of light-armed troops at the expense of the hoplites and this may contribute to our perplexity about what they were doing (Van Wees 2004, 61–65). I have argued that the Helots were part of the same formation as the Spartiates, which would not then have been composed entirely of hoplites (Hunt 1997; cf. Van Wees 2004, 69, 177–183). Herodotus' ratio of seven Helots to one Spartiate was an estimate based on the common model of the hoplite phalanx as eight ranks deep and his belief that only the front rank were Spartiates. This mode of calculation is consistent with Herodotus' methods elsewhere and closely parallel to Thucydides' method of calculating the size of the Lacedaemonian army at Mantinea (Thuc. 5.68). It also coheres with other evidence that suggests a strong distinction between the front rank and the rest of a Lacedaemonian phalanx.[16] Nonetheless, my theory requires us to radically rethink how much we actually know about the rear ranks of a hoplite formation and some historians have found it implausible.[17]

Third and finally, Helots served as hoplites in their own special detachments. These Helots, known as *neodamodeis* ("new citizens") or Brasidians (after their commander Brasidas) earned freedom from Helotage and sometimes even farmland as a reward for their service (e.g. Thuc. 4.80.5, 5.34.1). The Spartans relied on them because they could go on lengthy and distant campaigns, which the Spartans themselves were reluctant to undertake in large numbers. Helot soldiers made good this crucial weak point in Sparta's ability to acquire and control an empire. The *neodamodeis* are best attested during the late fifth and early fourth centuries, but must have existed before (Thuc. 5.34.1; cf. 4.80.3). For example, Spartan support for Syracuse, when the Athenians were besieging it, consisted of 1,000 *neodamodeis* and 2 Spartiate commanders. At certain points in the fourth century, there were more *neodamodeis* in active service than there were full Spartan male citizens in total (Hunt 1998, 170–175). Freeing Helots represented for Sparta a loss of agricultural labor, but these Helots could eventually be settled in places where they would bolster Spartan power, for example, at Lepreon, whose independence Sparta was asserting against the claims of Elis (Thuc. 5.31, 5.34.1).

The promises were made because no one expected Brasidas and his men to return alive. Two thousand helot volunteers were simply killed.

15 E.g. Rahe 2015, 309–323 with 384 fn. 55; Cartledge 2013b, 106–107.

16 Isoc. *Archidamus* 99. Compare also the original use of the Sacred Band in the front rank of the Theban phalanx (Plut. *Pel.* 19.3; Diod. Sic. 12.70).

17 E.g. Flower and Marincola 2002, 159; Lazenby 2004, 261 fn. 15; but cf. Van Wees 2004, 181; Sekunda 2013, 212.

Slaves in the Navy

Compared to land warfare, naval battles often required larger numbers of men and put them at a greater risk. Already at Salamis, the Athenians manned 180 ships requiring about 36,000 men (Hdt. 8.44.1)—and this without being able to draw on the Aegean and Ionian allies that helped man its Peloponnesian War navies, since these areas were under Persian control. Not only during the Persian invasions, but in general, cities were under great pressure to man as many ships as they could afford. Indeed, historians are often surprised at the large navies that islands like Chios, Corcyra, and Samos could muster. The casualties in naval warfare were also greater than in land warfare (Strauss 1986, 80, 179–182). As a result, Greek cities often used all available sources of manpower to man their navies. Slaves in particular provided a proportion of the rowers in classical navies ranging in reported cases from 20% to 80%.[18]

Greek naval warfare depended largely on the ability of triremes to move quickly, to maneuver, and to ram their enemies. In this sense, all rowers were combatants and we find frequent references to their morale or skill as crucial to the outcome of battles (esp. Thuc. 2.83–2.92). That rowers did not carry weapons or wear armor eliminated many of the dangers that the use of slaves in land warfare presented. In fact, Aristotle reassures his readers that it was not necessary that the rowers of triremes be of the citizen class—presumably in contrast to hoplites and cavalry who should be—because the rowers could be kept in line by the armed marines on board, who were indeed citizens (Arist. *Pol.* VII 1327b8–11). It might still seem difficult for the 30 or so officers, specialists, marines, and archers to control the 170 or so rowers on board a typical trireme, were they all slaves. But, although the proportion of slaves in a given navy varied depending on the city-state and on particular circumstances, the rowers were never all slaves. Often slaves served alongside equal or greater numbers of free rowers, either from the city manning the navy or foreigners rowing for pay. In many navies, the problems of control envisaged by Aristotle never came up. We are most definitely not talking about "slave galleys" with a crew consisting entirely of miserable slaves chained to the benches and motivated by whipping.[19] Now let us turn to the evidence for slave rowers and consider first non-Athenian navies.

Among the passages that mention the status of rowers in non-Athenian navies, all imply the participation of slaves. Before the Peloponnesian War, at the naval battle near Sybota, the Corinthians captured over 1,000 prisoners from the Corcyraean navy: fully 800 of these were slaves, whom they sold (Thuc. 1.55.1). This seems a high proportion of slaves, but in a description of the discontent in a large Peloponnesian navy late in the Peloponnesian War, Thucydides makes the remarkable statement that "insofar as most of the Syracusan and Thurian crews were free men, they most boldly rushed at [Astyochos the admiral] and demanded their pay" (Thuc. 8.84.2). From this we can infer that some members of these Syracusan and Thurian crews were not free men, that is, they were slaves. More remarkable is the additional implication that free men were not the majority on the other crews in this large Peloponnesian navy of 112 ships (Thuc. 8.79.1, 8.87.4). Other passages reveal the

---

18 Diod. Sic. 14.58.1 reports that the tyrant Dionysius of Syracuse freed slaves and manned 60 ships (requiring about 10,000 rowers) with them, but I do not believe that we can infer a navy manned entirely by ex-slaves from such a brief report.

19 Cf. Hale 2009, xxvi–xxvii.

presence of slaves in Corinthian and Chian navies.[20] In sum these passages attest to slave rowers, from the beginning of the Peloponnesian War to its end, and in the four most powerful navies after the Athenian: the Corcyraean, the Corinthian, the Syracusan, and the Chian.[21] Since no ancient source states or implies the contrary, the conclusion that slave rowers regularly served in non-Athenian navies is certain.

This point has two important consequences for the more controversial question of slave use in the Athenian navy.[22] First, those who argue against the use of slaves in the Athenian navy need to show that the Athenian navy was exceptional in this respect.[23] Second, there cannot have been any insurmountable practical difficulties that prevented the use of slaves in naval crews. Corinth, Corcyra, Syracuse, and Chios were great naval powers with major victories as well as defeats against the dominant Athenians, so we can rule out arguments such as "rowing was hard and skilled work, so slaves couldn't do it." There cannot have been any *a priori* reasons that navies with some slave rowers could not fight effectively. And, as it turns out, several pieces of evidence show that Athens was not exceptional.

First and most striking, our only surviving, relatively complete, Athenian crew list, an inscription on stone dating from somewhere between 410 and 390, lists slaves among the crews in an Athenian naval squadron of eight ships.[24] Indeed, on three of the four triremes whose crew composition can be reconstructed with some probability, most of the rowers were slaves (Osborne and Rhodes 2017, 546). Since slaves are identified by their names and then those of their owners—as on Athenian state work records inscribed on stone—scholars can link individual slaves with their owners among the trierarchs, marines, and other officers, and specialists: "twenty-two of the thirty-two trierarchs and officers whose names are preserved are found to own slaves who also serve."[25] The reason for honoring this squadron of eight ships on the Acropolis is unknown, but the list does provide important evidence for slave rowers in the Athenian navy. A second key piece of evidence is the inclusion of slaves as one of the subsets of the naval crews described by Nicias in his letter about deteriorating conditions in the Athenian navy before Syracuse (Thuc. 7.13.2). Although earlier commentators tried to explain away or emend the manuscript, in an important article A.J. Graham persuasively defended the manuscript reading and the implication that slaves were regular members of Athenian crews.[26]

For the use of slave rowers to be a practical and continuing practice, Athenian slaves had to be motived without the promise of freedom, since you can only free a slave once. At the most basic level, all members of a ship's crew had a good chance of dying if the ship was rammed and sunk—depending on the conditions of the particular battle, whether it was

20 Thuc. 2.103, 8.15.2. Robert 1935, 459.

21 See Myron *FGrH* 106 F1 and Xenophon, *Hell.* 5.1.11, 7.1.12 with Bertosa 2005 for Helot crews on Spartan ships.

22 See Potts 2008 for a thorough recent treatment.

23 The use of slaves as rowers does not vary with the system of government: Syracuse and Corcyra were democracies when slave rowers are attested; Chios and Corinth were oligarchies.

24 *IG* I3 1032, also in Osborne and Rhodes 2017, no. 190 538–949. Detailed recent treatments include Graham 1992, 263–269; Graham 1998; Bakewell 2008, all of whom argue for or accept the regular use of slave rowers in the Athenian navy. *SEG* LIV (2004) no. 226 is a fragment of an Athenian crew list, but it contains only a partial list of one ship's officers, who are of course free citizens. See also Isoc. *On the Peace* 48.

25 Osborne and Rhodes 2017, 548. This way of listing slaves prevents me from accepting Hale's theory (2009, 355) that the inscription commemorates the liberation of slaves to fight at Arginusae; the nomenclature is exactly that of the unambiguous slaves on Athenian work records.

26 Graham 1992, 257–263; accepted by Hornblower 1991–2008, III.563; cf. Morrison et al. 2000, 117–118.

fought near to land, whether the victors went around killing the survivors in the water (Thucydides 1.50.1), and whether the particular crew members could swim. This prospect, however dire, was a bit distant, so rewards played a role as well. Rowing was wage labor and the same wage was allotted to each member of the crew, whether slave, foreigner, or citizen. In fact, slaveholders, a powerful group of course, would not have accepted anything less as the Erechtheion work records confirm. On these, the legal status of a worker (citizen, foreigner, or slave) had no effect on the amount he was paid; that depended only on the type of work he did: for example, was he a sculptor or a mason (Randall 1953, 208)? Of course, masters had the legal right to the entirety of their slaves' pay and some may just have pocketed the whole amount, both in the navy and on public work projects. More likely, a slave rower would keep a fraction of his earnings as an incentive while the master took the rest.[27] If the slave saved up and bought his freedom, the master would get back the money he had "shared" and could buy another slave to replace the one he was freeing. Thus, the parallel of the Erechtheion work records suggests that some slaves could be motivated by the hope of eventual freedom—or just spending money—but without an immediate promise.

Two sources seem at first to suggest that slaves did not regularly serve as rowers in the Athenian navy, but these are easily explained. First, in Thucydides, Pericles reassures the Athenians that their naval dominance will be secure in the coming war:

> And if they tried to win over the foreign sailors by taking the money at Olympia or Delphi and offering a higher wage, that would be a terrible danger, were it not for the fact that we'd still be a match for them, even with just ourselves and the metics embarking. (Thucydides 1.143.1–2; cf. 1.121.3, 3.16.1)

Here Pericles has one aim only: to discount the fear that the Athenians will lose their crews if the Peloponnesians can pay more. Since Pericles does not even worry about metics leaving Athens, he *a fortiori* did not consider slaves a flight risk. They required no separate treatment, but were included among "we ourselves," a simplification for which unambiguous parallels exist.[28]

Second, slaves won their freedom and citizenship for serving the navy raised in an emergency to rescue a large Athenian squadron trapped on Lesbos. These slaves helped the Athenians win a great victory at Arginusae and constitute the most famous case of slave rowers in the Athenian navy. Is this the exception that proves the rule that slaves did not usually serve in the Athenian navy? No. The slaves of Arginusae were exceptional, not in the fact that they rowed in the navy, but for two other reasons. They were part of an unprecedented full mobilization of Athenian manpower including even cavalrymen; not just some slaves but according to our source, Xenophon, all (male) slaves were recruited.[29] And the Athenians gave them not only freedom but also citizenship, on the same terms as the refugees in Athens from its longstanding ally Plataea.[30] Specific factors favored this generosity: first, slaves enrolled *en masse* would not have been handpicked by their masters as in the previous system where slaveholders brought only a few slaves with them on a trireme. Rather, the slaves of

27 Trundle 2016, 76–78; [Xen.] *Ath. Pol.* 1.11 may refer to this practice.

28 Hunt 1998, 96 on Thuc. 1.90.3 and Diod. Sic. 11.40; Thuc. 1.32.5, 1.55.1.

29 Xen. *Hell.* 1.6.24 with Hunt 2001; Demont 2004, 185–191; Tritle 2010, 209–210; Hamel 2015, 44.

30 Ar. *Ran.* 33–34, 190–191, 693–694; Hellanicus in *FGrH* 323a25; cf. Dem. 59.104.

Arginusae came from types of slaves—for example, mine slaves and slaves on large farms—that were harder to motivate to train and fight hard for Athens. Finally, at this point in the war, both sides were competing for rowers and the Peloponnesians had been winning thanks to the deep pockets of their Persian allies. Granting slaves citizenship gave the Athenians a much better chance to retain such ex-slaves because their status was assured at Athens, but insecure everywhere else.

One passage mentions a restriction on the employment of slaves in the Athenian navy. Thucydides describes the crew of the Paralos, a "sacred trireme" assigned special missions for the state, as "all Athenians and free." In other words, by law or custom, the Paralos had no metics, foreigners, or slaves among its complement.[31] But of course, this is the exception that proves the rule. Thucydides mentions this feature of the Paralos' crew because it was special and not the regular practice. In the rest of the navy, there was no such restriction and so the proportions of citizens, foreigners, metics, and slaves among Athenian crews varied over time in complicated ways that we can only guess at. For example, during the Peloponnesian invasions of the early Peloponnesian War, displaced farmers are likely to have wanted jobs in the navy—but why not for their idle slaves too? (Cf. Rosivach 1985.) The presence of slaves in Athenian crews was not exceptional, but rather the regular practice—as in the rest of the Greek world.

Warfare as a Source of Slaves

Heracleitus famously stated that "War is father and king of everything; and it shows some to be gods, others men; it has made some slaves, other free" (Diels and Kranz 1951, B53). How war (or strife) shows some as gods is not at issue here, but war obviously resulted in the enslavement of some people while it left others free. Indeed, enslavement in war has been invoked to explain the references to foreign women in Linear B tablets, the slaves in the *Iliad* and *Odyssey*, and the foreign slaves that made up most of the slave population in Classical Athens. The hypothetical origin of all slaves as defeated enemies encouraged a militaristic justification of slavery that played a large role in popular thinking: slaves had failed in an important test of their basic worth and manhood, which the victors had manifestly passed. It was probably as important as the ethnocentric justification for the enslavement of "barbarians" and certainly more common than Aristotle's attempt to link slavery with incomplete rationality (Hunt 1998, 153–164; Kraut 2002, 284–290). But, although a host of passages makes it obvious that slavery was the fate of some of those captured in war, the overall proportion of war captives in the slave population is a complicated topic. For example, most slaves at Athens in the late fifth century seem not to have been Greek. So, if they were enslaved in war—which is likely enough—it was not in the wars between Greek states that we hear most about in our histories of that period. They may rather have originally been enslaved in wars beyond the borders of the Greek world and transported to Greek markets or in raids on the periphery, especially by the many Greek cities on the coasts of the Mediterranean and Black Sea on unfriendly terms with their non-Greek neighbors.

31 Thuc. 8.73.5 with Graham 1992, 259; Hornblower 1991–2008, III.973–974. The same restriction to free citizens presumably applied to the Salaminia, the other state trireme.

The enslavement of war captives was always one option that a victorious army could exercise (Pritchett 1971–1991, 5.203–312). Attested alternatives included simply killing people rather than taking prisoners, mass execution after capture, internment as prisoners of war to be traded back at the end of hostilities, and ransoming. The choice among these options varied depending how the victors felt—bitter feelings could lead to massacres—and what they thought was advantageous. Most of the time, Greek city-states did not expect to destroy their opponents and thus kept captured soldiers as prisoners of war and expected their enemies to do the same for eventual release at the end of hostilities. Even while conflicts continued, cities could ransom their prisoners. This was apparently regular enough a practice that we hear that two *mnai* was the standard ransom in the Peloponnese (Hdt. 6.79). The original status of the captives also played a big role in how they were treated. In one case, Phormion sold the slave members of some naval crews he had taken and brought the free members back to Athens as prisoners of war (Thuc 2.103). In contrast to this relatively benign treatment of captured soldiers, brutality marked the treatment of captured cities (Gaca 2010). In that case, there was nobody left to ransom the captives, no expectation of an eventual peace treaty, and women and children were as vulnerable as soldiers. Cities could surrender on terms, of course, but if they were taken by storm or surrendered unconditionally, their inhabitants had little reason to expect anything but death and enslavement. The treatment of Melos is an infamous example: the men were executed and the women and children enslaved (Thuc. 5.116.3–4). But there were several such cases every generation.

Women were often raped by soldiers when they were captured. Coerced sex was also a part of slavery for many women—and for some men—but it is hard to estimate its prevalence (Hunt 2018b, 106–111). Alcibiades supposedly found himself a Melian mistress from among the captives after Athens destroyed Melos (Plut *Alc.* 16.5; cf. 8.3). And in some cases during the Persian Wars, Greek soldiers seem to have acquired women slaves directly (e.g. Hdt. 9.81). During an ongoing campaign—which was most of the time—armies did not want to guard recently captured slaves, often familiar with the area and prone to escape, so we regularly hear of captives being sold and of slave traders who made it their business to follow armies. They could buy captives cheaply near the point of capture and then transport and sell them in distant markets. Once in possession of a slave trader a person could end up anywhere in the Greek world or even beyond—Carthage, for example, was a slave society and provided a large market for newly enslaved persons (Lewis 2018, 259–266). Direct capture in war and the slave trade dependent on war contributed substantially to the slave supply in both in Greece and in the wider Mediterranean.

Conclusion

Military historians do not typically spend much time thinking about slavery. To readers of this Companion, some relationship between warfare and slavery in a world with frequent wars and numerous slaves may seem likely enough, but only in theory. My goal in this brief treatment has been to flesh out this abstract likelihood with concrete cases of interaction between warfare and slavery in the Classical Greek world. Some of these had direct and important consequences for how the Greeks fought their wars: the role of Helots in the Spartan military and of slave rowers in Greek navies stand out. Other cases shaped, in less direct but still important ways, the whole Greek experience of warfare: the fear of enslavement, the omnipresence of slave attendants, and, on the other hand, the risk of slave flight or Helot rebellion. Without warfare, slavery in ancient Greece would have been unrecognizable if it existed at all; many aspects of Greek warfare would also have been profoundly different without slavery.

CHAPTER 22

Agriculture and Greek Warfare

Jeanne Reames and Ann Haverkost

Agriculture affected Greek warfare by way of crop destruction, the fortification of fertile regions, logistics often determining where troops marched or camped, and wars fought to acquire farmland or pasturage. Yet with the exception of Garlan's work on crop destruction in connection of with Philo of Byzantium, and of the wider-ranging monograph of Hanson, *Warfare and Agriculture in Classical Greece* (1983), modern writing on these topics began only in the late 1980s.[1] This eventually led Hanson to reissue his 1983 monograph in a revised form in 1998. He had already contributed *The Other Greeks* in 1995. Thorne and Hughes have offered recent contributions.[2]

An important obstacle for scholars in this field is that Greek literature, unlike Roman, has no treatise that deals with farming and animal husbandry, although Hesiod, Xenophon, and especially Theophrastus provide information.[3] The majority of our ancient accounts date only to the Peloponnesian War and later, yet, as we shall see, the Peloponnesian War changed the nature of Greek military operations, and thus changed how Greeks used and abused the local ecology.[4] This chapter thus begins by establishing the relation between farming and warfare in the Archaic period, well before this shift, and then turns to Classical and Hellenistic developments–devastation techniques, forms of invasion and defense, and merchant supply for armies.[5]

1 Garlan 1972, 1974; Ober 1985; followed by Osborne 1987; Munn 1987, 1993; Spence 1990, 1993.

2 Thorne 2001; Hughes 2013.

3 McHugh 2017, 13. Xenophon: *On Horsemanship* and *Oeconomicus* .

4 Monographs on agriculture: Amouretti 1986; Gallant 1991; Sallares 1991; Isager and Skydsgaard 1992; Burford 1993 Howe 2008; McHugh 2017.

5 In the interest of space, we will not deal with logistics at length, referring readers to Chapter 17 in this volume. Likewise, animal husbandry is only lightly touched on; the authors direct readers to Howe 2008, ch. 3.

A Companion to Greek Warfare, First Edition. Edited by Waldemar Heckel, F.S. Naiden, E. Edward Garvin, and John Vanderspoel.

Archaic Era Farming and Warfare

Little evidence exists for the use of marginal lands, or *eschatiai*, in the Archaic period, much less the late Geometric.[6] Instead, our archaeological record for those eras shows nucleated settlements next to good agricultural land.[7] Horden and Purcell accordingly describe rural land and small and larger settlements as a continuum, and argue that treating them as separate entities does violence to Mediterranean settlement ecology. While their work covers the *longue durée* of ancient and medieval Mediterranean life, their observations match the Archaic archaeological record.[8]

Farming conducted in these circumstances would have made systematic crop destruction problematic, given the unencumbered, rapid response of the local populace to any incursion.[9] In addition, Geometric and early Archaic conflicts often amounted to intermittent, haphazard raiding between settlements for livestock and booty; they are better termed landlord-sponsored feuding than *polis* warfare.[10] Although such conflicts might last for years, campaigns themselves were brief, for, outside Sparta, troops were often ill-equipped and ill-trained farmers who needed to return to maintain their own fields.

Some Archaic wars, however, were fought to relieve land shortages by obtaining additional cropland for the poor, or additional pasturage for the elite.[11] Villages might also reallocate land into more equal parcels, which could result in disputes.[12] Rather than force such redistribution, city-states might seek to expand, as Sparta did when it invaded Messenia.[13] Forced emigration and the creation of colonies was another way to deal with poor acreage, drought, or related farming obstacles. Last, border conflicts over scrubland or wetland likely represented attempts to expand potential pasturage rather than to acquire more cropland.[14]

Whatever the aim, Archaic wars occurred largely according to demands of the agricultural calendar. Of the two chief crops, barley was ready for harvest in May, while wheat began in June. Once cut, grain needed to be dried before threshing and winnowing. Fruits came much later, in autumn, with figs in August, grapes in September, and olives in October and November. All this made June and July optimal times for warfare. The need to return before August, in order to process one's own grain, as well as to prepare for autumn fruit picking, implied a short war season.

6 Jameson 1994, 57–58; Krasilnikoff 2008; Foxhall 2013, 215–218. A contrary view: Hanson 1983/1998, 47ff., 217–218 (references are to the 1998 edition).

7 Foxhall 2013, 215–216. Cf. Hanson (1983/1998: 215–217), but the bulk of his extensive evidence is Classical, or at best, late Archaic. McHugh 2017, makes a solid case for Greek farmsteading as well, but again, much of what she discusses is Classical and later.

8 Horden and Purcell 2000, 96–112; Foxfall 2013, McHugh 2017.

9 As Hanson theorizes, minus the homesteads (1989, 4, 1995, 219ff.). Connor 1988 agrees, as do various others, such as Cartledge 2013b.

10 Singor 2009, 586.

11 Even if sources such as Thuc. 3.50 speak of agriculture, not animal husbandry.

12 Morris 2009, 68. While typically idealistic, Plato explains how such "overstepping" should be handled, *Lg.* 843.

13 Van Wees 2009, 463.

14 Howe 2008, 79–81, with a number of possible specific examples in the pages following.

Complications always made themselves felt. Xenophon says crops should be sown three times—early, middle, and late in the season—so as to provide a staggered harvest (*Oecon.* 17.4–6). Additionally, field elevation could affect when grain would be ripe, as could the weather. Many farmers doubled as pastoralists, and were thus subject to sundry seasonal dictates.

In sum, Archaic wars were economic and political ventures intimately tied to agriculture. Yet agricultural destruction was not the goal. Agricultural usurpation was.

Military Depredations

Methods of agricultural devastation came to maturity during the Classical era, with invaders trampling, slashing, burning, and cutting for use. The choice of method involved several factors, first, the state of the fields themselves, especially grain crops: still green, ready for harvest, or already harvested? The goal of the invaders mattered as well. Was it a brief raid, an extended occupation, or something in between? The speed of response to the invasion played a role, as did the presence of field walls or other fortifications. Finally, the supplies the invaders brought or could purchase affected the length of their stay.

Methods of destruction: trampling and slashing

Trampling and slashing were preferred when time was of the essence, either for fear of counterattack or because the incursion was intended only to harry. These two methods worked best with green crops that could not be burned, or with grape vines and younger fruit trees.[15] While vines are more delicate than the virtually unkillable olive, they are also notoriously stubborn (Hanson mentions uprooted vines coming back the next year[16]). Similarly, fruit trees are most vulnerable only in their first decade or so. As figs regularly live over 100 years, olives for centuries, and even apples can bear fruit at age 50, these species are hardy for most of their lifespan. In addition, all fruit and nut trees would be good candidates for terraced hillsides rather than flat land, complicating attempts to damage them.[17] Would-be devastators would have to scramble around the rocky landscapes with several strains of prickly weeds.[18] For the most part, troops would have stuck to the more accessible, level fields, and left to the farmer whatever crops remained in the *eschatiai*.

The results may have provided psychological satisfaction to attackers, yet not done systematic damage. Crops would not have been trampled uniformly, so portions of a field might revive later.[19] Doing more serious economic damage would require a large army to

15 Aristoph. *Ach.* 231; Hanson 1983/1998, 49–55, 68–71; Foxhall 1993, 88–89.

16 Hanson 1983/1998, 224, with 55–68, 222–223; Foxhall 2007.

17 Foxhall 1993, 88.

18 Remote areas might also conceal dangerous fauna, including vipers. If most Greek viper venom is not deadly to humans, snakebites, or scorpion stings might incapacitate soldiers. An epitaph by lyric poet Anyte mentions a hound killed by a "spot-throated" viper, presumably while hunting in a thicket (*Lyra Graeca* Anyte 165).

19 Foxhall 1993, 87.

invade a small city-state and flood numbers of farms with platoons of men, a mere handful not being adequate to devastate even a typical small farm of only a few acres.[20]

Army encampments themselves could flatten fields, and any extended stay would have compacted the soil, stunting the next year's crops.[21] Yet battlefields might increase field production via a tragic form of fertilizer. Plutarch described the fields of Massalia as excessively fertile for years after a bloody battle there, affirming Archilochus's statement that such events "fatten" fields (*Mar.* 21.3).

Methods of destruction: burning

Fire produces far more extensive and comprehensive damage, but only in certain circumstances. First, crops need to be within a few weeks of harvest, or just harvested and still in the fields; green grain does not burn. Vines and living trees (green wood) do not burn easily, either. Second, while some wheat or barley might be ready for harvest and combustible, intercropping would make it more difficult for fire to spread over a wide area.[22] Third, invaders would seek a controlled burn, not a raging wildfire. Aside from all the conditions involving the crops, attention to the weather would be crucial. Winds in Greece can be fierce, and fire leaping from crops to dry maquis scrub might result in a rapidly moving, out-of-control blaze that threatened the invaders as much as the invaded.[23]

Contrary to Thorne's view, we should not assume the Spartans (or anyone else) used this problematic method unless the fact is plainly stated.[24] Incidents such as the firing of (stored?) grain in Sicily, or Philip II's burning of Elean colonies' lands, or the accidental immolation of shrine lands and sanctuary of Athena of Assesus by the Lydians, were so many exceptions to the rule, which was that harvestable or harvested grain was far more valuable to take than to destroy.[25]

Invaders were more likely to burn buildings. In fact, fire was virtually the only means invaders used to demolish looted houses and barns either in towns or in the countryside.[26] Although most of these buildings were constructed of stone or mudbrick, wood and straw, two flammable materials, were used for support and in the roof. To hamper fire-setting, evacuees might take removable wood with them, including doors, shutters, and even lintel frames (Thuc. 2.14.1). In timber-starved Greece, wood was an expensive material well worth preserving and transporting.[27] Perhaps for this reason, states on the Greek mainland seldom scorched the earth of territories occupied by their armies. The Laconians apparently burned the countryside to hamper the Theban occupation of 370 (Xen. *Hell.* 6.5.50),

20 Van Wees 2004, 122.

21 Hughes 2013, 130.

22 Intercropping consisted of planting legumes between grain crops, or of planting two types of grains together, or of planting cereals between olive trees or vine rows; Xenophon adds planting figs to support vines (*Oecon.* 19.18). Even if intended primarily to guard against devastation from disease, intercropping worked to reduce other sorts of damage.

23 The 2018 July wildfires in Attica, spread by summer winds, drove people into the surf for safety.

24 Thorne 2001, 231, citing Thuc. 2.21, who describes distraught Athenians watching the land laid waste in Acharnia, 7 miles away; Thorne supposes these Athenians see smoke, not people, but Spartans were invading at the time of the harvest in order to use it, not burn it.

25 Sicily: Thuc. 6.94.3. Eleans: Dem. 7.32. Lydians: Hdt. 1.19.

26 Hanson 1983/1998, 72–74.

27 Meiggs 1982, 208; Borza 1987.

yet Xenophon implies that some of the crops were harvested or stolen. To judge from the experience of the Ten Thousand and of Alexander, scorched-earth policy was more common in Asia Minor.[28]

Methods of destruction: cutting for use

Finally, we come to invaders cutting (*temno*) ripe crops, as with the Spartans during the Archidamian War in Thucydides.[29] That the Spartans depended on this grain for food emerges when, during the third of their annual invasions of Attica, they were forced to withdraw because they had arrived when the grain stood still green in the fields (Thuc. 4.6).[30]

The challenge in using cut grain was preparation. To be edible, both wheat and barley require about a month from field-cutting through winnowing. This is too long for armies needing an immediate food supply, or invading for only a short period, as was typical in Greece prior to the Peloponnesian War. Only when campaigns lengthened did harvesting grain for use become routine logistics. Xenophon thus mentions the burning of crops by the retreating Persians to deprive the Ten Thousand of food (*Hell.* 1.6.1, 5.2.3). When Alexander invaded Anatolia, he brought only limited stores, intending to eat off the land or buy supplies and fodder from local villages.[31]

Extent of devastation and recovery

Of the Mediterranean triad of foodstuffs—grain, grapes, and olives—grain is the most vulnerable to devastation, but also the most easily replaced. Cold-crop vegetables, such as beans and peas, were also easily replaced, legumes being significant secondaries to grains, even as fodder. Diversification of crops may have also protected the food supply against devastation, but we do not know how much, or where and when, this topic being one neglected by scholars.[32]

Hanson notes that damage done by trampling and slashing may be offset by various methods, while Foxhall points out that the loss of higher-yield crops with longer growing periods can be alleviated by substituting short-growth barley and vegetables.[33] Small city

28 Hanson 1983/1998, 118–120. Thorne 2001, 232, says water was used as a contaminating agent, and little would be needed to cause grain to rot, but if grain were kept in granaries, fire would be as easy to use as water, assuming that the attackers did not take the grain for themselves. Greece, of course, is a dry country where rivers run dry; for water management on farms, see McHugh 2017, 17–18.

29 Thuc. 2.19.2, 2.20.1, 2. 4, 2.21.2, 3; see Hanson's discussion 1983/1998, 190–193, where he discusses the term's use more widely.

30 Thucydides neglects to say whether the army was too early to purchase grain from the markets. Although some wheat can be eaten green by first drying and then roasting it (*freekah*), this is possible only if the grain is at a particular stage of unripeness.

31 Plut. *Alex.* 15.2). Hence Memnon's proposing a scorched-earth policy to Darius, who rejected it (Arr. *Anab.* 1.12.9–10, Diod. Sic. 17.18.2–4).

32 In his assessment of crop destruction, Thorne 2001, 228, overestimates grain losses, while calling vines and olives "relatively minor components." In times of severe deprivation, however, high-caloric foodstuffs with a relatively long shelf-life—oil and wine being foremost—become crucial. In any event, olives and grapes, not to mention dried figs and beans, are not minor Greek components.

33 Hanson suggests, "special manuring and irrigation of damaged plants, grafting, prompt replanting and resowing, application of mud seals to chopped and broken limbs, pruning back of hacked vines or trees, and additional propping" (1983/1998, 177; Foxhall 1993, 89.

plots with a fig, pomegranate, or even olive tree could augment the rural supply; on the other hand, wild plants, fish, and game would help sustain the food supply as new crops grew.[34] If invading armies departed by July, vegetables could be replanted in August or September for an October and November harvest.

As a further safeguard against crop failure, Greeks distributed crops across several plots. Even in the event one or more fields were destroyed, a family might thus scrape by on the produce from plots elsewhere.[35] An unintended safeguard for crops took the form of the common features of any rural neighborhood—walls, gullies, streambeds, and hills that broke up land and thus impeded destruction.[36] Before an invasion, stored grain might have been transferred to two kinds of hiding places or *ambaria*, underground, airtight pottery containers, or local caverns. Foxhall has drawn attention to this modern Greek practice, and argued that it was a likely way for ancient Greeks to avoid the labor-intensive process of evacuating grain, as described by Thorne.[37]

Destruction of expensive and immovable farm equipment (such as olive presses) would have had a greater long-term impact than these seasonal losses. Likewise, the loss of slaves would have cut into the labor needed to work the farm. Although our sources say nothing about the former, they are voluble about the latter: states at war encouraged slave desertion.[38] In Aristophanes's *Peace* (450–451), Trygaio references slaves escaping to the enemy, and in the *Clouds* (6–7), Strepsiades bemoans that war means he cannot punish slaves (for fear they might run). Helots escaped to the Athenians at Pylos.[39] And on Chios, the advice of runaways, including agricultural slaves, helped the invading Athenians do greater damage.[40]

By Foot, Horse, Ship, and Fort: Offensive and Defensive Strategies

Thorne usefully identified three chief types of invasions responsible for crop destruction:

> All ravaging operations can be categorised as one of three types: "general invasion," in which an overwhelming force invaded a territory; "amphibious raiding," in which relatively small forces of ravagers went ashore for short periods at distantly separate points to devastate a littoral; and "fortified occupation," or *epiteichismos*, in which a fortified post was set up, from which (again, in relatively small numbers) troops could sortie and ravage the surrounding area on a permanent, or at least an open-ended, basis.[41]

We know most, of course, about destruction due to general invasions by large armies. By contrast, small armies struck quickly and focused on looting or pilfering rather than on

34 Foxhall 1993: 87; Hanson 1983/1998: 80; Gallant 1991, 115–120.
35 Gallant 1991, 39–42.
36 Hanson 1983/1998, 218.
37 Foxhall 1993, 89; Thorne 2001, 242–245.
38 Hunt 2018b, 157–159.
39 Thuc. 5.35.6–7, 7.26.3; Kagan 2003, 138–139; Hunt 2018b, 158–159.
40 Thuc. 8.40.2; Hunt 1998, 102–115.
41 Thorne 2001, 233.

destroying crops. Similarly, amphibious raids were in and out. The occupation of a fortified site, albeit harmful to agriculture, was rare and typically reserved as a last resort during long, inconclusive wars.

General invasion

Who engaged in crop destruction? Hanson argues that light-armed troops were primarily responsible, while Spence points to hoplites, and Foxhall suggests both.[42] Besides foot soldiers of either type, cavalry might serve, and Hanson, Spence, and Willekes likewise highlight the importance of cavalry for cropland defense.[43] Given their speed of movement, they are beneficial when the defenders do not know from which direction enemy troops will attack. As Willekes says, "The speed and mobility provided by the horse ensured that the only way to stop an incursion of mounted raiders was to send horsemen out against them." Both invaders and attackers would have relied on small bodies of mounted troops, leading to skirmishes as opposed to organized combat.[44]

The number of troops used for looting and devastation during invasions may have been small, if for no other reason than that some men needed to remain behind and guard the camp. How efficient these troops were is a matter of educated guesswork, Halstead and Jones estimating that a man could devastate a Greek *stremma* of cereals in one to three days, and a *stremma* of pulses in half the time, or, in English terms, an acre per week.[45] Farmers who had been forewarned—and that would be common in the case of large invasions—would lower this rate by harvesting what they could and taking it to safety.[46] As wars grew more lengthy and crop destruction became more frequent, farmers might build obstacles or even fortifications, causing more difficulties.[47]

Among the first, and certainly best-known, instances of crop destruction were the Spartan invasions during the Archidamian War. The Periclean response was to evacuate the countryside when Sparta arrived with troops (Thuc. 1.143.5). In Books 2–4, Thucydides is vague about what forms the depredations took, using simple terms such as "ravage" or "destroy," "cut" being the most specific. Yet the effects were surely widespread, the second and fourth invasions being the most severe.[48] The Persians of 480, expecting only a short struggle, may have wrought less harm (Hdt. 8.50–2, 9.13).

Owing to her wealth, as well as to the protection afforded by the long walls that linked the city to her port, Athens was able to withstand these by importing the needed grain supplies. Thucydides also makes clear that while the Athenians did not come out in force, they did defend themselves, by way of forays by cavalry and light troops.[49]

Potidaea, in contrast, demonstrated how enemy occupation of the countryside could devastate a city that lacked Athens's wealth or navy. The Athenian siege of Potidaea lasted

42 Hanson 1983/1998, 20–21; Spence 1990, 93 with Xen. *Anab.* 1.2.25; Foxhall 1993, 87.

43 Memnon suggests using cavalry to trample fodder and crops (Arr. *Anab.* 1.12.9). Speed: Thuc. 3.1.2, Xen. *Hell.* 2.4.4. Secondary literature: Spence 1990, 93–101; 1993, 121–133; Hanson 1983/1998, 122–125, 230–231.

44 Willekes 2016, 180, and generally, 178–181; and 173–176.

45 Halstead and Jones 1989: 47. A *stremma* was 1,000 m2.

46 Thorne 2001, 234.

47 Ober 1985; Munn 1993, discussed later.

48 They occurred in 431, 430, 429, 427, 425. For a thorough discussion, see Hanson 1983/1998, 132–143.

49 Spence 1990, 106.

two-and-a-half years, driving the inhabitants to the point of cannibalism. Meanwhile, the Athenians cultivated Potidaean farmlands for their own benefit. The invaders' success was short-lived, however, as attacks on Potidaea's allied neighbors ultimately resulted in an Athenian defeat at Spartolus (Thuc. 2.70).

In the next century, threatening to burn crops was a negotiating ploy for Agesilaus, who made Acarnania surrender this way. Plutarch reports that he had even allowed the Acarnanians to plant crops in the first place, in hopes of making this threat once the grain was ripe. (Plut. *Ages.* 22.5). An occupying army could also prevent the sowing of crops altogether, rather than destroy them at a later date.[50]

In all, how much damage was done to crops, and how much impact did threats make? Hanson questioned extensive, long-term damage. Thorne replied by calling devastation as "a formidable threat," arguing that the Athens of the Peloponnesian War was an exception to the rule illustrated by Potidaea during the same war.[51] Nonetheless, most crop devastation would have remained incomplete for reasons ranging from the hardiness of the olive to intercropping and to counterstrokes that would compel raiders to turn from pillage to self-defense.

Naval predations

If cavalry lent speed to the defenders, naval raids lent speed, and unpredictability, to the attackers. Ships could move faster than troops on land, and so, with numerous locations to observe and protect, the defenders were often at a loss. The attackers felt a compelling impetus to strike, for if provisions were not being transported by a following merchant ship, they would have to make successful raids in order to avoid starving.[52]

The numbers employed in amphibious raids were not large, and they commonly attacked targets well outside cities, not urban locations.[53] Yet even limited raids served to instill fear and terror in the local population, as well as to cause agricultural damage. As Thorne observes, these raids sought to avoid pitched battle, not incite it. The example of Sparta and Athens in the Peloponnesian war bears him out: although each side raided the other, they never fought a pitched battle, save at Mantinea in 418, when they were nominally at peace with one another and no raiding was taking place. The Peloponnesian War ended in an altogether different fashion, as a result of naval battles that gave Sparta control of the Dardanelles and thus curtailed the Athenian grain supply.[54]

Fortifications

Walls and forts of several kinds increased in numbers during and after the Peloponnesian War.[55] Earlier, these would have been largely private affairs, like the wall 5 cubits high to be

50 Hanson 1983/1998, 54–55.

51 Hanson 1983/1998; Thorne 2001, 252.

52 Furthermore, the design of the trireme, the most popular naval battleship of the time, made it difficult for troops to sleep aboard ship. As a result, the men would beach for the night and make camp. There was also little room for storing provisions.

53 Thorne 2001, 236–237.

54 Kagan 2003, 402–404, Tritle 2010, 181–182.

55 Ober 1985, Hanson 1983/1998, 79–102.

built around a farm according to an inscription found on Thasos.[56] Such walls, fences, and terraces, although not intended originally or wholly to deter agricultural devastation, would work to that end (Hanson 1983/1998, 214).

Larger walls expressly to prevent invasion functioned best in bottlenecks such as at Gallipoli.[57] By the early fourth century, these walls had become more elaborate, as with the Theban stockade and the Athenian Dema Wall, both apparently constructed to ward off Spartan incursions during the Spartan–Boeotian Wars (378–362).[58] Xenophon describes Agesilaus' surprise upon encountering the stockade, although he managed to circumvent it and ravage the cropland.[59] The much larger Dema Wall, protecting Athens from incursions via Eleusis, may have had the additional military purpose of denying armies of hoplites easy access to Attica.[60]

Forts guarding key roads and passes might provide incidental protection for cropland, but these facilities were not always reliable. Along the Boeotian border, the Athenians had several forts that occasionally changed hands because of treachery.[61] Sometimes light troops guarded Athenian border forts, and sometimes mercenaries; by the fourth century, sometimes ephebic trainees did.[62] Perhaps Laconia's formidable border forts were better manned (Xen. *Hell.* 6.5.24). One of the military improvements Thucydides (2.100.1–3) attributes to the Macedonian King Archelaus was riverine fortifications, as along the Axius, with wide farmlands and pasturage above the swamps at its mouth.

Invaders might use forts and walls to put both agricultural and social pressure on the countryside. During the Peloponnesian War, the Athenians constructed fortifications at Pylos from local stone reportedly in six days (Thuc. 4.3–5). The Spartans felt threatened enough by this move to recall troops from Attica, albeit this decision may have been made easy because that was the year the Spartans invaded Attica only to find the grain still green (Thuc. 4.6).

At the advice of Alcibiades, the Spartans erected a fort at Decelea in 413 (Thuc. 7.27.3–28.2). This post allowed them to occupy and pillage the country year round, preventing Athens from planting crops. Decelea also interfered with mining operations at Laurium by harboring more than 20,000 escaped slaves, some of them surely miners, and thereby cutting into Athenian funds for purchasing grain. Thucydides refers to these runaways as "artisans," or *technitai*: "They were deprived of their whole country: more than twenty thousand slaves had deserted, a great part of them artisans, and all their sheep and beasts of burden were lost."[63]

In addition, Decelea cut supplies coming to Athens from Oropus.[64] This fort, which Thucydides called a severe hindrance to Athens, would eventually contribute to Athens' vulnerability and isolation after her naval defeats in the Hellespont.

56 McHugh 2017, 20.

57 Hdt. 6.36–37; Plut. *Per.* 19.1–2; Xen. *Hell.* 3.2.10–11.

58 Munn dates the Dema Wall to c.378 (1993, 42–46).

59 Xen. *Hell.* 5.4.4, *Ages.* 2.22; Munn 1987.

60 Munn 1993, 47–62; van Wees 2004, 129.

61 Hanson 1983/1998, 89–91; van Wees 2004, 129–130.

62 McCredie 1966; Pritchett 1974, 59–125; Ober 1985, 90–95.

63 Thuc. 7.27.5 with Tritle 2010, 116. Concerning Laurium, Hanson 1992 and Harvey 1986 argue that the runaway slaves were more likely farmhands. Sallares 1991, 294–389, suggests there was more farming going on in Attica than scholars suppose, notably in the southern part of the peninsula, farthest from Decelea.

64 Kagan 2003, 299–300; Tritle 2010, 165.

Additional Food Supplies on Campaign

Troops typically brought some rations with them, then sought supplies via plunder or purchase. At Pylos, the Athenians arrived with only 20 days' rations. Then they would need to make other arrangements (Thuc. 4.39). Timomachus took some unspecified amount of grain to Stryme, along with some Thasian peltasts. Once he captured the city, it would feed his men (Dem. 50.21). Alexander the Great brought a supply train with him across Asia. When morale was low, he even offered extra rations to soldiers' families (Diod. Sic. 17.94.4).

Grain merchants accompanying armies appear in the fourth century, when soldiers were commonly paid 2–6 obols per day, or enough to buy provisions.[65] We learn of these merchants incidentally. For example, Dion, son of Hipparinus of Syracuse, hired two Corinthian merchants to carry armor back home to wage war against the tyrant Dionysius (Diod. Sic. 16.6.6). Dionysius, in turn, hired merchants to bring him grain as he pillaged the countryside (Diod. Sic. 16.13.3).[66] A scene in Xenophon suggests these merchants were familiar figures: When the Greeks in the service of Cyrus display their charging skills during a parade, the merchants following Cyrus are the ones to flee (Xen. *Anab.* 1.2.18).

Soldiers also bought grain at local markets as they passed through, or paid to have it shipped. Xenophon's *Anabasis* reports a half dozen examples of the former practice.[67] Such grain was often milled.[68] One of the few examples of the latter: after the Thebans paid 10 talents for grain to be sent from Pagasae, the Spartans captured the triremes carrying it (Xen. *Hell.* 5.4.56). If grain proved unobtainable, soldiers ate meat as a last resort (Xen. *An.* 1.5.6).

Absent rations, merchants, or transport, armies sometimes turned to pillage.[69] Yet this alternative was less tempting than mere convenience might suggest. It alienated neutrals or allies, and it encouraged revenge by enemies. If the troops were mercenaries, it might alienate potential future employers. If they were part of an army of conquest, like Alexander's, it alienated future subjects.

Warfare's Benefits for Agriculture

Aside from the gruesome notion of decomposing corpses "fattening" fields, warfare did benefit agriculture. New crops entered Magna Graecia in the Hellenistic period as a result of conquest. In particular, Alexander's campaigns brought agricultural cross-pollination—east to west and west to east—with the arrival of non-native plant species such as apricot trees.[70]

---

65 Trundle 2020, 21.

66 Dionysius was devastating this land as punishment for rising up against him, not to feed his men.

67 Xen. *Anab.*1.4.19, 1.5.5, 2.3.24, 2.3.27–28, 2.4.9, 2.4.28. Xenophon's accounts may not reflect warfare in Greece itself, but there is no reason to think Greek towns would have refused help, for fear of being raided.

68 Lee 2007, 215 explains that the typical hand mill during the era of Xenophon weighed too much for soldiers to carry, and required a table at which to work, another drawback.

69 Thorne 2001, 235; Foxhall 1993, 88.

70 E.g. Alexander reportedly imported Greek ivy to Babylon for the gardens there, as at Horden and Purcell 2000, 259.

Admittedly, these new plants spread in an erratic fashion "largely dependent on the whimsical communications of members of the elite."[71]

The citron, usually assumed to have arrived in the Mediterranean via Alexander, may rather have come to Greece via Italy. The Romans, in turn, might have obtained it from Palestine as a result of the Jewish diaspora, for the citron is indispensable to the Sukkot festival.[72] However this fruit reached Greece, it became a staple of Greek cuisine.

Agriculture's Contribution to Greek Warfare

The recruitment of Greek armies lies outside the scope of this chapter, but it is worth observing that many Greek soldiers were farmers.[73] By the Classical Age, if not earlier, many hoplites were farmers; at every period, pastoralists served in the light infantry. According to the laws of Solon, rich landowners served in the cavalry. The military role of hoplites is another topic lying outside the scope of this chapter, but the importance of this heavy infantry to Greek armies runs parallel to the importance of agricultural produce, especially grain, for Greek campaigning.[74] Again and again, Greeks farmers under arms trampled and burned crops, or defended crops against other farmers who were invading partly in order to destroy them.

It would follow that Greek terms for soldiers' efforts would be the same as the term for a farmer's, and one such term, *ponos*, applied to both. Homer described warriors as "toiling" (*ponos*) in combat, a term later employed of farming by Hesiod. Garlan states that by Xenophon's day, soldiers, farmers, and hunters are presented as engaging in *ponos*, whereas craftsmen are excluded.[75] In his *Oeconomicus* (4.2), Xenophon has Socrates claim that in some states, especially those with a military reputation, only non-citizens are permitted to pursue crafts.[76] Further on he links farming and warfare intimately (5.1–9): land is the prize for valor, and farmers are cast as "robust" of body, able to run, throw a javelin, and leap with the best. He assumes farmers make good warriors. This attitude need not have been universal. Plutarch records a tradition that the ancestors of the Athenians, the four sons of Ion, were the forebears of tribes known as warriors (*Hoplitai*), farmers (*Geleontai*), craftsmen (*Ergadeis*), and goatherds (*Aigikoreis*).[77] In this tradition, farmers had no military vocation.

Yet groups such as craftsmen and goatherds figure much less prominently in Greek warfare than farmers do. In the late Geometric and Archaic age, wars, or mere border scuffles, occurred at least partly to acquire additional land for farming. Extended campaigns that took

71 Horden and Purcell 2000, 259.

72 Zohary 2013 discusses the etymology and archaeology.

73 For recruitment, see Chapter 24 in this volume.

74 For the role of hoplites, see Chapter 6 in this volume.

75 Garlan 1972, 8–91.

76 Almost certainly Sparta is meant, although other states may be meant as well.

77 The four sons: Hdt 5.66. The list of occupational tribes: Plut. *Sol.* 23.4. Cf. Hindu castes named for occupational categories, priests (*Brahmins*), warriors (*Kshatriya*), farmers and merchants (*Vaishya*), and laborers (*Shudra*). The four classes of Solon are much more agricultural: horsemen (*Hippeis*), 500 bushel men (*Pentakosiomedimnoi*), yoke men (*Zeugetai*), and laborers (*Thetes*). Attempts to describe the Athenian groups in concrete, agricultural terms: Osborne 1987; Gallant 1991; Hanson 1983/1998, 1999, 2003; van Wees 2003, 2013; Rosivach 2005; Guia and Gallego 2010; Samaras 2012. For further bibliography, see Chapter 6 in this volume.

the farmer far from his homestead were rare. Citizen-soldiers brought their own food with them, and did not serve long enough to require much in the way of logistics. Those serving overseas typically did so as mercenaries.[78] With the notable exception of the Spartan conquest of Messenia for her farmland and slave labor, these early wars were such rural affairs that agriculture was never severely harmed.

By the time of the Peloponnesian Wars, invading armies began to target agriculture, especially ripe crops. Invasions now focused on the harvest, not only to save money by living off the land but also to ensure crop destruction by trampling, burning, and cutting. Advancements in sea travel let troops assail coastal farmland at will. Yet crop destruction took time, and invaders' harvesting of crops for use, another new practice, took even longer. These delays and other complications have led recent scholars to question the effectiveness of wartime depredations against agriculture.

Longer wars, and campaigns farther from home, compelled invaders to supplement their supplies by frequenting local markets or by allowing merchants to follow armies and thus create a new type of occasional market. As for those being invaded, they could make up agricultural losses and deficiencies by purchasing foreign grain; the most famous, but by no means only, example of this being Athens during the Decelean War. Invaders responded to these displays of commercial self-sufficiency by redoubling their efforts to cut off grain supplies. One method was to enter enemy territory and occupy fortified places year round, bringing economic life in the countryside to a halt. The defenders' counter-method was to build walls blocking invaders. Athens now built walls to protect its western approaches, just as it had earlier built walls to protect communications with the Piraeus. By the fourth century, these field walls and fortifications became the norm for controlling roadways and farmland. Yet for all these changes, smash-and-grab tactics never became obsolete. When the Ten Thousand found themselves without food or leadership on their way back from Mesopotamia, they raided and pillaged their way home.

Extended, year-round campaigning for conquest of new territory developed in the second half of the fourth century with Philip of Macedon (Dem. 9.50). These campaigns required far more sophisticated logistics. Frontinus tells us (*Strat.* 4.1.16) that Philip insisted his men carry a 30-day supply of flour, but would also have urged them to eat locally where possible, either by confiscation or purchase, especially by army-supply merchants. Long campaigns might also lead to crop devastation for purposes other than to deprive the enemy of foodstuffs. Sometimes it was to alleviate troop boredom, as happened in the case of Alexander's army in India (Diod. Sic. 17.94).

From warfare intended to acquire land for farming and pasturage, to land devastation as military strategy, to an increasing dependence on logistics for long campaigns, agronomics remained bound up with ancient Greek warfare.

78 Hale 2013.

PART VI

Social and Political Context of Greek Warfare

CHAPTER 23

Battle Trauma in Ancient Greece

Lawrence Tritle

In the midst of the battle of Marathon (490), the Athenian soldier Epizelus watched a fellow Athenian killed beside him. In an instant Epizelus was blind, and remained so the rest of his life (Hdt. 6.117). In the middle years of the Peloponnesian War (c. 420 BCE), the Sicilian Greek rhetorician Gorgias delivered a demonstration speech defending the innocence of Helen of Troy as the cause of the Trojan War. As part of his defense, Gorgias utilized an argument based on the power of sight. In explaining this he referred obliquely to soldiers so traumatized by what they had seen on the battlefield that some suffered from unusual illnesses, while others were no longer able to work (Gorg. *Hel.* 16–18).

How do we today, in the first decades of the twenty-first century, explain such pieces of evidence? A century ago, W.W. How and J. Wells, writing their commentary on Herodotus, noted Epizelus' experience, and placed it in a context of myth while noting the experience of St. Paul on the road to Damascus.[1] Only a few years later they would have learned many times over of similar responses to battlefield terrors on the Western Front. In short, with the knowledge of modern science, we should be able to improve our understanding of what happened to Epizelus and others like him who suffered psychological trauma on the battlefield. Today the medical community would explain Epizelus' response to battle as an example of "conversion disorder," what is more popularly called "hysterical blindness."[2] In the same way, the battlefield reactions that Gorgias relates are well known to us today not only from the twentieth-century World Wars and Vietnam, but from more recent conflicts in Iraq and Afghanistan.[3]

1 How and Wells 1912, 1:114, 354, 356.
2 Tritle 2000, 64; see further Weinstein 1995, I, 4: 383–407.
3 See further Tritle 2019.

A Companion to Greek Warfare, First Edition. Edited by Waldemar Heckel, F.S. Naiden, E. Edward Garvin, and John Vanderspoel.

Since the publication of Jonathan Shay's *Achilles in Vietnam* (1994), and my own contribution to the experience of battle trauma, *From Melos to My Lai* (2000), a growing acceptance has emerged that combatants in the ancient Greek world experienced the trauma of war no less than their modern counterparts.[4] Evidence from the Greek world is certainly limited and hardly compares with the mass of material from the modern one. Of greater significance perhaps is the knowledge gap, the reality that those in the ancient world had little idea, as we do today, of the workings of human physiology.

While different by today's standards, attempts—usually invoking mythology—were made in the Greek world to explain traumatic responses to the experience of battle. A notable example occurs in Euripides' story of Heracles, the greatest of heroes, who on returning home slaughters his children and wife. Euripides makes clear that Heracles has returned not from his famous labors, but from "war"—*polemos*—(Eur. *HF* 1135, 1273, note also the reference to *dory*, spear, l.1176) only to find his family threatened by a local tyrant whom he kills. But as so many times in war, there is no on/off switch and Heracles continues to kill, this time his own family. The poet explains the "madness" of Heracles as a response to divine intervention (Hera ordering Iris and Lyssa into action), but this should be seen as a rationalizing effort to explain that which could not otherwise be explained.[5] Euripides told of Heracles' homecoming from battle in the midst of the Peloponnesian war when, as Gorgias also relates, men were returning home from war variously shaken by their experiences.

That Euripides places Heracles' actions in a context of human and divine mischief and jealousy should hardly be surprising—how else could he explain it? His understandable ignorance stands apart from similar gaps in understanding today. Pulitzer Prize winning scientist Edward O. Wilson observes that most people nowadays respect science but find it baffling, preferring instead science fiction, fantasy, and pseudoscience, "like stimulants to jolt their cerebral pleasure centers."[6] This attitude extends as well to some scholars in the humanities who reject the application of modern notions of trauma studies and psychology. For example, J.C. Couvenhes has dismissed my investigations into wartime trauma in the Greek world as an idiosyncratic and distinctly American pursuit.[7] His critique suffers from the same lack of imagination K.J. Dover rejected when describing research into pre-Christian religious attitudes. The subject, he said, "requires a great imaginative effort, and those who make it are commonly regarded as imposters by those who cannot."[8]

This investigation explores the nature of battle trauma in the Greek world from a perspective that takes into account present-day science and psychology. Extensive historiographical or methodological explanation is impractical here, but two primary arguments may be presented in support of this analysis of ancient battle trauma. First, postmodern thought posits that psychology is a social construct and so without merit in the consideration of human motivations. This is not a cogent argument. Society and culture are the creations of people and so biological. While societies and cultures do differ greatly on account of the nature of their experiences, they are ultimately of human and biological origin.[9]

4 See especially the introductory remarks of Brice and Roberts 2011.
5 See further Tritle 2010, 127–128.
6 Wilson 1998, 293.
7 Couvenhes 2005, 431–454.
8 Dover 1987, 195. See also the review by J.D. Galla of Meineck-Konstan 2014 in *Michigan War Studies Review* (2015–097), 1, calling attention to this issue in the humanities.
9 Wilson 1998, 136–177.

This brings us to the second consideration, namely the "Universalist" and "Presentist" debate. "Universalists" emphasize the primacy of the human condition through time and place as a result of biology, while "Presentists" assert that society and culture are particular to the here and now. As the foregoing should have made clear, my view is "Universalist."

In a 2013 symposium at Loyola Marymount University and dedicated to the "Many Faces of War," I had occasion to relate to one of the participants, a medical doctor, the seemingly strange story of Epizelus.[10] The doctor was surprised to learn of this, as many others who may also imagine that "hysterical blindness" is a response to the battle in the World Wars, not ancient Greece. Yet such a "modern" response as Epizelus' is true and other similar responses could be cited.[11] His story illustrates the dilemma over the universality of war and its traumas, particularly in assessing the reality of battle trauma and what is popularly known as post-traumatic stress (disorder), or PTS(D).[12]

Essentially, Epizelus' physiological response to the terror of battle and its repetition in other places and times, and other peoples, argues for the "universalist" approach. Today our knowledge of biology and medicine, science generally, increases almost daily. Not so long ago schizophrenia, for example, was believed to be "an intrusion of the dream state into the waking state." Today schizophrenia is known to be about chemical imbalances in the brain, possibly the result of stress, which can be corrected.[13] The "modern" human brain is some 50 to 100,000 years old and pretty much functions in the same way. Only recently has science begun to understand the brain and how it regulates the body, as our understanding of schizophrenia today makes clear.[14]

On the other side the "presentists" argue for the primacy of culture and society, seemingly unaware of the origins of society. As Edward Wilson reminds us, "history makes little sense without prehistory, and prehistory makes little sense without biology" (Wilson 2014, 9). In a sense to characterize PTS(D) as a cultural construct is simplistic: all cultures are the products of their societies, and societies—people—are the products of biology. To define PTS(D) then as a social-cultural construct is not an especially remarkable conclusion.

Today in the early twenty-first century, the importance of uniting history and science is becoming increasingly clearer if for no other reason than telling the story of mankind from today's vantage point and not that of the nineteenth century. In reviewing Daniel Snail's thought provoking *On Deep History and the Brain*, University of Reading archaeologist Steven Mithen observes that in the future students of history will need to become "scientifically literate," learning "a new set of impersonal actors"—chemicals such as dopamine, serotonin, and oxytocin. A dose of evolutionary biology and psychology, I think, would be no

10 Organized by Tritle, and available in Tritle and Warren 2018.

11 An example from the Vietnam War came to light in 2014 when actual combat footage from the 1967 Dak To campaign (not lost but forgotten in archives) emerged. "Raw War: The Lost Film of Dak To," details an incident of hysterical blindness and deafness by a Fourth Infantry Division soldier coming to the aid of others hit by North Vietnamese mortar fire. Though he recovered, the soldier related how he was struck blind and mute at the sight of his mangled friends. Film available on YouTube.

12 Note here that "Disorder" is deliberately set off. Many psychologists and trauma counselors today (as at the West Los Angeles Veterans Hospital) believe that the condition is a natural response to traumatic events and in no way a "disorder".

13 Eagleman 2015, 60–61 and id., 2011.

14 See also Wilson 1998, 145 (brain development and age in *Homo sapiens*), 141–142 (culture and society).

less useful. At the same time, Mithen adds, biologists and physiologists will have to become "historically minded," appreciating how "human bodies and brains are the products of society and culture."[15]

Terror on the Battlefield

Classicist Michael Flower once asked me, rather skeptically it seemed, if Homer's accounts of ghastly head trauma were credible. My non-committal response was appropriate to the question's implicit tone. Readers of what follows should look closely at the skull of "Gamma 16," the Theban soldier killed at Chaeronea by a Macedonian cavalryman's sword, most likely the heavy, single-edged *kopis*, or "chopper." Readers might also consider the Waterloo story of Life Guardsman Corporal Webster: he tells how he watched fellow Guardsman John Shaw attack a French cuirassier, cutting through the cuirassier's helmet so that "his face fell off him like an apple."[16] Battle trauma in the ancient, medieval, and modern world of war is just that—traumatic for the body and mind. The consequences of such actions linger, sometimes for life, as Webster relates.

In today's western democracies, less than 1% of the population serves in the military. An even smaller figure actually serves in ground combat roles and experiences the cruelties of ground combat in battlefield encounters. Drones and cyber warfare, snipers killing from a distance of a mile, have made hand-to-hand combat seemingly rare, though it certainly occurs. Such realities are an obstacle, not an impassable one perhaps, but an obstacle still, in understanding the fear induced by the prospect of battle: of standing in a hoplite phalanx facing sharp, foot-long spearheads, or charging headlong into battle on a horse without stirrups, and all against others ready to kill.

A few examples of the terror. A passage from Polyaenus describes soldiers awaiting battle to begin, their teeth chattering audibly (*Strat.* 9.1, 8). This finds echoes in Homer's description of Alexandros (Paris) facing Menelaus: "the heart was shaken within him … he draws back and away, cheeks seized with a green pallor."[17] Chattering teeth and terrified faces may be the closest Greek authors can come to expressing the reality, known from other places and times, that some men facing battle lose control of their bowels and bladders. This may in fact be the inspiration for the Aristophanic characters, Dionysus in *Frogs* (l.479), the Sausage-Seller in *Knights* (l.995), and Strepsiades in *Clouds* (l.1389), shitting themselves. It must be remembered that the vast majority of the Aristophanic audience were men, most of whom had fought and knew well this stressful reality of war. K.J. Dover notes that "Old Comedy exploits to the utmost the humorous potentialities of the bowels," and military service might well have played a part in creating this metaphor.[18]

The sights of battle are no less fearsome. Homer again refers many times to men decapitated, or the severing of other limbs. In the prelude to his fateful duel with Hector, Achilles quickly kills 14 Trojans, 5 of whom are decapitated or suffer ghastly head wounds.[19] These

15 Mithen 2008, 24–25.

16 Haydon 1877/2010, 232.

17 *Il.* 3.30–35, trans. R. Lattimore 1951/2011, 118.

18 Dover 1968, 93.

19 Hom. *Il.* 20.382–489. Separated in time by a millennium, but in terms of weaponry little, a Viking grave mound (c. CE 750) from the Baltic uncovered in 2008 revealed one warrior the top of whose head had been sheared away by a sword (Pringle 2017, 39).

poetic references find their counterparts in real life at Marathon and the mortal wounding of Cynegirus, brother of the playwright Aeschylus, his hand severed by an axe; the Theban warrior identified as Gamma 16, his face sliced off by a Macedonian sword at Chaeronea in 338; the arm of the Persian satrap Spithridates severed by Clitus the Black at the Granicus, thereby saving the life of Alexander the Great.[20] The blades of a scythed chariot no doubt inflicted similarly severe wounds.[21]

"Detached limbs," as wounds such as these are defined in today's military terminology, are not easy to stomach. Nor are wounds to the torso as would be caused by a hoplite's spear thrust, easily capable of penetrating a shield (such a spearhead would be about 30 cm in length). Brasidas, the famous Spartan commander, was mortally wounded in this way at Amphipolis (Thuc. 5.10). Another Spartan killed in the fighting to restore the Athenian democracy in 403 also suffered such a fatal wound (Xen. *Hell.* 2.4.33). His remains were found in the Spartan tomb in the Athenian Kerameikos, the spearhead still embedded in his chest.[22]

Sometimes it is claimed that modern wartime wounds are more horrific than those of previous times, but this is debatable. "Pink-misting," the obliteration of a body from bombs or rockets, does not leave much behind. Wounds produced by artillery or shell-fire remain the nastiest in conventional warfare, as the body is shredded on the impact of fragments; wounds from high-powered rifles are often terrible, particularly on exiting the body. All this is a matter of degree: evidence from the ancient world onward suggests that hand weapons as well as "gunpowder" weapons are variously lethal, and each type imprints fearful images of death on contemporaries. Too often military veterans and scholars alike engage in a fruitless debate over the nature and depth of human suffering. Such attitudes obscure the horror of war: seeing the face of another human being like Gamma 16 sheared off is hardly different than seeing a body burned by napalm.

Home After Battle

The experience of the returning combatant in ancient Greece was little different than today. Socrates, returning to Athens from the fighting around Potidaea, found himself responding to the many questions of his friends (Pl. *Char.* 153b–c). Interestingly, his responses were brief and ambiguous, as so often the case of soldiers returning home from war today. Returning combatants in Athens, at least by c. 460, would also have been accompanied by the boxes containing the ashes of those killed in battle and whose remains were cremated, packed, and then returned home (date provided by the reference in Aesch. *Ag.* 434–436).

Alongside the living and the dead would also have been the walking wounded. Gorgias makes clear that life after battle was no easier for them than the survivors of modern or

20 Hdt. 6.118; Ma 2008a, 72–127; Arr. *Anab.* 1.15.8. See also Salazar 2013, 297–299, citing other types of wounds.

21 Scythed chariots were difficult to employ effectively, but when they were they could inflict severe losses, as a group of Greek mercenaries discovered in 395 (Xen. *Hell.* 4.1.17–19). Cf. also Curt. 4.15.14–17 and Diod. Sic. 17.58.2–5.

22 Camp 2001, 133, 134 (tomb details); Salazar 2000, 233–234 (detailed treatment of the wounds suffered by these Spartans).

contemporary wars. A number of the extant testimonies to combat trauma in ancient Greece come from crippled and injured veterans who sought relief from the sanctuary of Asclepius in Epidaurus.[23] Like the Athenian Epizelus, some of these testimonies tell of others who were struck blind in the midst of battle.[24] Both Anticrates of Cnidus and a former soldier named Timon suffered from blindness but unlike Epizelus, both found healing at Epidaurus.[25] The testimony left by Anticrates is clear in its reference to the blindness that temporarily affected him:

> This man had been struck with a spear through both his eyes in some battle, and he became blind and carried around the spearhead with him, inside his face. Sleeping here, he saw a vision. It seemed to him the god pulled the dart and fitted the so-called girls (i.e., the pupils of the eyes) back into his eyelids. When day came he left well.[26]

The Greeks certainly knew of cases of blindness resulting from witnessing the horrors of battle. The sounds of battle were no less frightful to some, as in the case of Nicanor whose fear of flute players the Hippocratic writings record: it was well known that the Spartans marched into battle to the tune of flutists.[27] Another physiological-psychological response to battlefield sights and sounds was mutism, what the Greeks sometimes referred to as *aphonia*. Treatises belonging to the Hippocratic Corpus discuss medical conditions in which *aphonia* occurs, as also the Epidaurian miracle inscriptions, though not of a soldier.[28] Herodotus provides an opposite but complementary example of a mute moved to stress-induced speech: the son of Croesus speaking for the first time warning off a Persian soldier from killing his father (Hdt. 1.87.1).

A fragment of the Hellenistic era poet Posidonius of Pella preserves a literary example of mutism. The text relates how the Cretan mercenary soldier Arcas, described as mute (*kōphos*), made an offering to Asclepius, which healed him (Voutiras 1994, 27–31). Taciturn as well as braggart soldiers are a favorite trope of Greek authors, ranging from Aristophanes, who caricatured the Athenian general Lamachus, to Menander, who created the stereotype of the blustering soldier. Posidonius' examples may now be added to this literary form.

Blind and mute survivors of battle would not have been rare sights in ancient Greece, just as other survivors trying to cope with life after battle. The soldier Euhippos carried a

---

23 First published in Edelstein and Edelstein 1945/1998, more recently in LiDonnici 1995. The latter presents 70 testimonies, of which eight explicitly or implicitly refer to wounds suffered in war as the cause of their visit to the sanctuary. LiDonnici 1995, 15–23, also provides a technical discussion of the texts and their composition.

24 Hdt. 7.229 tells that the Spartans Eurytus and Aristodemus also suffered from blindness. The former ordered his helot attendant to lead him into battle and death, while Aristodemus chose to return home where his fellow Spartans shunned him as a coward.

25 As in antiquity, there are in modern times cases of battle blindness being cured, and sometimes quickly, e.g. World War II soldier Albert Blithe (Ambrose 1992, 98) and Vietnam soldier Robert Walkowiak (see n. 11). This applies as well to mutism, as may be seen in the John Huston film *Let There Be Light* (filmed 1947; released 1981).

26 LiDonnici 1995, 109. Just to be clear: Anticrates' wound is psychological (or more accurately perhaps, neurological). A foot-long spearhead inside one's face or head would cause a surely mortal wound. This caveat applies to Timon as well (LiDonnici 1995, 115) and possibly to another anonymous survivor (LiDonnici 1995, 125).

27 Sources in Salazar 2013, 299, n. 3.

28 Edelstein and Edelstein 1945/1998, 230–231.

spearhead in his jaw for six years before his body finally expelled it; Gorgias of Heraclea returned home with an arrowhead buried in a lung, and suffered for over a year before his body too rejected the missile; Hermodicus of Lampsacus, perhaps a combatant, suffered from paralysis before recovering the ability to walk.[29] Finally, one of the great dramas of Sophocles, *Philoctetes*, features an injured veteran whose foul smelling and painful wound leads his comrades to abandon him on a remote island.[30] Greek physicians knew about sepsis and gangrene, injuries to bones and tissues that, in the case of the former, could only be stopped (one hesitates to say cured) by amputation.[31] As in the tragic case of Philoctetes, gangrene is characterized by "putrefaction and mortification of tissues."[32] This means nasty odors, the sort of thing that would drive people away, just as Sophocles describes.[33] Such wounds could easily result not only from ground combat, but also naval warfare, crushed hulls and shattered oars.

Such survivors as these would have encountered various hardships regards working and living. Imagine, as in the case of Euhippus, trying to manage any occupation with a spearhead (or a piece of one) embedded in the jaw, or the possible pain in extracting it or reducing it in size so that you could work. Or Gorgias of Heraclea with his festering lung wound, which at the time of its healing left 57 basins of pus.[34] Their dedications at Epidaurus point to the physical and psychological stresses, and these latter are reported clearly by Gorgias of Leontini in his *Encomium on Helen*. The passage is valuable as it reveals plainly the psychological dimensions of battle trauma, telling that:

> when warriors put on their armor and weapons, some for defense, others for offense, the mere sight of these terrifies the souls of some, who flee panic-stricken even from anticipated dangers as if these were really present. ... And [later] some of these [men] lose presence of mind, and others become unable to work, suffering terrible diseases or incurable madness.[35]

What are we to make of this testimony? Gorgias is defending Helen's virtue so in this part of his speech argues a point based on sight and how what people see influences their actions. The reference to soldiers is incidental, which makes it all the more valuable as a reflection of what both Gorgias and his listeners knew was happening in the midst of bitter fighting in the Peloponnesian War.

29 References in Edelstein and Edelstein 1945/1998, 232–237 and LiDonnici 1995, 109. The miracle texts record other cases of paralysis, also cured as in the case of Hermodicus.

30 Sources for Sophocles' *Philoctetes*: Tritle 2010, 191–192; Edwards 2000, 55–69. Additional references in "Sayings of the Spartans" (= Plut. *Mor.* 210F), refer to disabled Spartan veterans wishing to fight despite their injuries (e.g. Androclides); others also disabled are encouraged to do so by their mothers (= "Sayings of Spartan Women" = Plut. *Mor.* 331B). Passages also available in Talbert 1988, 128, 161.

31 See Salazar 2000, 32–34.

32 Salazar 2000, 32. Gangrene can kill and in a matter of weeks (or sooner) or months. The elderly Louis XIV survived from 10 August to 1 September 1715, fighting off a gangrenous foot wound (or injury), lately depicted in the 2017 film entitled after him.

33 Edwards 2000, 61–64, also discusses injuries that fail to heal and then become gangrenous. While Sophocles no doubt has complex intellectual issues in mind for this drama's tale, wartime realities, as the setting of the plague in *Oedipus Tyrannos*, provides the inspiration. As a former Athenian general, Sophocles would have been no stranger to the perils of battle.

34 On the nature of pus, identified by physicians as good and bad, see Salazar 2000, 26–27.

35 Text and translation in MacDowell 1982/1993, 16–17, with translation slightly adapted. See Tritle 2010, 158–160 for further discussion.

We might ask, though a definitive answer is elusive, about the numerical value of this reference. How many trauma-stricken soldiers must there have been to prompt Gorgias' statement? A handful, five or six, seems unlikely, but if we multiply this several times, we might have a critical mass. The larger demographic context for our question is uncertain. Population figures for ancient Greece are famously obscure and hotly debated.[36] Combatant numbers, and especially casualty figures, are no less vague. Another issue is how many badly wounded men survived. Authorities estimate that in modern times the ratio of dead to wounded has decreased from approximately 1 to 3 to 1 to 16.[37] This shift underscores the limitations of ancient medicine, which could not prevent wounded men from dying of sepsis weeks or months after being injured.

Gorgias tells us that there were psychiatric casualties, but again we have no figures.[38] Nonetheless, the Epidaurian miracle texts substantiate Gorgias' testimony about mental trauma, or what we know today as PTS(D).[39]

Understanding the Trauma of War

In ancient Greece, and for that matter until the later years of the twentieth century, there was little understanding of how the mind responded to the traumas related in the foregoing discussion. The Homeric explanation of divine intervention and threats of imaginary giant warriors as in Herodotus will plainly not do for the early twenty-first century. Thanks today to the invention of (functional) magnetic resonance imaging, or (f)MRI), increased knowledge of biochemistry and neurology, we can explain much better—though much remains to be learned—of what happens to the mind when humans experience horrific events, as common in war.

Psychiatrist Theodore George compares the human mind to a child's slinky toy—a flexible coiled spring that can take many forms—as it makes its way across a floor or down stairs. But once one of the metal coils is bent out of shape it does not work as before. The traumatized mind is like a bent slinky—yes, it can function but not as well as in its original perfect order. The trauma of war is more consequential in its effects than the ordinary events that happen in life, the death of an aged relative or parent or an accidental injury. Wartime trauma is different in that it "is a threat to personal security or survival," something that "triggers profound, intense feelings of fear, helplessness, and even horror."[40] Such events trigger a response in the nervous system that distorts neurotransmitter activity in the brain so that misfortunes are continually re-experienced. A key part of this nervous system breakdown is damage or stress to that part of the brain called the periaqueductal gray that receives signals from the amygdala, a memory center (among other things), and

---

36 Evidence from Athens is most extensive; Gomme 1933 is foundational to the debate, with contributions from Hansen 1988 and Cohen 2000, 17, 65.

37 Glasser 2011, 33.

38 In this case, however, the modern situation may be only a little better; for discussion see Tritle 2019.

39 Full discussion is impossible, but note at the battle of Delium (424) an Athenian force of some 7,000 suffered nearly 1,000 dead (Thuc. 4.93.3); the wounded (from minor to critical) are unrecorded, but a one to three ratio yields some 3,000 (cf. Krentz 1985b: 16, silent about any wounded, as if that were possible). Several men might well have been ruined by the experience, just like Anticrates, Arcas, or Epizelus.

40 George 2013, 168.

from the cortex, the more advanced part of the brain. In effect, the amygdala assumes control over the body's entire survival system, placing the body in a constant state of alert.[41]

It is perhaps easy enough to see the impact of wartime experience in the physiological responses—blindness, mutism, paralysis—as seen in the individuals recorded on the Epidaurian inscriptions. These are clear examples of how the brain and the nervous system respond to traumatic events in order to protect the body. But there are also less noticeable reactions than blindness, as in depression and emotional shutdown. Such responses as these seem apparent in what Gorgias refers to as those sick and unable to work, on the one hand, and, at the other extreme perhaps, flat emotional and predatory responses, as seen most clearly in the life and career of Clearchus, one-time Spartan commander, later renegade and mercenary general.

A Contrasting Pair of Veterans: Clearchus and Socrates

It is Xenophon, the Athenian soldier of fortune, companion of Socrates, and author who tells us that Clearchus was a man who loved war, who loved a good fight, and who thought, evidently, of little else. Xenophon would know. He spent months in close company with Clearchus as part of the Greek mercenary army organized by the Persian prince Cyrus to overthrow his brother Artaxerxes and become the Great King himself. That venture failed in the battle of Cunaxa: though the Greeks won the battle over Artaxerxes' forces, Cyrus was killed as he rashly charged against his brother (Xen. *An.* 1.8). Left leaderless and unemployed, the victorious Greek soldiers soon lost their generals, including Clearchus, to Persian treachery. Invited to a parley, the Greek commanders were seized, tortured, and killed (Xen. *An.* 2.5).

This in short is the story of Xenophon's "Ten Thousand," thrilling in itself and investigated many times over since its writing. But it is in Xenophon's mini-biography of Clearchus that we find a picture of a man who loved fighting. After a snapshot account of Clearchus' service before the expedition, Xenophon writes:

> This seems to me to be the actions of a man who loved war.[42] He could have lived in peace without dishonor and harm, but he chose to make war. He could have lived a life of ease, but he preferred a hard life in order to make war. He could have had money and security, but he chose to have less money by engaging in war. Just as some might spend money on love affairs or other pleasures, he liked spending money on war. (Xen. *An.* 2.6.6)

Elsewhere Xenophon says that Clearchus was fond of adventure and loved to lead an attack. While this encouraged those he led into battle, afterward many abandoned him as quickly as they could. His tough and savage ways, Xenophon adds, were not endearing qualities

41 George 2013, 69, 36. It is important to note that many or most combatants do not succumb to wars' traumas, as noted by George 2013, 169. While researchers believe that environment and heredity play a part in enabling some to withstand and transcend the horrors of battle, the issue continues to be an intense subject of contemporary research.

42 Translators (e.g. Warner in the Penguin, Waterfield in the Oxford) continue to render ***philopolemos*** as "devoted to war" when the words clearly identify someone who "likes" or "loves" (so ***philos***, as in philosophy, "lover of philosophy") "war" (so ***polemos***). The distinction may seem slight, but loving to kill—because this is what war is about, killing—is not some harmless, unproblematic avocation.

(Xen. *An.* 2.6.11). Xenophon's portrait of confrontational behavior and brute rage suggests that Clearchus is a survivor of violence; in particular, Clearchus' maniacal outbursts of anger correspond to a typical symptom of PTS(D). As psychiatrist Jonathan Shay observes, such outbursts reflect the "persistent mobilization of the body and mind for lethal danger, *with the potential for explosive violence*" (italics added). Clearchus' constant preparation for war reflects the "persistence and activation of combat skills *in civilian life*" (italics added).[43]

For other ancient veterans, exposure to war and violence was transformative in a different way. After his distinguished wartime service, Socrates resumed his inquiries into the meaning of life. It is clear from surviving testimonies reflected in Plato that ethics were of great concern to him and led him to one of his great conclusions, that it is better to suffer wrong than do wrong (Pl. *Cri.* 46b–50a). This is the response of a veteran soldier, weary of killing, who believes that there must be a better way to live. Yet the friends and acquaintances who welcome him home from the front do not grasp this response, and Socrates does not refer to it. Although they smother him with questions about battles and the fates of others, he answers only briefly. Socrates exhibits the veteran's typical reluctance to talk about combat, especially with civilians.[44]

Conclusions

Greek literature and the story of war and violence begins with Homer. The Homeric *Iliad*, so influential in the foundations of western literature, is replete with references to the horror and trauma of war, though only lately has this been recognized.[45] Many of his literary successors echoed this theme: Pindar writes "war is sweet to the inexperienced" (Pind. Fr. 110); Herodotus adds that the Greeks wage war stupidly, with massive losses for the victors while the vanquished are all but wiped out (Hdt. 7.8–9); Thucydides tells that war is a violent teacher, bringing people down to the level of their circumstances (Thuc. 3.82.2). While it may be easy to discount the opinions of intellectuals, their verdicts are more than authenticated by the shattered face of Gamma 16 and by walking wounded like Euhippos and Gorgias, and others like them who attempted to find relief from the trauma of war at Epidaurus and surely other sacred precincts as well.[46]

Appendix: Summary List of Traumatized and Wounded Soldiers

The following individuals suffered either psychic injury (e.g. blindness, mutism) or a physical wound (sources have been cited for both types of wounds). Those identified with an asterisk also have in common a visit or consultation of the healing god Asclepius, suggesting a need for what today might be called psychological relief or counseling. This list cannot be

43 Shay 1994, xx.

44 For this reluctance, see Monoson 2014, 131–162.

45 Homer's impact as the poet of war is complex and would require a long, separate discussion. For useful comments see Rutherford 1989, 13; Vandiver 2010, 228–280. Shay 1994 presents the "real" Homer's take on the horrors and consequences of battle.

46 As an example, note that the Athenians moved the cult of the healer god Asclepius to Athens from Epidaurus in the midst of the Peloponnesian War—a response not only to the plague but the trauma of war. See further Tritle 2010, 115–116; Wickkiser 2008, 65–72.

complete owing to space limitations. It also omits those killed: as philosopher and historian George Santayana said of wartime survivors in post-World War I Oxford, only the dead see the end to war.[47]

Alexander, Macedon: wounded in battle, multiple times[48]
Alcibiades, Athens: wounded in battle[49]
*Anticrates, Cnidus: conversion disorder[50]
Aristodemus, Sparta: conversion disorder
*Arcas, Crete: conversion disorder
Anonymous 1: PTS(D) (unspecified number alluded to, but unnamed by Gorgias)
Anonymous 2: conversion disorder (cf. Posidonius of Pella)
*Anonymous 3: wounded (cf. Euhippus)
Brasidas, Sparta: wounded, later killed in battle
Epaminondas, Thebes: wounded, later killed in battle[51]
Epizelus, Athens: conversion disorder
*Euhippus: wounded in battle, spearhead in jaw
Eurytus, Sparta: conversion disorder
*Gorgias, Heraclea: wounded in battle, infected lung wound
*Hermodicus, Lampsacus [?]: paralysis
Clearchus, Sparta: PTS(D)
*Climenes, Argos, paralysis
Leosthenes, Athens: wounded, (head), later dying[52]
Philip, Macedon: wounded, multiple times[53]
*Timon: conversion disorder

47 Often attributed to Plato, but Palaima and Tritle 2013, 734–735, reveals the link to Santayana.
48 Alexander, wounds: Arr. 2.27.2 (shoulder), 3.30.11 (leg), 6.10.1–2 (chest), 7.9.2 (general reference); Curt. 4.6.23 (leg).
49 Alcibiades, wounded: Pl. *Symp.* 220 E, Plut. *Alc.* 7.4.
50 As noted above, conversion disorder includes blindness, mutism, and paralysis.
51 Epaminondas, wounds: Plut. *Pel.* 4.5 (chest); later killed by the same type of wound, Diod. Sic. 15.87.1.
52 Leosthenes, fatal wound, Diod. Sic. 18.13.1–2.
53 Philip, wounds: Diod. Sic. 16.34.5 (head), 13.3–5 (leg).

CHAPTER 24

Religion and Warfare

F.S. Naiden

Speaking before the battle of Munychia in the winter of 404–403, the Athenian general Thrasybulus summarized the role of the gods in warfare:

> The gods are obviously fighting alongside us. When we need it, they cause storms, and when we come to blows with the enemy, they let us prevail and erect a trophy, even though we are few in number and the enemy is stronger. (Xen. *Hell.* 2.4.14–5)

The general does not need to say why the gods help the Athenians: The Athenians have prayed, made vows and offerings, and hearkened to omens and oracles. They fight for a cause the gods can endorse, and after the battle they will give thanks and erect dedications enriching the gods' shrines.

This chapter will explore the aspects of religion that Thrasybulus omits as well as those he mentions, while noting challenges he overlooks. Previous treatments of this subject go back only about a century, for early handbooks, and even later ones, down through *The Cambridge History of Greek and Roman Warfare*, minimize religion. These works, and others such as biographies of Alexander the Great, assign it the role of morale booster.[1] In this chapter religion plays not only this role but several others—social cement, stress antidote, and planning aide.

Many gods made war, or helped men make it, especially Zeus, Hera, Apollo, Demeter, Artemis, Ares, Dionysus, and Athena. So did Heracles, the Dioscuri, and other heroes.

1 Case 1915; Schwenn 1920–1922; Pritchett 1971–1991, v. 3; Lonis 1979; Jacquemin 2000; Brisch 2008. Alexander the atheist: Lane Fox 1973; Worthington 2014, preceded by Droysen 1833; Tarn 1948. Articles on particulars such as pre-battle sacrifice are cited later.

A Companion to Greek Warfare, First Edition. Edited by Waldemar Heckel, F.S. Naiden, E. Edward Garvin, and John Vanderspoel.

The position of Ares as the only or main god of war is a fiction of Homer and other poets, not a cult phenomenon. Only one leading military power, Thebes, honored Ares as a municipal patron or as a progenitor of the city's armed men. Just as Roman Mars developed out of an agricultural divinity, Greek Ares developed out of a local god worshiped in Thebes and elsewhere. Two other local gods, Enyalius and the female Enyo, became doublets for Ares.[2]

One lesser god, Phobos, the fear that makes men run in battle, deserves notice because of Spartan devotion to him. Spartan worshippers asked him to visit others and not themselves, an exclusively apotropaic cult. In peacetime, the god's shrine was closed.[3] Phobos also resembled a ghost, or demon, as when painted on a warrior's shield.[4] The cult of Phobos reflected the character of hoplite warfare, in which causing an enemy to turn and flee often proved decisive. The Spartans, masters of this kind of warfare, worshiped their own chief military goal.

A second lesser god, Nike, deserves notice because victory inspired most religious rituals occurring after battle. Nike, Zeus, and Athena form a post-battle triad, just as Zeus, Athena, and Heracles were most important before and during battle.

Just as many familiar gods were military, many familiar rituals were staples of warfare. Commanders and soldiers did not need liturgical instruction, and by the same token this survey will take the basics of prayer and sacrifice for granted, just as it will take Zeus and Heracles for granted.

Pre-battle Rituals and Norms

Religion figured in several Greek preparations for war. One of the first preparations, especially in Greek *poleis*, was to enroll young citizen males for military service, an occasion Sparta turned into a mock battle. Spartan recruits divided into two camps, and each let loose a boar against the other. The winning boar provided a good omen to one side. Then the youths assailed one another using boxing gloves loaded with iron weights like brass knuckles. At the start of each campaigning season, Macedonian soldiers fought similar mock battles against each other, but with sticks.[5]

Another preparation for war was to purify the army at the start of the campaigning season, a precaution much like purifying public places such as courts and assemblies. Spartan recruits sacrificed puppies and in Macedon soldiers marched between the severed parts of a dog. Other armies may have performed this sort of purification, too, but the sources do not say whether the celebrants were soldiers as opposed to male citizens. On all these occasions, Hecate and Heracles were mistress and master of ceremonies.[6]

---

2 So, after prevailing at Tanagra, the Lacedaemonians and their allies thanked Zeus, Demeter, Apollo, Athena, Poseidon, several heroes, and two abstractions, *Phobos* and *Pasikrateia*, "Power of every kind" (Paus 5.10.4).

3 Sicyon: *SIG3* 1122. Closed: Plut. *Cleom.* 8–9.

4 *Il.* 11.37, Hes. *Scut.* 144; Heysich. s.v. γοργῶ.

5 Sparta: Paus. 3.4.10 with Baege 1911, 223. Macedon: Liv. 40.6.5–6. A mock mock-battle fought by servants: Plut. *Alex.* 31.2. A brief survey of sacrifices before departure on campaign: Szymanski 1908, 9–20.

6 Macedonian lustration: Liv. 40.6.1; Curt. 10.9.12; Arr. *Succ.* 1.4. A Boeotian example for citizens and not just soldiers: Plut. *QR* 290d. Heracles in Macedon: Hatzopoulos 1994, 92.

These two kinds of preparation initiated recruits, washed away sins, rallied troops, channeled aggressiveness, and provided rudimentary training. They also illustrated the tendency of Greek sources to regard successful armies as pious—in modern terms, bonded through ritual.

Thus purified and invigorated, a recruit deserved sacred weapons, and Greek religion provided them in the symbolic numbers suitable for phalanx warfare. The tyrant Alexander of Pherae carried a spear imbued with divine power, and the two Theban Boeotarchs each carried a sacred or magical spear (Plut. *Pel.* 29). Many if not all major shrines had their sacred weapons, which soldiers were expected to protect, not use. (The sacred weapons at Delphi moved against the Persians of their own accord.[7]) Perhaps no sacred weapon made as much of an impression as the apotropaic devices found on shields and sometimes greaves. These images frightened the enemy by warning him that gods and demons such as Phobos were fighting for the other side. Only a few armies foreswore these devices in favor of uniform shields lacking them.

Mockeries finished and sacred weapons accounted for, an army might face ill omens. An eclipse, lightning bolt, or comet trail would serve, and if Zeus did none of these things Poseidon might cause an earthquake. Lesser, sometimes unidentified gods burned or wrecked houses, or sent plagues. If an eagle picked up a stray javelin near headquarters, the tyrant in command was doomed. Or a bird might fly to the sinister left, not to the auspicious right. Someone might sneeze at a crucial moment, an omen that, unlike most others, is missing from Roman or Near Eastern military sources.[8] Not even the best armies could prevail against the fears aroused by these events. When Agesilaus asked his Peloponnesian allies to contribute to his Asia Minor expedition for restoring Greek liberties there, the Corinthians, the most important ally, refused because Zeus's Corinthian temple had caught fire.[9] On other occasions he and other generals convinced his men omens were positive.[10]

The contrast between positive and negative omens was characteristic of Greek and other ancient military encounters with the supernatural. Commanders had to be sure to distinguish between them, as Agesilaus did, and for help they often turned to seers, or *manteis*. An omen that said "no" might not be final, and ambiguity was interpretable.

Once equipped like magic and fortified by omens, an army needed an enemy and a strategy, and gods provided both through oracles. The military use of oracles by great powers was so common that the Spartan kings had Delphic envoys called *pythioi*, and the Athenians had envoys called *theopropoi*.[11] The typical military consultant assumed who the enemy was, and how to fight him, and asked only whether he and his army would prevail.[12] The oracle thus served most of the same purposes as omen interpretation, but gave more power to two

7 Hdt. 8.37; Rhodes and Osborne 2003, no. 88i.

8 Lonis 1979, 45, especially Plut. *Dion* 24.4–5.

9 As at Lonis 1979, 43–4, especially Paus. 3.9.1. The longest treatment of the subject: Pritchett 1971–1991, 3.91–154.

10 Naiden 2013, 175–81, closely followed below in the section on "Epiphanies" as well.

11 We should dismiss the Elian claim that using an oracle to learn whether to attack fellow Greeks was impious (Xen. *Hell.* 3.2.22, without any parallel for other oracular shrines).

12 As noted at Plut. *E apud Delph.* 386c, and shown by the compilations of Parke and Wormell 1956, and Fontenrose 1978. The only Delphic oracle saying when to fight (Hdt. 5.5), has been reckoned as *post eventum* by How and Wells 1928, ad loc. The only oracle saying where: Plut. *Arist.* 21.3.

parties, the envoys from belligerent states and oracular priests beholden to the amphictyonies that controlled the shrine.[13]

A general who could not get a favorable oracle in one place might get it in another, or might have his publicists describe favorable signs.[14] Although Alexander lacked an oracle from Delphi, he got one from Siwa, where heaven-sent crows, his court historians said, got the king through the Sahara. Themistocles was another commander who excelled at manipulating oracles and omens.[15]

Since no hierarchy among oracles compelled a consultant to accept one over another, shrines competed for business and sometimes *poleis* competed to control shrines. The oracle at Delphi was the occasion for four "sacred" wars that, whatever their geopolitical complications, were partly wars for a valuable military and diplomatic resource—but wars that, once they began, did not differ from other wars, except in length, since two of them lasted ten years.[16]

Before making war, Greek *poleis* usually declared it, and like states in other times and places they justified themselves partly by invoking moral and legal commonplaces supposed to be of divine provenience. One such commonplace was the sanctity of shrines and shrine property, an excuse to fight the Delphic Sacred Wars. The duties of hospitality and supplication were commonplaces dear to Zeus, and the former justified the Achaean attack on Troy, the home of the perfidious Paris. Related to the duty of hospitality was the duty to protect heralds under the protection of Zeus' son Hermes. Religious commonplaces were rhetorical tools. In the Hellenistic Period, these commonplaces found their way into the language of treaties that belligerents would cite in order to claim divine support when defending their independence or frontiers.[17]

War having been justified and declared, a Greek or Macedonian army participated in *diabateria*, sacrifices conducted at the time of departure from home. Next came sacrifices when the army crossed the frontier, amply reported in Herodotus, Thucydides, and Xenophon.[18] (Meanwhile, the invaded made a sacrifice asking the god to ward off the invaders.[19]) Armies would sacrifice daily while on the march; Agesilaus did so at dawn. Armies would also sacrifice before undertaking enterprises such as foraging and building forts.[20] When crossing a river, for example, armies observed the traveler's custom of giving sacrifices to local gods.[21]

13 Hdt. 7.138–44, the oracles being defended by Robertson 1987.

14 E.g. before Leuctra, when Thebes consulted Apollo at Delphi, and also Trophonius at Lebadaea (Paus. 4.32.5; Xen. *Hell.* 4.6.7).

15 For Alexander's religious skills, see Naiden 2018, ch. 5.

16 The first from 595 to 586 and the third from 356 to 346, as at Hammond 1949.

17 Supplication: Naiden 2009, 158–160. Territorial integrity: Chaniotis 2004. General remarks: Momigliano 1961, ch. 1. For other examples of these commonplaces, see Chapter 31 in this volume.

18 Departure: Paus. 4.22.5, X. *HG* 6.5.49, 7.2.21. Frontier: Hdt. 9.10.3, the frontier of the Peloponnesian League; Thuc. 5.54.2, 5.55.3, 5.116.1; Xen. *Hell.* 3.4.3, 5.3.14, 5.4.7, 6.4.19, 6.5.12 *polis* frontiers. Lonis 1979, 97–98 gives fewer examples without drawing the distinction between *diabateria* and other types of sacrifice; so also Pritchett 1971–1991, 1.113–115. A brief survey: Szymanski 1908, 20–35.

19 Paus. 9.13.6, an act of *enagismos* by Epaminondas.

20 Before foraging Xen. *Hell.* 4.6.6, An. 6.4.9. Erecting fort: Xen. *Hell.* 4.7.7; *An.* 5.6.15; Arr. *Anab.* 3.1.5.

21 Daily: Xen. *Hell.* 4.6.6, 5.4.49, 6.5.18. River: Hdt. 6.76, Arr. *Anab.* 1.4.5, 5.8.2. Voyage: Xen. *An.* 6.6.35. Local gods: Xen. *Hell.* 3.1.21 (Dercylidas at Skepsis); Plut. *Pyrr.* 12 (Pyrrhus at the acropolis of Athens); Arr. *Anab.* 1.18.2 (Alexander at Ephesus), 2.5.8 (Soli), 3.1.4 (Memphis), 4.22.6 (Nicaea); Plut. *Cleom.* 30 (Antigonus Gonatas at Sparta); Plb. 4.73.3 (Philip V at Olympia).

Onward to battle? Religion prohibited fighting during sacred truces and festivals or months sacred to important gods. The most frequent sacred truces occurred just before and during leading festivals with athletic competitions, especially the circuit of the Olympian, Delphic, Nemean, and Isthmian games.[22] At Olympia, it covered the games and the month before; at Delphi it eventually covered a year. Prohibited were attacks on the host shrine and interference with participants and spectators en route. Violations were rare, and individual violators might be fined and whipped. In contrast, the punishment imposed on Sparta, a city found in violation of the truce at Olympia, backfired on the games' sponsor, Elis.[23]

Manipulating these truces might prove fruitless. Xenophon reports that when the Argives claimed an extra month for the Nemean truce the Spartans balked, consulted Delphi, and attacked anyway (*Hell.* 4.7.2).

All these were Pan-Hellenic truces. A single state also could declare a truce and suspend fighting for some religious purpose. Every year, at the time of the festival of the Hyacinthia, Spartiates from the village of Amyclae, site of the festival, marched home and left their comrades behind.[24] This festival in honor of the hero Hyacinthus, killed by a discus, lasted five days during which girls rode chariots and even slaves disported themselves.[25] The Hyacinthia was home leave at its most generous, but like much religious custom, it was irregular. Other Spartiates never got their turn, and neither did the *perioikoi* serving with them, the helots serving under them, or the allies serving alongside them.

With these exceptions, and after these various sacrifices, an army could turn to the acts of divination that preceded any siege, battle, or invasion.[26] These acts, accomplished by slaying a sheep or goat and inspecting the entrails, are the best recorded if not most common ritual performed by armies, and for this reason deserve a section of their own.

Military Divination

Pre-combat divination was, in Greek terms, a kind of sacrifice, or *thusia*, but it differed from most *thusiai* on several important counts. Although the victim needed to be in good fettle, it did not need to meet the special requirements for age, color, or gender found in many sacrifices. The commander could use any altar he could find, whereas many civic sacrifices had to take place at one particular location. He could select any day he liked, and so no sacred calendar constrained him. He did not have to have a *mantis*, and sometimes could not find one. He did not have to feed his men afterwards or distribute honorific portions of the animal to a *mantis* or others.[27]

---

22 Olympia: Thuc. 5.49. Nemea: Pind. *N.* 3.1–5; Xen. *Hell.* 4.7.2–3. Isthmia: Thuc. 8.9.1. Delphic: *IG* ii2 1126.44–48.

23 Whipping: Thuc. 5.49–50. Fine: Dem. 19 *hypoth.* 2. Elis: Xen. *Hell.* 3.2.21–31.

24 Xen. *Hell.* 4.5.11; Paus. 3.10.1. Unanswered by the handbooks and others (as in "Organization," n. 4) is the question of how these men were distributed among the *morai.*

25 Truce: Paus. 4.19.3. Festival details: Ath. 4.139e–f.

26 Xenophon alone supplies numerous examples: *An.* 1.8.15, 2.2.3–4, 4.3.9, 6.4.9, 5.6.29, 6.4.9, 6.4.14, 20, 7.8.23; *Hell.* 2.4.15, 3.1.17, 19, 3.4.15, 4.2.18. A complete list would be several times as long.

27 Like *thusia*, *thuein* and *thuesthai* were both used of divination. Casabona 1966, 101, warns against the view that *thuesthai* was the *vox propria* for divinatory sacrifice.

All this contributed to making the commander central. He and he alone put forward the alternatives between which the god chose, typically between attacking and standing pat, but also whether or not to obtain supplies, join forces, or build a wall.[28] Then he or the ***mantis*** slew the animal and inspected the entrails, especially the liver. Unlike the rituals described so far, extispicy was very largely the private concern of commanders, their seers, and any officers invited to attend. Only to quell doubts among disaffected soldiers would a commander invite them to the ceremony.

The sources treat this task as a special version of inspecting entrails during many other acts of ***thusia*** portrayed in vase paintings and adumbrated in Homeric passages allocating innards and other cuts of meat to kings and priests. Positive results were ***kala***, and negative were ***ou kala***, with a bad omen perhaps added.[29] On some occasions the commander and his ***mantis*** responded to sacrifices that were ***ou kala*** by interpreting the altar fire or the contours of other parts of the sacrificial animal.[30] Extispicy was just one way for the gods to say "yes" or "no" to a military plan.

If the commander did not like the result of an act of divination, he could repeat the ritual twice more in the course of one day.[31] Among 36 known cases of negative results, seven commanders tried again. In Xenophon's ***Anabasis***, the army's leaders tried twice to learn whether the gods would favor their return to the harbor of Calpe.[32] On another day, the leaders tried three times to learn whether they should forage.[33] Another general, Cleander, commanding a Peloponnesian garrison at Byzantium, tried repeatedly, over three days, to learn whether to advance. So did another Spartan, Dercylidas, when in command in Asia Minor.[34]

Three tries being made, an act of divination was final. Of the 36 negative results, only seven led to a commander's disobeying divine warnings, and three of these seven involved Alexander the Great. Two more involved generals more or less subject to literary reworking: Mardonius in Herodotus and Adrastus in Euripides' ***Suppliants***. Only two instances involved commanders who were historical Greeks: the Spartans Anaxibius and Callicratidas. Anaxibius and Callicratidas disobeyed and met with defeat, the same as Mardonius and Adrastus.[35] Only Alexander disobeyed successfully—his father, Philip, never disobeyed—and even Alexander did not always succeed fully.[36] Victorious at Gaza, he was frustrated, if not defeated, at the Tanais River in Scythia, and wounded at the siege of the Malli capital.[37] A divine warning was to be heeded.[38] Or so say the historical sources. Perhaps some commanders had

28 Naiden 2013, App. B nos.16–18, 21, 25.

29 Naiden 2013, App. B. Omen: no. 28. Signal: no. 13. Words to that effect: nos. 11, 21. Manteis: nos. 24, 35. The only other list of failures: Pritchett 1971–1991, 3.73–80.

30 Liver: Naiden 2013, App. B, nos. 25, 34. Fire: no. 6; E. *Ph.* 1255–1259 (a successful sacrifice). Head: no. 27. Bladder and tail: schol. E. *Ph.* 1255.

31 Xen. ***An.*** 6.4.13, 6.4.15. This section of the chapter largely follows Naiden 2013, 175–181. A survey of sacrifices of this type: Szymanski 1908, 48–53.

32 Naiden 2013, App. B, nos. 14–15.

33 Naiden 2013, App. B, nos. 16–18.

34 Naiden 2013, App. B, nos. 20, 22.

35 Alexander: Naiden 2013, App. B, nos. 29, 32, 31 (in two of three sources). Mardonios: no. 3. Adrastos: no. 6. Anaxibios: no. 26. Kallicratidas: no.2 8.

36 Philip: Naiden 2013, App. B, no. 27.

37 Gaza: Naiden 2013, App. B, no. 28. Frustration: no. 30, with escape of the enemy (Arr. ***Anab.*** 4.4.3) and sickness in camp (*It. Alex.* 85). Wound: 21. Other views of no. 30: Pritchett 1971–1991, 3.80 with refs.

38 In one case, in Xenophon, a subordinate disregarded negative results, as is discussed in what follows.

a different opinion, but kept it to themselves. Even men such as Themistocles and Alexander mostly preferred manipulation to defiance.

Like some Greek magistrates, a commander might pick a particular *mantis* to assist him over the course of his term of office or longer.[39] Unless the commander was Alexander, he and his *mantis* rarely quarreled. In any event, the commander had the last word. He regarded the skill of the *mantis* as he would that of the physicians or craftsmen whom Xenophon says accompanied a Spartan army on campaign (*Lac.* 13.7, 11.2). Without drawing on these skills, the commander would find victory harder to achieve, but responsibility for achieving it fell to him, not on subordinates. He was the one who would be reelected if successful, and perhaps put on trial if unsuccessful. In contrast, no *mantis* is said to have suffered for being wrong; some were rewarded for being right.[40]

Generations of scholars have recognized that generals manipulated this rite. An early expression of this view came from the historian of religion Arthur Darby Nock, who compared divination to oratory. Both, he said, were means of persuasion.[41] Other historians of religion went farther in the same direction. Conceding that divination was sometimes a means of persuasion, these scholars said that at other times divination was a trick commanders or *manteis* played on soldiers.[42] Still others have dissented, especially in recent years, and called divination a kind of "imaginative interpretation" rather than a rhetorical device or a trick.[43] In this view, the commander did not persuade or deceive others, but might explain events for his own ends. Some scholars, including Walter Burkert, have held that divination allowed the worshiper to avoid indecision and resolve disputes. The rite served a cognitive purpose.[44] Military historians have mostly followed Nock. They have not addressed the question of why divination by way of extispicy is absent from Homer, and why it is more prominent in the Classical than Hellenistic period.

Did Greek troops find the rite persuasive, as Nock claimed? The celebrant persuaded the men to follow his lead, and he did so partly by consulting the gods, just as he did by consulting oracles, but the god, not the celebrant, persuaded the army to take some course of action. If we ask whether the celebrant tricked his men, we must answer "no." In none of the 36 cases of rejection did the celebrant suppress the news and report the opposite to his troops. This evidence outweighs any vague, general statements that accuse generals of this kind of deception, partly because the statements are few, and partly because they form a motif in late authors. According to Plutarch, Agesilaus sometimes gave orders to place letters signifying good omens on a sacrificial liver. (The liver would hold the markings and the king would invite his soldiers to see for themselves, *Apophth. Lac.* 215d.) Polyaenus picks up the motif but attributes it to Attalus of Pergamum (Polyaen. 4.20.1). Frontinus assigns it to that likely target, Alexander the Great (*Str.* 1.11.14).

In contrast, the one early source for deception, Xenophon, explains how a general would deal with a false accusation of manipulating results. In the *Anabasis*, his men suspected him

39 Cities: Pritchett 1971–1991, 3.61–62. Commanders: Lonis 1979, 105.

40 E.g. Tisamenus of Sparta, rewarded with citizenship for being right in five major battles (Hdt. 9.33, 35).

41 Nock 1942, 476, "Divination was comparable with oratory." So also Popp 1959, 39–74.

42 Trickery: Meyer 1925, 190–191 regarding Xen. *An.* 6.4 .12–5.2.

43 Imaginative if pious interpretation: Jameson 1991, 220–221.

44 Burkert 1985, 109–114; Flower 2008, 105: "it aided decision-making, circumvented indecision, and arbitrated disputes." An account stressing *manteis*, not commanders: Pritchett 1971–1991, 3.73–91.

of this, and to quiet their suspicions he invited them to attend a sacrifice.[45] Xenophon knew that the celebrant must not appear to contradict the god. This obligation accounts for the unscrupulous Alexander's action on another occasion when the results of extispicy proved unfavorable. Told he must take Tyre within a month, or abandon the siege, Alexander added two days to the length of the month.[46]

If we ask whether the commander shaped the process, the view of recent scholars, we must agree. Besides deciding whether to try again, the commander timed the rite. Some few commanders did more.[47] Alexander induced his *mantis* to find negative results positive, and Iphicrates delayed in obeying the god. Yet these exceptions appear in the Alexander historians and in Polyaenus. Such sources show not what most did, but what the boldest or most devious did. The power to shape, like the powers to persuade or deceive, was circumscribed.

All of these scholarly views tend to justify rejected sacrifices of this kind—to make them persuasive, imaginative, or useful. Yet these rejections cost lives. At Plataea, rejected acts of divination left the Spartans vulnerable to showers of arrows from the Persians.[48] If we ask one last question, which is why celebrants ran the risk of this sort of suffering, our sources give two implicit answers.

First, negative results were unavoidable. In Xenophon's *Anabasis*, the biggest source, 12 rejections rival 14 successes.[49] The second biggest source, his *Hellenika*, presents one-quarter rejections.[50] Thucydides and Herodotus present more rejection than success, including rejection motivated by authorial bias against Spartans and Persians.[51] (In contrast, Xenophon shows Spartans both succeeding and failing, the same as others.[52]) Only tragedians are optimistic about extispicy.[53]

Second, commanders risked rejection in the hope that, as the army waited for a good sign, the situation might improve. When Dercylidas approached Cebren in Asia Minor, just after taking nearby places without a fight, the town's refusal to surrender made him angry, and he sacrificed in order to give battle, only to be rejected. Meanwhile the townspeople convinced

---

45 Naiden 2013, App. B, no. 14, repeating no. 13, at which they did not attend.

46 Naiden 2013, App. B, no. 33.

47 Alexander: Naiden 2013, App. B, no. 31 (but cf. Arr. *An.* 4.4.3, *It. Alex.* 85). Iphicrates: no. 36.

48 Unless this act was not *thusia*, but *sphagia*, a kind of pre-battle sacrifice discussed below. Hdt. 9.72 is one of only two cases of any pre-battle sacrifice (the other, Xen. *An.* 6.4.8) being described as σφαγία χρηστά vel sim.

49 Twelve rejections: Xen. *An.* 2.2.3, 5.5.2, 5.6.16, 6.1.22–4, 6.4.13, 6.4.15, 6.4.19, 6.4.20, 6.4.22, 6.4.25, 6.6.35, 7.6.44, or nos. 10–21 in Naiden 2013, App. B. Fourteen successes, Xen. *An.* 1.8.15, 2.2.4, 3.5.17–8, 4.3.9, 4.6.23, 5.4.22, 5.6.29, 6.4.9, 6.5.2, 6.5.8, 6.5.21, 7.2.15–6, 7.8.10, 7.8.20. Just one of these 26 instances is described as *sphagia* (6.5.8).

50 Five rejections: Xen. *Hell.* 3.1.7, 3.4.15, 4.4.5, 4.7.7, 4.8.36, or nos. 22–26 in Naiden 2013, App. B. Fifteen successes: Xen. *Hell.* 3.1.19, 3.2.16 (unless interrupted at 3.2.18), 3.5.7, 4.2.18, 4.5.10, 4.6.6, 4.7.2, 5.1.33, 5.3.14, 5.4.37, 5.4.49, 6.4.19, 6.5.12, 6.5.17, 6.5.49. 6.5.18 is ambiguous: the Spartans must retreat, but do so safely.

51 Thucydides, three out of five, with rejections starred: 4.92.7, 5.54.2*, 5.55.3*, 5.116.1*, 6.69.2; these three are nos. 7–9 in Naiden 2013, App. B. Herodotus, four out of seven with rejections starred: 6.76*, 9.19 bis, 9.36–7.1* bis, *9.61, 9.62; these four are nos. 2–5 in App. B. Spartan rejection in Thucydides: all three starred examples, vs. no examples of success. Persian rejection in Herodotus: 9.36, again vs. no example of success.

52 Spartan success: Xen. *An.* 4.6.23 (Cheirisophos); *HG* 3.1.19, (Dercylidas), 3.5.7 (Pausanias), 4.5.10, 4.6.6 (Agesilaus), 4.7.2 (Agesipolis), 5.1.33, 5.3.14, 5.4.37, 5.4.49 (Agesilaus), 6.4.19 (Archidamus), 6.5.12, 6.5.17, (Agesilaus). Rejection: Naiden 2013, App. B, no. 10 (Clearchus), 20 (Cleander), 22 (Dercylidas), 23 (Agesilaus), 25 (Agesipolis), 26 (Anaxibius), 27 (Calllicratidas). Others' success: X. *Hell.* 4.2.18 (Thebans), 6.5.49 (Iphicrates). Other failures: Naiden 2013, App. B, nos. 11, 14 (commanders of the Ten Thousand), 18 (Cleanor of Arcadia).

53 Tragic rejection: Naiden 2013, App. B, nos.1, 6. Success: E. *Heracl* 399–402, 672-3; *Ph.* 1109–1011.

their commander that it was wrong to side with the brigand who now ruled the region and better to side with Dercylidas. At last the sacrifices turned favorable, and he advanced, only to find that the city would surrender.[54] The Ten Thousand benefited, too. When they reached the Tibarenians, the generals at first refused to accept the locals' hospitality. Plunder might provide more supplies. After the sacrifices proved negative for war, the generals reversed themselves, accepted hospitality, and found themselves in a Greek colony just two days later. The negative sacrifice had saved them from needless fighting.[55]

The explanation for the appearance of this ritual in the Classical period, and its disappearance only a few centuries later, may be two changes in the position of the supreme commander. In the *Iliad* Agamemnon scarcely controls his army, and in the *Odyssey* Odysseus scarcely controls his sailors. These commanders did not try to ensure control by resorting to a private rite of sacrifice. In the Classical period, *polis* commanders exerted some control, yet also feared political and legal challenges, and a private rite might buttress their position. In the Hellenistic Period, royal commanders exerted much more control, and the political and legal challenges (rather than the military ones) had vanished. A ruler who received worship in his own right did not need the measure of legitimacy that extispicy provided.

Perhaps another change that devalued extispicy was the shift from personal combat leadership to battle planning and management. The general hoping to plan and control military encounters could leave less up to the good luck that extispicy procured. In this respect, the skills of the general competed with those of the *mantis*, and the general's *techne* came to prevail. In later antiquity, the *mantis* gave way to diviners with more abstruse skills, such as the Chaldaean astrologers favored by Roman emperors.

Even after divination, a final sacrifice remained. When in full view of the enemy, Spartans sacrificed a nanny goat to Artemis, and Athenians made some customary sacrifice, too, as did the Ten Thousand; other Greeks probably did. The commander slew the victim but did not skin it, let alone dismember it. If possible, the sacrifice occurred at the last convenient moment.[56] A positive result was virtually guaranteed, for the animal need only bleed visibly and copiously.[57] That propitiated Artemis and also the chthonic spirits that were perhaps addressed by the Athenians and the mercenaries. The choice of a goat is unexplained, but these animals were cheap and easy to feed. A herd accompanied every Spartan army.[58]

If a commander had no time for this ritual, he would try to perform a sacrifice later, during a lull in the fighting.[59] If he did have time, this ritual was *de rigueur*.[60] Of all kinds of military sacrifice, it was surely the most common.

The Greeks termed this sacrifice *sphagia*, or bloodletting, and not *thusia*, which fundamentally meant "making smoke."[61] In all of Greek literature, only one sacrifice blurs this distinction, and

54 Naiden 2013, App. B. no. 23 with X. *HG* 3.1.18–19.

55 Naiden 2013, App. B, no. 11 with X. *An.* 5.5.3. Similar: no. 25, where Agesipolis abandons his plan to build a wall, but only after he has succeeded in his campaign against Argos.

56 Interruptions that could not be avoided this way: Plut. *Arist.* 17. One case of sacrifice at a distance: Xen. *An.* 6.5.8, because enemy cavalry have been spotted approaching.

57 Another apparent exception, Plut. *Phoc.* 13.1, is avowedly doubtful.

58 Chthonic addressees: Pritchett 1971–1991, 3.67–68. Similar sacrifice at Syracuse: Jameson 1991, 212.

59 No time for sacrifice: Thuc. 5.11.2. Later: Thuc. 6.69.6.

60 Sparta: Hdt. 9.33, Xen. *Hell.* 3.4.23, 4.2.20; Plut. *Lyc.* 22.4–5; Diod. Sic. 15.85.1. Athens: Hdt. 6.112; Plut. *Them.* 13.1, *Phoc.* 13. The Ten Thousand: Xen. *An.* 1.8.15 (*sphagia*).

61 Neither eaten nor inspected: Xen. *An.* 1.8.15 and 6.5.21, where *sphagia* refers to killing without inspecting, and *hiera* refers to both.

a Persian, Cyrus the Great, performs it in Xenophon's romantic biography. The Persian king even tells his hungry men to eat the flesh of an animal killed in a pre-battle sacrifice.[62] Yet the difference between the two kinds of sacrifice did not keep them from being alike in one respect: both assured the army of divine favor. Divination and propitiation shared this aim.

After this sacrifice, the troops prayed, and then the flutists Sparta brought into battle began to play the tune known as "Castor" as the men raised the Paean. Other Greeks sang the Paean, too, but minus the flutists. The noise was loud enough that singing enemies hidden from view by trees or hillocks could be heard before they were seen.[63] After singing what was surely this very short hymn, the soldiers raised a war cry Xenophon terms "alalai," the meaning of which is unknown.[64]

No text of this kind of Paean (for there were several other kinds) survives, and neither does any vase painting or picture that is sure to portray soldiers singing it. Only scholiasts say that it was sung to Ares or Enyalius.[65] What was it for, besides boosting morale and frightening the enemy? Along with the flute music, the Paean kept men marching in order. It also gave Ares a prominent place that this god lacked during pre-battle sacrifices. The war god was, in effect, a god of pitched battle, a kind of Olympian field officer to be contrasted with the higher-ranking Zeus. The blending of Ares and Enyalius, a mere spirit, confirmed this status. Ares thus came to the fore only after other large forces, figures, and emotions sent him into action. The soldiers, in turn, surrendered themselves to Ares at the very moment, the onset of combat, when their officers would lose some or any control over them. During lulls in the action, the officers might regain control, make men regroup, as Agesilaus did at Coronea, and order a new attack (Xen. *Hell.* 4.3.18–19). Then the troops raised the Paean and war cry again. In this sort of case, Ares reinvigorated them.

This last pre-combat rite is the only one for which Greek writers provide an origin (but unsurprisingly provide it only for Sparta). In early Sparta, a scholiast says, the priests of Ares stepped between the Spartans and their enemies as soon as the goat had been slaughtered. They then threw lighted torches toward the enemy (Schol. Eur. *Phoen.* 1186). The use of flutes and marching music came later. The early custom makes sense for the period, perhaps down to the seventh century, when the Spartans did not fight in formation. Fire would intimidate the enemy, and it would not matter whether smoke sometimes obscured them. A melee would ensue, not a set-piece battle.

Only then, in the frantic confusion of combat, did the Greek gods toss aside the panoply of ritual or rhetoric and manifest themselves directly. These manifestations, or epiphanies, were diverse, just as sacrifices were.

Epiphanies in Battle

Once combat began, a god or hero could appear in any of several guises, or appear first in one guise, then another, or appear one way in one source and another in the rest. The observers of the epiphany might be civilians, messengers, officers, or soldiers. These variables

62 Xen. *Cyr.* 6.4.13, 7.1.1.

63 Lonis 1979, 117–118.

64 Xen. *An.* 4.3.19, 1.8.17.

65 Schol. Thuc. 1.50.5, 4.44.3. The meter is unknown, but other military marching songs were iambic or anapestic, as at Page 1962, 856–857, 870.

made it possible for commanders to fake epiphanies, a stratagem that has interested historians of religion more than military historians.[66]

A god or hero might appear in his own person, but to judge from the sources at our disposal, which are almost all literary, he did not do so most of the time. Outside of Homer, Zeus, Poseidon, and Hermes appeared only once apiece, Apollo, Athena, and especially Artemis more often. Sundry heroes appeared, but only the Dioscuri frequently did. A number of gods and heroes are never reported to appear in this form, and instead appear in a dream, and in these cases the sources mostly report visitations to commanders. Access to gods depended on rank, the same as access to sacrifice.[67]

Less fulsomely, a god or hero might appear as a social or natural phenomenon. When panic struck an army, Greeks detected Phobos or Pan. After the Greeks invoked Boreas, this god swooped down on the Persian fleet and scattered them. From Homer onwards, plague was a divine manifestation, too. The gods and heroes mentioned in the preceding paragraph all made themselves known in this fashion, but Artemis and Athena were less important, and the Dioscuri, shining at sea for sailors, were more so.

A god could also appear as a statue or other image an army brought with it on campaign, and then switch guises, and appear on the battlefield, giving aid, even though the source may not say any soldiers saw him. The Spartans made a practice of bringing statues of the Dioscuri with them. In these cases, the god was a public, liturgical phenomenon, a matter of civic policy rather than personal experience.

Rather than appear as a statue, the god could become an anonymous human being and the hero could become a snake, two guises recognized after the fact (Demeter appears only in the former way). The supernatural being was thus employing a stratagem or a sort of proleptic epiphany. By the same token, a god could appear near a battlefield, and not on it, or appear in some place connected to a campaign, and either way a civilian might be the only person to see him. For the soldiers told of this event, there was an epiphany via hearsay. Pan appeared this way to the Athenian messenger Pheidippides before the battle at Marathon.

Last of all, the god might remain invisible, and show himself by the unexplained movement of objects, especially weapons. At Delphi, gods or heroes did this twice, warding off the Persians and the Gauls from the most important Greek oracle.[68] Epiphanies during these two great crises were axiomatic.

To all these possibilities, generals added another, stratagems of their own. Epaminondas either had sacred arms moved from a shrine, or exploited the accidental disappearance of the weapons, or, less boldly, had them cleaned and polished, and thus convinced his men Heracles was fighting for them. That stratagem worked on the best infantry in Greece, the Boeotians. The tyrant Pisistratus worked a bold political stratagem on the Athenians, convincing them that a country girl dressed in bright armor in a chariot was Athena, coming to town to endorse him.[69]

66 There being no military monograph on the subject, nor even a long article devoted to it. Cf. Wheeler 2009, on Roman battlefield epiphanies.

67 For this and the next four paragraphs, see Pritchett 1971–1991, 19–39, to which add *POxy* 3965, with an epiphany of the Dioscuri and perhaps of two other heroes, and Archilochus 94, 98 *PLG*.

68 Persians: Hdt. 8.38–39. Gauls: *SIG*3 398 vs. Diod. Sic. 22.9.5; Justin 24.8.5–12.

69 Epaminondas: Diod. Sic. 15.53.4; Polyaen. 2.3.8. Sceptical of Epaminondas' role: Xen. *Hell.* 6.4.7. Pisistratus: Hdt. 1.60. Herodotus objects to the credulity of the Athenians, but does not claim that Greeks in other cities would have detected the ruse.

Scholars of the nineteenth and early and mid-twentieth centuries responded to this array of evidence by focusing on the superstitiousness of ancient soldiers. In recent decades, however, scholars have viewed epiphanies as an effect of traumatic stress disorder during combat.[70] The older theme, superstitiousness, misses certain semiotic aspects of the evidence. Gods and heroes entered battle as variously as they responded to sacrifices, prayers, vows, and hymns. Just as livers spoke, so did dust clouds. Statues came to life, and so did the weather. Epiphanies depended on ontological slips—the wind-borne god who was only wind, the goddess who was only an anonymous woman, the panic that was also Pan. The Greek pantheon made these slips easy. It included anthropomorphic beings, personifications, and ghosts.

The newer theme, stress, explains cases like that of Epizelus, the Greek who saw a giant fighting beside him at Marathon. Temporary blindness such as Epizelus suffered may be regarded a symptom of conversion disorder, in which the victim, rather than evaluate a threat, displaces and then ignores it.[71] Stress, however, cannot explain the resemblances between combat epiphanies and epiphanies in other circumstances, whether among civilians or before battle. It cannot explain Pisistratus' success in using a country girl as a goddess, and it is not the only explanation for Phidippides encountering Pan at an opportune moment. Seeing Pan out in the countryside was commonplace in rural life.

The remarks of Thrasybulus quoted at the start of this chapter show the importance of epiphanies, a religious phenomenon not to be underestimated. By the same token, the rituals that preceded and followed epiphanies should not be overestimated. Fundamentally, ritual and epiphany should not be segregated from each other. Every ritual aimed to make a god manifest in some sense. Epiphanies were partly an indirect reward for ritual piety, partly one more proof of divine power and unpredictability. They made violent death supernatural, and thus made it bearable yet unavoidable.

Post-battle Rituals and Norms

Once the fighting stopped, three religious duties fell upon the victor—to erect a trophy, receive suppliants, and make thanks offerings. Another duty, conducting funerals, fell on both sides.

Erecting a trophy was the first and simplest of the three. The victors took the armor from one of the enemy dead or from a prisoner—sometimes the armor of the enemy commander—and hung it from a crucifix made of wood or some other perishable material. After erecting the trophy the victors sang a Paean and poured libations. The trophy belonged to Zeus *Tropaios*, the Zeus who had caused the turning point in the battle. No other god received this epithet save for Artemis and Poseidon, and she received it only in her home ground of the Tauric Chersonese, while he received it just once, after sending waves that brought about the turning point in a battle fought on a beach.[72]

70 Superstitious and irrational, and increasingly so in the Hellenistic Period: Chaniotis 2005, 145. Stress disorder: Tritle 2014. Resemblance between military epiphanies and others, and between Greek and Roman: Pfister 1924. Bibliography: Platt 2015. The longest list of Greek military epiphanies: Pritchett 1971–1991, 3.19–39.

71 For a different view of these instances, see Chapter 23 in this volume.

72 Zeus and Artemis: *SEG* 47.177. Poseidon: Poseidonios *FGrH* 87 F 28.220.

Before the fifth century, apparently, the trophy stood at the turning point itself. Later, trophies built of bronze and stone were put in some prominent place on the battlefield. These trophies were monuments as well as offerings.[73]

The trophy's history was brief. Missing from Homer, it began with hoplite fighting, which provided both the armor and the turning point, and lasted until the end of ancient Greek warfare in the second century BCE.[74] The one perverse or unusual use of this custom came before the battle at Leuctra, when Epaminondas erected a trophy using the arms of an anti-Spartan hero. This turning point was meant to be prophetic, for the hero was the Messenian Aristodemus, who had defeated the Spartans in the Second Messenian War.[75]

Why honor Zeus *Tropaios* with a kind of human effigy? Answers involving magic fail to explain why the trophy is missing from Homer. Answers involving the underworld fail to explain why the libations were of an Olympian kind, and were evidently poured chiefly to Zeus and Nike. Answers involving purification fail because Zeus did not regard killing in combat as wrongful.[76] Trophies remain an uncanny element of Greek warfare, but one that confirms the role of offerings and of Zeus rather than Ares.

Next, victors disposed of any suppliants. The captured commonly resorted to supplication in order to increase their chances of being spared and ransomed. On the battlefield, they surrendered with more or less ceremony. In cities, they often took refuge in shrines and pled with victorious supplicandi from the safety of an altar. All acts of supplication, however, had a religious aspect, for Zeus *Hikesios* watched over them, just as sundry gods watched over altars.

Homer provides the most examples of the situation in which the individual soldier acts on his own and takes suppliant captives. Achilles, Diomedes, and Odysseus all deal with suppliants, the same as Agamemnon.[77] In later sources, supplication by individual soldiers gives way to supplication by large numbers, even whole armies. Defeated Greeks supplicate Greek victors, throwing down their arms and raising their hands and asking to be spared. Victors commonly comply with this request, the reason being that both parties are Greek. Thucydides calls this practice a *nomos*—in others words, an irrefragable appeal that leaves the supplicandus no choice but to accede or face accusations of flouting the norms of warfare.[78] Other sources agree with Thucydides, and instances in which the appeal meets with rejection are few and notorious. The Theban suppliants who appealed to Alexander's men after the siege of the city were hauled away and slain, a scandal that some of Alexander's generals thought caused his eventual death.[79]

After being spared, these Greeks might be ransomed or sold into slavery. The *nomos* kept them alive, but did not liberate them. That depended on the choice of the victor and the efforts of the relatives or *polis* of the captives.

73 Athens: *IG* ii2 1006.28–29, 100817–100818, 1028.26–27, 10780.2; *SEG* 5.104.22, 19.108.71; Thebes: *IG* vii 2462. Illyria: *SEG* 35.680, 38.527. Thrace: *JÖAI* 23 (1926) Bb 119.3. Pergamon: *IvP* I 237 (cult alone mentioned). A full list of relevant sources for trophies: Rabe 2008.

74 Thus also Lonis 1979, 137. Yet there is no secure evidence for *tropaia* erected at battlefields before the fifth century BCE, as noted by Rabe 2008, 12–15.

75 Paus. 4.32.5–6, saying that Epaminondas obeyed a Delphic oracle—a very unlikely oracle considering Delphi's cautious attitude towards the political claims of Messenia, discussed at Naiden 2009, 210–212.

76 A survey: Lonis 1979, 130–136; Rabe 2008, 15–22.

77 Agamemnon: *Il.* 11.130–144. Achilles: 20.463–469, 21.64–119, 22.337–360. Diomedes: 10.454–456. Odysseus: *Od.* 22.310–356, a quasimilitary situation.

78 Th. 3.58.2; also D.S. 13.23.5, 26.3, and Sopat. Rh. *RG* 8.241, with Naiden 2009, 87.

79 Naiden 2018, App. 2, no. 1.

Both sparing suppliants and enslaving them made strategic sense. Mercy would encourage future enemies to surrender and enslavement would create a labor force at the victor's disposal, and make rebellion among the captives more difficult. That paragon of piety, Agesilaus, was one of many commanders who sold shrine suppliants into slavery; Alexander sold the most, and Achilles is the first figure in Greek literature to be reported to do it.[80]

If the surrendering enemy controlled a city, an invading army would welcome supplication as a way of avoiding a siege. The advantage of this sort of supplication was a quick, painless victory, and the disadvantage was that the suppliants might seek to negotiate terms, and the invaders might find the terms onerous but costly to reject. For example, Asian cities confronted by the ruthless Alexander would send emissaries to entreat him. Their gesture of choice was to wave palms, and their hope was to achieve a settlement falling well short of abject surrender.[81] The priests and people of Babylon used this method with great success; Indian cities with less success. Supplication, in short, allowed for negotiations, just it allowed for a *nomos* protecting Greek combatants.

On the battlefield, supplication was a kind of victor's justice. In shrines, it was a kind of justice dispensed by those in charge. If it was unclear who was in charge, ambiguous situations might arise, like the one after the battle of Delium. Athenian soldiers captured the temple of Apollo and took refuge in it. The Theban victors did not want to attack what had been, and would be, a Boeotian shrine, but said the Athenians had no right to remain there.[82]

Resorting to shrines would seldom be possible in captured camps, a third venue for supplication. Suppliants caught in camp reverted to a Homeric style of supplication, clasping the knees of the victors and especially the commander. One of Mardonius's camp followers, a Greek courtesan, saved herself by appealing this way to King Pausanias. Strategic considerations aside, mercy was becoming to a conqueror, and a century later Agesilaus knew it as well as Pausanias did.[83]

Receiving suppliants was just one of a victor's privileges. Holding thanksgivings was another, whereas giving fallen comrades a funeral was a solemn duty. These military postscripts were as important as the pre-battle divination on which scholars have concentrated.

Like much else in Greek religion, thanksgivings followed acts of *thusia*. These elaborate sacrifices, all ending in a feast, were called *epinikeia* or *niketeria*, or *euangelia*. Mere deliverance merited a sacrifice called *soteria*.[84] Following the sacrifice came festivities that might include games and other contests; Alexander punctuated most of his great victories and conquests with these festivities.[85] Next came the dedication of thanks offerings, including those made in fulfillment of vows. One form of thanks offering was the establishment of a new, regular sacrifice in honor of the victory. In Athens, the largest of all annual sacrifices, 500 goats given to Artemis *Agrotera*, began in honor of Marathon.[86] If an army did not know what to offer, or to whom, an oracle might advise them. After Marathon, the Athenians

80 Xen. *Hell.* 4.4.3–5 with Naiden 2009, 120–122, 2018, App. 2.

81 Naiden 2018, App. 2 no. 4 (Miletus), 14 (Babylon, 24–27 (Indian cities)).

82 Thuc. 4.98 reports the Athenians' sophistic reply.

83 Hdt. 9.76.2; Xen. *Hell.* 4.3.20, Nep. *Ag.* 4.6–8.

84 Two surveys: Pritchett 1971–1991, 3.189–196, concentrating on Athens; and Szymanski 1908, 63–71.

85 Naiden 2018, App. 1a nos. 15, 28, 33, 39, 43. *Epinikeia* and *soteria* both: no. 49 (siege of Malli town).

86 Xen. *An.* 3.2.12. The goat Spartans offered just before battle also was for Artemis *Agrotera*. The goddess thus received *thusia* in one city, and *sphagia* in another.

wanted to thank the giant seen by Epizelus, and asked Delphi who this being was. The oracle identified him and told the city to worship him as a hero.[87]

The most common form of thanks offering was the dedication of spoils of war. As a rule, armies dedicated a tithe of the spoils, but often after selling them and commissioning the creation of some suitable, often spectacular, object, like the 18'-high statue dedicated by the Greeks after Salamis. Aside from the tithe, the victors dedicated certain impressive spoils to the gods—after Salamis, three captured Phoenician triremes.[88] To make sure a god had received his due, the victors might ask an oracle, and so, when Delphi said that the Aeginetans had been stingy after the Salamis victory, the Aeginetans commissioned and dedicated three gold stars atop a bronze mast at Delphi[89] (Hdt. 8.122). Often several gods shared the spoils; after Plataea, Zeus, Poseidon, and Apollo did.

Most victories provided the gods much less. "Religion," volume three of Pritchett's *Greek State at War*, provides an inventory of dedications of mostly mundane military items—shields, corselets, greaves, swords, spears, spear-butts, helmets, bows, and trumpets, to which prisoners' chains should be added.[90] Individual soldiers contributed some of these items, but commanders contributed more. Like many things religious, dedications centered on commanders and the *poleis* that appointed them.

All these dedications were peculiar to warfare, but victors also made dedications of the same kinds as those made in peacetime. The most common of these was a statue of a god. The grandest was a temple, followed by a portico, a treasury, an altar, a bronze statue of animals, a crown, or a tripod.[91] War's profits were an important source for public works of these kinds.

In a different category was a statue of a victorious general. Dedications of this kind resembled the statues of athletic victors erected at Olympia and elsewhere. Such statues conferred honor, but unlike statues of gods they did not generally elicit worship. Those few statues of generals that could or did receive worship formed part of the context of Hellenistic ruler cult, not the older practice of conferring honor on victors.[92] Perhaps some statues of generals formed part of the context for hero worship such as the cult of Brasidas at Amphipolis.[93] Statues of other soldiers, such as those of Alexander's cavalrymen killed at the Granicus River, received honor rather than worship.[94]

The Greeks did not always burn their dead, but Greek armies did, partly because they lacked the means to preserve any large number of corpses from decay on a long journey

87 Paus. 1.32.5; other examples at Lonis 1979, 150.

88 I.e. *akrothinia* as opposed to the *dekate*, which was a kind of *aparche*, or "first fruits." In some cases, however, a dedication cannot be identified as being one or the other. *Akrothinia* means "tip-top" or "foremost." None of these terms appear in Homer in a military context.

89 Hdt. 9.81. Athena and Apollo after a battle in Phocis: Hdt. 8.27.

90 I.e. *skula*, as at Pritchett 1971–1991, 3.240–296. Chains: Hdt. 1.66.3. Lonis 1979 is longest treatment in French, Popp 1959 the longest in German. Rabe 2008, 26–37, 115–116, shows that some of these dedications were called *tropaia*, the word for "trophies".

91 Lonis 1979, 162, 165–166, 170.

92 Memorials: *IG* i3 683, the oldest, for a general who fell at Marathon. The statues paraded at Aegae in the parade before the assassination of Philip II (Diod. Sic. 16.93.4–5) are often thought to belong in a gray area between honor and worship, as at Fredricksmeyer 1979.

93 Statues of generals: Lonis 1979, 164. There is, however, no known example of a heroic statue of a dead general in a *heroon* where this object would receive worship. Brasidas: Thuc. 5.11.1.

94 Plut. *Alex.* 19.8, saying the statues were erected at Dion, but not specifying that they were erected somewhere in the shrine of Zeus. Arr. *An.* 1.16.4 and Justin 11.6.12 do not locate the statues.

home from the battlefield. (There were thus no ancient controversies about bringing back bodies—for example, the controversies about bringing them back from Vietnam to the United States after the Vietnam War, or bringing them back from France to Germany after World War I.) At the pyre the comrades of the fallen made offerings or *enagismata* epitomized by the cattle, sheep, horses, dogs, and captives Achilles slaughtered for Patroclus, and his libations of honey, oil, and wine. Unlike most offerings, these were burned whole (*Il.* 23.166–176). The common soldier of later times, buried with others in a *polyandrion*, received personal items such as strigils and weapons. At most these soldiers received small animals such as birds, as did the Lacedaemonians buried in Attica after fighting there in 404–403.[95]

Athens differed from most cities in bringing the bones and ashes of the dead home for public burial. Beginning in the early fifth century, an annual ceremony filled three days, during which mourners placed bones and ashes in a tent in the agora. Then the mourners carried the remains to the public tomb in the Kerameikos. After Marathon and Plataea, however, the Athenians left their dead in a *polyandrion* on the battlefield, the same as the rest of the Greeks.

Near the tomb stood monumental lists of casualties. These lists did not mention personal exploits, and until the fourth century they even omitted branch of service and rank. An annual festival did take place at the tomb, but perhaps on a different date than the three-day burial. Sacrifices accompanied this festival, but apparently not sacrifices to the dead rather than to the gods.[96]

If the Athenians could not recover the body of a dead comrade, they erected a cenotaph. Not far from the public tomb, an empty bier commemorated the Athenian "Unknown Soldier."[97] Enemies, too, had a right to be buried, but if they left behind their dead, these soldiers received short shrift: at Marathon, the Athenians piled the Persians into a ditch and buried them without ceremony (Paus. 1.32.5).

Various religious notions motivated these ceremonies. A desire to avoid the deterioration of the corpse explained cremation. The belief that a soldier who died after being wounded kept his wounds in the Underworld reflected assumptions associated with inhumation, not cremation. Belief in an afterlife explained burying tools and instruments for the dead.[98] (For further treatment of post-mortem ceremonies, especially secular practices such as the payment of pensions and the delivery of funeral orations, see Chapter 29 in this volume.)

Aside from the "Unknown Soldier," other cities may have equaled, or nearly equaled, this network of practices. Alexander did so during his conquest of the Persian Empire.[99] No

95 Kurtz and Boardman 1971, ch. 13, describes this and other mass burials of soldiers. Soldiers dying in bed were sometimes inhumed, and buried beside their weapons and small food offerings, or cremated with these offerings in separate pits, as at Kurtz and Boardman, 62, 74–75.

96 Another view: honor given to some dead soldiers verged upon and often turned into worship, as at Loraux 1986, 39–40, and Parker 1996, 135–137. *Enagismata* for historical persons were rare: *SEG* 13 312.13 (Megara), Plut. *Arist.* 21.3–5 (Plataea). Prohibited in Athens: Plut. *Sol.* 21.6. The term *heros* for historical persons was also rare: Pi. fr. 133.5–6; Ar. *Ra.* 1039 (Lamachus). Harmodius and Aristogiton, however, made the transition from honor to worship: Dem. 19.280 with MacDowell 2000 ad loc., Arist. *Ath.* 58.1 with Rhodes 1981 ad loc.

97 The Unknown Soldier: Thuc. 2.34.3. Similarly, the Athenians erected cenotaphs for the soldiers buried at Marathon and Plataea.

98 Bremmer 1983, 83–84.

99 Naiden 2018, App. 1.

doubt his Successors did, too, but we know little about the religious practices of these armies, save to say that the introduction of ruler cult did not cause the Macedonian king to lose his traditional religious duties such as pre-battle sacrifice and post-battle thanksgiving. One characteristic of Greek and Macedonian religion was that some lesser gods, such as Hermes, poured libations, and the god-kings of the Successor kingdoms followed this hallowed example, but added animal sacrifice to the liturgical menu.

Concluding Remarks

As Frederick the Great said in his "Instructions to his Generals," religion could be "a valuable military arm."[100] Was it more valuable to some cities or generals than to others, or more valuable to the Greeks than to their foreign foes? The regrettably prudent answer is that we cannot tell whether a reputation for piety made winning easier or whether winning inspired a reputation for piety. The piety of the Spartans, like that of the Romans, was exemplary, but not necessarily explanatory. Piety was important in ancient warfare, for it overlapped with patriotism and propaganda, but the effect of piety is difficult to measure, as is true of patriotism and propaganda in modern warfare.[101]

We can surely say that Greek and Macedonian armies never faced major opponents to whom they attributed a religion greatly different from their own. The *interpretatio graeca*, which attributed Greek gods to foreign peoples, prevented that. Greek wars were thus never conceived as wars between one religion and another, or one sect and another. The religious impulses felt in Greek wars did not lead to martyrdom for the dead, conversion of the defeated, or crusades by the aggressors. Wars with these features appeared only *Annis Domini*.

100 Phillips 1940, 357, cited in, and explored by, Naiden 2019.

101 Cf. Napoleon's famous but vague dictum that "in war three-quarters turns on personal character and relations; the balance, which is manpower and materials, counts only for the remaining quarter," as at Vaillant 1858, 17.471–472. Which relations does Napoleon mean, and among whom?.

CHAPTER 25

Women and War in the Greek World

Elizabeth D. Carney

After the Thebans revolted and Alexander defeated them despite their strong resistance, 6,000 of them were dead and the king decided to sell the remaining citizens, some 30,000, into slavery. The city was to be razed. Plutarch twice (*Alex.* 12.1–3; *Mor.* 259e–260d) tells the story of one of these captive survivors of the fall of Thebes, Timoclea, whom he describes as a chaste woman of good repute. While Thracian troops (part of Alexander's forces) plundered Timoclea's house, their commander forced her to talk to him, raped her, and then interrogated her about the location of any gold or silver she might have. She conceded that she did have some hidden and led him into her garden to show him the well in which she had concealed these valuables. He bent over the well to look at what she had placed there; Timoclea pushed him in and then killed him by throwing stones on top of him. When the angry soldiers took Timoclea in front of Alexander, though her hands were bound, she showed herself a dignified and high-minded person by her countenance and her carriage: she was undaunted and fearless as she followed her captors. When the king asked who she was, she told him that she was the sister of Theagenes, a Theban general who had fallen at Chaeronea. Alexander, impressed by her answer and her actions, commanded that she depart in freedom, with her children.

The story of Timoclea introduces many (though not all) of the issues and difficulties that encompass any attempt to grasp the role, or rather roles, of women in warfare in the Greek world. As we shall see, enslavement and rape were indeed the common if not universal lot of women taken captive when a city fell, but aggressive revenge by women was not. In both his accounts, but particularly in his essay in the *Moralia* (*Mulierum Virtutes*), Plutarch stresses the social and economic status of Timoclea, specifically her family's importance (significantly, it is the rank of the family of her birth, not of her marriage, that dominates Plutarch's account). Alexander frees her because she is brave, but in Plutarch's narrative, in effect, she

A Companion to Greek Warfare, First Edition. Edited by Waldemar Heckel, F.S. Naiden, E. Edward Garvin, and John Vanderspoel.

is brave because she is the sister of Theagenes. In the *Moralia* (260c–d), Plutarch specifically tells us that Alexander could tell by her face and walk that she was of high repute and well born and the king pardons all those related to her (not just her children) and tells his officers not to let such an insult happen again to this sort (i.e. this class) of family. Class did sometimes (though hardly regularly) play a role in the treatment of captives, male and female, and female members of elites sometimes took a more aggressive part in war and its aftermath than other women.

A story as good as Timoclea's need not be untrue. In its general outlines, particularly in the shorter version in the *Alexander*, there is nothing overtly ahistorical about this account, although it obviously furthers Plutarch's desire to demonstrate Alexander's *sophrosyne* (moderation, particularly sexual moderation), but the incident is fable-like, as are many other war narratives of the bravery of individual women or groups of women. In *Mulierium Virtutes*, the tale of Timoclea becomes a parable about the excess of the commander of the Thracians and of Timoclea's moderation, circumspection, and good birth (in this version, the officer goes down the well himself—Timoclea does not push him in—and she is assisted in her revenge by her servants, as is appropriate for an upper class woman).

Virtually all of the material relevant to analysis of the role of women in war was created by men. Two well-known early modern paintings depicting the story of Timoclea illustrate how much difference the gendered perspective of the "narrator" can make. The first painting, "Timoclea before Alexander the Great," by Domenchino c. 1615, fairly faithfully reproduces the tone and content of Plutarch's narrative(s): Alexander the magnanimous ruler makes a gesture of pardon to Timoclea while she, tall and calm, stands in the middle of the painting, followed by soldiers dragging in her screaming children. Elisabetta Sirani's 1659 "Timoclea killing Alexander's Captain" shows a gleeful Timoclea upending the officer into the well; Alexander does not appear. Whereas the first painting tells the story from the point of view of the male ruler and of the male narrator, Plutarch, the second is uninterested in Alexander and focuses on Timoclea, narrating the episode from her point of view. My point here is not the gender of the writers or artists (though our Greek sources are indeed nearly all male),[1] but rather that our ancient accounts are told from the point of view of men, not women. Thucydides (2.45.2), for instance, tells us what Pericles thought women who had lost male relatives in war should do (not be mentioned), but does not tell us what those grieving women thought about his pronouncements.

How Women Participated in War

The male point of view of nearly all extant sources is not the only source problem for attempts to understand the relationship between women and war in the Greek world. Most of our sources are uninterested in women and thus fail to mention them. Even when a male writer is interested—and Plutarch, for instance, is very interested in heroic women—he may mention women or a woman for didactic purposes, whether the lesson is taught by means of a bad or a good model. Nonetheless, the story of Timoclea warns us that we should not assume that if a story has a lesson of some sort that it is false: George Washington did not cut down a

1 Golahny 2011, 39 suggests that the model for Sirani's version was the work of an earlier male artist, Matthaeus Merrian.

cherry tree and then admit it but Abraham Lincoln really was born in a log cabin. Didactic sources need to be read with skepticism but not discounted. Often it is possible to read against the texture of a moral tale and find useful material in what is assumed to be normative and thus is not meant to be the "point." Unfortunately, because of long-lasting positive and negative reads on the Spartan system and way of life, it is particularly hard to parse material about Spartan women. For instance, Aristotle (*Pol.* 2.9.9–10) may be correct in asserting that Spartan women panicked after the massive defeat at Leuctra—he says they were less useful than women in other cities—but he embeds his assertion in an attack on the Spartan system's tendency to allow women more freedom and less control than elsewhere.[2]

Genre and purpose are often relevant. Sometimes an author mentions in passing the situation of women in a given city; although one wishes there were more information, the mere fact that the author is not as interested as we are can signify that the information conveyed in passing is relatively accurate. On the other hand, even a "good" source or major writer can be wrong. A source many centuries removed in time from the events it mentions may preserve important and relatively correct information in spite of anachronisms. Compared to other written sources, inscriptions and contemporary speeches have obvious advantages, but there are relatively few of them.

The majority of our written sources other than inscriptions remain Athenian in origin, fifth or fourth century, and upper class in point of view. These sources used to be read for universal truths: the worlds of Greek men and women were completely divided and separate, a split found in all periods of Greek history, in all areas in which Greeks lived, and in all classes. Yet we cannot reject something as false simply because Plato or Aristotle or Xenophon thought it was true. Nor should we conclude that material is false because it is at odds with generalizations about Greek women; what we think is exceptional is not necessarily untrue and indeed may not have been exceptional.[3]

Some generalizations seem valid: women (apart from the occasional royal one) were not generally participants in offensive military action. Greek artists and writers imagined women who were—the Amazons—but were these fictions created from scattered information about the warrior role of women in some nomadic cultures on the edge of the Greek world, and from Greek fears about people they considered "barbaric."[1] Once in a while a woman or group of women of non-Greek origin, or women who originated in a culture and place that combined Greek and non-Greek features, played a military role, but they were exceptions. Authors occasionally report that a group of women did fight or participate in battle; such reports should be treated with skepticism but not rejected out of hand; Pausanias' account (2.20.8–10) of Telesilla, an Argive poet, who, after the defeat of the Argive army, led women and other traditional non-combatants against the attacking Spartans, is a good example.[5] We

2 For instance, Polyb. 12.6b.8. On the problem of Spartan women and the sources, see Napolitano 1987; Ducat 1999; Powell 2005. Schaps' point (1982, 211) that most of our Spartan women stories are of Roman date remains an important one. Ducat 1999 and Napolitano 1987 stress the great inconsistency about the military tradition of Spartan women. Still, see Penrose 2016, 57–64 for a discussion of ways to "use" the story of Telesilla and other stories about Spartan women.

3 Ducrey 2015 provides a useful recent overview of the topic of women in Greek warfare. See also Chaniotis 2005, 102–114. Schaps 1982 is still helpful, if dated in some respects. See also Loman 2004.

4 See Mayor 2014, 17–83; Penrose 2016, 67–118.

5 See also Plut. *Mul. Virt.* 245d–f.; Polyaen. 8.33. Schaps 1982, 207 n. 115 rejects this story partly because Herodotus doesn't tell it and sees it as an "aetiological invention." Graf 1984b, 246–249 rejects it, essentially because of perceived chronological problems and then rejects all stories of women fighting victoriously against men.

need not agree with Xenophon (*Ec.* 7.23) that women's bodies were unsuited for outside work such as campaigns, but we should concede they were unlikely combatants. Women were not generally trained for war as young men were, and did not usually even have athletic training; lacking these skills, the likelihood that women would choose to engage in battle was small, but it cannot be ruled out. Desperate circumstances easily produce atypical behavior.[6] Fear of rape and enslavement must have been powerful motivating factors for all the actions of women in besieged cities.[7]

Although women did not participate in combat, they did support the men in their families and their cities, especially cities under siege. Although Hector saw fighting as men's work (*Il.* 6.490–493), he also rejects a more cautious approach to warfare because he fears the shame he would feel before Trojan men and women if he did so (*Il.* 441–443).[8] Stories support the notion of the importance of women's gaze.[9] Vase paintings and some other documentary evidence indicate that women—perhaps primarily the mothers of soldiers—were part of two ceremonies or practices relating to a departing warrior. These were pouring libations and the actual arming of the soldier.[10] They do not seem to have participated in religious activity connected to battle itself, but they certainly made offerings to the tutelary gods of the city. Female deities were sometimes understood as defenders of cities, most famously Athena. Women played a critical role in lament for any member of their family and were primarily responsible for the care of the grave and offerings to the dead (Thuc. 2.34.4). If possible, widows ought quickly to remarry, while older women had to hope that their sons, if they survived, would care for them;[11] unmarried girls, without perhaps a father or brother to help with a dowry, could be in difficult straits.

Until the Hellenistic period, men on military service were not often gone for very long periods of time, but they were absent, and if killed or wounded might never return. In practice, this meant that women and old men had to manage the household and somehow support it. If the city itself were to be threatened, non-combatants—women, the elderly, children, and sometimes slaves—might be evacuated (Hdt. 8.36.2, 41.1; Thuc. 2.6.4, 2.14), a frightening hardship in itself. On some occasions, women refused to leave (Polyaen. 8.49), perhaps out of affection for their male family members or loyalty to their city, and also because they considered this risk preferable to what they would face as refugees without male kin and with children and old people to support.

Women along with the entire community might turn out to build or rebuild defenses, bring up weapons, help get ammunition and food to the defenders of the city, cook food, and care for the wounded.[12] We also hear of women and slaves and the elderly throwing tiles and stones from rooftops if invaders broke into a city.[13] Reports of this activity once were doubted,

6 Paus. 10.22.5–6 says that when attacked by the Gauls, male and female Aetolians fought together against the invader. Plutarch (*Apophth. Lac.* 227d), discussing the alleged athletic training of Spartan women, says that one of the purposes was so that they could, if necessary, defend themselves, their children, and their country.

7 Penrose 2016, 60–61.

8 Rousseau 2015 discusses this and many other passages in Homer; she reads Hector's remarks more narrowly than some, but her discussion is illuminating. See also Nappi 2015.

9 See Fuhrer 2015; Nappi 2015 on women as spectators, commenters, and mourners of war.

10 Lissarrague 2015 observes that it is usually the woman who presents the arms, possibly because of the mythological model of Thetis. He notes (2015, 71) that the focus is on the mothers of warriors rather than wives.

11 See McGinn 2008, 22 for discussion and references.

12 Plut. *Pyrrh.* 27.2–5; Polyaen. 8.49.9.70: Thuc. 5.78.3, 82.6.

13 E.g. Thuc. 2.4.2; Plut. *Mul. Virt.* 245c.

but the last century of war, one in which street fighting has played such a part, reminds us that this kind of warfare, though not a way to defeat an entire army, could channel enemy forces into streets and areas targeted by the defenders and could eliminate commanders, as famously happened in the case of Pyrrhus.[14] Granted that women were closely associated with the house/household (*oikos*), such domestic-based defense enabled them to remain in a domestic situation.[15] Moreover, even if such efforts ultimately failed to protect the city, they would still have been satisfying for people who knew what fate to expect when the enemy finally took the city. From the fourth century on, siege warfare grew more common and more successful.[16] In the Hellenistic period, a woman might pay for a defensive tower, and women as well as men might contribute to defense funds or help with decorative military armaments.[17]

What did happen if a city was taken? In some cases, usually because of surrender, the surviving population, including men and women, were allowed to depart, typically with little more than the clothes on their backs (Thuc. 2.70.3). While it is true that cities were occasionally allowed to be reconstituted at some later date and that some allies took refugee families in and found some place for them, departure generally meant becoming a permanent exile.[18] Being a refugee, then as now, made one vulnerable to disease and starvation and, if a woman's male kin were dead, quite possibly rape and/or prostitution.

If the city did not surrender and fell to the attackers, the defenders were often killed and the women enslaved.[19] Women were sometimes killed, quite possibly more often than our sources specifically indicate, especially if a woman belonged to someone important in the opposing faction, and enmity against him was redirected against her.[20] At other times, the entire population was simply enslaved.[21] In Homer, capturing, raping, and enslaving the wives and daughters of your enemy is not only part of victory, it in a sense is victory (*Il.* 6.447–465, 20.193–194). Since personifications of cities were female, besieging and storming a city was a kind of figurative as well as literal rape.[22] In modern times we have seen rape used as an instrument of war and terror; in varying degrees it seems to have been so employed in antiquity as well.[23] Funerary inscriptions survive in which women committed suicide rather than be captured and raped, narratives such as that of Pausanias describe similar situations, and Timoclea, as we have seen, avenged her own rape, but most women were probably

14 Plut. *Pyrrh.* 34; Paus. 1.13.8; Polyaen. 8.68. See Barry 1996 *contra* Schaps 1982, 195.

15 Barry 1996, 69–72.

16 Ducrey 2015, 195, who gives as an example Eretria which was captured four times in 400 years.

17 Chaniotis 2005, 109–110. See especially Bielman 2012, 92–95.

18 Reconstituted cities: Diod. Sic. 13.89.4. Permanent exile: Thuc. 1.103.1, 2.27.2.

19 Thuc. 5.32.1. There are reports that have the men of a city kill the women and children and destroy their possessions rather than let the enemy acquire them: Polyb. 16.32.1–2; Paus. 10.1.6–10, Plut. *Mul. Virt.* 244a–d.

20 Women killed: Diod. Sic. 13.57.1–3, Hdt. 8.3. Women killed for factional reasons: Diod. Sic. 18.37.2. A second, gruesome example of this kind of killing: After the knifing of Hermias, the women of Apamea stoned his wife to death and the boys of the city did the same to his sons (Polyb. 5.56.15). Stories sometimes involve stripping women and exposing them to the crowd before they were killed; rape may be implied (Polyb. 15.33.7–12).

21 Ducrey 2015, 195 provides some rough statistics about how common each alternative was and argues that of his sampling of 100 cities over nearly four centuries, 25 ended with massacres of males and enslavement of females, 34 with the entire population's enslavement, and 41—those that surrendered—with the forced departure of the population.

22 Chaniotis 2005, 103.

23 Thuc. 7.68.2. See Gaca 2015, 2016 for discussion and further references.

raped, enslaved, and raped again while having to spend the rest of their lives in drudgery.[24] When describing the fall of Olynthus, Demosthenes tells a grim story about a newly enslaved woman taken captive and nearly beaten to death because, when forced to attend a drinking party and asked to sing (both things expected of courtesans), she asserted that she did not know how to sing.[25] Timoclea's rank did not save her from rape, even though it apparently meant that she was raped by an officer, whose property she had become. Sometimes elite women were to be raped and killed, but captors allowed them to commit suicide, a dignified alternative.[26]

Still, some captives were freed; those freed were likely to be members of elites whose ransom was paid by friends or family members. Civic decrees thanking well-off inhabitants for paying for the freeing of captive women, men, and children indicate that this was a civic duty. At least one person who ransomed captive citizens was a woman, Timessa of Arcesine.[27] Obviously, though, people of wealth, or with important connections, were more likely to be freed. The survival of family members or friends was critical for the chance of ransom.[28] Demosthenes (19.194–195) says that the actor Satyrus, a friend of a man who had been involved in the assassination of Philip II's brother Alexander II, managed to persuade Philip to free the assassin's daughters (they had been captured during the fall of Olynthus); he promised to find them dowries and husbands.

Although most women and children stayed at home while the men went off to war, this was not true of the courtesans, prostitutes, and flute girls who accompanied armies, or women and children in the families of mercenaries. The use of mercenaries in warfare began to increase at the end of the fifth century and by the era of the Successors—when it was often hard to distinguish national troops from mercenaries—they were commonplace.[29] The wives of mercenaries often began as captives but, over time, made a transition, as apparently happened with the men of Xenophon's force.[30] Alexander's troops, many of whom never returned from Asia, seem to have acquired wives in the same fashion. On the other hand, some Hellenistic mercenary forces were in effect the consequence of mass migration (from Crete, for instance) and their established families traveled with them.

In any case, the wives and children of mercenaries typically followed along with the baggage train—they literally were baggage—and were therefore often in proximity to battle.[31] Generals could endear themselves to opposing troops by leaving the baggage train alone and

24 *Pal. Anthol.* 7.492, 493; Paus. 10.22.4. Penrose 2016, 55 makes the point that Timoclea's bravery, like other women in *Mulierum Virtutes*, is admired because it was expressed in the context of defense/revenge of chastity.

25 Dem. 19.196–199. The literal truth of this story, part of Demosthenes' case against Aeschines, is not my point; it is the picture of the woman who has not yet learned to cope with her apparent prostitution.

26 The wife of Aristotimus, after his death, locked the doors of her chamber and killed herself; the angry crowd seized and dragged out his two unmarried daughters, but Megisto convinced the crowd to let them hang themselves in private (Plut. *Mor.* 253b–e). After the death of King Nicocles of Paphos, his wife persuaded her sisters-in-law to kill themselves; she killed her own unmarried daughters to prevent an enemy from mastering them and then herself (Diod. Sic. 20.21.2–3).

27 Bielman 1994, no. 39 = *IG* XII.7 36. See Bielman 1994, 324–327 on the factors leading to the freeing of captives, including women, and the importance of class and connections in this process, particularly family.

28 See discussions and references to inscriptions in Bielman 1994, 236–237, 324–325, nos. 32, 38; Ducrey 1999, 283–288; Chaniotis 2005, 111–113.

29 Chaniotis 2005, 104–106; Roisman 2012, 57–60 has a useful discussion of mercenaries in the early Hellenistic period and the role of their families and how that role could be manipulated.

30 Lee 2007, 65–71.

31 The word *aposkeue* included both human and other kinds of baggage. See Loman 2004, 51.

not enslaving their children or "shaming" their wives (Plut. *Pyrrh.* 9.6). Troops might even change sides because the enemy forces had come to control their wives and children (Plut. *Pyrrh.* 18.1; *Pelop.* 27.4). Philo of Byzantium (5.94.26–29) advises the authorities in besieged cities to pay attention to the widows of mercenaries as a way to increase the loyalties of other mercenary families. Women traveling with the mercenaries helped to care for and even decorate their arms and armor.[32] Diodorus (17.84.5–6) recounts that when Alexander's forces surrounded rebellious mercenaries, the mercenaries put the women and children inside their circle and that when the men began to fall, the women replaced them, most fighting to the death. Whatever the truth of this tale, it suggests how much women were part of the mercenary life.

Royal Women at War

We have already noted that social rank sometimes dramatically shaped women's experience of warfare and its aftermath, but in the case of women connected to the ruling dynastic families of Macedonia (the Argeads) and the Hellenistic kingdoms (Ptolemaic and Seleucid rulers), a connection to power led to military roles for some royal women (the mothers, wives, sisters, and daughters of kings).[33] So far as we know, before the reign of Philip II, no such role existed, but one soon developed.[34] His daughter Cynnane (by his Illyrian wife Audata) went into battle and supposedly killed another warrior woman (Polyaen. 8.60). She married another Argead, whom Alexander killed soon after the death of Philip II. Cynnane apparently trained her daughter Adea Eurydice in military matters. Elite Illyrian women sometimes engaged in combat: one reason that this alteration in the role of royal women in war happened has to do with the influence of a non-Hellenic culture on Macedonia.[35]

But this particular influence of Illyrian culture on Macedonia might not have mattered long term had a scarcity of adult male Argeads not developed after the death of Alexander the Great in 323. His infant son, Alexander IV, and his half-brother, the mentally limited Philip Arrhidaeus, became co-kings, but because neither could rule, a series of regents ruled for them. In response, Cynnane left for Asia, despite the military efforts of Antipater to prevent her departure. She planned to marry her daughter to Philip Arrhidaeus. Cynnane had some sort of military force with which she successfully opposed Antipater's men. When she arrived in Asia, Perdiccas, the current regent, opposed her plans and there was some sort of confrontation (possibly another military one) that led to her death. However, the Macedonian troops present, incensed at the violent death of the daughter of Philip and sister of Alexander, forced Perdiccas' faction to allow Adea Eurydice's marriage to go forward.[36]

---

32 Loman 2004, 51–53.

33 Zaidman 2015, discussing Euripides' *Phoenician Women* and role of Iocasta, points to the ways in which Euripides' version of the Theban saga gives her a military aspect: Iocasta acts as commander in chief, imposing a truce (in this version Iocasta is a member of the dynasty by birth as well as marriage). Rather than hang herself after the deaths of her sons, she chooses to stab herself with a sword and dies on the battlefield. The play in some ways prefigures the roles royal women would play in similar situations in Hellenistic dynasties.

34 See discussions in Carney 2004; Pillonel 2008; Penrose 2016, 184–222.

35 Carney 2000, 69–70.

36 Arr. *FGrH* 156F 9.22–23; Polyaen. 8.60. Penrose 2016, 184 attributes Macedonian acceptance of the military roles of royal women to ethnic difference; I believe that the centrality of dynastic monarchy and loyalty to it was more important, though one could argue that this, in itself was a kind of ethnic difference.

Despite her youth, Adea Eurydice was soon addressing the troops, apparently trying to wrest command from them into her own hands. After some initial success, she was forced to submit to Antipater and returned with her husband and Alexander IV to Greece.[37] Subsequent to the death of Antipater, she escaped the control of the current regent, Polyperchon, reached Macedon with her husband and—acting as though she were regent for him—formed some sort of alliance with Cassander (Just. 14.5.1–3). While Cassander was in the Peloponnese, the army of Polyperchon, along with Olympias and Alexander IV, crossed the border into Macedon and Adea Eurydice and her husband led the Macedonian army out to meet it. However, when the home army saw the mother of Alexander, Olympias, it switched sides. Adea Eurydice and Philip Arrhidaeus were captured and then executed.[38]

It is unclear whether Adea Eurydice actually went into battle or commanded troops or whether she was simply present with the troops, but it is very likely that she addressed them and wore military attire (Duris *ap.* Ath. 13.560f). As we have seen, the Macedonian troops supported this mother and daughter pair against the male elite. Evidently, the military role of these women (whatever combination of overt and indirect it was) did not trouble assorted bodies of troops. Not coincidentally, the royal woman buried in the antechamber of Tomb II at Vergina/Aegae, whether Adea Eurydice or a wife of Philip II, was buried as a warrior, with weapons and armor.[39]

Olympias had no known military training, yet she developed a military role, albeit a smaller one than that of Cynnane and Adea Eurydice. Olympias may not have commanded the army that opposed Adea Eurydice and Philip Arrhidaeus—Polyperchon probably did that—but she accompanied it, faced danger, and acted as the symbolic leader. Like Cynnane, Adea Eurydice, and subsequent royal women, she probably rode horseback.[40] After the deaths of the royal pair and the departure of Polyperchon, Olympias, perhaps by accident, came to exercise administrative military leadership: she dispatched troops and gave orders to commanders. She was in charge when Cassander besieged Pydna, which was in her control. She attempted to escape, but ultimately made the decision to surrender. Cassander then had her killed, but her career served as a precedent for royal women to come.[41] Her connection by birth and marriage to the dynasty mattered more, particularly as a focus for loyalty, than the fact that she was a woman, granted the absence of adult males and a role, whether official or not, that functionally was that of regent.

A number of factors enlarged the military role of royal women during the period of the Successors: dynastic women generally functioned as legitimators of the new dynasties,[42] but royal women began to control cities, often by means of mercenary troops, and thus functioned as commanders of small forces, including besieged garrisons, as in the case of Olympias.[43] Often this was possible because of the wealth these women commanded. The Successor's realms were far-flung, and royal women could act as alter egos—handling shipping, equipment, and troop morale—for their often distant husbands. Typically, they

37 Diod. Sic. 18.39.1–4; Arr. *FGrH* 156F. 9.30–33.

38 Diod. Sic. 19.11.1–7; Just. 14.5.8–10.

39 See Carney 2016.

40 Pillonel 2008, 126–127. The woman buried with armor in Tomb II at Vergina apparently rode horses: see Carney 2016, 114 for references.

41 Diod. Sic. 19.50.5, 51.2–6; Just. 14.6.5, 6.6–12.

42 Carney 2000, 203–233.

43 Carney 2004, 192–193.

exercised symbolic and administrative military leadership, less often battlefield command, although the sources often do not make such a distinction certain.[44]

Soon after the last of the Successors had died, Ptolemy II married his full sister Arsinoë II, beginning what became the institutionalized sibling marriage policy of the Ptolemaic dynasty. This policy generated a power base for royal women that did not necessarily depend on the absence of adult royal males: Arsinoë II, along with her brother, was part of a lifetime cult; she was associated with various pharaonic titles and symbols either in her lifetime or soon after her death, and one of her cults linked her to the Ptolemaic navy. She and her brother traveled together to the frontier to inspect defenses, a critical part of what was a primarily defensive Ptolemaic military policy.[45] Her brother (and many subsequent male Ptolemies) did not himself command in battle, but rather entrusted the task to professionals, a decision that blurred the distinction between male and female Ptolemies.

The manipulation of pharaonic precedent—presumably playing to the Ptolemies' Egyptian audience—helps to explain Ptolemy II's practices.[46] Yet it also made possible the further development of a military role for Ptolemaic women.[47] Several Ptolemaic women had teams that won chariot races, and Hyginus (*Astr.* 2.24) claims that Berenice II was so excellent a horsewoman that she rallied fleeing troops on horseback and led them to victory.[48] Arsinoë III, sister-wife of Ptolemy IV, accompanied him to the battle of Raphia in 217, may have had a role in command, and addressed the troops when the battle seemed to turn against the Ptolemies.[49] The support of Egyptian troops was critical to the Ptolemies from this point forward, and, as noted, Egyptians may have had a welcoming attitude to royal women.

At the beginning of the second century, a series of marriage alliances between the Ptolemaic and Seleucid dynasties increased the institutional power of royal women by a kind of cross-fertilization. Ptolemaic women began to be co-rulers, starting with Cleopatra I Syra, the daughter of a Seleucid king. Cleopatra IV raised an army in Cyprus as part of a Ptolemaic dynastic dispute, and then took her troops to Syria where she involved herself and them in a Seleucid dynastic dispute, only to be killed in its course (Justin 39.3). Cleopatra III dispatched troops (Justin 39.4); later she put the army under two generals and herself headed forces invading Palestine (Josephus. *AJ* 13.13.1). Cleopatra VII made use of a mercenary army in an attempt to regain her throne; when Julius Caesar arrived she and her brother Ptolemy XII had armies facing each other.[50] Arsinoë IV escaped her sister's control and went off to join an opposing army, sharing command with one of its generals.[51] Later, Cleopatra attempted, if with little success, to employ the Egyptian fleet in the round of civil wars that followed Caesar's death (App. *B Civ.* 5.8). Her fleet and policy were essential for Mark Anthony, though not capable, of course, of saving either one of them.[52]

---

44 *Contra* Pillonel 2008, 125.

45 Ager 2005; Carney 2013.

46 Carney 2013, 90; Penrose 2016.

47 Penrose 2016, 197–206 points to the pharaonic precedents for female power and rule that were often appropriated by the Ptolemies—e.g., the title of "pharaoh" given to Arsinoe II and Berenice II—and to the association of some pharaonic women with warlike images. See further Carney 2013, 7–9.

48 See Clayman 2014, 142–158 for a discussion of Ptolemaic royal women and equestrian victories. On the Hyginus passage, see Penrose 2016, 211.

49 *Maccabees* 3.1–4; Polyb. 5.83.

50 App. *B Civ.* 2.84.1; Caes. *BCiv.* 3.103, 107.

51 Arsinoë IV: Dio. 42.39. Wars after Caesar's death: Caes. *B Civ.* 3.112. Anthony and Cleopatra: Plut. *Ant.* 58.5.

52 On women and military command in this period, see discussion and references in Penrose 2016, 216–219.

Often we cannot tell whether a queen was exercising command or whether some professional officer was; indeed the same is often true for kings. We can determine that royal women were often present on the battlefield and in danger, whatever the command structure. For ordinary women, as well as for those who were members of royal dynasties, defense of the *oikos* (household), whether the literal house (or a city full of houses) or the dynastic house, was essential. Uncommon on battlefields, women were important participants in sieges, whether throwing tiles or supporting troops; indeed, they were in a sense what was besieged, for they were the symbols and spoils of victory and defeat. Epic or legendary stories of how women caused wars or guided heroes preserve a literary version of a fundamental truth about Greek warfare, and about warfare in many times and places.

CHAPTER 26

Piracy, Brigandage, and Terrorism

Sheila L. Ager

In the context of a volume on warfare, one might expect a discussion of piracy, brigandage, and terrorism to be somewhat tangential. Contemporary definitions of "war" privilege the legalistic description of "an armed conflict which has been formally instituted by a declaration of war or the expiry of an ultimatum" (Berridge and Lloyd 2012, 388). The parties to war are "independent political units" (Weigall 2002, 237), or, in the case of a civil war, the formally constituted governing group and a coherent and recognizable insurgent group. Just war theories in particular require that a warring party must have *ius ad bellum*, part of which is that it must have the legitimate authority to make war, legitimate authority chiefly being defined as sovereign states. Under such an understanding, acts of piracy and terrorism are neither just nor are they even war.

In the world of the twenty-first century, it is clear that the legalistic definitions of the previous paragraph are outdated and simplistic. Armed conflict continues between and within states regardless of labels and public declarations; sovereign states may support insurgencies within the borders of another state; and non-state actors have gained access to weapons or other materiel of such destructive capacity that from the perspective of the victims, it might as well be war. The terrorist attacks of 9/11 resulted in the deaths of roughly 3,000 people, three times the number of battle-related deaths David Singer and Melvin Small set as a threshold for considering an armed conflict a "war."[1] True, the victims were non-combatants; but perhaps the distinction between combatants and non-combatants as a criterion of war (especially "just" war) is another metric that should be abandoned.

1 Singer and Small, COW http://www.correlatesofwar.org/; 8 October 2018.

A Companion to Greek Warfare, First Edition. Edited by Waldemar Heckel, F.S. Naiden, E. Edward Garvin, and John Vanderspoel.

Acts of terrorism, not to mention piracy, are not considered "legitimate" forms of violence in the modern world. The same might be said of antiquity, though once again our definitional categories are slippery: then as today, labels mattered, because wars could be fought through rhetoric as well as arms. The Athenian Demosthenes labeled Philip II of Macedon a pirate, despite the fact that he was the acknowledged ruler of a sovereign state. As for Philip, he returned the compliment in his accusations against Athens. From the point of view of their victims, however, it mattered little who the real pirate was. Matters of perspective and propaganda will therefore play a large part in the discussion that follows.

Piracy and Brigandage

Minos, according to tradition, was the first person to organize a navy. He controlled the greater part of what is now called the Hellenic (Aegean) Sea; he ruled over the Cyclades, in most of which he founded the first colonies, putting his sons in as governors after having driven out the Carians. And it is reasonable to suppose that he did his best to put down piracy (τὸ ληστικόν) in order to secure his own revenues. For in these early times, as communication by sea became easier, so piracy became a common profession both among the Hellenes and among the barbarians who lived on the coast and in the islands. The leading pirates were powerful men, acting both out of self-interest and in order to support the weak among their own people. They would descend upon cities that were unprotected by walls and indeed consisted only of scattered settlements; and by plundering such places they would gain most of their livelihood. At this time, such a profession, so far from being regarded as disgraceful, was considered quite honorable. It is an attitude that can be illustrated even today by some of the inhabitants of the mainland among whom successful piracy is regarded as something to be proud of; and in the old poets, too, we find that the regular question always asked of those who arrive by sea is "Are you pirates (λησταί)?" It is never assumed either that those who were so questioned would shrink from admitting the fact, or that those who were interested in finding out the fact would reproach them with it.[2]

Thucydides remarks on the apparent nonchalance and even sneaking admiration with which the early population of the Greek world was supposed to have viewed the profession of piracy, somewhat similar to the glamorizing idealization of the dashing highwaymen of eighteenth-century Britain, captured in Alfred Noyes' famous poem.[3]

Thucydides expands on his account of early piracy by pointing out that such behaviors took place both at sea and on land:

> The same system of armed robbery prevailed by land (ἐλῄζοντο δὲ καὶ κατ' ἤπειρον ἀλλήλους); and even up to the present day much of Hellas still follows the old way of life—among the Ozolian Locrians, for instance, and the Aetolians and the Acarnanians and the others who live on the mainland in that area. Among these people the custom of carrying arms still survives from the old days of robbery (τῆς παλαιᾶς ληστείας). (Thuc. 1.5)

2 Thuc. 1.4–5; this and all subsequent Thucydides quotations are from the Rex Warner translation.

3 *The Highwayman*, first published in *Blackwood's Magazine* (August 1906).

The same terminology is used to describe "armed robbery," whether by land or sea. Typically, modern scholars have settled on the term "pirate" for seaborne activity, while employing some other term such as "bandit" or "brigand" for land-based raiders (de Souza 1999/2002, 1). This chapter will follow the same convention, but it is important to note that the ancients made little linguistic distinction between the two, which suggests that they saw little qualitative difference (it seems possible that one reason for that is that a great number of piratical attacks in antiquity actually took place on land, i.e. raids on a coastal settlement; de Souza 2014, 24). There was what we might call a "quantitative" difference, however: piracy was more demanding in terms of resources (ships and a base from which to operate), and also brought greater rewards through its greater operating range and capacity for transport.[4]

In the aforementioned quotation, Thucydides is referring to passages from the *Odyssey*. When Odysseus' son Telemachus begins his search for his father by visiting the court of Nestor, he is greeted with the following questions:

> Strangers, who are you? From where do you come sailing over the watery ways? Is it on some business, or are you recklessly roving as pirates do (οἷά τε ληιστῆρες), when they sail on the salt sea and venture their lives as they wander, bringing evil to alien people? (Homer Odyssey 3.68–74; Lattimore translation)

The same questions arise in Book 9, where Odysseus is recounting his adventures with the Cyclops; this time, it is the Cyclops who interrogates Odysseus and his men as to their livelihood and intentions. In some respects, of course, the identical phrasing is to be attributed to the repetitive style of epic poetry—yet it is hard not to think that the poet was having some fun by putting these trepidatious and accusatory words into the mouth of an anthropophagous monster.

The earliest appearance of the term *leistai* or *leisteres* for "pirates" is in these passages from the *Odyssey*.[5] Contrary to Thucydides' view that the question "Are you pirates?" implies no reproach, the Homeric passages do suggest that the viewpoint of the victim is not without merit ("bringing evil [*kakon*] to alien people"). Moreover, there would appear to be a scale of virtue on which piratical plundering was measured: in a later passage in the *Odyssey*, where Odysseus is spinning a tale that he once engaged in such activities, he dissociates himself from the greedy and brutal actions of his pirate companions. Yet his description of the wealth and status associated with maritime raiding seems very much in keeping with the goals and the behaviors of Homeric heroes in general.[6]

Hans van Wees, in his study of Homeric warriors, argues that it is heroes who engage in "status warfare," while "predatory warfare" is the realm of "freebooters" (*leïsteres*).[7] But the activities of fighting and plundering and the taking of booty, both animate and inanimate, are virtually identical in both modes of warfare (de Souza 1999/2002, 21). Moreover, van Wees acknowledges that successful pillaging of a lot of wealth does raise a hero's status, more so if he takes it by violence than if he takes it by trade (1992, 238–258). This ambiguity between piratical raiding and warfare (or, as van Wees puts it, two different types of warfare) is there

4 De Souza 1999/2002, 11; Gabrielsen 2003, 390–391.

5 De Souza 2014, 25; see also the *Homeric Hymn to Apollo* 3.452–455. The term *peiratai* does not appear until the Hellenistic period.

6 *Od.* 14.229–234, 17.424–444; see de Souza 1999/2002, 17–21, 2008, 72–73, 2014, 25–26.

7 Van Wees 1992, 207–217; Rawlings 2007, 151–157.

from the very beginning of our written sources, and it remains a consistent characteristic of accounts of piracy and brigandage throughout Greek history.

The Homeric poems reflect to a large extent the milieu of the eighth century rather than providing a faithful depiction of the Bronze Age Mycenaeans. Contemporary documentary evidence does suggest that the Greeks of the Bronze Age may have engaged in the kind of plundering expeditions, both on land and by sea, that in later years were largely indistinguishable from piracy. The correspondence between the Hittite kingdom and the ruler(s) of "Ahhiyawa" from the fifteenth through the thirteenth centuries speaks of Ahhiyawan sea raids on the island of Cyprus and of Ahhiyawa providing support and a safe haven for rebellious Hittite vassals and freebooters.[8] Nevertheless, the contemporary evidence is not voluminous, and should not be stretched beyond its capacity.

The epic poems do not suggest that Agamemnon and the rest were aiming at expansionistic occupation and territorial acquisition when they assaulted Troy. Rather, the war is motivated by considerations of honor and vengeance, and the material outcome is booty, not land. So, although the Hittite documents certainly suggest that the Ahhiyawans aimed at territorial control of places in western Anatolia such as Wilusa,[9] it is perhaps indicative of the more constrained world of the eighth century that Homer's Achaeans plunder Troy—and then depart with their booty and their slaves, leaving the ruined city behind.

The historians Herodotus and Thucydides provide us with glimpses of Archaic piracy and brigandage, though the distinction between mercenaries, pirates, bandits, and "legitimate" state actors is often blurred. This distinction, or lack thereof, is perhaps best illustrated by the Archaic colonization movement. Thucydides states specifically that the founders of Sicilian Zancle were pirates from (Italian) Cymae (6.4.5). Presumably, he is singling them out as somehow different from other colonizers, but the experience of the earlier populations in the regions colonized by the Greeks would have justified the term "pirates" in most of these cases. Indeed, worse than pirates, since they came to settle and to dispossess. Indigenous populations may have been incorporated, particularly through intermarriage, but like the women of Miletos—who, in memory of their murdered kinfolk, neither dined with their husbands nor addressed them by name (Hdt. 1.146.2–3)—they may have carried a legacy of violence down through the generations.

The longstanding tradition of self-help in Greek culture is an important factor in any consideration of piracy and brigandage. Vendetta had a lengthy history—it is the root of the Trojan War—and in the absence of an international court, self-help is all there is. Specifically, *rhysia* or *sylē*—the custom of declaring reprisals against another community for alleged theft or other wrongdoing—gave the right of raiding and plundering to any and all members of the declaring community.[10] In the *Iliad*, Nestor reminisces fondly about the *rhysia* of his youth (11.670–683). From the first half of the fifth century, we have an inscription recording an agreement between two small neighboring *poleis*, Chaleion and Oeantheia, whereby they

8 Beckman et al. 2011, AhT 3–4, AhT 11; see pp. 1–6 for the identification of the Ahhiyawans with the Mycenaeans; Dickinson 2006, 28–30. The archaeological record shows that there was significant material exchange between the Bronze Age Greeks and the contemporary Near Eastern kingdoms, but the nature of this exchange (trading, raiding, diplomatic gifting) is not easily discerned from the material record alone; see Dickinson 2006, 30–35, and cf. de Souza 1999/2002, 22.

9 (W)Ilios?; Beckman et al. 2011, 121–122, 276.

10 Bravo 1980; Garlan 1989, 185–189; Rigsby 1996; de Souza 1999/2002, 29, 34–36; Buraselis 2003.

regulated the degree and amount of *sylē* that could go on between them.[11] Thucydides tells us that in 416/415, while Athens and Sparta were still technically at peace, the Spartans granted to any of their citizens the right of reprisal (in this case, λῄζεσθαι) against Athens to offset the significant plundering the Athenian forces at Pylos were doing (5.115).

While not war in and of itself, *rhysia* could clearly be used as a weapon of war: it "often represents a stage in the growing tension between two states which may ultimately end in war" (Walbank 1957–1979, 1. 474). In the Hellenistic age we find similar examples, such as Philopoemen's grant of *rhysia* to Achaeans against Boeotia, "which produced a by no means insignificant quarrel between the two nations" (Polyb. 22.4.13–14). The legitimacy of a declaration of reprisals is, as are all other issues of legitimacy explored in this chapter, a matter of perspective. The letters of marque issued to privateers such as Francis Drake would not have prevented his Spanish victims from seeing him as a pirate, and such parajudiciary practices could devolve into brigandage pure and simple (Garlan 1989, 189). In his account of the Peloponnesian War, Thucydides provides multiple examples of both sides engaging in behaviors indistinguishable, at least lexically, from piracy and brigandage.[12]

I have used the term "piratical" repeatedly here in order to underline the unity of Thucydides' vocabulary (which applies to attacks both on land and at sea).[13] But Thucydides does not appear to use the *leist-* compounds in order to challenge the legitimacy of the actors by dismissing them as robbers and brigands and their actions as mere piracy. Rather, he appears to use these terms to indicate actions that are a normal and regular part of warfare, and that act as a complement to set battles. Lightning raids, destruction of crops, seizure of livestock and other goods: all of these—and perhaps even more, the fear of these—played a role in harming an enemy's interests and hampering his ability to act. The Spartans' hands were tied with the dual occupations of Pylos and Cythera, and they were thereby driven to make peace. The *leist-* terms employed by Thucydides, then, seem for the most part to be a description of military tactics, rather than a political judgment (Roth 2016). The ongoing reality simply seems to be that some of the modes of warfare—small-scale raiding and plundering—were still much the same for pirate-brigands and for states.[14] Even the regular assaults on Attica by the Peloponnesian armies during the Archidamian War—the chief Spartan tactic of the first few years of the war—were essentially large-scale robber raids.

Even though Thucydides' use of the term in these instances may be neutral, denouncing one's enemy as a pirate could be employed as a weapon. Accusations and counter-accusations of piracy (or of harboring pirates) reached a height in the rhetorical war between Athens and Philip II. Demosthenes repeatedly accused Philip of supporting pirates, acting

11 *IG* IX.1^2 3 717; Zunino 2005. De Souza points out that these communities were located in precisely the region (West Locris) Thucydides describes as still backward in his day (1999/2002, 29); nevertheless, it is clear that the agreement was seeking to replace all-out *sylē* with judicial practices.

12 Melesander sent against Peloponnesian pirates (τὸ λῃστικὸν) to prevent them from attacking shipping from the eastern Mediterranean (2.69); Nicias occupying Minoa to prevent Megarian piracy (λῃστῶν; 3.51); Corcyrean piratical attacks (ἐλῄζοντο, ἐλῃστεύοντο; 3.85, 4.2); a Messenian pirate ship at Pylos (λῃστρικῆς Μεσσηνίων τριακοντόρου; 4.9); more Messenian piratical raids (ἐλῄζοντό, λῃστείας; 4.41); piratical attacks from an Athenian garrison on Methana (ἐλῄστευον; 4.45); Cythera as an Athenian base for piratical raids (4.53–54) that incline the Spartans toward peace in 422 (5.14; cf. 7.18, 7.26); ships under the Spartan Gylippus taken for pirates (λῃστικώτερον; 6.104); Chian fear of piracy during the siege by Athenian forces (λῃστείαις; 8.40).

13 In the vast majority of these cases, translators of Thucydides choose to employ a variety of other words: raids, plundering, guerilla warfare, privateering, and so on; see Warner 1954; Lattimore 1998.

14 Gabrielsen 2001; Rawlings 2007, 67–68; de Souza 2008, 2014, 27.

like a pirate, or simply being a pirate, "the plunderer of the Greeks (ληστοῦ τῶν Ἑλλήνων)."[15] Philip for his part accused the Athenians of condoning piracy and of engaging in acts of piracy against him and his allies.[16] And if accusations of piracy against others served as useful propaganda, so too did the claim to be a champion of the defenseless against the scourge of piracy.

> Suppression is the response most readily associated with states in relation to piracy—not necessarily because states historically, as a general rule, have tended to act forcefully to suppress piratical activity but, rather, because states produce official histories in which the state's success in clearing the seas of pirates is often emphasised and used as a means of legitimising the state's monopoly on legal violence. (Amirell and Muller 2014b, 15–16)

Making war on pirates was a regular motif in Greek history. Thucydides' account of the thalassocracy of Cretan Minos—which the historian himself acknowledges is conjectural—indicates that he considered it the natural role of a naval hegemon to suppress piracy, if only for the benefit of the hegemon himself.[17] But while the Aegean Sea saw many hegemons and would-be hegemons over the centuries, there was not always a direct correlation between the presence of powerful navies and the suppression of piracy.

States themselves could behave in piratical fashion. Raphael Sealey argued that "the original purpose of the Delian League was piratical," and that Thucydides is stating no less than the truth when he says that the object of the allies was "to compensate themselves for their losses by ravaging the territory of the King of Persia"; in other words, a declaration of reprisals.[18] Philip de Souza rejects Sealey's position, but does acknowledge that the brutal and extortionate behavior of many Athenian commanders was, from the perspective of the victims, little different from that of outlaws.[19] Vincent Gabrielsen points to a number of factors that complicate the picture, such as the naval protector's own political ambitions, possible alliance with mercenary-pirates, and the advantages of running a protection racket (2003, 395–398). "The connection between state power, piracy, and predatory behaviour is not straightforward" (Manning 2018, 258).

In the Hellenistic period, it was the trading state of Rhodes that had the leading reputation of protecting the Aegean from pirates.[20] Rhodes falls into the category of Hellenistic middle powers identified by Clemens Koehn, which also includes the Achaeans, the Aetolians, and Attalid Pergamum (Koehn 2007). All of these powers portrayed themselves as fighting

15 [Dem.] 7.14–15; 10.34; de Souza 1999/2002, 36–41, 2008, 73–74; Gabrielsen 2003, 398–399. Even if some speeches preserved in the Demosthenic corpus were not by Demosthenes (or in the case of Philip's letter, by Philip), they are likely to preserve the spirit of the contemporary propaganda.

16 [Dem.] 12.2, 12.5. De Souza 2014, 28–29.

17 Cf. Thuc. 1.13, on the Corinthians, and de Souza 1999/2002, 23.

18 Thuc. 1.96.1; Sealey 1966, 237.

19 De Souza 1999/2002, 26–28; we will return to this point in the section on terrorism. We do have evidence from the fourth century that speaks specifically to an Athenian squadron dedicated to guarding against pirates (*IG* II^2 1623 ll. 276–282; Bielman 1994, 21–22).

20 See Ducrey 1983; Gabbert 1986; Reger 1994, 44–46; de Souza 1999/2002, 48–53, 2008, 74–76; Chaniotis 2005, 8–9; Gabrielsen 2013, 73–76; Manning 2018, 247. Wiemer's massive 2002 study examines the interplay between war, commerce, and piracy in the history of Hellenistic Rhodes.

against the natural enemies of civilization: Gauls, tyrants, and pirates.[21] The Rhodian reputation permeates our sources and may result in an exaggerated picture of the degree of naval policing that actually went on. Scholars thus question the extent, purpose, and duration of the Rhodian *prostasia* of the sea.[22] Gabrielsen accepts the image of a Rhodian "protection route" that reduced "transaction costs" for merchants plying the route between Rhodes and Alexandria.[23] Gary Reger, on the other hand, argues that after the mid-third century, the Aegean islanders had no anti-pirate champion to rely on, pointing to a rise in the number of individual *asylia* treaties—interstate covenants providing guarantees of exemption from raids and plundering—with the Aetolians and the Cretan cities.[24] The First Cretan War, fought between Rhodes and some of the Cretan *poleis* at the end of the third century, may well have been instigated by the piratical activity with which the Cretans were so commonly associated; on the other hand, this war extended well beyond these players, and the involvement of the Macedonian ruler Philip V indicates that the stakes for Rhodes went well beyond the suppression of piracy.[25]

Polybius' persistently negative view of all things Cretan inevitably colors our reading of events in the Hellenistic period.[26] He is equally persistent in the negativity with which he approaches the Aetolians, and so it is not surprising to hear that the Aetolians too were a pirate nation.[27] In Hellenistic Asia Minor, it was the Galatians who were the demonic stereotype of the barbarian brigand, with his penchant for indiscriminate massacre, mass rape, cannibalism, and necrophilia.[28] These pictures are clearly exaggerated and propagandistic, for reasons too numerous to cover in detail here. Aetolian willingness to grant the right of reprisals and plundering to their own citizens was notorious, according to Polybius:

> The Aetolians have a custom not only to make booty of the persons and territory of those with whom they are themselves at war, but if any other peoples are at war with each other who are friends and allies of theirs, it is permissible nevertheless to the Aetolians without any public decree to help both belligerents and pillage the territory of both, so that with the Aetolians there is no precise definition of friendship and enmity, but they promptly treat as enemies and make war on all between whom there is a dispute about anything. (18.5.1–3)

The words are tendentious, spoken by Philip V and recorded by Polybius.

---

21 Cf. Mitchell 2003, 284–287. It may be somewhat ironic that the Aetolians figure in this list (see further in what follows), but their role in repelling that Gallic invasion of the early third century (and the subsequent establishment of the Soteria festival at Delphi) enabled them to claim this mantle.

22 For the term, see Polyb. 4.47.1, 5.90.5.

23 Gabrielsen 2013; see also Manning 2018, 247. For honors to Rhodian citizens for keeping the sea and the islands safe, see *IG* XI 4 596, 751.

24 Reger 1994, 45–46. The term *asylia* derives from *syle*: *a-sylia* was intended to guarantee that the grantee was not subjected to *syle*.

25 Diod. Sic. 28.1; Brulé 1978, 29–56; de Souza 1999/2002, 80–84, 2008, 74–75; Perlman 1999; Wiemer 2002, 143–176.

26 Polyb. 4.53, 6.46–7, 7.11, 8.16, 8.19, 24.3, 28.14, 33.16.

27 On Cretan and Aetolian piracy and brigandage, see Brulé 1978; de Souza 1999/2002, 48; Perlman 1999; Scholten 2000, 114–116; Gabrielsen 2003, 403; Chaniotis 2005, 118, 135–136; Criscuolo 2013, 162–163; de Souza 2014, 37. Rigsby comments that Aetolian activity in the Aegean in the third century may have been no more piratical than Athenian activity there in the fifth (1996, 17).

28 Paus. 10.22.2–3; Bielman 1994 nos. 21, 22; Mitchell 2003.

Nevertheless, Thucydides too identified the inhabitants of central Greece, and specifically the Aetolians, with a brigandish kind of life in his day (1.5–6).[29] Moreover, the epigraphic record—most notably the *asylia* treaties cited earlier—suggests that there was a genuine concern that gave rise to these compacts.[30] The third century saw a much increased Aetolian presence in the Aegean, and inscriptions inform us that at least some of this activity included raiding: a mid-third century inscription from Naxos speaks of the ransoming of numerous persons carried off by Aetolian raiders.[31] A lengthy inscription from Delphi records a compact between the island state of Chios and the Aetolian League whereby the Aetolians agreed to outlaw any Aetolian plundering of Chian territory, in all likelihood at the Chians' request.[32] Much Aetolian activity during the Social War (220–217) is presented by Polybius as piratical. We should not necessarily suppose, however, that piracy was an official state policy in the Aetolian League; private adventurers were no doubt behind much of this activity (Scholten 2000, 115). Moreover, what was said above in the context of the Peloponnesian War holds true for the Hellenistic period: much warfare, especially warfare involving large alliances where allies fought separate engagements, looked a good deal like raiding.

The great powers of the Hellenistic age—the Hellenistic monarchs and the Romans—at times assisted in the suppression of piracy, or at least claimed to do so. The pact of the Hellenic League created by Antigonus Monophthalmus and Demetrius Poliorcetes may have contained an injunction against piracy, but the inscription is both lacunose and propagandistic (*IG* IV^2 1 68 ll. 38–39). The Ptolemaic garrison on Thera was effective in defeating a night attack by pirates on the island, and a Ptolemaic official seems to have played a role in repatriating Theraean citizens who had been seized by pirates from Allaria in Crete.[33] By the first century, it was Rome that had taken on the mantle of war against the pirates, "the common enemies of humankind."[34] The *lex de provinciis praetoriis*, promulgated in 100, was a public statement calling on the peoples of the eastern Mediterranean (specifically the Ptolemaic and Seleucid rulers) to prevent pirates from using their territories as a base of operations.[35] Expeditions were made against the Cilician and Cretan pirate bases, expeditions that provided "a cause and legitimation of Roman imperialism and territorial expansion."[36] "I pacified the sea, freed it of pirates" was Augustus' proud proclamation in his *Res Gestae* (25.1).

But in truth, and in spite of the propaganda, pirate forces are remarkable for showing up rather frequently as (no doubt mercenary) allies of the Hellenistic kings and later of Roman commanders during the civil wars. Anne Bielman discusses the use of pirates and brigands as

29 The marginalization and ethnic prejudices that Perlman 1999 detects in the case of the Cretans may lie behind Thucydides' remark that the Aetolians could barely speak intelligible Greek (3.94.5).

30 See Bravo 1980; Rigsby 1996; Dreher 2003; Gabrielsen 2003, 401.

31 *IG* XII 5 36; Bielman 1994 no. 26; Scholten 2000, 105–116; cf. de Souza 1999/2002, 70–76.

32 *FD* III 3 214; *Syll*^{3} 443; Austin 2006 no. 64. For other Aetolian grants of *asylia*, see Rigsby 1996 nos. 54, 67, 132, 178. Rigsby also includes dozens of grants of *asylia* from the cities of Crete; see Brulé 1978, 70–105. See Rigsby passim; the essays collected in de Souza 1999/2002, 69–76; Scholten 2000, 106–107; Dreher 2003.

33 *IG* XII.3 1291, 328; Bielman 1994 no. 54; Ager 1998; de Souza 1999/2002, 53–57; Chaniotis 2005, 70; Criscuolo 2013.

34 Cicero *Off.* 3.107; cf. Polyb. 8.12 (on the Illyrians).

35 For text, translation, and commentary, see Crawford 1996, 231–270; the Greek inscriptions: *FD* III.4 37; *IK Knidos* I 31.

36 Braund 1993, 203; cf. Chaniotis 2005, 12; de Souza 2008, 2014; Tröster 2009.

additional armed forces, and suggests too that their modes and methods, such as assaults on civilian populations, may have been deliberately terroristic (2003). Crete, as Michel Austin points out, was "an important recruiting ground for mercenaries" and the demarcation between mercenary service and piracy might be only a matter of dates (and, again, perspective).[37] Aetolians and Galatians also figured commonly as mercenaries in the Hellenistic period, though the phenomenon of the pirate-mercenary can be traced much earlier than that.[38] The Antigonids (Poliorcetes, Gonatas, and Philip V) had a habit of recruiting pirates to fight for them, as did at a later date Mithridates, Sertorius, and Sextus Pompey.[39] Strabo states that the great scourge of Cilician piracy in the late second and early first century was the direct result of Seleucid monarchic rivalry:

> The first place in Cilicia, then, to which one comes, is a stronghold, Coracesium, situated on an abrupt rock, which was used by Diodotus, called Tryphon, as a base of operations at the time when he caused Syria to revolt from the kings and was fighting it out with them, being successful at one time and failing at another. Now Tryphon was hemmed up in a certain place by Antiochus, son of Demetrius, and forced to kill himself; and it was Tryphon, together with the worthlessness of the kings who by succession were then reigning over Syria and at the same time over Cilicia, who caused the Cilicians to organize their gangs of pirates; for on account of his revolutionary attempts others made like attempts at the same time, and thus the dissensions of brethren with one another put the country at the mercy of any who might attack it. The exportation of slaves induced them most of all to engage in their evil business, since it proved most profitable; for not only were they easily captured, but the market, which was large and rich in property, was not extremely far away, I mean Delos, which could both admit and send away ten thousand slaves on the same day.[10]

Cilician piracy may qualify as the kind of "enduring and large-scale venture" of persistent piracy that actually begins to evolve into a state (Amirell and Müller 2014b, 1).

Strabo links the growth of Cilician piracy with the slave market on Delos. It has long been recognized that piracy, and specifically though not solely the slave trade, was an integral part of the ancient Mediterranean economy; the authors of *The Corrupting Sea* call it "a systemic epiphenomenon of connectivity."[41] Booty, both animate and inanimate, was the reward of the pirate's labors, and ransoming or selling that booty his source of income. The fact that booty was frequently the reward of fighting forces as well, and that the booty might take exactly the same form as the pirate's plunder (goods, livestock, and human beings) is another thing that

37 Austin 2006, 216. On pirates/mercenaries, see also Shaw 1984; Garlan 1989, 161–163, 194–198; McKechnie 1989, 86–87, 91, 101–141; de Souza 1999/2002, 43–48; Bielman 2003; Trundle 2004, 23–24; van Wees 2004, 41–42; Criscuolo 2013. On the lasting nature of the use of "illegitimate" forces by "legitimate" authorities, see Tilly 1985, 173.

38 Herodotus (2.152) speaks of Ionian and Carian sea-raiders who plundered Egypt, and were then enrolled as mercenaries by Psammetichus (Psamtik I).

39 Diod. Sic. 20.82, 83, 97, 110, 28.1; Polyb. 4.28.7; Polyaenus 4.6.18, 5.19; Bielman 1994 no. 24 (decree of Rhamnous for Epichares = Austin 2006 no. 62). Gabbert 1986; Garlan 1989, 195; Reger 1994, 29; de Souza 1999, 43–48; Gabrielsen 2003, 395; Chaniotis 2005, 134.

40 Strabo 14.5.2. On Diodotos Tryphon, see Garlan 1989, 195; de Souza 1999/2002, 98–99. On the ambiguities around the equation of "Cilicians" with "pirates," see Avidov 1997; Gabrielsen 2003, 400.

41 Horden and Purcell 2000, 387. See also Garlan 1989, 75–79, 173–201; Reger 1994, 30–31, 261–263; de Souza 1999/2002, 22–23, 56–65; Horden and Purcell 2000, 157, 388; Gabrielsen 2001, 2003; Chaniotis 2005, 128–137.

makes the line between pirate, mercenary, and soldier fuzzy.[42] Hellenistic kings themselves reaped astonishing amounts of booty with their military labors, such as Alexander's plundering of the Persian treasuries; this booty became a significant part of the Hellenistic economy.[43]

> It was an elegant and truthful reply that was made to Alexander the Great by a certain pirate he had captured. When the king asked the fellow, what did he think he was doing when he tormented the sea, he replied with defiant outspokenness: "The same as you when you torment the world! I do it with a little ship, so I am called a pirate (*latro*). You do it with a large fleet, and so you are called a king (*imperator*)."[44]

This is not the place for a nuanced discussion of the role of piracy and brigandage in the ancient Mediterranean economy. The impact on its victims, however, is relevant here, in part because that impact was largely indistinguishable from the consequences of war. Damage to and loss of property, crops, and livestock would be bad enough; completely devastating would be the loss of loved ones through death and kidnapping. As for those kidnapped, they could only hope that ransoming would be possible; otherwise, the only way for their abductors to benefit financially was to sell them as slaves. Ransoming might be the more profitable option in the case of older captives of high social standing, who could be expected to fetch a high ransom, no doubt higher than the amount they could command if sold as a slave.[45]

For the poor, however, and the young, who might command a higher price at the slave market, the consequences of captivity, whether through war or brigandage by land or sea, were dire. In his account of war in the Hellenistic period, Angelos Chaniotis devotes some words to the anonymous victims: "the women and children who were carried away from sacked and destroyed cities to be sold in distant places were usually lost for ever."[46] The lost child, stolen by pirates and rediscovered in the nick of time many years later, was a romantic device employed by Menander and the Roman playwrights, as well as the ancient novelists; but that recovery would be a most rare occurrence in reality. An inscription from Aegiale on Amorgos suggests that women and girls might be specifically targeted by raiders.[47] For the women—and no doubt some of the children and some of the men—sexual assault was to be expected. Chaniotis cites Plutarch's account of an Aetolian assault on the *polis* of Pellene in 241:

> For as soon as they had entered the city, the common soldiers had scattered themselves among the houses, jostling and fighting with one another over the booty, while the leaders and captains were going about and seizing the wives and daughters of the Pellenians, on whose heads they put their own helmets, that no one else might seize them, but that the helmet might show to whom each woman belonged.[48]

---

42 Trundle 2004, 98–101 distinguishes between pirates and mercenaries, but points out the central importance of booty to mercenary service.

43 Austin 1986; Chaniotis 2005, 13, 129–130.

44 Augustine *City of God* 4.4; translation Austin 2006 no. 23.

45 De Souza 1999/2002, 65–69; Gabrielsen 2003, 392–393; Chaniotis 2005, 113. On ransoming of prisoners, see the thorough study in Bielman 1994.

46 Chaniotis 2005, 113. See Bielman 1994, 324–325; Bielman Sánchez 2011 for a discussion of the female captives of war and piracy.

47 *IG* XII.7 386; Austin 2006 no. 105. Cf. the story of the Ephesian slaughter of a Chian military contingent that was mistakenly supposed to be a gang of brigands after the Ephesian women (Hdt. 6.16).

48 Plut. *Aratos* 31; Chaniotis 2005, 133.

The fate of the Pellenian women pales in comparison to Pausanias' account of a Gallic assault on Aetolian Callium at the time of Brennus' invasion in 279 (10.22.3–4): every living being was slaughtered, the children cannibalized, and the women raped and re-raped even after they were dead. One hopes that this is an exaggerated account of the "hysterical demonization" of the Gauls (Mitchell 2003), but it is not without parallel in either the ancient or the modern record.[49]

The epigraphic record offers numerous examples of honorary decrees for those who helped out in the wake of a land or sea assault, chiefly in ransoming prisoners.[50] The Cretan Eumaridas, from the state of Cydonia, was awarded a gold crown and a statue by Athens for the benefactions he performed when one Bucris (perhaps an Aetolian) raided Attica and kidnaped a number of Athenians whom he then took to Crete.[51] Eumaridas ransomed the Athenian captives out of his own pocket, and furthermore advanced loans to them to pay for their travel home. An unfortunately fragmented inscription from Delos tells us that a Delian citizen by the name of Semus leant his assistance to some women and children from Theangela; perhaps they had shown up at the slave market on Delos?[52] An honorary decree from Teos for benefactors who helped out in the wake of a piratical assault contains detailed measures for the payment of the ransom.[53] While in some instances it seems that the benefactor acted without any expectation of reward or payback, it is probable that in others those who paid the pirates their ransom expected to recoup their loss in some way or other (Gabrielsen 2003, 394).

It has been argued here that piracy can be indistinguishable from "legitimate" warfare not only in its methods, but also in its impact on victims. Modern scholars also point to a nexus between contemporary maritime piracy and terrorism: the ultimate goals may differ, but tactics and targets can merge.[54] Piracy, moreover, is one means by which terrorist groups can fund themselves. But the ineluctable purpose of terrorism is political, not economic, and the terrorist may even have the moral edge over the pirate, as we see in this exchange from the 1988 film *Die Hard*:

Takagi to Gruber: "You want money? What kind of terrorists are you?"
Gruber to Takagi (chuckle): "Who said we were terrorists?"

Terrorism

An exploration of terrorism in the ancient world presents a host of methodological challenges. The task of defining even modern terrorism is complicated by the fact that this is not purely an academic consideration, though academics have had much to say on the subject; politicians, policymakers, and the public at large have all contributed to the interchange of

49 For further examples, see Chapters 21 and 25 in this volume.
50 For examples, see Bielman 1994 *passim*.
51 *IG* II^2 844; Bielman 1994 no. 31; Scholten 2000, 109; Austin 2006 no. 106.
52 *IG* XI 4 1054; *SEG* 3 666; Bielman 1994 no. 32; Bielman Sánchez 2011, 78.
53 *SEG* 44 949; Meier 2017.
54 Banlaoi 2007; Chen 2007; Snoddon 2007.

views.[55] As Alex Schmid states, "terrorism may well be the most politicized term in the political vocabulary these days" (Schmid 2011c, 40).[56]

Brian Jenkins' 1975 essay on international terrorism is inevitably somewhat dated now, but his construction of the basic parameters of terrorism still has merit:

> The threat of violence, individual acts of violence, or a campaign of violence designed primarily to instil fear—to terrorise—may be called terrorism. Terrorism is violence for effect, not only, and sometimes not at all, for the effect on the actual victims of the terrorists. In fact, the victim may be totally unrelated to the terrorist's cause. Terrorism is violence aimed at the people watching. Fear is the intended effect, not the by-product of terrorism. (Jenkins 1975, 14)

Scholars have continued to emphasize the core theatrical character of terrorism. "Most victims of terrorism … are mere props for the staging of a violent spectacle meant to influence the perception and behavior of one or several other audiences—the ultimate targets in the macabre spectacle of terror."[57]

In addition to emphasizing the performative qualities of terrorism, most modern definitions point to the (frequently) non-military nature of terrorism's targets: the innocent civilian (Berridge and Lloyd 2012, 365). This statement, however, requires further nuance: most terrorist attacks in recent decades have not been completely indiscriminate in their choice of object. The "actual victims of the terrorists" are very frequently members of a specifically targeted ethnic or religious group, or are in a locale chosen by the terrorists for its symbolic character (another aspect of the theatrical nature of terrorist acts).

Finally, all acts of terrorism have a political end, whether it is sharply defined or not.[58] The term *terrorism* and the concept of terror as a tool to achieve political goals originated in the French Revolution, when Maximilien de Robespierre associated *la Terreur* positively with his own deeply held values around virtue and democracy. In England, Edmund Burke took up the term and applied it pejoratively to Robespierre and his colleagues ("those hell-hounds called terrorists"), the first use of the term in English.[59] Freedom fighters, patriots, revolutionaries, holy warriors: these are all appellations adopted sincerely by those condemned as terrorists, now that "terrorist" has become a pejorative term. "Among members of the international community, the definition of terrorism has still been held hostage to the relativism embedded in the well-worn cliché that one person's terrorist is another's freedom fighter."[60]

A consideration of terrorism in the ancient Greek world, especially as a potential component of warfare, leads inevitably to the question of state (as opposed to revolutionary or sectarian) terrorism. The term "state terror" was coined in the early 1960s in order to distinguish the use of terror by the official organs of the state—as was the case during the French

55 For an in-depth treatment of the application of the term and concept of terrorism to the events of ancient history, see Howe and Brice 2016, especially the Introduction; see also Wheeler 1991b.

56 See also Schmid 2011b; Easson and Schmid 2011 for massive surveys of the struggle around defining terrorism.

57 Schmid 2011c, 80. See also Weigall 2002, 217; Hoffman 2006, 32; Easson and Schmid 2011; McAllister and Schmid 2011, 246–247; Law 2015a, 4.

58 Law 2015a, 6; cf. Hoffman 2006, 2.

59 Laqueur 2001, 6; Hoffman 2006, 4; Law 2015a, 1. See Rapport 2015 for a discussion of both the "top-down" state terror of the Montagnards and the "bottom-up" popular terror of the *sans-culottes*. Burke: *Fourth Letter on a Regicide Peace* 358.

60 Law 2015a, 4; cf. Jenkins 1975, 21; Wheeler 1991b; Hoffman 2006; Ciment 2011, xix; Schmid 2011b, 19–23. The same can be said for modern "pirates": see Kamola 2018.

Revolution—from the increasing use of terror tactics by "clandestine or expatriate organizations."[61] Some scholars, however, such as Bruce Hoffman, reject the notion of state terrorism; they are willing to apply the term "terrorism" only to non-state actors (Hoffman 2006).[62] Hoffman's view aligns with the official American position, but others such as Alex Schmid oppose it.[63] Noam Chomsky has been a particularly vehement critic of the view that states by definition cannot be terrorists; he challenges the preferential treatment given to powers such as the United States, whose terror tactics are called "coercive diplomacy" rather than terrorism.[64] In making this point, Chomsky pointed to Augustine's story of Alexander the Great and the pirate cited earlier; the story holds as true for the ambiguity of terrorists as it does for pirates.

Since the word "terrorism"—in the sense of the highly debated and politically charged term it is today—did not exist in antiquity, we are not bound here by the need to assess certain players and actions as *either* terrorist *or* not-terrorist. Nor do we, I think, need to adopt a rigid stance on the question of the legitimacy of the actors. For one thing, in circumstances of extreme civil strife—the Corcyrean *stasis* in the 420s, the last century of the Roman Republic, or the French Revolution—it can be difficult to determine just who *is* the legitimate state actor at any given time. We may therefore explore here the question of the use of terror as a means to a political end, while maintaining fairly broad parameters for our discussion.

The manifestation of terrorism in antiquity, then, is chiefly detectable in the form of state terrorism (Brice 2016, 14–16). Our sources may be in part responsible for this: terroristic behaviors by non-state actors (and some state actors for that matter) are typically represented as criminal rather than political, as we saw in the previous section. The motivations of these individuals and groups are dismissed as greed and lust for violence rather than ideologically charged performance. Technology—or the lack thereof—may also have a role to play here: without a weapon capable of sudden and massive destruction, such as a hijacked airliner or a nuclear bomb, ancient insurrectionists with a terroristic bent were restricted in their capacity to cause the kind of harm, shock, and attendant terror that follows on such an act today (and the lack of immediate media coverage available in antiquity is also significant). Thus, aside from a few exceptions, theatrical acts intended to instill fear in observers seem mostly to have been the purview of the already powerful in antiquity, who relied on their reputations to frighten and intimidate. Scholars who have tackled the subject are in general agreement that most terrorism in the ancient world was state-endorsed, and more akin to the terror inspired by Nazi Germany and Stalinist Russia than that of Al-Qaeda and Osama bin Laden.[65]

---

61 *OED*, 2nd ed. See Stohl and Lopez 1984b, 4; Weigall 2002, 210; Hoffman 2006, 3.

62 Hoffman 2006 *passim;* Hoffman argues that violence against one's own citizens should be designated as "terror," while "terrorism" should be reserved for non-state actors (2006, 15–16); cf. Schmid 2011c, 89.

63 Schmid 2011c, 44–49, 68–70; see also Stohl 2006, 6. Title 22 of the United States Code 2656 f: "the term "terrorism" means premeditated, politically motivated violence perpetrated against non-combatant targets by subnational groups or clandestine agents".

64 Chomsky 2002, 119, citing Stohl 1988.

65 The one example of revolutionary terrorism in antiquity most frequently cited in modern scholarship actually lies outside the parameters of this volume: the instigation of an anti-Roman insurrection by the Sicarii in the Jewish Revolt/War of CE 66–73, one of the very few examples of ancient insurgency that combined the political and religious ideals that motivate modern groups such as Al-Qaeda and Isis (Taylor and Gautron 2015, 33).

The ancient Greek model that continually rises to the surface of any discussion of state terror is that of classical Sparta. A recent essay by Ellen Millender (2016) has thoroughly examined the evidence from ancient Sparta from the perspective of state terror. She cautiously accepts some of the more contested claims made by ancient authors about the treatment of the helots by their Spartan masters, arguing that the processes adopted shaped the identity not only of the helots, but also of the Spartans themselves. This "identity formation," as Millender calls it, was a "key element of state terrorism," and Millender believes that it was ramped up in the late fifth and early fourth centuries as the decline in Spartiate citizen numbers made them increasingly dependent on the helot population (121). She argues that threats and rumors of violence would have been as effective as actual deeds in instilling terror. Thus, for example, the alleged anti-helot brutality of the mysterious *krypteia*—famously, if inaccurately, known as the "secret service" of Sparta—might have been much more random and much less systematic than the sources suggest, but still could have been effective as a tool of terror.

Bradley McAllister and Alex Schmid point out that "minority elites in highly stratified societies are likely to use terror routinely as an instrument of rule" (2011, 207). While their focus is on terrorism in the modern world, this analysis fits the Spartan paradigm well, and Millender emphasizes the efficacy of terror tactics not only in keeping the helots submissive but also in solidifying Spartiate identity and maintaining a gulf between them and the helots. This brings us to one of the most significant features of terrorism, whether ancient or modern: the "othering" of the target. The helots were, as Millender states, a "demonized out-group," and the dogskin caps they were said to have worn (whether the story is true or not; *BNJ* 106 F2) may be compared with the yellow Star of David worn by the Jews in Nazi-occupied Europe.

As we see in the case of Sparta, designating an individual or group as "other" is particularly important in the context of conflict *within* the state, whether potential or actual. One way of doing so was the public declaration that one's opponent was no longer a legitimate member of the state: unilaterally stripping a man (and by extension his family) of his citizenship was an instant means of removing him from all the consideration due to a fellow citizen, or even a fellow human. It was a means by which Robespierre was able to defend the reign of terror: "social protection is due only to peaceful citizens … there are no citizens in the Republic but the republicans" (Rapport 2015, 67). In the ancient world, the proscriptions during the last century of the Roman Republic come to mind, but one of the most dramatic single examples of this type of other-ing in antiquity is the trial and death of the Athenian Theramenes in 404.

In the wake of its defeat by Sparta in the Peloponnesian War, Athens acquiesced in the institution of a narrow oligarchic government that came to be known as the Thirty Tyrants, of which Theramenes was a member. A list of 3,000 of the top property-owners in Athens was drawn up; these men were to be considered citizens, but real power lay with the Thirty (backed by the Spartan garrison), definitely a minority elite.[66] A moderate, Theramenes soon came into conflict with his more extremist colleagues, particularly Critias. The Thirty had instituted a veritable reign of terror in Athens (often compared with that of Robespierre), arresting and judicially murdering anyone they saw as a potential opponent to their rule.

66 See Lopez 1984 on the particular association of state terror with regime change. On terrorism in Athens in 404, see Bearzot 2006.

Theramenes' protests against these actions resulted in Critias' denunciation of him as a traitor before the *boulē* (council of 500). If terrorism is seen as primarily theatrical—the victimization of one or a few in order to instill fear in the many—Critias' next actions were the epitome of terrorism:

> When Theramenes had concluded his speech, the council made it clear by their applause that they were on his side and Critias realized that, if he allowed them to vote on the case, Theramenes would be acquitted. Not being able to bear the thought of this, he went and held a brief consultation with the Thirty. He then went out and told the men with daggers [supporters whom he had previously brought into the chamber] to stand at the railing separating the council from the public. He then rejoined the council and said: "Members of the council, ... these men who are standing by the railings say that they simply will not allow us to let go a man who is so obviously doing harm to the oligarchy. Now in the new laws it is provided that without your vote no one on the list of the 3,000 may be put to death, but that the Thirty has power of life and death over all not on the list. And therefore, with the full approval of the Thirty, I am striking this man Theramenes from the list. And we, the Thirty, now condemn him to death."[67]

Although Xenophon does not say so explicitly, it would not be surprising if Critias enhanced the theatricality of his *tour de force* by holding up the actual list and drawing a line through Theramenes' name, thereby demonstrating to everyone present how easy it would be to delete their names as well. Fearful of the armed men, the council made no protest. Critias' strategy may be likened to that of the practice of creating "the disappeared" in Latin American dictatorships: "this strategy ... seems to have two objectives: to directly eliminate an important person opposed to the government; and indirectly to communicate to the *potential* opposition more generally ... that opposition is dangerous" (Stohl 2006, 10).

Tyrants and dictators, like other minority elites, are indeed apt to use terror tactics. Yet some authors searching for phenomena in antiquity that could be qualified as terrorism oddly pinpoint political assassination, and in particular, the assassination of tyrants. Julius Caesar is given as a prime example, while in the Greek world, the murder of the Athenian tyrant Hipparchus in 514 has been identified as an act of terrorism (Szuscikiewicz 2011). But the characterization of political assassination as terrorism is problematic: the performative quality of terrorism means that we need to think about target audience. As Schmid puts it, "an assassination of a tyrant is not terrorism, as victim and target are the same" (2011b, 35). While Caesar's friends and supporters might indeed have been badly frightened by the events of the Ides of March—Antony played for time, while Cleopatra got out of town—the main goal of the conspirators was to get rid of Caesar, not to instill terror (though terror of course ensued on both sides as the civil wars heated up again).

Likewise, in the case of Hipparchus, we are told that the assassins' chief objective was to rid Athens of its tyrants (plural); Hipparchus' brother Hippias survived the attempt only because the plot miscarried (Thuc. 6.53–59). There is no indication that Harmodius and Aristogeiton (and their co-conspirators) intended to kill Hipparchus with a view to terrorizing Hippias and his supporters. As it turned out, though, that is exactly what happened: Thucydides tells us that after the death of Hipparchus, Hippias became much more oppressive, executing numerous citizens. If we are to seek terrorism in these events, it seems more

67 Xen. *Hell.* 2.3.50–51; trans. Rex Warner (slightly adapted).

reasonable to see it in the tyrant's reaction, rather than in the initial assassination plot.[68] This brings us back again to the model of state rather than revolutionary terrorism.

In focusing on the question of terror specifically as a weapon of war (internal and external), we are again faced with analytical and theoretical challenges. In the seventeenth century, Hugo Grotius noted that "wars, for the attainment of their objects … must employ force and terror (*vis ac terror*) as their most proper agents."[69] It is such an obvious point—particularly with the greater sensitivity in the contemporary world to the psychological impact of violence—that it seems almost fatuous to assert that terror is a core component of warfare. But terror*ism* is another matter, and there is significant debate about whether the term can be applied to acts of war.[70] This is in part connected to the reluctance to label acts taken by legitimate state actors as terrorism, since war is seen by many as something that can only be waged by such actors (Hoffman 2006, 25–26). If violence is committed by a non-state actor it is *ipso facto* terrorism, if it is committed by a state actor it is *ipso facto* war, and never the twain shall meet. As with the larger picture of terrorism in antiquity, however, it seems unhelpful to create or debate finely tuned legalistic categories, given that so many acts of modern terrorism—and war, for that matter—resist legalistic categorization. This is particularly so if we allow that civil war is, in fact, war. What the evidence suggests is that in antiquity it is in times of war that we see the most consistent and widespread use of terror tactics.

Brian Jenkins argues that a distinguishing mark of international terrorism is that it is "waged outside the accepted rules and procedures of international diplomacy and war," and cites as an example, civilian targets (1975, 20). But while the death of civilians—especially those deemed incapable of defending themselves, such as children—might have been considered tragic and regrettable, even despicable, in antiquity, it would not necessarily have been seen as unusual or as transgressing norms of international behavior. Moreover, given that most ancient Greek states had no professional military, and that all able-bodied citizens served when necessary, the modern notion of a "civilian" population has less applicability in the ancient paradigm (Brice 2016, 13). Grotius acknowledged that civilian immunity was not a regular feature of warfare: "whenever war is declared against any power, it is at the same time declared against all the subjects of that power … Even women and children are frequently subject to the calamities and disasters of war" (3.4.8–9).

Among the ancient examples Grotius cited was the Roman commander Germanicus' campaign against the Germanic tribes in CE 14. A night attack took each of the separate Marsic villages by surprise and unarmed, and, as Tacitus tells it, "no pity was shown to age or sex" (*Ann.* 1.51). Scholars exploring the question of terrorism in the ancient world often look to the Roman example. As Polybius remarked, when discussing the ruthlessness of the Roman sack of Carthago Nova, "They do this, I think, to inspire terror" (10.15.5).[71] Citing similar comments made by Diodorus about the Romans (34.2, 34.4), Francisco Simón states

68 Dillinger 2015. For an analogy between Robespierre and Peisistratos, the father of Hippias and Hipparchus and the creator of the Athenian tyranny, "already in everyone's minds" in 1794, see Jones and MacDonald 2018.

69 *De iure belli ac pacis* (1625) 3.1.6; trans. A.C. Campbell.

70 Brice, for example, distinguishes between "terrorism" and the "terror of war" (2016). Wheeler argues that terrorism can be seen as a stratagem of war, the use of "trickery, deceit, employment of indirect means and avoidance of pitched battle," which he labels the "Odysseus ethos" (1991b, 24).

71 Polyb. 10.15.5. See Wheeler 1991b; Ziolkowski 1993; Campbell 2002, 169–170; Thornton 2006.

the following: "The reason for this cruelty ... is the terror instilled in the enemy, paralyzed in the face of the horror experienced and impeded from undertaking any new action against troops that practiced such extreme cruelty."[72] The pageantry and terror of Robespierre's guillotine might be compared to Crassus' crucifixion of thousands of rebellious slaves along the Appian Way (App. *Civ.* 1.120).

The paradigm of the excessively brutal Roman, however, has been challenged, most extensively by Arthur Eckstein (2006, 2008), who argues that the Romans were no more innately violent than their contemporaries. Grotius also cited Thucydides' description of the fate of the small *polis* of Mycalessus in Boeotia in 413.[73] The Athenians, strapped for cash in the wake of the Sicilian disaster, laid off a group of Thracian mercenary soldiers, and sent them on their way. Escorting them on their northward journey was the Athenian commander Diitrephes, who was under orders to harry the Boeotian enemy where possible. After some (piratical?) raiding and plundering, Diitrephes and the Thracian troops assaulted Mycalessus:

> The Thracians burst into Mycalessus, sacked the houses and temples, and butchered the inhabitants, sparing neither the young nor the old, but methodically killing everyone they met, women and children alike, and even the farm animals and every living thing they saw. For the Thracian race, like all the most bloodthirsty barbarians, are always particularly bloodthirsty when everything is going their own way ... Among other things, they broke into a boys' school, the largest in the place, into which the children had just entered, and killed every one of them ... Thus disaster fell upon the entire city, a disaster more complete than any, more sudden and more horrible.

Thucydides highlights the events at Mycalessus as particularly horrifying, and is able to scapegoat the Thracians for it; but in truth, his own account provides example after example of warfare that wiped out entire communities. Greek history is liberally strewn with such incidents. The custom of *andrapodismos*—the enslavement of an entire population, frequently accompanied by the massacre of all males of military age—was on a large scale what the stripping of Theramenes' citizenship was on the small: it constituted the wiping out of civic and corporate identity, a literal "disappearance" of those now made non-persons.[74]

The knowledge of what might happen in times of war—the suffering entailed in being under siege and the dread of the outcome—would have inspired terror in the entire population of a *polis*. This kind of fear could be exploited in order to control the population's response, and was, it seems, useful to Athens in its drive to control the members of its empire during the Peloponnesian War.[75] In 416, when the Athenians invested the small island of Melos, they responded to the Melians' suggestion that they (the Melians) could simply pursue a course of peaceful neutrality with the stark statement that their purpose in subjecting Melos was to send a message to others (Thuc. 5.95). Thucydides is of course shaping these words, but there is no reason to suppose that the argument itself—that brutality inspires terror, which in turn can be used to the terrorist's advantage—was not commonly held. The

72 Diod. Sic. 34.2; 34.4; Simón 2016, 221, 224.

73 Thuc. 7.29–30; Bettalli 2006.

74 For the brutal treatment of the defeated enemy, cf. the injunctions in Deuteronomy 20.10–18.

75 As it was centuries earlier to the Assyrians, whom Jonathan Fine identifies as the first people known to have explicitly articulated a policy of instilling terror (Fine 2010); see also Wheeler 1991b, 14.

Athenians might have hoped that the Melians would have been able to take the lesson from earlier examples of ***andrapodismos***, such as at Scione earlier in the war (Thuc. 5.32); they certainly were convinced that the ***andrapodismos*** they subsequently imposed on Melos itself would send that lesson to others.

Particularly disturbing to the populations of the Greek states, especially but not exclusively in the context of the Peloponnesian War, was the danger of insurrection from within. Covert support of a fifth column in an enemy state has been qualified as international state terrorism, and the decades of the Peloponnesian War were rife with democratic and oligarchic *stasis*.[76] The more threatened a regime feels itself to be, the more apt it is to employ terroristic tactics against its own population (McAllister and Schmid 2011, 207). In regard to the revolution at Corcyra in the 420s, Thucydides gives us a horrific description of the brutality and terror and degradation of morale (and morals) that both sides experienced, and outlines the progress of this phenomenon through the Greek world:

> So savage was the progress of this revolution (*stasis*), and it seemed all the more so because it was one of the first which had broken out. Later, of course, practically the whole of the Hellenic world was convulsed, with rival parties in every state—democratic leaders trying to bring in the Athenians, and oligarchs trying to bring in the Spartans. In peacetime there would have been no excuse and no desire for calling them in, but in time of war, when each party could always count upon an alliance which would do harm to its opponents and at the same time strengthen its own position, it became a natural thing for anyone who wanted a change of government to call in help from outside. (Thuc. 3.82)

When the *polis* of Abydus was besieged by Philip V in 200, the fear of what might happen to their families led the citizens to resolve on killing the women and children before they could be captured by Philip's troops (Polyb. 16.31). Polybius considered this to be a noble and courageous decision, lamenting that the moral weakness of certain citizens undermined it and prevented the mass killing of the women and children (16.32–33). We do not know what the women and children thought, beyond the fact that arrangements had been made to kill them, not to assist them in committing suicide; when the city finally fell, Polybius says that the surviving Abydenes took every opportunity to kill themselves, their children, and their wives (16.34.9–12).[77]

Josephus tells a particularly gruesome story of terror tactics in the midst of war. During the Syrian–Egyptian–Judaean conflict of 103–101, Ptolemy IX ordered his troops to kill the women and children they discovered in some Judaean villages, cut up their bodies and stew them, and then eat the stew. This, so that "those who had escaped from the battle and had returned to their homes might get the notion that the enemy were eaters of human flesh, and so might be the more terrified by this sight."[78] The story strains credulity somewhat, and Josephus seems to think so too, since he stresses that both Strabo and Nicolaus of Damascus record these events. Even if untrue, however, such a rumor could be highly effective at intimidating the enemy. Josephus' account shows that he clearly understood the purpose of such

76 Jenkins 1975, 31, who refers to this brand of terrorism as "surrogate warfare"; Stohl 1984.

77 16.34.9–12. Cf. Plut. *Mor.* 244, where the women of Phocis vote to set fire to themselves if the men are killed; the Phocian children vote separately to do the same thing (!). See Schaps 1982, 200–201.

78 Josephus *AJ* 13.345–347; cf. Hdt. 3.11.

actions: to terrify and to demoralize, to fight the war through emotional manipulation. While there may not have been an articulated concept of terrorism in antiquity, our sources indicate that the ancients understood very well indeed the tactics and purpose of terror:

> Those whose object is to gain dominion over others use courage and intelligence to get it, moderation and consideration for others to extend it widely, and paralyzing terror to secure it against attack ... Because of their surpassing humanity, therefore, kings, cities, and whole nations went over to the Roman standard. But once they held sway over virtually the whole inhabited world, they confirmed their power by terrorism and by the destruction of the most eminent cities. Corinth they razed to the ground, the Macedonians (Perseus for example) they rooted out, they razed Carthage and the Celtiberian city of Numantia, and there were many whom they cowed by terror. (Diod. Sic. 32.2.1, 32.4.5)

Conclusion

The pirate and the brigand were well-recognized figures in the ancient world, and featured in the popular mind as both the Robin Hoods and the Snidely Whiplashes of antiquity. But when examined through the lens of war, where states engaged in or sanctioned acts of piracy, the concepts of piracy and brigandage become much less simplistic. In fact, banditry, piracy, and aggressive violence in general have historically been at the heart of much state development: the "organized crime" of state making, as Charles Tilly calls it, exemplified in the forceful Greek colonization of Sicily.[79] As for the terrorist, he was not for the most part a recognized personage in antiquity, just as terrorism was not an articulated concept. But here too, when we contemplate terrorism in the context of war, we find multiple examples of the theater of violence, the use of extreme ruthlessness in order to intimidate and control the audience of such actions. And for the populations captured, assaulted, killed, or enslaved through an act of war, the physical and emotional experience was little different from what they might expect of a pirate raid. It is in the circumstances of the victims—their pain, their grief, and their terror—that we see most clearly the interconnectedness of piracy, brigandage, terrorism, and war. Whether acts of violence were committed by legitimate or non-legitimate actors would make little difference to those at the receiving end: the well-disciplined Roman army at Carthago Nova, the bloodthirsty Thracian mercenaries at Mycalessus, and the Gallic thugs at Callium all functioned at the same level of brutality.

79 Tilly 1985; cf. Amirell and Müller 2014b.

CHAPTER 27

Civil War in Greece: Forgetting as the Path to Reconciliation

Gordon Shrimpton

In this chapter, I examine civil conflict in two Greek cities, Athens and Corcyra. I have chosen these because they are famous, because we have a good deal of information about them from antiquity, and because the comparison of the two is instructive. Civil conflict in these cities and elsewhere took several forms: conflicts short of war (commonly called *stasis*) and those escalating into outright fighting, whether in the streets or in pitched battles (commonly called *polemos*). Even a brief study of these conflicts should help to clarify the conditions under which conflict or war tended to break out, and, especially in the case of Athens, the circumstances that made reconciliation possible.

Civil strife is a bitter thing that rarely ends "civilly." The cruel memories of relatives and neighbors killing each other are hard to erase, even with the help of the passage of time. "All wars are bad," said Charles de Gaulle (speaking of the Spanish Civil War of 1936–1939), "[but] civil wars, in which there are brothers in both trenches, are unforgivable, because peace is not born when war concludes."[1] In Xenophon, an Athenian herald puts this sentiment in ancient terms when speaking of conflict in Athens under the rule of the Thirty. The Thirty and their democratic foes have just fought in the streets of Piraeus, in 404:

> And when we might live in peace as fellow citizens, [says Cleocritus, the herald] these men [the Thirty tyrants, so called] bring upon us war (*polemos*) with one another, a war most utterly shameful and intolerable, utterly unholy and hated by both gods and men. Yet for all that, be well assured that for some of those now slain by our hands not only you, but we also, have wept bitterly. (Xen. *HG* 2.4.22, Brownson 1922, tr.)

1 Charles de Gaulle, quoted in Armitage 2017, 9.

A Companion to Greek Warfare, First Edition. Edited by Waldemar Heckel, F.S. Naiden, E. Edward Garvin, and John Vanderspoel.

In this case—unlike the wars evoked by de Gaulle—bitterness and loss led to regrets, and thus to reconciliation, even within a few short years. A well-chosen detail of Xenophon's foreshadows the coming reconciliation: although the victorious democrats take the weapons of their fallen enemies, they do not strip their tunics, and return the bodies to the other side under truce. Thus, reconciliation begins.

Concordia was the word the Romans gave to that highly desirable condition in which citizens foreswore conflict and chose to cooperate harmoniously. For the Romans, the worst sort of conflict was *bellum civile*; Thucydides speaks of ὁμόνοια (*homonoia*) replacing *stasis.*[2] These and other Greek terms deserve some further remarks, however, for they cannot always be translated by the same words in English. *Stasis* means primarily "the taking of a stand," but Andrew Lintott always translates it as "civil strife."[3] Mogens Hansen, however, gives this account of its use:

> The word *stasis* actually means stance but it underwent shifts of meaning as follows: (1) stance, (2) standpoint, (3) group of people with the same standpoint, (4) in the plural: two or more groups with opposing standpoints, (5) the split between groups, and (6) civil war. *Stasis* always means a group that wants to preserve or obtain power by deceit or violence, i.e. a revolutionary group, never a political group operating within the constitutional framework of the city-state, i.e. what we call a political party.[4]

Stasis can bleed over into *polemos.*

Greek historians supply further distinctions. Herodotus uses the expression στάσις ἐμφυλος (8.3.1), which looks as if it could mean civil war, or at least "civil conflict," because ἐμφυλος means "between the tribes" (i.e. internal tribes or groups within a state) but, in this case, that is not how Herodotus uses the expression. The subject is the preparation of the alliance of individual Greek cities to resist the Persian invasion of Greece in 480. The Athenians chose to submit their fleet to Spartan command, and to avoid being contentious over leadership of the alliance. They could thus prevent στάσις ἐμφυλος among the allies.

Thucydides is more complex. In Book VIII, Athens finds itself divided between supporters and opponents of the new oligarchic regime of the Four Hundred. The Four Hundred had come to power through a *stasis* that Thucydides narrates at some length (8.63–98). The opposing sides had murdered and expelled some of their enemies, "but not many" (8.70.2). Then comes news of a threat from the Spartans. The Athenians suspend what Thucydides calls their "internal war" (*idios polemos*) to rally together against the menace of Spartan forces nearby, considering the external threat to be more serious. They stop short of an outright civil war in order to prevent a foreign invasion.

In Book III, Thucydides speaks of *oikeios* polemos, meaning domestic or household war. The context: in 426, the Athenians sailed to Sicily, ostensibly to support Athens' allies against Syracuse, but evidently to prevent grain from being shipped from Sicily to the Spartan Peloponnese, and also to explore the possibility of bringing the island under Athenian control (Thuc. 3.86.5). In the summer of 424, the Greek cities in Sicily became

2 Thuc. 8.91–93. Armitage 2017, 41–44 identifies the Romans as the true inventors of civil war (*bellum civile*), and argues that the Greeks did not experience it on anything like the Roman scale. They did, however, call it "war" (*polemos*), both in Xenophon and in Thucydides 8.94.3.

3 Lintott 1982.

4 Hansen 2004, 125.

concerned about the threat posed by the Athenians, and sent delegates to Gela to discuss a general peace. Thucydides says little about the conference beyond the fact that the delegates brought forward sundry complaints, but does provide a lengthy account of the decisive speech delivered by Hermocrates of Syracuse (4.59–64). Toward the end of his speech, Hermocrates says:

> We shall go to war again, I suppose, when the time comes, and again make peace among ourselves by means of future congresses; but the foreign invader, if we are wise, will always find us united against him, since the hurt of one is the danger of all; and we shall never, in future, invite into the island either allies or mediators. By so acting we shall at the present moment do for Sicily a double service, ridding her at once of the Athenians, and of civil war (*oikeios polemos*), and in future shall live in freedom at home, and be less menaced from abroad. (Dutton 1910, tr.)

The word *oikeios* alludes to the political and cultural identity shared by Sicilian Greeks—an identity that should exclude Athens, even though some Sicilians were Ionians, as the Athenians were. Hermocrates is thus declaring one sort of civil conflict illegitimate, but not conflict among those supposedly in the same tribe, as in Herodotus, or those actually in the same city, but instead those in the same region.

In one respect, *oikeios polemos*, *idios polemos*, and *stasis emphylos* are alike: they characterize potential civil conflict as a self-destructive failure to meet the danger posed by an outsider, be it Persia, Sparta, or Athens. The two historians are contrasting civil war with foreign war, and implying that civil war may be unpatriotic.

In his account of the civil strife on the island of Corcyra, Thucydides again works with the theme of foreign foes affecting a domestic or local conflict, but also shows how *stasis* gives way to *polemos*. The foreign factor on Corcyra was the popular desire for an alliance with Athens and the contrary oligarchic wish to keep out of that alliance and avoid the wrath of Corinth (supported by Sparta); but the upshot of this political disagreement was a civil war marked by acts of treachery and savage warfare committed by citizen against citizen. The strife on Corcyra thus became an extension of the existing Peloponnesian War. I shall treat it, however, as domestic violence that repeatedly flared up as outright war.

In 433, the Corcyraeans sent an embassy to Athens seeking an alliance against Corinth (Thuc. 1.31–44). The Corinthians sent an embassy at the same time to urge the Athenians, who had a formidable navy, to reject the Corcyraean request. The Athenians decided to conclude a defensive alliance with Corcyra. By doing so, they hoped to avoid a war against Corinth, a Spartan ally, but this war could not be avoided. When Corinth attacked Corcyra, Athenian ships happened to be present, and they came to the aid of their ally, with the result that they engaged Corinthians (Thuc. 1.52–54). Corinth complained to Sparta, and owing to this and other provocations, Sparta declared war on Athens in the spring of 432 (1.87). Meanwhile, the Peloponnesian War commenced with the Theban attack on Plataea in 431 (2.1–7).

So much for the foreign background; Thucydides traces the domestic, local cause of the conflict on Corcyra to the release of 800 Corcyraean prisoners captured in earlier sea battles. Their release was sponsored by Corinthians, who expected them to work for the dissolution of the Corcyraean alliance with Athens once they were home (Thuc. 3.70). Their efforts came to naught. Next, the oligarchs put Pithias, leader of the Corcyraean *demos* (Thuc. 3.70.3), on trial for supporting the alliance with Athens; they claimed that he was enslaving the Corcyraeans to Athens. Acquitted, Pithias went on the offensive, putting five of his aristocratic opponents on trial for alleged acts of sacrilege. He won that case, and had them assessed a fine that they claimed was too heavy. Their pleas for a reduction of the fine fell on deaf ears. Believing now that Pithias was promoting a full alliance with Athens—one over and above the existing defensive one—oligarchic assassins armed with daggers burst in on a meeting of the Council and murdered him and some 60 others.

The oligarchs now called an assembly of the citizens in which they argued that they had acted for the best. As Thucydides reports (3.71.1), they rammed through a decree to the effect that Corcyra renounce alliance with Athens and agree to regard anyone who approaches in more than one ship as an enemy.

One trireme did arrive with a Spartan delegation on board and "the people in control" (the few, or *oligoi*) now attacked the *demos* and defeated it in battle (3.72.2). The *demos*, however, occupied some high ground and won support from many of the field slaves. The oligarchs responded by hiring 800 mercenaries. They met defeat, however, in the ensuing battle that evidently took place in the city streets. To cover their retreat, they set fire to the buildings around the agora. A dozen Athenian ships now arrived under Nicostratus, who attempted to impose peace: Hostilities were to end provided that ten of the "most responsible" were to be put on trial (but instead of standing trial they went into hiding), Corcyra to be allied to Athens both offensively and defensively. As Nicostratus prepared to leave, the democrats persuaded him to leave behind five ships. They would make up his numbers by giving him five of their own, manned by Corcyraeans.

The *demos* was now in power, and used the opportunity to name oligarchs as members of the crews to leave with Nicostratus. When they refused to go, Nicostratus was barely able to restrain the *demos* from murdering them. At least 400 of them took refuge as suppliants in the temple of Hera. They were removed to an island and evidently held under arrest. Four or five days later, a Peloponnesian fleet of 53 ships arrived and the Corcyraeans hastily manned 60 ships of their own. As they were about to close with the enemy, two of their ships deserted to the other side and the crews of some others began fighting among themselves. In the ensuing battle, they lost 13 ships, but the skill and experience of the Athenian crews, though greatly outnumbered, saved them and Corcyra from disaster. Now, however, fearing lest the Peloponnesians follow up their victory, the Corcyraeans removed the men from the island back to the temple of Hera. Here, "desiring to save their city," they persuaded some of the oligarchs to go aboard ship and serve as crewmen. The Corcyraeans expected another attack from the Peloponnesians, but it never came. The Peloponnesians had reached Corcyra, but withdrew after despoiling some outlying territory.

Now came the climax. Having stayed in power, how would the *demos* treat the oligarchs who had supplicated in the temple of Hera, and should have been doubly safe, since the *demos* had put them back in the shrine and effectively acknowledged that the oligarchs had a right to remain there? Greek unwritten law, or *agraphos nomos*, let political authorities remove suppliants under certain circumstances, but only if the suppliants were given some process of law. Greek custom never allowed those who had taken refuge to be slain without any ado.[5]

5 Naiden 2009, 147–153, 166–167, reporting earlier scholars, such as J. Gould, according to whom harming or merely removing suppliants from shrines would be even less justifiable according to *agraphos nomos*.

Thucydides' description of the people's behavior is chilling. Perhaps Thomas Hobbes was thinking of it when he wrote *Leviathan* during the English Civil War of 1642–1651. He had translated Thucydides in 1629. I give his translation of the last two sections of the passage:

> [3] But the most of the sanctuary men, that is, all those that were not induced to stand to trial by law, when they saw what was done [the others were condemned], killed one another there right in the temple; some hanged themselves on trees; everyone as he had means made himself away. [4] And for seven days together that Eurymedon stayed there with his sixty galleys, the Corcyraians did nothing but kill such of their city as they took to be their enemies, laying to their charge a practice to have everted the popular government. Amongst whom some were slain upon private hatred and some by their debtors for the money which they had lent them. All forms of death were then seen; [5] and (as in such cases it usually falls out) whatsoever had happened at any time happened also then, and more. For the father slew his son; men were dragged out of the temples and then slain hard by; and some immured in the temple of Bacchus [Dionysus] died within it. So cruel was this sedition and seemed so the more because it was of these the first. (Thuc. 3. 81.2–5)[6]

In this fashion, the Corcyraeans brought the civil war in their city to an end, but it was a fashion that only Joseph Stalin and his like could envy—a massacre of the losing side.[7]

Besides adumbrating the excesses of future authoritarian states, starting with a Hobbesian Leviathan or *Behemoth* (the title of his other work inspired by the Civil War) and continuing until the dictatorships of Stalin and the like, Thucydides was also providing a cultural warning. The protocols that made communication possible in a Greek city-state, such as respect for the conventions of religion, had broken down in the atmosphere of chronic distrust. The consequence was that political language no longer conveyed the meaning intended for it:

> Words had to change their ordinary meaning and to take that which was now given them. Reckless audacity came to be considered the courage of a loyal ally; prudent hesitation, specious cowardice; moderation was held to be a cloak for unmanliness; ability to see all sides of a question inaptness to act on any. Frantic violence, became the attribute of manliness; cautious plotting, a justifiable means of self-defence. The advocate of extreme measures was always trustworthy; his opponent a man to be suspected. To succeed in a plot was to have a shrewd head, to divine a plot still shrewder; but to try to provide against having to do either was to break up your party and to be afraid of your adversaries. In fine, to forestall an intending criminal, or to suggest the idea of a crime where it was wanting, was equally commended, until even blood became a weaker tie than party, from the superior readiness of those united by the latter to dare everything without reserve; for such associations had not in view the blessings derivable from established institutions but were formed by ambition for their overthrow; and the confidence of their members in each other rested less on any religious sanction than upon complicity in crime. (Thuc. 3.82.4–7; Dutton, 1910, tr.)

Somehow, the bloodletting must cease short of extermination, but how? That is our next topic.

6 Cf. most famous words in *Leviathan*, ones describing life without civil government: "which is worst of all, continual fear, and danger of violent death; and the life of man, solitary, poor, nasty, brutish, and short".

7 On Stalin's ruthlessness see, *inter multos*, Westwood 1987, 309–317. On mass graves in Spain after the Franco regime, see Smith 2009, 279–288. The American Civil War offers an ambiguous example. Grant and Lee achieved a peaceful transition, as observed by Brands 2012, 362–368, but this agreement did not end bloodshed elsewhere in the South, as observed by Blight 2001, 394–397.

Me mnesikakein

How then does a community heal itself? In 403, after the conflict with which this chapter began, the Athenians took an oath not to remember, or call to mind past evils (*mnesikakein*), the so-called amnesty decree. This happened after the Spartans had brokered a settlement between the combatants. The terms of peace: the parties, the men of the Piraeus (democrats) and the men of the city (supporters of the Thirty), were to live in peace with each other and return to their homes. Excluded were the Thirty and their immediate appointees. Anyone who felt fearful could remove to Eleusis (Xen. *HG* 2.4.38). The people who removed to Eleusis did not remain quiet, however. When they began to hire mercenaries, the Athenians marched against them in full force, put their generals to death, and imposed renewed peace conditions. These terms included the oath not to call evils to mind (Xen. *HG* 2.4.43).

This settlement was very successful. It sustained the Athenian democracy until its defeat at the hands of Philip and the Macedonians at Chaeronea in 338. Does this mean that the Athenians did the impossible: put the bitterness of nearly two centuries of sometimes violent civil conflict behind them, "forgot" the brutality, and lived in harmony (*homonoia*) after 403? To judge from the surviving record, they divided *stasis* into two main categories, conflict between the aristocrats (*gnorimoi*) and the people (*plethos*) from which it appears that they suppressed or downplayed the memories of brutality; and struggles against (would-be) tyrants in which the murder of tyrants and the deaths of citizens who fought them were memories to celebrate. When citizens made common cause with tyrants, however, as in 403, then the need to expel tyrants compelled the opponents of tyranny to fight fellow citizens, however reluctantly. The speech of Cleocritus captured the citizens' ambivalence when they found themselves in this situation.

Ambivalence could explain a peculiar feature of Herodotus' description of the battle of Pallenis (546) in which the Athenians faced the Pisistratids, who wished to return to power. The Pisistratids advanced from Marathon, where they had landed, and met the force sent out from the city at Pallenis (a temple of Athena). When the invaders attacked, the Athenians were snoozing or playing dice after breakfast. They scattered in flight, and Pisistratus sent his sons ahead on horseback to tell the Athenians to take heart and return to their homes. He succeeded in disarming the Athenians and established himself in the tyranny. Herodotus (1.63) reports this "battle," but fails to mention any casualties or describe the fighting. Next, he gives a substantial description of the way Pisistratus rooted his tyranny in Athens, purifying Delos and forging alliances. After a paragraph, he concludes: "some Athenians died in the battle and others went into exile with the Alcmaeonids" (1.64.3). This separation of the occasion from its consequences removes the battle from the class of heroic victories like Marathon, Thermopylae, and Plataea where Herodotus proceeds directly to an enumeration and celebration of the fallen.[8] The historian wishes to ignore the civil strife of 546, but cannot quite do so.

Later, two significant battles resulted in the final expulsion of the Pisistratus' son, Hippias (Hdt. 5.63.4–64). Hippias had summoned aid from Thessaly, while his opponents, the Alcmaeonids, had tricked the Spartans into supporting them against Hippias. Anchimolius, the Spartan commander, arrived with a force by sea, and disembarked, but was attacked by Thessalian cavalry. The Thessalians killed many Spartans, including Anchimolius. Later, King

8 Marathon: Hdt. 6.114–117; Thermopylae: 7.225–228; Plataea: 9.67–73.

Cleomenes arrived from Sparta by land with a larger force. The Thessalian cavalry, sent to block their entry into Attica, attacked and was routed with the loss of more than 40 men (5.64.2). It was remarkably generous of the Thessalians and Spartans to come to Attica and fight and kill each other to help the Athenians settle a political dispute. Did the Athenians merely cheer from the sidelines, or did some fight with their allies? If so, were there no casualties? In the second battle, the Thessalians presumably attacked before the Spartan army could enter Attica. Were there no Athenians with Cleomenes on that occasion, or with the Thessalians in either battle? It is a curious dilemma. No one saw the need either to explain the improbability of Athenian absences from the two battles or record the casualties they did (or did not) receive. Only foreign losses are noticed.[9] Once again, Herodotus is minimizing what he does not ignore.

With the Pisistratids gone, Cleisthenes the Alcmaeonid and Isagoras, friendly to Sparta, were poised to challenge each other for leadership of the state. The Athenians meanwhile rallied to expel the Spartans and restore the exiles. Herodotus says that "the rest" (presumably Athenian supporters of Isagoras) were imprisoned and condemned to die, but he names no Athenians in this regard, just a certain Timasitheus of Delphi.[10]

Aristotle's *Constitution of the Athenians* generally takes an even milder line regarding violent Athenian *stasis*. This work was found as a papyrus late in the nineteenth century and appears to be intact but for the opening paragraphs, but the Cylonian affair of c.630 was evidently among the first important events.[11] Evidently, it began with the Cylonian affair. Cylon, an Olympian victor, attempted to establish himself as tyrant of Athens, but he was driven out and his supporters, having taken refuge on an altar of Athena, were dragged away and killed. This act of sacrilege brought a curse on the perpetrators, led by the Alcmaeonid Megacles, who was a magistrate. In expiation, the perpetrators themselves were expelled from the city and their descendants, who remained, were henceforth held to be accursed by their political enemies. Three other sources, Herodotus, Thucydides, and Plutarch provide damning yet arguable details of the affair. Herodotus says that the magistrate and his associates made a promise not to put them to death; Thucydides says the promise was to do no harm; Plutarch says they promised a trial.[12] All this weighed against the magistrates, but Cylon was attempting to establish a tyranny, and the one exception to the rule that suppliants should not be put to death on the spot was supplication by a tyrant. In this case, death was deserved, and never promises or trials. Herodotus supplies one instance and Plutarch's *Timoleon* supplies another.[13]

We lack Aristotle's account, but he describes the next few decades as a period of *stasis* between what he calls the *gnorimoi* (the nobles) and the *plethos* (the people). Ignoring the issue of sacrilege and murder, and of the possible perversion of legal process, he attributes this stasis to economic causes. The impoverished *plethos* was obliged to negotiate loans on the security of their own persons (*Ath. Pol.* 4.4) and then, when they could not meet their obligations, they fell into slavery. The *stasis* became violent "for a long time" (5.2), and the desperate Athenians appointed Solon as an arbitrator. Among other improvements, Solon

9 Indeed, the sole "heroic" notice is the burial of Anchimolius: "at Alopeke, near a temple of Heracles in Cynosarges" (5.63.4).

10 [Arist.] *Ath. Pol.* 20.1, 20.3; Hdt. 5.72.4.

11 Rhodes 1981, 1ff.

12 Hdt. 5.71.1, Thuc. 1.126.11, Pl. *Sol.* 12.1.

13 Hdt. 5.56, Plu. *Tim.* 4.5–6 with Naiden 2009, 157.

forbade loans on the security of one's person (9.1). Most memorable was his *seisachtheia*, probably an abolition of some kinds of debts (10.1). After sanitizing the origin of this period of stasis, Aristotle has simplified its outcome.

Aristotle goes on to say that Solon extracted an oath from the Athenians not to change any of his reforms without his consent. They agreed, and he departed for ten years. Peace prevailed for four years, but there was discontent, because some people were not able to collect debts owed to them, and *stasis* erupted once more (13.1). At about this time, the Athenians were fighting the Megarians for the control of the island of Salamis, and we are told that Pisistratus distinguished himself in the fighting (14.1). Thanks to this achievement and to his support of the *demos* (a.k.a. the *plethos*; 13.4, 14.1), Pisistratus became a tyrant. He did so by violence—first, by deliberating wounding himself so that he might be granted a bodyguard, and then by seizing the acropolis, and the tyranny along with it. He held power for five years before two leading families combined to drive him out (14.3). Throughout this account of extended, bitter *stasis*, Aristotle, who has time to recall the name of the mover of the motion to grant Pisistratus a bodyguard, does not record a single murder or other violent death.

Stasis continues, but to believe Herodotus, the other main source for these events, it now became unsanitizable. Megacles, a disgruntled noble, invited Pisistratus to return from exile, and proposed to cement a political alliance through the union of Pisistratus with his daughter, who would bear the tyrant a successor. Pisistratus thwarted Megacles by not having sexual relations with her "in the usual way" (Hdt. 1.60.1), in other words, by having anal rather than vaginal intercourse. Aristotle, however, says that he declined to have intercourse (*Ath. Pol.* 15.1). This evasive formulation did not keep Aristotle from having to report that the neglectful husband fled for his life, and that Pisistratus now turned to outright war in order to regain power, first hiring mercenaries and recruiting allies in Thebes, Eretria, and on Naxos, and then fighting the battle of Pallenis, as already described (15.3). Characteristically, Aristotle makes no mention of the Athenians who died at Pallenis.

Although Aristotle omits murders, assassinations, casualties resulting from *stasis*, he seems to have no compunction regarding violent acts perpetrated by tyrants and assassins of tyrants. He records how Hippias murdered "many" and drove others into exile (19.1). Aristotle (19.3) quotes a *skolion* that sings of the lives lost at the siege of Lipsydrium, in which Hippias unsuccessfully tried to dislodge political opponents who had taken refuge there. Aristotle even overlooks executions ordered by the opponents of tyrants. In narrating the events of 403, Aristotle fails to report the fate of the generals in command of some supporters of the Thirty. As already noted, Xenophon (*HG* 2.4.43) says that this group had withdrawn to Eleusis, where the Athenian democrats attacked them and put their generals to death. Aristotle describes this scene (40.4), but omits the executions.

I would explain Aristotle's distaste for reporting violent acts of Athenian by proposing that fourth-century Athens had little appetite for remembering these evils.[14] The city sometimes censored or reproved public mention of unwelcome events. The first recorded example comes after the fall of Miletus to the Persians in 494. The Athenians were devastated by the fall of the city, and when the tragedian Phrynichus staged a play about it, *The Capture of Miletus*, the audience burst into tears. The Athenian assembly fined Phrynichus heavily, and outlawed production of the play because it called to mind a domestic disaster (Hdt. 6.21.2).

---

14 Cf. Rhodes 1981, 15–31, speculating on what literary sources were available to him; more important, I think, is the evasiveness he and the sources might share.

At the end of the same century, a magistrate in Aristophanes' *Lysistrata* told the heroine *me mnesikakein* after she complained that the magistrates bereave women of their sons by sending them away to die in battle (Ar. *Lys.* 591). These are literary examples, but the amnesty decree provides a legal one.[15] This cultural exercise was not exclusively Athenian: The Thessalians, who had grievances against the Phocians, informed them that they would not remember evil (*mnesikakein*) for a fee of 50 silver talents (Hdt. 8.29). In Athens, however, a long history of civil strife gave the city reason to make use of this exercise in order to facilitate reconciliation among political enemies. Not only dramatists and legislators, but also historians giving public readings, such as Herodotus, would feel some pressure to oblige the Athenian public.

Conclusion

In 1900, George Grote quoted the Tory William Mitford's condemnation of democratic Athens as "giddy, irascible, unjust and greedy," and refuted Mitford by referring to the *me mnesikakein* clause of the settlement of 403.[16] Grote was right to point out this theme in Athenian politics, but he omitted two qualifications. First, as Thucydides says about the events of 411, the Athenians were sometimes forgiving toward one another in the face of a foreign threat. Their motive was fear as much as magnanimity. Second, the practice of forgetting past differences was just that, a practice, and it required both censorship and self-censorship. This kind of forgetting was, after all, self-negating. Whenever the Athenians urged one another to do it, the event they wished to forget returned to mind, and impressed itself on the city's memory. Controlling public discourse was a struggle.

Sometimes it required scapegoating, as in the case of the Peloponnesian War. Did Athens fail owing to the bad advice of the *gnorimoi* or to the reckless impulsiveness of the *plethos*? Andrew Wolpert shows how the two factions found a way between the horns of the dilemma by blaming the Thirty.[17] As I have written elsewhere, renewed dreams of empire unified the Athenians of this era.[18] Imperialism was another way to displace civil strife.

15 For other examples, see Carawan 2012, who canvasses the Attic orators.

16 Grote 1900, 301, starting a chorus of praise: Hignett 1952, 298; Finley 1962, 183; Fine 1983, 523. Cf. Hornblower 1991, 182–183.

17 Wolpert 2002, 125.

18 Shrimpton 2006, 308.

PART VII

Warfare in Art and Literature

CHAPTER 28

The Iconography of War

Olga Palagia

This brief survey concentrates on the iconography of historical battles in Greek art. Fights of gods and giants or amazons and Greeks or even episodes of the Trojan War are not treated here as they are not expected to reflect contemporary military practices. Only a small selection of historical battles was represented in art, usually not long after the events. The Battle of Marathon, a pivotal moment in Athenian history, was depicted twice in the monumental art of Athens, once in the lifetime of the battle's veterans, and again during the Peloponnesian War, in a temple glorifying the Athenians' military prowess. Anonymous battles of Greeks and Persians were also depicted in Attic red-figure vase-paintings produced shortly after the events. Battles against Persians reappear after Alexander the Great's conquest of the Persian Empire. His Successors commissioned paintings of Alexander's fights against Darius; anonymous battles against Persians can also be found in the funerary monuments of the veterans of Alexander's campaign.

Among the battles between Greeks we single out the *Battle of Oenoe* between Athenians and Lacedaemonians, dedicated by the Athenians in the Painted Stoa in tandem with the *Battle of Marathon*. In the fourth century four rather insignificant military events were memorialized by major paintings. Later, the defeat of Perseus of Macedon at Pydna was immortalized by the Roman commander, Aemilius Paullus, in a sculptured relief frieze that decorated the base of his equestrian statue at Delphi.

A Companion to Greek Warfare, First Edition. Edited by Waldemar Heckel, F.S. Naiden,
E. Edward Garvin, and John Vanderspoel.

Unidentified Early Battles

The development of the Greek phalanx is reflected in Corinthian painting and vase painting around the middle of the seventh century. We do not know if the battles depicted are historical or mythological but the military equipment is based on contemporary practices. The polychrome Chigi olpe represents two Greek hoplite armies coming to blows, each soldier carrying a shield and two spears (Fig. 28.1).[1] A single flute player encourages the hoplites marching to right. Behind the two armies, groups of men run forward holding spears. On an aryballos attributed to the same painter, the phalanx formation has disintegrated into duels, which no doubt happened in battle.[2] Fragments of a wall-painting excavated in the seventh-century temple at Kalapodi depict two opposing armies in a style similar to the Chigi vase.[3] We do not expect to see the depiction of a historical battle at this early date, but the contemporary arms and armor are remarkable.

Historical Battles

The Persian wars

The Battle of Marathon (490)

The Battle of Marathon became embedded in the collective memory of the Athenians as their greatest triumph. The Athenian general Miltiades who led them into battle was commemorated in a panel painting dedicated in the Painted Stoa, which was erected in the Athenian Agora at the instigation of Miltiades' son, Cimon, in the 460s.[4] The painting has perished but we get glimpses of its contents from the literary sources.[5] The most comprehensive description is by Pausanias (1.15.3), who saw it surrounded by three other panels, two of them depicting mythological battles, while the third, which he names the *Battle of Oenoe*, also illustrated a historical battle involving the Athenians.[6] The Painted Stoa thus contained two of the earliest depictions of historical battles in Greek art.

According to Pausanias, the *Battle of Marathon* painting showed three phases of combat: first, the Athenians and Plataeans fighting the Persians in an equal fight, then a rout, with the Persians chased into the swamp, and finally the battle at the ships, where the Athenians attempted to stop the Persians from boarding their ships and fleeing. We do not know if

1 Middle Proto-Corinthian. Rome, Villa Giulia 22679, from Veii. Boardman 1998, fig. 178, 1–2; Wheeler and Strauss 2007, 197, fig. 7; Mugione and Benincasa 2012, pls. 1–6; Hurwit 2014, 80, pl. 2.9.

2 Middle Proto-Corinthian. Macmillan aryballos, London, British Museum 1889.4–18.1, from Thebes. Boardman 1998, fig. 176, 2.

3 Niemeier et al. 2012.

4 On the Battle of Marathon and its aftermath, see Krentz 2010. On the date of the Painted Stoa, see Palagia 2019. On the painting of the Battle of Marathon, see Harrison 1972; Hölscher 1973, 50–68; Castriota 1992, 28–32; Boardman 2005, 66–71; Di Cesare 2015, 188–190; Duffy 2018, 86–87.

5 On the sources, see Wycherley 1957, 31–45; Harrison 1972, 370–378.

6 On the Battle of Oenoe, see n. 35.

Fig. 28.1 Hoplites on the Chigi Vase. From Veii. Mid-Seventh-Century Rome, Villa Giulia 22679. Photo from Boardman 1998, fig. 178, 2.

Pausanias "read" the painting from left to right or from right to left. In any event, the painting seems to tally with Herodotus' account of the battle (6.109–115). It very likely predated Herodotus, in which case it was based on accounts of the battle by survivors.

Pausanias adds that the figures were named, and that gods and heroes attended the battle. Callimachus and Miltiades were among the most prominent Athenians, with Miltiades shown urging the Athenians to fight. Pliny (*HN* 35.57) adds the names of Cynegirus and Datis and Artaphernes, while Aelian (*NA* 7.38) adds Epizelus. One of the warriors was accompanied by his dog (Ael. *NA* 7.38), which may be evidence of the participation of dogs in battles, at least until the fifth century. The Athenians were supported by their patron goddess Athena; their national hero, Theseus; Heracles, who had a cult in Marathon; the eponymous hero of Marathon; and the hero Echetlus. Theseus was shown as a half-figure, emerging from the ground. It has been suggested that the divine figures were shown separately, in an upper register, by analogy with contemporary Attic red-figure vase paintings dividing the action up and down by means of groundlines.[7] But the gods can also be shown alongside the mortals, as in the north metopes of the Parthenon depicting the Fall of Troy.[8] Considering the fact that the Marathon dead were heroized (Paus. 1.32.4), we could readily imagine the warriors commingling with gods.

7 Harrison 1972; Boardman 2005, fig. 6.1, cites the Attic red-figure calyx-krater by the Niobid Painter in Paris, Louvre G 341, Beazley, *ARV2* 601, 22.

8 Aphrodite stands between Helen and Menelaos in North metope 25: Berger 1986, pl. 22.

During the Archidamian War the Athenians rebuilt the temple of Athena Nike on the Acropolis; its sculptured decoration was a celebration of victory in its various manifestations.[9] It is generally acknowledged that the south frieze represents the Battle of Marathon.[10] It was carved some 40 years after the painting of the battle in the Stoa Poikile and follows a rather different pattern. Unless their names were painted on the frieze and are now lost, no gods or heroes are involved. There is no attempt to differentiate the individual figures Greek hoplites fight against the Persian cavalry and a handful of foot soldiers (Figs. 28.1–28.2). The presence of Persian horsemen had raised doubts as to the identification of the battle[11] but Herodotus is silent on the issue and in fact the presence of horses on the frieze could be taken as evidence that the Persian cavalry *was* involved. The Greeks are shown in heroic nudity with the occasional billowing cloak, while the Persians wear trousers and headgear with lappets, and carry bow cases. The Greeks have the upper hand as the ground is littered with Persian bodies. The battle is generic save for a distinctive scene where two Greek hoplites drag a Persian down from his collapsing horse (Fig. 28.3).[12] Even though we have no record of a notable Persian perishing in the Battle of Marathon, this episode probably reflects a historic moment in the battle which lingered in the collective memory of the Athenians.

In addition to the two state-sponsored depictions of the Battle of Marathon, the Athenian repulse of the Persian invasion is illustrated in private settings such as Attic vase paintings, mainly from 490 to 450.[13] No cavalry is depicted; the opponents fight in single combat and there is a high degree of accuracy in the costumes and equipment of the Persians, especially in vase paintings produced soon after the events. Persians holding tall rectangular

Fig. 28.2 Block from the South Frieze of the Temple of Athena Nike. *The Battle of Marathon.* London, British Museum.
Photo Olga Palagia.

9 On the temple of Athena Nike, see now Shear 2016, 341–358.

10 The frieze is divided between the Acropolis Museum and the British Museum. Harrison 1972; Leventi 2014, 73–76; Shear 2016, 352–353. For an illustration of the entire frieze, see Felten 1984, pl. 47.2.

11 The presumed lack of cavalry at the Battle of Marathon is based on Suda, s.v. χωρὶς ἱππεῖς. The alternative interpretation of the frieze would have been the Battle of Plataea but the Athenians would not have wished to commemorate their collaboration with Sparta in the middle of the Peloponnesian War. See Harrison 1972, 353–354; Palagia 2005, 184.

12 Athens, Acropolis Museum 18145. Palagia 2017, 178–179.

13 See Muth 2008, 239–267; Miller 2011; Zournatzi 2019.

Fig. 28.3 Block from the South Frieze of the Temple of Athena Nike. *The Battle of Marathon.* Athens, Acropolis Museum 18145.
Photo Olga Palagia.

shields, wielding spears, and wearing body armor fight Greek hoplites on an Attic red-figure cup by the Painter of the Oxford Brygus, painted not long after the *Battle of Marathon*.[14] On a somewhat later Attic red-figure oenochoe by the Chicago Painter, a Greek hoplite, nude save for greaves and helmet, spears his Persian opponent, who is dressed in Persian costume and wields sword and bow.[15] The Battle of Salamis was commemorated in art by an allegory painted on one of the screens between the legs of the throne of Phidias' Zeus at Olympia (Paus. 5. 11.5): a personification of Salamis holding a ship's figurehead was depicted alongside a personification of Hellas. Pausanias (5.11.6) attributes this painting to Panaenus, Phidias' brother, to whom he also attributes the painting of the Battle of Marathon in the Painted Stoa, whereas other sources ascribe the *Battle of Marathon* painting to Micon (Ael. *NA* 7.38).

Around 200, Attalus I of Pergamon dedicated on the Athenian Acropolis an under-life-size bronze group of the Battle of Marathon along with groups of an Amazonomachy, a Gigantomachy, and a battle of Pergamenes and Gauls (Paus. 1.25.2).[16] Three Roman copies of defeated Persians are extant: they depict the Persians nude or in partial nudity, indicating that the clothing and equipment of the Persian Empire had long been forgotten.[17]

---

14 Oxford, Ashmolean Museum 1911.615. Beazley, *ARV2* 399, P 580; Muth 2008, 246–247, figs. 161A–B; Miller 2011, 131, fig. 3; Zournatzi 2019, 6, fig. 2.

15 Boston, Museum of Fine Arts 13.196. Beazley, *ARV2* 631.38; Muth 2008, 12, 258, figs. 10, 167; Miller 2011, 133, fig. 5; Zournatzi 2019, 122, fig. 1.

16 Stewart 2004, 204–205.

17 Copies: Vatican Museums 2794, Stewart 2004, 296, fig. 3; Picón and Hemingway 2016, 178–179, no. 99. Aix-en-Provence, Musée Granet, Stewart 2004, 296–297, fig. 4. Naples, Museo Archeologico Nazionale 6014, Stewart 2004, 297, fig. 21.

The Asian campaign of Alexander the Great

Panel paintings illustrating Alexander's battles with the Persians were commissioned after his death, presumably executed in the lifetime of his veterans. Three such paintings are recorded: a) a battle between Alexander and Darius by Philoxenus of Eretria, a member of the Athenian School of painting, commissioned by Cassander;[18] b) the battle of Issus, painted by Helena of Egypt;[19] and c) a multi-figured composition of [Alexander?] fighting the Persians, painted by Aristides of Thebes and bought by Mnason, tyrant of Elatia.[20]

The Battle of Issus (333)

All three paintings are now lost but one of them may be reflected in a floor mosaic, known as the Alexander mosaic, from the House of the Faun in Pompeii, created in the late second century (Fig. 28.4).[21] This evidently draws on a monumental painting of the late fourth century, given the verisimilitude of the equipment of Alexander the Great (Fig. 28.5), which resembles the arms and armor excavated in Tomb II in Vergina. Alexander's articulated cuirass with wide shoulder straps is akin to the iron cuirass of Vergina, while his sword is similar to the Vergina sword.[22] The mosaic is fragmentary and was repaired in antiquity, and most of the Macedonian troops are missing, but it represents a cavalry battle telescoped within a narrow frame. Alexander leads the charge from the left in pursuit of Darius. He thrusts his spear into a young Persian horseman whose horse is collapsing under him (Fig. 28.6). Alexander's sword hangs from a shoulder-strap at his left side, and he rides a chestnut horse, presumably Bucephalas. He is bareheaded and wears a cuirass with *pteryges* over a long-sleeved chiton. It has been argued that his chlamys was originally purple, indicating royalty, and that it has lost its original coloring.[23] The Persians retreat to the right; a line of sarissas in the background indicates that the Macedonian phalanx surrounds them. Darius in his chariot towers above everyone else in the picture (Fig. 28.7). He wears an upright tiara and royal tunic with white centerpiece. He appears to be in panic at Alexander's approach; his horses are in disarray as their path is littered with dead bodies (Fig. 28.8). The scene seems to reflect a pivotal moment in the battle of Issus as described by Diodorus (17.34.1–7): Darius' brother, Oxathres, threw himself and his noblemen in Alexander's path to stall his charge, thus enabling Darius to escape from the field.[24]

The prototype of the Alexander mosaic is usually attributed to Philoxenus of Eretria and identified with the painting commissioned by Cassander.[25] There is, however, a distinctly

18 Plin. *HN* 35.110. Stewart 1993, 134, 376.

19 Photius, *Bibliotheca* 190.149b.29–33 Bekker. Stewart 1993, 376. This painting was later removed to Rome and exhibited in Vespasian's temple of Peace.

20 Plin. *HN* 35.99. Alexander's presence is conjectural. Stewart 1993, 378.

21 Naples, Museo Archeologico Nazionale 10020. Stewart 1993, 130–150; Cohen 1997; Moreno 2000; Palagia 2017, 177–180 with earlier references.

22 Palagia 2017, 180. Iron cuirass from Tomb II in Vergina: Andronicos 2004, figs. 95–96. Sword from Tomb II: Andronicos 2004, fig. 99. An iron cuirass, iron greaves and an iron helmet were also excavated in Agios Athanasios, Macedonian Tomb III, of ca. 300: Tsimbidou-Avloniti 2011.

23 Guimier-Sorbets 2016, 227–228.

24 Stewart 1993, 134–139; Palagia 2017, 179–181.

25 Stewart 1993, 147–150. Moreno (2000) identifies it as the Battle of Gaugamela and attributes it to Apelles.

Fig. 28.4 The Alexander Mosaic. From Pompeii. Second-Century Naples, Museo Archeologico Nazionale 10020.
Photo Hans R. Goette

Egyptian flavor in the mosaic decoration of the House of the Faun, as other mosaics depict Nilotic scenes with crocodiles and cobras.[26] The association with Egypt supports the attribution of the Alexander mosaic to Helena of Egypt (whoever she was), indicating that the

Fig. 28.5 Alexander as He Appears in the Mosaic. Naples, Museo Archeologico Nazionale 10020.
Photo Hans R. Goette.

26 De Caro 2019, 100–102, figs. 60–61.

Fig 28.6 Alexander on the Attack in the Mosaic. Detail. Naples, Museo Archeologico Nazionale 10020.
Photo Hans R. Goette.

original was probably commissioned by Ptolemy I, who could easily have wished to commemorate his participation in the battle.[27]

The episode of Alexander piercing a Persian rider reappears in the battle scene between Macedonians and Persians on one of the long sides of the so-called Alexander sarcophagus from the royal cemetery of Sidon, dated toward the end of the fourth century.[28] Alexander is recognized by his lion-head helmet, a type that he is not known to have worn in life; it seems to be inspired by the head of Heracles with Alexander's features reproduced on posthumous Alexandrines, notably the coins of Philip III Arrhidaeus.[29] The sculptor of the sarcophagus was not familiar with Alexander's features; however, the appearance of Alexander dispatching the Persian rider suggests that the battle scene refers to the victory at Issus. The

27 Palagia 2017, 180.
28 Istanbul Archaeological Museum 370. Stewart 1993, 299–301, 422–423, figs. 102–103; Palagia 2017, 180, 182, fig. 11.4.
29 See Palagia 2017, 182.

Fig. 28.7 Darius as He Appears in the Mosaic. Naples, Museo Archeologico Nazionale 10020. Photo Hans R. Goette.

sarcophagus is usually attributed to Abdalonymus, who was appointed by Alexander king of Sidon in 332, after the battle of Issus.[30]

Alexander's pursuit of Darius also appears on four Apulian vases of c.330, all attributed to the Darius Painter (Fig. 28.9).[31] Alexander's proximity to Darius' chariot on most of the vases is unhistorical. Moreover, Alexander is bearded and wears the wrong cuirass (muscle cuirass) and a Corinthian helmet (outdated by the fourth century), while Darius and his charioteer wear theatrical costumes; Darius rides a Greek chariot. The painter obviously did not know what Alexander looked like, and the inaccuracy of the costumes and military equipment indicates that the scene drew on hearsay, not any contemporary prototype. The surrounding divinities, heroes, and personifications raise the episode into the realm of myth.

---

30 Palagia 2017, 180, 182. For an alternative attribution of the sarcophagus to Mazaeus, satrap of Babylon, see Heckel 2016b. *Contra* Stewart 2014, 258 n. 2; Palagia 2017, 182.

31 Apulian volute-krater, Naples, Museo Archeologico Nazionale 3256; Apulian amphora, Naples, Museo Archeologico Nazionale 3220; Apulian amphora fragment, Copenhagen, National Museum 13.320; lost vase, ex Hamilton collection, drawn by Tischbein. Moreno 2000, 84–87, figs. 17–20; Stewart 1993, 150–157, 431–432; Palagia 2017, 179.

Fig. 28.8 Darius' Horses in the Mosaic. Naples, Museo Archeologico Nazionale 10020. Photo Hans R. Goette.

Cavalry battles of Greeks and Persians are also represented in the funerary monuments of Alexander's veterans in Macedonia. The battles seem generic though they may have had special significance to the tombs' owners. Good examples are offered by the plaster relief frieze on the façade of the Judgment Tomb at Leukadia and the painted marble couch in a Macedonian tomb at Dion.[32]

Greeks against Greeks

Oenoe (506)

When Pausanias visited the Painted Stoa (1.15.1), the first panel painting he saw was the *Battle of Oenoe* between the Athenians and the Lacedaemonians. He described the painting

32 Leukadia: Petsas 1966, 159–170, pls. 27–30. Dion: Pandermalis 1999, fig. on p. 268.

Fig. 28.9 The Pursuit of Darius on an Apulian Vase, ex Hamilton Collection, Now Lost. C. 330 Alexander Pursues Darius.
Drawing by Tischbein. From Stewart 1993, text Fig. 30.5.

as depicting two armies facing each other before coming to blows. This, along with the *Battle of Marathon*, would have been the earliest painting of historical battles in Greek art. It is reasonable to suggest that the Battle of Oenoe would have been as significant to the Athenians at the time as the Battle of Marathon, so one must look for a major historical event before the 460s. Where was the Battle of Oenoe? Pausanias located Oenoe in the Argolid and this has generated controversy, since no battle between Athens and Sparta in the Argive Oenoe appears in the historical sources. As a consequence, scholars have changed either the battle venue or the ethnicity of the opponents. The answer to the question has come from a newly excavated votive inscription from Thebes, dated c.500.[33] It mentions a single campaign, which included Oenoe, Phyle, Eleusis, and Chalcis, and this corresponds to the victorious campaign of the newly established Athenian democracy against the attack of the combined forces of Sparta, Thebes, and Chalcis in 506.[34] Herodotus (5.75) relates how the armies of

33 Thebes Museum 35900. Aravantinos 2006.
34 Hdt. 5.74–77.

Athens and Sparta and her allies faced one another on the plain of Eleusis but did not fight because the Peloponnesian allies fell out among themselves and retreated. The Athenians then marched off and drove the Thebans out of another Oenoe, in Attica, and crushed the Chalcidians in Euboea. The first victory of the Athenian democracy was commemorated with a bronze quadriga dedicated on the Athenian Acropolis (Hdt. 5.77), which was removed by Xerxes in 480/479 and replaced, probably in the 460s.[35] Pausanias' *Battle of Oenoe* thus refers to the campaign of Oenoe, which was the first significant military campaign launched by the Athenian democracy.

The Peloponnesian War (431–404)

No particular battles of the Peloponnesian War were commemorated in the state art of Athens. What we have instead are generic representations of battles between Athenian cavalry and enemy foot soldiers on the crowning reliefs of the knights' casualty lists that the state erected in the Athenian *demosion sema* after 431.[36] The best preserved example was found reused in a cemetery of the Roman period.[37] The inscription contains two separate casualty lists in two different scripts, with the more recent one written on top, and an epigram. The fallen knights are listed by tribes. The inscriptions name several battles, some of which are not mentioned by Thucydides. The earlier list dates from the 420s, which also dates the relief, while the second may be from around 409. The relief depicts two Athenian horsemen to the left, battling two enemy foot soldiers, one of whom is lying on the ground.

The iconography of state burials is also reflected in private monuments of the same period. Although individuals are named in them, the battle scenes remain generic. A fine example is offered by a relief dedicated to Demeter and Kore at Eleusis by the hipparch Pythodorus sometime in the decade 425–415 (Fig. 28.10).[38] Only a fragment of the relief is preserved, comprising a two-tiered scene of horsemen attacking foot soldiers from the right. One of the horsemen can be identified with Pythodorus, who clearly wished to commemorate his victory in a particular battle while he served as cavalry chief. Some conflicts commemorated by official artworks were not particularly significant to posterity but captured the political mood of the moment. Three minor engagements of the fourth century were commemorated by monumental paintings.

The Battle of Plataea (c. 379)

The earliest known painting of a cavalry engagement was painted by Androcydes of Cyzicus to commemorate a victory of the pro-Spartan oligarchy of Thebes (382–379) (Plut. *Pelop.* 25.5–6).[39] The painting was seized after the fall of the oligarchy and dedicated in commemoration of a minor skirmish at Plataea sometime before the Battle of Leuctra (371) with the addition of the Theban commander's name, Charon. This probably indicates that the battle piece was generic and could be used as a memorial to any event.

35 See Palagia 2019. For the association of the *Battle of Oenoe* painting with the events of 506, see also Stewart 2018, 305–306.

36 Goette 2009; Arrington 2015, 99–102.

37 City of Athens Ephorate storerooms in Plato's Academy M 4551. Parlama and Stambolidis 2000, no. 452; Goette 2009, 189; Arrington 2015, 100–101, fig. 3.2.

38 Eleusis Museum 5101. Clinton 2005, no. 39, pl. 16; Goette 2009, 198–199, color pl. 4; Arrington 2015, 194–195, fig. 5.5.

39 Palagia 1980, 52; Kansteiner et al. 2014, no. 1789.

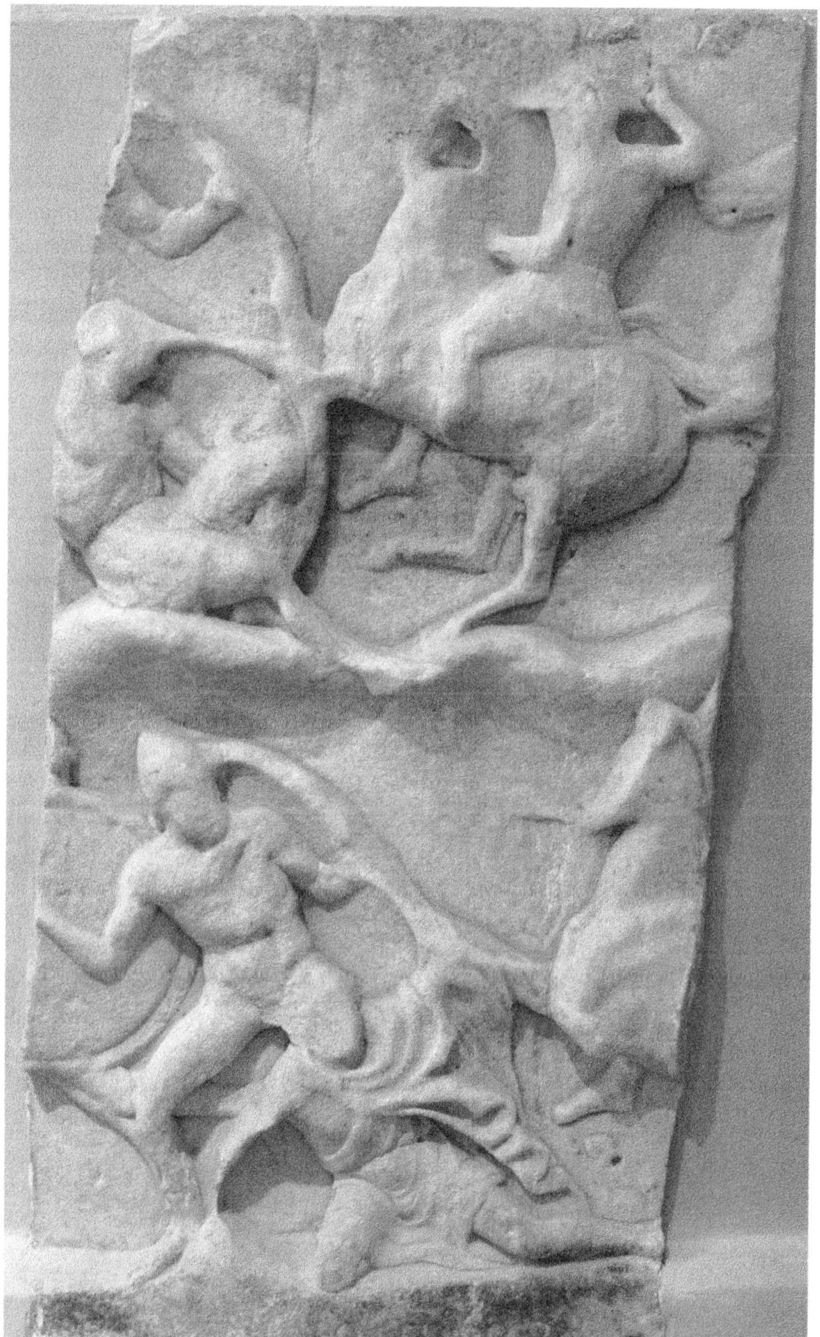

Fig. 28.10 A Cavalry Attack According to a Funeral Relief, Dedicated by Pythodorus to Demeter and Kore. About 425–415 Eleusis Museum 5101.
Photo Hans R. Goette.

The Battle of Phlious (second quarter of the fourth century)

The Athenians commissioned a battle piece from Pamphilus of Amphipolis to commemorate an Athenian victory in Phlious in the fourth century (Plin. *HN* 35.76). We do not know which battle was involved, or the whereabouts of the picture.[40]

The Battle of Mantinea (362)

The Athenian state commissioned a panel painting of the cavalry battle between Athenians and Thebans before the battle of Mantinea. The battle was painted by Euphranor and was dedicated in the Stoa of Zeus in the Athenian Agora.[41] Considering that the battle was indecisive, the main point of the painting was the glorification of Gryllus, Xenophon's son, who fell in that preliminary engagement. The painting was unhistorical, for Gryllus was shown killing Epameinondas even though Epameinondas perished the following day. It was, however, one of the earliest battle pieces showing a cavalry engagement and it was so successful that a Roman copy was erected in a chapel in the gymnasium of Mantinea, probably in the time of Hadrian (Paus. 8.9.8). Like Leonardo da Vinci's painting the *Battle of Anghiari*,[42] Euphranor's *Battle of Mantinea* immortalized an obscure historical event on account of its great artistic merit.

Macedonians against Romans

The Battle of Pydna (168)

After the defeat of King Perseus of Macedon, at the battle of Pydna, the Roman commander, Aemilius Paullus visited Delphi, where he saw an unfinished marble pillar that Perseus had been preparing to support his equestrian statue near the temple of Apollo.[43] Paullus appropriated the pillar, erected his own equestrian statue and dedicated it in his name from the spoils of Perseus. The dedicatory inscription is in Latin, forming a sharp contrast with the Greek inscriptions on all the earlier votive monuments at Delphi.[44] The image of Paullus has perished but we have the statue base, decorated with a relief frieze of the battle of Pydna, no doubt carved by a Greek sculptor (Fig. 28.11).[45] The Macedonians are distinguished from the Romans by means of their arms and armor. They carry round shields, some decorated with stars placed within concentric circles (Fig. 28.11),[46] others with a horizontal spine, while the Romans hold the *scutum*, a large oval shield.[47] The Macedonians wear corselets

---

40 Palagia 1980, 52.

41 Pausanias 1.3.4, 8.9.8, 8.11.6, 9.15.5. See also Palagia 1980, 50–54; Dillery 2017, 211. On the Battle of Mantinea: Xen. *Hell.* 7.5.16–27.

42 Frank 2019.

43 Plut. *Aem.* 28.2. Livy (45.27.1) and Polybius (30.10.2) were under the impression that Perseus had erected multiple pillars.

44 Jacquemin 1999, 350, no. 424; Taylor 2016, 560.

45 Delphi Museum. On the relief frieze: Hammond and Walbank 1988, 613–617; Boschung 2001; Taylor 2016 (with earlier bibliography). On the Battle of Pydna: Polyb. 29.14–18; Livy 44.33–42; Plut. *Aem.* 13–22. See also Hammond and Walbank 1988, 547–557.

46 On this very distinctive Macedonian design, see Liampi 1998.

47 For the *scutum*, see Polyb. 6.23.2–6.

with *pteryges* (Fig. 28.11),[48] while the Romans wear chain mail[49] or muscle cuirasses. The warriors' metallic weapons were attached separately and are now lost. The Romans wear their swords on the right side of their belts,[50] the Macedonians on the left. There have also been attempts to distinguish the allies of the two armies.[51] The battle represents a melee of cavalry and infantry, where the Macedonians are consistently defeated; it probably evokes the final phase of the fight, after the Macedonian phalanx had been routed.[52] Neither Paullus nor Perseus can be identified and are probably not shown on the frieze. A Macedonian cavalry commander, riding a collapsing horse, can be distinguished by his animal skin saddle-cloth (Fig. 28.11), which was introduced to the Macedonian cavalry by Alexander the Great after the conquest of Asia.[53] Alexander himself uses a panther skin saddle-cloth on the Alexander mosaic (Fig. 28.5). It is interesting that the frieze shows Macedonian horsemen in distress even though the Macedonian cavalry escaped from the battlefield almost unscathed. It is also interesting that the frieze does not include Aemilius Paullus' elephant contingent.

Fig. 28.11 The Battle of Pydna According to a Relief on a Statue Base. Detail. Delphi Museum. Photo Olga Palagia

48 These have been described as linen corselets (Taylor 2016, 562) though the Macedonians also wore iron armor, as attested by the corselets excavated in Macedonian tombs, see n. 20. Compare Alexander's cuirass in the Alexander mosaic (fig. 5).
49 On Roman chain mail corselets, see Polyb. 6.23.15.
50 Polyb. 6.23.6. Taylor 2016, 562.
51 Taylor 2016.
52 Thus, Taylor op. cit.
53 Palagia 2003, 148.

CHAPTER 29

War Monuments and Memorials

Lawrence A. Tritle

Perhaps every human society has commemorated acts of wartime heroics and triumph, and sometimes defeat and disaster. The third-millennium Mesopotamian *Epic of Gilgamesh* speaks of memorial statues and of the names of famous royal commanders, these names sometimes having been obliterated and thus lost to memory.[1] The Circle Graves excavated by Heinrich Schliemann at Mycenae in the 1870s mark the resting places of elite warriors, some engaged in battles portrayed on sculptured and (presumably once painted) stelae that stand beside the dead. Do the stelae remind the viewer of hard-fighting warriors, and thus constitute a monument to these men, or do the stelae let the viewer personally remember them, and thus constitute a memorial? Both readings of the stelae seem possible.

"Monument" derives from the Latin *monere*, to remind, warn, or teach; "memorial" from the Latin *memoria*, designating remembrance or recollection. In either case, viewers bring their own perspective and experience to monuments and memorials, a process that may evoke complementary or conflicting thoughts and emotions.[2] One example from mid-twentieth-century America is the refusal of President Dwight Eisenhower to speak at the dedication of the United States Marine Corps Memorial (better known perhaps as the "Iwo Jima" Monument), as he believed it represented "crass self-promotion" that demeaned the far greater sacrifices of the United States Army.[3] An example from both ancient and modern Greece is the shocked response of the 1908 *Macmillan Guide to Greece* to the exposed

1 Gilgamesh: George 1999, 56 (removing names), 65 (commemorative statues). Commemoration of the dead is, of course, far older. Neanderthal etchings: Zimmer 2018. Early burials: Wilson 1979/2004, 169, Dunbar 2004, 6 (and canine burial, c. 7,000: Roman-German Museum, Cologne, Germany). Pyramids: Borg 1991, 52–53.

2 Arthur Danto, cited in Sturken 1997, 47, observes that the two may overlap.

3 Haas 1998, 140 n. 69.

A Companion to Greek Warfare, First Edition. Edited by Waldemar Heckel, F.S. Naiden, E. Edward Garvin, and John Vanderspoel.

remains of the Theban dead of Chaeronea in the National Archaeological Museum in Athens. Just as Eisenhower objected to a kind of vainglory, the guidebook objected to a kind of desecration, saying that the remains had nothing to do with art and were better left in the ground.[4]

Surveying this long history, *The Dominion of the Dead* by Robert Pogue Harrison observes that civilization itself owes much to the habit of burial and remembrance.[5] Words that continue to define us—humanity (*humanitas*) and human (*humanus*)—derive from *humus*, the earth, and so, Harrison says, putting the dead to rest in the ground (*humo*, *humatio*, a burying) as a way of remembering them is a fundamental human act. Sophocles knew this, too, as the story of Antigone and Creon feuding bitterly over burial rituals for the torn body of Polynices makes clear. Sophocles' *Antigone* is the most celebrated example of this obligation. Yet the obligation was not always observed: the Thebans denied the Athenian request to recover their dead at Delium, leaving the bodies exposed for 17 days before relenting (Thuc. 4.101.1). The condition of the bodies may be imagined by referring to Menander. Some lines in his *Aspis* (69–79) tell of a soldier's body left exposed for three days, the face so bloated that it could not be recognized. The Athenians acted similarly in suppressing the Samian Revolt of c. 440.[6]

This vengeful practice resembled Roman *damnatio memoriae*, which insulted some of the dead but in a different way, by excising names from monuments. It casts a shadow over the story of commemoration that this chapter will discuss—the perennial attempt, through art, memorials, cemeteries, and mere casualty lists, to see to it that the "heroic" dead shall not die.[7]

Greek Remembrance Begins

While the origins of the Greek city state, or *polis*, are matters of debate, the *polis* is present in the pages of Homer, and it was not long after the Homeric poems took shape that heroic remembrance in the Greek city-state took a new form—the *kouros* sculptures first appearing around 650.[8] Egyptian influenced and meaning literally "young man," the earliest of these life-sized marble *kouroi* honored gods like Apollo and heroes like the legendary Cleobis and Biton, immortalized in Apollo's sanctuary at Delphi. Similar status and honors were soon bestowed on dead contemporaries, especially for the purpose of preserving the names and memory of those who died heroically in battle. Perhaps the best example of this is the *Anávissos Kouros* erected in honor of a noble Athenian youth named Croesus, whose story accompanies the preserved name. It tells how Croesus died in battle: "Destroyed one day as

4 Ma 2008a, 72.

5 Harrison 2003, x–xi.

6 Tritle 2010, 13–14.

7 Many may believe that this part of the mythology of war–*dulce et decorum est pro patria mori*—lasted until the slaughter of World War I led poets like Wilfred Owen and novelists like Erich Maria Remarque to demolish it. Yet Homer, Archilochus, Thucydides, and Euripides make clear that even ancient Greece questioned the idea that that death in combat was sweet and proper.

8 Rise of the *polis*, democracy and freedom: Morris 2001, 68–76; Raaflaub 2004; Arnason et al. 2013. Cf. also Hom. *Od.* 15.412 and the numerous references to hoplite arms and battle in the *Iliad*.

he fought in the foremost ranks."[9] So began a tradition of honorific sculpture that we know well from modern examples like the Iwo Jima sculpture to which Eisenhower objected.

Aristocratic families such as that of Croesus contributed greatly to the development of the *polis*, which over time became more democratic, Athens being the exemplar. Sharing political rights and responsibilities, non-elites soon learned, included the hazards and consequences of battle. The citizen holding his shield and spear became the basic fighting man of the Greek world, and navies became powerful forces employing thousands in Athens and in island communities such as Lesbos and Chios. Service as naval rowers brought the *thetes*, Athens' lowest census class, not only employment but also pay and a sense of self-importance that they would manifest in meetings of the Athenian assembly. These rowers earned recognition for heroic deaths, just as soldiers did. So did some slaves, for they rowed as well, and less often fought. The community would honor them, too, as happened after the battles of Marathon, Arginusae, and Chaeronea.[10]

Marathon was but one of a series of battles and sieges, beginning with the Ionian Revolt of 499–494, that brought home the costs of war to Athens and other Greek cities: cities such as Athens and Miletus being burned to the ground, others such as Eretria being "netted," their populations taken away into captivity, and the intense, titanic struggles at Thermopylae and Salamis in 480, and at Plataea in 479. The Greek response to these events strengthened and transformed practices of commemoration in ways that persisted through later antiquity. The heroic dead were variously interred (as at Marathon), memorialized by literary tributes (at Thermopylae, Simonides' tribute to the seer Megistias and his famous verses, "Go Tell the Spartans," inscribed on the cenotaph of the dead Three Hundred), and honored with shrines and festivals (at Plataea, the Eleuthera, a festival that began with the burial of the Greek dead, and took place annually for centuries).[11]

These commemorations and sanctifications were agonistic. So was the custom that the victor marked the spot where the losing side halted and turned back by a trophy (Greek *tropaion*) that was a pile of arms and armor stripped from the bodies of the dead. This tradition lasted centuries also, but with time the ephemeral battlefield trophy took the form of a memorial in stone. Even today, travelers may behold the Theban victory monument that marked the spot where the Thebans repelled and defeated the Spartans at the battle of Leuctra in 371. The Thebans may have erected this bronze trophy at the very spot where the commander of the Spartan army, King Cleombrotus, died after being attacked by the elite Theban unit, the Sacred Band. Although permanent trophies had been erected before, this was the first to commemorate a victory over fellow Greeks.[12] It advertised the victors' prestige and power.[13]

---

9 *Kouros* style (and Croesus translation): Stewart 2008, 28–31. The Athenian Croesus was named after the Lydian potentate (or *tyrannos*) Croesus (fl. c. 560–540), providing an approximate date for the hero and his statue. Discovered in 1936 (base in 1938), the statue was dismantled and buried possibly during the fifth-century Persian invasion in order to protect it, only for it ironically to be forgotten.

10 Paus. 1.29.7, with Pritchett 1971–1991, vol. 4, 146–147.

11 Mound at Marathon: Thuc. 2.34.1 with Gomme 1956, 94. Tributes at Thermopylae: Hdt. 7.228. Tombs, shrines, and the Eleuthera festival at Plataea: Thuc. 3.58.4–5 with Gomme 1956, 344, Plut. *Arist.* 21.1, Paus. 9.2.4.

12 Munn 1997, 86.

13 See Pritchett 1971–1991, v.2, 246–275, for extensive discussion of the battlefield trophy and a list of known trophies (pp. 264–269).

At Chaeronea, the Macedonians dedicated a great burial mound (discussed later) to those of their comrades who died, but this is the only known example of a Macedonian commemoration of this kind. Alexander appears not to have commemorated his victories in such fashion, although like other commanders past and present he recovered his dead even in difficult circumstances and with little doubt buried them.[14] Alexander also erected statues to remember friends and intimates who died heroically. At Dion, the chief Macedonian shrine of Zeus, honorific statues commemorated the 25 troopers in the companion cavalry, or royal horse guard, who died at the Granicus (Arr. *Anab.* 1.16.4). The musician Aristonicus, who died fighting bravely in Afghanistan, was remembered in bronze at Delphi, a harp in one hand, a spear in the other (Plut. *Fort. Alex.* 334e–f). Spoils of war also made for influential political capital. Persian arms taken at the Granicus (Arr. *Anab.* 1.16.7) were sent to Athens engraved with a bit of propaganda, "Alexander son of Philip and the Greeks, except for the Lacedaemonians, set up these spoils from the barbarians dwelling in Asia."

Nearly constant warfare through the Hellenistic age that followed Alexander inspired many more military memorials. These appear to have followed the Alexandrian precedent of dedicatory statues and related commemorative monuments. The famous statues of the "Dying Gauls" in Rome today, once part of the great commemorative complex honoring the victories of the Attalid kings of Pergamum, is probably the best known of these (discussed further in later).

Battlefield Graveyards

We should bear in mind that some battlefields, even the most famous, were in reality open graveyards, at least for some period of years, and were cultivated by local farmers. "Fields fattened by corpses" was Archilochus' description, one picked up by Plutarch in his *Life of Marius* (21) and *Table Talk* (658b). At Chaeronea, the Macedonian dead were cremated and interred so that they might escape this fate. As for the two armies fighting on the other side, the ashes of the Athenians were returned home for the same purpose, while the Theban Sacred Band, who died virtually to the last man, were buried en masse. Yet many of the Chaeronea dead, allied Greeks from distant communities and mercenaries, were not spared the plow. Theophrastus reports that the region around the battlefield later suffered from a period of disease, evidently the result of the decomposing corpses that littered the site.[15]

Invading enemies, such as the Gauls who invaded Greece in 279, suffered this fate too, perhaps because the ferocious conduct of the Gauls earned them little sympathy from the Greeks who defeated them. As Pausanias reports, after the Gallic defeat at Thermopylae it was impossible to discover the number of their dead "since a large number of them vanished into the mud" (Paus. 10.22.4).

Sometimes burial came late. At the great Macedonian defeat at Cynoscephalae in 197, the phalanx were slaughtered, thousands being left unburied where they fell. Six years later, they finally received an honorific mound, not on the orders of their king and commander, Philip V, still less on the orders of the victorious Romans, but thanks to the pretender Philip of

14 See Diod. Sic. 17.25.6 for Halicarnassus, 68.4 for the Susian Rocks.

15 Theophr. *HP* 4.11.3, noted by Ma 2008a, 74.

Megalopolis, who was seeking to ingratiate himself with the Macedonian people.[16] Two thousand men were reportedly needed to collect the bones and build the mound that covered them.[17]

"Twas a famous victory," concluded Robert Southey in *The Battle of Blenheim* (1796), but not because of the defeat Britain administered to France back in 1704. Farmers had been plowing up bones ever since. In this way, too, a Greek military tradition lived on.

Military Cemeteries

At the great fifth-century battlefields of Marathon and Plataea, the dead were interred where they fell. This remained the Spartan tradition, one that would serve to intimidate rivals or remind others of the aid given to friends.[18] The Athenians, however, may have been the first to bring the dead home for burial in military cemeteries. Sometime after the Persian Wars, and certainly before c. 460, they commenced the practice of cremating the dead where they fell and returning the remains to Athens in caskets.[19]

Located outside the walls of Athens in the district known as the Ceramicus—what Thucydides believed the most beautiful part of the city—was the *demosion sema*, the "Public Cemetery."[20] This became the hallowed ground of Athenian heroes, in American terms, the Athenian Arlington National Cemetery.[21] An elaborate ritual developed: the display of the caskets according to the Athenian tribe to which the dead man belonged (and thus according to his infantry regiment, since these units were tribal), then a period of public mourning, and last an oration or eulogy offered by a leading citizen.[22] (Thucydides relates all this in respectful detail.). Similar memorials served to remember the dead of the Corinthian War (395–387/386), the war with Philip of Macedon that ended with Chaeronea (338), and the Lamian War (322/321).

The speeches given by Demosthenes for the Chaeronea dead and Lycurgus for those of the Lamian War survive. They make calls to civic duty and commemoration that echo the oration that Thucydides reports Pericles giving on the same occasion.[23]

In addition, the Athenians recorded the names of the fallen in lists organized by tribe and by rank and noting the year and place of death. Many of these inscriptions have survived.

---

16 The actions of his Megalopolitan ally drove Philip V to support the Romans and thus help bring out the defeat of Rome's remaining enemy in the region, the Seleucid king Antiochus III.

17 Liv. 36.8 with Hammond and Walbank 1988, 451.

18 Plut. *Apoph. Lac.* 233c notes a Spartan retort to an Argive boast that many Spartans lay buried in Argive soil: yes, the Spartan replied, but no Argives lay in Sparta. Cf. the Spartan tomb in the Athenian Ceramicus in honor of Spartans fallen to revive the Athenian democracy.

19 Date provided by the reference in Aesch. *Ag.* 434–436.

20 See also Pritchett 1971–1991, v.4, 145–158, with a table of the Athenian monuments, pp. 148–149.

21 Thuc. 2.34.5, translated variously as "public tomb" (Lattimore), or "public cemetery" (Hammond). Patterson 2006, 54–56, objects to the notion of a "public cemetery," but the critique is unpersuasive. Ceramicus as Arlington: Wills 1992, 296 n. 10; Tritle 2000, 166–167.

22 Thuc. 2.34–46 provides a full description of the rituals performed by the Athenians, and the famous "Funeral Oration" written for Pericles and given over the Athenian dead of the first year's fighting in the Peloponnesian War.

23 Discussion of the Athenian funeral orations is vast. Especially useful are Loraux 1986; Pritchett 1971–1991, vol. 4, 106–124; Raaflaub 2001.

Although mostly fragmentary or otherwise damaged, they have provided scholars valuable evidence regarding the Athenian citizen community.[24]

Similarly, Sparta erected a stele with the names of the Spartans who died with Leonidas at Thermopylae. The occasion for this memorial seems to have been the repatriation of Leonidas' remains by the Spartan commander Pausanias c. 440 (Paus. 3.14.1). Herodotus evidently saw the stele, and thus could claim he learned the names of the Spartans who fought at Thermopylae.[25]

Other Greek communities, Corinth and Megara, for example, had set up dedications to the memory of citizens who died fighting the Persians at Salamis or Plataea but evidently did not erect a list of names.[26] The difference between Sparta and Athens, on the one hand, and these additional cities, on the other, raises the question of the motive for permanently recording the names of fellow citizens killed in battle. This motive may be deep-seated, as argued by Harrison, who finds origins as remote as Stone-Age burials by Neanderthals; others point to a deep-seated spiritual impulse.[27] Whatever the explanation, in Classical Greece this practice became prominent without being universal.

Commonly buried with the dead were white *lekythoi* (the plural of *lekythos*), perhaps the most beautiful type of Greek pottery. The paintings on these *lekythoi* depict fifth-century wartime realities of death and dying, remembrance and memorializing, and not only in Athens. Findings of this type of funerary vase in locales stretching from the Greek islands to Italy demonstrate the widespread appeal of these images, and also show how Greek attitudes toward war and commemoration changed over centuries. The earliest paintings featured creatures from myth and legend. In the fifth century, the older, black-colored vase gave way to the white *lekythos*, which told stories of contemporary Greeks suffering from the scourges of death and war. Seen now were visits to graves by those left behind, offerings of possessions once prized by the dead, and scenes that reveal the emotional toll of war—tearing one's hair and other gestures of mourning.[28]

The great size, wealth, and power of Athens provided this city the means to bestow lavish honors on their fallen citizens. Poorer cities, including Megara, Tanagra, and Thespiae, also organized soldier's cemeteries and preserved the names of the fallen. Thespiae, whose men fought to the death against the Persians at Thermopylae, joined their neighbors in battle a generation later against the Athenians at Delium (424). The Thespians again suffered heavily. These losses may have sparked the dedication of a memorial precinct for their dead. It included a wall inscribed with the names of the fallen and surmounted by a crouching lion.[29]

Thebes, the powerful and ambitious leader of the central Greek states, had led its neighbors to victory against Athens at Delium. The victors took great spoil from the defeated Athenians, and this evidently financed monuments in both Thebes and Thespiae. Thespiae built a modest structure, but the more powerful and influential Thebans built a great stoa or

24 ML 125–128, include a list from 447(?) that gives locations where Athenians died (two, plus an ambiguous "other wars") and rank, namely that one general (*strategos*) was among the fallen. Additional inscriptions collected and discussed in Bradeen 1969; Arrington 2010; Pritchett 1971–1991, vol. 4, 139–140, 145–151, 153–228.

25 Hdt. 7.224.1. Presumably the names of the infamous survivors, Aristodamas and Pantites, were omitted (Hdt. 7.229–231, 232), yet their stories survived.

26 Corinth, 480 BCE: ML 52–53; Megara, 479 BCE: Fornara 1977, 58–59.

27 Wilson 1979/2004, 169 ff.

28 Full discussion: Oakley 2004.

29 Thespiae: Low 2003, 104–111. Other communities: Pritchett 1971–1991, vol. 4, 140–145.

colonnade decorated with bronze statues and displays of captured armor nailed to the columns (Diod. Sic. 12.70.4). The Thebans also financed a festival celebrating the victory, the Delia (Diod. Sic. 12.70.5). Some black stelai portraying fallen warriors—among the few military remains of ancient Thebes—may also commemorate the victory at Delium.[30]

One-upmanship may have prompted the Theban festival. The Eleuthera festival celebrated in nearby Plataea, a community long hostile to Thebes, commemorated another great victory, over Persia. Since Thebes had been Persia's ally, Thebans could scarcely participate in these Plataean celebrations. Now they had a comparable festival of their own. The name, Delia, was the same as one of an Athenian festival, and that would make for useful posturing, especially as the Theban festival was celebrated on or near the battle site, which was on the boundary between Boeotia and Attica.

While the Thebans were building and decorating their new stoa with wealth taken from dead Athenians, the Athenians were doing something similar, displaying captured shields taken from the Spartans in the island battle at Sphacteria near Pylos.[31] One of these shields survived into the modern era, to be found by archaeologists in the Athenian agora. (The inscription reads, "The Athenians from the Lacedaemonians [i.e. Spartans] at Pylos."[32]) Displays of captured arms and armor were common among the Greeks, not only in particular cities but also in the Panhellenic sanctuaries at Delphi and Olympia.[33]

Theban fortunes declined in the mid-fourth century, and at Chaeronea in the summer of 338 the Thebans and other members of a Greek coalition suffered a crushing defeat at the hands of Philip of Macedon. The Sacred Band of Thebes, a picked force 300 strong, was annihilated, at least 255 dying on the field and then being buried nearby. Owing most likely to the August heat, the ceremony was so hurried that the dead were buried with their boots on. Some years later, as a result of the Macedonian warlord Cassander's political intrigues, the famous Lion of Chaeronea was set up on the site. Reconstructed in its present form in 1902, this is the famous lion visible to tourists today. Historian John Ma notes that modern excavation reports suggest that when the Lion and its *peribolos* (precinct boundary) were established, the exact location of the Theban dead may have been forgotten, as some remains were found outside the boundary. If correct, this suggests that a number of remains, perhaps as many as 45, remain to be recovered, and that the Sacred Band died nearly to a man.[34]

A mile away from the Theban burial, in the plain of Chaeronea, the Macedonians organized their own tribute. Whereas the Theban site was individualized, the Macedonian one was communal. Archaeologists uncovered a thick layer of cremated remains (about 2.5' thick

30 The small archaeological museum at Thebes ascribes these stelai to Aristeidis, a student of the noted artist/sculptor Polyclitus, and thus assigns a date in the later fourth century [https://www.mthv.gr/en/permanent-exhibition/clasical-period/]. Perhaps, but the attribution is unattested: the triumph at Delium seems a likelier occasion to celebrate Theban glories than the slump that begins in mid-fourth century and leads to Chaeronea.

31 This Athenian parallel strengthens the case for the Theban Delia, which some scholars have doubted. See Gomme et al. 1945, 3, 567; Hornblower 1991–2008, 2, 308, expressing caution; and Pritchett 1971–1991, vol. 3, 206, regarding the Theban Delia as probable.

32 Paus. 1.16.5, with Camp 1986, 71–72 (Spartan shields); the surviving example in Athens today was recovered from a well in the Athenian agora dating to the third century; Pausanias would have seen others from the group preserved in the Stoa Poikile, the "Painted Porch".

33 Displays of armor at Athens: Paus. 1.15.2 (trophies from the Macedonians), 1.27.2 (Persian spoils). Delphic displays: Paus. 10.9.3. Tegean spoils from Sparta: op. cit. 10.10.2, 10.11.4. Treasuries of Thebes and Athens deriving from military triumphs: op cit. 10.10.5. Olympic displays: op. cit. 5.27.11. See Pritchett 1971–1991, vol. 3, 240–295, for detailed discussion and many more examples.

34 Ma 2008a, 80–84.

in the middle, 100 yards^2 in area) and various metal objects (remains of swords, arrow heads, and javelin and sarissa heads) suggesting that Macedonian losses were not slight.[35]

As for the other important participant in the battle, the Athenians, their dead, only 1,000 in number, were cremated on site and returned to Athens in a somber and politically calculated gesture.[36]

The Chaeronea burials show how both the battleground and subsequent commemorations became contested. The Theban memorial expressed tardy, calculated defiance, and the Macedonian one tacitly rebuked the Thebans for their opposition and their decision to support Athens against Macedon.[37] Athens ducked the conflict, commemorating the dead in its own, well-established way.

Archaeology brings its own weapons to this contested ground. The excavators and scholars studying the remains have documented the horrific wounds that some, perhaps many, of the fallen suffered. One example, the soldier known to us today as Gamma 16, reveals a ghastly head wound—the face nearly sliced away by the downward stroke of a cavalryman's sword. Other types of wounds are multiple sword-cuts to the lower legs or to the skull. One soldier received a *coup de grâce* to the head from a Macedonian spear butt—a circular hole in the skull surrounded by a depressed fracture.[38] These wounds make abundantly clear the ferocity of ancient battle and the psychological toll inflicted upon combatants. (For further remarks on the toll of combat, see Chapter 23 in this volume.)

Classical and Hellenistic Developments

The statue of Croesus described earlier is not, of course, authentic or naturalistic. Instead it is a type, that of the "god-like" warrior, valiant in death. Not until the fifth century, and perhaps only at mid-century, do images, like that of Socrates for example, begin to be naturalistic.[39] Celebrated examples such as the *Riace Bronzes*, two slightly larger than life-size warriors recovered off the coast of Italy in 1972, may be likenesses of generals of the Peloponnesian or Persian War.[40] The numerous identifications with multiple personalities of the time reveal the limits of our knowledge.

On the other hand, portraits of famous figures of the Hellenistic age, above all Alexander the Great, are somewhat better attested and may be regarded as authentic, always allowing for more than a touch of artistic license.[41] All images of Alexander are more or less heroic.

Fallen foes, however, receive unheroic treatment. Numerous examples of memorial sculpture, such as the well-known funeral relief of Dexileus, a young Athenian cavalryman killed in battle in the Corinthian War (Fig. 29.1), and of grave reliefs, such as that of an Athenian hoplite c.400, depict fallen foes being put to death.[42] Similar scenes appear in the west frieze

35 Ma 2008a, 74–75; also Pritchett 1971–1991, vol. 4, 138.

36 Diod. Sic. 16.87.3, Plut. *Dem.* 21.2, Polyb. 5.10.4.

37 Cf. Buckler and Beck 2008, 259–276.

38 See further Ma 2008a, 75, with photographs.

39 Two fundamental texts: Xen. *Mem.* 3.10.1–5 a discussion between Socrates and Parrhasius on painting and portraiture; Plin. *NH* 35.61–66, a report of the celebrated competition between Parrhasius and Zeuxis.

40 Osborne 1998, 160–161.

41 See Stewart 2008, 282–290 for full discussion.

42 Dexileus "memorial" (so Stewart 2008, 236–237), whereas Osborne 1998, 14 refers to it as a "monument".

of the Athena Nike temple in Athens (c. 415?) as well as in a commemorative relief from Eleusis, c.411, depicting a fight between Athenian cavalry and Spartan hoplites.[43] To modern viewers these scenes may be repugnant, but to readers of Homer they cannot be surprising. On numerous occasions the poet relates a hero such as Hector or Agamemnon finishing off a wounded opponent (Patroclus by Hector: *Il.* 16.816–830), or refusing an offer of surrender and then slaughtering an adversary (Adrastus by Agamemnon, *Il.* 6.42–60). In these instances, literature and art converge, confirming the ferocity of ancient Greek battle.

Other works, especially from the Hellenistic period, evoke pathos. Perhaps the greatest surviving monument of this type was built by the Attalid kings of Pergamum to celebrate their triumphs over the Gauls or Celts in a series of struggles in the third century.[44] *The Dying Gaul* (Rome's Capitoline Museum, a.k.a. the *Trumpeter*), and the (*Ludovisi*) *Gaul and Wife* (a.k.a., *Dying Gaul and Wife* in the Palazzo Altemps of Rome's National Museum) combine heroic death and self-sacrifice with Hellenistic ideals of pathos and melodrama.[45]

Another fourth-century and Hellenistic trait, vividness, was exemplified by the honorific statue of the general Chabrias that stood in the Athenian agora. Chabrias died in battle in 356.[46] To commemorate his bravery, the sculptor adopted an "at-ease" stance that Chabrias once took when daring another distinguished commander, the Spartan King Agesilaus, to attack him in 378 (Diod. Sic. 15.32–33). The king withdrew rather than fight, and the Athenian general added this unfought triumph to his laurels. Possibly listed on the statue's base were names of the campaigns in which Chabrias led the Athenians to victory.[47] The Spartan memorial to Leonidas was a comparable but less striking tribute.

Aside from these works of art in changing styles, the Greeks of the Classical and Hellenistic periods remembered the dead by displaying military objects associated with heroic acts.[48] In Athens, for example, Pausanias saw shields taken from the Spartans in a battle fought some 650 years earlier and painstakingly preserved ever since as dedications to Athena. Pausanias describes other dedications that convey heroic sacrifice, such as the shield of the young Athenian Cydias, killed while defending Greece against the Gallic invaders of 279. His shield stood with others like it in the Stoa of Zeus Eleutherios ("the Liberator") in the agora. It is inscribed,

The shield of a brave man, Zeus's offering,
Pining away for the youth of Cydias:
The first shield his left arm ever put on,
When raging War went hottest at the Gauls.[49]

The pathos and sadness at life extinguished is transparent.

The Athenians displayed other shields in the same stoa, among them the shield of Leocritus, who won fame in battle against the Macedonians in the third century (Paus. 1.26.2). His

43 Eleusis battle relief in Rossiter 1981, 195; Athena Nike, in Stewart 2008, 194–196.

44 "Heroized" as the portrait of Gallic conduct in the invasion of Greece was hardly heroic, as Pausanias tells (10.22–23).

45 Cf. Pollitt 1986, 85–89.

46 At Chios (Diod. Sic. 16.7.3, Plut. *Phoc.* 6.2).

47 Buckler 1972.

48 See further Sturken 1997, 76–81; cf. white *lekythoi* variously decorated with mourning imagery and objects dedicated to the dead in Oakley 2004, 49, 63–65, 72, 175; pl. 5B.

49 Paus. 10.21.3 (trans. Levi). Pausanias may draw from the account of Hieronymus of Cardia, a younger contemporary of Alexander the Great, whose Greek history ended with events c. 272; see further Habicht 1985, 84–86, 97.

shield and that of Cydias and others were plundered during the sack of Athens that followed the city's fall to the Roman commander Sulla in 86.[50] The Romans evidently appreciated these talismans of acts of heroism.

Conclusions

Kurt Raaflaub argues that these Greek monuments and memorials of heroic death and communal sacrifice convey a message of what we might today imagine as "civic virtue" or responsibility.[51] As young Athenians, Spartans, and other Greeks walked through their cities, they confronted statues of seemingly brave generals and fearless soldiers, and long lists of names, many of which would have been familiar, as being those of relatives or neighbors, all of these sights conveying stories of sacrifice and service encouraging emulation. This practice continues today, although in muted and subdued form. Tourists flock to see the changing of the guard at the Unknown Soldiers' Tomb at Washington's Arlington National Cemetery, and similar ceremonies take place in Athens at the Memorial of the Unknown Soldier and in places around the world such as Egypt and Iraq. The message is powerful, even intoxicating, yet it remains as deadly as ever.

Fig. 29.1 The Funeral Relief in Honor of Dexileus. Fourth century. Ceramicus Museum. Photo Lawrence A. Tritle.

---

50 Paus. 10.21.3 with Camp 2001,105, 184–185, *pace* Habicht 1985, 85; although Athens suffered, destruction was limited, as Plut. *Sul.* 14.3–6 notes.

51 Raaflaub 2001, 307–356, preceded by Raaflaub 1998, 9–32.

CHAPTER 30

War in Greek Poetry

J. Vela-Tejada

This chapter deals with the literary treatment of Greek warfare, as opposed to warfare in general, and it is confined within two additional limits: first, the limits of the Archaic and Classical periods, which were decisive in forming and shaping Greek culture and in establishing a social setting for the *polis* as a political institution; and second, the limit imposed by the medium of verse, prose authors being the staple of the other chapters in this volume.[1] Prose authors and authors of the Hellenistic and Roman periods thus appear incidentally, by way of comparison or background.[2] Other than Homer, the chief topics are Archaic poetry about the hoplite phalanx and anti-war sentiments in Athenian drama. Whatever the topic, the literary sources for war are just that—literary. They contribute to cultural, not political, history.[3] If wars be regarded as a science, they do not form part of this science. If wars be regarded as an art, they have much to offer thoughtful students and practitioners.

1 Rawlings 2007, 223: "Any study of Greek warfare reveals but a fraction of the vast array of human experience at war".
2 This chapter thus omits military material in Middle and New Comedy, such as the topos of the *miles gloriosus*, who is better attested in Latin than in Greek sources. Hellenistic poets such as Apollonius Rhodius add little to Homer and other earlier authors where military matters are concerned, and most especially in regard to combat. The Alexander historians, like their subject, Alexander III of Macedon, made ample use of Homeric precedents, but the late Classical and early Hellenistic sources that inform their work are largely lost; for an introduction, see Heckel 2008, 5–13.
3 Cf. Hornblower 2007, 53: "Greek literature may indeed be an unsafe and rhetorical guide to reality." Connor 2007, 6 also criticizes the use of literary works as evidence: "Surely war was an extremely important fact for that culture and its writers. But to gauge its frequency by its prominence in literature would be to commit the historiographical fallacy—mistaking literary representation for historical fact".

A Companion to Greek Warfare, First Edition. Edited by Waldemar Heckel, F.S. Naiden, E. Edward Garvin, and John Vanderspoel.

Heraclitus observed that "war is the father and king of all: some he has made gods, and some men; some slaves and some free" (fr. 53 D–K, *polemos panton men pater*).[4] War, indeed, was so frequent that periods of peace were regarded as extraordinary. Thus, Plato (*Lg.* 626a) has Clinias say that war was normal and that peace was only a truce.[5] Aristotle explains why: war was "natural" as a "way of acquisition" (*Pol.* 1256b 20). Later, at the dawn of Roman hegemony, Polybius expatiated on this view: "for being all, by nature, inclined to impose their hegemony (*hegemonikoi*) and lovers of liberty (*phileleutheroi*), people are in a state of constant warfare, none being disposed to yield the first place to his neighbor" (5.106.5). In the words of Kurt Raaflaub (2007, 9), "war prompts action, momentous events, and change; it deeply affects the lives of individuals, communities, and states." Although preoccupied by war, and sometimes fascinated by it, Greeks could also be realistic about the aspects of human nature to which war appealed. Aristotle (*Pol.* 1253a6) says that "the 'clanless, lawless, hearthless man', reviled by Homer as antisocial, is also a lover of war."[6]

Homeric Epic

The Ancients attributed the beginning of military theory to Homer, but recent scholars prize the *Iliad* for the cultural strata being documented: from the Mycenaean world and the material culture of bronze, through the Dark Age—the period that seems to predominate—to the first vestiges of hoplite fighting in the eighth century. Unifying this material is a world of nobles who made raids, not wars, for the sake of booty. In this milieu, the fantasy of a national expedition to rescue a woman is simply the gods' will, while the true causes of the war remain hidden.[7] Homer sings not about war so much as about "the glorious deeds of the heroes" (*ta klea andron*); hence, he portrays combat as a struggle among individuals, as in the duel between Ajax and the Trojan Hippothoon for possession of Patroclus' corpse. At the same time, Homer gives a detailed description of their weaponry (*Il.* 17.293–298):

> The son of Telamon, sweeping in through the mass of fighters,
> struck him at close quarters through the brazen cheeks of his helmet
> and the helm crested with horsehair was riven about the spearhead
> to the impact of the huge spear and the weight of the hand behind it
> and the brain ran from the wound along the spear by the eyehole,
> bleeding.[8]

The singer is no strategist, as the Greeks mistakenly thought, but he is an acute observer of ancient infantry battle.

The army is a mere mass (*laos*) at the service of protagonists who display *aristeia*, or excellence, through their personal exploits. References to the hoplite phalanx, which had

4 See Millett 2013, 51; Rawlings 2013, 5.

5 Raaflaub 2007, 9 suggests, by contrast, "Peace seems static, uneventful, devoid of opportunities, and boring." It is thus unsurprising that, as Vilatte says, 1999, 207, we should not speak of *pacifism* among Greek thinkers: many of whom glorify the war, especially the war against Persia. e.g. Lys. 2.55–57; Plat. *Menex.* 241d–242a. See also van Wees 2004, 77–79, and the conclusion of this chapter.

6 *Il.* 9.63; see also 4.446–456, 16.210–220, and 772–775.

7 Cf. van Wees op. cit., 249–252.

8 All translations of Homer are by Lattimore 1951.

developed in the eighth century, are clear but few.[9] References to earlier weapons appear willy-nilly amid hoplite gear: next to the short spear comes the Homeric *dolichoskion enchos*; while a concave shield—the *aspis*; *hoplon* is later—differs from its leather predecessor, the *talaurinon* of Ares.

In the *Iliad* even the Olympian gods fight, as in the *theomachia* of Book XXI (ll. 330–520).[10] A prelude to this encounter comes in Book XX, 31–74, when the goddess of love, Aphrodite, takes part in the struggle and the antithetical forces, Ares–Aphrodite (war–love), show up together in Homer. The gods fight much as the heroes do, but without the stimulus of fearing their own deaths. Immortality renders them both superior and inferior to their worshippers.

There are many approaches to Homer, but the military heart of the epic is the twin theme of heroic combat and human mortality.[11] Thus, together with the military greatness of the hero in his *aristeia*, Homer describes and praises another kind of greatness, that of the "beautiful death" (*kalos thanatos*) that confers imperishable glory on fallen heroes.[12] This sort of death, not Helen or booty, is what Achilles will die for.

Thanks to two episodes, Hector is paradigmatic no less than Achilles, but in his own way: first, the farewell of Hector and Andromache, who holds their infant Astyanax in her arms, and second, Hector's duel with Achilles. In Book VI, Hector bids farewell to his infant son with the prayer to Zeus that Astyanax "may be as I am, pre-eminent among the Trojans" (ll. 477). After leaving him in the arms of his wife, Hector addresses her, foretelling his imminent death (ll. 486–489):

> Poor Andromache! Why does your heart sorrow so much for me?
> No man is going to hurl me to Hades, unless it is fated,
> but as for fate, I think that no man yet has escaped it
> once it has taken its first form, neither brave man nor coward.[13]

Since death is inevitable, Hector reasons, what could be better for a hero than to die doing his duty. As a result, women and children must yield to the ideal of the hero's beautiful death.

Although Achilles will win the contest with the Trojan leader, the fate that awaits the "Best of the Achaeans" (e.g. *Il.* 1.91, 2.82) will be no different.[14] His mother, the divine nymph Thetis, foretells it at the beginning of the poem (*Il.* 1.414–418):

> Ah me, my child. Your birth was bitterness. Why did I raise you?
> If only you could sit by your ships untroubled, not weeping.

9 According to van Wees 2004, 165, the *Iliad* reveals the hero to be a less exotic figure, different from the hoplite by degree rather than by category.

10 Cf. Rawlings 2007, 179, who does not contrast human suffering and divine inanity: "The intervention of the gods in Homeric warfare might be rationalized as poetic embellishment".

11 As Hornblower 2007, 25 points out, "there is not a consistent perspective on war in Homer: dislike of violence is combined with sordid details of that very thing". See also van Wees 2007, 165.

12 The theme of beautiful death has been propounded by Loraux (2008, 37–95) in connection with the Athenian funeral oration. Vernant, 2001, 45 ff. argues that the epic endows the hero's death with a kind of eroticism.

13 Redfield 1994, 122–124, says that the poet dramatizes the negative side of the warrior's role, which is that he must leave his family.

14 Cf. Vernant 2001, 46.

> since indeed your lifetime is to be short, of no length.
> Now it has befallen that your life must be brief and bitter
> beyond all men's. To a bad destiny I bore you in my chambers.

Achilles will choose a premature death in combat because it promises eternal glory. In this version of the beautiful death, the hero binds himself to posterity rather than to his own family and community.

Consequently, the combat between Hector and Achilles in Book XXII is the poem's climax. After Athena tricks him, making him believe he has Deiphobus at his side, Hector understands that his time has come and he accepts the ineluctable end of his life (ll. 297–301):

> No use. Here at last the gods have summoned me deathward.
> I thought Deiphobos the hero was here close beside me,
> but he is behind the wall and it was Athena cheating me,
> and now evil death is close to me, and no longer far away,
> and there is no way out.

Athena's trickery aside, Zeus and far-shooting Apollo have planned this outcome. They superintend the beautiful death (and the Fates, or *Moirai*, superintend the entire action, as at ll. 303–305):

> But now my death is upon me.
> Let me at least not die without a struggle, inglorious,
> but do some big thing first, that men to come shall know of it.

Here Hector thinks as Achilles does. The mode of death links the two opponents.[15]

Achilles prevails and satisfies his thirst to avenge the death of Patroclus at Hector's hands. Then anger gives way to piety—and this enhances him as a hero—when Achilles returns Hector's corpse to Priam and allows the celebration of an 11-day funeral (*Il.* 24.669–670). This last stage, or ritualistic fulfillment, of the beautiful death brings the poem to a natural conclusion, the final line being "Such was their burial of Hector, breaker of horses" (*Il.* 24.804).

Thanks to Homer, his heroes lived for future generations of Greeks as paradigms of the past and inspirations for Hellenic culture, especially for the genre of tragedy, which Aeschylus reportedly described as leavings from Homer's banquet.[16] A different view emerges from Hesiod, who, in contrast to Homer's approach, shows concern for the disastrous consequences of war for agriculture in his *Works and Days*.[17] Hesiod offers a reminder of how much Greek poetry fails to say about war, including its destructive effects on the environment as well as on civilians and on economic life.

15 Cf. Redfield 1994, 128–130, saying that Hector's death is the consequence of an error due to the adverse conditions in which the hero finds himself, but adding that Hector is humanized by his error. For Redfield, the *Iliad* lies juxtaposes and contrasts the two heroes (223).

16 Aeschylus: Ath. 8.347e. Foundations of Hellenic culture: Vernant 2001, 85. For Homer's influence on Greek prose, see Hornblower 2007, 30–34; for historiography in particular, 39–47.

17 According to van Wees 2007, 299, the *Iliad* addresses and overcomes the ancient audience's reflexive, Hesiodic preference for peace. See also Rawlings 2007, 5–6.

Archaic Poets and Hoplites

Some poetic forms known to us from Archaic examples are far older, and thus co-existed with early epic poetry. They often form part of rituals. One of these forms, the *paian*, the invocation to Apollo that pertains to war, appears in the *Iliad*, when Achilles urges the Myrmidons to return home after killing Hector (*Il.* 22.391–392):

> But now, you young men of the Achaeans, let us go back, singing
> A victory song, to our hollow ships ...[18]

Similarly, choral poetry, another old form, can raise a war while echoing the theme of a beautiful, divinely ordained death:

> Hearken, Alala, daughter of War,
> Thou prelude of clashing spears, thou to whom are offered
> Heroes in the holy sacrifice of death. (Pi. fr. 78)[19]

The language of sacrifice, rendering the death both sacred and routine, marks a departure from Homer. Choral poetry, moreover, assigns this message not to an epic singer or an aristocratic audience but to choirs drawn from the *polis*, as in Pindar's athletic odes and other songs, and those of rival poets such as Bacchylides.

Again, similarly, lyric fragments of the seventh to sixth centuries link communal singing, Homeric ethos and the emergence of the hoplite phalanx, notably in the poems of Callinus, Tyrtaeus, and Archilochus.[20] These poems express the ideal of a heroic death, but subordinate the desire for *kleos aphthiton* to the interests of the community. The poet thus exhorts his fellow citizens to fight in order to attain *arete*, human excellence, and makes use of a *parainesis*, or warlike exhortation, rather than the *mythoi*, or authoritative stories, of the epic singer.

Callinus, for instance, calls on his fellow citizens to defend the homeland, most likely in response to the Cimmerian attack on Ephesus around 650:

> Have you no shame for your sloth amid those that dwell round about you?
> Do you propose to sit in peace although the land is full of war?[21]

---

18 According to Lonis 1979, 124, "Le péan est donc, de tous les rites, celui qui s'adapte sans doute le mieux aux différents états d'âme du guerrier. Avant le combat, il a un effet d'exaltation: Arès préside alors à ces transports. Après le combat, c'est l'exultation placée sous le signe de Dionysos, puis le calme, l'hésychia que procure enfin Apollon." See also Xen. *HG* 5.4.10; Plut. *Lyc.* 22.4, and Chapter 24 in this volume. Connor 2007, 94 links the *paian* to the choral performance of *komos*, or revel, by victorious armies returning home; cf. Ar. *Av.* 1763.

19 Plut. *Glor. Ath.* 349c, as translated by F. Babbit, *Loeb Classical Library* no. 305.

20 Millet 2013, 49 stresses that the performance of lyric poetry abetted the social and political processes that led both to creating constitutional government and to establishing armies of citizen-soldiers. The emergence of the phalanx remains controversial: Sekunda, 2013, 199 considers the relationship between the *polis* and the "hoplite revolution" to be indirect, while van Wees 2000, 156, argues that hoplite infantry developed slowly, over almost four centuries. Other views: Hanson 2000, 219; Rawlings 2000, 250; Krentz 2002, cited from the reprint in Wheeler 2007b, 113–122; Chapter 6 in this volume.

21 1 Gentili-Prato = 1 West, ll. 2–4. The context: Millet, 2013, 50.

The hero who fights for his country, wife, and children attains honor (*time*) and glory, (*aglaon*, ll. 6–8). On the assumption that the death is inevitable, he has no reason to be afraid and to avoid the dangers of the combat (ll. 9–13):

> Nay, so soon as war is mingled let each go forward
> spear in poise and shield before stout heart;
> for by no means may a man escape death,
> nay not if he come of immortal lineage.
> Hector said the same, but to his wife, not his fellow citizens.

Tyrtaeus' elegies provide profound insight into the hoplite spirit at the time of the Second Messenian War that concluded with the annexation of Messenian territory by Sparta.[22] The graphic description of the fight in the phalanx, where men stand shoulder to shoulder, is unparalleled in Greek literature (8G-P = 11W, ll. 21–26):

> So let each man bite his lip and abide
> firm-set astride upon the ground,
> covering with the belly of his broad
> buckler thighs and legs below and breast and shoulders above;
> let him brandish the massive spear in his right hand,
> let him wave the dire crest upon his head.

Shield, sword, helmet, and spear form the panoply (27–34):

> Let a man learn how to fight by first daring to perform mighty deeds,
> Not where the missiles won't reach, if he is armed with a shield,
> But getting in close where fighting is hand to hand, inflicting a wound
> With his long spear or his sword, taking the enemy's life,
> With his foot planted alongside a foot and his shield pressed against shield,
> And his crest up against crest and his helm up against helm
> And breast against breast, embroiled in the action—let him fight man to man,
> Holding secure in his grasp the haft of his sword or his spear!

For the first time, a Greek writer clearly distinguishes between heavy, hoplite infantry and light troops (see ll. 35–38).

Other fragments emphasize that the soldier must be willing to sacrifice even his own life; the supreme achievement of a man is to fight and die in battle for the homeland (6G-P = 10W, ll. 1–2):

> For it is a beautiful (*kalon*) thing for a noble man (*agathon*) to fall
> in the front line and die fighting for the fatherland:

Thus, no honor or reward can match the glory the hoplite gains on the battlefield (9G-P = 12W, ll. 13–14):

> This is excellence (*arete*), the greatest (*ariston*) possession of mankind,
> The most virtuous prize that a youth can acquire.

22 Van Wees 2004, 172. This war, difficult to date, occurred sometime in the sixth century.

In sum, true *arete* (ll. 43–44) belongs to the warrior serving the community by not avoiding combat.[23]

Tyrtaeus' poems reflect the Homeric agonal ideal of searching for honor through combat. Yet the poet puts this longing for glory in the framework of the society of the *polis*, as Callinus does, and subordinates glory to the common good.[24] His verses also anticipate the consolidation of the military model for the Spartan society, one that emerged by the time of the Persian Wars. The cohesion in the ranks of the phalanx would eventually become a metaphor for Spartiate society. Tyrtaeus never mentions the *perioikoi*, or autonomous "neighbors," who later contributed much of the manpower to the phalanx. No doubt Spartiates provided most or all the officers. Tyrtaeus' message suited them especially, for they often fought in the front rank.

Archilochus, by contrast, uses the same elegiac medium to describe and satirize the war against the Saeans, a Thracian tribe. His actions are neither idealized nor glorious: he shrugs at losing his shield, something normally taken to be a sign of cowardice (5W):

> One of the Saeans now delights in the shield I discarded
> Unwillingly near a bush, for it was perfectly good,
> But at least I got myself safely out.
> Why should I care for that shield? Let it go.
> Some other time I'll find another no worse.

He ridicules and neglects his own "warlike deeds" by recounting in a serious measure, that of tetrameters, that (101W):

> Of seven that lie dead whom we overtook in the pursuit,
> we are the thousand slayers.

In other poems, Archilochus nevertheless acquires gravity, announcing in dystics the contest of Ares against the Euboeans (3W) or, in tetrameters, a call to save the imperiled ship of the state (105W). In other dystics, he calls himself servant of both Enyalius and the Muses (1W).[25] This formulation differs from Homer's invocation of the Muses alone. Archilochus is reshaping an epic, aristocratic heritage.

One of the most famous poets of Lesbian monody, Alcaeus, exhorts members of his political faction to fight in a civil war.[26] He describes armament housed in the temple of Ares in evocative, politically charged, detail, and then calls on his listeners to follow the example of their ancestors and remember their military duty:

> The great house gleams with bronze
> All the roof is well furnished with bright helms;
> White horse-hair plumes nod down from them,
> Ornaments for the heads of men.
> Bronze-shining greaves hang around and hide pegs,
> A fence against the strong dart

23 Rawlings 2007, 208.

24 In the formulation of Rawlings 2007, 101, Tyrtaeus balances concern for the individual's contribution to communal well-being with a second concern, the value of the collective effort made by the phalanx.

25 Rawlings 2007, 48 stresses another aspect of Archilochus—the mercenary who has quit the agricultural life typical of a citizen-soldier and served foreign masters for pay. See also Millet 2013, 49–50.

26 Rawlings 2007, 45 suggests the poet presents an image familiar to his audience, but this only consisted of his wealthy fellow aristocrats. See also Millett 2013, 50.

Corselets of fresh linen and hollow shields are thrown down;
Beside them are blades from Chalcis,
Beside them many a belt and tunic.
These we must not forget since we first undertook the task[27]

In these lines, the Homeric heritage serves a partisan as well as military purpose.

To sum up, these Archaic poets adopt the warrior ideal to sundry circumstances and purposes—to ritual forms like the *paian*, to political occasions like armed conspiracies, to satire as well as exhortation. The poet commonly transmits a message that brings together religious and political elements, even in Archilochus' satire, for the mocking verses serve as a means of ritual purification, as in apotropaic rites meant to expel evil. Even satire thus benefited the community or group. Many more variants are no doubt lost to us, but a few of these may be deduced from stray references, for example, to the *embaterion*, a marching song sung by young Spartans according to several sources (and also called a song to sing "while in armor," *enopilion*[28]). It survives only in a few fragmentary lines from a late source:

Come on, sons of citizen fathers
In Sparta with its good men,
Step lively and put yourselves forward.
Boldly brandish your spears
—
Life's nothing to be sparing with.
That's not the custom in Sparta (D. Chrys. *Or.* 2.59)

Perhaps other Greeks sang similar songs. As shown by the *paian*, they sang other marching songs. To judge from the source, the Roman-era orator Dio Chrysostom, these songs long outlasted the armies that sang them.

War on the Stage

The Persian Wars mark a turning point in Greek warfare, and military poetry changed accordingly, as shown by the epigrams and epitaphs attributed to Simonides of Ceos on the subjects of Artemisium, Salamis, and Plataea (see frgs. 532–536).[29] These poems developed the Homeric tradition of commemorating the noble deaths of citizen-soldiers by extending it to whole communities.[30] More important for this chapter is the role of warfare in a new genre, drama, that would sweep the Greek world in a matter of decades, attaining a sudden popularity comparable to that achieved by cinema in the twentieth century.

Of the hundreds of tragedies written and performed in the Classical period, just over 30 have survived, all authored by the three great tragedians of the Athenian theater. With the

27 Clay 2013, 24 sees a Homeric resonance, for the warriors in the *Iliad* are constantly being reminded to recall their valor by means of the description of the arms stored in Odysseus' Palace at Ithaca (cf. *Od.* XXI and XXII).
28 Brill's *New Pauly*, s.v. "Embaterion," and s.v. "Kastoreion" (E. Robbins), with Plut. *De mus.* 1140c, where the King sings it, and Polyaen. *Strat.* 1.101.
29 According to van Wees 2004, 183, the Persian Wars added two new dimensions: the notion that Greeks fought of their own free will; and the idea that a hoplite never gave ground but fought to the death (cf. Herodotus 7.103.4, 104.4, 223.3, 225.3). See also Krentz 2002, cited from the reprint in Wheeler 2007b, 122.
30 Thus Vernant 2001, 88–98, contrasting Simonides with Homer and Tyrtaeus.

exception of Aeschylus' *Persians*, all tell mythical or legendary stories rather than describe contemporary conflicts. The plays of Aeschylus and Euripides nonetheless describe warfare from a Classical, not Homeric, perspective; indeed, these plays more or less clearly allude to recent events.[31] Through myth and poetry, contemporary war and violence find their way onto the stage.

About half of the extant tragedies concern the Trojan War, the War of the Seven against Thebes, the rebellion of Eteocles against Thebes, or the immediate aftermath of these stories of besieged cities.[32] Stories about Trojan captives bulk large, a circumstance that reflects Greek suffering during and after sieges in the Persian and Peloponnesian Wars. Euripides concentrates on this theme; Aeschylus adds a new perspective on the Persian Wars, that of the court of the Great King of Persia. Herodotus explains the preference for a distant time and place, or, in the case of Aeschylus' *Persians*, a distant perspective. The historian reports (6.21.2) the impact of the performance of *The Sack of Miletus* by Phrynichus, another fifth-century playwright. This play described the capture of Miletus by the Persians during the Ionian Rebellion:

> The whole theater fell to weeping; he had to pay a thousand drachmas for bringing to mind this calamity. Athens forbade the performance of that play forever.[33]

Never again did a tragedian chose a contemporary subject.

Aeschylus witnessed the struggle against the Persians, having served as a hoplite at Marathon.[34] Yet when he diverged from myth into history in the *Persians* he did not adopt a chauvinistic perspective. The play's treatment of the battle of Salamis reveals Xerxes' ambition to dominate Greece, but it also extends some sympathy to him and to members of his court. Xerxes falls victim to what is a familiar human fault according to Greek religious thinking—*hybris*. Punishment by the gods must inevitably follow. Thus, we read at lines 454–455:

> Grievously did he misjudge the issue. For when some god
> had given the glory to the Hellenes in the battle on the sea.

This catastrophe has a political meaning: *hybris* is likely to befall a despot, such as Xerxes. Yet the play is mainly a religious and ethical drama about warfare, one in which the gods take part in the conflict, not for the sake of helping the Greeks, but because Xerxes has considered himself a man "like the gods" and overstepped human limits in the belief that his power was endless; then "Xerxes groaned aloud when he beheld the extent of the disaster"

---

31 An example of contemporary military details in tragedy: in Eur. *Heracl.*, two characters debate the use of light troops. Lycus (ll. 157–164) deprecates archers and mobile troops: "Archery is no test of manly bravery; no! he is a man who keeps his post in the ranks and steadily faces the swift wound the spear may plough." Amphitryon (190–192) replies that "A man who fights in line is a slave to his weapons, and of his fellow-comrades," and praises light infantry (ll. 201–203). Connor 2007, 109, draws a parallel between this debate about arms and tactics and other Euripidean and Athenian debates on subjects such as politics, religion, economics, and philosophy. One play of Sophocles that has a military setting, *Antigone*, alludes to timeless Greek burial customs, if not to any fifth century war.

32 For drama and sieges, see Rawlings 2007, 142; Millett 2013, 53.

33 Hdt. 6.21.2. See Millett 2013, 52.

34 Millett loc. cit.

(465).[35] Aeschylus has inverted Homer's theme of the gods being responsible for beautiful deaths, and made them righteously responsible for hideous suffering.

Euripides, however, is the author who has bequeathed the most tragedies concerned with war, as is in keeping with the fact that most of his productions took place during or just before the three decades of the Peloponnesian War, a conflict that brought plague, privation, and defeat to Athens. The anti-war sentiments in *Hecuba, The Suppliants, The Trojan Women, Heracleidae, Helen*, and *Andromache* have prompted some critics to call these plays "political works." *The Suppliants* condemns war as cruel and senseless, with Adrastus telling the Chorus:

> O wretched sons of men!
> Why do you get weapons and bring slaughter on one another?
> Cease from that, give over your toiling,
> and in mutual peace keep safe your cities. (ll. 949–952)

The Trojan Women condemns war even more forcefully, leading some critics to consider Euripides "the champion of peace."[36] In this play, Troy has been taken and sacked; only women survive, waiting to be allotted as war prizes. The Trojan queen, Hecuba—as pious as Xerxes is impious—leads a funeral threnody for the city and for their dead men. Of the royal family, there remain only her, Cassandra, the virgin devoted to Apollo, and Andromache, Hector's widow. Polyxena has been immolated on Achilles' tomb. Other surviving women form the Chorus.

The messenger Talthybius calls for Cassandra to be taken to the ship of Agamemnon, who has chosen her as war booty. This prompts her appearance on the stage (l. 308), where she sings an ironic *Hymenaios*, or marriage hymn (ll. 308–340), that she treats as a song of victory for her upcoming wedding with Agamemnon. When Cassandra emerges from her trance, however, she recovers all her reasoning powers and employs rational if rhetorical arguments: her royal nuptials to Agamemnon (l. 354) will cause the destruction of his dwellings (l. 359), and therefore she will exact revenge for all that her country and her family have suffered, even paying with her own life.[37] In the epilogue of this elaborate intervention (ll. 400–402), Cassandra transmits her final reflection, through which Euripides condemns not war in general—a safe theme—but the transgression of codes of military conduct. The warning given in the following lines evokes the worst consequences of the Peloponnesian War, which had begun 19 years before the play's premiere:

> Whoever is wise should fly from making war;
> but if he come to this, a noble death will crown his city
> with glory, a coward's end with shame.[38]

---

35 In the words of Sommerstein 1996, 96, "such is the nemesis of Persian power." A different view: Podlecki 1999, 26 suggests a defense of Themistocles, who was ostracized in 471. In fact, the piece was performed a decade after the battle of Salamis.

36 Vela-Tejada 2008, 219, considers the literature on Euripides' politics and on his pacifism (if that is the word). See my remarks in the Conclusion section of this chapter.

37 Aside from the religious theme of a virgin prophetess being degraded and becoming a concubine, Cassandra's situation presents one more instances of what Deacy and McHardy 2015, 268, call the characteristic gender violence visited on enemy females by conquering warriors. See Chapters 25 and 26 in this volume.

38 The 415 production date might suggest a response to the genocide at Melos or to the imminent expedition to Sicily. Vela-Tejada 2008, 220, compares Euripides' attitude to the pessimism of Thucydides.

The execution of little Astyanax, son of Hector (ll. 1118 ff.), marks the climax of the play. It is, perhaps, the most barbarous act in any poem or play about the Trojan War, and the son of Achilles has committed it, without remorse.[39] Perhaps the verdict that Euripides was expecting from the public—his fellow citizens—appeared in the prologue, when Poseidon declares (ll. 95–97):

> A fool is he who sacks the towns of men,
> with shrines and tombs, the dead man's hallowed home,
> for at the last he makes a desert round himself and dies.

The war of conquest transgresses both the divine and human order. Euripides has demystified the Achaean victory at Troy, and objected to idealized death even more pointedly than Aeschylus did in the *Persians.*

An earlier play, *Andromache*, produced between 428 and 421, describes the eventual rather than immediate harm done by war. This play may be topical, too—the Amphipolis disaster may explain anti-war sentiment in the fourth *stasimon* (ll. 1009–1046)—but Euripides concentrates on the plight of war widows and orphans in a society in which survivors may find themselves enslaved. Lines 1037–1041 describe both the lamentations of bereaved parents who have sons in the struggle and war widows who have left home to become the slaves, and often concubines, of a foreign conqueror.[40] The most powerful and prestigious of the bereaved parents, Achilles' father Peleus, reproaches a war hero who is often Euripides' target, the Spartan Menelaus, for having destroyed so many (*Andr.* 611–613):

> ... rather, you lost lives, many and brave,
> and left old women at home bereft of their sons
> and robbed grey-headed fathers of their noble children.

Menelaus proceeds to bring home the evils of the Trojan War in another way, by tricking Andromache off an altar where she has taken refuge as a suppliant. No such misdeed occurs in Homer, Aeschylus, Sophocles, or Thucydides. Then Andromache supplicates Peleus, who rights the wrong at the expense of the Spartan villain. This play warns the public that barbarized victors may make war at home—but softens the warning by giving a Spartan and not an Athenian example. Attributing misconduct to other Greeks is one of tragedy's distancing devices.

Aristophanes, the chief author of the other dramatic genre, Old Comedy, often condemns wars of conquest. Well known are the messages of his early plays—*Acharnians, Knights, Lysistrata,* and *Peace* (performed in 421, days before the Peace of Nicias), which all aim political and social satire at statesmen and generals, among other targets. Even more than Euripides, Aristophanes objects to unjust war, but unlike any tragedian he specifies leaders and policies. Perhaps *Lysistrata* is the most pointed of these protests, for it allows women, war's largest class of victims, to compel armies to stop fighting.[41] In all the plays, a range of comic techniques, from ridicule to pratfalls, wordplay to obscenities, underscores the absurdity of war.

39 Vela-Tejada 2008, 211.

40 See a similar argument by Cassandra in *Tr.* 380–381. See also van Wees 2007, 33–35.

41 See also the remarks of Rawlings 2004, 4–16.

Acharnians may best represent the perspective from which the poet writes. In this play a Chorus of rural Attic farmers complain about the protagonist, Dicaeopolis (Just City), who wants peace with Sparta:

> Father Zeus and all you gods!
> Let's prosecute anyone who has made
> peace with our foes,
> against whom, on my side,
> "There waxes hate-filled war"
> because of my lands. (ll. 221–226)[42]

The farmers want vengeance because the war has harmed their vineyards. Aristophanes happened to be a native of a rural deme, and he readily expresses the frustration of country dwellers who find themselves the victims of a strategy of letting the Spartans invade Attica and devastate the fields. He uses the play to show this group, a large part of his audience, that peace, not war, will relieve them of their suffering. The poet wants peace, but not pacifism. He sympathizes with the vengeful farmers.

Conclusion

Major military events and trends in every historical period drew the attention of Greek poets: the legendary raids and piracy romanticized in epic; the hoplites, mercenaries, and self-sacrificing citizen-soldiers of Archaic poetry; the bigger wars and sharper criticisms found in fifth-century Athens, where drama emerged. All these genres linked warfare to myth and ritual, yet preserved some of Homer's directness, bluntness, and frankness. Greek war poetry met the standard set by Victor Davis Hanson when writing about modern narratives of battles—and thus about prose, not verse:

> Battle history [should] bring a much needed reality, a morality, to the whole time-honored notion of an "art of warfare", that obscene phrase which operational historians employ when investigating the "science" of killing and maiming faceless mobs of humanity.[43]

Greek poetry does not, however, adopt a pacifist attitude toward war. For that, it would need to make general objections to violence in political and social relations. These objections were never forthcoming in a society that featured commonplace animal slaughter and widespread slavery.

42 This comedy was staged at the rural Dionysia of December, 425. In the Dionysia one year later, the *Farmers* (*Georgoi*) was performed as a peace-play on the theme of return to the countryside. See Sommerstein 1980, 61 along with Millett 2013, 52–53.

43 Hanson 1991, 255.

CHAPTER 31

War and Propaganda

Sabine Müller

The term "propaganda," in the sense of influencing public opinion, or, more fully, of circulating political and social norms, values, and symbols according to the circulator's interest, is first attested during the French Revolution.[1] For the study of antiquity, doubt exists about this term's possibly anachronistic character and its negative connotation, yet most scholars regard propaganda as part of ancient political representation in the widest sense—*Sprachregelungen* that aim at acceptance and support, often in a manipulative and suggestive way.[2] The striving for legitimacy, here defined as the official justification of political acts by spreading ideologically coded messages, lies at the core of propaganda. War, above all, needed to be justified by the propagation of new claims of legitimacy and even by the establishment of new definitions of legitimacy.

The ancient sources for Greek and Macedonian warfare thus record numerous official justifications for making war. One audience for these justifications was the citizenry or public, but another was soldiers and sailors. A shared, communal legitimization of their acts served to mobilize their energy and team spirit. For example, Thucydides reports that at the end of the eighteenth year of the Peloponnesian War, the Spartans, who were preparing another invasion of Attica, were particularly eager to go to war, for they were convinced that this time the Athenians "had been the first to infringe the truce. In the former war, they considered, the offence (*paranomema*) had been more on their own side .... For this reason they thought that they deserved their misfortunes," while now, they were encouraged.[3]

1 Cf. Maletzke 1972, 157; Burke 2001, 13; Bussemer 2008, 26–27. All translations are those of LCL.

2 Weber 1993, 35, 60, 123–124. See the title of the volume by Baynham and Walsh 2018: *'Selling Alexander': Alexander the Great and Propaganda*.

3 Thuc. 7.18.2–3. Cf. Hornblower 2008, 572–576.

A Companion to Greek Warfare, First Edition. Edited by Waldemar Heckel, F.S. Naiden, E. Edward Garvin, and John Vanderspoel.

This passage sheds light upon the widespread belief that warfare without some justification was a hazardous, even impious, task. The fourth-century Athenian politician and peripatetic philosopher Demetrius of Phalerum says:

> A declaration of war...that can be labelled as just (*dikaios*) makes both victories appear greater and failures less dangerous while one that leaves a bad impression as feeble serves the opposite. (*BNJ* 228 F 29)

The question of what was "just," however, presented complications. In this connection, Polybius zeroes in one element, pretexts or excuses for making war (*prophaseis*, or official, if untrue, justifications, such as the claim by Philip II to avenge the injuries inflicted on the Greeks by Xerxes' troops).[4] Prextexts differ from causes (*aitiai*, or events that influence decisions) and beginnings (*archai*, or initial attempts to carry out a plan). Polybius mentions many such pretexts, and so do other ancient sources.

These sources often depict the enemy as aggressors waging war out of a thirst for fame, empire, or money, and thus as counter-images of ideal warriors and leaders who are just and honorable. Through this kind of moralizing, the sources isolate war from its political and often commercial context, and serve to create the impression of an enemy waging an unjust, ungodly, tyrannical campaign. For example, the Attic orator Demosthenes characterizes Macedonian expansion as a series of crimes (raids, robberies, deceits, enslavements, and bribes) committed by Philip II, "the plunderer of the Greeks" (10.34),[5] driven by an insatiable, wicked greed for power without even an attempt to circulate a pretext.[6] The orator furthermore says a Macedonian "barbarian ought to be subject to Greeks" (3.24), not vice versa.[7] From the delegitimization of the enemy sprang ethnic prejudice.

Key Themes in Propaganda: Helping Others

One frequent propaganda theme is the claim to come to the aid of others (with or without their invitation), be they allies, kin, or anyone proclaimed needy. The protectors act out of a sense of justice, a love of order, and a spirit of helpfulness untainted by ambition. Using this theme, the Attic orators idealized the history of Athens' relations with other Greek *poleis* in the fifth and fourth century. Demosthenes claims that both Athenians and Spartans "thought it our duty to fight on account of wrongs which we saw the others suffering" (9.25). In his *Epitaphios* honoring the Athenian *strategos* Leosthenes and his soldiers as fallen heroes of the Lamian War 323/322, Hyperides dwells upon Athens' role as Greece's protector, using a metaphor:

---

4 Polyb. 3.6.3–14; cf. Walbank 1957–1979, 1.305–309. See Thuc. 1.23.6 for another use of *aitia* and *prophasis*.

5 Philip's official reason to attack Olynthus—the Olynthians had given shelter to his two surviving rebellious half-brothers and refused to surrender them—is mentioned by Just. 8.3.10, not by Demosthenes.

6 E.g. Dem. 1.3; 2.7–9, 3.16, 3.20. Cf. Wirth 1985, 68, 74, 171; Pownall 2005, 270–271; Müller 2010, 173–174; Worthington 2012, 71–254.

7 On the "barbarous" Macedonians, see Dem. 3.16; 19.327; MacDowell 2000, 348.

> Compare her with the sun that visits the whole world and duly separates the seasons, disposing all things for the best ... so our city never fails to punish the wicked, help the just, mete out to all men fairness in place of wrong, and at her individual peril and expense assure the Greeks a common safety.[8]

Another example comes from fourth-century Thessaly. After the assassination of Jason, tyrant of Pherae, leaders in the rival Thessalian city of Larisa, particularly the Aleuads, sought to win back the hegemony Larisa had lost to Pherae. They appealed to their long-standing ally, Macedon, for aid.[9] The new Macedonian king, young Alexander II, needed a military success to prove himself, and lost no time in responding favorably to this call (Diod. Sic. 15.60.3). He scooped Pherae's new tyrant, his namesake Alexander, and seized Larisa and Crannon (Diod. Sic. 15.61.4). Then he dropped the mask of being a protector. Instead of handing the cities back to his Thessalian allies, as they expected, he installed Macedonian garrisons. The situation changed when rival "helping hands" arrived—the Thebans under Pelopidas, who was gratefully using an invitation from other Thessalians in order to go against Pherae. This invitation provided Thebes a pretext to expand northwards. Pelopidas forced Alexander II to withdraw his garrisons.[10] The Theban protectors were the next to expand their influence at the expense of Thessalian freedom.

The Liberation Theme

Often connected with the protection theme is the claim to liberate people subjected to foreign rule. Greek states commonly invoked this mainly illusory theme. After the Persian Wars, the Athenians convinced other Greek *poleis* to join their First Naval Confederacy by proclaiming that Athens was the only power capable of securing its members from future Persian attacks (Hdt. 7.138–140). This postwar propaganda obscured the role played by Athens in triggering the Persian Wars in the first place. Otherwise, Greeks could have recalled that Athens first submitted to Persia in 507/506 in exchange for protection against Sparta (Hdt. 5.73), then became a rebel to the Great King by supporting the Ionian revolt (Hdt. 5.97.3), and finally, when confronted with the prospect of Persian punishment, involved other Greek *poleis* in the war through alliances. (See Chapter 2 for the political difficulties caused by the increasingly severe policy of the Athenian Naval Confederacy, and for discontent among its members.) During the Peloponnesian War, the Spartan commander Brasidas used the freedom theme to damage Athens' position in northern Greece. He encouraged members of Athens' league to revolt by promising to liberate them from Athenian enslavement. Thucydides says he told the people of Acandus:[11]

> And for myself, I have come here not to hurt but to free the Hellenes. Witness the solemn oaths by which I have bound my government that the allies that I may bring over shall be independent; and besides my object in coming is not by force or fraud to obtain your alliance, but to offer you

8 Hyp. 6.4. The Lamian War: Wirth 2004.

9 Diod. Sic. 15.60.5–61.3; Xen. *Hell.* 6.4.28–32. Cf. Sprawski 2005, 33–36.

10 Diod. Sic. 15.67.3. Cf. Wirth 1985, 24; Georgiadou 1997, 190–193; Müller 2016b, 219–220.

11 Thuc. 4.84.1, 4.88.1–2. He calls them untrue. Cf. Hornblower 1996, 280–284; Tritle 2010, 106.

> mine to help you against your Athenian masters ... Endeavor, therefore, to decide wisely, and strive to begin the work of liberation for the Hellenes, and lay up for yourselves endless renown, while you escape private loss, and cover your commonwealth with glory. (4.86.1; 4.87.6)

Spartan leaders favored this theme because it dovetailed with Sparta's record (as they saw it) of being opposed to tyranny.

At the start of the next century, the Spartan king, Agesilaus II, posed as a savior of the whole of Greece from the threat of a Persian despotism. Campaigning in Asia Minor in 396, he announced his intention: "That the cities in Asia shall be independent."[12] In fact, he and his troops left Asia Minor two years later (and, according to Xenophon, left sad Ionians behind). The Spartan army had plundered Phrygia, Lydia, Paphlagonia, Mysia, and Cappadocia in the name of freedom for the Asiatic Greeks. Agesilaus wished to finance further Spartan wars.[13] In 387/386, Sparta, still steered by Agesilaus, accepted Persian control over the Ionian Greeks. (Chapter 2 explores the role of the Persian king in these diplomatic developments, which led to the King's Peace.)

By the end of their hegemony (which lasted from 404 to 371), the Spartans came to be widely perceived as oppressive.[14] Athens responded by stylizing itself as Sparta's opposite. It founded a Second Athenian Confederacy that guaranteed *eleutheria* to its members.[15] The liberation theme colors the decree of 378/377 proposed by the Athenian diplomat and logographer Aristotle, for it invites cities to join the new *symmachia* "so that the Spartans shall allow the Greeks to be free and autonomous and to live at peace, possessing securely all their own (territory)."[16]

In the West, Dionysius I legitimized his seizure of power in Syracuse by styling himself as the liberator of the Sicilian Greeks from the Carthaginian foe (Diod. Sic. 14.45.4–5), announcing: "The Syracusans declare war upon the Carthaginians unless they restore freedom to the Greek cities that they have enslaved" (Diod. Sic. 14.46.5). His claim to wage war on behalf of the freedom of the Sicilian Greeks was, however, a pretext for the consolidation of his position. Frances Pownall shows that while posing as a guardian of Greek freedom against the "barbarians," Dionysius treated Greeks and non-Greeks alike.[17]

A Hellenistic example of the theme is the Panhellenic coloring given to the Macedonian war against Persia initiated by Philip II and led by Alexander III.[18] Traces of this coloring survive in the fragmentary official report, the *Praxeis Alexandrou*, written by the court historiographer Callisthenes, Alexander's "non-combatant mercenary."[19] He repeatedly characterized Alexander as the divinely chosen liberator of the Greeks. Yet a long history belied this pretext for the Macedonian enterprise. In 335, the revolting Thebans refused to surrender to Alexander and proclaimed that their own aim was to free the Greeks from the Macedonian

---

12 The plan: Xen. *Ages.* 1.6–7; *Hell.* 3.4.1; Nep. *Ages.* 2.1–2. The citation: Xen. *Hell.* 3.4.5.

13 Nep. *Ages.* 3.1–2; *Hell. Oxy.* 14.1–16.2, 24.1–25-4. Cf. Xen. *Ages.* 1.34. Plut. *Ages.* 19.3 mentions rich booty. Sad Ionians: Xen. *Ages.* 1.38.

14 As predicted by Thuc. 1.77.6.

15 Dreher 1995.

16 Rhodes and Osborne 22, ll. 9–12 = *IG* II^2 43 = *SIG*^{3} 147.

17 Cf. Pownall 2018. See also Günther 1983, 65–71.

18 Polyb. 3.6.3–14: Diod. Sic. 16.89.3, 17.4.9; Just. 11.2.5; Curt. 3.10.8–9, 5.6.1; Plut. *Alex.* 16.8. Cf. Flower 2000; Heckel 2008, 36–38; Squillace 2010.

19 Golan 1988, 101.

tyrant.[20] After Alexander invaded Asia Minor the next year, cities such as Miletus and Halicarnassus resisted alleged Macedonian "liberation."[21] After Alexander's death, resistance to Macedonian hegemony in the Lamian War, under a coalition headed by Athens, also invoked the liberation theme. According to Hyperides, Athens responded to the humiliated and enslaved Greeks' need for a military leader to put an end to Macedonian arrogance, despotism, and insolence and bring back *koine eleutheria* and the rule of law.[22]

In spite of this history, Alexander's successors tried to exploit the theme of Greek freedom for their own purpose, which was to court Greek support in their wars against one another. In 315, Antigonus Monophthalmos took a firm propagandistic stance by championing Greek *eleutheria*. In the name of the Macedonian assembly as well as in his own, he decreed, "All the Greeks should be free, exempt from garrisons, and autonomous" (Diod. Sic. 19.61.3). He aimed at winning Greek allies to provide him manpower and resources, while denouncing his rival Cassander, whom he wished to isolate.[23] Diodorus reports that Ptolemy imitated Antigonus, for the new ruler of Egypt "published a similar decree himself, since he wished to inform the Greeks that he was no less interested in their autonomy than was Antigonus" (19.62.1–2). Yet in 308, when Ptolemy set out to "liberate" Greece by campaigning in the Peloponnese, the slogan failed to serve its purpose: The self-appointed champion of Greek *eleutheria* did not win much support, perhaps because of his former association with Cassander.[24] Disappointed, Ptolemy dropped the mask of liberator and installed garrisons in the allegedly free *poleis* Corinth, Sicyon, and Megara. He also joined forces with Cassander.[25]

Antigonus eventually proved more successful. According to the manifesto he sent to several Greek *poleis* under his control, and probably to other cities as well, he took credit for an agreement that the Greeks should be autonomous. Cassander, Ptolemy, Lysimachus, and Antigonus himself all were parties to this Peace of 311.[26] This claim of being the protector of Greek *eleutheria* may have been a device to justify further wars against his Macedonian rivals, particularly Cassander, whose Greek policy was widely perceived to be oppressive. The fragmentary inscription of a copy of Antigonus' letter to one Greek city, Scepsis in the Troad, sheds light on his self-fashioning as a zealous "martyr" for Greek *eleutheria*: he alleges he was forced to "make no small concessions and distribute money besides" because others, notably Cassander, blackmailed him with burdensome demands.[27] In 307, when Antigonus' son Demetrius Poliorcetes expelled Cassander and his officials from Athens, the Athenians honored father and son as benefactors and saviors who restored the *patrios politeia* to them.[28] That did not prevent Demetrius from occupying the Piraeus. The Macedonian garrison Demetrius put in Athens' port made it impossible for the city to

---

20 Diod. Sic. 17.9.5; Plut. *Alex.* 11.4. Thereby, reportedly, they referred to the terms of the King's Peace (Xen. *Hell.* 5.1.3) characterizing the Persian king as one of their warrantors. See also Dein. 1.20.

21 Strab. 14.1.7; Arr. *Anab.* 1.18.3–19.4, 1.20.2–23.8; Plut. *Alex.* 17.1. Cf. Bosworth 1988, 47–49; Heckel 2008, 43. In the Aegean, there was also resistance by Mytilene.

22 Hyp. 6.10–11, 13, 16–17, 20, 24, 40. Cf. Wirth 2004.

23 Austin 2006, 80–81.

24 Diod. Sic. 20.37.1–2; Suda s.v. *Demetrius Antigonou*. Cf. Simpson 1959, 390; Billows 1990, 314; Lund 1992, 118.

25 Diod. Sic. 20.37.1–2; Polyaen. 8.58; Diog. Laert. 2.115; Plut. *Demetr.* 15.1.

26 Diod. Sic. 19.105.1–2. Cf. Kralli 2000, 117.

27 *OGIS* 5 (= Austin 2006, no. 38). See Simpson 1959, 394; Billows 1990, 311. The Scepsians thanked him by decreeing cultic honors for him, a *temenos* with an altar, and a cult image (*OGIS* 6 = Austin 2006, no. 39).

28 Plut. *Demetr.* 8.1–12.2; Athen. 12.536a. Cf. Kralli 2000, 117; Paschidis 2013, 123–126.

regain freedom altogether, and so the restoration of the "ancestral constitution" ranked as one more propagandistic slogan.[29]

A century later, the liberation theme proved to be an attractive ideological tool in Greece and Anatolia. During the First Mithridatic War, Mithridates VI of Pontus presented himself as a savior who would free these regions from the Roman yoke, and even as a new Dionysus guaranteeing prosperity in lieu of oppressive Roman taxes.[30] In the end, however, Mithridates and his supporters failed to convince Greece and Anatolia that they provided an alternative to Rome. Partly because of the ongoing wars, they could not fulfill their promises of prosperity. They, too, may have been regarded as oppressors.[31]

The Revenge Theme

Another theme concerns acts of revenge for misdeeds and injustice. There are several "avengers" in the history of Greek and Macedonian warfare.[32] While campaigning in Asia Minor, Agesilaus II pretended to "pay back the Persian in his own coin for the former invasion of Greece" (Xen. *Ages.* 1.8). To ennoble his propaganda with heroic symbolism, Agesilaus associated his campaign with the Trojan War and himself with Agamemnon. He even set out to perform a pre-voyage sacrifice at Aulis, in Boeotia, where Agamemnon sacrificed before sailing to Troy. Sparta's tense relations with Thebes, the head of the Boeotian League, spoiled Agesilaus' self-fashioning: The Boeotarchs arrived at Aulis and forbade him to sacrifice.[33]

Philip II of Macedon became involved in the Third Sacred War (355–346) through his Thessalian connections, especially with Larisa, but he officially proclaimed that he was avenging the insults to Delphic Apollo caused by Phocian sacrileges.[34] He ostentatiously ordered his soldiers to wear wreaths of laurel, the god's sign, when they battled the Phocians at the Crocus Field (352).[35] Later, Philip commemorated this victory by a series of gold staters showing Apollo's head with a wreath of laurel on the obverse.[36] In fact, the campaign served to improve Philip's position in Thessaly. He now became master of Thessaly's famous cavalry and appropriated Thessaly's vote in the Delphic Amphictyony. Apollo's self-appointed avenger even tried to invade central Greece on the pretext of pursuing the Phocians, but this time the Athenians were on their guard and blocked him at Thermopylae.[37]

Philip's successor Alexander was also a self-appointed avenger. Serving as the new king's foil was Greece's ultimate bogeyman, Xerxes. As Diodorus reports, Alexander had a meeting of Greek delegates at Corinth pass a resolution "to join in an expedition against Persia seeking satisfaction for the offences which the Persians had committed against Greece"—an

29 Cf. Thonemann 2005, 71–72; Paschidis 2008, 153–154; Lund 1992, 93.
30 Just. 38.7.8; App. *Mithr.* 21. Cf. McGing 1986, 88–108.
31 Erciyas 2006, 165, 176.
32 Cf. Hdt. 5.105.1–2: When the Persian king Darius I heard about the burning of Sardis in the Ionian Revolt, he swore vengeance upon the Athenians.
33 Xen. *Hell.* 3.4.3–4: Plut. *Ages.* 6.4–6.
34 Buckler 1989.
35 Just. 8.2.3; Diod. Sic. 16.35.3–6.
36 Cf. Le Rider 1996, 50–53; Ritter 2002, 139, 143.
37 Diod. Sic. 16.38.1–2. Cf. Wirth 1985, 50; Müller 2016b, 254.

echo of the official claim of the Naval Confederacy of 478/477.[38] This allusion to the era of the Persian Wars was illusory, and ironic, since Alexander's own ancestor Alexander I had fought on Xerxes' side in Greece.[39] Macedonians tried to ignore this embarrassing fact (starting with Alexander I's own rehabilitation campaign),[40] but Athenian leaders remembered it.[41]

Alexander's propaganda was flexible. After the assassination of Darius III in 330 by the Persian king's own officials,[42] Alexander ceased to pose as the avenger of the Greeks and instead began to pose as the avenger of the former royal enemy he had failed to eliminate in battle. This switch was meant to appeal to Alexander's new Iranian subjects, and it let him act as Darius' successor, duly pursuing Bessus, the leader of the assassins of Darius, and then capturing and executing him.[43] In fact, Bessus severely threatened the Macedonians. Probably an Achaemenid himself, he was proclaimed as king, bearing the throne name Artaxerxes V. More legitimate in Persian eyes than the Macedonian invader, Bessus became the focus of indigenous resistance.

The *diadochos* Lysimachus was a self-appointed Macedonian avenger who also played the role of a devoted, chivalrous lover. When invading northwestern Anatolia in 302/301, he had taken the profitable step of marrying the Achaemenid Amastris. As the widow of the tyrant of prosperous Heraclea Pontica, she served as regent for two minor sons, and so she provided Lysimachus control over army supplies and a strategically important port.[44] After the polygamous Lysimachus married Ptolemy's daughter Arsinoë in about 300, Amastris took up residence at her eponymous Paphlagonian foundation, Amastris, while governing Heraclea under Lysimachus' supervision.[45] After she died under obscure circumstances in 285/284, he exploited rumors about her having been assassinated by her sons. Recalling "the hidden spark of his former love" (Memnon *BNJ* 434F 1.5.3), he claimed to be avenging her death, executed the sons, annexed her realm, and plundered Heraclea.[46] To remind the public of his devotion to Amastris, he continued to mint her coins, which bore the legend "Queen Amastris," ΑΜΑΣΤΡΙΟΣ ΒΑΣΙΛΙΣΣΗΣ.[47] This avenger was both a sentimentalist and a murderer.

As a pretext to set out against Lysimachus in the last War of the Successors, Seleucus posed as the avenger of a victim of Lysimachus. This victim was Lysimachus' own son, Agathocles, killed on orders from his father, probably for treason.[48] Seleucus could claim that Agathocles'

38 The quotation: Diod. Sic. 17.4.9. Cf. Just. 11.2.5; Curt. 3.10.8–9, 5.6.1. The official claim: Thuc. 1.96.1.

39 Hdt. 7.25.2, 7.143.3, 7.185.2, 8.34, 9.31.5, 8.126.2, 9.89.4.

40 Hdt. 9.44–46, 5.18.2–20.5; Just. 7.3; Plut. *Arist.* 15.2. Cf. Müller 2016b, 129–134.

41 Dem. 6.11; Lycurg. *Leoc.* 71. Contra Speus. *Phil.* 3–4; [Dem.] 12.21.

42 Diod. Sic. 17.73.2; Just. 11.15.1–14; Curt. 5.3.15–25; Arr. *Anab.* 3.20–21.

43 Arr. *Anab.* 4.7.3–4; Curt. 7.5.43. Cf. Jacobs 1992, 183; Heckel 2006, 72.

44 Strab. 12.3.10; Memnon *BNJ* 434F 1.4.9. Achaemenid descent: Arr. *Anab.* 7.4.5; Diod. Sic. 20.109.7; Memnon *BNJ* 434 F 1.4.4. Cf. Burstein 1976, 81, 143 n. 43; Landucci Gattinoni 1992, 146–147; Müller 2009, 33.

45 Memnon *BNJ* 434F 1.4.9; Strab. 12.3.10.

46 Just. 16.3.2–3; Trog. *Prol.* 16; Memnon *BNJ* 434F 1.5.2–3. Cf. Burstein 1976, 85; Lund 1992, 98, 105; Carney 2000, 208; Heckel 2006, 21.

47 Forrer 1969, 2–4.

48 Memnon *BNJ* 434F. 1.5.6; Cf. Müller 2009, 45–46.

widow, Lysandra, had fled to his court with her children, and begged him to take revenge.[49] In fact, Lysimachus' influence in Asia Minor threatened Seleucus.[50]

The next self-appointed Macedonian avenger turned this useful propagandistic device against Seleucus. It was now 281. That year, Lysimachus died in the battle of Corupedium and Seleucus seized his throne. Ptolemy I's elder son, Ptolemy Ceraunus, eliminated Seleucus. He was acting, he claimed, in the name of Lysimachus.[51] Besides, an oracle had warned Seleucus to leave Asia Minor (App. *Syr.* 56.53). Just before, Ceraunus had probably fought with Seleucus against Lysimachus, a contradiction Ceraunus tried to cover up, first by involving the Macedonian assembly in his accession and then by marrying Lysimachus' widow, Arsinoë. Ceraunus did not scruple at her being his half-sister.[52] He then tried to kill Lysimachus' surviving sons, who were his rivals to the throne.[53] By this démarche, Ceraunus unmasked himself. Lysimachus' sons were also Ceraunus' own nephews. This avenger was a would-be parricide as well as a hypocrite.

One after another, the Successors invoked the theme of spurious vengeance, then died at the hands of avengers.

The Theme of Safeguarding the Peace

The need to wage war to safeguard the peace formed another major theme of Greek war propaganda. So, after the establishment of the *koine eirene* (common peace) by the treaty of 387/386, the Spartans used their position as protector (*prostates*) of the Peace as a vehicle for their own dominion.[54] For example, they interpreted the autonomy clause in the treaty for all Greek *poleis* very strictly, in order to prevent rival alliances.[55] In 371, after the reaffirmation of the Peace, Sparta marched against Phocis in the name of preserving the *koine eirene*. Phocis was an ally of the recently refounded Boeotian League, and the Spartans claimed that the League violated the autonomy clause.[56] In fact, they aimed to crush the League, which they regarded as an instrument of their opponent, Thebes, the League's most powerful member.[57]

After defeating the Greek allies at Chaeronea in 338, Philip II used the *koine eirene* to control Greece. He forced a general peace upon the Greek representatives assembled at Corinth, who responded by electing Philip *hegemon*.[58] After his death, Alexander III succeeded him. In 319, during the Wars of Alexander's Successors, the Macedonian general Polyperchon, then in control of much of Greece, justified waging a new war as a way to preserve the Peace:

---

49 Polyaen. 6.12; App. *Syr.* 10.64; Paus. 1.10.4.

50 App. *Syr.* 62, Phrygia; Just. 17.1.11, *imperii cupiditas;* Nep. 21.3.2–3; Memnon *BNJ* 434F 1.5.7.

51 Nep. 21.3.4; App. *Syr.* 62–63; Paus. 1.16.1, 10.19.7; Just. 17.2.4; Trog. *Prol.* 17; Memnon *BNJ* 434F 1.8.3; Plin. *NH* 6.31; Plut. *Ser. num.* 555b–c; Strab. 13.4.1.

52 Acclamation by the assembly: Nep. 21.3.4; App. *Syr.* 62; Paus. 1.16.1; Trog. *Prol.* 17; Memnon *BNJ* 434, F 1.8.3; Marriage: Just. 17.2.6–9, 24.3.2–3. Cf. Müller 2009, 67–81.

53 Just. 24.3.3–10. He killed the two younger sons. The eldest had escaped.

54 Xen. *Hell.* 5.1.25–26; Diod. Sic. 14.110.2–4. See Cawkwell 1981a.

55 Buckler 2003, 184–189.

56 Xen. *Hell.* 6.3.18–20; Diod. Sic. 15.50.4; Plut. *Ages.* 27.4–28.2; Paus. 9.13.2.

57 Xen. *Hell.* 6.4.1–2; Plut. *Pelop.* 23; Diod. Sic. Sic. 15.51.3. Cf. Georgiadou 1997, 160–161.

58 Just. 9.5.1–7; Diod. Sic. 16.89; *StV* 3.403.

> We believed it necessary to restore all to peace and to the forms of government that Philip our father established … preparing peace for you and such governments as you enjoyed under Philip and Alexander. (Diod. Sic. 18.55.2–3)

In fact, he strove to weaken his rival Cassander by overthrowing the oligarchic regimes established by Cassander's father, Antipater, in several Greek cities.[59] His ploy did not succeed. Cassander eventually prevailed, in spite of the appeal to Peace and to "Philip, our father."

The Theme of Fighting for One's Rights

Another theme frequently used to justify military campaigns is the claim to assert unjustly ignored rights, be they rights to territory, legacies, rewards, freedom, autonomy, or even the right to live under an ancestral constitution or other established laws. For example, Demosthenes tried to rouse the Athens against Macedon, quickly expanding under Philip II, by saying that the city should fight for freedom and autonomy. The city was literally compelled to fight (4.10, ἀνάγκη). Amphipolis affords an example of a fight for territorial rights. The strategically and economically important *polis* in the Strymon valley, founded by Athens in 437/436, was lost to her in 424/423, never retaken despite several attempts, and conquered by Philip II in 357.[60] Defending Athens' age-old rights to the lost *apoikia*, the Attic orator Aeschines claimed that Amphipolis originally fell to Athens as the dowry of the wife of a son of Athens' mythical king Theseus. Philip therefore did not take Amphipolis from its rightful owners, making his claim void.[61] This case illustrates the often pivotal role of forefathers (*progonoi*) in war propaganda. These ancestors might be heroes, as Theseus was, a fillip giving some religious value to the claim.

In the same spirit, Greeks often invoked divine support for their rights. Ceraunus evidently claimed that the oracle warning Seleucus not to leave Asia Minor put god on Ceraunus' side. Not just oracles, but also prophecies, miracles, signs, and dreams would serve this purpose, and the party doing a god's will might be a *polis* as well as a king or a prince. For example, the Athenians justified defending their freedom and independence against Xerxes by invoking Athena, the city's patroness. The goddess revealed her will through a holy olive tree on the Acropolis. According to legend, the Persians who sacked the Acropolis did not succeed in destroying the holy tree:

> It happened that the olive tree was burnt by the barbarians with the rest of the sacred precinct, but on the day after its burning, when the Athenians ordered by the king to sacrifice went up to the sacred precinct, they saw a shoot of about a cubit's length sprung from the stump, and they reported this. (Hdt. 8.55.1)

---

59 Diod. Sic. 18.55.2–56.8. Cf. Simpson 1959, 388–389. On the oligarchies: Diod. Sic. 18.57.1, 18.68.3, 18.69.3; Polyb. 9.29.1–30.4.

60 Foundation: Thuc. 1.100.3, 4.102.3, 4.108.1; Diod. Sic. 12.32.3, 12.68.2; Polyaen. 6.53. Loss: Thuc. 4.103.1–3, 4.107.2–3. Macedonian Conquest: Diod. Sic. 16.8.2; Polyaen. 4.2.17.

61 Aeschin. 2.31–33. Cf. Dem. 7.27–29: Those who inhabited Amphipolis, before Philip took it, were holding Athenian territory. See Harris 1995, 42–43; Squillace 2011.

Since the olive had been Athena's gift to the city, this story showed that she had signaled her refusal to abandon her people. The story thus predicted "the eventual revival of Athens, an interpretation so obvious that Herodotus need not to give it."[62] The olive leaves that soon began to decorate her crested helmet on Athenian coins were a symbol of predestined triumph.[63]

Another example is a tradition about a portent that appeared during Macedonian conquest of Lycia after the battle of Granicus in 334. Plutarch reports that a spring suddenly bubbled up at Alexander's arrival "and cast forth a bronze tablet bearing the prints of ancient letters, in which it was made known that the empire of the Persians would one day be destroyed by the Greeks and come to an end."[64] Both Macedonian and local spectators were thus invited to identify Alexander as a savior whose arrival was welcomed by the forces of nature itself. He was predestined to rule Asia Minor. To make this report unmistakable, the age-old tablet was added, providing more proof of Alexander's timeless destiny to establish "Greek" (in fact, Macedonian) rule.

Conclusions

In sum, the chief themes of Greek and Macedonian war propaganda are help and protection, liberation, revenge, safeguarding the peace, and fighting for one's rights, often by invoking heroes and gods. Ideas of justice, lawfulness, and divine approval may blend, this last idea being crucial, each being defined in accordance with the political aims and intentions of ancient leaders. Like propaganda at other times and places, the war propaganda of Greece and Macedon tends toward black and white, not shades of gray.

62 Mikalson 2003, 73. Cf. Bowie 2007, 141.

63 Cf. Kraay 1976, 62; Hollmann 2011, 84.

64 Plut. *Alex.* 17.2. Cf. Hamilton 1969, 43; Hammond 1993, 46.

Bibliography

Abdullaev, K. 2012. "Baktriskii katafraktarii." *Istoriia material'noi kul'tury Uzbekistana* 37: 74–91.

Adam, Jean-Pierre. 1982. *L'architecture militaire grecque.* Paris.

Adcock, Frank E. 1957. *The Greek and Macedonian Art of War.* Berkeley and Los Angeles; reprinted 1962.

Adornato, G. 2011. *Akragas arcaica: Modelli culturali e linguaggi artistici di una città greca d'Occidente.* Milan.

Afshar, A. and J. Lerner. 1979. "The Horses of the Ancient Persian Empire at Persepolis." *Antiquity* 53: 44–47.

Ager, Sheila L. 1998. "Thera and the Pirates: An Ancient Case of the Stockholm Syndrome." *AHB* 12: 83–95.

Ager, Sheila L. 2005. "Familiarity Breeds: Incest and the Ptolemaic Dynasty." *JHS* 125: 1–34.

Agre, Daniela. 2011. *The Tumulus of Golyamata Mogila near the Villages of Malomirovo and Zlatinitsa.* Sofia.

Akrigg, Ben and Rob Tordoff (eds.). 2013. *Slaves and Slavery in Ancient Greek Comic Drama.* Cambridge.

Albanese Procelli, R.M. 1996. "Greeks and Indigenous People in Eastern Sicily: Forms of Interaction and Acculturation." In R. Leighton (ed.), *Early Societies in Sicily: New Developments in Archaeological Research*, London, 167–176.

Albracht, Franz. 2005. *Battle and Battle Description in Homer. A Contribution to the History of War*, trans. P. Jones. London.

Alexandrova, Sirma. 2017."Спасително археологическо проучване на ранноримски временен военнен лагер край с. Поленица, общ. Сандански, обект №15 от АМ 'Струма' лот 3.3." *Археологически открития и разкопки през 2016 s.* София: 279–281.

Alonso Troncoso, Victor. 2013. "The Diadochoi and the Zoology of Kingship: The Elephants." In Alonso Troncoso and Anson, 254–270.

A Companion to Greek Warfare, First Edition. Edited by Waldemar Heckel, F.S. Naiden, E. Edward Garvin, and John Vanderspoel.

Alonso Troncoso, Victor and Edward M. Anson (eds.). 2013. *After Alexander. The Time of the Diadochoi (323–281 BC)*. Oxford.

Alonso Troncoso, Victor and Mauricio Álvarez Rico. 2017. "Alexander's Tents and Camp Life." In Müller et al., 113–124.

Alusik, T. 2008. "Defensive Architecture in Crete in the Late/Final Neolithic and Bronze Age." In J. Czebreszuk, S. Kadrow, and J. Müller (eds.), *Defence Structures from Central Europe to the Aegean in the 3rd and 2nd millennia BC. Studia nad Pradziejami Europy Środkowej/Studien zur Archäologie in Ostmitteleuropa 5*, Poznań – Bonn, 11–27.

Alusik, T. 2016. "Fortifications of Prehistoric Crete: The Current State of Research." In Frederiksen et al., 53–65.

Álvarez, Mauricio G. 2002. "The Greek Military Camp in the Ten Thousand's Army." *Gladius* 22: 29–56.

Álvarez, Mauricio G. 2010. *El campamento militar griego de Homero a Jenofonte* [Ph.D.]. Universidad Autónoma de Madrid.

Álvarez, Mauricio G. 2013. *El campamento militar griego en época clásica*. Madrid.

Ambrose, Stephen E. 1992/2001. *Band of Brothers*. New York.

Amirell, Stefan Eklöf and Leos Müller (eds.). 2014a. *Persistent Piracy. Maritime Violence and State-Formation in Global Historical Perspective*. Basingstoke, UK and New York.

Amirell, Stefan Eklöf and Leos Müller (eds.). 2014b. "Introduction: Persistent Piracy in World History." In Amirell and Müller 2014a, 1–23.

Amouretti, Marie-Claire 1986. *Le pain et l'huile dans la Grèce antique. De l'araire au moulin*. Paris.

Anderson, John K. 1960. "Notes on Some Points in Xenophon's ΠΕΡΙ ΗΙΠΠΙΚΗΣ." *JHS* 80: 1–9.

Anderson, John K. 1961. *Ancient Greek Horsemanship*. Berkeley.

Anderson, John K. 1970. *Military Theory and Practice in the Age of Xenophon*. Berkeley and Los Angeles.

Anderson, John K. 1991. "Hoplite Weapons and Offensive Arms." In Hanson, 15–37.

Andrewes, Antony. 1956. *The Greek Tyrants*. New York.

Andrewes, Antony. 1992. "The Peace of Nicias and the Sicilian Expedition." In *CAH* V, 433–463.

Andronicos, M. 2004. *Vergina*. Athens.

Ansani, Fabrizio. 2019. "'This French artillery is very good and very effective.' Hypotheses on the Diffusion of a New Military Technology in Renaissance Italy." *Journal of Military History* 83: 347–378.

Anson, Edward M. 1985a. "The Hypaspists: Macedonia's Professional Citizen-Soldier." *Historia* 34: 246–248.

Anson, Edward M. 1985b. "Macedonia's Alleged Constitutionalism." *CJ* 80: 303–316.

Anson, Edward M. 1988. "Hypaspists and the Argyraspids after 323." *AHB* 2: 131–132.

Anson, Edward M. 2008a. "Macedonian Judicial Assemblies." *CP* 103: 135–149.

Anson, Edward M. 2008b. "Philip II and the Transformation of Macedonia: A Reappraisal." In Howe and Reames, 17–30.

Anson, Edward M. 2010a. "The Asthetairoi: Macedonia's Hoplites." In Carney and Ogden, 81–90.

Anson, Edward M. 2010b. "The Introduction of the SARISA in Macedonian Warfare." *AncSoc* 40: 51–68.

Anson, Edward M. 2013. *Alexander the Great: Themes and Issues*. London, New Delhi, New York, and Sydney.

Anson, Edward M. 2014. *Alexander's Heirs: The Age of the Successors*. Chichester and Malden.

Anson, Edward M. 2015. *Eumenes of Cardia: A Greek Among Macedonians*, 2nd ed. Leiden and Boston.

Anson, Edward M. 2017. "The Rise of Macedon, 359–336 BC." In Michael Whitby and Harry Sidebottom (eds.), *The Encyclopedia of Ancient Battles*, Chichester and Malden, 480–495.

Aravantinos, V.I. 2006. "A New Inscribed kioniskos from Thebes." *BSA* 101: 369–377.

Archibald, Zosia H. 1994. "Thracians and Scythianss." In CAH^2 VI, 444–75.

Archibald, Zosia H. 1998. *The Odrysian Kingdom of Thrace: Orpheus Unmasked*. Oxford.

Archibald, Zosia H. 2002. "A River Port and Emporion in Central Bulgaria: An Interim Report on the British Project at Vetren." *ABSA* 97: 309–351.

Archibald, Zosia H. 2007. "Археологически разкопки на британския екип в Пистирос (Ветрен), сезон 2006." *Археологически открития и разкопки през 2006 г.* София: 78–82.

Armitage, David. 2017. *Civil Wars. A History in Ideas.* New York.

Arnason, Johann P., Kurt A. Raaflaub, and Peter Wagner (eds.). 2013. *The Greek Polis and the Invention of Democracy. A Politico-cultural Transformation and Its Interpretations.* Malden.

Arrington, N.T. 2010. "Topographic Semantics. The Location of the Athenian Public Cemetery and Its Significance for the Nascent Democracy." *Hesperia* 79: 499–539.

Arrington, N.T. 2015. *Images and Memories. The Presence of the War Dead in Fifth-century Athens.* Oxford and New York.

Asheri, D. 1988. "Carthaginians and Greeks." In *CAH* IV, 739–90.

Asheri, D. 1992. "Sicily, 478–431 B.C." In *CAH* VI, 147–70.

Asheri, D., A. Lloyd, and A. Corcella 2007. *A Commentary on Herodotus, Books 1–4.* Oxford.

Ashley, James R. 1998; 2004. *The Macedonian Empire. The Era of Warfare under Philip II and Alexander the Great.* Jefferson.

Asmonti, Luca. 2015. *Conon the Athenian: Warfare and Politics in the Aegean, 414–386 B.C.* Stuttgart.

Atanasov, Georgi. 1995. "Тракийско въоръжение от фонда на Исторически музей-Шумен." В: Въоръжение от Древна Тракия. Каталог на изложба. Шумен, 29–91.

Atanasov, Georgi. 2003. "Групи от железни върхове на стрели от крепостта в местността "Градището" при село Драгоево IV–II в. пр. Хр." Есенни четения Сборяново. На ползу роду, том 1, Варна, 15–23.

Atanasov, Georgi. 2005. "Един нов тип стрели от елинистическата епоха в колекцията на РИМ–Шумен." Есенни четения Сборяново, том 2, Варна, 55–63.

Atanasov, Georgi. 2006. "Шпори от късножелязната епоха, открити в Шуменския регион." *Известия на историческия музей-Шумен* 13: 25–48.

Atanasov, Georgi, Stanimir Stoychev and Yuri Yorgov. 2006. "Резултати от разкопките на "Градището"край с. Драгоево, Шуменско през 1986 г." Есенни четения Сборяново. Силищен модел в Североизточна България, том 4, София, 66–73.

Atanasov, Georgi and Todor Todorov. 1999. "Оловни тежести за хвърляне с прашка от фонда на Исторически музей-Шумен." *Известия на Историческия музей-Шумен* 9: 61–66.

Austin, Michel M. 1986. "Hellenistic Kings, War, and the Economy." *CQ* 36: 450–466.

Austin, Michel M. 2006. *The Hellenistic World from Alexander to the Roman Conquest.* Cambridge.

Avidov, A. 1997. "Were the Cilicians a Nation of Pirates?" *MHR* 12: 5–55.

Avram, Alexandru 2011. "Marginalien zu griechisch beschrifteten Schleudergeschossen (I)." In L. Ruscu et al. (ed.), *Scripta classica. Radu Ardevan sexagenario dedicate*, Cluj-Napoca, 195–199.

Avram, Alexandru, Costel Chiriac and Ionel Matei. 2013. "Balles de fronde grecques en pays gète et ailleurs. Sur les traces de Zopyrion dans le bas Danube." *Revue archéologique* 2: 227–303.

Azzaroli, A. 1985. *An Early History of Horsemanship.* Leiden.

Badian, E. 1991. "The King's Peace." In Michael A. Flower and Mark Toher (eds.), *Georgica. Greek Studies in Honour of George Cawkwell*, London, 25–48.

Badian, E. 1993. *From Plataea to Potidaea: Studies in the History and Historiography of the Pentecontaetia.* Baltimore.

Badian, E. 1996. "Alexander the Great between Two Thrones and Heaven: Variations on an Old Theme." In A. Small (ed.), *The Cult of the Ruling Power in Classical Antiquity: Papers Presented at a Conference Held in the University of Alberta on April 13–15, 1994, to Celebrate the 65th Anniversary of Duncan Fishwick.* Ann Arbor, 11–26; reprinted in *Collected Papers on Alexander the Great*, ed. R. Stoneman, London, 2008, 365–386.

Badian, E. 2004. "Ionian Revolt." *Encyclopaedia Iranica* 13(2): 188–195.

Baege, W. 1911. *De Macedonum Sacris.* Halle.

Baitinger, Holger. 2001. *Die Angriffswaffen aus Olympia. Olympische Forschungen XXIX.* Berlin.

Baitinger, Holger. 2004. "Hellenistisch-frühkaiserzeitliche Reitersporen aus dem Zeusheiligtum von Olympia." *Germania* 82: 351–379.

Baker, Patrick. 2003. "Warfare." In Erskine, 373–386.

Bakewell, Geoffrey. 2008. "Trierarchs' Records and the Athenian Naval Catalogue (*IG* i^3 1032)." In Mackay, 143–162.

Bakhuizen, Simon C. 1986. "La Grande Batterie de Gorítsa et l'artillerie défensive." In Leriche and Tréziny, 315–321.

Balandier, C. 2016. "Étudier l'organisation défensive d'une région et son évolution: pour une archéologie historique des fortifications. Question de méthode." In Frederiksen et al., 417–434.

Balcer, Jack M. 1976. "Imperial Magistrates in the Athenian Empire." *Historia* 25: 257–287.

Balcer, Jack M. 1977. "The Athenian *episkopos* and the Achaemenid king's eye." *AJPh* 98: 252–263.

Balcer, Jack M. 1984. *Sparda by the Bitter Sea. Imperial Interaction in Western Anatolia.* Chico, CA.

Ball, Joanne E. 2014. "Small Finds and Roman Battlefields: The Process and Impact of Post-Battle Looting." In H. Platt, J. Pearce, C. Barron, J. Lundock, and J. Yoo (eds.), *TRAC 2013: Proceedings of the Twenty-Third Annual Theoretical Roman Archaeology Conference.* London 2013, 90–104.

Ballmer, Ariane, Manuel Fernández-Götz, and Dirk P. Mielke (eds.). 2018. *Understanding Ancient Fortifications. Between Regionality and Connectivity.* Oxford and Philadelphia.

Bankel, Hansgeorg (ed.). 1986. *Carl Haller von Hallerstein in Griechenland, 1810–1817. Architekt, Zeichner, Bauforscher.* Berlin.

Banlaoi, Rommel C. 2007. "The Abu Sayyaf Group: Threat of Maritime Piracy and Terrorism." In Lehr 2007, 121–137.

Bar-Kochva, B. 1976. *The Seleucid Army: Organization and Tactics in the Great Campaigns.* Cambridge.

Barringer, J.M. and J.M. Hurwit (eds.). 2005. *Periklean Athens and Its Legacy. Problems and Perspectives.* Austin.

Barry, William. 1996. "Roof Tiles and Urban Violence in the Ancient World." *GRBS* 37: 55–74.

Bauer, A. 1893. *Die Griechischen Kriegsaltertümer*, 2nd ed. Munich.

Baynham, E. and J. Walsh (eds.). 2018. *"Selling Alexander": Alexander the Great and Propaganda.* London and New York.

Bayrakov, Dimitar. 2015. *Планински светилища в Южна Тракия.* Дисертация за присъждане на образователната и научна степен "Доктор". София.

Bearzot, Cinzia. 2006. "Atene nel 411 e nel 404. Tecniche del colpo di stato." In Urso 2006, 21–54.

Bearzot, Cinzia and Franca Landucci (eds.). 2016. *Alexander Legacy. Atti del convegno, Milano, Università Cattolica del sacro cuoro, Settembre 2015.* Rome.

Beckman, Gary M., Trevor R. Bryce, and Eric H. Cline. 2011. *The Ahhiyawa Texts.* Atlanta GA.

Beloch, H. 1906. "Griechische Aufgebote, II." *Klio* 6: 34–78.

Beloch, Julius. 1897. *Griechische Geschichte, vol. II.* Strasbourg.

Berent, Moshe. 2000. "Anthropology and the Classics: War, Violence and the Stateless Polis." *CQ* 50: 257–289.

Berger, E. 1986. *Der Parthenon in Basel. Dokumentation zu den Metopen.* Mainz.

Berger, S. 1992. *Revolution and Society in Greek Sicily and Southern Italy.* Stuttgart.

Berridge, G.R. and Lorna Lloyd (eds.). 2012. *The Palgrave Macmillan Dictionary of Diplomacy*, 3rd ed. Basingstoke, UK and New York.

Bertosa, Brian. 2005. "The Social Status and the Ethnic Origins of the Rowers of Spartan Triremes." *War and Society* 23: 1–20.

Bertrand, J.-M. (ed.). 2005. *La violence dans les mondes grec et romain.* Paris.

Bessac, Jean-Claude. 1986. "Approche des problèmes posés par la construction des remparts grecs en pierre." In Leriche and Tréziny, 273–282.

Bessac, Jean-Claude. 1997. *La Construction des fortifications hellénistiques en pierre de Doura-Europos (Syrie).* 3 vols. Lille.

Bessac, Jean-Claude and Pierre Leriche. 1992. "L'analyse des techniques de construction en pierre et en brique crue." *Dossiers d'Archéologie* 172: 70–81.

Best, J. 1969. *Thracian Peltasts and Their Influence on Greek Warfare*. Groningen.

Beste, Heinz-Jürgen. 2016. "The Castle Euryalos of Syracuse." In Frederiksen et al., 193–206.

Beste, Heinz-Jürgen and Dieter Mertens. 2015. *Die Mauern von Syrakus: das Kastell Euryalos und die Befestigung der Epipolai*. Wiesbaden.

Beston, Paul. 2000. "Hellenistic Military Leadership." In van Wees, 315–336.

Bettalli, Marco. 1990. *Enea Tattico, La difesa di una città assediata*. Pisa.

Bettalli, Marco. 2006. "Hoi ton Hellenon aporoi: i mercenari del mondo greco classico tra violenza, emarginazione e integrazione." In Urso 2006, 55–64.

Bielman, A. 1994. *Retour à la liberté: Libération et sauvetage des prisonniers en Grèce ancienne*. Athens and Lausanne.

Bielman, A. 2003. "Piraterie et brigandage hellénistiques: razzias pour le seul profit ou actes de nature terroriste?" *Droit et cultures* 45: 127–141.

Bielman, A. 2012. *Femmes en public dans le monde hellénistique, IVe–Ie s. av. J.-C.* Paris.

Bielman Sánchez, Anne. 2011. "Des victimes entre silences et allusions: les captives en Grèce ancienne." In Francesca Prescendi and Agnes A. Nagy (eds.), *Victime: substantif féminin*, Genève, 67–82.

Billows, Richard A. 1990. *Antigonos the One-eyed and the Creation of the Hellenistic State*. Berkeley.

Billows, Richard A. 1995. *Kings and Colonists. Aspects of Macedonian Imperialism*. Leiden.

Bingen, Jean. 2007. *Hellenistic Egypt. Monarchy, Society, Economy, Culture*. (ed. with introduction by Roger S. Bagall). Edinburgh.

Bissa, E.M.A. 2009. *Governmental Intervention in Foreign Trade in Archaic and Classical Greece. Mnemosyne Suppl.* 312. Leiden, Boston.

Bittner, S. 1985. *Tracht und Bewaffnung des persischen Heeres zur Zeit der Achaimeniden*. Munich.

Blastenbrei, Peter. 1987. *Die Sforza und ihr Heer: Studien zur Struktur-, Wirtschafts- und Sozialgeschichte des Söldnerwesens in der italienischen Frührenaissance*. Heidelberg.

Blight, David W. 2001. *Race and Reunion. The Civil War in American Memory*. Cambridge, MA and London.

Boardman, J. 1998. *Early Greek Vase Painting*. London.

Boardman, J. 2005. "Composition and Content on Classical Murals and Vases." In Barringer and Hurwit, 63–72.

Boardman, J. and M. Price 1980. "The Late Geometric Pottery." In M.R. Popham, L.H. Sackett and P.G. Themelis (eds.), *Lefkandi 1: The Iron Age. The Settlement. The Cemetries*, London, 57–79.

Bol, Peter C. 1989. *Argivische Schilde. Olympische Forschungen XVII*. Berlin.

Bondi, M. 2006. "Obiettivi e modalità dell'azione militare di Cartagine in Sicilia." In Ampolo, 131–138.

Bopearachchi, O. and C. Sachs 2001. "Armures et armes des Indo-Scythes d'après leurs émissions monétaires et les données archéologiques." *Topoi* 11: 321–355.

Borg, Alan. 1991. *War Memorials. From Antiquity to the Present*. London.

Borza, Eugene N. 1987. "Timber and Politics in the Ancient World: Macedon and the Greeks." *TAPA* 131: 32–52.

Boschung, D. 2001. "Überlegungen zum Denkmal des Aemilius Paullus in Delphi." In C. Evers and A. Tsingarida (eds.), *Rome et ses provinces: genèse et diffusion d'une image du pouvoir*, Brussels, 59–72.

Bosworth, A. Brian. 1980–1995. *A Historical Commentary on Arrian's History of Alexander*. Oxford.

Bosworth, A. Brian. 1988. *Conquest and Empire. The Reign of Alexander the Great*. Cambridge.

Bosworth, A. Brian. 1996. *Alexander and the East. The Tragedy of Triumph*. Oxford.

Bosworth, A. Brian. 2003. "The Indian Satrapies under Alexander the Great." In Ian Worthington (ed.), *Alexander the Great. A Reader*, London and New York, 170–177.

Bottini, Angelo. 1988. *Antike Helme*. Mainz.

Bouzek, Jan. 2007. "Celts at Pistiros." In J. Bouzek, L. Domaradzka and Z.H. Archibald (eds.), *Pistiros III. Excavations and Studies*, Prague, 249–253.

Bouzek, Jan. 2013. "Addenda." In Bouzek et al., 288.

Bouzek, Jan and Mieczyslaw Domaradzki. 2010. "Fibulae and Other Bronze Objects." In J. Bouzek, L. Domaradzka and Z.H. Archibald (eds.), *Pistiros IV. Excavations and Studies*, Prague, 223–225.

Bouzek, Jan and Jiri Musil. 2002. "Adžijska Vodenica II, 1994–1998: Excavations Report by the Czech Mission." In J. Bouzek, L. Domaradzka, and Z.H. Archibald (eds.), *Pistiros II. Excavations and Studies*, Prague, 37–111.

Bouzek, Jan, Jiri Musil, and Iva Ondrejova. 2013. "Metal Objects." In Bouzek et al., 224–237.

Bouzek, Jan, L. Domaradzka, A. Gotzev, and Z.H. Archibald (eds.). 2013. *Pistiros V. Excavations and Studies*. Prague.

Bowden, Hugh. 1993. "Hoplites and Homer: Warfare, Hero Cult, and the Ideology of the Polis." In Rich and Shipley, 45–63.

Bowie, Angus M. 2007. *Herodotus: Histories Book VIII*. Cambridge.

Bradeen, David W. 1947. "The Lelantine War and Pheidon of Argos" *TAPA* 78: 223–241.

Bradeen, David W. 1969. "The Athenian Casualty Lists." *CQ* 19: 145–159.

Brands, H.W. 2012. *The Man Who Saved the Union. Ulysses Grant in War and Peace*. New York.

Brandt, A.L., Y. Hagos, Y. Yacob, V.A. David, N.J. Georgidis, J. Shoshani, and A.L. Roca. 2014. "The Elephants of Gash-Barka, Eritrea: Nuclear and Mitochondrial Genetic Patterns." *Journal of Heredity* 105: 82–90.

Brasse, Christiane and Silke Müth. 2016. "Mauerwerksformen und Mauerwerkstechniken." In Müth et al., 75–100.

Braun, K. 1970. "Der Dipylon-Brunnen B1, die Funde." *Mitteilungen des deutschen archäologischen Instituts (Athenische Abteilung)* 85: 129–269.

Braund, David. 1993. "Piracy under the Principate and the Ideology of Imperial Eradication." In Rich and Shipley 1993, 195–212.

Bravo, Benedetto. 1980. "*Sulân*. Représailles et justice privée contre des étrangers dans les cités grecques." *Annali della Scuole Normale Superiore di Pisa, Classe di Lettere e Filosofia* III 10: 675–987.

Bremmer, J. 1983. *The Early Greek Concept of the Soul*. Gröningen.

Bresson, Alain. 2016. *The Making of the Ancient Greek Economy: Institutions, Markets, and Growth in the City-States*, trans. S. Rendall. Princeton.

Briant, Pierre. 1994. "Sources gréco-héllenistiques, institutions perses et institutions macédoniennes: continuités, changements et bricolages." In H. Sancisi-Weerdenburg et al., 283–310.

Briant, Pierre. 1996. *Histoire de l'empire perse de Cyrus à Alexandre*. Paris.

Briant, Pierre. 2002. *From Cyrus to Alexander: A History of the Persian Empire*, trans. P.T. Daniels. Winona Lake.

Brice, Lee L. 2013. "The Athenian Expedition to Sicily." In Campbell and Tritle, 623–641.

Brice, Lee L. 2016. "Insurgency and Terrorism in the Ancient World, Grounding the Discussion." In Howe and Brice, 3–27.

Brice, Lee L. and Jennifer T. Roberts (eds.). 2011. *Recent Directions in the Military History of the Ancient World*. Claremont.

Bridges, Emma. 2015. *Imagining Xerxes. Ancient Perspectives on a Persian King*. London.

Brisch, N. (ed.). 2008. *Religion and Power: Divine Kingship in the Ancient World and Beyond. Chicago Oriental Institute Seminars 4*. Chicago.

Briscoe, J. 1981. *A Commentary on Livy Books XXXIV–XXXVII*. Oxford.

Brown, F.E. 1937. "A Recently Discovered Compound Bow." *Seminarium Kondakovianum* 9: 1–10.

Brownson, Carleton L. (trans.). 1922. *Xenophon*. Cambridge, MA.

Brulé, Pierre. 1978. *La piraterie crétoise hellénistique*. Paris.

Brunt, P.A. 1963. "Alexander's Macedonian Cavalry." *JHS* 83: 27–46.

Bryant, Joseph M. 1990. "Military Technology and Socio-Cultural Change in the Ancient Greek city." *Sociological Review* 38: 484–516.

Buck, Robert J. 1994. *Boiotia and the Boiotian League, 423–371 B.C.* Edmonton.

Buckler, John. 1972. "A Second Look at the Monument of Chabrias." *Hesperia* 41: 466–474.

Buckler, John. 1980. *The Theban Hegemony 371–362 BC*. Cambridge, MA.
Buckler, John. 1989. *Philip II and the Sacred War*. Leiden.
Buckler, John. 2003. *Aegean Greece in the Fourth Century BC*. Leiden.
Buckler, John and Hans Beck. 2008a. "Sphodrias' Raid and the Evolution of the Athenian League." In Buckler and Beck 2008b, 79–84.
Buckler, John and Hans Beck (eds.). 2008b. *Central Greece and the Politics of Power in the Fourth Century BC*. Cambridge.
Budiansky, Stephen. 1997. *The Nature of Horses: Exploring Equine Evolution, Intelligence, and Behavior*. New York.
Bugh, Glenn. 1988/2016. *The Horsemen of Athens*. Princeton.
Buraselis, Kostas. 2003. "Zur Asylie als auβenpolitischem Instrument in der hellenistischen Welt." In Dreher 2003, 143–158.
Burford, Alison M. 1960. "Heavy Transport in Classical Antiquity." *Economic History Review* 13: 1–18.
Burford, Alison M. 1993. *Land and Labor in Ancient Greece*. Baltimore: Johns Hopkins Univ. Press.
Burke, Peter. 2001^2. *Ludwig XIV. Die Inszenierung des Sonnenkönigs*. Berlin.
Burkert, W. 1985. *Greek Religion*, trans. J. Raffan. Cambridge, MA.
Burrer, F. 2008. "Sold und Verpflegungsgeld in klassischer und hellenistischer Zeit." In Burrer and Müller, 74–90.
Burrer, F. and H. Müller (eds.). 2008. *Kriegskosten und Kriegsfinanzierung in der Antike*. Darmstadt.
Burstein, Stanley M. 1976. *Outpost of Hellenism. The Emergence of Heraclea on the Black Sea*. Berkeley.
Burstein, Stanley M. (trans.). 1989. *Agatharchides of Cnidus on the Erythraean Sea*. London.
Bussemer, Thymian. 2008. *Propaganda. Konzepte und Theorien*. Heidelberg.
Buykliev, Hristo. 1991. "Използването на лъкът и стрелата в древна Тракия през ранножелязната епоха." *Известия на музеите югоизточна България* 14: 7–22.
Buykliev, Hristo. 1994. "Към проблема за тракийското въоръжение в хинтерланда на Севтополис." В: *Първи международен симпозиум "Надгробните могили в Югоизточна Европа", Казанлък, 4–8 юни 1993*, том 1. Велико Търново, 157–164.
Buykliev, Hristo. 1998. "Състояние на проучванията на въоръжението в древна Тракия от бронзовата епоха до установяването на римското владичество в българските земи." *Епохи* 3–4: 25–42.
Cahill, Nicholas. 2002. *Household and City Organization at Olynthus*. New Haven and London.
Cairns, F. 1984. "*ΧΡΕΜΑΤΑ ΔΟΚΙΜΑ*: IG XII 9, 1273f. and the Early Coinage of Eretria." *ZPE* 54: 145–155.
Caliò, Luigi M. 2012. "Dalla *polis* alla città murata. L'immagine delle fortificazioni nella società ellenistica." *ArchClass* 63: 169–221.
Callebat, Louis and Philippe Fleury. 1986. *Vitruve*, De L'Architecture, *livre X*. Paris.
Camp, John M. 1986. *The Athenian Agora. Excavations in the Heart of Classical Athens*. London.
Camp, John M. 1991. "Notes on the Towers and Borders of Classical Boiotia." *AJA* 95: 193–202.
Camp, John M. 2001. *The Archaeology of Athens*. New Haven.
Campbell, Brian. 2002. "Power without Limit: 'The Romans Always Win'." In Chaniotis and Ducrey, 167–180.
Campbell, Brian and Lawrence A. Tritle (eds.). 2013. *The Oxford Handbook of Warfare in the Classical World*. Oxford.
Campbell, Duncan B. 2003a. *Greek and Roman Siege Machinery 399 BC–AD 363*. Oxford.
Campbell, Duncan B. 2003b. *Greek and Roman Artillery 399 BC–AD 363*. Oxford.
Campbell, Duncan B. 2005. *Ancient Siege Warfare: Persians, Greeks, Carthaginians and Romans 546–146 BC*. Oxford.
Campbell, Duncan B. 2006. *Besieged: Siege Warfare in the Ancient World*. Oxford.
Cantrell, Deborah. 2011. *The Horsemen of Israel: Horses and Chariotry in Monarchic Israel (Ninth-Eighth Centuries B.C.E.)*. Winona Lake.
Carawan, Edward. 2012. "The Meaning of Mê Mnêsikakein." *CQ* 62: 567–568.

Carlier, P. 2000. "Homeric and Macedonian Kingship." In Roger Brock and Stephen Hodkinson (eds.), *Alternatives to Athens. Varieties of Political Organization and Community in Ancient Greece*, Oxford, 259–268.

Carney, Elizabeth D. 2000. *Women and Monarchy in Macedon*. Norman, OK.

Carney, Elizabeth D. 2004. "Women and Military Leadership in Macedonia." *AncW* 35: 184–195.

Carney, Elizabeth D. 2013. *Arsinoë of Egypt and Macedon: A Royal Life*. Oxford and New York.

Carney, Elizabeth D. 2016. "Commemoration of a Royal Woman as a Warrior: The Burial in the Antechamber of Tomb II at Vergina." *SyllClass* 27: 109–149.

Carney, Elizabeth D. and Daniel Ogden (eds.). 2010. *Philip II and Alexander the Great. Father and Son, Lives and Afterlives*. Oxford.

Cartledge, Paul. 1977. "Hoplites and Heroes: Sparta's Contribution to the Technique of Ancient Warfare." *JHS* 97: 11–27.

Cartledge, Paul. 1979. *Sparta and Lakonia*. London.

Cartledge, Paul. 1996. "La nascita degli opliti e l'organizzazione militare." In Salvatore Settis (ed.), *I Greci: Storia, Cultura, Arte, Societá, vol. II.I. Formazione*, Torino, 681–714.

Cartledge, Paul. 2001. "The Birth of the Hoplite: Sparta's Contribution to Early Greek Military Organization." In id. (ed.), *Spartan Reflections*, London, 153–166.

Cartledge, Paul. 2002. *Sparta and Lakonia*, 2nd ed. London and New York.

Cartledge, Paul. 2013a. "Hoplitai/Politai: Refighting Ancient Battles." In Kagan and Viggiano, 74–84.

Cartledge, Paul. 2013b. *After Thermopylae: The Oath of Plataea and the End of the Graeco-Persian Wars*. Oxford.

Casabona, J. 1966. *Recherches sur le vocabulaire des sacrifices en Grèce des origines à la fin de l'époque classique. Publication des annales de la faculté des lettres n.s. 56*. Aix-en-Provence.

Case, S. 1915. "Religion and War in the Graeco-Roman World." *AJTh* 19: 179–199.

Casson, Lionel. 1971. *Ship and Seamanship in the Ancient World*. Princeton.

Casson, Lionel. 1993. "Ptolemy II and the Hunting of African Elephants." *TAPA* 123: 247–260.

Castleden, R. 2005. *The Mycenaeans*. London.

Castriota, D. 1992. *Myth, Ethos, and Actuality. Official Art in Fifth-Century Athens*. Madison.

Caubert, A. 2002. "Animals in Syrio-Palestinian Art." In Collins, 211–238.

Caven, Brian. 1990. *Dionysius I. War-Lord of Sicily*. New Haven.

Cawkwell, George L. 1972. "Epaminondas and Thebes." *CQ* 22: 254–278.

Cawkwell, George L. 1981a. "The King's Peace." *CQ* 31: 69–83.

Cawkwell, George L. 1981b. "Notes on the Failure of the Second Athenian Confederacy." *JHS* 101: 40–55.

Cawkwell, George L. 1989. "Orthodoxy and Hoplites." *CQ* 39: 375–389.

Cawkwell, George L. 1993. "Sparta and Her Allies in the 6th c." *CQ* 43: 364–376.

Cawkwell, George L. 1997. *Thucydides and the Peloponnesian War*. London.

Cawkwell, George L. 2005. *The Greek Wars: The Failure of Persia*. New York.

Champion, J. 2014. *Antigonus the One-Eyed*. Barnsley.

Chandezon, Christophe. 2003. *L'élevage en Grèce (fin Ve-fin Ier s. a.C.). L'apport des sources épigraphiques*. Bordeaux.

Chaniotis, Angelos. 2004. "Justifying Territorial Claims in Classical and Hellenistic Greece: The Beginnings of International Law." In E. Harris and L. Rubenstein (eds), *The Law and the Courts in Ancient Greece*, London, 185–213.

Chaniotis, Angelos. 2005. *War in the Hellenistic World*. Oxford and Malden, MA.

Chaniotis, Angelos. 2008. "Policing the Hellenistic Countryside: Realities and Ideologies." In Cédric Brélaz and Pierre Ducrey (eds.), *Sécurité collective et ordre public dans les sociétés anciennes*, Entretiens sur l'antiquité classique, 54, Geneva, 103–153.

Chaniotis, Angelos and Pierre Ducrey (eds.). 2002. *Army and Power in the Ancient World*. Stuttgart.

Chankowski, Andrzej S. 2010. *L'Éphébie hellénistique. Étude d'une institution civique dans les cités grecques des îles de la Mere Égée et de l'Asie Mineure*. Paris.

Chapman, George A.H. 1972. "Herodotus and Histiaeus' Role in the Ionian Revolt." *Historia* 21: 546–568.

Charles, Michael B. 2008. "Alexander, Elephants and Gaugamela." *Mouseion* 8: 9–23.

Charles, Michael B. 2009. "Turrets, Gaugamela and the Historian's Duty of Care." *Scholia. Studies in Classical Antiquity* 18: 29–36.

Charles, Michael B. 2010. "Elephants, Alexander and the Indian Campaign." *Mouseion* 10: 327–353.

Charles, Michael B. 2011. "Immortals and Apple Bearers: Towards a better Understanding of Achaemenid Infantry Units." *CQ* 61: 114–133.

Charles, Michael B. 2012a. "Herodotus, Body Armour and Achaemenid Infantry." *Historia* 61: 257–269.

Charles, Michael B. 2012b. "The Persian Κάρδακες." *JHS* 132: 7–22.

Charles, Michael B. 2014. "Elephants in Vegetius' 'Epitoma Rei Militaris' (3.24.5–16)." *MH* 71: 189–203.

Charles, Michael B. 2015a. "Achaemenid Elite Cavalry: From Xerxes to Darius III." *CQ* 65: 14–34.

Charles, Michael B. 2015b. "The Chiliarchs of Achaemenid Persia: Towards a Revised Understanding of the Office." *Phoenix* 69: 279–303.

Charles, Michael B. 2016a. "Two Notes on Darius III." *CCJ* 62: 52–64.

Charles, Michael B. 2016b. "The Achaemenid Chiliarch *par excellence*." *Historia* 65: 392–412.

Charles, Michael B. 2016c. "Elephant Size in Antiquity-DNA Evidence and the Battle of Raphia." *Historia* 65: 53–65.

Charles, Michael B. and E. Anagnostou-Laoutides. 2016. "Athenaeus, Clearchus and the Dress of the Persian Apple Bearers." *Iranica Antiqua* 51: 149–164.

Chaumont, M.L. 1973: "Études d'histoire parthes. II. Capitales et residences des premiers Arsacides." *Syria* 50: 197–222.

Chen, Jeffrey. 2007. "The Emerging Nexus between Piracy and Maritime Terrorism in Southeast Asia Waters: A Case Study on the Gerakan Aceh Merdeka (GAM)." In Lehr 2007, 139–154.

Chichikova, Maria. 2016. "The Royal Palace in Seuthopolis." *Thracia* 21: 31–47.

Childs, William A.P. 1991. "A New Representation of a City on an Attic Red-figured Kylix." In Marion True (ed.), *Greek Vases in the J. Paul Getty Museum*, vol. 5, Malibu, 27–40, no. 84.

Chomsky, Noam. 2002. *Pirates and Emperors, Old and New. International Terrorism in the Real World.* Toronto.

Chykerda, C.M., M.J. Haagsma, and S. Karapanou. 2014a. "From City to Sea: Kastro Kallithea (Thessaly) and Its Role in the Defense of Southern Achaia Phthiotis." In D. Rupp and J. Tomlinson (eds.), *Meditations on the Diversity of the Built Environment in the Aegean Basin and beyond. Proceedings of the colloquium in commemoration of Dr. Frederick Winter, Canadian Institute in Greece, June 22–23,2012.* Athens, 285–311.

Chykerda, C.M., M.J. Haagsma, and S. Karapanou. 2014b. "Landscapes of Defense: Kastro Kallithea and Its Role in Fourth-Century BCE Achaia Phthiotis." *Backdirt December 2014*: 10–23.

Clay, Diskin. 2013. "Lesbian Armour: Alcaeus fr. 140 Voigt." *Prometheus* 39: 18–24

Clay, Jenny S. 2011. *Homer's Trojan Theater. Space, Vision and Memory in the Iliad.* Cambridge.

Clayman, Dee L. 2014. *Berenice II and the Golden Age of Ptolemaic Egypt.* Oxford and New York.

Clinton, K. 2005. *Eleusis. The Inscriptions on Stone.* Athens.

Clouet, A. 2015. *La cité de Téos à l'époque hellénistique.* Mémoire de Master 1. l'Université Lumière Lyon.

Cobet, J. 1997. "Milet 1994–1995. Die Mauern sind die Stadt: zur Stadtbefestigung des antiken Milet." *Archäologischer Anzeiger* 1997: 249–284.

Cohen, A. 1997. *The Alexander Mosaic.* Cambridge.

Cohen, Edward. 2000. *The Athenian Nation.* Princeton.

Collins, A.W. 2001. "The office of Chiliarch under the Successors." *Phoenix* 55: 259–283.

Collins, A.W. 2012. "Alexander and the Persian Court Chiliarchy." *Historia* 61: 159–167

Collins, B.J. (ed.). 2002. *A History of the Animal World in the Ancient Near East.* Leiden.

Connor, W. Robert. 1988. "Early Greek land Warfare as Symbolic Expression." *P&P* 119: 3–29; reprinted in Wheeler 2007b, 83–109.

Cool-Root, Margaret. 2002. "Animals in the Art of Ancient Iran." In Collins, 169–210.

Cooper, Frederick A. 2000. "The Fortifications of Epaminondas and the Rise of the Monumental Greek City." In J.D. Tracy (ed.), *City Walls: The Urban Enceinte in Global Perspective*, Cambridge, 155–191.

Cornell, T. (ed.). 2013. *The Fragments of the Roman Historians*. Oxford.

Corsi, Lucia and Paola Angeli Bernardini. 2008. "A Survey of Theban and Athenian Relations Between 403–371 BC." In Buckler and Beck, 33–43.

Coskun, Altay. 2012. "Deconstructing a Myth of Seleucid History: The So-Called 'Elephant Victory' Revisited." *Phoenix* 66: 57–73.

Coulson, W.D.E. 1985. "The Dark Age Pottery of Sparta." *ABSA* 80: 29–84.

Coulston, J.C. 1985. "Roman Archery Equipment." In M.C. Bishop (ed.), *The Production and Distribution of Roman Military Equipment*, Oxford, 220–366.

Coulston, J.C. 1986. "Roman, Parthian and Sassanid Tactical Development." In P. Freeman and D. Kennedy (eds.), *The Defence of the Roman and Byzantine East (BARIS 297/1)*, Oxford, 59–75.

Coulton, James J. 2002. See Sapouna-Sakellarikis et al.

Couvenhes, J.-C. 2005. "De disciplina graecorum: les relations de violence entre les chefs militaires grecs et leurs soldats." In Bertrand, 431–454.

Couvenhes, J.-C. 2006. "La place de l' armée dans l' économie hellénistique." In J. Andreau et al., (eds.), *Approches de l' économie hellénistique* (Saint-Bertrand-de-Comminges, Entretiens 7), Paris, 397–436.

Couvenhes, J.-C. 2014. "Les kryptoi spartiates." *DHA* 11: 45–76.

Crawford, Michael H. 1996. *Roman Statutes I*. London.

Criscuolo, Lucia. 2013. "Ptolemies and Piracy." In Buraselis et al. 2013, 160–171.

Crouch, Dora P. 1993. *Water Management in Ancient Greek Cities*. Oxford.

Cruickshank, William W. 1954. *Topography, Movement and Supply in the Warfare of Ancient Greece South of Thessaly and Epirus*. Ph.D. thesis, University of London.

Cuillandre, Joseph. 1944. *La droite et la gauche dans les poèmes homeriques en concordance avec la doctrine pythagorienne et avec la tradition celtique*. Paris.

Curtis, J. and N. Tallis. 2012. *The Horse: From Arabs to Ascot*. London.

Curtius, Ernst. 1851–1852. *Peloponnesos. Eine historisch-geographische Beschreibung der Halbinsel*. 2 vols. Gotha.

Damyanov, Margarit, Maria Reho and Krastina Panayotova. in print a. "Cold Case Reopened: A Late Classical Tomb in Apollonia Pontica." In G. Tsetskhladze (ed.), *Proceedings of Sixth International Congress on Black Sea Antiquities "The Greeks and Romans in the Black Sea and the Importance of the Pontic Region for the Graeco-Roman World (7th c. BC–5th c. AD): 20 Years On (1997–2017)," Constanța, September 18–22, 2017.*

Dana, Dan, Maria Kamisheva and Nicola Theodossiev. 2016. "Une signature d'artisan sur le fourreau d'une epee trouve pres de Beroia de Thrace." *ZPE* 199: 60–64.

Davies, J.K. 1967. "Demosthenes on Liturgies." *JHS* 87: 33–40.

Dawson, Doyne. 1996. *The Origins of Western Warfare. Militarism and Morality in the Ancient World*. Boulder.

De Angelis, F. 2016. *Archaic and Classical Greek Sicily: A Social and Economic History*. New York.

De Caro, S. 2019. *I mosaici e la casa del Fauno*. Milan.

De Luna, M.E. 2009. "Camarina sub-colonia di Siracusa: dalla fondazione al conflitto." In M. Lombardo and F. Frisone (eds.), *Colonie di colonie: le fondazioni sub-coloniali greche tra colonizzazione e colonialismo*, Lecce, 75–86.

de Rochas d'Aiglun, Albert. 1881. *Principes de la fortification antique. Précis des connaissances techniques nécessaires aux archéologues pour explorer les ruines des anciennes forteresses*. Paris.

de Souza, Philip. 1999/2002. *Piracy in the Graeco-Roman World*. Cambridge.

de Souza, Philip. 2008. "Greek Warfare and Fortification." In Oleson, 673–690.
de Souza, Philip. 2014. "Piracy in Classical Antiquity: The Origins and Evolution of the Concept." In Amirell and Müller 2014a, 24–50.
De Staebler, Peter D. 2016a. "The Building Experience." In Müth et al., 2016, 61–74.
De Staebler, Peter D. 2016b. "Physical Surroundings, Technique and Building Experience: The State of Research in the Network Fokus Fortifikation." In Frederiksen et al., 123–128.
De Ste. Croix, G.E.M. 1972. *The Origins of the Peloponnesian War*. London.
De Vido, S. 2009. "Selinunte. Gli ultimi anni." In C. Antonetti and S. De Vido (ed.), *Temi selinuntini*, Pisa, 111–128.
De Vido, S. 2013. *Le guerre di Sicilia*. Rome.
De Vido, S. 2016. "Guerra e battaglia nella storia di Sicilia: uno sguardo d'insieme." In M. Congiu, M. Micciché and S. Modeo (eds.), *Le grandi battaglie della storia antica di Sicilia. Atti del XII Convegno di Studi sulla Sicilia antica*, Roma, 7–17.
Deacy, Susan and Fiona McHardy. 2015. "Ajax, Cassandra and Athena: Retaliatory Warfare and Gender Violence at the Sack of Troy." In Geoff Lee, Helene Whittaker and Graham Wrightson (eds.), *Ancient Warfare. Introducing Current Research*, vol. I, Cambridge, 252–272.
Debord, Pierre and Raymond Descat (eds.). 1994. *Fortifications et défense du territoire en Asie Mineure occidentale et méridionale, Table Ronde CNRS, Istanbul 20–27 mai 1993 = REA* 96 (published 1995).
Dedyulkin, Anton. 2015. ""Пращевые" камни в погребениях скифской эпохи." В: (ред. С. И. Лукьяшко) *Война и военное дело в скифо-сарматском мире. Мат-лы междунар. науч. конф. памяти А. И. Мелю-ковой (с. Кагальник, 26–29 апреля 2014 г.)*. Ростов-на-Дону, 89–97.
Deger-Jalkotzy, S. 1978. *E-QE-TA. Zur Rolle des Gefolgschaftswesens in der Sozialstruktur mykenischer Reiche*. Vienna.
Delatte, Armand. 1933. *Le troisième livre des souvenir socratiques de Xénophon: Étude critique*. Liège and Paris.
Delbrück, Hans. 1887. *Die Perserkriege und die Burgunderkriege*. Berlin.
Delbrück, Hans. 1900. *Geschichte der Kriegskunst im Rahmen der Politischen Geschichte, 4 vols*. Berlin (originally published 1900–1908; with new editions to 1920; often reprinted).
Delev, Peter. 2014. "Елинистически тюрзис на Зайчи връх над античния град Кабиле." *Studia Archaeologica Universitatis Serdicensis, Tome* 4 (2008): 189–200.
Delev, Peter. 2015. "Thrace from the Assassination of Kotys I to Koroupedion (360–281 BCE)." In Valeva et al., 48–58.
Demont, Paul. 2004. "La peur et le rire: le perception de l'esclavage dans les Grenouilles d'Aristophane." In Anastasia Serghidou (ed.), *Fear of Slaves – Fear of Enslavement in the Ancient Mediterranean. Peur de l'esclave – Peur de l'esclavage en Mediterranee ancienne (Discours, représentations, pratiques)*, Rethymno, 179–192.
Desborough, V.R.d'A. 1975. "The End of the Mycenaean World. The Protogeometric Period." In *CAH*[3] *2.2*, 658–675.
Descat, Raymond. 1993. "La loi de Solon sur l'interdiction d'exporter les produits attiques." In Alain Bresson and Pierre Rouillard (eds.), *L'emporion*, Paris, 145–161.
Detienne, Marcel. 1968. "La phalange. Problèmes et controversies." In Vernant, 119–142.
Devine, A.M. 1986. "Demythologizing the Battle of the Granicus." *Phoenix* 40: 265–278.
Devlin, F. Roger. 2003–2004. "The Case of Victor Davis Hanson: Farmer, Scholar, Warmonger." *The Occidental Quarterly* 3: 37–54.
Di Cesare, R. 2015. *La città di Cecrope. Ricerche sulla politica edilizia cimoniana ad Atene*. Athens.
Dickinson, Oliver. 1994. *The Aegean Bronze Age*. Cambridge.
Dickinson, Oliver. 2006. *The Aegean from Bronze Age to Iron Age. Continuity and Change between the Twelfth and Eighth Centuries BC*. London and New York.
Diels, Hermann A. and Walther Kranz. 1951. *Die Fragmente Der Vorsokratiker*, 6th ed. Berlin.
Diels, Hermann A. and Erwin Schramm. 1919 [1920]. *Exzerpte aus Philons Mechanik B.VII und VIII (vulgo fünftes Buch), Griechisch und Deutsch*. Berlin.

Dillery, J. 1996. "Reconfiguring the Past. Thyrea, Thermopylai and Narrative Patterns in Herodotus." *AJPh* 117: 217–254.

Dillery, J. 2017. "Xenophon: The Small Works." In M.A. Flower (ed.), *The Cambridge Companion to Xenophon*, Cambridge, 195–219.

Dillinger, Johannes. 2015. "Tyrannicide from Ancient Greece and Rome to the Crisis of the Seventeenth Century." In Law 2015a, 15–27.

Dimitrov, Stanimir. 2002. "Два шлема от древна Тракия." *Известия на Регионален исторически музей—Велико Търново* 17–18: 179–187.

Dimitrov, Stanimir. 2005a. "Шлемове "тракийски" тип. Произход и разпространение." *Известия на Регионален исторически музей—Велико Търново* 20: 56–79.

Dimitrov, Stanimir. 2005b. "За тракийската ромфея." В: *МИФ 9. Тракия и околният свят. Научна конференция Шумен, 2004.* София, 285–294.

Dimitrov, Stanimir. 2006. "Беотийски шлем от Русенския музей." *Известия на Регионален исторически музей—Русе* 10: 243–250.

Dimitrov, Stanimir. 2007. "Картиране на антични шлемове от България." *Известия на Регионален исторически музей—Велико Търново* 22: 52–63.

Dimitrov, Stanimir. 2010. "Бронзови брони от Тракия." В*: Великотърновският университет "Св. Св. Кирил и Методий" и българската археология -1.* Велико Търново, 217–224.

Dimitrova, Diana. 2015. *The Tomb of King Seuthes III in Golyama Kosmatka Tumulus.* Sofia.

Dmitriev, S. 2005. *City Government in Hellenistic and Roman Asia Minor.* Oxford.

Dobson, Mike. 2008. *The Army of the Roman Republic. The Second Century B.C. Polybius and the Camps at Numantia.* Oxford.

Domaradzki, Mieczysław. 1991. "Западна крепостна стена на Кабиле." В: *Кабиле*, т. II. София, 54–82.

Domaradzki, Mieczysław. 1995. *Трако-гръцки търговски отношения (Емпорион Пистирос 1).* Септември.

Domaradzki, Mieczysław. 1998. "Тракийската култура в прехода към елинистическата епоха." В: М. Домарадски, В. Танева (ред.) *Тракийската култура в прехода към елинистическата епоха. (Емпорион Пистирос 2).* Септември, 11–76.

Donaghy, Thomas. 2014. *Horse Breeds and Breeding in the Greco-Persian World: 1st and 2nd Millennium BC.* Newcastle-upon-Tyne.

Donlan, Walter. 1970. "Archilochus, Strabo and the Lelantine War." *TAPA* 101: 131–142.

Donlan, Walter. 1997a. "The Homeric Economy." In Morris and Powell, 649–667.

Donlan, Walter. 1997b. "The Relations of Power in the Pre-State and Early State Polities." In Mitchell and Rhodes, 36–41.

Donlan, Walter. 1999. *The Aristocratic Ideal in Ancient Greece and Other Papers.* Wauconda, IL.

Dorion, Louis-André. 2006. "Xenophon's Socrates." In Sara Ahbel-Rappe and Rachana Kamtekar (eds.), *A Companion to Socrates*, Oxford and Malden, 93–109.

Dover, Kenneth J. (ed.). 1968. *Aristophanes Clouds.* Edited with Introduction and Commentary. Oxford.

Dover, Kenneth J. 1987. *Greek and the Greeks.* Oxford.

Dreher, Martin. 1995. *Hegemon und Symmachoi. Untersuchungen zum zweiten athenischen Seebund.* Berlin and New York.

Dreher, Martin. (ed.). 2003. *Das antike Asyl: Kultische Grundlagen, rechtliche Ausgestaltung und politische Funktion.* Cologne.

Drews, Robert. 1972. "The First Tyrants in Greece." *Historia* 21: 129–144.

Drews, Robert. 1993. *The End of the Bronze Age. Changes in Warfare and the Catastrophe of ca. 1200 BC.* Princeton.

Droysen, Hans. 1889. *Heerwesen und Kriegführung der Griechen.* Freiburg.

Droysen, J. 1833. *Geschichte Alexanders des Grossen.* Düsseldorf; reissued Berlin, 1966.

Droysen, J. 1877. "Alexanders des Grossen Armee." *Hermes* 12: 226–252.

Ducat, Jean. 1999. "La femme de Sparte et le guerre." *Pallas* 51: 159–171.

Ducrey, Pierre. 1983. "Les Cyclades à l'époque hellénistique: la piraterie, symptome d'un malaise économique et sociale." In *Les Cyclades. Matériaux pour une étude de géographie historique*, Paris, 143–148.

Ducrey, Pierre. 1985. *Guerre et guerriers dans la Grèce antique*. Fribourg.

Ducrey, Pierre. 1986. "La fortification des cités grecques: rôle, fonction et efficacité." In Leriche and Tréziny, 133–142.

Ducrey, Pierre. 1995. "La muraille est-elle un élément constitutif d'une cité?" In Mogens Herman Hansen (ed.), *Sources for the Ancient Greek City-State, Symposium August 24– 27,1994, Acts of the Copenhagen Polis Centre* 2, Copenhagen, 245–256.

Ducrey, Pierre. 1999^2. *Le traitement des prisonniers de guerre dans la Grèce antique*. Paris (first ed. 1968).

Ducrey, Pierre. 2015. "War in the Feminine in Ancient Greece." In Fabre-Serris and Keith, 181–199.

Duffy, X. 2018. *Commemorating Conflict. Greek Monuments of the Persian Wars*. Oxford.

Dunbar, Robin. 2004. *The Human Story*. London.

Dutton, E.P. (trans.). 1910. *Thucydides. The Peloponnesian War*. London and New York.

Eadie, J.W. 1967. "The Development of Roman Mailed Cavalry." *JRS* 57: 161–173.

Eagleman, David. 2011. *Incognito. The Secret Lives of the Brain*. New York.

Eagleman, David. 2015. *The Brain. The Story of You*. New York.

Easson, Joseph J. and Alex P. Schmid. 2011. "Appendix 2.1: 250-plus Academic, Governmental and Intergovernmental Definitions of Terrorism." In Schmid 2011a, 99–157.

Echeverría, Fernando. 2008. *Ciudadanos, Campesinos y Soldados. El Nacimiento de la Polis Griega y la Teoría de la 'Revolución Hoplita'*. Madrid.

Echeverría, Fernando. 2010. "Weapons, Technological Determinism and Ancient Warfare." In Fagan and Trundle, 21–56.

Echeverría, Fernando. 2011. "*Taktikè téchne*: The Neglected Element in Classical Hoplite Battles." *AncSoc* 41: 45–82.

Echeverría, Fernando. 2012. "Hoplite and Phalanx in Archaic and Classical Greece: A Reassessment." *CP* 10: 291–318.

Echeverría, Fernando. 2015. "Heroic Fiction, Combat Scenes, and the Scholarly Reconstruction of Archaic Greek Warfare." *BICS* 58: 33–60.

Eckstein, Arthur M. 2006. *Mediterranean Anarchy, Interstate War, and the Rise of Rome*. Berkeley.

Eckstein, Arthur M. 2008. *Rome Enters the Greek East. From Anarchy to Hierarchy in the Hellenistic Mediterranean, 230–170 BC*. Malden.

Edelstein, Emma J. and Ludwig Edelstein. 1945/1998. *Asclepius. Collection and Interpretation of the Testimonies*. Introduction by Gary B. Ferngren. Baltimore.

Edwards, Martha. 2000. "Philoctetes in Historical Context." In David A. Gerber (ed.), *Disabled Veterans in History*, Ann Arbor, 55–69.

Ehrenberg, Victor. 1932. *Der griechische und der hellenistische Staat*. Leipzig.

Ehrenberg, Victor. 1937. "When Did the Polis Rise?" *JHS* 57: 147–59.

Ehrenberg, Victor. 1974. *The People of Aristophanes*, 3rd ed. Leipzig.

Elkman, E. 1933. *Zu Xenophons Hipparchikos*. Inaugural dissertation, Uppsala.

Ellinger, P. 2010. "Artémis de tous les dangers à Épidamne-Dyrrhachion." *BCH* 134: 470–476.

Engels, David. 2017. *Benefactors, Kings, Rulers. Studies on the Seleukid Empire Between East and West*. Leuven.

Engels, Donald W. 1978. *Alexander the Great and the Logistics of the Macedonian Army*. Berkeley and Los Angeles.

Engels, Donald W. 1980. "Alexander's Intelligence System." *CQ* 30: 327–340.

Epplett, Chris. 2007. "War Elephants in the Hellenistic World." In Waldemar Heckel, Lawrence Tritle and Pat Wheatley (eds.), *Alexander's Empire: Formulation to Decay*, Claremont, 209–232.

Erciyas, Deniz B. 2006. *Wealth, Aristocracy and Royal Propaganda Under the Hellenistic Kingdom of the Mithridatids*. Leiden.

Erdkamp, Paul. 1998. *Hunger and the Sword: Warfare and Food Supply in Roman Republican Wars (264–30 B.C.)*. Amsterdam.

Errington, R. Malcolm. 1990. *A History of Macedonia*. Berkeley.

Erskine, Andrew (ed.). 2003. *A Companion to the Hellenistic World*. Malden.

Evans, J.A.S. 1976. "Herodotus and the Ionian Revolt." *Historia* 25: 31–37.

Evans, R.J. 2016. *Ancient Syracuse: From Foundation to Fourth Century Collapse*. London.

Fabre-Serris, Jacqueline and Alison Keith (eds.). 2015. *Women and War in Antiquity*. Baltimore.

Fachard, Sylvian. 2012. *La défense du territoire. Étude de la chôra érétrienne et de ses fortifications. Eretria 21*. Gollion.

Fachard, Sylvian. 2016a. "Studying Rural Fortifications: A Landscape Approach." In Müth et al., 207–230.

Fachard, Sylvian. 2016b. "Regional Fortifications and the Fortification of Regions." In Frederiksen et al., 413–416.

Fagan, G.G. and M. Trundle (eds.). 2010. *New Perspectives on Ancient Warfare*. Leiden and Boston.

Fallows, N. 2010. *Jousting in Medieval and Renaissance Iberia*. Woodbridge.

Felten, F. 1984. *Griechische Tektonische Friese archaischer und klassischer Zeit*. Waldsassen.

Ferraz, Vicente. 1801. *Tratado de castrametación*. Alcalá de Henares.

Feyel, Ch. 1999. "Aperçu sur le financement de la guerre dans la cité classique." In F. Prost (ed.), *Armées et sociétés de la Grèce classique (V^e^ et IVe s. av. J.-C.)*, Paris, 209–221.

Field, D. 1984. *Mycenaean Fortifications on the Mainland of Greece*. Ph.D. diss., University of Cambridge.

Figueira, T.J. 1977. *Aegina and Athens in the Archaic and Classical Periods*. Ph.D. diss., University of Pennsylvania.

Figueira, T.J. 2006. "The Spartan Hippeis." In S. Hodkinson and A. Powell (eds.), *Sparta and War*, Swansea, 57–85.

Filanovich, M. 1974. "Gjaur-kala." *Trudÿ Yuzhno-Turkmenskoï arkheologicheskoï kopleksnoï ékspeditsii* 15: 15–139.

Filimonos-Tsopotou, Melina. 2004. *Ρόδος, 1. Η ελληνιστική οχύρωση της Ρόδου. Δημοσιεύματα του "Αρχαιολογικού Δελτίου"* 86. Athens.

Fine, John V.A. 1983. *The Ancient Greeks. A Critical History*. Cambridge, MA.

Fine, Jonathan. 2010. "Political and Philological Origins of the Term 'Terrorism' from the Ancient Near East to Our Times." *Middle Eastern Studies* 46: 271–288.

Finley, Moses I. 1962. "Athenian Demagogues." *P&P* 21: 3–24. Reprinted in id., *Studies in Ancient Society*, London, 1974, 1.25; revised and reprinted in id. *Democracy Ancient and Modern*, London, 1985, 38–75 (with notes 177–179). Reprinted in P.J. Rhodes (ed.), *Athenian Democracy*, Oxford, 163–184.

Finley, Moses I. 1970. *Early Greece. The Bronze and Archaic Ages*. London.

Finley, Moses I. 1980. *Ancient Slavery and Modern Ideology*. New York.

Finley, Moses I. 1981. "Sparta and Spartan Society." In Moses I. Finley, *Economy and Society in Ancient Greece* (edited with an introduction by Brent D. Shaw and Richard P. Saller). New York 24–40.

Finnegan, R. 1977. *Oral Poetry*. Cambridge.

Fischer-Bovet, C. 2014a. *Army and Society in Ptolemaic Egypt*. Cambridge.

Fischer-Bovet, C. 2014b. "Est-il facile de conquérir l'Égypte? L'invasion d'Antiochos IV et ses conséquences." In C. Feyel and L. Graslin (eds.), *Le projet politique d'Antiochos IV*, Nancy, 209–259.

Fisher, Nick and Hans van Wees (eds.). 1998. *Archaic Greece: New Approaches and New Evidence*. London.

Flower, Michael A. 2000. "Alexander the Great and Panhellenism." In A. Brian Bosworth and Elizabeth Baynham (eds.), *Alexander the Great in Fact and Fiction*, Oxford, 96–135.

Flower, Michael A. 2008. *The Seer in Ancient Greece*. Berkeley.

Flower, Michael A. (ed.). 2017. *The Cambridge Companion to Xenophon*. Cambridge.

Flower, Michael A. and J. Marincola (eds.). 2002. *Herodotus, Histories, Book IX*. Cambridge.

Fol, Alexander. 1969. *Тракийско военно изкуство*. София.

Fontenrose, J. 1978. *The Delphic Oracle, Its Responses and Operations with a Catalogue of Responses*. Berkeley.

Forbes, C. 1930. "Peripoloi at Sicyon." *CP* 25: 75–77.
Fornara, Charles W. (trans. and ed). 1977. *Translated Documents of Greece and Rome. 1. Archaic Times to the End of the Peloponnesian War.* Cambridge.
Forrer, Leonard. 1969. *Portraits of Royal Ladies on Greek Coins.* Chicago.
Forrest, W.G. 1956. "The First Sacred War." *BCH* 80: 33–52.
Forster, E.S. 1941. "Dogs in Ancient Warfare." *G&R* 10: 114–117.
Foulon, E. 1995. "Μισθοφόροι et Ξένοι hellénistiques." *REG* 108: 211–218.
Foxhall, Lin. 1993. "Farming and Fighting in Ancient Greece." In Rich and Shipley, 134–145.
Foxhall, Lin. 1997. "A View from the Top. Evaluating the Solonian Property Classes." In Mitchell and Rhodes, 113–136.
Foxhall, Lin. 2007. *Olive Cultivation in Ancient Greece: Seeking the Ancient Economy.* Oxford.
Foxhall, Lin. 2013. "Can We See the 'hoplite revolution' on the Ground? Archaeological Landscapes, Material Culture, and Social Status in Early Greece." In Kagan and Viggiano, 194–221.
Frank, L. 2019. "La bataille d'Anghiari." In V. Delieuvin and L. Frank (eds.), *Léonard de Vinci*, Paris, 234–257.
Frasca, M. 2009. *Leontinoi: archeologia di una colonia greca.* Rome.
Fraser, A.D. 1942. "The Myth of the Phalanx Scrimmage." *Classical Weekly* 36: 15–16.
Frederiksen, Rune. 2011. *Greek City Walls of the Archaic Period, 900–480 BC.* Oxford.
Frederiksen, Rune, Eric Laufer. and Silke Müth. 2016. "Source Criticism: Fortifications in Written Sources and the Visual Arts." In Müth et al., 2016b, 173–195.
Frederiksen, Rune, Silke Müth, Peter Schneider, and Mike Schnelle (eds.). 2016. *Focus on Fortifications: New Research on Fortifications in the Ancient Mediterranean and the Near East. Fokus Fortifikation Studies*, 2. Oxford.
Fredricksmeyer, E. 1979. "Divine Honors for Philip II." *TAPA* 109: 39–61.
Frost, Frank J. 1984. "The Athenian Military before Cleisthenes." *Historia* 33: 283–294.
Fuhrer, Therese. 2015. "Teichoskopia: Female Figure Looking on Battles." In Fabre-Serris and Keith, 52–81.
Fuller, J.F.C. 1960/1998. *The Generalship of Alexander the Great.* Ware, Hertfordshire.
Funari, P.P.A and A. Pollini. 2005. "Greek Perceptions of Frontier in Magna Graecia: Literature and Archaeology in Dialogue." *Studia Historica. Historia Antigua* 23: 331–344.
Funke, Peter. 2006. "Western Greece (Magna Graecia)." In Kinzl, 153–173.
Fustel de Coulanges, Numa D. 1864. *La Cité Antique.* Paris.
Gabbert, Jan. 1986. "Piracy in the Early Hellenistic Period: A Career Open to Talents." *G&R* 33: 156–163.
Gabriel, Richard A. 2010. *Philip of Macedonia. Greater than Alexander.* Washington, DC.
Gabrielli, Marcel. 2006. *Le Cheval das L'Empire Achéménide.* Istanbul.
Gabrielsen, Vincent. 2001. "Economic Activity, Maritime Trade and Piracy in the Hellenistic Aegean." *REA* 103: 219–240.
Gabrielsen, Vincent. 2003. "Piracy and the Slave-Trade." In Erskine, 389–404.
Gabrielsen, Vincent. 2008. "Die Kosten der athenischen Flotte in klassischer Zeit." In Burrer and Müller (eds.), 46–73.
Gabrielsen, Vincent. 2013. "Rhodes and the Ptolemaic Kingdom: The Commercial Infrastructure." In Buraselis et al. 2013, 66–81.
Gaca, Kathy L. 2010. "The Andrapodizing of War Captives in Greek Historical Memory." *TAPA* 140: 117–161.
Gaca, Kathy L. 2015. "Ancient Warfare and the Ravaging Martial Rape of Girls and Women: Evidence from Homeric Epic and Greek Drama." In M. Masterson, Nancy S. Rabinowitz, and James Robson (eds.), *Sex in Antiquity: Exploring Gender and Sexuality in the Ancient World*, New York, 278–297.
Gaca, Kathy L. 2016. "Continuities in Rape and Tyranny in Martial Societies from Antiquity Onward." In Stephanie L. Budin and Jean MacIntosh Turfa (eds.), *Women in Antiquity: Real Women Across the Ancient World. Rewriting Antiquity*, London and New York, 1041–1056.

Gaebel, Robert. 2002. *Cavalry Operations in the Ancient Greek World*. Norman.
Gaebler, H. 1935. *Die antiken Münzen von Makedonia und Paionia 2*. Berlin.
Gagoshidze, I. 2008. "Weapons." In A. Furtwängler et al. (eds.), *Iberia and Rome*, Langenweißbach, 123–140.
Galla, Jan D. 2015. "Review, Peter Meineck and David Konstan (eds.). 2014. *Combat Trauma and the Ancient Greeks*." *Michigan War Studies Review*, 2015–097.
Gallant, Thomas W. 1991. *Risk and Survival in Ancient Greece: Reconstruction the Rural Domestic Economy*. Stanford.
Gardner, C.A.M and R.M. Seifried. 2016. "Euboean Towers and Aegean Powers: Insights into the Karystia's Role in the Ancient World." *Journal of Greek Archaeology* 1: 149–176.
Garlan, Yvon. 1972. *La Guerre dans l'Antiquité*. Paris. (= *War in the Ancient World: A Social History*, trans. Janet Lloyd, 1975. London).
Garlan, Yvon. 1973. "Cités, armées et stratégie à l'époque hellénistique d'après l'oeuvre de Philon de Byzance." *Historia* 22: 16–33.
Garlan, Yvon. 1974. *Recherches de poliorcétique grecque*. Athens.
Garlan, Yvon. 1988. *Slavery in Ancient Greece*, trans. Janet Lloyd. Ithaca, NY.
Garlan, Yvon. 1989. *Guerre et Économie en Grèce Ancienne*. Paris.
Garlan, Yvon. 1992. "L'enceinte hellénistique d'Apollonia." *Dossiers d'Archéologie* 167: 64–69.
Gatto, Maurizio. 2010. *Il peri mechanematon di Ateneo Meccanico*. Rome.
Gauss, Walter. 2018. "Fortification Systems of the Aegean Bronze Age. The Early and Middle Bronze Age." In Ballmer et al., 43–61.
George, Andrew (trans. and ed.). 1999. *The Epic of Gilgamesh*. London.
George, D. Theodore, M.D., with L. Berger. 2013. *Untangling the Mind. Why We Behave the Way We Do*. New York.
Georgiadou, Aristoula. 1997. *Plutarch's Pelopidas: A Historical and Philological Commentary*. Stuttgart and Leipzig.
Georgieva, Rumjana. 2016. "Riders' Burials in Thrace." In O. Henry and U. Kelp (eds.), *Tumulus as Sema: Space, Politics, Culture and Religion in the First Millennium BC*. Berlin, 269–280.
Gera, Deborah L. 1993. *Xenophon's Cyropaedia: Style, Genre and Literary Technique*. Oxford.
Gerolymatos, A. 1986. *Espionage and Treason: A Study of the Proxenia in Political and Military Intelligence Gathering in Ancient Greece*. Amsterdam.
Giese, Jürgen. 2010. "Bautechnische Beobachtungen am nördlichen und nordwestlichen Mauerabschnitt in Messene." In Lorentzen et al., 85–95.
Gillis, Daniel. 1979. *Collaboration with the Persians*. Wiesbaden.
Glasser, Ronald J., M.D. 2011. *Broken Bodies, Shattered Minds. A Medical Odyssey from Vietnam to Afghanistan*. Palisades, NY.
Glotz, Gustave. 1928. *La Cité Grecque*. Paris.
Glover, Richard. 1944. "The Elephant in Ancient War." *CJ* 39: 257–269.
Glover, Richard. 1948. "The Tactical Handling of the Elephant." *G&R* 17: 1–11.
Goette, Hans Rupprecht. 2000. *Ο αξιόλογος δήμος Σούνιον. Landeskundliche Studien in Südost-Attika*. Rahden/Westfalen.
Goette, Hans Rupprecht. 2009. "Images in the Athenian 'Demosion Sema'." In O. Palagia (ed.), *Art in Athens During the Peloponnesian War*. Cambridge, 188–206.
Golahny, Amy. 2011. "Elisabetta Sirani's 'Timoclea' and Visual Precedent." *Source: Notes in the History of Art* 30: 37–42.
Golan, David. 1988. "The Fate of a Court Historian: Callisthenes." *Athenaeum* 66: 99–120.
Goldsworthy, Adrian K. 1997. "The *Othismos*, Myths and Heresies: The Nature of Hoplite Battle." *War in History* 4: 1–26.
Goldsworthy, Adrian K. 2003. *The Complete Roman Army*. London.
Gomme, Arnold W. 1933. *The Population of Athens in the Fifth and Fourth Centuries B.C.* Oxford.
Gomme, Arnold W., Anthony Andrewes and Kenneth J. Dover. 1945–1981. *A Historical Commentary on Thucydides*, 5 vols. Oxford.

González, Francisco J. and Pedro López-Barja. 2012. "Neocon Greece: V.D. Hanson's War on History." *IJCT* 19.3: 129–151.

Goodall, D.M. 1977. *A History of Horse Breeding*. London.

Goody, Jack 2008: *The Theft of History*. Cambridge.

Gordon, E.I. 1958. "Sumerian Animal Proverbs and Fables: 'Collection Five.'" *Journal of Cuneiform Studies* 12: 1–21.

Graf, David. 1979. *Medism: Greek Collaboration with Achaemenid Persia*. Ph.D. diss., University of Michigan.

Graf, Fritz. 1984a. "Medism: The Origin and Significance of the Term." *JHS* 104: 15–30.

Graf, Fritz. 1984b. "Women, War, and Warlike Divinities." *ZPE* 55: 245–254.

Graham, A.J. 1982. "The Western Greeks." In *CAH* III.3: 163–195.

Graham, A.J. 1992. "Thucydides 7.13.2 and the Crews of Athenian Triremes." *TAPA* 122: 257–270.

Graham, A.J. 1998. "Thucydides 7.13.2 and the Crews of Athenian Triremes: An Addendum." *TAPA* 128: 89–114.

Grainger, John D. 2010. *The Syrian Wars*. Leiden.

Grandjean, Yves. 2011. *Le rempart de Thasos*. Études Thasiennes 22. Athens.

Gray, Vivienne J. 2011. *Xenophon's Mirror of Princes: Reading the Reflections*. Oxford.

Green, Peter. 2006. *Diodorus Siculus, Books 11–12.37.1*. Austin.

Greenhalgh, Peter A.L. 1973. *Early Greek Warfare. Horsemen and Chariots in the Homeric and Archaic Ages*. Cambridge.

Griffith, G.T. 1935. *The Mercenaries of the Hellenistic World*. Cambridge.

Griffith, G.T. 1980. "Philip as a General and the Macedonian Army." In Miltiades B. Hatzopoulos and Louisa D. Loukopoulos (eds.). *Philip of Macedon*, Athens, 58–77.

Griffith, Mark. 2006. "Horsepower and Donkeywork: Equids and the Ancient Greek Imagination." *CP* 101: 185–246.

Grote, George. 1846–1856. *A History of Greece*, 12 vols. London. often reprinted (also cited as 1900 = vol. 8).

Grotius, Hugo. 1625. *De iure belli ac pacis libri tres (Three Books on the Law of War and Peace)*. Paris. Trans. A.C. Campbell, *The Rights of War and Peace*, Washington and London (1901).

Grundy, George B. 1911. *Thucydides and the History of His Age*. London.

Guettel, Susan 1988. "The Use of Water in Greek Sanctuaries." In Robin Hägg (ed.), *Early Greek Cult Practice*, Stockholm, 161–165.

Guía, M. and J. Gallego. 2010. "Athenian 'Zeugitai' and the Solonian Census Classes: New Reflections and Perspectives." *Historia* 59: 257–281.

Guimier-Sorbets, A.-M. 2016. "Ajouter de la couleur aux mosaïques à l'époque hellénistique: Employ de peinture et de mortiers colorés." In L. Neira Jimenez (ed.), *Estudios sobre antiguos y medievales*, Rome, 222–228.

Günther, Linda-Marie. 1983. *Karthago und Sizilien*. Hildesheim.

Haas, Kristin Ann. 1998. *Carried to the Wall. American Memory and the Vietnam Veterans Memorial*. Berkeley.

Habicht, Christian. 1985. *Pausanias' Guide to Ancient Greece*. Berkeley, Los Angeles, London.

Habicht, Christian. 1997. *Athens from Alexander to Antony*. Cambridge.

Hagerman, Chris. 2014. "Weapons: Catapult Bolts, Arrowheads, Javelin and Spear Heads, and Sling Bullets." In G.P. Schaus (ed.), *Stymphalos: The Acropolis Sanctuary*, vol. 1. Toronto, Buffalo, and London, 79–102.

Hale, John R. 2009. *Lords of the Sea: The Epic Story of the Athenian Navy and the Birth of Democracy*. New York.

Hale, John R. 2013. "Not Patriots, Not Farmers, Not Amateurs: Greek Soldiers of Fortune and the Origins of Hoplite Warfare." In Kagan and Viggiano, 176–193.

Halstead, Paul. 2007. "Toward a Model of Mycenaean Palatial Mobilization." In M.L. Galaty and W.A. Parkinson, (eds.), *Rethinking Mycenaean Palaces*, revised and expanded 2nd ed., Los Angeles, 66–73.
Halstead, Paul and Glynis Jones. 1989. "Agrarian Ecology in the Greek Islands: Time Stress, Scale and Risk." *JHS* 109: 41–55.
Hamel, Debra. 1995. "*Strategoi* on the *Bema*: The Separation of Political and Military Authority in Fourth-Century Athens." *AHB* 9: 25–39.
Hamel, Debra. 1998. *Athenian Generals: Military Authority in the Classical Period.* Leiden.
Hamel, Debra. 2015. *The Battle of Arginusae.* Baltimore.
Hamilton, J.R. 1969. *Plutarch, Alexander. A Commentary.* Oxford.
Hammond, N G. . 1949. "Sacred War." In *Oxford Classical Dictionary*, 1st ed. Oxford, 806.
Hammond, N.G.L. 1975. "The End of the Mycenaean Civilization and the Dark Age, b) The Literary Tradition for the Migrations." *CAH*^{3} II.2:, 678–712.
Hammond, N.G.L. 1980a. "Training in the Use of a Sarissa and Its Effect in Battle, 359–333 B.C." *Antichthon* 14: 53–63.
Hammond, N.G.L. 1980b. *Alexander the Great: King, Commander and Statesman.* Park Ridge, NJ.
Hammond, N.G.L. 1983. "Army Transport in the Fifth and Fourth Centuries." *GRBS* 24: 27–31.
Hammond, N.G.L. 1987. "A Papyrus Commentary on Alexander's Balkan Campaign." *GRBS* 28: 331–347.
Hammond, N.G.L. 1993. *Sources for Alexander the Great. An Analysis of Plutarch's Lives and Arrian's* Anabasis Alexandrou. Cambridge.
Hammond, N.G.L. and G. Griffith 1979. *A History of Macedonia, vol. 2.* Oxford.
Hammond, N.G.L. and Frank W. Walbank. 1988. *A History of Macedonia. Vol. 3: 336–167 B.C.* Oxford.
Hansen, Mogens H. 1988. *Three Studies in Athenian Demography.* Copenhagen.
Hansen, Mogens Herman. 2006a. *Polis: An Introduction to the Ancient Greek City-State.* Oxford.
Hansen, Mogens Herman. 2006b. *The Shotgun Method: The Demography of the Ancient City-State Culture.* Columbia, MO.
Hansen, Mogens Herman and Th. H. Nielsen (eds.). 2004. *An Inventory of Archaic and Classical Poleis.* Oxford.
Hanson, Victor D. 1983/1998^2. *Warfare and Agriculture in Classical Greece.* Berkeley.
Hanson, Victor D. 1988. "Epaminondas, the Battle of Leuctra and the Revolution in Greek Battle Tactics." *ClAnt* 7: 190–207.
Hanson, Victor D. 1989/1998^2/2000^2/2009^2. *The Western Way of War: Infantry Battle in Classical Greece.* New York, Berkeley and Los Angeles (2nd ed. variously reprinted).
Hanson, Victor D. 1991a. "The Ideology of Hoplite Battle, Ancient and Modern." In Hanson, 3–11.
Hanson, Victor D. 1991b. "Hoplite Technology in Phalanx Battle." In Hanson, 63–84.
Hanson, Victor D. (ed.) 1991. *Hoplites. The Classical Greek Battle Experience.* London and New York.
Hanson, Victor D. 1992. "Thucydides and the Desertion of Attic Slaves during the Decelean War." *ClAnt* 11: 210–228.
Hanson, Victor D. 1995; 1999^2. *The Other Greeks. The Family Farm and the Agrarian Roots of Western Civilization.* Berkeley.
Hanson, Victor D. 1996. "Hoplites into Democrats: The Changing Ideology of Athenian Infantry." In Ober and Hedrick, 289–312.
Hanson, Victor D. 2000. "Hoplite Battle as Ancient Greek Warfare: When, Where, and Why?" In van Wees, 201–232.
Hanson, Victor D. 2005. *A War Like No Other: How the Athenians and Spartans Fought the Peloponnesian War.* New York.
Hanson, Victor D. 2007. "The Modern Historiography of Ancient Warfare." In Sabin et al., 3–21.
Hanson, Victor D. 2013. "The Hoplite Narrative." In Kagan and Viggiano, 256–275.

Harari, Yuval Noah. 2000. "Strategy and Supply in Fourteenth-Century Western European Invasion Campaigns." *Journal of Military History* 64: 297–333.

Harl, Kenneth W. 2008. "Legion over Phalanx: A Reassessment of the Battle of Magnesia, 190 B.C." In Howe and Reames, 257–282.

Harris, Edward M. 1995. *Aeschines and Athenian Politics.* Oxford and New York.

Harris, Edward M. 2015. "Military Organization and One-Man Rule in the Greek *Polis.*" *Ktema* 40: 83–93.

Harris, Edward M. and David M. Lewis. 2016. "Introduction: Markets in Classical and Hellenistic Greece." In Edward M. Harris, David M. Lewis, and Mark Woolmer (eds.). *The Ancient Greek Economy: Markets, Households and City-States.* Cambridge, 1–38.

Harrison, E. B. 1972. "The South Frieze of the Nike Temple and the Marathon Painting in the Painted Stoa." *AJA* 76: 353–378.

Harrison, Robert Pogue. 2003. *The Dominion of the Dead.* Chicago.

Harrison, Thomas. 2006. "The Greek World, 478–432." In Kinzl, 509–525.

Harthen, David. 2001. *The Logistics of Ancient Greek Land Warfare.* Ph.D. thesis, University of Liverpool.

Harvey, P. 1986. "New Harvests Reappear: The Impact of War on Agriculture." *Athenaeum* 74: 205–218.

Hatzopoulos, Miltiades B. 1994. *Cultes et rites de passage en Macédoine.* Athens.

Hatzopoulos, Miltiades B. 2001. *L'organisation de l'armée macédonienne sous les Antigonides. Problèmes anciens et documents nouveaux. Meletemata 30.* Athens.

Hauser, S. 2006. "Was There No Paid Standing Army?: A Fresh Look on Military and Political Institutions in the Arsacid Empire." In Mode et al., 295–320.

Haydon, Benjamin Robert. (1877/2010). *Correspondence and Table-Talk*, vol. II. Boston.

Haztopoulos, Miltiades B. and P. Juhel. (2009). "Four Hellenistic Funeral Stele from Gepyra, Macedonia." *AJA* 113: 423–437.

Heckel, Waldemar. 1992. *The Marshals of Alexander's Empire.* London and New York.

Heckel, Waldemar. 2006. *Who's Who in the Age of Alexander the Great.* Oxford.

Heckel, Waldemar. 2008. *The Conquests of Alexander the Great.* Cambridge.

Heckel, Waldemar. 2013. "The Three Thousand: Alexander's Infantry Guard." In Campbell and Tritle, 162–178.

Heckel, Waldemar. 2016a. *Alexander's Marshals. A Study of the Makedonian Aristocracy and the Politics of Military Leadership.* London and New York.

Heckel, Waldemar. 2016b. "Mazaeus, Callisthenes and the Alexander Sarcophagus." *Historia* 55: 385–396.

Heckel, Waldemar. 2017a. "Dareios III's Military Reforms before Gaugamela and the Alexander Mosaic: A Note." *AHB* 31: 65–69.

Heckel, Waldemar. 2017b. "Geography and Politics of Argead Macedonia." In Müller et al., 67–78.

Heckel, Waldemar. 2020. *In the Path of Conquest: Resistance to Alexander the Great.* Oxford.

Heckel, Waldemar and Ryan Jones. 2006. *Macedonian Warrior. Alexander's elite infantryman.* Oxford and New York.

Heckel, Waldemar and Jolene L. McLeod. 2015. "Alexander the Great and the Fate of the Enemy: Quantifying, Qualifying, and Categorizing Atrocities." In Heckel et al., 233–267.

Heckel, Waldemar, Sabine Müller, and Graham Wrightson (eds.). 2015. *The Many Faces of War in the Ancient World.* Cambridge.

Heckel, Waldemar and Lawrence A. Tritle (eds.). 2009. *Alexander the Great. A New History.* Chichester.

Heckel, Waldemar, C. Willekes and G. Wrightson. 2010. "Scythed Chariots at Gaugamela: A Case Study." In Carney and Ogden, 103–109.

Heinrichs, Johannes. 1989. *Ionien nach Salamis. Die kleinasiatischen Griechen in der Politik und politischen Reflexion des Mutterlands.* Bonn.

Heinrichs, Johannes. 2015. "Military Integration in Late Archaic Arkadia." In Heckel et al., 1–89.

Heinrichs, Johannes. 2017. "Coins and Constructions. Origins and Developments of Achaemenid Coinage under Alexander I." In Müller et al., 79–98.

Heinrichs, Johannes. 2018. "Antiochos III and Ptolemy, Son of Thraseas, on Private Villages in Syria Koile around 200 BC: The Hefzibah Dossier." *ZPE* 206: 272–311.
Heinrichs, Johannes and Sabine Müller. 2008. "Ein persisches Statussymbol auf Münzen Alexanders I. von Makedonien. Ikonographie und historischer Hintergrund des Tetrobols SNG ABC, Macedonia I, 7 und 11." *ZPE* 167: 283–309.
Helbig, Wolfgang. 1909. "Ein homerischer Rundschild mit eine Bugel." *JÖAI* 12: 1–70.
Helbig, Wolfgang. 1911. "Über die Einführungszeit der geschlossenen Phalanx." *Sitzungsberichte der Königlich Bayerischen Akademie der Wissenschaften* 12: 3–41.
Hellmann, Marie-Christine. 2010. *L'architecture grecque. Vol. 3, Habitat, urbanisme et fortifications.* Paris.
Hermary, Antoine. 2010. "Les vases et leur décor à l'époque classique: transfert de formes et d'images entre grecs et Thraces (Ve s. av. J.-C.)." In H. Treziny (ed.) *Grecs et indigènes de la Catalogne à la Mer Noire*, Paris, 481–486.
Herrmann, G. 1989. "Parthian and Sasanian Saddlery. New Light from the Roman West." In L. De Meyer and E. Haerink (eds.), *Archaeologia Iranica et Orientalis. Miscellanea in honorem Louis Vanden Berghe*, vol. II. Gent, 757–809.
Hignett, Charles. 1952. *A History of the Athenian Constitution.* Oxford.
Hobbes, Thomas. 1651. *Leviathan.* London.
Hobbes, Thomas (trans.). 1843 (first published 1629). *Thucydides. The English Works of Thomas Hobbes of Malmesbury.* London.
Hodkinson, S. 1983. "Social Order and the Conflict of Values in Classical Sparta." *Chiron* 13: 241–265; reprinted in M. Whitby (ed.), *Sparta*, London (2002), 104–130.
Hoffman, Bruce. 2006. *Inside Terrorism.* New York.
Hoffmann, Adolf and Claudia Bührig. 2000. "Die Stadtmauern der hellenistisch-römischen Dekapolisstadt Gadara. Zusammenfassender Bericht über die seit 1991 durchgeführten Ausgrabungen und Untersuchungen." *Archäologischer Anzeiger* 2000: 175–233.
Holladay, A. James. 1977. "Sparta's Role in the First Peloponnesian War." *JHS* 97: 54–63.
Holladay, A. James. 1982. "Hoplites and Heresies." *JHS* 102: 94–103.
Hollmann, Alexander. 2011. *The Master of Signs: Signs and Interpretation of Signs in Herodotus.* Cambridge, MA and London.
Hölscher, T. 1973. *Griechische Historienbilder des 5. und 4. Jahrhunderts v. Chr.* Würzburg.
Hölscher, T. 1998. *Öffentliche Räume in frühen griechischen Städten.* Heidelberg.
Holt, Frank. 2003. *Alexander the Great and the Mystery of the Elephant Medallions.* Berkeley.
Hoover, O. D. 2011. *Handbook of Coins of the Peloponnesos.* Lancester, London.
Hope Simpson, R. 1981. *Mycenaean Greece.* Park Ridge, NJ.
Hope Simpson, R. 2002. "The Mycenae Roads and Mycenaean Chariots." *Mouseion* 2: 125–133.
Horden, P. and N. Purcell. 2000. *The Corrupting Sea. A Study of Mediterranean History.* Malden.
Hornblower, Simon. 1991–2008. *A Commentary on Thucydides.* 3 vols. Oxford.
Hornblower, Simon. 1991^2; 2011^4. *The Greek World 379–323 BC.* London and New York.
Hornblower, Simon. 2007. "Warfare in Ancient Literature: The Paradox of War". In Sabin et al., 22–53.
Hornblower, Simon and Christopher Pelling. 2017. *Herodotus Histories Book VI.* Cambridge.
How, W. and J. Wells. $1912/1928^2$. *A Commentary on Herodotus.* Oxford.
Howard, Michael. 1976. *War in European History.* Oxford.
Howe, Timothy. 2008. *Pastoral Production: Animals, Agriculture and Society in Ancient Greece.* Claremont.
Howe, Timothy. 2016. "Plutarch, Arrian and the Hydaspes: An Historiographical Approach." In Bearzot and Landucci, 25–39.
Howe, Timothy and Lee L. Brice (eds.). 2016. *Brill's Companion to Insurgency and Terrorism in the Ancient Mediterranean.* Leiden and Boston.
Howe, Timothy, E. E. Garvin and G. Wrightson (eds.). 2015. *Greece, Macedon and Persia: Studies in Social, Political and Military History in Honour of Waldemar Heckel.* Philadelphia.

Howe, Timothy, S. Müller and R. Stoneman (eds.). 2017. *Ancient Historiography on War and Empire.* Oxford and Philadelphia.

Howe, Timothy and Jeanne Reames (eds.). 2008. *Macedonian Legacies: Studies in Ancient Macedonian History and Culture in Honor of Eugene N. Borza.* Claremont, CA.

Hoyos, B. D. 2015. *Mastering the West: Rome and Carthage at War.* Oxford.

Hristov, Ivan. 2011. "Metal Finds." In *Kozi Gramadi. Studies of an Odryssian Ruler's Residence and Sanctuaries in Sredna Gora Mt. 8th–1st centuries BC*, vol. I. Sofia, 116–118.

Hristov, Ivan. 2012. "Military Operations before the Walls of the Residence in 341 BC in Accordance with the Evidence of the Lead Sling-balls from the Site." In *Kozi Gramadi. Studies of an Odryssian Ruler's Residence and Sanctuaries in Sredna Gora Mt. 8th–1st centuries BC*, vol. II, Sofia, 79–89.

Hristov, Ivan. 2014. *A Sanctuary of Zeus and Hera on Kozi Gramadi Peak in Sredna Gora.* Sofia.

Hristov, Ivan and Metodi Manov. 2011. "Sling Bullets from the Area of a Thracian Rulers' Residence near the Peak of "Kozi Gramadi." *Archaeologia Bulgarica* 1: 21–33.

Hughes, J. Donald. 2013. "Warfare and Environment in the Ancient World." In Campbell and Tritle, 128–139.

Hülden, Oliver. 2000. "Pleistarchos und die Befestigungsanlagen von Herakleia am Latmos." *Klio* 82: 382–408.

Hülden, Oliver. 2018. "Greek Fortifications before the Persian Wars. An Overview." In Ballmer et al., 93–109.

Hunt, Peter. 1997. "The Helots at the Battle of Plataea." *Historia* 46: 129–144.

Hunt, Peter. 1998. *Slaves, Warfare, and Ideology in the Greek Historians.* Cambridge.

Hunt, Peter. 2001. "The Slaves and the Generals of Arginusae." *AJP* 122: 359–380.

Hunt, Peter. 2007. "Military forces." In Sabin et al., 108–146.

Hunt, Peter. 2009. "War and Society." In George Boys-Stones, Barbara Graziosi and Phiroze Vasunia (eds.), *Oxford Handbook of Hellenic Studies*, Oxford, 226–237.

Hunt, Peter. 2011. "Slaves in Greek Literary Culture." In Keith Bradley and P.A. Cartledge (eds.), *The Cambridge World History of Slavery. Vol. I: The Ancient Mediterranean World*, Cambridge, 22–47.

Hunt, Peter. 2016. "Slaves or Serfs?: Patterson on the Thetes and Helots of Ancient Greece." In John Bodel and Walter Scheidel (eds.), *On Human Bondage: After Slavery and Social Death*, Malden, 55–80.

Hunt, Peter. 2018a. "Ancient Greece as a Slave Society." In Catherine M. Cameron and Noel Lenski (eds.), *What Is a Slave Society? The Practice of Slavery in Global Perspective*, Cambridge, 61–85.

Hunt, Peter. 2018b. *Ancient Greek and Roman Slavery.* Malden.

Hurwit, J. M. 2002. "Reading the Chigi Vase." *Hesperia* 71: 1–22.

Hurwit, J. M. 2014. "The Lost Art: Early Greek Wall and Panel Painting, 760–480." In J.J. Pollitt (ed.), *The Cambridge History of Painting in the Classical World*, Cambridge, 66–93.

Hutchinson, Godfrey. 2000. *Xenophon and the Art of Command.* London.

Hutton, William E. 2005. *Describing Greece: Landscape and Literature in the* Periegesis *of Pausanias.* Cambridge.

Hyland, A. 2003. *The Horse in the Ancient World.* Westport.

Hyland, John O. 2018. *Persian Interventions: The Achaemenid Empire, Athens, and Sparta, 450–386 BCE.* Baltimore.

Illinois Greek Club (= W.A. Oldfather and others). 1923. *Aeneas—Asclepiodotus—Onasander.* Cambridge, MA and London.

Isager, Signe and Mogens Herman Hansen, trans. Judith Hsiang Rosenmeier. 1975. *Aspects of Athenian Society in the Fourth Century: A Historical Introduction to and Commentary on the Paragraphe-Speeches and the Speech Against Dionysodorus in the Corpus Demosthenicum B.C.* Odense.

Isager, Signe and Jens Erik Skydsgaard. 1992. *Ancient Greek Agriculture: An Introduction.* London and New York.

Ivanov, Dimitar. 2000. *Колекция "Арес."* Veliko Tarnovo.

Ivanov, Teofil. 1948. "Предпазно въоръжение на тракиец от Асеновград." *Разкопки и проучвания* I: 99–108.

Ivanov, Yavor. 2017. "Златен предмет от Шивачева могила—идентификация и функция." В: Попов, Х., Цветкова, Ю. (ред.) *ΚΡΑΤΙΣΤΟΣ. Сборник в чест на професор Петър Делев*. София, 483–499.

Jackman, T. 2006. "T. Ducetius and Fifth-Century Sicilian Tyranny." In S. Lewis (ed.), *Ancient Tyranny*, Edinburgh, 33–48.

Jacobs, Bruno. 1992. "Der Tod des Bessos." *Acta praehistorica et archaeologica* 24: 177–186.

Jacquemin, A. 1999. *Offrandes monumentales à Delphes*. Paris.

Jacquemin, A. 2000. *Guerre et religion dans le monde grec (490-322 av. J.-C.)*. Paris.

James, S. 2004. *The Excavations at Dura-Europos Conducted by Yale University and the French Academy of Inscriptions and Letters 1928 to 1937: Final Report and The Arms and Armour and Other Military Equipment*. London.

Jameson, Michael. 1991. "Sacrifice before Battle." In Hanson, 197–227.

Jameson, Michael. 1992. "Agricultural Labor in Ancient Greece." In Berit Wells (ed.), *Agriculture in Ancient Greece*, Stockholm, 135–146.

Jameson, Michael H. 1994. "Class in the Ancient Greek Countryside." *Collection de l'Institut des Sciences et Techniques de l'Antiquité* 508: 55–64.

Jarva, Eero. 1995. *Archaiologia on Archaic Greek Body Armour*. Rovaniemi.

Jarva, Eero. 2013. "Arms and Armor. Part I. Arming Greeks for battle." In Campbell and Tritle, 395–418.

Jeffery, L.H. 1962. "The Campaign between Athens and Aegina in the Years before Salamis (Hdt. 6.87–93)." *AJPh* 83: 44–54.

Jeffery, L. H. 1976. *Archaic Greece. The City-States c. 700-500 B.C.* London.

Jehne, Martin. 1994. *Koine eirene*. Stuttgart.

Jenkins, Brian M. 1975. "International Terrorism: A New Mode of Conflict." In David Carlton and Carlo Schaerf (eds.), *International Terrorism and World Security*, London, 13–49.

Jobst, W. 1978. "Hellenistische Außenfortifikationen um Ephesos." In Sencer Şahin, Elmar Schwertheim and Jörg Wagner (eds.), *Studien zu Religion und Kultur Kleinasiens. Festschrift für Friedrich Karl Dörner zum 65. Geburtstag am 28. Februar 1976*, Leiden, 446–456.

Jones, Colin and Simon MacDonald. 2018. "Robespierre, the Duke of York, and Pisistratus During the French Revolutionary Terror." *The Historical Journal* 61: 643–672.

Juhel, P. and N.V. Sekunda 2009. "The *Agema* and 'The Other Peltasts' in the Late Antigonid Army, and in the Drama/Cassandreia Conscription *Diagramma*." *ZPE* 170: 104–108.

Kagan, Donald. 1991. *Pericles of Athens and the Birth of Democracy*. New York.

Kagan, Donald. 2003. *The Peloponnesian War*. New York.

Kagan, Donald and Gregory F. Viggiano (eds.). 2013a. *Men of Bronze. Hoplite Warfare in Ancient Greece*. Princeton.

Kagan, Donald and Gregory F. Viggiano. 2013b. "The Hoplite Debate." In Kagan and Viggiano, 1–56.

Kahn, Dan'el and Oded Tammuz. 2008. "Egypt Is Difficult to Enter: Invading Egypt—A Game Plan (Seventh–Fourth Centuries BCE)." *Journal for the Society for the Study of Egyptian Antiquities* 35: 37–66.

Kamola, Isaac. 2018. "Pirate Capitalism, or the Primitive Accumulation of Capital Itself." *Millennium: Journal of International Studies* 47: 3–24.

Kansteiner, S., K. Hallof, B. Seidensticker, and S. Prignitz 2014. *Der Neue Overbeck*. Berlin.

Karlsson, Lars. 1992. *Fortification Towers and Masonry Techniques in the Hegemony of Syracuse, 401–211 B.C.* Stockholm.

Karunanithy, David 2013. *The Macedonian War Machine*. Barnsley, UK.

Katinčarova, Daniela. 2007. "Survey along the Main Street—From the Eastern Gate of the Emporion Pistiros Inwards and in the Sector of the Easter City Wall." In J. Bouzek, L. Domaradzka, and Z. Archibald (eds.) *Pistiros III. Excavations and Studies*. Prague, 35–61.

Katzarov, Gavrail. 1913. "Битът на старите траки според класическите писатели." *Сборник на БАН*, 1: 1–72.

Kayumov, Ildar and Alexander Minchev. 2013. "The καμβέστριον and other Roman military equipment from Thracia." In M. Sanader, A. Rendić-Miočević, D. Tončinić, and I. Radman-Livaja (eds.). *Proceedings of the XVIIth Roman Military Equipment Conference Zagreb 2010. Weapons and Military Equipment in a Funerary Context.* Zagreb, 327–345.

Keegan, John. 1976. *The Face of Battle.* London.

Kent, J.H. 1941. "A Garrison Inscription from Rhamnous." *Hesperia* 10: 342–350.

Kent, R.G. 19532. *Old Persian.* New Haven, CT.

Kern, Paul B. 1999. *Ancient Siege Warfare.* Bloomington, IN.

Khazanov, A.M. 1968. "Katafraktarii i ikh rol' v istorii voennogo iskusstva." *Vestnik drevnej istorii* 1968: 180–191.

Khazanov, A.M. 1971. *Ocherki voennogo dela sarmatov.* Moscow.

Kienast, Hermann J. 1978. *Die Stadtmauer von Samos.* Samos 15. Bonn.

Killen, J.T. 2006. "Conscription and *corvée* at Mycenaean Pylos." In M. Perna (ed.), *Fiscality in Mycenaean and Near Eastern Archives*, Napoli, 73–87.

Killen, J.T. 2007. "'Followers' and 'Watchers' at Pylos and Knossos." In E. Alram-Stern and G. Nightingale (eds.), *Keimelion: Elitenbildung und elitärer Konsum von der mykenischen Palastzeit bis zur homerischen Epoche*, Wien, 263–267.

Kinzl, Konrad H. (ed.). 2006. *A Companion to the Classical World.* Oxford and Malden; reprint 2010.

Kirk, G.S. 1975. "The Homeric Poems as History." In *CAH*^{3}, 2.2, 820–850.

Kistler, John. 2007. *War Elephants.* Lincoln, Nebraska.

Kleu, M. 2015. *Die Seepolitik Philipps V. von Makedonien.* Bochum.

Klöppel, U. 2014. "Athener Eulen in Not. Notgeld und Münzfälschungen am Ende des Peloponnesischen Krieges." *Geldgeschichtliche Nachrichten* 49: 65–75, 125–134.

Knigge, Ursula. 1988. *Der Kerameikos von Athen. Führung durch Ausgrabungen und Geschichte.* Athens.

Knoepfler, D. 1993. "Les *kruptoi* du stratège Épicharès à Rhamnonte et le début de la guerre de Chrémonidès." *BCH* 117: 327–341.

Koehn, Clemens. 2007. *Krieg – Diplomatie – Ideologie: zur Außenpolitik hellenistischer Mittelstaaten.* Stuttgart.

Kolev, Panto. 1975. "Проблеми на тракийското военно дело." *Векове* 6: 15–22.

Konijnendijk, Roel. 2012. "'Neither the Less Valorous Nor the Weaker.' Persian Military Might and the Battle of Plataia." *Historia* 61: 1–17.

Konijnendijk, Roel. 2016. "Mardonios' Senseless Greeks." *CQ* 66: 1–12.

Konijnendijk, Roel. 2018. *Classical Greek Tactics. A Cultural History.* Leiden.

Koshelenko, G.A. 1963. "Parfianskaia fortifikatsiia." *Sovetskaia arkheologiia* 1963: 57–73.

Koshelenko, G.A. 1980. "Les cavaliers parthes. Aspects de la structure sociale de la Parthie." *DHA* 6: 177–199.

Koukouli-Chrysanthaki, Ch. 1990. "The '*metalla*' of the Thasian *peraia*." In id. and Olivier Picard (eds.), *Μνήμη Δ. Λαζαρίδη. Πόλις καὶ Χώρα στὴν ἀρχαία Μακεδονία καὶ Θράκη*, Thessaloniki, 493–515 (in Greek, with French summary).

Kourinou, Eleni. 2007. "The Representation of Means of Transport on Reliefs in the Collection of the National Archaeological Museum in Athens." In Colin Adams and James Roy (eds.), *Travel, Geography and Culture in Ancient Greece, Egypt and the Near East*, Oxford, 88–103.

Koutsoukou, A. and C. Kanellopoulos. 1990. "Towers from North-West Andros." *ABSA* 85: 155–174.

Kraay, C. M. 1976. *Archaic and Classical Greek Coins.* London.

Kraay, C. M. 1984. "Greek Coinage and War." In W. Heckel et al., (eds.), *Ancient Coins of the Graeco-Roman World.* Waterloo, 3–18.

Kralli, Ioanna. 2000. "Athens and the Hellenistic Kings (338–261 B.C.): The Language of the Decrees." *CQ* 50: 113–132.

Krasilnikoff, J.A. 1992. "Aegean Mercenaries in the Fourth to Second Centuries BC. A Study in Payment, Plunder and Logistics of Ancient Greek Armies." *C&M* 38: 229–262.

Krasilnikoff, J. A. 2008. "Attic φελλεύς. Some Observations on Marginal Land and Rural Strategies in the Classical Period." *ZPE* 167: 37–49.

Kraut, Richard. 2002. *Aristotle: Political Philosophy.* Oxford.

Kremydi-Sicilianou, S. 1999. "The Financing of Alexander's Asian Campaign." *NomKhron* 18: 61–68.

Krentz, Peter. 1985a. "The Nature of Hoplite Battle." *ClAnt* 4: 50–61.

Krentz, Peter. 1985b. "Casualties in Hoplite Battles." *GRBS* 26: 13–20. reprinted in Wheeler 2007b, 349–356.

Krentz, Peter. 1991. "The Salpinx in Greek Warfare." In Hanson, 110–120.

Krentz, Peter. 1994. "Continuing the *Othismos* on *Othismos*." *AHB* 8: 45–49.

Krentz, Peter. 2000. "Deception in Archaic and Classical Greek warfare." In van Wees, 167–200.

Krentz, Peter. 2002. "Fighting by the Rules: The Invention of the Hoplite Agôn." *Hesperia* 71: 23–39; reprinted in Wheeler 2007b, 111–127.

Krentz, Peter. 2007a. "Warfare and Hoplites." In H. Alan Shapiro (ed.). *The Cambridge Companion to Archaic Greece*, Cambridge, 61–84.

Krentz, Peter. 2007b. "Part I: Archaic and Classical Greece: War." In Sabin et al., 147–185.

Krentz, Peter. 2010. *The Battle of Marathon.* New Haven.

Krentz, Peter. 2013a. "Hoplite Hell: How Hoplites Fought." In Kagan and Viggiano, 134–156.

Krentz, Peter. 2013b. "Marathon and the Development of the Exclusive Hoplite Phalanx." In Christopher Carey and Michael Edwards (eds.), *Marathon—2,500 Years. Proceedings of the Marathon Conference 2010.* London, 35–44.

Krentz, Peter and Everett L. Wheeler. 1994. *Polyaenus,* Strategems of War. Chicago.

Krischen, Fritz. 1922. *Die Befestigungen von Herakleia am Latmos. Milet* 3,2. Berlin.

Krischen, Fritz. 1941. *Die Stadtmauern von Pompeji und griechische Festungsbaukunst in Unteritalien und Sizilien.* Berlin.

Kroll, J. H. 1977. "An Archive of the Athenian Cavalry." *Hesperia* 46: 83–140.

Kromayer, J. and G. Veith. 1903–1907. *Antike Schlachtfelder in Griechenland.* Berlin.

Kromayer, J. and G. Veith. 1928. *Heerwesen und Kriegführung der Griechen und Römer. Handbuch der Altertumswissenschaft 4.3.2.* Munich.

Krotscheck, U. 2010. "Going with the Grain: Athenian State Formation and the Question of Subsistence in the 5th and 4th Centuries BCE." Princeton/Stanford Working Papers in Classics, No. 010603 (available online).

Kuhrt, Amélie. 1995. *The Ancient Near East c. 3000–330 BC,* 2 vols. London.

Kuleff, Ivelin, Ilia Iliev, Ernst Pernicka, and Diana Gergova. 2006. "Chemical and Lead Isotope Compositions of Lead Artefacts from Ancient Thracia (Bulgaria)." *Journal of Cultural Heritage* 7: 244–256.

Kurtz, D. C. and J. Boardman. 1971. *Greek Burial Customs.* London.

Lacey, Jim. 2011. *The First Clash: The Miraculous Greek Victory at Marathon and its Impact on Western Civilization.* New York.

Laffineur, R. (ed.) 1999. *Polemos. Le contexte guerrier en Égée à l'Âge du Bronze. Actes de la 7e Rencontre égéenne internationale, Univ. de Liège 1998.* Liège.

LaForse, Bruce. 2006. "The Greek World, 371–336." In Kinzl, 544–559.

Lambert, S. 2018. "Fitzwilliam Museum, Cambridge." *Attic Inscriptions in UK Collections* 3: 1–54.

Lammert, E. 1899. "Die geschichtliche Entwicklung der griechischen Taktik." *Neue Jahrbücher für das Klassische Altertum, Geschichte und Deutsche Literatur und für Pädagogik* 3: 1–29.

Lammert, E. and F. 1921. "Schlachtordnung," RE Supp. A.1, col. 435–494.

Landucci Gattinoni, Franca. 1992. *Lisimaco di Tracia nella prospettiva del primo ellenismo.* Milan.

Lane Fox, R. 1973. *Alexander the Great.* London.

Lang, Franziska. 1996. *Archaische Siedlungen in Griechenland.* Berlin.

Lang, M. 1990. "The *oka* Tablets Again." *Kadmos* 29: 113–125.

Laqueur, Walter. 2001. *A History of Terrorism.* New Brunswick, NJ and London.

Latacz, Johachim. 1977. *Kampfparänese, Kampfdarstellung und Kampfwirklichkeit in der Ilias, bei Kallinos und Tyraios.* Munich.
Lattimore, Richmond (trans.). 1951/2011. *Homer. The Iliad* (with introduction and notes by Richard Martin). Chicago.
Lattimore, Steven (trans.). 1998. *Thucydides. The Peloponnesian War.* Indianapolis.
Laufer, Eric. 2016. "Die Befestigung als historische Quelle." In Müth et al., 196–206.
Lauffer, Siegfried. 1979. *Die Bergwerkssklaven von Laureion*, 2nd ed. Wiesbaden.
Launey, Marcel. 1950; 1987^2. *Recherches sur les armées hellénistiques.* Paris (2nd ed. with addenda and postscript by Y. Garlan, Ph. Gauthier and Cl. Orrieux).
Lauter, Hans. 1986. *Die Architektur des Hellenismus.* Darmstadt.
Law, Randall D. 2015a. "Introduction." In Law 2015b, 1–11.
Law, Randall D. (ed.). 2015b. *The Routledge History of Terrorism.* London and New York.
Lawrence, Arnold Walter. 1979. *Greek Aims in Fortification.* Oxford.
Lazär, Simona. 2009. "Helmets of the Chalcidian Shape Found in the Lower Danube Area." *Dacia* 53: 13–26.
Lazaridis, Dimitrios. 1986. "Les fortifications d'Amphipolis." In Leriche and Tréziny, 31–38.
Lazaridis, Dimitrios. 2003. *Amphipolis*, 2nd ed. Athens.
Lazenby, John F. 1985. *The Spartan Army.* Warminster.
Lazenby, John F. 1991. "The Killing Zone." In Hanson, 87–109.
Lazenby, John F. 1994. "Logistics in Classical Greek Warfare." *War in History* 1: 3–18.
Lazenby, John F. 2004. *The Peloponnesian War: A Military Study.* London and New York.
Lazenby, John F. and David Whitehead. 1996. "The Myth of the Hoplite's *Hoplon.*" *CQ* 46: 27–33.
Le Bohec, Yann 2004. *El ejército romano.* Barcelona.
Le Rider, Georges. 1977. *Le monnayage d' argent et d' or de Philippe II frappé en Macédoine de 359 à 294.* Paris.
Le Rider, Georges. 1996. *Monnayage et finances de Philippe II.* Paris.
Le Rider, Georges. 1999. "La date des premiers statères d'or de Philippe II." In id. (ed.), *Études d' histoire monétaire et financière du monde grec 2*, Athènes, 485–493.
Leake, William. Martin. 1830. *Travels in the Morea.* London.
Lee, John W.I. 2001. "Urban Combat at Olynthus, 348 BC." In Ph. Freeman and A. Pollard (eds.) *Fields of Conflict: Progress and Prospect in Battlefield Archaeology* (BAR International Series 958). Oxford, 11–22.
Lee, John W.I. 2006. "Warfare in the Classical Age." In Kinzl, 480–508.
Lee, John W.I. 2007. *A Greek Army on the March: Soldiers and Survival in Xenophon's Anabasis.* Cambridge.
Lee, John W.I. 2013. "The Classical Greek Experience." In Campbell and Tritle, 143–161.
Lee, John W.I. 2017. "Xenophon and His Times." In Flower, 15–36.
Lee, John W. I. 2020. "Daily Life in Classical Greek Armies, c. 500–330 BCE." In Lee L. Brice (ed.), *New Approaches to Greek and Roman Warfare.* Hoboken, NJ, 39–51.
Lehmann, Gustav A. 2008. *Perikles. Staatsmann und Stratege im klassischen Athen.* Munich.
Leighton, R. 2000. "Indigenous Society Between the Ninth and Sixth Centuries BC: Territorial, Urban and Social Evolution." In Ch. Smith and J. Serrati (eds.), *Sicily from Aeneas to Augustus. New Approaches in Archaeology and History.* Edinburgh, 15–40.
Leighton, Richard M. 2017. "Logistics." *Encyclopedia Britannica* (pages not numbered, web publication accessed August 7, 2018 at https://www.britannica.com/topic/logistics-military)
Lendle, Otto. 1975. *Schildkröte: antike Kriegsmaschinen in poliorketischen Texten.* Wiesbaden.
Lendle, Otto. 1981. "Antike Kriegsmaschinen." *Gymnasium* 88: 330–356.
Lendle, Otto. 1983. *Texte und Untersuchungen zum technischen Bereich der antiken Poliorketik.* Wiesbaden.
Lendon, J.E. 1994. "Thucydides and the 'Constitution' of the Peloponnesian League." *GRBS* 35: 159–177.

Lendon, J.E. 2005. *Soldiers and Ghosts: A History of Battle in Classical Antiquity*. New Haven and London.

Lenfant, Dominique (ed.). 2004. *Ctésias de Cnide: La Perse. L'Inde. Autres Fragments*. Paris.

Lengauer, Włodzimierz. 1979. *Greek Commanders in the 5th and 4th Centuries B.C. Politics and Ideology. A Study of Militarism*. Warsaw.

Lentini, M.C. 2012. "Recent Investigation of the Early Settlement Levels at Sicilian Naxos." *MedArch* 25: 309–315.

Leriche, Pierre. 1993. "L'extrême orient hellénistique: le monde de la brique crue." *Dossiers d'Archéologie* 179: 75–83.

Leriche, Pierre. 1995. "L'Étude archéologique des fortifications urbaines grecques." In Debord and Descat, 9–27.

Leriche, Pierre. 1996. "Le Chreophylakeion de Doura-Europos et la mise en place du plan Hippodamien de la ville." In Marie-Françoise Boussac and Antonio Invernizzi (eds.), *Archives et sceaux du monde hellénistique—Archivi e sigilli nel mondo ellenistico. Torino, 13-16 gennaio 1993. BCH Supplement* 29, Athens, 157–169.

Leriche, Pierre and Henri Tréziny (eds.) 1986. *La fortification dans í histoire du monde grec. Actes du Colloque International: La fortification et sa place dans l'histoire politique, culturelle et sociale du Monde Grec. Valbonne, décembre 1982*. Paris.

Leshtakov, Lyuben. 2015. *Типология и хронологияна бронзовите върхове за копия от късната бронзова и началото на ранножелязната епоха в Югоизточна европа*. София.

Leshtakov, Petar, Emil Nankov, Alexandrina Tsoneva, Jordanka Ilieva and Deyan Dryanovski. 2016. "Издирване на археологически обектив землищата на селата Сотиря и Тополчане, община Сливен." *Археологически открития и разкопки през 2015*, София, 869–872.

Lesquier, J. 1911. *Les institutions militaires de l'Égypte sous les Lagides*. Paris.

Leventi, I. 2014. *Πόλη σε κρίση. Αρχιτεκτονική γλυπτική της Αθήνας στην περίοδο του Πελοποννησιακού πολέμου*. Athens.

Lewis, David M. 1992a. "Mainland Greece, 479–451 B.C." In *CAH* V^2, 96–120.

Lewis, David M. 1992b. "The Archidamian War." In *CAH* V^2, 370–432.

Lewis, David M. 1994. "Sicily, 413–368 B.C." In *CAH* VI.2, 120–155.

Lewis, David M. 1997. "The Origins of the First Peloponnesian War." In id. (ed.), *Selected Papers in Greek and Near Eastern History*, ed. P.J. Rhodes, Cambridge, 9–21. Originally published in G.S. Shrimpton and D.J. McCargar (eds.), *Classical Contributions: Studies in Honour of Malcolm Francis McGregor*. Locust Valley, NY, 1981, 71–78.

Lewis, David M. 2018. *Greek Slave Systems in their Eastern Mediterranean Context*. Oxford.

Ley, Judith. 2009. *Stadtbefestigungen in Akarnanien: ein bauhistorischer Beitrag zur urbanen Entwicklungsgeschichte einer antiken Landschaft*. Berlin.

Ley, Judith, Elke Richter and Brita Jansen. 2016. "Die Dokumentation vor Ort: Methodik und Organisation." In Müth et al., 24–39.

Lezzi-Hafter, Adrienne. 1997. "Offerings Made to Measure: Two Special Commissions by the Eretria Painter for Apollonia Pontica." In J. Oakley, W. Coulson, and O. Palagia (eds.) *Athenian Potters and Painters*. Oxford, 353–369.

Liampi, K. 1998. *Der makedonische Schild*. Bonn.

LiDonnici, Lynn R. 1995. *The Epidaurian Miracle Inscriptions*. Atlanta.

Lindenlauf, Astrid. 2003. "Constructing the Memory of the Persian Wars in Athens." In Ann Brysbaert, Natasja de Bruijn, Erin Gibson, Angela Michael and Mark Monaghan (eds.), *Symposium on Mediterranean Archaeology (SOMA): Proceedings of the 6th annual meeting of postgraduate researchers, University of Glasgow 2002*. Oxford, 53–62.

Lintott, Andrew. 1982. *Violence, Civil Strife and Revolution in the Classical City*. London.

Lipka, Michael. 2002. *Xenophon's Spartan Constitution. Introduction, Text, Commentary*. Berlin and New York.

Lippelt, O. 1910. *Die griechischen Lichtbewafftneten bis auf Alexander den Grossen*. Jena.

Lissarrague, François. 2015. "Women Arming Men: Amor and Jewelry." In Fabre-Serris and Keith, 71–81.

Litvinskii, B. A. 1966. "Slozhnosostavnyi luk v drevnei Srednei Azii." *Sovetskaia arkheologia* 1966: 51–69.

Lloyd, Alan B. (ed.). 1996. *Battle in Antiquity*. London.

Lloyd, Matthew. 2017. "Unorthodox Warfare? Variety and Change in Archaic Greek Warfare (ca. 700–ca. 480 BCE)." In Brian Hughes and Fergus Robson (eds.), *Unconventional Warfare from Antiquity to the Present Day*, Cham, Switzerland, 231–252.

Loader, N.C. 1995. *The Definition of Cyclopean: An Investigation into the Origins of the LH III Fortifications on Mainland Greece*. Ph.D. diss., Durham University.

Lochée, Lewis 1778. *An Essay in Castrametation*. London.

Lohner-Urban, Ute and Peter Scherrer. 2016. "Hellenistische Prunktore—ein wissenschaftlicher Irrtum? Vorläufige Grabungsergebnisse vom Osttor von Side aus der Kampagne 2012." In Frederiksen et al., 232–243.

Loman, Pasi. 2004. "No Woman No War: Women's Participation in Ancient Greek Warfare." *G&R* 51: 34–54.

Loman, Pasi. 2005. "Mercenaries, their Women and Colonisation." *Klio* 87: 346–365.

Lomas, K. 1993. *Rome and the Western Greeks, 350 BC–AD 200: Conquest and Acculturation in Southern Italy*. London.

Lonis, Raoul. 1979. *Guerre et religión en Grèce a l'époque classique. Recherches sur les rites, les dieux, l'ideologie de la victoire*. Paris.

Loomis, W.T. 1992. *The Spartan War Fund. IG V1,1 and a New Fragment. Historia Einzelschriften* 74. Stuttgart.

Lopez, George A. 1984. "A Scheme for the Analysis of Government as Terrorist." In Stohl and Lopez 1984a, 59–81.

Loraux, Nicole. 1986. *The Invention of Athens: The Funeral Oration in the Classical City*, trans. A. Sheridan. Cambridge, MA. (= *La invención de Atenas*. Barcelona, 2008 [= *L'invention d'Athènes*. Paris, 1993; English ed. Cambridge, 2006]).

Lord, L.E. 1939. "Watchtowers and Fortresses in Argolis." *AJA* 43: 78–84.

Lorentzen, Janet. 2010. "Die Stadtmauern des hellenistischen Pergamon." In Lorentzen et al., 107–139.

Lorentzen, Janet, Felix Pirson, Peter Schneider, and Ulrike Wulf-Rheidt (eds.) 2010. *Aktuelle Forschungen zur Konstruktion, Funktion und Semantik antiker Stadtbefestigungen, Kolloquium 9/10. Februar 2007 in Istanbul*. Byzas 10. Istanbul.

Lorimer, Hilda L. 1903. "The Country Cart in Ancient Greece." *JHS* 23: 132–151.

Lorimer, Hilda L. 1947. "The Hoplite Phalanx with Special Reference to the Poems of Archilochus and Tyrtaeus." *ABSA* 42: 76–138.

Loukopoulou, L. 2004. "Thrace from Strymon to Nestos." In Hansen and Nielsen, 854–869.

Low, Polly. 2003. "Remembering War in Fifth-Century Greece: Ideologies, Societies, and Commemoration beyond Democratic Athens." *World Archaeology* 35: 98–111.

Luginbill, Robert D. 1994. "*Othismos*: The Importance of the Mass-Shove in Hoplite Warfare." *Phoenix* 48: 51–61.

Lund, Helen S. 1992. *Lysimachus. A Study in Early Hellenistic Kingship*. London and New York.

Luraghi, Nino. 1994. *Tirannidi arcaiche in Sicilia e Magna Grecia*. Florence.

Luraghi, Nino. 2002. "Helotic Slavery Reconsidered." In Anton Powell and Stephen Hodkinson (eds.), *Sparta: Beyond the Mirage*. Swansea, 227–248.

Luttwak, Edward N. 1993. "Logistics and the Aristocratic Idea of War." In Lynn, 3–7.

Lynn, John A. (ed). 1993. *Feeding Mars: Logistics in Western Warfare from the Middle Ages to the Present*. Boulder.

Lynn, John A. 1993a. "The History of Logistics and *Supplying War*." In Lynn, 9–29.

Lynn, John A. 1993b. "Food, Funds, and Fortresses: Resource Mobilization and Positional Warfare in the Campaigns of Louis XIV." In Lynn, 137–159.

Ma, John. 2000. "Fighting Poleis of the Hellenistic World." In van Wees, 337–376.
Ma, John. 2008a. "Chaironeia 338: Topographies of Commemoration." *JHS* 128: 72–91.
Ma, John. 2008b. "The Return of the Black Hunter." *Cambridge Classical Journal* 54: 188–208.
Ma, John. 2008c. "Mysians on the Can Sarcophagus? Ethnicity and Domination in Achaemenid Military Art." *Historia* 57: 243–254.
Ma, John. 2010. "Autour des balles de fronde 'camiriennes'." *Chiron* 40: 155–173.
MacDowell, Douglas M. (ed.). 1982/1993. *Gorgias,* Encomium of Helen. Edited with Introduction, Notes, and Translation. Bristol.
MacDowell, Douglas M. 1986. *Spartan Law*. Edinburgh.
MacDowell, Douglas M. 2000. *Demosthenes: On the False Embassy (Oration 19)*. Oxford.
Machiavelli, N. 1965. *The Art of War*, trans. E. Farneworth. Indianapolis.
Mackay, E. Anne (ed.). 2008. *Orality, Literacy, Memory in the Ancient Greek and Roman World*. Leiden.
Madreiter, Irene. 2012. *Stereotypisierung—Idealisierung—Indifferenz. Formen der Auseinandersetzung mit dem Achaimeniden-Reich in der griechischen Persika-Literatur*. Wiesbaden.
Maenchen-Helfen, O.J. 1973. *The World of the Huns: Studies in Their History and Culture*, ed. M. Knight. Berkeley, Los Angeles and London.
Maher, M. 2018. "The Defense Network in the Chora of Mantineia." *Hesperia* 87: 451–495.
Maier, Franz-Georg. 1959–1961. *Griechische Mauerbauinschriften*, 2 vols. Heidelberg.
Maier, Franz-Georg. 2008. *Nordost-Tor und persische Belagerungsrampe in Alt-Paphos. Grabungsbefund und Baugeschichte*. Ausgrabungen in Alt-Paphos auf Cypern 6. Mainz.
Maletzke, Gerhard. 1972. "Propaganda. Eine begriffsgeschichtliche Analyse." *Publizistik* 17: 153–164.
Malitz, J. 2008. "Thukydides und die Finanzen Athens." In Burrer and Müller (eds.), 28–45.
Mallett, Michael E. 1974. *Mercenaries and Their Masters: Warfare in Renaissance Italy*. London.
Mallett, Michael E. 1984. "Introduction: The European Context 1400–1525." In Michael E. Mallett and John R. Hale. *The Military Organization of a Renaissance State: Venice c. 1400 to 1617*, Cambridge.
Manning, Joseph G. 2018. *The Open Sea. The Economic Life of the Ancient Mediterranean World from the Iron Age to the Rise of Rome*. Princeton and Oxford.
Manov, Metodi and Narcis Torbov. 2016. "Inscribed Lead Sling Bullets with the Name of Alexander the Great and with Other Names and Symbols Found in Thrace." *Archaeologia Bulgarica* 2: 29–43.
Manville, Philip B. 1977. "Aristagoras and Histiaios: The Leadership Struggle in the Ionian Revolt." *CQ* 27: 80–91.
Marangou, Lila I. 2005. *Αμοργός*. vol. 2, *Οι αρχαίοι πύργοι*. Athens.
Marazov, Ivan. 2005. *Thracian Warrior*. Sofia.
Marazov, Ivan. 2006. "Богинята Атина в царската идеология на Древна Тракия." *Проблеми и изследвания на тракийската култура* I: 38–73.
Marazov, Ivan. 2011a. *Тракия и древният свят XV-I в. пр. Хр. "Колекция Васил Божков."* София.
Marazov, Ivan. 2011b. "Philomele's Tongue: Reading the Pictorial Text of Thracian Art." In L. Bonfante (ed.), *The Barbarians of Ancient Europe. Realities and Interactions*, Cambridge, 132–189.
Marazov, Ivan, Gergana Kabakchieva, Gavril Lazov, and Tatyana Shalganova. 2005. *Колекцията Васил Божков*. София.
Marazov, Ivan, Tatyana Shalganova and Nartzis Torbov. 2012. *Тракийско въоръжение в български частни колекции. Кратка енциклопедия*. Велико Търново.
Markle, M.M. 1974. "The Strategy of Philip in 346 B.C." *CP* 24: 253–268.
Markle, M.M. 1977. "The Macedonian Sarissa, Spear, and Related Armor." *AJA* 81: 323–339.
Markle, M.M. 1978. "Use of the Sarissa by Philip and Alexander of Macedon." *AJPh* 82: 483–497.
Marksteiner, Thomas. 1994. "Kastell oder Herrensitz?" *JÖAI* 63: 95–120.
Marsden, Eric W. 1969. *Greek and Roman Artillery: Historical Development*. Oxford.

Marsden, Eric W. 1971. *Greek and Roman Artillery: Technical Treatises.* Oxford.

Marsden, Eric. W. 1977. "Macedonian Military Machinery and Its Designers under Philip and Alexander." *Ancient Macedonia II: papers read at the second international symposium held in Thessaloniki, 19–24 August 1973*, 211–223.

Marshall, J. 1951. *Taxila*, vols I–III, Cambridge.

Masaracchia (ed. and trans.). 1990^2. *Erodoto: La battaglia di Salamina. Libri VIII delle Storie.* Milan.

Martin, Roland. 1947. "Les enceintes de Gortys d'Arcadie." *BCH* 71–72: 81–147.

Martini, Wolfram. 2016. "Form, Funktion und Bedeutung der Stadtmauern von Perge in Pamphylien." In Frederiksen et al., 220–231.

Matthew, Christopher A. 2009. "When Push Comes to Shove: What Was the *othismos* of Hoplite Combat?" *Historia* 58: 395–415.

Mayer, Walter 1995. *Politik und Kriegkunst der Assyrer.* Münster.

Mayor, Adrienne. 2014. *The Amazons: Lives and Legends of Warrior Women Across the Ancient World.* Princeton.

McAllister, Bradley and Alex P. Schmid. 2011. "Theories of Terrorism." In Schmid 2011a, 201–271.

McCredie, James R. 1966. *Fortified Military Camps in Attica. Hesperia Supplement* 11. Athens: ASCSA.

McGing, Brian. 1986. *The Foreign Policy of Mithridates VI Eupator, King of Pontus.* Leiden.

McGinn, T.A. J. 2008. *Widows and the Patriarchy: Ancient and Modern.* London.

McHugh, Maeve. 2017. *The Ancient Greek Farmstead.* Oxford and Philadelphia.

McHugh, Maeve. 2019. "Going the Extra Mile: Travel, Time and Distance in Classical Attica." *ABSA* 114: 1–34.

McIntire, R. 1965. *The Training of Dogs for Field Reconnaissance.* College Park, MD.

McKechnie, Paul. 1989. *Outsiders in the Greek Cities in the Fourth Century BC.* London and New York.

McKechnie, Paul R. and Stephen J. Kern (eds.). 1988. *Hellenika Oxyrhynchia.* Warminster.

McNicoll, Anthony W. 1986. "Developments in Techniques of Siegecraft and Fortification in the Greek World ca. 400–100 B.C." In Leriche and Tréziny, 305–313.

McNicoll, Anthony W. 1997. *Hellenistic Fortifications from the Aegean to the Euphrates.* With revisions and an additional chapter by N.P. Milner. Oxford.

Meeus, A. 2009. "Some Institutional Problems Concerning the Succession to Alexander the Great: Prostasia and Chiliarchy." *Historia* 58: 287–310.

Meier, Ludwig. 2017. "Der sogenannte Piratenüberfall auf Teos und die Diadochen: Eine Neuedition der Inschrift SEG 44, 949." *Chiron* 47: 115–188.

Meier, M. 1998. *Aristokraten und Damoden. Untersuchungen zur inneren Entwicklung Spartas im 7. Jh. v. Chr. und zur politischen Funktion der Dichtung des Tyrtaios.* Stuttgart.

Meiggs, R. 1982. *Trees and Timber in the Ancient Mediterranean World.* Oxford.

Meineck, Peter and David Konstan (eds.). 2014. *Combat Trauma and the Ancient Greeks.* New York.

Meister, K. 1984. "Agathocles." In *CAH* VII.1, 384–411.

Meister, K. 2008. "Die finanzielle Ausgangssituation Athens zu Beginn des Peloponnesischen Krieges." In Burrer and Müller (eds.), 19–27.

Meritt, B.D., H.T. Wade-Gery and M.F. McGregor. 1949. *The Athenian Tribute Lists 2.* Princeton.

Mertens, Dieter. 1999. "Die Landschaftsfestung Epipolai bei Syrakus." In Schwandner and Rheidt, 143–149.

Mertens, Dieter. 2003. *Selinus. Vol. 1, Die Stadt und ihre Mauern.* Mainz.

Meyer, Eduard. 1884–1902/1925^2–1958^2. *Geschichte des Altertums.* Stuttgart; subsequent editions and reprints.

Mielczarek, Mariusz. 1993. *Cataphracti and Clibanarii: Studies on the Heavy Armoured Cavalry of the Ancient World.* Łódź.

Migeotte, L. 1992. *Les souscriptions publiques dans les cités grecques.* Geneva and Québec.

Migeotte, L. 2000. "Les dépenses militaires des cités grecques." In Jean Andreau, Pierre Briant and Raymond Descat (eds.), *La guerre dans les économies antiques, Entretiens d'archéologie et d'histoire* 5, Saint-Bertrand-de-Comminges, 145–209.

Migeotte, L. 2008. "Kriegs- und Verteidigungsfinanzierung in den hellenistischen Städten." In Burrer and Müller (eds.), 151–160.
Mihailov, Georgi. 2015. *Траките. Второ допълнено и преработено издание* (ред. М.Василева, В.Маринов), София.
Mikalson, Jon P. 2003. *Herodotus and Religion in the Persian Wars*. Chapel Hill.
Mikov, Rumen. 2010. "Ромфеята от Кабиле в контекста на проблемите, свързани с датирането на оръжията от тази категория. Разпространение и производство." В: *Stephanos Archaeologicos in honorem Professoris Stephcae Angelova* (Studia Archaeologica Universitatis Serdicensis, V). Велико Търново, 207–227.
Millender, Ellen G. 2016. "Spartan State Terror: Violence, Humiliation, and the Reinforcement of Social Boundaries in Classical Sparta." In Howe and Brice 2016, 117–150.
Miller, Martin. 1995. *Befestigungsanlagen in Italien vom 8. bis 3. Jahrhundert vor Christus*. Hamburg.
Miller, M.C. 2011. "Imaging Persians in the Age of Herodotus." In R. Rollinger, B. Truschnegg and R. Bichler (eds.), *Herodot und das persische Weltreich—Herodotus and the Persian Empire*, Wiesbaden, 123–157.
Millett, P.C. 2013. "Winning Ways in Warfare." In Campbell and Tritle, 46–73.
Milns, R.D. 1966. "Alexander's Seventh Phalanx Battalion." *GRBS* 7: 159–166.
Milns, R.D. 1976. "The Army of Alexander the Great." In E. Badian (ed.), *Alexandre le Grand: image et realité*, Fondation Hardt, Entretiens 22, Geneva, 87–130.
Minchev, Alexander. 2013. "Нови данни за бронзолеярството в североизточна Тракия през V–I в. пр. Хр." В: *Известия на народния музей Варна* XLVI (LXI) 2011, Варна.
Mitchell, Lynette G. and Peter J. Rhodes (eds.). 1997. *The Development of the Polis in Archaic Greece*. London.
Mitchell, Peter. 2018. *The Donkey in Human History: An Archaeological Perspective*. Oxford.
Mitchell, Stephen. 1996. "Hoplite Warfare in Ancient Greece." In Lloyd, 87–105.
Mitchell, Stephen. 2003. "The Galatians: Representation and Reality." In Erskine, 280–293.
Mithen, Steven. 2008. "When We Were Nicer," *London Review of Books*, January 24, 24–25.
Momigliano, A. 1961. *Secondo Contributo alla Storia degli Studi Classici*. Rome.
Monoson, S. Sara. 2014. "Socrates in Combat: Trauma and Resilience in Plato's Political Theory." In Meineck and Konstan, 131–162.
Moore, R. 2013. "Generalship: Leadership and Command." In Campbell and Tritle, 457–473.
Moreno, P. 2000. *Apelle. La battaglia di Alessandro*. Milan.
Morgan, Catherine. 2001. "Symbolic and Pragmatic Aspects of Warfare in the Greek World of the 8th to the 6th Centuries BC." In Bekker-Nielsen and Hannestad, 20–44.
Moritz, L.A. 1949. "Ἄλφιτα—A Note." *CQ* 12: 113–117.
Morris, Ian. 1987. *Burial and Ancient Society. The Rise of the Greek City-State*. Cambridge.
Morris, Ian. 1996. "The Strong Principle of Equality and Archaic Origins of Greek Democracy." In Ober and Hedrick, 19–48.
Morris, Ian. 2001. "The Use and Abuse of Homer." In Douglas L. Cairns (ed.), *Oxford Readings in Homer's Iliad*, Oxford, 57–91.
Morris, Ian. 2009. "The Eighth-Century Revolution." In Raaflaub and van Wees, 64–80.
Morris, Ian and Barry Powell (eds.). 1997. *A New Companion to Homer*. Leiden.
Morris, S.P. and J.K. Papadopoulos. 2005. "Greek Towers and Slaves: An Archaeology of Exploitation." *AJA* 109: 155–225.
Morrison, John S. and Roderick T. Williams. 1968. *Greek Oared Ships (900–322 B.C.)*. Cambridge.
Morrison, John S., J. F. Coates, and N. B. Rankov. 2000. *The Athenian Trireme: The History and Reconstruction of an Ancient Greek Warship*, 2nd ed. Cambridge.
Mossé, Claude. 1968. "Le role politique des armées dans le monde grec a l'époque classique." In Vernant, 221–230.
Mugione, E. and A. Benincasa (eds.). 2012. *L'Olpe Chigi. Storia di un agalma*. Paestum.
Muth, S. 2008. *Gewalt im Bild*. Berlin.

Müller, Sabine. 2009. *Das hellenistische Königspaar in der medialen Repräsentation. Ptolemaios II. und Arsinoë II.* Berlin and New York.

Müller, Sabine. 2010. "Philip II." In Roisman and Worthington, 166–185.

Müller, Sabine. 2016a. "Arrian, the Second Sophistic, Xerxes, and the Statues of Harmodios and Aristogeiton." In Robert Rollinger and Saana Svärd (eds.), *Cross-Cultural Studies in Near Eastern History and Literature*, Münster, 173–202.

Müller, Sabine. 2016b. *Die Argeaden. Geschichte Makedoniens bis zum Zeitalter Alexanders des Großen.* Paderborn.

Müller, Sabine. 2017. *Perdikkas II.—Retter Makedoniens.* Berlin.

Müller, Sabine, Tim Howe, Hugh Bowden, and Robert Rollinger (eds.). 2017. *A History of the Argeads—New Perspectives.* Wiesbaden.

Munn, Mark. 1987. "Agesilaos' Boiotian Campaigns and the Theban Stockade of 378-77 B.C." *ClAnt* 6: 106–138.

Munn, Mark H. 1993. *The Defense of Attica. The Dema Wall and the Boiotian War of 378–375 B.C.* Berkeley.

Munn, Mark. 1997. "Thebes and Central Greece." In L. Tritle (ed.), *The Greek World in the Fourth Century*, London and New York, 66–106.

Munson, Rosaria B. 2001. *Telling Wonders: Ethnographic and Political Discourse in the Work of Herodotus.* Ann Arbor.

Murray, O. 1980. *Early Greece.* Cambridge, MA.

Murray, William M. 2012. *The Age of Titans: The Rise and Fall of the Great Hellenistic Navies.* Oxford.

Mussche, H. 1961. "La forteresse maritime de Thorikos." *BCH* 85: 176–205.

Musti, D. 2005. *Magna Grecia: il quadro storico.* Roma.

Müth, S. 2008. *Gewalt im Bild.* Berlin.

Müth, S. 2010. "Fortifikationskunst und Repräsentation an der Stadtmauer von Messene." In Lorentzen et al., 57–83.

Müth, S. 2016a. "Urbanistic Functions and Aspects." In Müth et al., 2016, 159–172.

Müth, S. 2016b. "Functions and Semantics of Fortifications: An Introduction." In Frederiksen et al., 183–192.

Müth, S. and U. Ruppe. 2016. "Regional begrenzte Phänomene." In Müth et al., 231–248.

Müth, S., E. Laufer, and C. Brasse. 2016a. "Symbolische Funktionen." In Müth et al., 126–158.

Müth, S., A. Sokolicek, B. Jansen, and E. Laufer. 2016b. "Methods of Interpretation." In Müth et al., 1–23.

Müth, S., P. Schneider, M. Schnelle, and P. De Staebler (eds.). 2016. *Ancient Fortifications: A Compendium of Theory and Practice. Fokus Fortifikation Studies*, 1.

Nafissi, Massimo. 2009. "Sparta." In Raaflaub and van Wees, 117–137.

Naiden, F.S. 2007. "The Invention of the Officer Corps." *Journal of the Historical Society* 1: 35–60.

Naiden, F.S. 2009. *Ancient Supplication*; revised ed. Oxford.

Naiden, F.S. 2013. *Smoke Signals for the Gods: Greek Sacrifice from the Archaic Through Roman Periods.* Oxford.

Naiden, F.S. 2014. "Finley's War Years." In Naiden and Talbert, 243–266.

Naiden, F.S. 2017a. "Military Communications: The Case of the Classical Battlefield." In F.S. Naiden and Richard J.A. Talbert (eds.), *Mercury's Wings: Modes of Communication in the Ancient World*, Oxford, 62–84.

Naiden, F.S. 2017b. "Moses Finley's Communist Party Membership." *AJPh* 138: 739–742.

Naiden, F.S. 2018. *Soldier, Priest, and God: A Life of Alexander the Great.* Oxford and New York.

Naiden, F.S. 2019. "War Aims of Alexander the Great." In F. Naiden and Daniel Raisbeck (eds.), *Reflections on Macedonian and Roman Grand Strategy*, Bogota, Colombia, 57–79.

Naiden, F.S. and Richard J.A. Talbert (eds.). 2014. *Moses Finley in America: The Making of an Ancient Historian = AJPh* 135.2.

Naiden, F.S. and Richard J.A. Talbert (eds.). 2017. *Mercury's Wings: Modes of Communication in the Ancient World*, Oxford.

Nakas, Yannis D. 2016. "Isolated Towers in the Fortification Network of Ancient Molossia: A Case Study." In Frederiksen et al., 446–455.

Nankov, Emil. 2007. "An Ivory Scabbard Chape from Seuthopolis Rediscovered: Evidence for a Xiphos from Early Hellenistic Thrace?" *Archaeologia Bulgarica* 1: 37–46.

Nankov, Emil. 2008. "The Fortifications of the Early Hellenistic City of Seuthopolis: Breaking the Mold." *Archaeologia Bulgarica* 3: 15–56.

Nankov, Emil. 2009. *Phrouria Lokrika: Aspects of Military Presence in Hellenistic Opountian Lokris.* Ph.D. diss. Cornell.

Nankov, Emil. 2010. "Why One Needs 'The Odd Man Out'? The Deer Hunter with Lagobolon from the Frescoes in the Thracian Tomb near Alexandrovo." *Archaeologia Bulgarica* 1: 35–55.

Nankov, Emil. 2015a. "The Mobility of Macedonian Army in Thrace during the Reign of Philip II and the Inscribed Lead Sling Bullets from Kozi Gramadi." *Bulgarian E-Journal of Archaeology* 5: 1–13.

Nankov, Emil. 2015b. "Urbanization." In Valeva et al., 399–411.

Nankov, Emil. 2016. "Inscribed Lead Sling Bullets from the Regional Museum of History in Shumen. New Data on the Macedonian Campaigns in the Lands of the Getae in the Time of Philip II and Alexander III." В: И. Маразов, Д. Аладжова, Б. Хараланова, Д. Руменов, С. Стойчев (ред.). *Тракия и околният свят. Сборник с доклади от Национална научна конференция, 27–29 октомври 2016 г. Шумен*, 282–293.

Nankov, E. forthcoming. *Markers of Mobility: Toward an Archaeology of the Sling in Ancient Thrace.*

Napolitano, M.L. 1987. "Le donne spartane e la guerra: Problemi di tradizione." *Aion (archeol)* 9: 127–144.

Nappi, Marella. 2015. "Women and War in the *Iliad*: Rhetorical and Ethical Implications." In Fabre-Serris and Keith, 34–51.

Nefedkin, A.K. 2006. "The Tactical Development of Achaemenid Cavalry", *Gladius* 26: 5–18.

Nicholls, R.V. 1958–1959. "'Old Smyrna': The Iron Age Fortifications and Associated Remains on the City Perimeter." *ABSA* 53: 35–137.

Niemeier, W.-D., B. Niemeier, and A. Brysbaert. 2012. "The Olpe Chigi and New Evidence for Early Archaic Greek Wall-Painting from the Oracle Sanctuary of Apollo at Abai (Kalapodi)." In E. Mugione and A. Benincasa (eds.), 2012. *L'Olpe Chigi. Storia di un agalma, Atti del Convegno Internazionale, Salerno, 3–4 giugno 2010*, Salerno, 79–86.

Nikoloudis, S. 2008. "The Role of the *ra-wa-ke-ta*." In A. Sacconi et al., (eds.), *Colloquium Romanum. Atti XII Coll. Int. Micenologia, Roma 2006, II*, Pisa, 587–594.

Nikonorov, V.P. 1987. *Vooruzhenie i voennoe delo v Parfii.* Ph.D. thesis abstract, Leningrad.

Nikonorov, V.P. 1994. "Sredneaziatskie katafraktarii kak produkt vzaimodeistviia voennykh shkol Zapada i Vostoka v epokhu rannego ellinizma." In V. M. Masson (ed.), *Vzaimodeistvie drevnikh kultur i tsivilizatsii i ritmy kulturogeneza*, Saint-Petersburg, 47–51.

Nikonorov, V.P. 1995. "K voprosu o parfianskoi taktike (na primere bitvy pri Karrakh)." In A. M. Iliushin (ed.), *Voennoe delo i srednevekovaia arkheologiia Tsentralnoi Azii*, Kemerovo, 53–61.

Nikonorov, V.P. 1997. *The Armies of Bactria (700 BC-450 AD)*, vols. I–II. Stockport.

Nikonorov, V.P. 1998. "Cataphracti, Catafractarii and Clibanarii: Another Look at the Old Problem of Their Identifications." In G. V. Vilinbakhov and V. M. Masson (eds.), *Military Archaeology*, Saint-Petersburg, 131–138.

Nikonorov, V.P. 2000. "The Use of Musical Percussions in Ancient Eastern Warfare: Parthian and Central Asian Evidence." In E. Hickmann, Ingo Laufs and Ricardo Eichmann (eds.), *Musikarchäologie früher Metallzeiten*, Rahden, 71–81.

Nikonorov, V.P. 2002. "On the Saddles of the Parthian Cavalry." In Iu. S. Khudiakov and S.G. Skobelev (eds.), *Voennoe delo nomadov Severnoi i Tsentral'noi Azii*, Novosibirsk, 21–27.

Nikonorov, V.P. 2005. "K voprosu o parfianskom nasledii v sasanidskom Irane: voennoe delo." In id. (ed.), *Tsentral'naia Aziia ot Akhemenidov do Timuridov*, St. Petersburg, 141–179.

Nikonorov, Valerii P. 1997. *The Armies of Bactria, 700 BC-450 AD*, 2 vols. Stockport.

Nikonorov, V.P. 2013a. "The Parade Hatchet-Klevets from Old Nisa." *Anabasis. Studia Classica et Orientalia* 4: 179–232.

Nikonorov, V.P. 2013b. "More About Western Elements in the Armament of Hellenistic Bactria: The Case of the Warrior Terracotta from Kampyr-Tepe." In G. Lindström et al., (eds.), *Zwischen Ost und West. Neue Forschungen zum antiken Zentralasien*, Darmstadt, 187–204.

Nilsson, Martin. 1928. "Die Hoplitentaktik und das Staatswesen." *Klio* 22: 240–249.

Nilsson, Martin. 1929. "The Introduction of Hoplite Tactics at Rome: Its Date and Its Consequences." *JRS* 19: 1–11.

Nock, A. 1942. "Religious Attitudes of the Ancient Greeks." *PAPhS* 85: 472–482 = Arthur Darby Nock, *Essays on Religion and the Ancient World*, ed. Zeph Stewart, Oxford, 1972, 534–551.

Notopoulos, James. 1941. "The Slaves at the Battle of Marathon." *AJP* 62: 352–354.

Nussbaum, G. 1959. "The Captains in the Army of the Ten Thousand." *C&M* 20: 16–29.

O'Connor, Stephen. 2013. "The Daily Grain Consumption of Classical Greek Sailors and Soldiers." *Chiron* 43: 327–356.

O'Connor, Stephen. 2015. "Private Traders and the Food Supply of Classical Greek Armies." *Journal of Ancient History* 3: 173–219.

Oakley, John H. 2004. *Picturing Death in Classical Athens.* Cambridge.

Ober, Josiah. 1985. *Fortress Attica: Defense of the Athenian Land Frontier, 404-322 B.C.* Leiden.

Ober, Josiah. 1987. "Early Artillery Towers: Messenia, Boiotia, Attica, Megarid." *AJA* 91: 569–604.

Ober, Josiah. 1991. "Hoplites and Obstacles." In Hanson, 173–196.

Ober, Josiah. 1992. "Towards a Typology of Greek Artillery Towers: The First and Second Generations (c. 375–275 B.C.)." In van de Maele and Fossey, 1992, 147–169.

Ober, Josiah. 1994. "The Rules of War in Classical Greece." In Michael Howard, George J. Andreopoulos and Mark R. Shulman (eds.), *The Laws of War. Constraints on Warfare in the Western World*, New Haven, 12–26, 227–230.

Ober, Josiah. 1996. "The Rules of War in Classical Greece." In id., *The Athenian Revolution: Essays on Ancient Greek Democracy and Political Theory*, Princeton, 53–71.

Ober, Josiah and Charles Hedrick (eds.). 1996. *Demokratia: A Conversation on Democracies, Ancient and Modern.* Princeton.

Oehler, J. 1905. "Εφηβία." *RE* 5.2, col. 2741–2742.

Ognenova, Lyuba. 1961. "Les cuirasses de bronze trouvées en Thrace." *BCH* 85: 501–538.

Ognenova-Marinova, Lyuba. 1984. "Дребни находки, теракота, скулптура." В: *Севтополис*, т. 1. София, 159–223.

Ognenova-Marinova, Lyuba and Totko Stoyanov. 2005. "The Chalkidian Helmets and the Origin of the North Thracian Ceremonial Armour." In *Stephanos Arcaeologicos in honorem Professoris Ludmili Getov* (Studia Archaeologica Universitatis Serdicensis, Suppl. IV). Sofia, 519–539.

Olbrycht, M.J. 1993. "Some Remarks on Hellenistic Influence upon the Fortification of Northeastern Iran in the Arsacid Period." *Folia Orientalia* 29: 131–151.

Olbrycht, M.J. 1998. *Parthia et ulteriores gentes. Die politischen Beziehungen zwischen dem arsakidischen Iran und den Nomaden der eurasischen Steppen*, Munich.

Olbrycht, M.J. 2003. "Parthia and Nomads of Central Asia. Elements of Steppe Origin in the Social and Military Developments of Arsakid Iran." In I. Schneider (ed.), *Militär und Staatlichkeit. Mitteilungen des SFB "Differenz und Integration" 5*, Halle (Saale), 69–109.

Olbrycht, M.J. 2004. *Alexander the Great and the Iranian World.* Rzeszów.

Olbrycht, M.J. 2010a. "K voprosu o proiskhozhdenii konnitsy katafraktov v Irane i Srednei Azii." In I.V. Erofeev et al., (eds.), *Rol' nomadov evraziiskikh stepei v razvitii mirovogo voennogo iskusstva*, Almaty, 66–85.

Olbrycht, M.J. 2010b. "Central Asian, Achaemenid and Parthian Cavalry Developments." In V.A. Alekshin (ed.), *Drevnie kul'tury Evrazii*, Saint-Petersburg, 196–200.

Olbrycht, M.J. 2010c. "Macedonia and Persia." In Roisman and Worthington, 342–369.

Olbrycht, M.J. 2012. "Cultural Transfer in the Central Eurasian Steppes." In *Cultures of the Steppe Zone of Eurasia and Their Interactions with Ancient Civilizations*, vol. 2, Saint-Petersburg, 505–509.

Olbrycht, M.J. 2013. "Iranians in the Diadochi Period." In Alonso Troncoso and Anson, 159–182.

Olbrycht, M.J. 2014. "Parthian Cities and Strongolds in Turkmenistan." *International Journal of Eurasian Studies* 2: 117–125.

Olbrycht, M.J. 2015. "Arsacid Iran and the Nomads of Central Asia—Ways of Cultural Transfer." In J. Bemmann and M. Schmauder (eds.), *Complexity of Interaction Along the Eurasian Steppe Zone in the First Millennium CE*, Bonn, 333–390.

Olbrycht, M.J. 2016. "Manpower and Army Organisation in Parthia." *AncSoc* 46: 291–338.

Oleson, John P. (ed.). 2008. *The Oxford Handbook of Engineering and Technology in the Classical World*. Oxford.

Oliver, Graham J. 2007. *War, Food, and Politics in Early Hellenistic Athens*. Oxford.

Olson, D.R. and N. Torrance (eds.). 1991. *Literacy and Orality*. Cambridge.

Olson, S. Douglas. 1998. *Aristophanes*: Peace. Oxford.

Ortbals, Candice and Lori Poloni-Staudinger. 2018. "How Gender Intersects with Political Violence and Terrorism." *Oxford Research Encyclopedias: Politics*, accessed online 03/09/18: http://politics.oxfordre.com/view/10.1093/acrefore/9780190228637.001.0001/acrefore-9780190228637-e-308.

Osborne, Robin. 1986. "Island Towers: The Case of Thasos." *ABSA* 81: 167–178.

Osborne, Robin. 1987. *Classical Landscapes with Figures*. London.

Osborne, Robin. 1996/2009^2. *Greece in the Making, 1200–479 BC*. London.

Osborne, Robin. 1998. *Archaic and Classical Greek Art*. Oxford.

Osborne, Robin. 2010. "The Economics and Politics of Slavery at Athens." In id., *Athens and Athenian Democracy*, Cambridge, 27–43.

Osborne, Robin and Simon Hornblower (eds.). 1994. *Ritual, Finance, Politics: Democratic Accounts Presented to David Lewis*. Oxford.

Osborne, Robin and P. J. Rhodes (eds.). 2017. *Greek Historical Inscriptions 478-404 BC*. Oxford.

Page, D. 1962. *Poetae Melici Graeci*. Oxford.

Paillet, Jean-Louis and Henri Tréziny. 2000. "Le rempart hellénistique et la porte charretiée de Glanum." In Jean Chausserie-Laprée (ed.), *Le temps des Gaulois en Provence*, Marseille, 189–190.

Palagia, O. 1980. *Euphranor*. Leiden.

Palagia, O. 2003. "The Impact of *Ares Macedon* on Athenian sculpture." In O. Palagia and S. V. Tracy (eds.), *The Macedonians in Athens 322–229*, Oxford, 140–151.

Palagia, O. 2005. "Interpretation of Two Athenian Friezes: The Temple on the Ilissus and the Temple of Athena Nike." In Barringer and Hurwit, 177–192.

Palagia, O. 2017. "Alexander's Battles Against Persians in the Art of the Successors." In Howe et al., 177–187.

Palagia, O. 2019. "The Painted Battle of Oenoe in the Stoa Poikile and the Events of 506." In J. Fouquet, S. Herzog, K. Meese, and T. Wittenberg (eds.), *Argonautica, Festschrift für Reinhard Stupperich*, *Boreas* Beiheft 12, 61–63.

Palaima, Thomas G. 1995. "The Last Days of the Pylos Polity." In R. Laffineur and W.-D. Niemeier (eds.), *Politeia, society and state in the Aegean Bronze Age. Proceedings of the 5th International Aegean Conference, Heidelberg 1994*, vol. 2, Liège and Austin, 623–633.

Palaima, Thomas G. and Lawrence A. Tritle. 2013. "Epilogue: The Legacy of War in the Classical World." In Campbell and Tritle, 726–742.

Palmer, L.R. 1963. *Mycenaean Greek Texts*. Oxford.

Palmer, L.R. and J. Boardman. 1963. *On the Knossos Tablets*. Oxford.

Panagiotakis, N. 2004. "Contacts Between Knossos and the Pediada Region in Central Crete." *British School at Athens Studies* 12: 177–186.

Panagiotakis, N., M. Panagiotaki, and A. Sarris. 2013. "The Earliest Communication System in the Aegean." *Electryone* I: 13–27.

Pandermalis, D. 1999. *Δίον. Η ανακάλυψη*. Athens.

Panovski, Stefan and Vojislav Sarakinski. 2011. "Memnon the Strategist." *Macedonian Historical Review* 2: 7–26.

Papadopoulos, John and Sarah Morris. 2005. "Greek Towers and Slaves: An Archaeology of Exploitation." *AJA* 109: 155–225.

Parke, H. 1933. *Greek Mercenary Soldiers from the Earliest Times to the Battle of Ipsus.* Oxford.

Parke, H.W. and P. Wormell. 1956, *The Delphi Oracle.* Oxford.

Parker, R. 1996. *Athenian Religion: A History.* Oxford.

Parker, V. 1997. *Untersuchungen zum Lelantischen Krieg und zu verwandten Problemen der frühgriechischen Geschichte.* Stuttgart.

Parker, Geoffrey. 1988. *The Military Revolution. Military Innovation and the Rise of the West, 1500–1800.* Cambridge.

Parlama, L. and N. Stambolidis (eds.). 2000. *The City Beneath the City.* Athens.

Parry, M. 1987. *The Making of Homeric Verse.* Oxford.

Parvin, Meglena. 2008. "Параден нагръдник от инвентара на погребението в могила Голяма Косматка." В: Д. Димитрова (съст. и ред.). *По пътя на миналото. Сборник научни статии по повод 65-годишнината на д-р Георги Китов.* София, 163–166.

Parvin, Meglena. 2010. "Биметални брони от Тракия." В: *Великотърновският университет "Св. Св. Кирил и Методий" и българската археология -1.* Велико Търново, 225–230.

Paschidis, Paschalis. 2008. *Between City and King.* Athens.

Paschidis, Paschalis. 2013. "Agora XVI 107 and the Royal Title of Demetrius Poliorcetes." In Alonso Troncoso and Anson, 121–141.

Patterson, Cynthia. 2006. "'Citizen Cemeteries' in Classical Athens?" *CQ* 56: 48–56.

Patterson, Orlando, 1982. *Slavery and Social Death: A Comparative Study.* Cambridge, MA.

Paunov, Evgeni. 2005. "Ромфея – новооткрит тракийски меч от Западните Родопи." В: *Stephanos Archaeologicos in honorem Professoris Ludmili Getov. (Studia Archaeologica Universitatis Serdicensis, Suppl.* IV). София, 368–375.

Paunov, Eugeni and Dimitar Dimitrov. 2000. "New Data on the Use of War Sling in Thrace (4th-1st century BC)." *Archaeologia Bulgarica* 3: 44–57.

Pedersen, Poul. 2010. "The City Wall of Halikarnassos." In Riet van Bremen and Jan-Mathieu Carbon (eds.), *Hellenistic Karia. Proceedings of the First International Conference on Hellenistic Karia, Oxford, 29 June–2July 2006*, Pessac, 269–316.

Penrose, Walter Duvall. 2016. *Postcolonial Amazons: Female Masculinity and Courage in Ancient Greek and Sanskrit Literature.* Oxford and New York.

Perjés, Geza. 1970. "Army Provisioning, Logistics, and Strategy in the Second Half of the 17th Century." *AHistHung* 16: 1–51.

Perlman, Paula. 1999. "*Krētes aei leistai?* The Marginalization of Crete in Greek Thought and the Role of Piracy in the Outbreak of the First Cretan War." In Vincent Gabrielsen, Per Bilde, Troels Engberg-Pedersen, Lise Hannestad, and Jan Zahle (eds.), *Hellenistic Rhodes: Politics, Culture, and Society,* Aarhus, 132–161.

Petrakos, Vasileios Ch. 1999.*Ο Δήμος του Ραμνούντος. Σύνοψη των ανασκαφών και των ερευνών (1813–1998) vol. 1, Τοπογραφία.* Athens.

Petsas, Ph. 1966. *Ο τάφος των Λευκαδίων.* Athens.

Pettersson, M. 1992. *Cults of Apollon at Sparta. The Hyakinthia, the Gymnopaidiai and the Karneia.* Stockholm.

Pfister, F. 1924. "Eiphanie." *RE* Supp. 4, col. 277–323.

Phillips, T. (ed.) 1940. *The Roots of Strategy: A Collection of Military Classics.* Harrisburg, PA.

Picón, C.A. and Hemingway, S. 2016. *Pergamon and the Hellenistic Kingdoms of the Ancient World.* New Haven and London.

Pilipko, V.N. 2006. "Arms and Armours from Old Nisa." In Mode et al., 259–294.

Pilipko, V.P. 2015: *Stanovlenie i razvitie parfianskoi kul'tury na territorii Iuzhnogo Turkmenistana.*, Saint-Petersburg.

Pillonel, C. 2008. "Les reines hellénistiques sur les champs de bataille." In F. Bertholet, A. Bielman Sánchez, and R. Frei-Stolba (eds.), *Egypte-Grèce-Rome: les différents visages des femmes antiques: travaux et colloques du séminaire d'épigraphie grecque et latine de l'IASA 2002–2006*, Bern, 117–145.

Pimouguet-Pédarros, Isabelle. 2000. *Archéologie de la défense. Histoire des Fortifications antiques de Carie (Époques classique et hellénistique)*. Paris.

Platt, V. 2015. "Epiphanies." In E. Eidinow and J. Kindt (eds.), *The Oxford Handbook of Ancient Greek Religion*, Oxford, 491–505.

Podlecki, Anthony J. 1999 [=1966]. *The Political Background of Aeschylean Tragedy*. London.

Polignac, François. 1984. *La Naissance de la Cité Grecque. Cultes, espace et société, VIIIe-VIIe siècles avant J.-C.* Paris.

Pollitt, J.J. 1986. *Art in the Hellenistic Age*. Cambridge.

Popov, Hristo. 2017. "За одриските резиденции в Средна гора." В: Попов, Х., Цветкова, Ю. (ред.) *ΚΡΑΤΙΣΤΟΣ. Сборник в чест на професор Петър Делев*. София, 469–482.

Popp, H. 1959. *Die Einwirkung von Vorseichen, Opfern, und Festen auf die Kriegfhürung der Griechen im 5.und 4. Jahrhundert v. Ch.* Würzburg.

Potts, Samuel. 2008. *The Athenian Navy: An Investigation into the Operations, Politics and Ideology of the Athenian Fleet Between 480 and 322 BC*. Ph.D. thesis, Cardiff.

Powell, Anton. 2005. "Les femmes de Sparte (et d'autres cités) en temps de guerre." In Bertrand, 321–335.

Powell, B.B. 1991. *Homer and the Origin of the Greek Alphabet*. Cambridge.

Pownall, Frances. 1998. "What Makes a War a Sacred War?" *EMC* 17: 33–55.

Pownall, Frances. 2005. "The Rhetoric of Theopompus." *Cahiers des Études Anciennes* 42: 255–278.

Pownall, Frances. 2018. "Liberation Propaganda as a Legitimizing Principle in Warfare: Dionysius I as Antecedent to Philip and Alexander of Macedon." In Kai Ruffing, Sebastian Fink and Kerstin Droß-Krüpe (eds.), *Societies at War*, Münster (forthcoming).

Pretzler, Maria. 2007. *Pausanias: Travel Writing in Ancient Greece*. London.

Pretzler, M. and N. Barley (eds.). 2017/18. *Brill's Companion to Aeneas Tacticus*. Leiden.

Price, M.J. 1991. *The Coinage in the Name of Alexander the Great and Philip Arrhidaeus*. Zürich and London.

Pringle, Heather. 2017. "New Visions of the Vikings." *National Geographic* March 2017: 30–51.

Pritchett, W. Kendrick. 1971–1991. *The Greek State at War*, 5 vols. Berkeley and Los Angeles.

Pugachenkova, G.A. 1966. "O pantsirnom vooruzhenii parfianskogo i baktriiskogo voinstva." *Vestnik drevnei istorii* 2.96: 27–43.

Purcell, N. 1994. "South Italy in the Fourth Century." In *CAH* VI.2, 381–403.

Raaflaub, Kurt A. 1993. "Homer to Solon. The Rise of the Polis. The Written Sources." In Mogens Herman Hansen (ed.), *The Ancient Greek City-State. Symposium on the occasion of the 250th Anniversary of The Royal Danish Academy of Sciences and Letters, July, 1–4 1992*, Copenhagen, 41–105.

Raaflaub, Kurt A. 1996. "Equalities and Inequalities in Athenian Democracy." In Ober and Hedrick, 139–174.

Raaflaub, Kurt A. 1997. "Soldiers, Citizens and the Early Greek Polis." In Mitchell and Rhodes, 49–59.

Raaflaub, Kurt A. 1998. *Leaders in War and Bravery. The Ideology of War in Late Fifth-Century Athens*. The Gail Burnett Lectures in Classics. San Diego.

Raaflaub, Kurt A. 2001. "Father of All, Destroyer of All: War in Late fifth-Century Athenian Discourse and Ideology." In David McCann and Barry S. Strauss (eds.), *War and Democracy. A Comparative Study of the Korean War and the Peloponnesian War*, Armonk, NY, 307–356.

Raaflaub, Kurt A. 2004. *The Discovery of Freedom in Ancient Greece*. Chicago.

Raaflaub, Kurt A. 2005. "Homerische Krieger, Protohopliten und die Polis: Schritte zur Lösung alter Probleme." In Burkhardt Meissner, Oliver Schmitt and Michael Sommer (eds.), *Krieg, Gesellschaft, Institutionen. Beiträge zu einer vergleichenden Kriegsgeschichte*, Berlin, 229–266.

Raaflaub, Kurt A. 2007. "Introduction: Searching for Peace in the Ancient World." In id. (ed.), *War and Peace in the Ancient World*, Oxford, 1–33.

Raaflaub, Kurt A. 2008. "Homeric Warriors and Battles: Trying to Resolve Old Problems." *CW* 101: 469–483.

Raaflaub, Kurt A. 2009. "Learning from the Enemy: Athenian and Persian 'Instruments of Empire'." In John Ma, Nikolaos Papazarkadas and Robert Parker (eds.), *Interpreting the Athenian Empire*, London, 89–124.

Raaflaub, Kurt A. 2013a. "Early Greek Infantry Fighting in a Mediterranean Context." In Kagan and Viggiano, 95–111.

Raaflaub, Kurt A. 2013b. "Homer und die Agonie des Hoplitenkampfes." In Cedric Brélaz and Sylvian Fachard (eds.), 2013. *Pratiques militaires et art de la guerre dans le monde grec antique: Études offertes à Pierre Ducrey à l'occasion de son 75e anniversaire*, Paris, 17–33.

Raaflaub, Kurt A. and Hans van Wees (eds.). 2009. *A Companion to Archaic Greece*. Oxford and Malden.

Rabadzhiev, Kostadin. 2014. *Конят, колесницата и конникът. Към интерпретацията на образа в тракийската култура*. София.

Rabe, B. 2008. *Tropaia: tropē und skola. Entstehung, Funktion und Bedeutung des griechischen Tropaions*. Rahden, Westphalia.

Rackham, Oliver. 1990. "Ancient Landscapes." In Oswyn Murray and Simon R.F. Price (eds.), *The Greek City from Homer to Alexander*, Oxford, 85–111.

Radoslavova, Galena, Georgi Dzanev, and Nicolay Nikolov. 2011. "The Battle at Abritus in AD 251: Written Sources, Archaeological and Numismatic Data." *Archaeologia Bulgarica* 3: 23–49.

Radt, Timm. 2011. "The Ruins on Mount Karasis in Cilicia." In Anna Kouremenos, Sujatha Chandrasekaran and Roberto Rossi (eds.), *From Pella to Gandhara – Hybridisation and Identity in the Art and Architecture of the Hellenistic East*, Oxford, 49–64.

Radt, Timm. 2016. "Fortified Palaces and Residences in Hellenistic Times: The Upper Castle on Mount Karasis and Other Examples." In Frederiksen et al., 263–276.

Raepsaet, Georges. 2002. *Attelages et techniques de transport dans le monde gréco-romain*. Brussels.

Raepsaet, Georges. 2008. "Land Transport, Part 2: Riding, Harnesses, and Vehicles." In Oleson, 580–605.

Ragone, G. 2006. "Riflessioni sulla documentazione storica di Fidone di Argo." In Bearzot and Landucci, 27–103.

Rahe, Paul A. 2015. *The Grand Strategy of Classical Sparta: The Persian Challenge*. New Haven.

Randall, R.H. 1953. "The Erechtheum Workmen." *AJA* 57: 99–210.

Rangarajan, L.N. (trans.). 1992. *The Arthashastra*. Haryana, India.

Rapport, Mike. 2015. "The French Revolution and Early European Revolutionary Terrorism." In Law 2015b, 63–76.

Rawlings, Louis. 2000. "Alternative Agonies: Hoplite Martial and Combat Experiences beyond the Phalanx." In van Wees, 233–259.

Rawlings, Louis. 2007. *The Ancient Greeks at War*. Manchester and New York.

Rawlings, Louis. 2013. "War and Warfare in Ancient Greece." In Campbell and Tritle, 3–28.

Redfield, James M. 1975/1994. *The Tragedy of Hector. Nature and Culture in the Iliad*. Durham and London.

Reger, Gary. 1994. *Regionalism and Change in the Economy of Independent Delos, 314–167 B.C.* Berkeley, Los Angeles, and London.

Rhodes, P.J. 1981. *A Commentary on the Aristotelian* Athenaion Politeia. Oxford.

Rhodes, P.J. 1984. "Notes." In *The Athenian Constitution*, New York.

Rhodes, P.J. 2006; 2010^2. *A History of the Classical Greek World 478–323 BC*. Oxford and Malden.

Rhodes, P.J. 2011. *Alcibiades. Athenian Playboy, General and Traitor*. Barnsley.

Rhodes, P.J. 2018. *Periclean Athens*. London and New York.

Rhodes, P.J. and Robin Osborne (eds.). 2003. *Greek Historical Inscriptions: 404–323 BC*. Oxford.

Rich, John and Andrew Wallace-Hadrill (eds.). 1991. *City and Country in the Ancient World*. London.

Rich, John and Graham Shipley (eds.). 1993/1995. *War and Society in the Greek World*. London.

Richmond, J.A. 1998. "Spies in Ancient Greece." *G&R* 45: 1–18.

Ridley, Ronald T. 1979. "The Hoplite as Citizen: Athenian Military Institutions in Their Social Context." *AC* 48: 508–548.

Rigsby, Kent. 1996. *Asylia. Territorial Inviolability in the Hellenistic World.* Berkeley.

Rihll, Tracey. 2001. "Making Money in Classical Athens." In David J. Mattingly and John Salmon (eds.), *Economies beyond Agriculture in the Classical World*, Oxford and New York, 115–142.

Rihll, Tracey E. 2007. *The Catapult: A History.* Yardley, PA.

Ritter, Stefan. 2002. *Bildkontakte. Götter und Heroen in der Bildsprache griechischer Münzen des 4. Jahrhunderts v. Chr.* Berlin.

Robert, L. 1935. "Sur des inscriptions de Chios." *BCH* 59: 453–470.

Roberts, Jennifer T. 2017. *The Plague of War: Athens, Sparta, and the Struggle for Ancient Greece.* Oxford.

Robertson, N. 1987. "The True Meaning of the 'Wooden Wall'." *CP* 82: 1–20.

Robinson, David. 1941. *Olynthus, X. Metal and Minor Miscellaneous Finds, (an original contribution to Greek life).* Baltimore.

Roisman, Joseph. 1993. *General Demosthenes and His Use of Military Surprise.* Stuttgart.

Roisman, Joseph. 2012. *Alexander's Veterans and the Early Wars of the Successors.* Austin.

Roisman, Joseph. 2017. *The Classical Art of Command. Eight Generals Who Shaped the History of Warfare.* Oxford.

Roisman, Joseph and Ian Worthington (eds.). 2010. *A Companion to Ancient Macedonia.* Oxford and Malden.

Rollinger, Robert 1999. "Zur Lokalizierung von Parsu(m)a(š) in der Fārs und zu einigen Fragen der frühen Persischen Geschichte." *Zeitschrift für Assyriologie* 89: 115–139.

Rollinger, Robert 2008. "Dareios, Sanherib, Nebukadnezar und Alexander der Große: die Organisation großköniglicher Projekte, deren Infrastruktur sowie der Einsatz fremder Arbeitskräfte." In *Iranistik. Deutschsprachige Zeitschrift für iranistische Studien* 9–10 (Festschrift Kettenhofen), 147–169.

Rollinger, Robert 2014a. "Das teispidisch-achaimenidische Großreich: Ein 'Imperium' avant la lettre?" In Michael Gehler and Robert Rollinger (eds.), *Imperien und Reiche in der Weltgeschichte: Epochenübergreifende und globalhistorische Vergleiche*, Wiesbaden, 149–192.

Rollinger, Robert 2014b. "Thinking and Writing about History in Teispid and Achaemenid Persia." In Kurt A. Raaflaub (ed.), *Thinking, Recording, and Writing History in the Ancient World*, Malden, MA, 187–212.

Rollinger, Robert 2017. "Asssyria and the Far West: The Aegean World." In Eckart Frahm (ed.), *A Companion to Assyria*, Malden, MA, 275–285.

Rop, Jeffrey. 2019a. *Greek Military Service in the Ancient Near East, 401–330 BCE.* Cambridge.

Rop, Jeffrey. 2019b. "The Phocian Betrayal at Thermopylae." *Historia* 68: 413–435.

Rosenfield, J. 1967. *The Dynastic Art of the Kushans*, Los Angeles.

Rosivach, Vincent. 1985. "Manning the Athenian Fleet, 433–426." *AJAH* 10: 41–66.

Rosivach, Vincent. 2005. "Notes on the Pentakosiomedimnos' Five Hundred Medimnoi." *CQ* 55: 597–601.

Rossiter, Stuart. 1981. *Athens and Environs.* London.

Rostovtzeff, M.I. 1935. "Dura and the Problem of Parthian Art." *YCS* 5: 155–304.

Rostovtzeff, M.I. 1941. *The Social and Economic History of Hellenistic World.* Oxford.

Roth, Jonathan P. 1999. *The Logistics of the Roman Army at War (264 B.C.–A.D. 235).* Leiden.

Roth, Jonathan P. 2007. "War." In Sabin et al., 368–398.

Roth, Jonathan P. 2016. "Epilogue: Looking Ahead." In Howe and Brice 2016, 344–364.

Rousseau, Philippe. 2015. "War, Speech, and the Bow are Not Women's Business." In Fabre-Serris and Keith, 15–33.

Rousset, D. 1994. "Les frontières des cités grecques. Premières réflexions à partir du recueil des documents épigraphiques." *Cahiers du Centre Gustave Glotz* 5, 97–126.

Rubel, Alexander. 2000. *Stadt in Angst: Religion und Politik in Athens während des Peloponnesischen Krieges*. Darmstadt.

Ruberto, Antonella. 2010. "Il demos, gli aristocratici e i Persiani." *Historia* 59: 1–25.

Ruppe, Ulrich. 2007. "Neue Forschungen an der Stadtmauer von Priene—Erste Ergebnisse." *Istanbuler Mitteilungen* 57: 271–322.

Ruppe, Ulrich. 2010. "Die Stadtmauer von Priene—Zweckbau, Identifikationsobjekt oder Machtsymbol?" In Lorentzen et al., 141–163.

Russell, Frank S. 1999. *Information Gathering in Ancient Greece*. Ann Arbor.

Russell, Frank S. 2013. "Finding the Enemy: Military Intelligence." In Campbell and Tritle, 474–492.

Rüstow, Wilhelm and Hermann Köchly. 1852. *Geschichte des Griechischen Kriegswesens von der Ältesten Zeit bis auf Pyrrhos*. Aurau.

Rutherford, Andrew. 1989. *The Literature of War. Studies in Heroic Virtue*, 2nd ed. London.

Ruzicka, Stephen. 1988. "War in the Aegean, 333–331 B.C.: A Reconsideration." *Phoenix* 42: 131–151.

Ruzicka, Stephen. 1992. *Politics of a Persian Dynasty: The Hecatomnids in the Fourth Century BC*. Norman and London.

Ruzicka, Stephen. 2010. "The 'Pixodarus Affair' Reconsidered Again." In Carney and Ogden, 3–11.

Ruzicka, Stephen. 2012. *Trouble in the West. Egypt and the Persian Empire 525–323 BCE*. Oxford.

Sabin, Philip. 2007. "Land Battles." In Sabin et al., 399–433.

Sabin, Philip, Hans Van Wees and Michael Whitby (eds.). 2007. *The Cambridge History of Greek and Roman Warfare. Volume I: Greece, the Hellenistic World and the Rise of Rome*. Cambridge.

Sachs, G. 2014. *Phokaia und seine Kolonien im Westen*. Hamburg.

Saglio, Edmond. 1962. *Dictionnaire des antiquités grecques et romaines*. Graz.

Salazar, Christine. 2000. *The Treatment of War Wounds in Graeco-Roman Antiquity*. Leiden.

Salazar, Christine. 2013. "Treating the Sick and Wounded." In Campbell and Tritle, 294–311.

Sallares, Robert. 1991. *The Ecology of the Ancient Greek World*. New York.

Salmon, John B. 1977. "Political Hoplites?" *JHS* 97: 84–101.

Salmon, John B. 1984. *Wealthy Corinth. A History of the City to 338 BC*. Oxford.

Samaras, T. 2012. "Leisured Aristocrats or Warrior-Farmers? Leisure in Plato's Laws." *CP* 107: 1–20.

Sancisi-Weerdenburg, Heleen, Amélie Kuhrt and Margaret Cool Root (eds.). 1994. *Continuity and Change. Achaemenid History*, vol. 8. Leiden.

Santosuosso, Antonio. 1997. *Soldiers, Citizens, and the Symbols of War*. Boulder.

Sapouna-Sakellaraki, Efi, James J. Coulton and Ingrid R. Metzger (ed. James J. Coulton). 2002. *The Fort at Phylla, Vrachos: Excavations and Researches at a Late Archaic Fort in Central Euboea. The British School at Athens Supplementary Paper* 33, London.

Sawyer, R. (trans.). 1993. *The Seven Military Classics of Ancient China*. Boulder.

Schaps, D. 1982. "The Women of Greece in Wartime." *CP* 77: 193–213.

Schmid, Alex P. (ed.). 2011a. *The Routledge Handbook of Terrorism Research*. London and New York.

Schmid, Alex P. 2011b. "Introduction." In Schmid 2011a, 1–37.

Schmid, Alex P. 2011c. "The Definition of Terrorism." In Schmid 2011a, 39–98.

Schmitt, R. 1991. *The Bisitun Inscriptions of Darius the Great: Old Persian Text*. London.

Scholten, Joseph B. 2000. *The Politics of Plunder. Aitolians and Their Koinon in the Early Hellenistic Era, 279–217 BC*. Berkeley.

Schwandner, Ernst-Ludwig. 1977. "Die Böotische Hafenstadt Siphai." *Archäologischer Anzeiger* 1977: 513–551.

Schwandner, Ernst-Ludwig and Klaus Rheidt (eds.). 1999. *Stadt und Umland. Neue Ergebnisse der archäologischen Bau- und Siedlungsforschung. Bauforschungskolloquium in Berlin vom 7. bis 10. Mai 1997*. Mainz.

Schwartz, Adam. 2002. "The Early Hoplite Phalanx: Order or Disarray?" *C&M* 53: 31–64.

Schwartz, Adam. 2009. *Reinstating the Hoplite. Arms, Armour and Phalanx Fighting in Archaic and Classical Greece*. Stuttgart.

Schwartz, Adam. 2013. "Large Weapons, Small Greeks: The Practical Limitations of Hoplite Weapons and Equipment." In Kagan and Viggiano, 157–175.

Schwenn, F. 1920–1922. "Der Krieg in der griechischen Religion." *Archiv für Religionswissenschaft* 20: 299–322; 21: 58–71.

Schwertheim, Ute. 2010. "Monumentale Hoftore in Messene." In Lorentzen et al., 97–106.

Scott, Douglas D. and Andrew P. McFeaters. 2011. "The Archaeology of Historic Battlefields: A History and Theoretical Development in Conflict Archaeology." *Journal of Archaeological Research* 19: 103–132.

Scott, Lionel. 2005. *Historical Commentary on Herodotus Book 6. Mnemosyne Supplement* vol. 268. Leiden.

Scott, M. 2014. *Delphi.* Princeton.

Scranton, Robert L. 1941. *Greek Walls.* Cambridge, MA.

Scullard, H.H. 1974. *The Elephant in the Greek and Roman World.* Ithaca.

Sealey, Raphael. 1966. "The Origin of the Delian League." In E. Badian (ed.), *Ancient Society and Institutions. Studies Presented to Victor Ehrenberg*, Oxford, 233–255.

Sealey, Raphael. 1976. *A History of the Greek City States, ca. 700–338 B.C.* Berkeley.

Sears, Matthew. 2013. *Athens, Thrace, and the Shaping of Athenian Leadership.* Cambridge.

Sears, M. and C. Willekes. 2016. "Alexander's Cavalry Charge at Chaeronea, 338 BCE." *Journal of Military History* 80: 1017–1035.

Seifried, R. M. and W. A. Parkinson. 2014. "The Ancient Towers of the Paximadi Peninsula, Southern Euboia." *Hesperia* 83: 277–313.

Sekunda, Nicholas V. 1983. "The Rhomphaia: A Thracian Weapon of the Hellenistic Period." In A. G. Poulter (ed.). *Ancient Bulgaria*, vol. 1. Nottingham, 275–288.

Sekunda, Nicholas V. 1985. "Achaemenid Colonization in Lydia," *REA* 87: 7–30.

Sekunda, Nicholas V. 1988a. "Some Notes on the Life of Datames." *Iran* 26: 35–53.

Sekunda, Nicholas V. 1988b. "Achaemenid Military Terminology." *Archäologische Mitteilungen aus Iran* 21: 69–77.

Sekunda, Nicholas V. 1988c. "Persian Settlement in Hellespontine Phrygia." In Amélie Kuhrt and Heleen Sancisi-Weerdenburg (eds.), *Method and Theory. Proceedings of the Fifth Achaemenid History Workshop. Achaemenid History, vol. 3*, Leiden, 175–196.

Sekunda, Nicholas Victor. 1991. "Achaemenid Settlement in Caria, Lycia and Greater Phrygia." In Heleen Sancisi-Weerdenburg and Amélie Kuhrt (eds.), *Asia Minor and Egypt. Achaemenid History*, vol. 6, Leiden, 83–143.

Sekunda, Nicholas V. 1992. *The Persian Army, 560–330 BC.* London.

Sekunda, Nicholas V. 2007. "Land Forces." In Sabin et al., 325–356.

Sekunda, Nicholas V. 2010. "The Macedonian Army." In Roisman and Worthington, 446–471.

Sekunda, Nicholas V. 2013. "War and Society in Greece." In Campbell and Tritle, 199–215.

Sekunda, Nicholas V. 2017. "*Lochagoi* under Iphicrates: Forming a Mercenary Army in the Fourth Century." In T. Ñaco de Hoyo and F. Lopez-Sanchez (eds.), *War, Warlords, and Interstate Relations in the Ancient Mediterranean, Impact of Empire* 28. Leiden, 64–89.

Sekunda, Nicholas V. and Bogdan Burliga (eds.). 2014. *Iphicrates, Peltasts and Lechaeum.* Gdansk.

Sergeev, G. 1966. "Олонешткий античный клад." *Вестник древней истории* 2: 132–142.

Sergidis, K. 2012. *The Pursuit of Power and Security: The Influence of Natural Resources and Geography on Athenian Foreign Policy.* Ph.D. diss., University of Nottingham.

Sevinç, N., R. Körpe, M. Tombul, C.B. Rose, D. Strahan, H. Kiesewetter, and J. Wallrodt. 2001. "A New Painted Sarcophagus from Çan." *Studia Troica* 11: 383–420.

Shaw, Brent D. 1984. "Bandits in the Roman Empire." *P&P* 105: 3–52.

Shaw, Brent. 2014. "The Young Finley: Observations on Naiden, Perry, and Tompkins." In Naiden and Talbert, 267–280.

Shay, Jonathan. 1994. *Achilles in Vietnam. Combat Trauma and the Undoing of Character.* New York.

Shear, T. L., Jr. 2016. *Trophies of Victory. Public Building in Periklean Athens.* Princeton.

Shipley, Graham. 1987. *A History of Samos, 800–188 BC.* Oxford.

Shipley, Graham. 1993. "The Limits of War." In Rich and Shipley, 1–24.
Shrimpton, G. S. 2006. "Oh Those Rational Athenians! Civil War, Reconciliation and Public Memory in Ancient Athens." *Mouseion* 6: 293–311.
Sick, David H. 2002. "An Indian Perspective on the Graeco-Roman Elephant." *The Ancient World* 33: 126–146.
Sidnell, P. 2006. *Warhorse: Cavalry in Ancient Warfare*. London.
Simón, Francisco Marco. 2016. "Insurgency or State Terrorism? The Hispanic Wars in the Second Century BCE." In Howe and Brice 2016, 221–247.
Simpson, R. Hope. 1959. "Antigonos the One-Eyed and the Greeks." *Historia* 6: 371–373.
Simpson, R. Hope and D.K. Hagel. 2006. *Mycenaean Fortifications, Highways, Dams and Canals*. Sävedalen.
Singor, Henk. 2009. "War and International Relations." In Raaflaub and van Wees, 585–603.
Sismanidis, Konstantinos. 2003. *Das antike Stageira*. Athens.
Smith, Gavin. 2009. "Formal Culture, Practical Sense and the Structures of Fear in Spain." *Anthropologica* 51: 279–288.
Smith, J. 1992/1993. "The Pylos Jn Series." *Minos* 27.8: 167–259.
Snail, Daniel L. 2008. *On Deep History and the Brain*. Berkeley.
Snoddon, Robert. 2007. "Piracy and Maritime Terrorism: Naval Responses to Existing and Emerging Threats to the Global Seaborne Economy." In Lehr 2007, 225–240.
Snodgrass, Anthony M. 1964. *Early Greek Armour and Weapons from the End of the Bronze Age to 600 BC*. Edinburgh.
Snodgrass, Anthony M. 1965a. "The Hoplite Reform and History." *JHS* 85: 110–122.
Snodgrass, Anthony M. 1965b. "L'introduzione degli opliti in Grecie e in Italia." *RSI* 77: 434–444.
Snodgrass, Anthony M. 1967. *Arms and Armor of the Greeks*. London.
Snodgrass, Anthony M. 1971. *The Dark Age of Greece. An Archeaeological Survey of the Eleventh to the Eighth Centuries BC*. Edinburgh.
Snodgrass, Anthony M. 1980. *Archaic Greece. The Age of Experiment*. Berkeley.
Snodgrass, Anthony M. 1986. "Interaction by Design: The Greek City-States." In Colin Renfrew and John F. Cherry (eds.), *Peer Polity Interaction and Socio-Political Change*, Cambridge, 47–58.
Snodgrass, Anthony M. 1993. "The 'Hoplite Reform' Revisited." *DHA* 19: 47–61.
Snodgrass, Anthony M. 2013. "Setting the Frame Chronologically." In Kagan and Viggiano, 85–94.
Sokolicek, Alexander. 2009. *Diateichismata: zu dem Phänomen innerer Befestigungsmauern im griechischen Städtebau. Jahreshefte des Österreichischen Archäologischen Institutes in Wien, Ergänzungsheft* 11. Vienna.
Sommerstein, Alan H. 1980. *The Comedies of Aristophanes, vol. I.* Acharnians. Warminster.
Sommerstein, Alan H. 1996. *Aeschylean Tragedy*. Bari.
Spawforth, Anthony J.S. 2007. *Court and Court Society in Ancient Monarchies*. Cambridge.
Speidel, M. 1970. "The Captor of Decebalus: A New Inscription from Philippi." *JRS* 60: 142–153.
Spence, Iain G. 1990. "Perikles and the Defence of Attika during the Peloponnesian War." *JHS* 110: 91–109.
Spence, Iain G. 1993/1995²/1999². *The Cavalry of Classical Greece: A Social and Military History with Particular Reference to Athens*, 2nd ed. 1995, reprinted 1999. Oxford.
Spendel, A. 1915. *Untersuchungen zum Heerwesen der Diadochen*. Breslau.
Sprawski, Slawomir. 2005. "All the King's Men. Thessalians and Philip II's Designs on Greece." In D. Musiał (ed.), *Society and Religions. Studies in Greek and Roman History*, Toruń, 31–49.
Squillace, Giuseppe. 2010. "Consensus Strategies Under Philip and Alexander: The Revenge Theme." In Carney and Ogden, 69–80.
Squillace, Giuseppe. 2011. "Filippo e Anfipoli nella propaganda macedone e antimacedone." *Eirene* 47: 106–117.
Stadter, Philip A. 1989. *A Commentary on Plutarch's* Pericles. Chapel Hill.
Stadter, Philip A. 1991. "Fictional Narrative in the *Cyropaideia*." *AJP* 112: 461–491.

Stein-Hölkeskamp, E. 1999. "Lelantischer Krieg." *DNP* 7: 38–39 (English digitized version in Brill's *New Pauly*).

Stewart, A. 1993. *Faces of Power*. Berkeley, Los Angeles, and Oxford.

Stewart, A. 2004. *Attalos, Athens and the Akropolis. The Pergamene "Little Barbarians" and their Roman and Renaissance Legacy*. Cambridge.

Stewart, A. 2008. *Classical Greece and the Birth of Western Art*. Cambridge.

Stewart, A. 2014. *Art in the Hellenistic World*. Cambridge.

Stewart, A. 2018. "Notes on the Origins and Early Development of the 'Agora of the Kerameikos'." In C. M. Draycott, R. Raja, K. Welch, and W. T. Wootton (eds.), *Visual Histories of the Classical World*. Essays in Honour of R.R.R. Smith, Turnhout, 299–308.

Stohl, Michael. 1984. "International Dimensions of State Terrorism." In Stohl and Lopez 1984a, 43–58.

Stohl, Michael. 1988. "States, Terrorism and State Terrorism: The Role of the Superpowers." In Robert Slater and Michael Stohl (eds.), *Current Perspectives on International Terrorism*, Basingstoke, 155–205.

Stohl, Michael. 2006. "The State as Terrorist: Insights and Implications." *Democracy and Security* 2: 1–25.

Stohl, Michael and George A. Lopez (eds.). 1984a. *The State as Terrorist: The Dynamics of Governmental Violence and Repression*. Westport.

Stohl, Michael and George A. Lopez. 1984b. "Introduction." In Stohl and Lopez 1984a, 3–10.

Stoll, O. 2013. "Terror im Gebirge: Xenophon und die Anforderungen transkultureller Kriegführung. Der Rückzug des griechischen Söldnerkontingentes in Xenophons Anabasis und die Schilderung von Flussübergängen, Pässen und Bergbewohnern." *Göttinger Forum für Altertumswissenschaft* 16: 277–345.

Stoll, O. 2016. "Aus Wissen wird Können: 'Amplius prodest locus saepe quam virtus' (Veg. mil. III 26,11). Landschaft in der militärwissenschaftlichen Fachliteratur der Antike." *MBAH* 33: 87–130.

Storch, Rudolph H. 1998. "The Archaic Greek 'Phalanx', 750–650 BC." *AHB* 12: 1–7.

Stoyanov, Totko. 2003. "On the Spurs' Development in Thrace, Macedonia and Illyria during the Early Hellenistic Times." In ПЪТЯТ. *Сборник научни статии, посветени на живота и творчеството на д-р Георги Китов*, Sofia, 198–203.

Stoyanov, Totko. 2005. "Chalkidian-type Helmet from Gilyamo Shivachevo, Sliven District." In *Stephanos Arcaeologicos in honorem Professoris Ludmili Getov (Studia Archaeologica Universitatis Serdicensis, Suppl. IV)*. Sofia, 646–654.

Stoyanov, Totko. 2008a. "За обсадната и отбранителна техника в Ранноелинистическа Тракия." В: *Проблеми и изследвания на тракийската култура III*, Казанлък, 46–63.

Stoyanov, Totko. 2008b. "Бележки за въоръжението и снаряжението представени в люнета на царската гробница при Свещари." В: *Погребални практики и ритуали. Есенни четения Сборяново V*, Варна, 59–77.

Stoyanov, Totko. 2015. "Warfare." In Valeva et al., 426–442.

Stoyanov, Totko, Rumen Mikov and Tanya Dzhanfezova. 2013. "Надгробна могила от ранната елинистическа епоха край с. Кабиле, Ямболско." *Bulgarian E-Journal for Archaeology (Бе-JA)* 2: 245–314.

Stoyanova, Galina. 2015. "Богати гробни комплекси' на воини-конници от началото на IV—до средата на III в. пр. Хр. в Южна Тракия." В: Гюрова, М. (ред. и съст.) Рабаджиев, К., Стоянов, Т. (ред.). *Bulgarian E-Journal of Archaeology, Supplementum* 4 (Филови четения. Културно-историческото наследство: проучване и опазване. Доклади от докторантска конференция 15–16 декември 2014 г.), [online], 109–130.

Strauss, Barry S. 1986. *Athens After the Peloponnesian War: Class, Faction and Policy 403–386 B.C.* Ithaca.

Sturken, Marita. 1997. *Tangled Memories. The Vietnam War, the Aids Epidemic, and the Politics of Remembering*. Berkeley.

Stylianou, P. J. 1999. *A Historical Commentary on Diodorus Siculus, Book 15.* Oxford.

Szuscikiewicz, Paul. 2011. "Ancient Greece and Rome." In James Ciment (ed.), *World Terrorism: An Encyclopedia of Political Violence from Ancient Times to the Post-9/11 Era.* Armonk, NY, 43–46.

Szymanski, T. 1908. *Sacrificia Graecorum in bellis militaria.* Marburg.

Tacheva-Hitova, Margarita. 1971. "Извори за военното дело на траките." *Военноисторически сборник* 1: 51–63.

Tagliamonte G. 1994. *I figli di Marte: mobilità, mercenari e mercenariato italici in Magna Grecia e Sicilia.* Rome.

Talbert, Richard J.A. (ed. and trans.). 1988. *Plutarch on Sparta.* London.

Tancheva-Vasileva, Neli. 2007. "Въоръжение от хинтерланда на тракийския град Кабиле." *Известия на музеите от Югоизточна България* 33: 37–43.

Tänzer, Kurt. 1912. *Das Verpflegungswesen der griechischen Heere bis auf Alexander d. Gr.* Jena.

Tarn, W.W. 1930/1984. *Hellenistic Military and Naval Developments.* Cambridge; reprinted Chicago.

Tarn, W.W. 1948. *Alexander the Great.* Cambridge.

Tashkhodzhaev, Sh. S. 1963. "Razrez gorodskoi stenÿ Giaur-kalÿ." *Trudÿ Yuzhno-Turkmenskoï arkheologicheskoï kopleksnoï ékspeditsii* 12: 95–118.

Tausend, K. 1987. "Der Lelantische Krieg - ein Mythos?" *Klio* 69: 499–515.

Taylor, Donathan and Vannick Gautron. 2015. "Pre-modern Terrorism: The Cases of the Sicarii and the Assassins." In Law 2015b, 28–45.

Taylor, M.J. 2016. "The Battle Scene of Aemilius Paullus's Pydna Monument." *Hesperia* 85: 559–576.

Tedeschi, G. 1975. "La guerra lelantina e la cronologia esiodea." In Andrea Balanza and Paola Càssola Guida (eds.), *Studi triestini di antichità in onore di L.A. Stella*, Trieste, 149–167.

Teitler, G. 1977. *The Genesis of the Professional Officers' Corps*, trans. C.N. Ter Heide-Lopy. Beverly Hills.

Teleaga, Emilian. 2008. *Griechische Importe in den Nekropolen an der unteren Donau 6.Jh. – Anfang des 3. Jhs. v. Chr.* Rahden.

Teleaga, Emilian. 2015. *Die Anfänge der figuralen thrakischen Kunst in dem 5. Jahrhundert v. Chr.* Rahden.

Theocharaki, Anna Maria. 2011. "The Ancient Circuit Wall of Athens. Its Changing Course and the Phases of Construction." *Hesperia* 80: 71–156.

Thonemann, Peter. 2005. "The Tragic King: Demetrios Poliorketes and the City of Athens." In Olivier Hekster and Richard Fowler (eds.), *Imaginary Kings. Royal Images in the Ancient Near East, Greece and Rome*, Stuttgart, 63–86.

Thorne, James A. 2001. "Warfare and Agriculture: The Economic Impact of Devastation in Classical Greece." *GRB* 2: 225–253.

Thornton, John. 2006. "Terrore, terrorismo e imperialismo. Violenza e intimidazione nell'età della conquista romana." In Urso 2006, 157–196.

Tilly, Charles. 1985. "War Making and State Making as Organized Crime." In Peter B. Evans, Dietrich Rueschemeyer and Theda Skocpol (eds.), *Bringing the State Back In*, Cambridge, 169–191.

Todorov, Ivan. 2008. "Златен пръстен с желязно покритие от Рачева могила в околностите на град Мъглиж, Старозагорско." В: *PHOSPHORION. Studia in honorem Mariae Cicikova,* София, 219–224.

Topal, Denis. 2017. "Скифские мечи к югу от Дуная: болгарская группа памятников." *Добруджа* 32: 527–535.

Topouzi, S., A. Sarris, Y. Pikoulas, S. Mertikas, X. Frantzis, and A. Giourou. 2002. "Ancient Mantineia's Defence Network Reconsidered Through a GIS Approach." In G. Burenhult and J. Arvidsson (eds.), *Archaeological Informatics: Pushing the Envelope. CAA2001. Computer Applications and Quantitative Methods in Archaeology. Proceedings of the 29th Conference, Gotland, April 2001*, Oxford, 559–566.

Torbov, Narcis. 1998. "Spurs from the 2nd Century BC–1st Century AD Found in Northwestern Bulgaria." *Archaeologia Bulgarica* 1: 54–63.
Torbov, Narcis. 2016. *Антично въоръжение. От камъка до барута. Сбирка "Ватеви."* Пловдив.
Torelli, M. 1996. "The Encounter with the Etruscans." In G. Pugliese Carratelli (ed.), *The Western Greeks*, Venice, 567–576.
Tozzi, Pierluigi. 1978. *La rivolta ionica*. Pisa.
Tréziny, Henri. 1986. "Remarques sur la fonction du rempart hellénistique de Saint-Blaise." In Michel Bats and Henri Tréziny (eds.), *Le territoire de Marseille grecque: Actes de la Table-ronde d'Aix-en-Provence, 16 mars 1985*, Aix-en-Provence, 145–151.
Tréziny, Henri. 2006. "Les fortifications archaïques dans le monde grec colonial d'Occident." In Ampolo, 254–266.
Tréziny, Henri. 2016. "Archaeological Data on the Foundation of Megara Hyblaea. Certainties and Hypotheses." In L. Donnellan (ed.), *Conceptualising Early Colonisation*, Brussels, 167–178.
Tritle, Lawrence A. 1993. "Continuity and Change in Athenian Strategia." *AHB* 7: 125–129.
Tritle, Lawrence A. 2000. *From Melos to My Lai. War and Survival*. London.
Tritle, Lawrence A. 2010. *A New History of the Peloponnesian War*. Oxford.
Tritle, Lawrence A. 2014. "Ravished Minds in the Ancient World." In Meineck and Konstan, 87–103.
Tritle, Lawrence A. 2019. "Post-Traumatic Stress (Disorder) in the Post-9/11 World." In Jason W. Warren (ed.), *Landpower in the Long War: Projecting Force after 9/11*. Lexington, 307–321.
Tritle, Lawrence A. and J.W. Warren (eds.). 2018. *The Many Faces of War*. Los Angeles.
Tröster, Manuel. 2009. "Roman Hegemony and Non-state Violence: A Fresh Look at Pompey's Campaign against the Pirates." *G&R* 56: 14–33.
Trundle, Matthew. 2001. "The Spartan Revolution: Hoplite Warfare in the Late Archaic Period." *War and Society* 19: 1–17.
Trundle, Matthew. 2004. *Greek Mercenaries. From the Late Archaic Period to Alexander*. London and New York.
Trundle, Matthew. 2010. "Coinage and the Transformation of Greek Warfare." In Fagan and Trundle, 227–252.
Trundle, Matthew. 2016. "Coinage and Economics of the Athenian Empire." In Jeremy Armstrong (ed.), *Circum Mare. Themes in Ancient Warfare*, Leiden, 65–79.
Trundle, Matthew. (2020). "Wealth and the Logistics of Greek Warfare." In Lee L. Brice (ed.), *New Approaches to Greek and Roman Warfare*, Hoboken, 17–27.
Tuplin, Christopher. 1987. "The Leuctra Campaign: Some Outstanding Problems." *Klio* 69: 72–107.
Tuplin, Christopher 2010. "All the King's Horse." In Fagan and Trundle (eds.), 101–182.
Tsimbidou-Avloniti, M. 2011. "Agios Athanasios, Macedonian Tomb III: An Iron-Clad Companion of the King." In S. Pingiatoglou and T. Stefanidou-Tiveriou (eds.), *νάματα. Τιμητικός τόμος για τον καθηγητή Δημήτριο Παντερμαλή*, Thessaloniki, 351–363.
Tzvetkova, Julia. 2018. *ГИС и Тракология. Приложения на Географските информационни системи в изследванията на тракийската история*. София.
Urso, Gianpaolo (ed.). 2006. *Terror et pavor. Violenza, intimidazione, clandestinità nel mondo antico*. Pisa.
Vaillant, J.B.P. (ed.). 1858–1870. *Correspondence de Napoléon Ier: pupliée par ordre de l'empereur Napoleeon III*. Paris.
Valeva, Julia. 1992. "Изображения на шлемове и каузия в Казанлъшката гробница." *Проблеми на изкуството* 2: 19–26.
Valeva, Julia. 2005. *The Painted Coffers of the Ostrusha Tomb*. Veliko Turnovo.
Valeva, J., E. Nankov, and D. Graninger (eds.). 2015. *A Companion to Ancient Thrace*. London and Malden.
van Creveld, Martin. 1977. *Supplying War: Logistics from Wallenstein to Patton*. Cambridge.

van Creveld, Martin. 1989. *Technology and War: From 2000 B.C. to the Present.* New York and London.

van de Maele, Symphorien. 1992. "Le réseau mégarien de défence territorial contre l'Attique à l'époque classique (Ve et IVe s. av. J.-C.)." In van de Maele and Fossey, 95–99.

van de Maele, Symphorien and John M. Fossey (eds.). 1992. *Fortificationes Antiquae. Including the Papers of a Conference Held at Ottawa University, October 1988.* Amsterdam.

van de Mieroop, Marc. 2007. *A History of the Ancient Near East ca. 3000–323 BC.* Oxford.

van Wees, Hans. 1986. "Leaders of Men? Military Organization in the Iliad." *CQ* 36: 285–303.

van Wees, Hans. 1988. "Kings in Combat. Battles and Heroes in the *Iliad*." *CQ* 38: 1–24.

van Wees, Hans. 1992. *Status Warriors: War, Violence and Society in Homer and History.* Amsterdam.

van Wees, Hans. 1994a. "The Homeric Way of War: The *Iliad* and the Hoplite Phalanx—I." *G&R* 41: 1–18.

van Wees, Hans. 1994b. "The Homeric Way of War: The *Iliad* and the Hoplite Phalanx—II." *G&R* 41: 131–155.

van Wees, Hans. 1995. "Politics and the Battlefield: Ideology in Greek Warfare." In Anton Powell (ed.), *The Greek World*, London, 153–178.

van Wees, Hans. 1996. "Heroes, Knights and Nutters, Warrior Mentality in Homer." In Lloyd, 1–86.

van Wees, Hans. 1997. "Homeric Warfare." In Morris and Powell, 668–693.

van Wees, Hans. 2000a. "The Development of the Hoplite Phalanx. Iconography and Reality in the Seventh Century". In van Wees, 125–166.

van Wees, Hans (ed.). 2000b. *War and Violence in Ancient Greece.* London.

van Wees, Hans. 2001. "The Myth of the Middle-class Army: Military and Social Status in Ancient Athens." In Bekker-Nielsen and Hannestad, 45–71.

van Wees, Hans. 2002. "Tyrants, Oligarchs and Citizen Militias." In Chaniotis and Ducrey, 61–82.

van Wees, Hans. 2003. "Conquerors and Serfs: Wars of Conquest and Forced Labour in Archaic Greece." In Nino Luraghi and Susan Alcock (eds.), *Helots and Their Masters in Laconia and Messenia: Histories, Ideologies, Structures*, Cambridge, MA, 33–80.

van Wees, Hans. 2004. *Greek Warfare: Myths and Realities.* London.

van Wees, Hans. 2006. "Mass and Elite in Solon's Athens: The Property Classes Revisited." In Josine H. Blok and André Lardinois (eds.), *Solon of Athens. New Historical and Philological Approaches*, Leiden, 351–389.

van Wees, Hans. 2007. "War and Society." In Sabin et al., 273–299.

van Wees, Hans. 2009. "The Economy." In Raaflaub and Hans van Wees, 444–467.

van Wees, Hans. 2010. "'Those Who Sail Are to Receive a Wage': Naval Warfare and Finance in Archaic Eretria." In Fagan and Trundle, 205–226.

van Wees, Hans. 2013. "Farmers and Hoplites: Models of Historical Development." In Kagan and Viggiano, 222–255.

Vandiver, Elizabeth. 2010. *Stand in the Trench, Achilles.* Oxford.

Vandorpe, K. and M. Waelkens. 2007. "Protecting Sagalassos' Fortress of the Akra: Two Large Fragments of an Early Hellenistic Inscription." *Ancient Society* 37: 121–139.

Vansina, J. 1961. *Oral Tradition. A Study in Historical Methodology.* London.

Vansina, J. 1985. *Oral Tradition as History.* Nairobi and Oxford.

Vasilev, Vasil. 1980. "Производство на тракийски шлемове в Родопите." *Археология* 3: 1–18.

Vasilev, Vasil. 1983. "Повреди и поправки на шлемове от Тракия V–III в. пр. н.е." *Музеи и паметници на културата* 5–6: 5–12.

Vasileva, Miglena. 2015. "Нови перспективи пред археологията: археометричен анализ на предмети от медни сплави от VIII–III в. пр. Хр. от територията на югоизточна България." *Bulgarian E-Journal of Archaeology* 2: 137–161.

Vassileva, Maya. 2015. "Persia." In Valeva et al., 320–336.

Vela-Tejada, José. 2008. "El monólogo de Casandra: juicio y condena de la guerra injusta." *Faventia* 30: 209–222.

Velkov, V. and A. Fol. 1977. *Les Thraces en Egypte gréco romaine. Bulgarian Academy of Arts, Institute for Thracology, Thracian Studies*, vol. 4. Sofia.

Venco Ricciardi, R. 1998. "Pictorial Graffiti in the City of Hatra." *Electrum* 2: 187–205.

Venedikov, Ivan. 1948. "Разкопките в некропола на Аполония през 1946 г." *Разкопки и проучвания, 2. Аполония на Черно море*. София.

Ventris, M. and J. Chadwick. 1956/1973^2. *Documents in Mycenaean Greek*. Cambridge.

Vernant, Jean-Pierre (ed.). 1968. *Problèmes de la Guerre en Grèce Ancienne*. Paris.

Vernant, Jean-Pierre. 2001. *El individuo, la muerte y el amor en la antigua Grecia*. Barcelona [= *L'individu, la mort, l'amour*. Paris. 1989].

Vidal-Naquet, Pierre. 1968. "La tradition de l'hoplite athénien." In Vernant, 161–181.

Viggiano, Gregory F. 2013. "The Hoplite Revolution and the Rise of the Polis." In Kagan and Viggiano, 112–133.

Viggiano, Gregory and Hans van Wees. 2013. "The Arms, Armor, and Iconography of Early Greek Hoplite Warfare." In Kagan and Viggiano, 57–73.

Vilatte, Sylvie. 1999. "Les philosophes devant les guerres". In Patrice Brun (coord.), *Guerres et sociétés dans les mondes grecs*, Paris. 217–243.

Visonà, P. 2016. "Controlling the Chora. Topographical Investigations in the Territory of Locri Epizephyrii (Southeastern Calabria, Italy) in 2013–2015." *The Journal of Fasti Online* 351: 1–17.

Visser, E. 2006. "Homer and Oral Poetry." In S. Deger-Jalkotzy and I. S. Lemos (eds.), *Ancient Greece: From the Mycenaean Palaces to the Age of Homer*, Edinburgh, 427–438.

von Gall, H. 1990. *Das Reiterkampfbild in der iranischen und iranisch beeinflussten Kunst parthischer und sasanidischer Zeit*. Berlin.

von Gerkan, Armin. 1935. *Die Stadtmauer* (= Th. Wiegand (ed.) *Milet* 2.3). Berlin.

von Graeve, Volkmar. 2006. "Milet." In Wolfgang Radt (ed.), *Stadtgrabung und Stadtforschung im westlichen Kleinasien. Geplantes und Erreichtes*, Byzas 3, Istanbul, 241–262.

Voutiras, Emmanuel. 1994. "Wortkarge Söldner? Ein Interpretationsvorschlag zum neuen Poseidippos." *ZPE* 104: 27–31

Wade-Gery, H. 1944. "The Spartan *Rhetra* in Plutarch, *Lycurgus* VI.C: What Is the *Rhetra*?" *CQ* 38: 116–124.

Walbank, Frank W. 1957–1979. *A Historical Commentary on Polybius*, 3 vols. Oxford.

Walker, K.G. 2004. *Archaic Eretria. A Political and Social History from the Earliest Times to 490 BC*. London and New York.

Walser, Gerold. 1987. "Persischer Imperialismus und griechische Freiheit." *Achaemenid History* 2: 155–165.

Warner, Rex (trans.). 1954. *Thucydides. The Peloponnesian War*. New York.

Waters, Matthew. 2014. "Earth, Water, and Friendship with the King: Argos and Persia in the Mid-Fifth Century." In Michael Kozuh, Wouter Henkelman, Charles E. Jones and Christopher Woods (eds.), *Extraction & Control: Studies in Honor of Matthew W. Stolper = Studies in Ancient Oriental Civilization*, Number 68, Chicago, 331–336.

Webber, Christopher. 2011. *The Gods of Battle. The Thracians at War 1500 BC–150 AD*. Barnsley.

Weber, Gregor. 1993. *Dichtung und höfische Gesellschaft. Die Rezeption von Zeitgeschichte am Hof der ersten drei Ptolemäer*. Stuttgart.

Weber, Maximilian. 1922. *Wirtschaft und Gesellschaft*. Tübingen.

Weigall, David. 2002. *International Relations. A Concise Companion*. London.

Weinstein, E. A., M.D. 1995. "Conversion Disorders." In R. Zajtchuk (ed.), *Textbook of Military Medicine*, 4 Parts. Falls Church, 383–407.

Weiser, W. 1989. "Die Eulen von Kyros dem Jüngeren." *ZPE* 76, 267–296.

Welles, Charles B. 1970. *Alexander and the Hellenistic World*. Toronto.

Wellman, M. 1905. "Elefant." In *RE* V.2: col. 2248–2257.

Welwei, Karl-Wilhelm. 1974. *Unfreie im Antiken Kriegsdienst. Vol. 1. Forschungen zur Antiken Sklaverei* 5. Wiesbaden.

Welwei, Karl-Wilhelm. 2006. "The Peloponnesian War and Its Aftermath." In Kinzl, 526–543.
West, M.L. 2003. *Greek Epic Fragments From the Seventh to Fifth Century BC. Loeb Classical Library* 497. Cambridge, MA.
Westlake, H.D. 1936. "The Medism of Thessaly." *JHS* 56: 12–24.
Westlake, H.D. 1994. "Dion and Timoleon." In *CAH* VII.2, 693–722.
Westwood, J.N. 1987. *Endurance and Endeavour. Russian History 1812–1986*, 3rd ed. Oxford.
Wheatley, Pat and Elizabeth Baynham. 2015. *East and West in the World Empire of Alexander the Great: Essays in Honour of Brian Bosworth.* Oxford.
Wheeler, Everett L. 1987. "Ephoros and the Prohibition of Missiles." *TAPA* 111: 157–182.
Wheeler, Everett L. 1988. *Strategy and the Vocabulary of Military Trickery.* Leiden.
Wheeler, Everett L. 1991a. "The General as Hoplite." In Hanson, 121–170.
Wheeler, Everett L. 1991b. "Terrorism and Military Theory: An Historical Perspective." *Terrorism and Political Violence* 3: 6–33.
Wheeler, Everett L. 1996. "The Laxity of the Syrian Legions." In D. Kennedy (ed.), *The Roman Army in the East, JRA Supp. 18*, 229–276.
Wheeler, Everett L. 2007a. "Battle. A. Land Battles." In Sabin et al., 186–223.
Wheeler, Everett L. (ed.). 2007b. *The Armies of Classical Greece.* Aldershot.
Wheeler, Everett L. 2009. "Shock and Awe: Battles of the Gods in Roman Imperial Warfare, Part I." In Catherine Wolff and Yann Le Bohec (eds.), *L'armée romaine et la religion sous le Haut-Empire* (= *Collection d'Études Romaines et Gallo-Romaines*, n.s. 33). Lyon and Paris, 225–267.
Wheeler, Everett L. 2011. "Greece: Mad Hatters and March Hares." In Lee L. Brice and Jennifer T. Roberts (eds.), *Recent Directions in the Military History of the Ancient World*, *AAH* 10. Claremont, CA, 53–104.
Wheeler, E.L. and Strauss, B. 2007. "Battle." In Sabin et al., 186–247.
Whitehead, David. 1982. "Two Notes on the Old Oligarch." *LCM* 7: 119–120.
Whitehead, David. 1988. "*Klope polemou*: "Theft" in Ancient Greek Warfare." *C&M* 39: 43–53.
Whitehead, David. 1990/2001. *Aineias the Tactician, How to Survive Under Siege.* Oxford and (reissued with addenda and corrigenda) Bristol.
Whitehead, David. 2008. "Fact and Fantasy in Greek Military Writers." *AAntHung* 48: 139–155.
Whitehead, David. 2015. "Alexander the Great and the Mechanici." In Wheatley and Baynham, 75–91.
Whitehead, David. 2016. *Philo Mechanicus*, On Sieges. Stuttgart.
Whitehead, David and P. Henry Blyth. 2004. *Athenaeus Mechanicus*, On Machines. Stuttgart.
Wickkiser, Bronwen L. 2008. *Asklepios, Medicine, and the Politics of Healing in Fifth-Century Greece. Between Craft and Cult.* Baltimore.
Wiemer, Hans-Ulrich. 2002. *Krieg, Handel und Piraterie. Untersuchungen zur Geschichte des hellenistischen Rhodos.* Berlin.
Wiesehöfer, Josef. 2013. "Herodot und ein persisches Hellas." In Boris Dunsch and Kai Ruffing (eds.), *Herodots Quellen—Die Quellen Herodots*, Wiesbaden, 273–283.
Wilcox, P. 1986. *Rome's Enemies (3): Parthians and Sassanid Persians.* London.
Willekes, Carolyn. 2013. *From the Steppe to the Stable: Horses and Horsemanship in the Ancient World.* Ph.D. diss., University of Calgary.
Willekes, Carolyn. 2015. "Equine Aspects of Alexander the Great's Macedonian Cavalry." In Howe et al., 47–58.
Willekes, Carolyn. 2016. *The Horse in the Ancient World: From Bucephalus to the Hippodrome.* London.
Wills, Garry. 1992. *Lincoln at Gettysburg. The Words that Remade America.* New York.
Wilson, Edward O. 1979. *On Human Nature.* Cambridge; repr. with a new Preface, 2004.
Wilson, Edward O. 1998. *Consilience. The Unity of Knowledge.* New York.
Wilson, Edward O. 2014. *The Meaning of Human Existence.* New York.
Winkelmann, S. 2006. "Waffen auf parthischen Münzen." *Parthica* 8: 131–152.
Winter, Frederick E. 1971. *Greek Fortifications.* London.

Wirth, Gerhard. 1985. *Philipp II*. Stuttgart.

Wirth, Gerhard. 1999. *Hypereides, Lykurg und die autonomia der Athener. Ein Versuch zum Verständnis einiger Reden der Alexanderzeit*. Vienna.

Wirth, Gerhard. 2004. "Der Epitaphios des Hypereides und das Ende einer Illusion." In R. Kinsky (ed.), *Diorthoseis. Beiträge zur Geschichte des Hellenismus und zum Nachleben Alexanders des Großen*, Munich and Leipzig, 24–50.

Wiseman, D.J. 1953. *The Alalakh Tablets*. London.

Wójcikowski, R. 2014. "The graffito from Dura-Europos: Hybrid Armor in Parthian-Sasanian Iran." *Anabasis. Studia Classica et Orientalia* 4: 233–248.

Wolpert, Andrew. 2002. *Remembering Defeat. Civil War and Civic Memory in Ancient Athens*. Baltimore.

Wood, Ellen M. 1983. "Agricultural Slavery in Classical Athens." *AJAH* 8: 1–47.

Wood, Neal. 1964. "Xenophon's Theory of Leadership." *C&M* 25: 33–66.

Worley, L. J. 1994. *Hippeis: The Cavalry of Ancient Greece*. Boulder, San Francisco, and Oxford.

Worthington, Ian. 2008. *Philip II of Macedonia*. New Haven and London.

Worthington, Ian. 2012. *Demosthenes of Athens and the Fall of Classical Greece*. Oxford.

Worthington, Ian. 2014. *By the Spear. Philip II, Alexander the Great, and the Rise and Fall of the Macedonian Empire*. Oxford.

Wrede, Walther. 1933. *Attische Mauern*. Athens.

Wycherley, R. E. 1957. *Literary and Epigraphical Testimonia, Agora III*. Princeton.

Wylie, G. 1993. "Demosthenes the General—Protagonist in a Greek Tragedy." *G&R* 40: 20–30.

Yablonskii, L.T. 1999. *Nekropoli drevnego Khorezma. Moscow*.

Young, J.H. 1956. "Ancient Towers on the Island of Siphnos." *AJA* 60: 51–55; plates 38–39.

Zahrnt, Michael. 1992. "Der Mardonioszug des Jahres 492 v. Chr. und seine historische Einordnung." *Chiron* 22: 237–279.

Zaidman, Louise B. 2015. "Women and War: From the Theban Cycle to Greek Tragedy." In Fabre-Serris and Keith, 82–99.

Zavialov, V.A. 2007. "The Fortifications of the City of Gyaur Kala, Merv." In J. Cribb and G. Herrmann (eds.), *After Alexander: Central Asia before Islam*, Oxford and New York, 313–329.

Zhuravlev, Denis and Kirill B. Firsov (eds.). 2013. *Фракийское золото из Болгарии. Ожившие легенды. Выставка в ГИМ Москва*. Moscow.

Zimmer, Carl. 2018. "The Neanderthal, the Artist." New York Times, February 27: D1, D6.

Zimmermann, Konrad. 1980. "Thraker-Darstellungen auf griechischen Vasen." In R. Vulpe (ed.), *Actes du IIe Congrès International de Thracologie*, vol. I. Bucarest, 429–446.

Ziolkowski, Adam. 1993. "*Urbs Disrepta*, or How the Romans Sacked Cities." In Rich and Shipley 1993, 69–91.

Zohary, D., M. Hopf, and E., Weiss. 2013. *Domestication of Plants in the Old World: The Origin and Spread of Cultivated Plants in West Asia, Europe, and the Nile Valley*, 4th ed. Oxford.

Zournatzi, A. 2019. "Looking for Persians in Attic Vase Painting." In O. Palagia and E.P. Sioumpara (eds.), *From Hippias to Kallias: Greek Art in Athens and Beyond, 527–449*, Athens, 120–132.

Zunino, Maddalena Luisa. 2005. "La συνβολά fra Oiantheia e Khaleion (*IG* IX 1^2 717): il diritto di συλεν." *ZPE* 153: 113–126.

Index

Abantes, Thracian tribe, 215
Abdalonymus, King of Sidon, 377
Adea Eurydice, niece of Alexander the Great, 335–336
Aetolia, Aetolians, 134, 141, 344, 345–346
Abydus, 28, 356
Acarnania, 244, 248, 293
Achaea, Achaeans, 123, 134, 343, 344
Achaemenid, Achaemenids, see Persian Empire
Achilles, 138, 324, 327, 396–397
Acrae, 60
Acragas, Agrigentum, (see map of Sicily) 59, 60, 61, 62, 63, 64, 68, 69, 70, 71, 104, 106
Aegean Sea, 25, 27, 29, 39, 40, 101, 123, 163, 167, 264, 269, 340, 344, 346
Aegina (map 1), 20, 23, 33, 34, 35, 102, 326
Aegospotami, Battle of (405), 26, 38, 145, 166, 266
Aemilius Paullus, 382
Aetna, 63
Africa, 65
Agamemnon, mythical king of Mycenae, 138, 140, 342
Agathocles, Tyrant of Syracuse (-289), 64–65, 66, 70, 72
Agis III, Eurypontid King of Sparta (338–331), 29
  See also Megalopolis
Agesilaus II, Euripontid King of Sparta (398–360), 26–27, 273, 293, 294, 314, 315, 318, 321, 325, 392, 409, 411
Agrianes, Agrianian, 45, 48
Albania, 197
Alcibiades, 36–37, 67, 144, 240, 284, 294
Alexander I, King of Macedon (497–454), 23, 24, 154, 263, 412
Alexander II, King of Macedon (369–368), 334, 408
Alexander III, the Great, King of Macedon (336–323), 2, 27, 28, 41, 42, 79, 89, 109, 121, 123, 125, 145, 152, 197, 206–207, 216, 219, 252, 259, 268–269, 290, 295, 317, 324, 325, 327, 329
  Military reforms and organization, 45, 52, 54, 126–127, 129, 131, 133, 143, 150, 155, 156, 157, 245, 314, 317, 329–330, 348, 374–376, 387, 391, 409, 411–412, 415
  Chaeronea, Battle of (338), 46–47
  Granicus R., Battle of (334), 28–29, 168, 305
  Issus, Battle of (333), 29–30, 48
  Tyre, Siege of (332), 152, 319
  Gaza, 317
  And Zeus Amon, 135
  Gaugamela (Arbela), Battle of (330), 49, 180, 182, 202–203
  Hydaspes R., Battle of (326), 49, 145–146, 202, 207–208, 213

A Companion to Greek Warfare, First Edition. Edited by Waldemar Heckel, F.S. Naiden, E. Edward Garvin, and John Vanderspoel.

Alexander IV, 335
Alexander of Pherae, 314, 408
Alexandria, in Egypt, 345
Aluads, Aleuadae, 23, 263
Amorges, 166
Amorgos, Island in the Aegean, 242, 348
Amphictyonic Council, Amphictyonic Leauge, 40
See also Delphi
Amphipolis, city in Thrace (Map 1), 36, 110, 244, 266–268, 305, 326, 414
Amyntas I, King of Macedon (512/11–498), 23
Amyclae, Amyclaeans, see Sparta
Anatolia (Map 1), 15, 167, 168
Anaxander, Anaxandros, general under Philip II, 219, 221
Anaxilas of Rhegium, 66, 68
Andros, island in the Aegean (Map 1), 29, 102, 164, 242, 255
Antalcidas, 27, 39, 166
Antigonas Gonatas, 147, 212, 347
Antigonas Monophthalmus (the One-Eyed), Successor, 48, 49, 50–52, 153, 209, 210, 211, 346, 410
Antiochus I Soter, Seleucid King (281–261), 49, 199, 209
Antiochus I Theos, King of Commagene (70–31), 190
Antiochus II Theos, Seleucid King (261–246), 219
Antiochus III the Great, Seleucid King (222–187), 49, 53–54, 194, 195, 197, 198, 210, 212
Antiochus IV Epiphanes, Seleucid King (175–164), 54–55
Antiochus VII Sidetes, Seleucid King (138–129), 186, 198, 347
Antipater, Macedonian general, 123, 125, 127, 335–336, 414
Apamea, 91, 211, 212
Appius Claudius Caudex, 71
Arachosia, 199
Arbela, Battle of, See Gaugamela
Arcadia, Arcadian, 39
Archaic Greek Warfare, 76–77
Archelaus, King of Macedon (413–399), 294
Archidamian War (431–421), 35, 290, 292, 343, 372
See also Peloponnesian War; Archidamus II; Pericles
Archidamus II, Eurypontid King of Sparta (468–426), 35, 103, 105
Archidamus III, Eurypontid King of Sparta (360–338), 107
Archimedes, 59, 71, 112
See also Siege Warfare; Syracuse
Arcadia, 19, 229
See also Tegea
Archers, Archery, 13–14, 17, 107, 143, 182, 187
Argeads, Royal family of Macedon, 23, 24
Arginusae, Battle of (406), 37, 149, 166, 266, 282, 283, 386
Argos, Argives (map 1), 17, 19, 26, 33, 36, 38, 39, 102, 112, 125–126, 127, 133, 167, 245, 277, 278, 316, 331
Argyraspids, see Silver Shields
Aristagoras of Miletus, 22
See also Ionian Revolt
Aristobulus, 131
Aristomedes of Phae, 168
Ariston of Corinth, 153
Armenia, 91–92, 186, 195, 196
Armor, 13, 94
Arrow, see Archers
Arsaces I, King of Parthia (248–211/10), 186, 195, 198, 199
Arsacid(s), 185ff
Arsinoe, daughter of Ptolemy I, 412
Arsinoe II, wife of Ptolemy II, 337
Arsinoe III, wife of Ptolemy IV, 337
Arsinoe IV, Queen of Egypt (48–47), 337
Artabazus (Satrap of Phrygia), 27, 28, 29, 102, 167
Artagerses (Persian cavalry officer under Artaxerxes II), 180
Artaphernes, Persian general, 22, 23, 371
Artaxerxes I, King of Persia (465–424/3), 25
Artaxerxes II, King of Persia (404–359), 26–27, 151, 166, 180, 215, 266, 309
Artaxerxes III, King of Persia (359–338), , 27–28
Artemisium, Battle of (480), 164–165, 263
Asia Minor, 21, 25, 26, 28, 37
See also Ionia
Assyria, Assyrian(s) (Map 3), 92, 182, 239, 242
Atarneus, 239
Athens, Athenians, 17, 19, 22–24, 26, 27, 28, 31ff, 46–47, 58, 64, 65, 88, 101–102, 107, 110, 115, 125, 127, 129–130, 132, 134, 135, 138, 139, 140, 141–142, 144, 147, 155, 166, 167, 229, 238, 239, 244, 245, 250, 252, 255, 256–257, 261, 271, 274, 281–282, 284, 291, 292–293, 294, 305, 312, 314, 320, 322, 327, 331, 343, 349, 358ff., 369ff., 385, 389, 414
Athenian Acropolis, 24, 281, 371–372, 386, 414
Persian Wars, 163, 164, 165, 264, 277, 279, 325, 370–371, 388, 408
Athenian Empire 26, 31ff, , 122, 264, 344
Sicilian Expedition, 59, 60, 65, 66, 67, 72, 105, 122, 131, 153, 155, 158, 235, 258, 265, 276, 277, 410

Peloponnesian War, 104, 141, 264–265, 275, 293, 295, 351, 355, 390, 406
Second Athenian Confederacy, 27, 39, 409
and Philip II, 267–268, 343–344, 391
and Alexander the Great, 168
Attalus, Macedonian general, 28
Attalus I, King of Pergamon (241–197), 56, 269, 318, 373
Attica, see Athens
Autophratades, 29
Azes I (Sacan King), 196

Babylon, 26, 30, 45, 166, 185, 186, 198, 199, 203, 208, 211, 325
Bactra (map 3), 211Bactria, Bactrians (see map of Near East) 29, 30, 45, 165, 180, 187, 188, 189, 190, 194, 195, 198, 212
Barca, Siege of (512), 101
Battering Ram(s), 103, 105, 106
Berenice II, 337
Bessus, Satrap of Bactria, Artaxerxes V, , 30, 169, 412
Bianor of Acarnania, 168
Blockade, See Siege Warfare
Boeotia, Boeotians, 19–20, 23, 24, 34, 35, 38, 39–40, 46–47, 88, 122, 127, 130, 131, 132, 135, 140, 164, 165, 230, 238, 245, 254, 258, 322, 343, 355
See also Thebes
Boeotian League, 39, 411, 413
Bosporus, 25, 163
Brasidas, 36, 105, 122, 279, 305, 326, 408–409
Bronze Age, 11ff
Bronze Weapons and Armor, 11ff
Bubastis, 167
Bucephalus, Alexander the Great's horse, 374
Byzantium, 27, 28, 104, 110, 167, 268, 317

Caesar, Gaius Julius, 337, 353
Calabria, 255
Callias, Athenian general, 25
Callias, Peace of, 165
Callimachus, Athenian general, 370
Calinnicus, Battle of (171)
See also Perseus, King of Macedon
Camarina, 60, 62, 63, 64–65, 69, 70
Cambyses I, King of Anshan and Parsa (c.600–559), 228, 236
Cambyses II, King of Persia (530–522), 163
Camels, 93
Campania, 17, 65
Caphyae, Battle of (219), 52
Cappadocia, Cappadocian(s), 26, 55
See also Asia Minor
Caria (map 1), 22, 27, 28, 29, 245, 247, 255, 269, 340
Carrhae, Battle of (53), 187, 190, 196
Carthage, Carthaginians, 17, 42, 58, 59, 60, 62, 64, 65, 66, 67–71, 103, 104, 106–107, 142, 284, 409
Carystus, 163, 164
Casmenae, 60
Cassander, King of Macedon (305–297), 65, 123, 336, 373, 390, 410, 414
Catane, 61, 63, 67, 69, 155
Cataphract(s), 53, 55, 89, 94, 190 -
Catapult(s), 69, 106, 107, 110, 121, 142, 145, 233, 244–245
Cavalry 52–53, 57, 88ff., , 121, 124, 126–127, 130, 179–181, 235
Cecryphalea, 33
See also Aegina
Ceos, Keos island in the Aegean, 254, 255
Cephisus R., 46
See also Chaeronea
Chabrias, Athenian general, 134, 144, 167, 392
Chaeronea, Battle of (338), 41, 46–47, 96, 122, 304, 305, 329, 363, 384, 386, 387, 388, 390–391, 413
Chalcidice, 35, 105, 217, 263, 268
Chalcidian League, 39
Chalcis, Chalcidean, 16–17, 19–20, 115
Chandragupta, Mauryan King (c. 324–c.297), 205, 210
Chares, 29, 167
Charidemus, 144, 168
Chariot, Charioteer, 13, 15–16, 45
Charmande, 151
Chersonese, 25, 41
Chios, island in the Aegean, 27, 29, 155, 168, 242, 266, 272–273, 275, 280, 281, 291, 346, 386
Chorasmia, (map 3), 187, 189, 194
Chremonidean War (260s), 147
Cicones (Thracian tribe), 215
Cilicia (map 3), 29, 163, 167, 247, 346
Cimon, Athenian General, 25, 33, 101, 370
Clazomenae, 27
Cleander, Spartan general, 317
Clearchus, Spartan soldier, 151, 309
Cleisthenes, Athenian statesman, 126, 364
Cleobulus, 221
Cleombrotus, Agiad King of Sparta (380–371), 386
See also Battle of Leuctra
Cleomenes, Agiad King of Sparta (520–490), 278, 364
Cleon, 35–36, 142

Cleopatra I Syra, Queen of Egypt (193–176), Regent (180–176), 337
Cleopatra III, Queen of Egypt (142–131/127–116), 337
Cleopatra IV, Queen of Egypt (116–115), 337
Cleopatra VII Philopator, Queen of Egypt (51–30), 337, 353
Clitus, 135
Clitus 'The Black', 305
Cnidus (map 1), 27, 60, 166
Colophon, 245
Combined Arms tactics, 42, 45, 46, 96
Companion Cavalry, 45,46, 48, 53, 89, 121, 126–127
Conon, Athenian general, 27, 166
Corinth (map 1), 18, 19, 26, 33, 35, 38, 39, 61, 67, 70, 229, 244, 263–264, 280, 281, 314, 316, 357, 360, 370, 389, 410, 413
Corinth, League of, 28, 277, 411
Corinthian War, (395 – 387/6) 26, 38ff, 216, 388
Corcyra/Corfu (map 1), 18, 35, 65, 144, 154, 272, 274, 277, 280, 281, 351, 356, 360–362
Coronea, Battle of (394), 39, 321
Coronea, Battle of (447/6), 34
Corsica, 66, 243
Corupedium, Battle of (281), 413
Cos, Island in the Aegean (map 1), 27, 29, 269
Cossaei, Cossaeans, 45
Cotys, King of the Odrysian Thracians (383–359), 215
Crannon, Battle of (322), 123
See also Lamian War
Crenides, see Philippi
Crete, Cretan(s), 15, 48, 53, 139, 219, 254, 334, 344, 345, 346, 347, 349
Critias, Athenian statesman, 353
Crocus Field, Battle of (352), 41, 411
Croesus, King of Lydia (561–547), 306
Croton, 66
Cumae, Battle of (474), 66
Cunaxa, Battle of (Map 3), 26, 130, 150, 166, 167, 180, 309
See also Cyrus the Younger
See also Artaxerxes II
Cylon, Athenian statesman, 364
Cyme, Cymae, in Campania, 17, 342
Cynnane, daughter of Philip II, 335
Cynoscephalae, Battle of (197), 216, 387 See Philip V of Macedon
Cynossema, Battle of (411), 37
Cyprus (Map 3), 22, 25, 27, 110, 163, 164, 167, 168, 242, 243, 252, 337, 342
Cyrene, 29
Cyrus II, The Great, King of Persia (559–530), 174, 179, 183, 204, 228, 236, 321
Cyrus the Younger (the Pretender), 26, 38, 144, 151, 153, 155–156, 166, 180, 215, 266, 295, 309
Cythera island off S. coast of Laconia, 343
Cyzicus, Battle of (410), 65, 166

Dahae (Skythian tribe), 194, 197
Damascus (map 3), 29
Dara (Parthian city), 198
Darius I, King of Persia (522–486), 21, 23, 163
Succeeded by Xerxes
Darius II, King of Persia (423–405), 25, 175, 265–266
Darius III, King of Persia (338–330), 28, 29, 169, 175, 176, 179, 180, 194, 202–203, 374–375, 412
See also Issus, Gaugamela
Dark Age, Greek. ca 1100 – 750 BC, 15
Datis the Mede, Persian general, 22, 23, 371
See also Marathon, Battle of
Decelea, town in Attica, 37, 98, 274, 294
Decelian War, Ionian War (413–404), 37
See also Peloponnesian War
Delian League, 25, 31–32, 35, 165, 264, 344, 408
Delium (map 1), 105, 385
Delium, Battle of (424), 36, 325, 389–390
Delos, Island in the Aegean (map 1), 32, 347, 349, 363
Delphi (map 1), 34, 40, 65, 282, 314–315, 316, 322, 326, 346, 364, 382, 385, 387, 390, 411
Demades, 131
Demetrius Poliorcetes (the Besieger), 51, 110, 112, 121, 123, 145, 209, 346, 347, 410
Demosthenes (Athenian politician), 28, 41, 261, 268, 272, 340, 406, 414
Demosthenes (Athenian general), 35, 37, 141, 154, 275
Dercylidas, Hercylides (Justin) Spartan general, 26, 238, 317, 319–320
Diadochoi, See Successors
Dinocrates, 64–65
Diocles, 68
Diomedes, mythical Greek king, 258, 324
Dionysius I, the Elder, Tyrant of Syracuse (406–376), 59, 64, 66, 68–69, 72, 104, 106, 107, 140, 142, 145, 244, 295, 409
Dionysius II, the Younger, Tyrant of Syracuse (367–357/346–344)), 64
Dionysius of Phocis, 157
Dolope, 23

Dorieus, 60, 67
Doris (map 1), 18, 33, 34, 164
Drangiana (map 3), 199
Ducetius, 61, 64

Ecbatana, capitol of Media (map 3), 199
Egesta (map 2), 60, 67, 68
Egypt, Egyptian Warfare, 13, 15, 24, 29, 33, 53, 126, 144, 162, 165, 166, 167, 182, 210, 216, 337
Eion, 25, 102
Elba, island off Italy, 66
Elephantine, in Egypt, 162
Elephants in Warfare 45, 49, 50, 53–54, 55, 93, 144, 146, 202ff., , 268
Eleporus River, Battle of (399), 66
Eleusinian Mysteries, 36
Eleusis (map 1), 294
Elis, city in the Peloponnese, , 279, 316
  See also Olympic Games
Enia, 23
Enna (map 2), 65
Epaminondas, 40, 122, 124, 128, 130, 131, 132, 143, 275, 322, 324, 382
  See also Theban Hegemony
Ephesus (map 1), 245, 247
Ephialtes of Athens, 168
Ephialtes of Malis, 164
Epichares, 145, 147
Epidaurus (map 1), 33, 102, 155, 306
Epipolae, city on Sicily, 155, 246, 247
Eretria, city on Euboea (map 1), 17, 23, 121, 163, 243, 250, 365, 386
Erythrae, city in Asia Minor, 245
Eryx (map 2), 68
Etna, Mt. (map 2), 64
Etruria, Etruscans, 17, 66
Euboea (Map 1), 17, 23, 34, 164, 244, 255, 263
Eumenes of Cardia, 48, 49, 50–52, 54, 131, 153
Eumenes II, King of Pergamum (197–159), 49, 55–57
Euphrates R., 185
Euryalus, fortress guarding Syracuse, 246
Eurymedon River, Battle of, 25
European Scythians, see Scythia, Scythians
Evagoras II, King of Salamis on Cyprus (361–351), 167, 252

Foragers, Foraging, 97
Fortification(s) 101, 133, 154, 155, 241ff., 293–294

Gabene, Battle of (315), 51
  See also Second War of the Successors; Eumenes of Cardia; Antigonus Monophthalmus
Gaius Flaminius, 71
Galatia, Galatian(s), 55
Gaul, Gauls, 349
Gaza (map 3), 29
Gaza, Battle of (312), 49
Gaugamela, Battle of (331) (map 3), 30, 45, 49, 127, 129, 168, 176, 179, 180, 182, 194, 202–203
Gaul(s), Gallic, 49, 53, 209, 387, 392
Gela (map 2), 59, 60, 62, 63, 64–65, 68, 69, 70, 106, 360
Gelon, Tryrant of Syracuse, 62, 63, 65, 68, 122, 263
Glaucus of Aetolia, 168
Gorgias, Greek rhetorician, 301
Gorgus, 131
Granicus River, Battle of (334) (map 1), 28–29, 127, 168, 179, 305, 326, 387, 415
Greco–Persian Wars, see Persian Wars
Gryllus, Athenian, 382
Gylipus, 37, 65

Haliartus, 38
Halicarnassus (map 1), 28, 29, 410
Halicrnassus, Siege of (334), 110, 168
Halieis, 33
Halos, 110
Hamilcar, 68, 70, 106
Hannibal Mago, 68, 106
Hector, 138, 140
Hellenic League, 24
  See also Persian Wars
Hellenistic Warfare 42ff, 58, 94, 111–112, 126–127, 144 - 147, 202ff., , 348
Hellespont, 25, 28, 65, 104, 163, 164, 265
Helorus, Battle of, 62
Hephaestion, 131
Heraclea, Battle of (280), 49
  See also Pyrrhus of Eprirus
Heraclea Minoa (map 2), 62, 65, 71
Heraclea Pontica, 237
Herbessus (map 2), 65
Hermocrates of Syracuse, 65, 360
Herodotus, 16, 19
Hermocrates, 59, 67
  See also, Sicilian Expedition, Syracuse
Hesiod, 17
Hicetas, Tyrant of Leontini, 70
Hieron I, Tyrant of Syracuse (478–467), 63, 66
Hieron II, Tyrant of Syracuse (270–215), 71, 72
Hieronymus, King of Syracuse (215–214), 71
Himera (map 2), 62, 63, 64, 68, 71–72, 106
Himera, Battle of (480), 63, 68, 122
Himilco, 68–69
Hindu Kush (map 3), 206

Hipparchus, 353
Hippeis, See Cavalry
Hippias, Tyrant of Athens, 22, 353
See also Peisistratus
Hippocrates of Gela, 62, 63
Histiaeus, 22
See also Ionian Revolt
Hittite, Hittite Warfare, 13, 15, 92, 342
Homer, see Iliad and Odyssey see also Troy, Trojan War
Hoplite 43–44, 48, 52, 75ff., , 119, 122, 124, 133, 160, 231, 233, 276, 279, 292, 313, 398
Hybla (map 2), 59
Hydaspes R., Battle of (326), 45, 49, 145, 202, 207–208, 213
Hypaspists, 46, 48, 50, 52, 54, 129
Hyphasis River, 30
Hyrcania, Hyrcanians (map 3), 185, 197, 198
Hysiae, 36
See also Argos

Iaxartes, Jaxartes R. (map 3), 317
Iberia (map 3), 197
Idrieus of Caria, 167
Iliad, Homer, 13, 15
Ilium, See Troy
Illyria, Illyrians, 43, 266, 335
Immortals, the Ten Thousand, 174–177, 183
Inarus, 165
India, Indians, 123, 145–146, 165, 202, 204, 208, 210, 211, 325
Indus R., 202
Infantry Warfare, 42, 43–44, 57, 75–87, 121, 124 130, 133, 145, 183
See also Phalanx; Hoplite; Hypaspists; Peltasts
Ionia, Ionians, Ionian Revolt, 21, 22–23, 25, 26–27, 37, 102, 162, 163, 164, 165, 167, 242, 386, 408, 409
See also Persian Wars
Iphicrates, Athenian general, 122, 127, 132, 134, 142–143, 144, 156, 166–167, 168, 176, 215–216, 275, 318
Ipsus, Battle of (301), 50, 123, 209, 210, 212, 269
Iran, Iranians, 45
Isagoras, Athenian statesman, 364
Issus, Battle of (333), 29, 48, 168, 174, 176, 182, 194, 373–376
See also Alexander the Great, Darius III
Ister R. (Danube), 163
Isthmian Games, see Corinth
Isthmus of Corinth, see Corinth
Italy, , 66, 107
Ithome, Mt. (map 1), 33, 101, 102, 275
Italy, 36

Jason, Tyrant of Pherae, 88, 122, 127, 408
Javelin(s), 17, 95, 142
Judaea, 212

King's Peace (387), 27, 28, 39–40, 166, 409
See also Antalcidas, Artaxerxes II
Knossos, Crete, 13–4

Lacedaemon, Lacedaemonia, see Sparta
Laconia, see Sparta
Lade, Battle of (494), 22, 157, 163
Lamachus, Athenian general, 67, 306
Lamian War (322–321), 388, 410
Larisa (map 1), 23, 263
Laurium (map 1), 23
Lechadum, Battle of (390), 142
Lelantine War, Lelantine Plain, 16–17, 20, 121
Leonidas, Agiad King of Sparta (490–480), 24, 142, 175, 389, 392
See also Persian Wars, Thermopylae
Leontini (map 2), 59, 62, 63, 64–65, 67, 69, 70, 72, 106, 242, 243
Lesbos (map 1), 29, 37, 168, 266, 282, 386
See also Mytilene
Leuctra, Battle of (371), 27, 40, 130, 143, 324, 331, 380, 386
See also Sparta, Thebes
Levant, 15, 25, 29
Libya, Libyan(s), 53, 101
Lilybaeum, on Sicily, 60
Locri Epizephyrii (map 2), 66
Locris, Locrians, 23, 38, 164, 165, 243
Lycia, Lykia (South western Anatolia), 25, 29, 247, 415
Lycius, cavalry officer in the Ten Thousand, 259
Lycon, 165
Lydia (Map 1), 26, 29, 165, 242, 289
See also Asia Minor
Lysander, Spartan commander, 26, 38, 104, 144, 145, 266
Lysias, 272
Lysimachus, Successor, king of Thrace etc (306–281), 50, 222, 410, 412–413

Macedon, Macedonia, 21, 23, 24, 43, 88, 109–111, 123, 124, 125, 127, 134, 135, 139, 143, 165, 166, 167, 168–169, 194, 218, 220, 223, 245, 258, 263, 266, 316, 335, 382–383, 387
Magna Gracia, 295
Magnesia (map 1), 23, 194, 212
Magnesia, Battle of (190), 49, 54–57, 145
See also Antiochus III, the Great
Mago, 69
Malis, Malians (map 1), 165

Mantinea (map 1), 39, 236, 245
Mantinea, Battle of (418), 36, 125, 128, 234, 278, 279, 293
Mantinea, Battle of (362), 40, 47, 122
Mantinea Battle of (207), 52
Marathon, Battle of, 21, 22, 23, 131, 163, 165, 184, 277, 301, 305, 323, 325–326, 327, 370, 372, 373, 379, 386, 388, 402
Marcellus, Marcus Claudius, 112
Mardonius (Persian commander under Xerxes), 22–24, 174, 177–178, 179, 181, 183, 317, 325
See also Plataea, Battle of
Mark Antony (Marcus Antonius), 186, 337, 353
Masistius (Persian commander under Xerxes), 151–152, 182
Massagetae (Scythian tribe), 194, 200
Massalia, 289
Mausolus of Caria, 27
Media, Medes (map 3), 51, 55, 91–92, 165, 185, 195, 197
See also Persians
Median War, see Persian Wars
Megabazus, Persian commander, 163
Megabyzus, Persian commander, 165
Megacles, Athenian statesman, 364, 365
Megalopolis, Battle of (331), 169, 210, 212
Megara (map 1), 33, 34, 38, 59, 104, 165, 165, 212, 244, 245, 262, 263, 274, 365, 389, 410
Megara Hyblaea (Map 2), 59, 60, 62, 63, 243
Melos (map 1), 23, 36, 155, 164, 284, 355
Memnon of Rhodes, 28, 29, 167, 168
Memphis (Egypt), 211
Menon, 130, 151
Mentor of Rhodes, 167
Mercenaries 48, 50, 53, 54, 58, 63, 64, 65, 66, 69, 72, 119, 125, 134, 144, 151, 167, 168, 176, 256, 268, 269, 334, 336, 347, 355
Messana/Zancle (map 2), 64, 65, 66, 69, 70, 71, 72
Messene, 69, 245, 246
Messenia, Messenian Wars, 12, 19, 35, 40, 103, 275, 278, 287, 297, 324
Methone (map 1), 110, 266
Methymna, 104
Miletus, Miletos (Map 1), 22, 23, 29, 168, 242, 246, 247, 255, 342, 365, 386, 410
Miltiades, Athenian general, 23, 370–371
Mitanni, 13
Mithridates I, Arsacid King of Parthia (165–133/2), 196, 198, 199, 200
Mithridates II the Great, Arsacid King of Parthia (124–91), 187, 190, 196, 199, 200, 347
Mithridates VI, King of Pontus (120–63), 411
Mnasippus, Spartan general, 154, 274
Morgantina, 61
Motya, Motyon, 61, 64, 68, 69, 107, 142, 145
Munychia, Battle of (404/3), 312
Mycale, Battle of (479), 24, 102, 165
See also, Persian Wars
Mycenae, Mycenaean 11ff, , 254, 261, 384
Myae, 67
Myndus, 163
Myrmidons, 124
Myronides, 33, 34
Mysia, 26
Mysteries, see Eleusinian Mysteries
Mytilene, 29, 104, 155, 168
See also Lesbos

Naupactus, city on the gulf of Corinth, 102, 275
Navarino, 12
Naval Warfare, 12, 17, 31, 59, 69, 264, 271–72, 280–283, 293
Naxos (map 1), 59, 63, 102, 155, 163, 346, 365
Naxos, Siege of (500), 101
Nemea, Nemean Games, 316
Nemea, Battle of (394), 39
Nestor, mythological King of Pylos, 138, 341
Nestor, Palace of, at Pylos, 11ff
Nicanor, 129
Nicias, Athenian general, 36–37, 67, 122, 131, 134, 156, 252, 281
Nicias, Peace of (421), 36, 157, 275, 404
Nicostratus, Athenian general, 361
Nile River, 209
Nisaea, 33, 195, 244
Nobility, the role of in military affairs, 12
Notium, 165
Notium, Battle of (406), 37

Odrysian(s) Thracian tribe, 215
Odysseus, mythical Greek king, 138, 258, 324, 341
Oenidae, 102
Oenoe, Battle of (506), 105, 244, 369, 376-
Oenophyta, 34
Olympia (map 1), 66, 282, 316, 326, 390
Olympias, mother of Alexander the Great, 336
Olympic Games, 65
Olynthus, 39, 110, 218, 219, 266, 334
Olynthus, Second Battle of (381), 96
Onomarchus of Phocis, 110
Orcynia, Battle of (319), 48
See also Eumenes of Cardia, Antigonus Monophthalmus
Orodes II, Arsacid King of Parthia (58/7–37), 189, 199

Orontobates, 28
Oropus, 294
Ortygia, 64
Oxathres, brother of Darius III, 374

Paeonia, Paeonian, 48, 215, 259, 266
Pagondas, 131
Palestine, 337
Palisades, See Fortifications
Pallenis, Pallene, Battle of (546), 363
Pammenes, 124, 167
Pamphylia, 25, 29
Panion, 194
Panormus (map 2), 68
Paphlagonia, 26
Paphos, on Cyprus, 242, 243, 244
Paraetacene, Battle of (317), 48, 49, 50, 52
  See also Eumenes of Cardia, Antigonus Monophthalmus, War of the Successors, Second
Parmenio, Macedonian general, 28, 127, 129, 135
Paros, Siege of (490), 101, 102
Parthia (map 3), 51, 52, 185
Pasion, 130, 151
Patron of Phocis, 168
Pausanias, Spartan Regent (480–477/6), 39, 325, 389
Pellene, 348–349
Pelopidas, Theban general, 39, 131, 408
Peloponnese, 18, 63, 263
Peloponnesian League, 19, 32, 33, 36, 39, 121–122, 264–265
  See also Sparta
Peloponnesian War (431–404), 25, 26, 35–38, 41, 65, 89, 104, 122, 141, 150, 166, 215, 230, 236, 238, 239, 244, 264, 274, 275, 280, 281, 283, 286, 290, 292, 294, 297, 301, 343, 352, 355–356, 360, 380, 408
Peloponnesian War, First (460 – 446/5), 32–34, 406
Peltast(s), 45, 53, 122, 130, 142–143, 176, 207, 215
  See also Javelin(s)
Pelusium, Battle of (525), 162
Perdiccas Perdikkas (successor, Regent to Philip Arrhidaeus of Macedon), 209, 335
Perdiccas III, King of Macedon (365–359), 43, 266
Pergamum, Pergamon (map 1), 57, 195, 246, 344, 373, 387, 392
Pericles, 35, 102, 141, 244, 282, 292, 330
Perinthus (map 1), 28, 110, 167, 268
Perrhaebia, 23
Persepolis, royal seat of the Achaemenid Empire (map 3), 181, 187
Perseus, Antigonid King of Macedon (179–168), 56–57, 212–213, 216, 383
Persia, Persians, Persian Empire 19, 21ff, , 37, 65, 92, 101, 123, 144, 194, 252, 264, 265, 278
Persian Wars 19, 21ff, , 31, 41, 89, 155, 164, 238, 244, 261, 263, 280, 284, 292, 319, 322, 327, 359, 370, 388, 401, 408
  See also Marathon, Battle of; Thermopylae; Salamis, Battle of; Plataea, Battle of; Mycale, Battle of.
  See also Darius I; Xerxes
Phalanx, 43–44, 57, 78, 96, 126, 155
  See also Infantry Warfare
Phalaris, Tyrant of Acragas (6th c.), 60, 72
Phanes, 162
Pharnabazus, 27, 29
Pharsalus, city in Thessaly, 238
Philip II, King of Macedon (360/59–336), 27–28, 40, 42–43, 109, 110, 121, 131, 133, 140, 167, 216, 289, 297, 317, 334, 336, 340, 343–344, 363, 407, 409, 411, 413, 414
  Military reforms, 43–45, 52, 54, 143, 218, 219, 221, 266–269
  Chaeronea, 46–47, 390
Philip III Arrhidaeus, son of Philip II, 335, 376
Philip V, King of Macedon (221–179), 216, 219, 269, 345, 347, 356, 387
Philippi, Crenides, city in Thrace, 267–368
Philopoemen, Achaean general, 343
Philotas, Macedonian general, 28, 135
Phocaea, Phocaeans (Map 1), 17, 34, 242
Phocion, 167
Phocis, Phocians (map 1), 38, 40–41, 110, 164, 165, 167, 238, 263, 366, 411, 413
Phoenicia, Phoenicians, 15, 25, 29, 67, 106, 141, 163, 165, 167, 326
Phormio, Phormion, Athenian general, 284
Phraates II, Arsacid King of Parthia (133/32–127), 186, 198, 199
Phraates IV, Arsacid King of Parthia (37–2), 186, 199
Phrygia (map 3), 26, 167
Phthia, 23
Pike(s), Pikemen, 95, 143
  See also Sarissa
Piraeus, 33, 38, 110, 147, 244
  See also Athens
Pisistratus, Peisistratus, Tyrant of Athens, 135, 262, 322, 365
Pissuthnes, 165
Pithon, 51
Pixodarus, Satrap of Caria, 28
Plataea (map 1), 35, 39–40, 103, 104–105, 151–152, 154, 165, 370

Plataea, Battle of (479 BC), 23, 24, 88, 101, 121–122, 124, 155, 164, 165, 177, 181, 184, 238, 278, 282, 319, 326, 327, 388, 389
Plataea, Battle of (379), 380–381
Plunder (Booty), 152
Polycrates of Samos, 162
Polyperchon, 210, 212, 336, 413
Pompeii, 374
Porus, Indian King, 145–146, 207–208
See also Alexander the Great; Battle of the Hydaspes R.
Potidaea (map 1)
Persian Siege of (479), 101, 102
Athenian Siege of (432–430/29), 35, 104, 115, 155, 156, 240, 292–293, 305
Siege of by Philip II (356), 110
See also Chalcidice
Priene, 249, 262
Proxenus, 151
Psammetichus, King of Egypt (664–610), 262–263
Ptolemaic Egypt, See Egypt
Ptolemy I Soter, King of Egypt (305/4–283), 49, 131, 209, 376, 410, 413
Ptolemy II, Philadelphus, King of Egypt (283–246), 147, 337
Ptolemy IV Philopator, King of Egypt (221–204), 53–54, 210, 337
Ptolemy IX Soter, King of Egypt (116–107/88–81), 356
Ptolemy XII, King of Egypt (51–47), 337
Punic War, First, 71, 72
Punic War, Second, 212
Pydna (map 1), 110, 266, 336
Pydna, Battle of (168), 47, 50, 54, 56, 212–213, 216, 369, 382–383
See also Perseus, King of Macedon
Pylos (map 1) 11ff, , 35, 141–142, 154, 254, 261, 275, 291, 294, 295, 343, 390
Pyrrhus, King of Epirus (297–272), 49–50, 65, 70–71, 112, 211, 333

Quadrireme(s), 42
Quinquereme(s), 69

Ram, See Battering Ram
Raphia, Battle of (217), 53, 123, 127, 210, 212, 337
See also Antiochus III; Ptolemy IV; Arsinoe III
Reconnaissance, 97, 98, 150, 252ff
Rhamnus (map 1), 244, 256, 259
Rhegium (map 2), 63, 64, 66, 67
Rhodes (Map 1), 27, 28, 62, 112, 145, 243, 246, 247, 248, 269, 344–345
Rhoetium, Rhoeteium, 168
Rome, Romans, 47, 54–57, 59, 65, 71, 102, 211, 346, 357, 359

Sacae, 165, 196–197
Sacred Band, 39, 41, 46–47, 123, 130, 143, 386, 387, 390
See also Thebes; Pelopidas
Sacred War, Third (356–346), 40–41, 315
Sacred War, Fourth (339), 41
Salamis, Battle of (480 BC), 24, 165, 272, 280, 326, 373, 389
Salamis, Cypriot, 110, 112, 242, 252
Salamis (Map 1), 164, 264, 365
Samos, Samians (map 1), 35, 102–103, 141, 165, 168, 242, 245–246, 280, 385
Samos, Siege of (523), 101
Sapping, Sappers, 108
Sardis, Capitol of Lydia (map 1), 22, 23, 25
Sarissa, 44, 47, 48, 143, 146
See also Infantry Warfare
Sarmatia, 196
Saronic Gulf, 33
Scione, 356
Scouts, See Reconnaissance
Scyros, 102
Scythes of Cos, 63
Scythia, Scythians, 23, 29, 45, 163, 185, 215, 224
Sea Peoples, 15
Seleucia Pieria, 115
Seleucid Empire, 53, 192, 203, 210, 211, 337
Seleucus I, 50, 209, 210, 211, 269, 412–413
Seleucus II, 199
Selimbria, 104
Selinus, 59, 60, 68, 69, 71, 106, 246
Sellasia, Battle of (222), 52
Semiramis (semi-mythical Queen of Assyria), 204, 213
Sepia, Battle of (c. 494), 277, 278
Sestos, 102, 115
Seuthes III, King of the Odrysian Thracians (324–312), 219
Sicels, 60, 61, 63, 64, 69
Sicilian Expedition, See Athens
Sicily (Map 2) 36, 37, 58ff, , 106–107, 122, 134, 142, 242, 244, 247, 248, 265, 289, 342, 359, 409
Sicyon (map 1), 33, 255, 258, 410
Sidon (map 3), 167, 376
Siege Warfare, 59, 99–118, 121, 143, 145, 155, 232
Siege Tower(s), 107, 145
Silver Shields, 50–52, 53, 123
Siphnos (map 1), 29, 254

Sitalces, King of the Odrysian Thracians (431–424), 215
Siwa, oracular temple in Egypt, 315
Slaves, Slavery and Warfare, 121, 233, 271ff
Slingers, 13–14, 17
Smyrna, on Cyprus, 242, 247
Social War (357–355), , 27
Social War (220–217), 52, 346
Socrates, 305
Sogdiana, Sogdianians (map 3), 29, 30, 45, 189, 194
Soleis, 68
Soli, Siege of (498), 101
Solon, Athenian statesman, 89, 364–365
Sparilises (Sacan King), 196
Sparta, 18–19, 22, 23, 24, 25–26, 32ff, 60, 65, 67, 101–102, 104, 121, 122, 123, 124–125, 128, 131, 134–135, 138, 139, 140, 141–142, 142–143, 152, 155, 156, 164, 165, 166, 167, 183, 235, 236, 239, 259, 265, 272, 273, 275, 277, 278–279, 287, 289, 294, 306, 313, 316, 318, 319, 320–321, 331, 343, 352, 359, 364, 390, 406, 413
  See also Peloponnesian War
Spartan Hegemony, 26–27, 39–40
Spear, Spearheads, 14, 95
Sphacteria (map 1), 36, 142, 390
  See also Pylos
Spitamenes (Sogdian commander), 194
Spithridates, Persian satrap, 305
Stratobates (perhaps mythical, King of India), 204, 213
Successors, The, 48, 53, 54, 89, 94, 121, 123, 126, 127, 130, 134, 202, 208–213, 269, 336, 412–413
  See also Wars of the Successors; Hellenistic Warfare.
Sunium, 244
Susa, Shushan, Shush, capitol of Elam and a major hub of the Persian Empire (map 3), 181, 198, 203
  See also Persepolis
Susiana, 197
Sword(s), 95
Sybota, Battle of (433), 35, 280
Syria, 337
Syracuse, Syracusans (map 2), 37, 59–66, 68–69, 107, 112, 140, 153, 154, 244, 258, 263, 265, 276, 277, 279, 280, 281

Tanagra, 34, 102, 389
Tanais River (Don River, sea of Azov) (map 3)
Tanais River (Aral Sea) mistaken for Iaxartes River
Tapur, Tapurians (Scythian), 45
Tarentum, 51
Tauromenium (map 2), 65, 72, 115
Taxiles, Indian King, 207
  See also Alexander the Great
Tegea, Arcadian, 19, 165, 278
Tegyra, Battle of (375), 39
Teispis, Teispids, Royal family of Persis, 22
Tempe, Vale of, 24, 263
  See also Xerxes
Ten Thousand Immortals, See Immortals
Ten Thousand, March of, 122, 124–125, 127, 130, 134, 135–136, 150, 151, 155, 157, 168, 234, 237, 258–259, 270, 290, 297, 309, 320, 334
Tenos, 164
Teos, 257, 349
Thasos (Map 1), 33, 102, 104, 243, 246, 248, 245, 294
Thebes (map 1), 20, 23, 26, 34, 35, 38, 39, 40, 41, 46–47, 104, 112, 121, 123, 124, 134, 164, 165, 167, 168, 229, 239, 244, 275, 289, 294, 295, 313, 314, 324, 325, 329, 365, 379, 380, 382, 384, 389, 390, 409, 411
Theban Hegemony, 27, 40
Themistocles, Athenian statesman, 2, 244, 264, 318
Thera (map 1), 346
Theramenes, Athenian statesman, 352–353
Thermae, 69
Thermopylae, Battle of (480), 18, 24, 142, 164–165, 175, 183, 263, 278, 386, 389–390
Thermopylae, Battle of (279), 387
Thersites, 138
Thespiae, 23, 165, 389
Thessaly, Thessalians (map 1), 23, 24, 33, 40–41, 88, 121, 123, 127, 164, 165, 238, 239, 247, 255, 258, 263, 363, 366, 408, 411
Thessalian Cavalry, 48
Thibron, 26
Thirty Years' Peace (446/5), 34–35
Thrace, Thracian(s) (Map 1) 23, 24, 28, 48, 53, 55, 143, 163, 214ff., , 244, 259, 266–268, 329
Thrasybulus, Athenian general, 168, 312
Thrasydaeus of Acragas, 63
Thucydides, Athenian general, 16, 17, 32
Thurii, 66, 280
Thurium, Mt., 46
Tigranes II the Great, King of Armenia (95–55), 196
Tigranocerta, Battle of (69), 196
Timasion, cavalry commander in the Ten Thousand, , 259
Timoleon of Corinth, 64, 70
Timotheus, Athenian statesman, 122
Tissaphernes, Satrap of Ionia under Darius II, general under Artaxerxes II, 25, 26, 166, 234

Tithraustes, Satrap of Sardis, 167
Torone, 104
Tower(s), See Siege Towers
Trireme, 20, 42, 121, 280
Troy, Trojan War, 13, 137–139, 150, 154, 262, 301, 315, 332, 341–342
Tryphon (Diototus) usurper king of the Seleucid Empire (142–138), 347
Tyre (map 1), 29
Tyre, Siege of (332/1), 42, 110, 111, 152
Tyrtaeus, 18

Vergina, in Macedonia, 374

War of the Successors, Second (320–316), 50–52, 412–413
 See also Eumenes of Cardia; Antigonus Monophthalmus

Xanthos, 247
Xenias, 151
Xenodicus, 64–65
Xenophon, commander of the Ten Thousand, 259, 266, 309, 382
Xerxes I (486–465) King of Persia, 22, 24, 65, 164, 165, 175, 178, 180, 182, 183, 215, 263, 402–403

Zancle, 63, 66, 342
 See also Messana
Zopyrion, Governor of Thrace under Alexander the Great, 219, 222